Monsoon dynamics

Monsoon dynamics

EDITED BY
SIR JAMES LIGHTHILL &
PROFESSOR R. P. PEARCE

CAMBRIDGE UNIVERSITY PRESS
CAMBRIDGE
LONDON · NEW YORK · NEW ROCHELLE
MELBOURNE · SYDNEY

CAMBRIDGE UNIVERSITY PRESS
Cambridge, New York, Melbourne, Madrid, Cape Town, Singapore, São Paulo, Delhi

Cambridge University Press
The Edinburgh Building, Cambridge CB2 8RU, UK

Published in the United States of America by Cambridge University Press, New York

www.cambridge.org
Information on this title: www.cambridge.org/9780521104265

First published 1981
This digitally printed version 2009

A catalogue record for this publication is available from the British Library

Library of Congress Cataloguing in Publication data
Joint IUTAM/IUGG International Symposium on Monsoon Dynamics, Delhi, 1977
Monsoon dynamics
Consists of revisions of selected papers presented at the symposium which was sponsored by the Indian National Science Academy and the Indian University Grants Commission
Includes index
1. Monsoons–Asia–Congresses 2. Monsoons–India–Congresses 3. Dynamic meteorology–Congresses
I. Lighthill, M. J., Sir II. Pearce, Robert Penrose, 1924– III. Indian National Science Academy IV. India (Republic) University Grants Commission QC939.M7J64 1977 551.5′184 78–72091

ISBN 978-0-521-22497-0 hardback
ISBN 978-0-521-10426-5 paperback

Contents

Contributors

Bavadekar, S. N., Indian Institute of Tropical Meteorology, Ramdurg House, University Road, Poona-5, India.

Bedi, H. S., India Meteorological Department, Lodi Road, New Delhi 110003, India.

Brode, R. W., Laboratory for Atmospheric Research, University of Illinois at Urbana-Champaign, Urbana, Illinois 61801, USA.

Cadet, D., Laboratoire de Météorologie Dynamique, Ecole Polytechnique, Rue Departmentale 36, 91128 Palaiseau, France.

Chander, S., Indian Institute of Technology, Hauz Khas, New Delhi 110 029, India.

Charney, J. G., Department of Meteorology, Massachusetts Institute of Technology, Cambridge, Massachusetts 02139, USA.

Cox, M. D., Geophysical Fluid Dynamics Laboratory/NOAA, Princeton University, PO Box 308, Princeton, New Jersey 08540, USA.

Das, P. K., India Meteorological Department, Lodi Road, New Delhi 110003, India.

Datta, R. K., India Meteorological Department, Lodi Road, New Delhi 110003, India.

Dhar, O. N., Indian Institute of Tropical Meteorology, Ramdurg House, University Road, Poona-5, India.

Findlater, J., Meteorological Office, Met. O.11, London Road, Bracknell, Berkshire RG12 2SZ, UK.

Flather, R. A., Institute of Oceanographic Sciences, Bidston Observatory, Birkenhead, Merseyside L43 7RA, UK.

Fleer, H. E., Instut für Meterologie und Klimatologie, Universität Hannover, Herrenhäuser Str. 2, 3 Hannover 21, W. Germany.

Gadgil, Sulochana, Center for Theoretical Studies, Indian Institute of Science, Bangalore 560012, India.

Gangadhara Rao, L. V., National Institute of Oceanography, Dona Paula, Goa 403004, India.

Gilchrist, A., Meteorological Office, London Road, Bracknell, Berkshire RG12 2SZ, UK.

Goswami, B. N., Physical Research Laboratory, Navrangpura, Ahmedabad 380009, India.

Hurlburt, H. E., NOL/NORDA, Code 322, NSTL Station, Mississippi 39529, USA.

Johns, B., Department of Meteorology, University of Reading, 2 Earley Gate, Whiteknights, Reading RG6 2AU, UK.

Kanamitsu, M., Meteorological College, Japan Meteorological Agency, Asahi-cho, Kashiwa, Chiba-Pref. 277, Japan. (Present address: Electronic Computation Center, Japan Meteorological Agency, 1-3-4 Olemachi, Chiyoda-ku, Tokyo 100 Japan.)

Keshavamurty, R. N., Physical Research Laboratory, Navrangpura, Ahmedabad 380009, India.

Krishnamurti, Ruby, Department of Meteorology, Florida State University, Tallahassee, Florida 32306, USA.

Krishnamurti, T. N., Department of Meteorology, Florida State University, Tallahassee, Florida 32306, USA.

Kumar, A., Indian Institute of Technology, Hauz Khas, New Delhi 110 029, India.

Lighthill, Sir James, University College London, Gower Street, London WC1E 6BT, UK.

Lim, J. T., Department of Meteorology, University of Hawaii at Manoa, 2525 Correa Road, Honolulu, Hawaii 96822, USA.

Lin, L. B., 36 Northam Court, Slidell, LA 70458, USA

Luyten, J. R., Department of Physical Oceanography, Woods Hole Oceanographic Institute, Woods Hole, Massachusetts 02543, USA.

Mak, M. K., Laboratory for Atmospheric Research, University of Illinois at Urbana-Champaign, Urbana, Illinois 61801, USA.

Mandal, B. N., Indian Institute of Tropical Meteorology, Ramdurg House, University Road, Poona-5, India.

Marchuk, G. I., Computing Centre, Novosibirsk 6300 90, USSR. (Correspond via: VAAP–The Copyright Agency of the USSR, Export and

Import Department of Social and Scientific Publications, B. Bronnaja 6a, Moscow K-104, USSR 103104.)

Mooley, D. A., Indian Institute of Tropical Meteorology, Ramdurg House, University Road, Poona-5, India.

Murakami, T., Department of Meteorology, University of Hawaii at Manoa, 2525 Correa Road, Honolulu, Hawaii 96822, USA.

Murty, A. V. S., Central Marine Fisheries Research Institute, PB No. 1912, Cochin 682 018, India.

Olory-Togbé, P., Laboratoire de Météorologie Dynamique, Ecole Polytechnique, Rue Departmentale 36, 91128 Palaiseau, France.

Pant, P. S., Indian Meteorological Department, Poona-5, India.

Pearce, R. P., Department of Meteorology, University of Reading, 2 Earley Gate, Whiteknights, Reading RG6 2AU, UK.

Penenko, V. V., Computing Centre, Novosibirsk 6300 90, USSR. (Correspond via: VAAP – The Copyright Agency of the USSR, Export and Import Department of Social and Scientific Publications, B. Bronnaja 6a, Moscow K-104, USSR 103104.)

Pisharoty, P. R., Physical Research Laboratory, Navrangpura, Ahmedabad 380009, India.

Raghava, R. C., Indian Institute of Technology, Hauz Khas, New Delhi 110 029, India.

Rakhecha, P. R., Indian Institute of Tropical Meteorology, Ramdurg House, University Road, Poona-5, India.

Ramanadham, R., Department of Meteorology and Oceanography, Andhra University, Waltair 530 003, India.

Ramanathan, Y., India Meteorological Department, Lodi Road, New Delhi 110003, India.

Ramesh Babu, V., National Institute of Oceanography, Dona Paula, Goa 403004, India.

Rao, G. V., Department of Earth and Atmospheric Sciences, St Louis University, Box 8099 Laclede Station, St Louis, Missouri 63156, USA.

Rao, R. R., Naval Physical and Oceanographic Laboratory, Cochin 682004, India.

Rao, Y. P., India Meteorological Department, Lodi Road, New Delhi 110003, India.

Sadler, J. C., Department of Meteorology, University of Hawaii at Manoa, 2525 Correa Road, Honolulu, Hawaii 96822, USA.

Santa Devi, M. R., Naval Physical and Oceanographic Laboratory, Cochin 682004, India.

Satyan, V., Physical Research Laboratory, Navrangpura, Ahmedabad 380009, India.

Sharma, O. P., India Meteorological Department, Lodi Road, New Delhi 110003, India.

Shukla, J., Department of Meteorology, Massachusetts Institute of Technology, Cambridge, Massachusetts 02139, USA.

Sikka, D. R., Indian Institute of Tropical Meteorology, Ramdurg House, University Road, Poona-5, India.

Singh, M. P., Indian Institute of Technology, Hauz Khas, New Delhi 110 029, India.

Sinha, M. C., India Meteorology Department, Lodi Road, New Delhi 110003, India.

Somanadham, S. V. S., Naval Physical and Oceanographic Laboratory, Cochin 682004, India.

Spolia, S. K., Indian Institute of Technology, Hauz Khas, New Delhi 110 029, India.

Subbaramayya, I., Department of Meteorology and Oceanography, Andhra University, Waltair 530 003, India.

Sunderaramam, K. V., Naval Physical and Oceanographic Laboratory, Cochin 682004, India.

Sundqvist, H., Department of Meteorology, University of Stockholm, Arrhenius Laboratory, S 106 91 Stockholm, Sweden.

Swallow, J. C., Institute of Oceanographic Sciences, Brook Road, Wormley, Godalming, Surrey GU8 5UB, UK.

Van de Boogaard, H. M. E., National Center for Atmospheric Research, PO Box 3000; Boulder, Colorado 80307, USA.

Varadachari, V. V. R., National Institute of Oceanography, Dona Paula, Goa 403004, India.

Varma, Savita, Department of Mathematics, India Institute of Technology, Hauz Khas, New Delhi 110 029, India.

Verma, R. K., Indian Institute of Tropical Meteorology, Ramdurg House, University Road, Poona-5, India.

Washington, W. M., National Center for Atmospheric Research, PO Box 3000, Boulder, Colorado 80307, USA.

Webster, P. J., CSIRO, Division of Atmospheric Physics, PO Box 77, Mardialloc, Victoria, Australia 3195.

Editors' preface

The aim of this volume is to present an up-to-date survey of our state of knowledge of the physical and dynamical processes involved in the Asian monsoon. Although traditionally the main emphasis has been on the study of the atmospheric component, it has been known for a long time that the oceans play a vitally important role in determining the occurrence of this spectacular seasonal event. A scientific study of this phenomenon must involve a detailed investigation of the dynamical processes which occur in both the atmosphere and the ocean, on time scales of up to at least a year and on spatial scales from a few hundred kilometres or so up to that of the global atmospheric and oceanic circulations. Such problems are of considerable intrinsic interest not only to meteorologists and oceanographers but also to fluid dynamicists in general, and the Joint IUTAM/IUGG International Symposium on Monsoon Dynamics, held in Delhi from 5 to 9 December 1977 had as its major aim the bringing together of all of these groups of scientists. This volume serves as a Proceedings of this symposium. An attempt has also been made in selecting and editing the contributions to present a coherent survey of each of the meteorological, oceanographic and hydrological aspects, and of their implications for weather forecasting and flood prediction. Some of the most recent observations are presented, including those which have become available from satellites. On the theoretical side, there are several papers which describe the use of computerized models of the atmospheric and oceanic circulations, modern computers providing one of the most powerful tools for their simulation and study.

At the time of publication of this volume, the observational phase of the Monsoon Experiment (MONEX), a sub-programme of the Global Atmospheric Research Programme (GARP) organized by the World Meteorological Organization (WMO) and the International Council of Scientific Unions (ICSU), will have been completed. The extensive observational data from this experiment will enable many aspects of monsoon behaviour to be more accurately described and provide a much improved observational input into the study of monsoon dynamics. It is hoped that much of the work described in this volume will serve as a useful basis for those who will become involved in this highly important research activity.

The detailed arrangements for the symposium were taken care of by the authorities at the Indian Institute of Technology, Delhi, and the undoubted success of the symposium owes much to the interest and enthusiasm of the Director, Professor N. M. Swani, and of the Institute's noted fluid dynamicist, Professor M. P. Singh. The meeting was arranged through the International Union of Theoretical and Applied Mechanics (IUTAM) jointly with the International Union of Geodesy and Geophysics (IUGG) who provided financial support for the participants. The sponsorship of the Indian National Science Academy and the Indian University Grants Commission, together with the collaboration of the India Meteorological Department and the cooperation of the American Meteorological Society, is gratefully acknowledged.

JAMES LIGHTHILL
R. P. PEARCE

Keynote address to the 1977 Symposium

This address was delivered by Sir James Lighthill at the opening ceremony of the Joint IUTAM/IUGG International Symposium on Monsoon Dynamics, New Delhi, 5 to 9 December 1977.

Exactly two years ago, in the second week of December 1975, a few of us (who today are all present in this hall) consulted together and jointly planned to hold a meeting on '*Monsoon Dynamics, with applications to weather forecasting and flood prediction*' at IIT, Delhi in the second week of December 1977. Since then, all of us have been working strenuously to ensure the effectiveness of the Monsoon Dynamics Symposium. But it is a wonderful reward for all that hard work to look at this distinguished audience, comprising 46 participants from 16 countries outside India and another 131 participants from within India, who will be actively pushing forward together towards real progress in the understanding and knowledge of Monsoon Dynamics during the coming week, encouraged by the high honour that has been done to us by the Honourable Minister for External Affairs, Shri Atal Behari Vajpayee, graciously accepting our invitation to inaugurate the symposium. The Honourable Minister has indicated by his kind words the importance attached to the aims of our symposium by the Government of India, and this is further indicated by the honour being done to delegates on Wednesday when they will be received by the Prime Minister at his official residence.

Those few of us who formulated the original plans included yourself, Mr Chairman [Dr A. Ramachandran, Secretary of the Department of

Science and Technology], together with the Director of IIT, Delhi, Professor N. M. Swani, and the Director General of Observatories of the Indian Meteorological Department, Mr Y. P. Rao. Key members of the Indian National Committee for Theoretical and Applied Mechanics, such as Professor S. D. Nigam of IIT, Madras and the Committee's secretary, Professor M. P. Singh, were also closely involved. In addition, I myself was concerned as a member of the seven-man Bureau which administers the International Union of Theoretical and Applied Mechanics, IUTAM.

All the members of IUTAM's Bureau had for some time felt dissatisfied with the fact that such international scientific bodies hold almost all of their meetings in Europe or North America. We wanted, therefore, to hold more meetings in countries which are outside those areas but where the sciences of mechanics are strong. It was important, however, that such a meeting should attract to itself well-qualified participants from many other areas of the world so that it would have the full impact of a globally international meeting.

In 1974 the Bureau received from INSA, the Indian National Science Academy, a proposal by the Indian National Committee for Theoretical and Applied Mechanics for a symposium within the general area of fluid mechanics affected by Coriolis forces, to be held in India. In 1975 the Bureau requested me, since I was due to be in India in December, to discuss with the Indian National Committee how to formulate the detailed proposals in such a way that a scientifically strong and globally international attendance would be achieved, while at the same time the work of the symposium would be of real value to scientists in India and in neighbouring countries. I remember that a valuable discussion with Professor Nigam on this campus led to the formulation of the title and subtitle 'Monsoon Dynamics, with applications to weather forecasting and flood prediction'.

This title and subtitle included *two* syntheses which have continued to seem important to all concerned with organizing the symposium. First, there is the synthesis between 'basic' and 'applied'; between the fundamental mechanics of the monsoon in relation to global processes and the important applications of that mechanics to problems of great significance for India and her neighbours. Secondly, the subtitle indicated a synthesis between air problems and water problems; for example, between monsoon winds and the associated ocean currents; between monsoon depressions and the associated local storm surges; between monsoon rainfall and the associated water-resource and river-management problems.

Such a grouping of subject matter seemed to us to meet all of the requirements that had been stated to us. If our aim of holding a major international symposium in India was the stimulus to further and yet more fruitful researches which would be received by the many Indian scientists who would be able to attend, then what topic for the symposium could be more appropriate than one of such great importance for India herself as the monsoon with all its immense benefits and attendant dangers? To encourage the study of monsoon dynamics, with applications to weather forecasting and flood prediction, by a still wider base of scientific effort within India, comprising universities as well as research establishments, and comprising all the relevant scientific disciplines, seemed potentially of great value. In addition, no topic for a symposium to be held in India could be more attractive to the global scientific community than a topic like monsoon dynamics which combines great scientific interest with immense significance to India, and dependence on data accumulated in India.

Professor Singh reacted most positively to these ideas and arranged for me to be able to consult both the Director General of Observatories of IMD, Mr Y. P. Rao and the Director of IIT, Professor N. M. Swani. Both warmly welcomed the proposal and pledged their support, pledges which have been honoured most generously in the ensuing two years. The Director General pointed out that the proposed meeting in December 1977 would be usefully preceded by a meeting of considerable importance, but confined to the meteorology of monsoons, organized jointly by his Department and the American Meteorological Society. The Director of IIT indicated plans for an Indo-French Workshop which could equally have a most valuable function of preparing the ground for a globally international meeting of much wider scope. They both emphasized that the impact on the direction of scientific effort within India would be greatest if a succession of meetings devoted to the theme were held in an ascending sequence of scientific scope and width of international cooperation. Such a sequence would also fulfil best the aim of drawing established workers in geophysical fluid dynamics from many other parts of the world into active monsoon researches. All these views have since been supported by the American Meteorological Society cordially agreeing to sponsor *this* meeting too, with the generous help of the US National Science Foundation.

Another point of importance that emerged in my discussion with Mr Y. P. Rao and Professor Swani was the fact that their establishments are rather close neighbours here in the Hauz Khas area of New Delhi! In fact, the proximity between IIT and the Headquarters of the IMD here in Hauz Khas is most advantageous for the general aim of bringing about close cooperation between university research work and the work of operational departments in this field. This, together with the central position of Delhi in relation to the country as a whole, suggested the great advantage of IIT, Delhi as the location for the Monsoon Dynamics Symposium.

Next, I had the opportunity of discussing the proposal with you, Mr Chairman, and was delighted to receive a strong impression that support from the Department of Science and Technology would be possible. Furthermore, my discussions with yourself and with other leading members of INSA, the Indian National Science Academy, of which I have the honour to be a Foreign Fellow, indicated clearly that INSA would approve the proposed changes and, soon afterwards, INSA sent to the Bureau of IUTAM a formal request for a Monsoon Dynamics Symposium to be held at IIT, Delhi during the present week. I wrote to all the Bureau members explaining the proposal's merits at length and recommending its acceptance. I also recommended that IUTAM should invite IUGG, the International Union of Geodesy and Geophysics, with its extensive interests in meteorology and oceanography, to be a joint sponsor of the symposium. The Bureau fully concurred with my proposals and then IUGG, after extensive correspondence to determine the precise role of this symposium in relation to other monsoon meetings, enthusiastically accepted joint sponsorship, providing support from its central funds and additional support from both the International Association of Meteorology and Atmospheric Physics and the International Association of the Physical Sciences of the Ocean.

Our two sponsoring Unions are both members of the same family, ICSU (the International Council of Scientific Unions), and we have to recognize, of course, that ICSU, besides being an 'umbrella' for different Unions, has its own Global Atmospheric Research Programme (GARP), which includes an important Monsoon Experiment MONEX. However, careful study indicated that the GARP programme, while of immense value in its own area of relations between monsoon phenomena and global processes, could hardly meet the need we had recognized for a programme offering some sort of synthesis between information of that kind and local information of practical predictive value. Ultimately, ICSU itself was convinced by our arguments and gave a special grant for our meeting to IUTAM, to increase the money available to this symposium without, I hasten to say, in any way depriving GARP.

One great advantage of the sponsorship by two International Unions is that it emphasizes still more the basic interdisciplinary nature of the subject. The IUTAM sponsorship aims to ground the symposium firmly in the dynamics of *fluids*. This not only *includes* the dynamics of air and of water and of their mutual interactions, but also tends to emphasize all the many common features of fluid motion that are prominent whether the fluid is air or water. One good example is the branch of fluid dynamics treated in my own latest book which is to be published early in 1978 by Cambridge University Press, entitled *Waves in Fluids*. This gives a somewhat general treatment of all sorts of wave propagation phenomena that occur in fluids, and usefully views many different phenomena of real geophysical interest from the standpoint of a unifying body of general theory. The properties of *boundary layers*, in general, form another group of unifying concepts, and there are several other such groups. The basic background science of fluid dynamics is part of what *meteorologists* and *oceanographers* have in common and its emphasis at this meeting can help to prevent any compartmentalization between meteorologists and oceanographers.

Please note: the whole purpose of this meeting would be negated if the meteorologists among you went only to the meteorological lectures! It is essential that you attend the oceanographical lectures as well. The most obvious reason is known to you: the oceans, and especially ocean temperatures, have a strong influence on atmospheric phenomena. Equally important, however, is the need of the oceanographers for your advice when they are discussing the influence of the atmosphere on the ocean. Then they need your help in ensuring that they are discussing a realistic pattern of atmospheric inputs. All these remarks hold also in reverse, of course: oceanographers must attend meteorology lectures; and it is equally important for all of us, the basic fluid dynamicists, the meteorologists and the oceanographers, to go along and help discuss the hydrologists' problems of water resources and river management, whose importance provides the *raison d'être* for so much of monsoon research, and where our techniques are highly relevant. Again, the problem of generating effective predictions of disastrous coastal flooding demands a combination of the skills, both of our leading experts, whom we have brought together into this hall, on predicting the combined tide and surge effects generated by a given moving cyclonic pattern of winds, and of the meteorologists' skills at forecasting not just winds in general but those features of such cyclonic wind patterns that are most significant for wind-generated surges.

So the keynote of this symposium, which I have been asked to give you with my tuning fork as it were, is its interdisciplinary character, merging the disciplines of fluid dynamics, atmospheric physics, oceanography, estuary dynamics and hydrology; merging the basic sciences of global geophysical processes with the applied sciences of local weather forecasting and flood prediction; merging the professional activities of operational departments in the field and those of university research workers. All of us taking part, by learning from each other in the lectures and in informal exchanges throughout these five days, will be enabled to get an overview of basic and applied monsoon research as a whole, which can guide us to become more effective workers in this vitally important field.

Part I

The large-scale climatology of the tropical atmosphere

Introduction

R. P. PEARCE

It has been recognized for many years that studies of the monsoon must be intimately related to the study of the seasonal climatology of the general atmospheric and oceanic circulations. As far as the atmospheric component is concerned, the main surface feature of the climate of south Asia is the changeover in early summer from the winter regime with its dry north to northeasterly wind to the summer rain-bringing southwesterly monsoon. The winter regime becomes re-established again during the early autumn. With upper-atmospheric data becoming regularly available from the late 1940s, this surface flow transition has been found to be linked to an equally substantial change in the upper-level flow over the whole of the low-latitude belt. This manifests itself in an intensification and northward movement of the main high-pressure region over southern Asia with the onset of the summer monsoon. Particularly dramatic is the generation and maintenance of an easterly jet-like flow at 10° to 15° N across the whole region from the South China Sea to east Africa with a maximum speed of about 30 m s^{-1} at a height of about 13 km over Sri Lanka. Thus in the same way as the seasonal changes of the low-level features of the atmosphere on the global scale are largely dominated by the circulations corresponding to the winter and summer Asiatic monsoon regimes, the high-level large-scale features are similarly dominated by the associated monsoonal events. This undoubtedly mainly reflects the influence of the seasonal changes in heating of the extensive Asiatic and African land masses. During the past 20 years or so, sufficient observational material has become available to enable reliable and detailed climatologies of the large-scale atmospheric circulation to be derived, particularly for the tropics and subtropics. However, there are still large data-sparse areas in these regions and the climatologies of some areas, particularly over the Pacific and in the southern hemisphere, are still likely to be in error in some details. The recent volume *The General Circulation of the Tropical Atmosphere* by R. E. Newell, J. W. Kidson, D. G. Vincent and G. J. Boer (MIT Press, Vol. I 1972; Vol. II 1974) contains an extensive series of maps showing seasonal distributions of wind, temperature, moisture, sea-surface temperature and rainfall based on observational data collected during the period 1958–63.

The observational studies described below examine particular aspects of the global climatology of the tropical atmosphere, and should be read with a background knowledge of the main monsoon features briefly outlined above. Chapter 1 presents a modern statistical approach aimed at identifying links between the monsoonal cycle over Asia and large-scale changes over the rest of the globe. Although statistical studies cannot lead to an understanding of the physical causes of such links, they can establish their existence and enable the physical problem to be properly posed. The occurrence of the monsoonal cycle is an event which takes place every year with unremitting regularity. However, its intensity and precise date of onset in a given region in any particular year vary considerably and have enormous economic significance. These inter-annual variations pose for the meteorologist two of his most challenging problems – identifying the physical causes and providing reliable predictions.

Chapter 2 presents a study of the summer monsoonal characteristics of a particular year, 1972, and identifies certain features which set it apart from the 'average' monsoon year. Chapter 3 examines global temperature patterns and relates the heat source associated with the latent heat release in the monsoon rain areas to other heat sources and sinks over the globe. Chapter 4 examines the maintenance of planetary-scale disturbances by considering the kinetic energy transformations between the mean zonal flow, the 'long' monsoon waves and the transient disturbances, and finally Chapter 5 gives examples of upper-level flow patterns constructed using the recently developed technique of measuring winds at upper levels over the tropics using cloud displacements from satellite images. This technique will undoubtedly make a major impact on the study of large-scale tropical flow patterns, particularly with the imminent setting-up of a global coverage of the tropical atmosphere by a geostationary satellite system. This development could be as significant for the tropics as the earlier development in the 1940s when upper-air soundings were first made on a regular basis using balloon-borne sensors.

Developments in computer technology over the past 20 years or so have provided meteorologists with the opportunity to solve approximate forms of the mathematical equations expressing the physical laws which the atmosphere obeys. This process, which is referred to as numerical modelling of the atmospheric circulation, has opened up a new dimension in the study of the physical processes occurring in the atmosphere and has also provided a powerful tool for the weather forecaster. Since these models can be integrated at a much faster rate than the atmosphere itself can evolve, the computer products can be used to provide the forecaster with valuable guidance concerning future atmospheric developments at least on a timescale of up to a few days. Such techniques are widely used by weather services in midlatitude countries, but are only just being developed for use by weather services in the tropics. However, such models have been used extensively as a research tool in some institutions, and Chapters 7 and 8 provide up-to-date general accounts of the present status of numerical modelling insofar as the tropical atmosphere is concerned. It is against this background of numerical techniques that the fundamental question of atmospheric predictability in low latitudes is addressed in Chapter 6. Chapter 9 explores the energy transformations associated with monsoon circulations as depicted in such a numerical model. Chapter 10 considers a model which couples the atmosphere with the underlying ocean and examines the large-scale interactions in a context in which the land and ocean areas are highly idealized. Such an approach to numerical modelling is extremely valuable in being able to isolate basic processes which are often obscured in more complicated models. Finally, in Chapter 11 an example is provided of the use of mathematical analysis techniques to study the interactions of long waves on a background zonal flow near the Equator. An understanding of such interactions is of fundamental importance in the larger context of the theory of the maintenance of the general atmospheric circulation.

Part Ia

Observational studies

1

Teleconnections of rainfall anomalies in the tropics and subtropics

H. E. FLEER

Statistical relations between the monthly rainfall amounts for 290 stations in the tropics and subtropics (30° S to 35 °N) are investigated by means of autocorrelation, power-spectrum, band-pass-filter, cross-correlation, cross-spectrum, and coherence analyses with respect to space and time.

The main point of interest is the regional distribution of lagged serial correlation coefficients in India and in Africa and the teleconnections and phase relationships of long-term fluctuations between stations in the equatorial Pacific, Indonesia, India, Africa and South America.

1.1 Introduction

One of the most striking climatic variations in recent years occurred in 1972. Drought conditions were prevalent not only across Africa and northern India, but also in South America and Australia. The severe deficiency of rainfall in the subtropics was compensated by excessive rainfall closer to the Equator in southern India, the Phillippines, the western Pacific and in the highlands of southern Africa.

A preliminary survey by Flohn (1974) suggested that the occurrence of such anomalies in different parts of the tropics at almost the same time is not a particularly rare event.

This investigation centres on the question of whether droughts in the tropics and subtropics occur simultaneously, randomly or mutually exclusively. How dominant are quasi-periodic processes with periods of more than one year, and what kind of spatial and temporal coherence exists in the tropics and subtropics?

1.2 Observations

The statistical calculations are based on long series of monthly rainfall measurements carried out over periods ranging from 66 to 135 years. The main interest is focused on the regional anomalies and the long-term variations of rainfall, because rainfall is the most essential, yet perhaps the most highly variable parameter within the tropics.

The regional distribution of the 290 selected stations within the area from 30° S to 35° N is given in Fig. 1.1. Missing data have been replaced either by estimating rainfall values from records of the neighbouring stations or by the long-term monthly means.

1.3 Persistence of precipitation

Lag-autocorrelation coefficients were computed in order to investigate the influence of persistence on monthly rainfall amounts. For all time series, a maximum lag of 120 months was chosen.

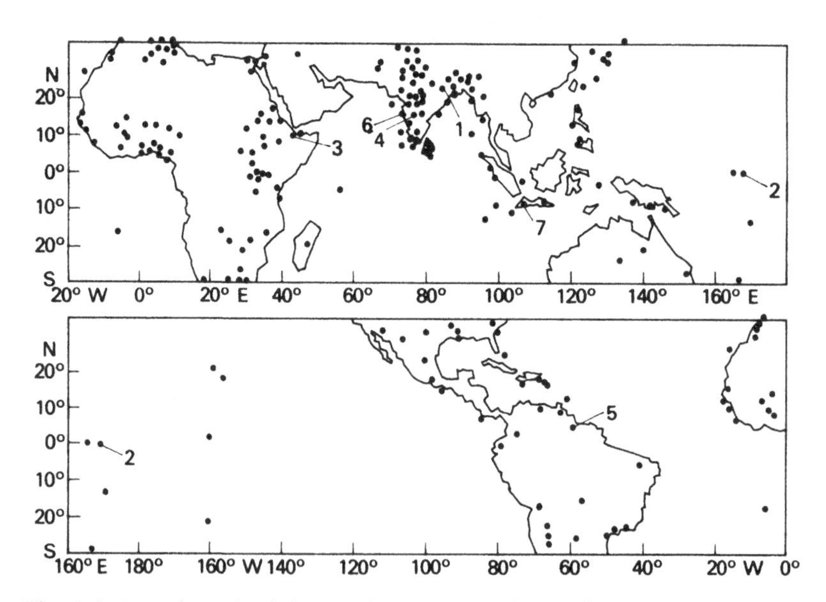

Fig. 1.1. Location of rainfall stations. The stations referred to specifically in this chapter are labelled: 1, Patna; 2, Nauru; 3, Djibouti; 4, Sholarpur; 5, Georgetown; 6, Poona; and 7, Djakarta.

The autocorrelation function of the monthly rainfall amounts of Patna (Fig. 1.2*a*) exhibits a highly significant one-month lag value and a well-marked annual cycle. Autocorrelation coefficients of deviations from the monthly averages (i.e., departure) generally approach zero (Fig. 1.2*b*).

The main property of the lagged-autocorrelation analysis is that persistence of monthly rainfall amounts is mainly determined by the annual cycle.

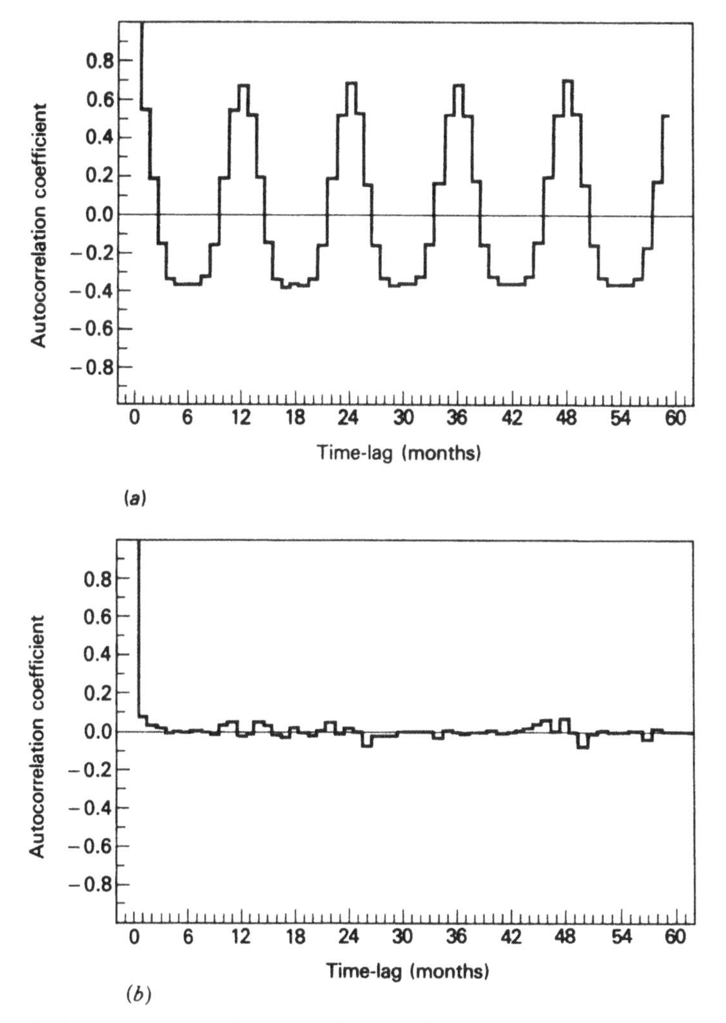

Fig. 1.2. Autocorrelation function of Patna (data from 1868–1960): (*a*) with annual cycle; (*b*) without annual cycle.

Only the autocorrelation function for Pacific island stations is a notable exception. In the case of Nauru (Fig. 1.3*a*), for example, the autocorrelation remains significant up to a lag of 6 months and the annual progression becomes negligible. It is interesting that the function for the departures (Fig. 1.3*b*) does not show a remarkable change. Fig. 1.3*a* shows a pattern similar to Fig. 1.3b, unlike the patterns for Patna (Fig. 1.2).

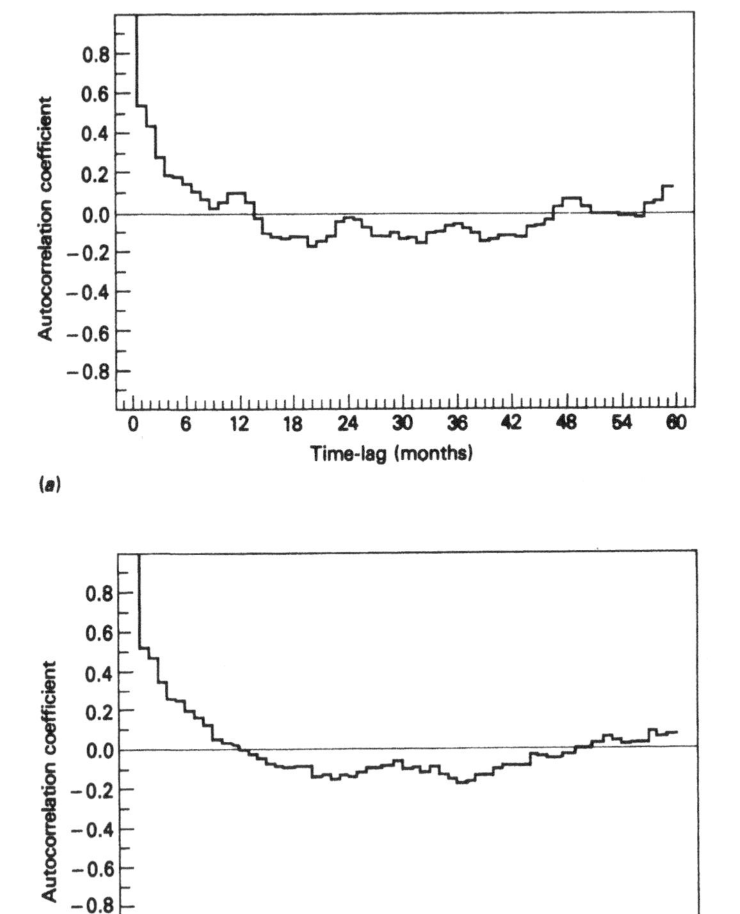

Fig. 1.3. Autocorrelation function of Nauru (data from 1892–1972); (*a*) with annual cycle; (*b*) without annual cycle.

1.4 Oscillations in the records

Several authors have, between them, found numerous cycles to be significant. Since such cycles may disappear or be replaced by other cycles over some time periods, the diagnostic and prognostic value of this type of investigation is restricted, unless the contribution of the individual cycles to the variance of the series is known.

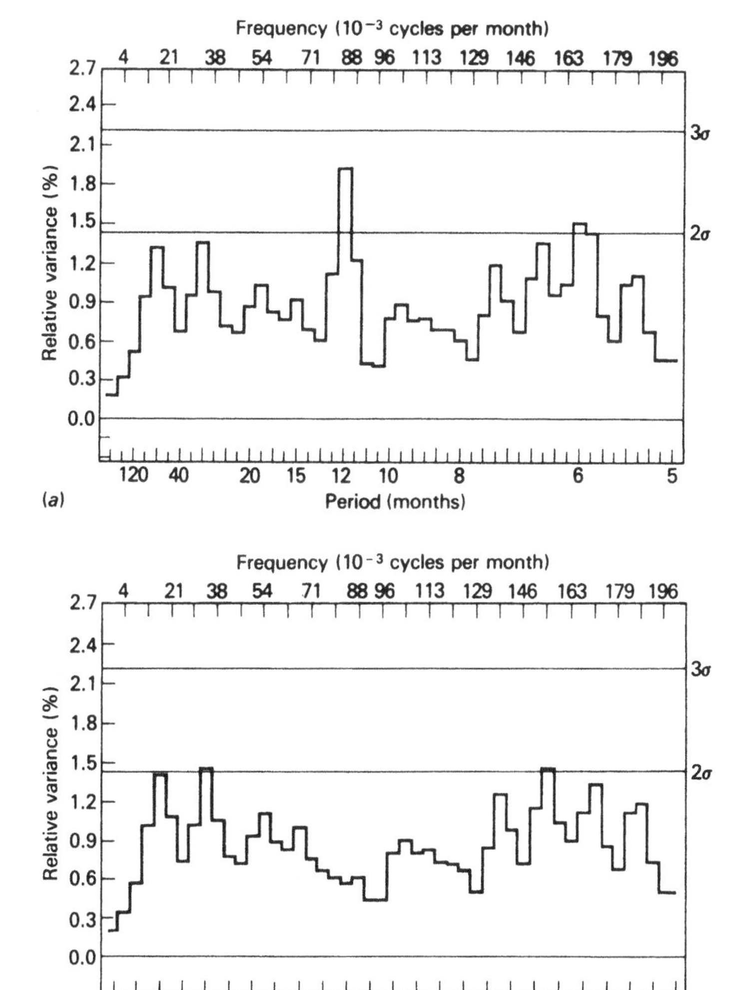

Fig. 1.4. Spectrum of monthly rainfall amounts at Djibouti (data from 1901–72): (*a*) with annual cycle; (*b*) without annual cycle.

Spectrum analysis avoids such difficulties and provides information on the relative importance of different periodicities. The spectra have been normalized to directly obtain the percentage of variance associated with particular frequency bands.

There are four basic spectrum types:

(i) The *random spectrum* (Fig. 1.4*a* and *b*) is represented primarily by some African series. The spectra do not show any frequency range significant above the 3σ level. Although the annual cycle may be

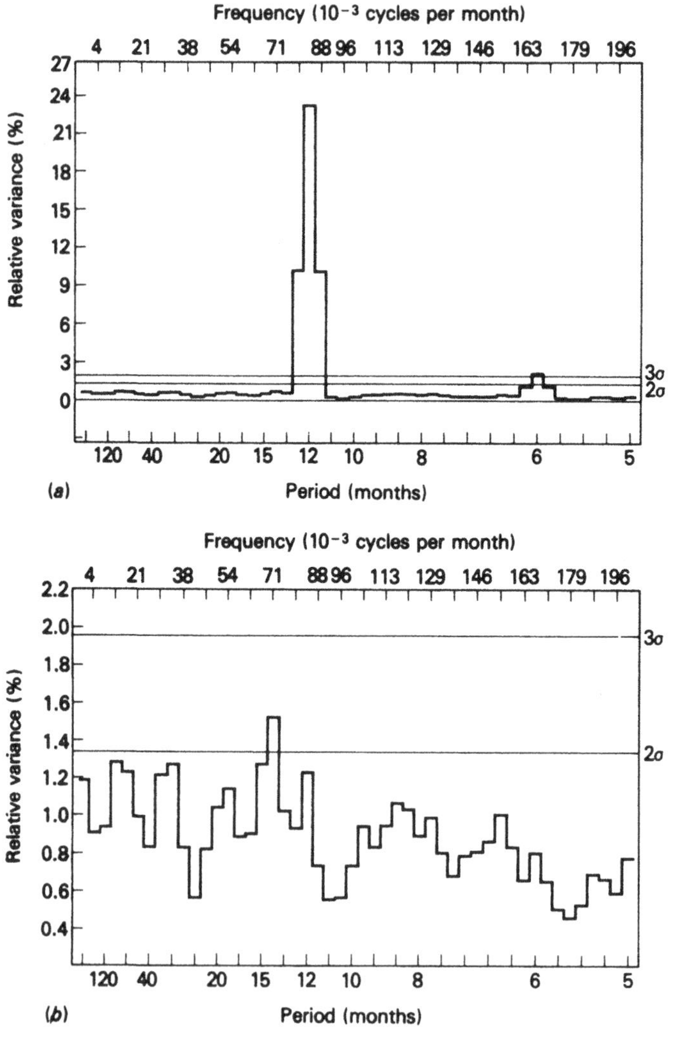

Fig. 1.5. Spectrum of monthly rainfall amounts at Sholarpur (data from 1875–1973): (*a*) with annual cycle; (*b*) without annual cycle.

significant at the 2σ level, this does not contribute much variance to the series.

(ii) Most of the Indian stations show a *white-spectrum* type (Fig. 1.5*a* and *b*). The spectrum of the monthly precipitation amounts exhibits a marked annual cycle. The anomaly spectrum has a white-noise character.

(iii) In the zone 10° S to 10° N a *mixed-spectrum* type is most prevalent (Fig. 1.6*a* and *b*). Both annual progression and periodicities longer than

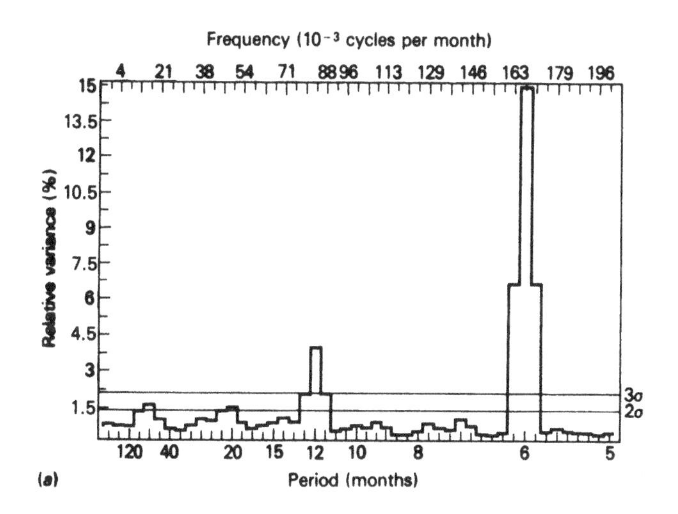

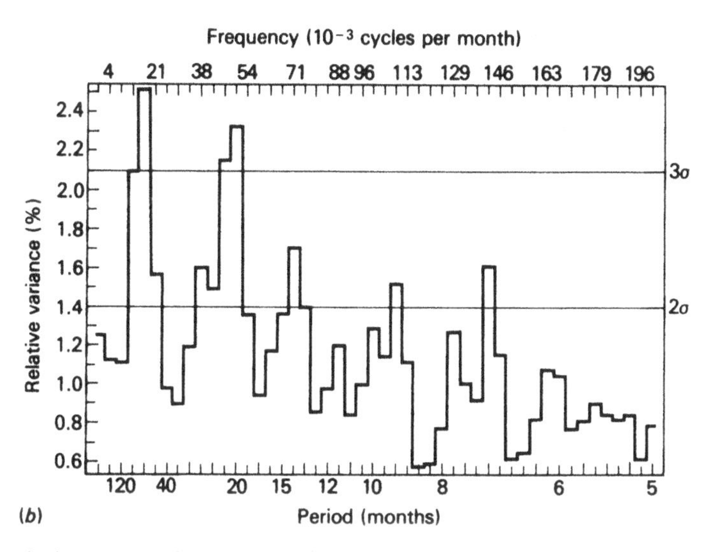

Fig. 1.6. Spectrum of monthly rainfall amounts at Georgetown (data from 1894–1973): (*a*) with annual cycle; (*b*) without annual cycle.

one year add significant variance to the spectrum; therefore, the anomaly spectrum is distorted towards low frequencies.

(iv) In the Pacific area, spectra are highly skewed (*red-spectrum* type) with dominant periodicities longer than one year (Fig. 1.7*a* and *b*), i.e. the annual cycle becomes negligible.

The distribution of the spectral peaks in the range of long-term

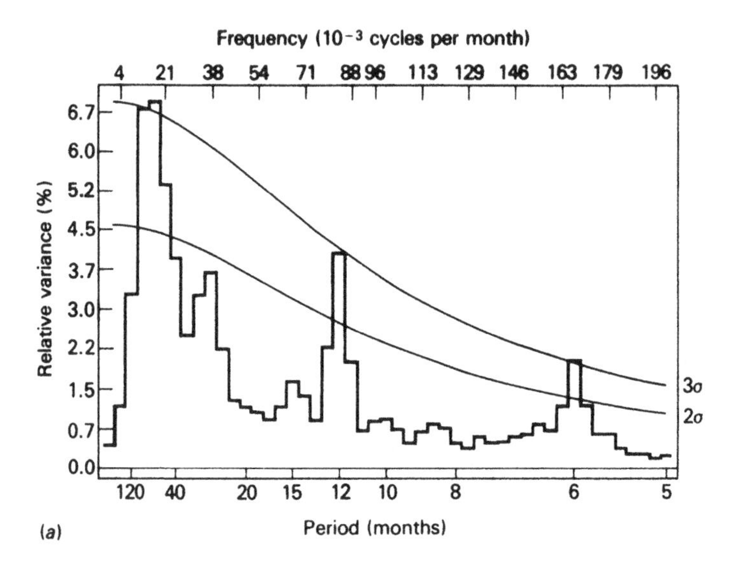

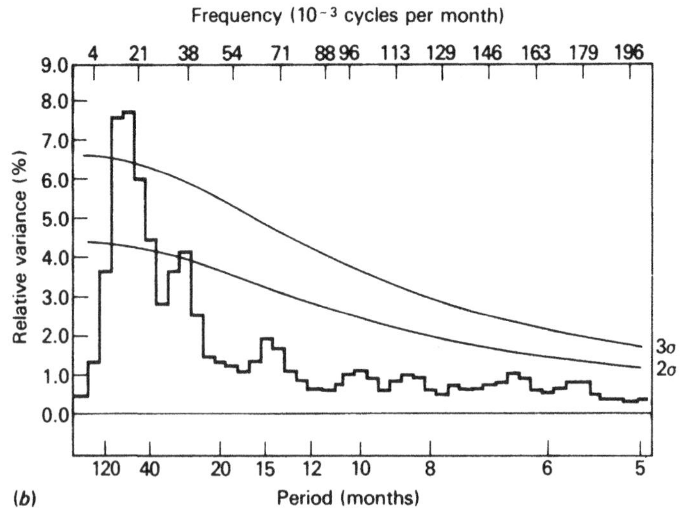

Fig. 1.7. Spectrum of monthly rainfall amounts at Nauru (data from 1892–1972): (*a*) with annual cycle; (*b*) without annual cycle.

oscillations shows its strongest maximum at 60 months. A second smaller maximum appears at around 27 months.

In order to obtain information about the role of the quasi-5-year and the quasi-biannual pulse all series have been subjected to the same band-pass-filtering technique. The two filters contain a 120-term moving average and preserve amplitude, phase and period for cycles of approximately 27 and 60 months, respectively.

Fig. 1.8*a* gives the filtered record of Nauru, and Fig. 1.8*b* that of Poona. The main properties of the filtered series are:

(i) The variance maxima in the power spectrum are in fact produced by

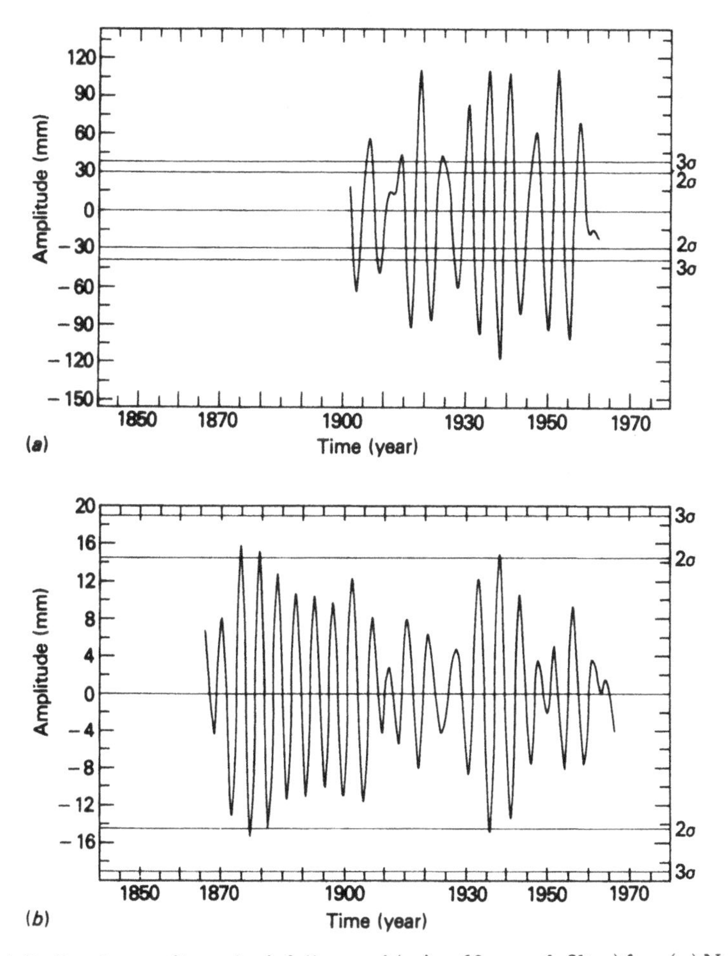

Fig. 1.8. Band-pass-filtered rainfall record (using 60-month filter) for: (*a*) Nauru (data from 1892–1972); and (*b*) Poona (data from 1856–1975).

an oscillatory behaviour of the time series in this frequency band. These oscillations are, however, largely unpredictable, because the amplitude, phase and period vary irregularly and even discontinuities in phase may occur.

(ii) The amplitude of the quasi-5-year oscillation is generally larger than that of the quasi-biannual pulse. The mean amplitude of the former is, in the Pacific, over 100% of the mean amplitude of the annual cycle (Fig. 1.8*a*), but in the Indian peninsula (Fig. 1.8*b*) the contribution of these fluctuations is about 10% of the mean annual amplitude.

1.5 Teleconnections

The term teleconnections refers to the statistically or empirically determined coupling of large-scale abnormalities of the atmospheric circulation in time and space. Such links were first discovered by Walker (1924), who found linear correlations between sea-level pressure in different parts of the world.

Investigations were carried out by Schütte (1969) and Doberitz (1967, 1968, 1969). Additional aspects were dealt with by Flohn and Fleer (1975). Even for such distant stations as Nauru and Djakarta (Fig. 1.9) there is a marked coherency (frequency-dependent correlation coefficient) over a period of 5 years. An inverse behaviour of the rainfall fluctuations in the Pacific and Indonesian area is verified by the negative cospectrum.

Along the north coast of New Guinea, a marked shift separates two contrasting areas, the equatorial Pacific and the Indonesian islands. These regions are internally consistent (i.e., all stations within a region are in phase with each other), but the oscillations of the two regions themselves are out-of-phase. The discontinuity coincides to some extent with Berlage's (1966) and Troup's (1965) pressure correlations.

In order to obtain a global survey of rainfall teleconnections, more than 1000 cross-spectra have been computed with several reference stations. Fig. 1.10 shows the regional distribution of phase relationships at the period of 5 years between Nauru and 289 selected stations in the zone 30° S to 35° N. Positive correlation (i.e., in-phase) and negative correlation (i.e., out-of-phase) are indicated in this figure.

Several interesting relationships are found. Between the equatorial Pacific and Indonesia there exists a negative correlation, as mentioned earlier. The equatorial African rainfall is in-phase with the equatorial Pacific rainfall, while the equatorial South American rainfall and the

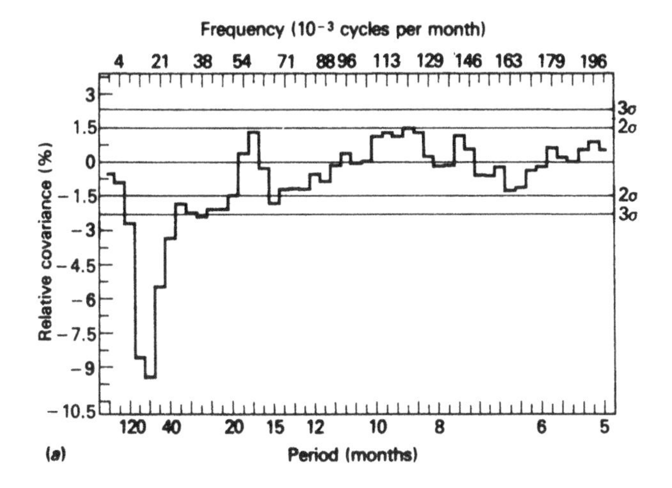

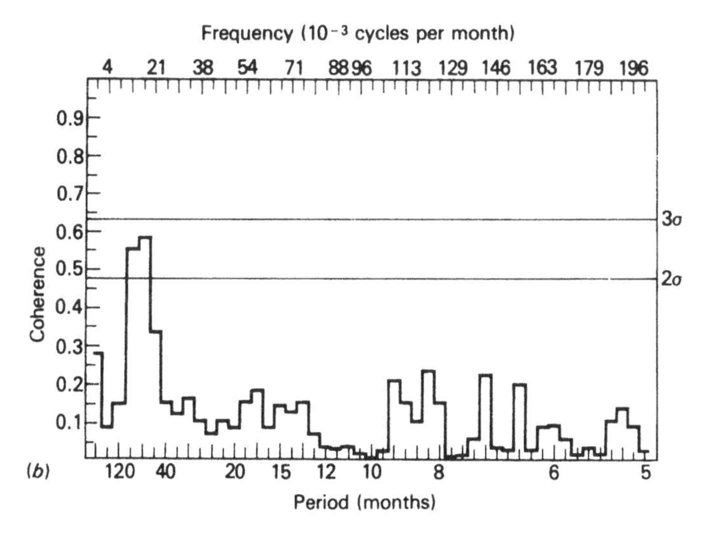

Fig. 1.9. (*a*) Cospectrum Nauru–Djakarta. (*b*) Coherency Naura–Djakarta. In both, data is from 1905–70.

majority of West African stations are out-of-phase with the equatorial Pacific.

A comparison with higher latitudes indicates that anomalies in the equatorial Pacific are in-phase with those in southern Japan, the Near East and the subtropics of North America, but out-of-phase with those in the Indian peninsula and most of the northern part of the African continent.

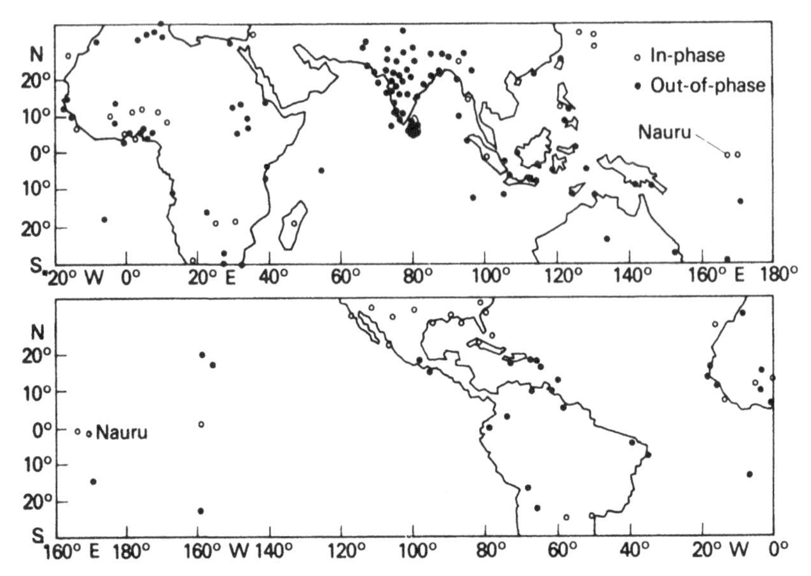

Fig. 1.10. Regional distribution of coherent relationships between rainfall anomalies; reference station Nauru.

The more detailed Fig. 1.11 shows the internally consistent region on the Indian peninsula, which is out-of-phase with Nauru. Along the southeast coast of India another discontinuity occurs, and there also exists an internally consistent area on Sri Lanka which is in-phase with Nauru.

1.6 Physical interpretation

A physical interpretation of such coincidence of rainfall anomalies in different parts of the tropics can be given in terms of a zonally-extended, thermally-driven Walker circulation, as first described by Bjerknes (1969).

In the tropics and subtropics the zonal temperature differences are not negligible in comparison with meridional differences. In the equatorial Pacific and Atlantic there exists a heat source on the western flank (the 'maritime continent' of Indonesia and the Amazon basin), together with a cooling area on the eastern side produced by upwelling off the western coast of South America and, to a lesser degree, off the west coast of Africa. A weak zonal thermal gradient exists also between the cool western Indian Ocean (including the area of upwelling off Somalia and southeast Arabia) and the unusually warm waters around Sumatra, which may create a Walker circulation, opposite to the circulations above the Pacific and Atlantic (Fig. 1.12; Flohn, 1975).

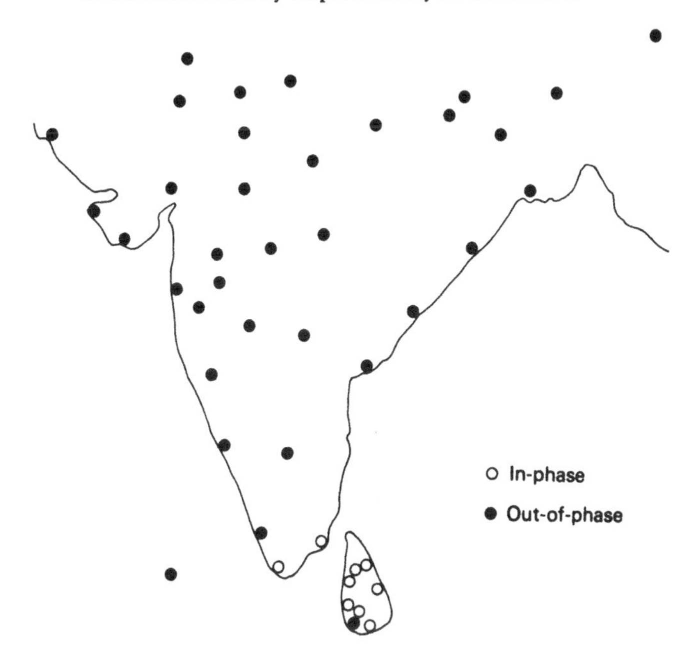

Fig. 1.11. Regional distribution of coherent relationships between Nauru and the Indian peninsula.

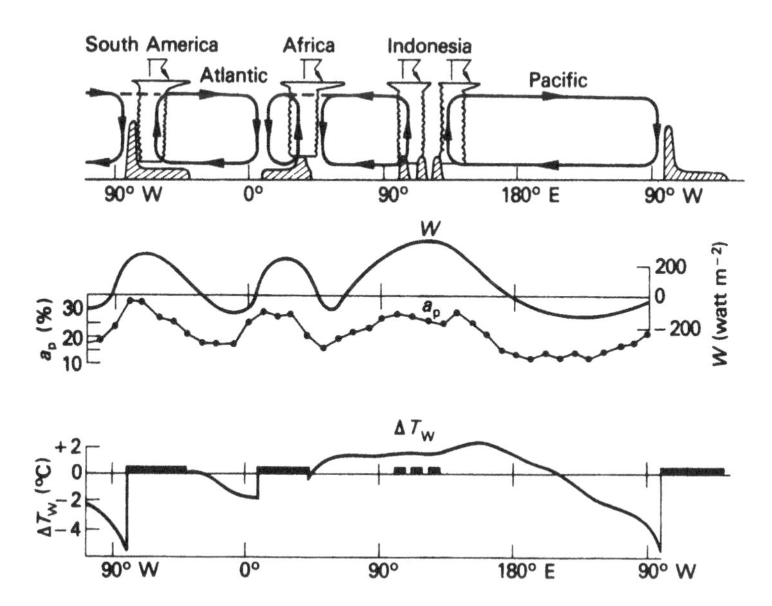

Fig. 1.12. Zonal ('Walker') circulation along the Equator. ΔT_W is the sea-surface temperature anomally in equatorial latitudes. W is the net heat input of an atmospheric column, and a_p is the planetary albedo.

1.7 Conclusions

Spectrum and band-pass-filter analyses verify that the occurrence of rainfall anomalies is not strictly periodic; instead it is quasi-periodic, with preferred frequencies at around 27 and 60 months, respectively. The amplitude of the 5-year oscillation in rainfall is not negligible in comparison with the amplitude of the annual cycle and is also more significant than the quasi-biannual pulse. Teleconnections across parts of the tropics and subtropics, possibly resulting from a Walker circulation, are clearly demonstrated.

References

Berlage, H. P. (1966) The southern oscillation and world weather. *Mededelingen en Verhandelingen*, Koninkijk, Nederlands Meteorologisch Institut, No. 88, 152 pp.

Bjerknes, J. (1969) Atmospheric teleconnections from the equatorial Pacific. *Mon. Wea. Rev.*, **97**, 163–72.

Doberitz, R. (1968) Kohärenzanalyse von Niederschlag und Wassertemperatur im tropischen Pazifischen Ozean. *Ber. Dtsch. Wetterdienstes*, **112**, cf. also, *Bonner Meteor. Abhandl.*, **8**, 62.

Doberitz, R. (1969) Cross spectrum and filter analysis of monthly rainfall and wind data in the tropical Atlantic region. *Bonner Meteor. Abhandl.*, **11**, 53.

Doberitz, R., Flohn, H. and Schütte, K. (1967) Statistical investigations of the climatic anomalies of the equatorial Pacific. *Bonner Meteor. Abhandl.*, **7**, 78.

Flohn, H. (1974) Instabilität und anthropogene Modifikation des Klimas. *Annalender Meteorologie* (New Series), **9**, 25–31.

Flohn, H. (1975) Tropische Zirkulationsformen im Lichte der Satellitenaufnahmen. *Bonner Meteor. Abhandl.*, **21**, 82.

Flohn, H. and Fleer, H. (1975) Climatic teleconnections with the equatorial Pacific and the role of ocean/atmosphere coupling. *Atmosphere*, **13**, 96–109.

Schütte, K. (1969) Untersuchungen zur Meteorologie und Klimatologie des El Nino Phänomens in Ecuador and Nordperu. *Bonner Meteor. Abhandl.*, **9**, 152.

Troup, A. J. (1965) The southern oscillation. *Quart. J. Roy. Meteor. Soc.*, **91**, 490–506.

Walker, G. (1924) Correlation in seasonal variations of weather: VIII; IX. *Mem. India Meteor. Dept.*, **24**, 75–131, 275–332.

2

Northern summer planetary-scale monsoons during drought and normal rainfall months

T. N. KRISHNAMURTI AND M. KANAMITSU

In this paper the 200 mb flow regimes during a drought year (1972) are contrasted with those during a normal rainfall year (1967) over the global tropics for the northern summer months. It is shown that the deficient rainfall over central India and western Africa during 1972 may be related to the following: (i) warm sea-surface temperatures over the equatorial Pacific; (ii) an excessive number of typhoon days over the equatorial Pacific; (iii) strong east-northeasterlies over the equatorial eastern Indian Ocean (related to upper-level outflows from the typhoons); (iv) a weaker tropical easterly jet; (v) a weaker meridional pressure gradient over India; (vi) a weaker Tibetan High; (vii) a south-eastward shift over the major circulation patterns as well as of several dynamical parameters; (viii) a weaker vertical wind shear and a weaker measure of the combined barotropic–baroclinic instability over west Africa; and (ix) weaker westward steering for rain-producing disturbances over India and a consequent stronger influence of the mountains.

A sequential interrelationship of the above aspects of the drought problem are discussed in this paper.

2.1 Introduction

The monsoonal rainfall over south Asia and west Africa undergoes interannual variations. Although some of the anomalous periods have more of a regional character, i.e., a period of drought over parts of a continent may not occur at the same period in other regions, there do exist periods of widespread drought that extend from west Africa to India. Such droughts occurred during the summers of 1877, 1899, 1918 and 1972. An examination of the circulations of the flow fields during different years has led us to believe that the planetary-scale circulations show considerable interannual variations. An attempt will be made to clarify the definition of the 'Planetary-scale monsoons'. Furthermore, circulation regimes during a drought year will be examined and contrasted with those for a 'normal' rainfall year.

The following are some of the important elements of the planetary-scale monsoons during the northern summer months for normal years that are characterized by a marked zonal asymmetry over the global tropics between land (around 80° E, i.e., the monsoonal region) and oceans (30° W and 170° W, i.e., the trade-wind belt) over a latitude belt between the Equator and 30° N.

(*a*) Surface pressure (land: low; oceans: high).

(*b*) Pressure in the upper troposphere (land: high; oceans: low).

(*c*) Lower-tropospheric zonal wind (land: westerlies; oceans: easterlies).

(*d*) Upper-tropospheric zonal wind (land: easterlies, oceans: westerlies).

(*e*) Lower-tropospheric meridional wind (land: southerlies, oceans: northerlies).

(*f*) Tropospheric mean temperature (land: warm; ocean: cold).

(*g*) Total moisture (land: humid; ocean: relatively dry).

(*h*) Rainfall (much larger over land than in the trade wind belts over the oceans).

Several of the above parameters were examined using standard space–time spectral-analysis techniques utilizing data sets for 1967 for the northern summer months. A large proportion of the total variance was found to be in the long waves (i.e., zonal wavenumbers 1 and 2) for most of these variables; furthermore, a large proportion of the total variance was accounted for by the standing- (or climatological-) wave components. The following are relevant references: Krishnamurti (1971*a* and *b*), Kanamitsu *et al.* (1972), Chang (1977), Kanamitsu and Krishnamurti (1978).

Divergent circulations. One of the prominent features of the tropical circulation is the divergent circulation on the planetary scale. The divergent part of the wind is defined as $\boldsymbol{v}_\chi = -\nabla\chi$ where χ is a velocity potential and is obtained from the observed motion field. The Hadley circulation can be defined by a zonal average of the meridional component of the divergent part of the wind, i.e.,

$$I_{\mathrm{H}} = -\oint \frac{\partial \chi}{\partial y} \mathrm{d}x. \tag{2.1}$$

The 'east–west circulation' can be defined by a meridional average of the zonal component of the divergent part of the wind, i.e.,

$$I_{\mathrm{E}} = -\int_{y_1}^{y_2} \frac{\partial \chi}{\partial x} \mathrm{d}y, \tag{2.2}$$

where the limits of integration include a tropical domain bounded by two latitude circles.

Fig. 2.1 shows the field of the velocity potential χ at 200 mb for the years 1967 and 1972. The streamlines of the divergent part of the flow field are illustrated in the same diagram. The time-averaged upper outflows from the monsoon belt have a planetary-scale character. This field implies rising motions in the monsoonal belt and sinking motions over the mid-oceans (the Atlantic and Pacific Oceans). It has, again, a large variance for the zonal wavenumbers 1 and 2. Fig. 2.1(*a*) is for the year 1967, a year of normal summer rainfall over south Asia and west Africa. Calculations of I_{H} and I_{E} show that the intensity of the Hadley and the east–west circulations are comparable. This implies that these east–west circulations may be an important aspect of tropical atmospheric dynamics. In order to explore this phenomenon, Krishnamurti, Kanamitsu *et al.* (1973) examined the generation of eddy kinetic energy by the east–west overturnings of mass. Calculations of vertical velocity show that warm air ascends over the land areas, and relatively colder air descends over the oceanic areas. This overturning has a pronounced influence on the energetics of this scale of motion. Thus, a large proportion of the eddy kinetic energy of the long waves (zonal wavenumbers 1 and 2) is accounted for by these overturnings, this also being confirmed by Tsay (1978). Kanamitsu and Krishnamurti (1978) have shown that a large proportion of the variance of the eddy kinetic energy of the long waves is furthermore accounted for by the quasi-stationary component (i.e., the standing eddies). The identifiable synoptic map features that define these long waves are: the Tibetan and Mexican Highs; the mid-Atlantic and mid-Pacific troughs; and a tropical easterly jet. The latter can be seen on most tropical maps spanning the monsoon belt from roughly the International Date Line to the Greenwich Meridian. Thus, an important role

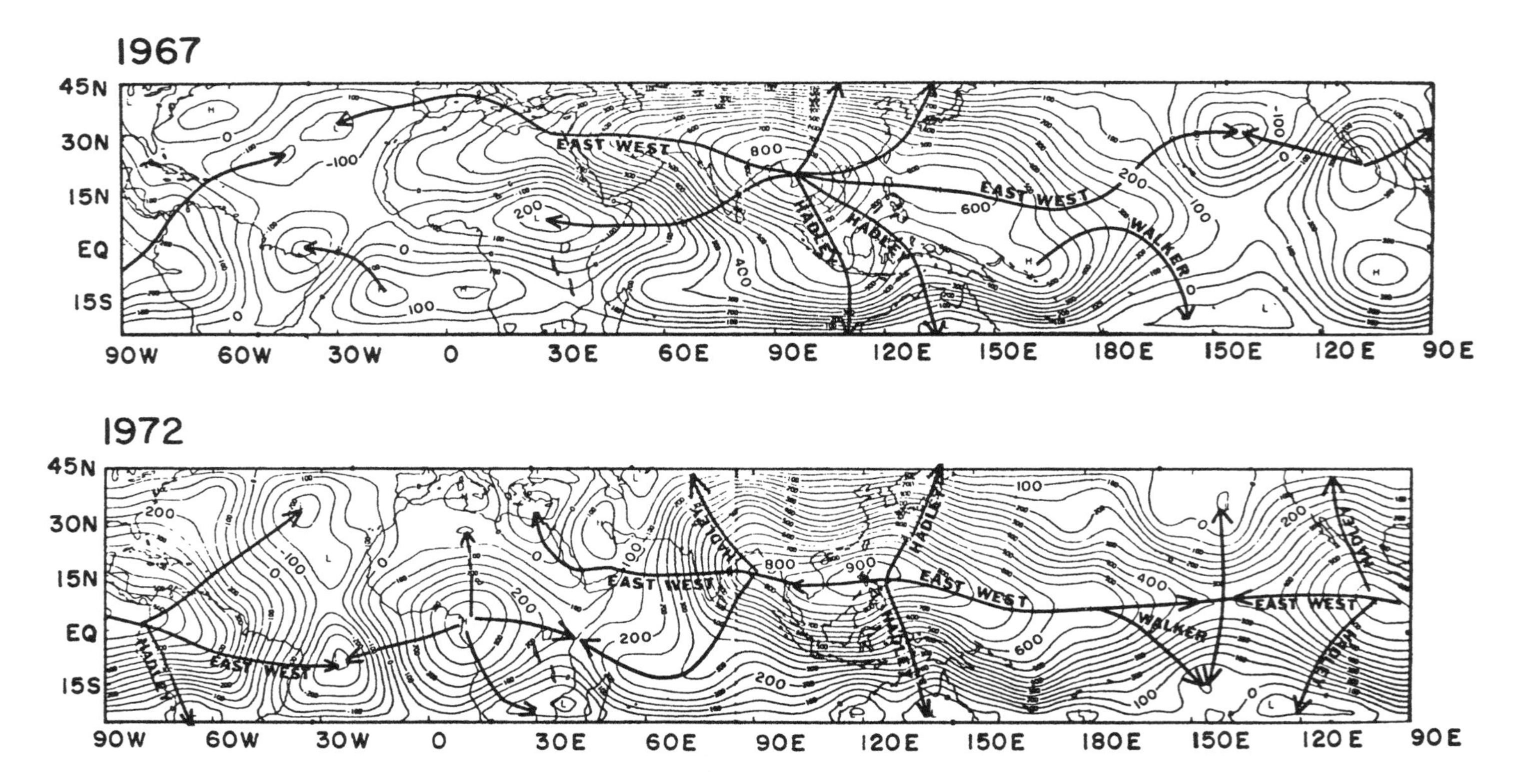

Fig. 2.1. Field of velocity potential based on the summer mean wind field at 200 mb for 1967 and 1972. Interval of analysis, $5 \times 10^5\ m^2\ s^{-1}$.

for the divergent east–west overturnings has been established, namely that they are a major source of energy for the planetary-scale quasi-stationary long waves of the tropical troposphere. For reasons not obviously apparent, the long waves have a very large tilt from southwest to northeast (Krishnamurti, 1971*a*). This feature is most striking between 5° N and 30° N during the normal monsoon months at 200 mb. A consequence of this tilt is the removal of westerly momentum poleward and away from the latitude of the tropical easterly jet located near 5° N. Viewed in terms of barotropic energetics, this implies 'barotropic stability' of the easterly jet with respect to these planetary-scale quasi-stationary long waves. The jet receives energy from the long waves which in turn are shown to receive energy from the east–west circulations. The zonal asymmetry between (i) land masses over Asia and Africa, and (ii) oceans (Atlantic and Pacific) provides a setting for an eventual differential heating that establishes itself on the scales of the land–ocean contrasts, i.e., zonal wavenumbers 1 and 2. In its fully-developed state the heating over monsoonal land areas is determined primarily by the release of latent heat, while the cooling over the mid-oceans is primarily a consequence of long-wave radiative flux divergence (Frank, 1970; Pellissier, 1972). This field of differential heating on the planetary scale is large near the Earth's surface as well as in the upper troposphere.

The evolution of the above-stated zonally-asymmetric parameters in the latitude belts between the Equator and 30° N during the northern summer months is here identified with the evolution (or onset) of the *planetary-scale monsoons*. The *definition* adopted here includes the asymmetries in the motion, temperature, humidity, divergent circulations, and differential heating.

At some time during the early part of June, the onset of the *planetary-scale monsoons* occurs. This is characterized by a sudden build-up in the variances of the zonal harmonics (wavenumbers 1 and 2). They account for roughly 70% of the total variance around 20° N.

Here, the evolution of the zonally asymmetric pressure field is included in the definition of the planetary-scale monsoons. In classical descriptions of the general circulation, the intensity of subtropical highs have long been related to the intensity of the Hadley cell (Lorenz, 1967). The observational studies of Newell *et al.* (1972) show that the intensity of the Hadley cell (of the northern hemisphere) is much stronger during northern winter than in the summer (see Fig. 2.2). However, when one examines the surface-pressure distribution (Fig. 2.2; Crutcher and Davis, 1969) it is apparent that the subtropical high is stronger during

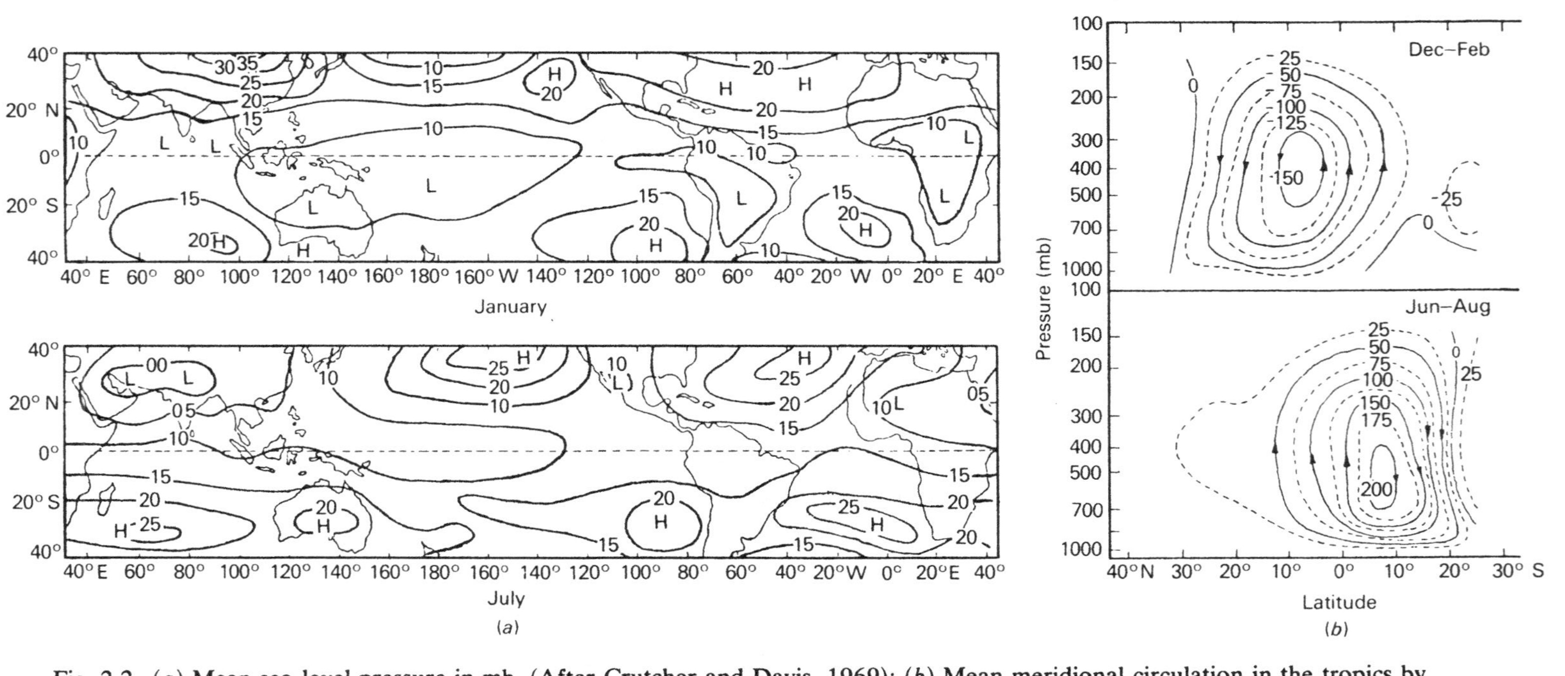

Fig. 2.2. (*a*) Mean sea-level pressure in mb. (After Crutcher and Davis, 1969); (*b*) Mean meridional circulation in the tropics by season, shown in the form of mass flow streamlines. Units, 10^{12} gm s^{-1}. (After Newell *et al.*, 1972.)

the northern summer than during the northern winter. Thus, it seems that the intensity of subtropical highs is related to the east–west as well as to the Hadley circulations in vertical planes. This is further supported by the studies of Krishnamurti (1971*b*) and Krishnamurti, Kanamitsu *et al.* (1973), where it is shown that the intensities of east–west circulations are somewhat stronger during the northern summer (normal monsoon years) than during the northern winter months.

Energetics analysis of monsoonal flows. Fig. 2.3 outlines the *gross-energetics of the planetary-scale monsoons.* The three components are (i) long waves (zonal wavenumbers 1 and 2), (ii) short waves (all other waves), and (iii) zonal flows. Fig. 2.3 illustrates what has been reviewed above. It also contains a link with the shorter waves. The shorter waves are maintained by three processes: (i) barotropic kinetic energy exchanges with longer waves via wave–wave interactions (Fig. 2.4); (ii) barotropic kinetic energy exchanges with the zonal flows via barotropic instability (Fig. 2.5); and (iii) heat released by cumulus convection in the smaller-scale disturbances. In the barotropic wave–wave interactions the long waves (planetary-scale monsoons) lose kinetic energy to other scales. The wave–zonal interactions also indicate that wavenumber 1 loses energy to the zonal flow; however, of interest here is the gain of energy from the zonal flows by several scales centred around wavenumber 8.

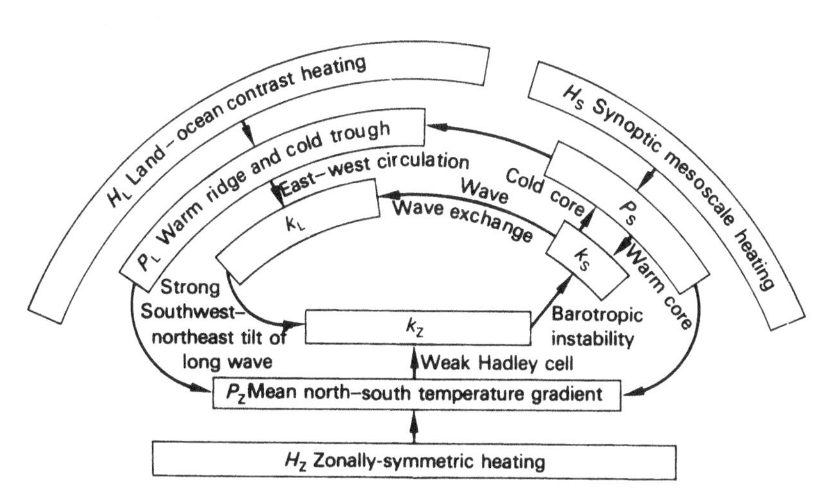

Fig. 2.3. A schematic outline of the energetics of the planetary-scale monsoons. Subscript L denotes the long waves, here identified as the planetary-scale monsoons. (After Krishnamurti, Daggupaty *et al.*, 1973.)

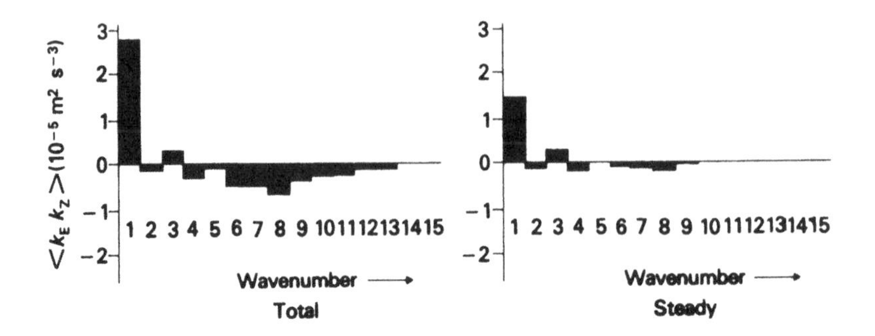

Fig. 2.4. Barotropic contributions to the wave–zonal (i.e. $\langle k_E k_Z \rangle$) energy exchanges during the northern summer, expressed as a function of zonal wavenumber.

Fig. 2.3 indicates potential energy exchanges among scales. Although the smaller-scale waves receive kinetic energy from longer waves as well as from the zonal flow, their contribution to the eddy available potential energy appears quite different. It has been known for some time from synoptic observations that the formation of an intense regional monsoon disturbance (such as a monsoon depression) leads to an intensification of the planetary-scale monsoons as well. This link was formally explored by Kanamitsu *et al.* (1977) and by Depradine (1978). They noted that the smaller-scale waves are a major source of available potential energy for the long waves and the zonal flow, while the long waves appear as the major source of kinetic energy for the other waves and the zonal flow. Thus the more intense smaller-scale disturbances intensify the planetary-scale monsoons by supplying available potential energy at the higher end of the wavenumber scale.

2.2 A drought year (1972) contrasted with a normal rainfall year (1967)

A brief review of some of the salient differences in the tropical circulations, rainfall, and sea-surface temperatures during the two years will be presented here. Some of the details of this comparison appear elsewhere (Kanamitsu and Krishnamurti, 1978). The year 1972 is well known for the major El Niño event off the coast of Peru. This relates to an absence of the well-known coastal upwelling and near-shore cold waters. During 1972 the water temperatures near the coast of Peru were roughly 10 °C warmer than in normal years. The El Niño events have been studied by many scientists, e.g., Wyrtki (1975) and Namias (1976). There probably

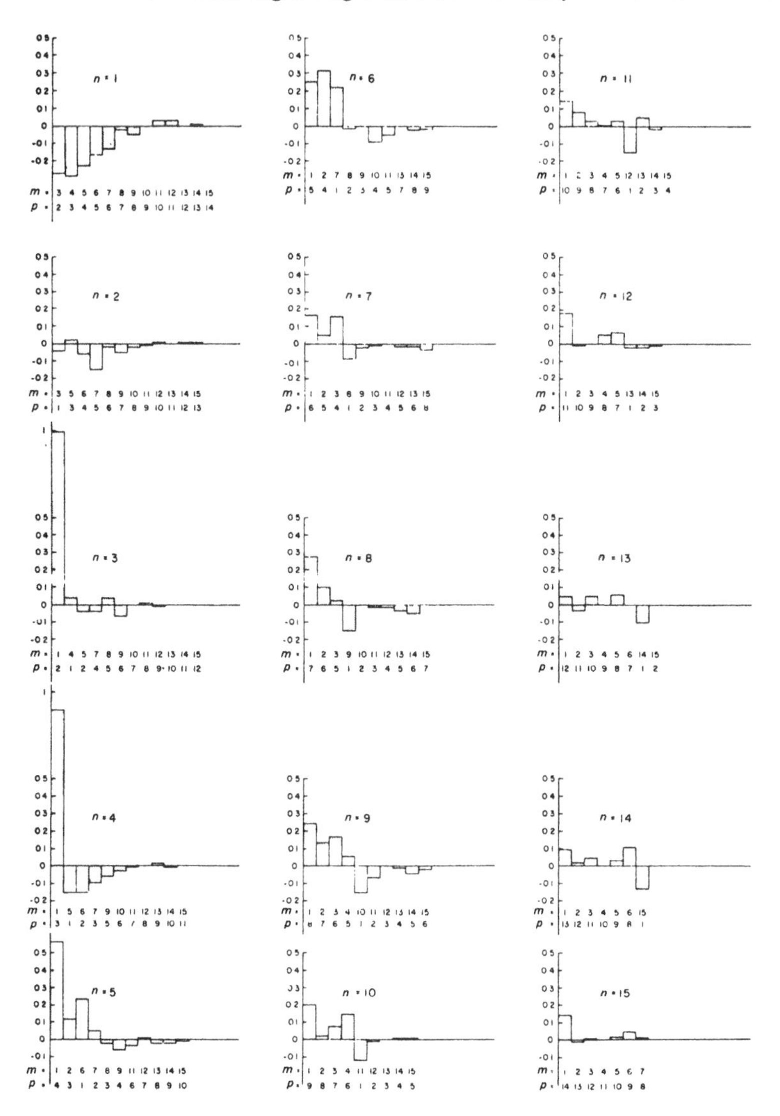

Fig. 2.5. Wave–wave energy exchanges between 15° N and 15° S. The gain of energy (along ordinate) by a wavenumber n due to its interactions with two other waves m and p indicated along the abscissa.

exist some remote teleconnections between the El Niño, southern oscillations, east–west circulations, Asian monsoons and the Pacific trade-wind systems. Walker (1923, 1924) and Berlage (1966) have shown that the long-term pressure variations at Jakarta are inversely correlated with the pressure changes at Easter Island in the eastern Pacific Ocean. This is an unexplained spectacular phenomenon whose period, according to Wyrtki, is of the order of 35 months (Fig. 2.6). It would be desirable to investigate any phase propagation that may be associated with the southern oscillations. Since the pressure rises and pressure falls would be associated with mass convergence and divergence fields, it is logical to assume that there must exist transient east–west circulations on the scales and periods of the southern oscillations. The mapping of such features of the east–west circulations remains to be carried out. The pressure oscillations over south Asia that are a part of the southern oscillation must in some way reflect the intensity of the monsoons (summer as well as winter). On the other hand, the pressure oscillations over the eastern Pacific that are a part of the southern oscillation must in some way reflect the intensity of the trade winds over the southern Pacific Ocean. It is only by exploring these teleconnections that one can bring into perspective the El Niño phenomenon in the tropical general circulation.

During 1972 the sea-surface temperatures over the central Pacific ocean were 2 to 3 °C warmer than in 1967. This is shown in Figs. 2.7*a*, *b* and *c*. The anomalous sea-surface temperatures were not restricted to the Peruvian coast. The reasons for this warming are not altogether clear. Starting with this feature, an attempt will be made to construct a sequential list of events that might provide a qualitative explanation for the anomalous rainfall during 1972. It should be recognized that this explanation is largely based on a phenomenological argument and is less formal. In a highly nonlinear global circulation problem of this complexity, even if one were able to provide a realistic simulation of the

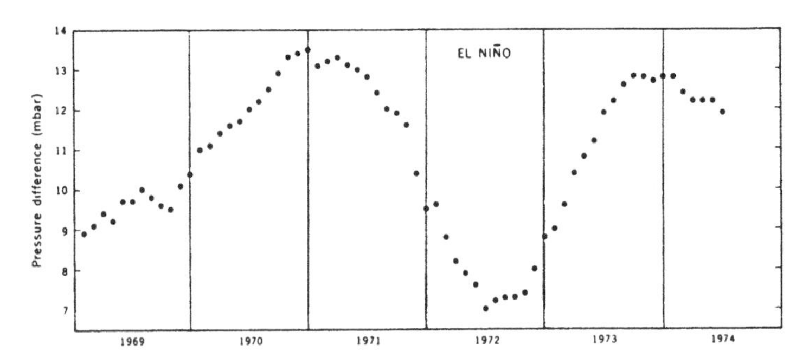

Fig. 2.6. Southern oscillation index for 1969–74 in relation to the 1972–3 El Niño and the 1975 prediction. The index is given by the 12-month running mean of the difference of atmospheric pressure between the Easter Islands and Darwin, Australia. (From Wyrtki, 1975.)

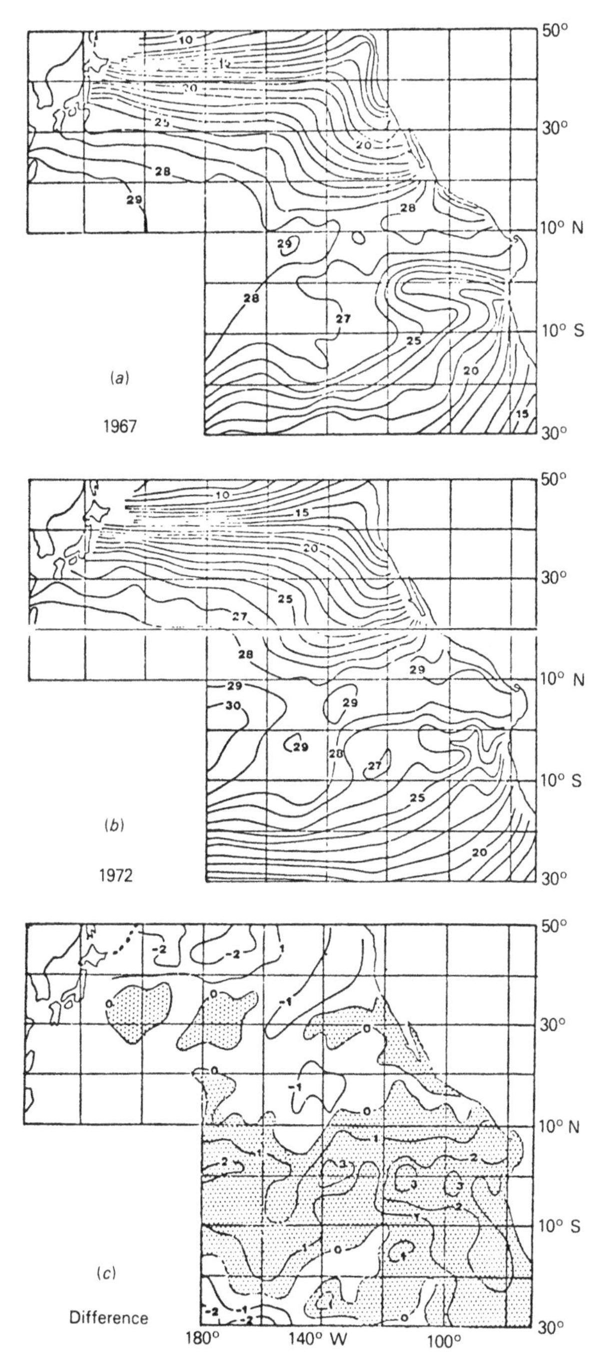

Fig. 2.7. Mean sea-surface temperature over the Pacific Ocean for summer seasons (June, July and August): (*a*) 1967; (*b*) 1972; and (*c*) the difference, i.e. 1972 minus 1967. Contours are labelled in degrees centigrade and the interval is 1 degree. In the difference chart, the area of positive values (areas warmer during 1972) are shaded. Over the western Pacific south of 10° N, reliable data are not available.

circulations, one is still left with the equally difficult task of providing an explanation.

In normal rainfall years there exists a preponderance of cold-cored tropical wave disturbances in the general region of the intertropical convergence zone (ITCZ) over the western Pacific Ocean (Reed and Recker, 1971). The year 1972 was characterized by a *large frequency of warm-core* disturbances, Chang and Miller (1977) attributing this feature to the anomalously warm sea-surface temperatures (SST) over the central Pacific Ocean. The warm-core disturbances generally contain more cumulus activity than the cold-cored easterly waves. The warmer SST is considered favourable for the occurrence of enhanced cumulus activity. The difference between a cold- and a warm-core disturbance may be regarded as resulting from mechanisms that cool, e.g., evaporative cooling of rain water, adiabatic ascent, radiative processes and those leading to heating, i.e. condensation and downdrafts around clouds. If a large fraction of a synoptic scale area is covered by deep cumulus convection, as in a hurricane, the warm core is more easily realized.

During 1972 a large number of these warm-core ITCZ disturbances became hurricanes. Statistics of what are called 'typhoon days' (one typhoon day counted as one typhoon per map per day) are regularly prepared by the Joint Typhoon Warning Center in Guam. For the months of June, July and August during 1972 there were 67 typhoon days contrasting with 23 days during 1967 (Kanamitsu and Krishnamurti, 1978).

Typhoons are usually associated with anticyclones in the upper troposphere. A high frequency of anticyclones was found during 1972. Their presence influences the climatology of the flows. Some of these aspects are next examined.

Non-divergent circulations. The rotational part of the wind is computed from the streamfunction field. This is illustrated for the two years in Fig. 2.8. It should be noted that the rotational part of the wind is about an order of magnitude larger than the divergent part. The fields for 1967 (normal) and 1972 (drought) are very similar for the most part. Some differences follow:

(i) The anticyclone to the northeast of Australia is very pronounced during 1972; also, large meridional flows from the north occur around this anticyclone, a substantial part of the meridional flow in this region being non-divergent.

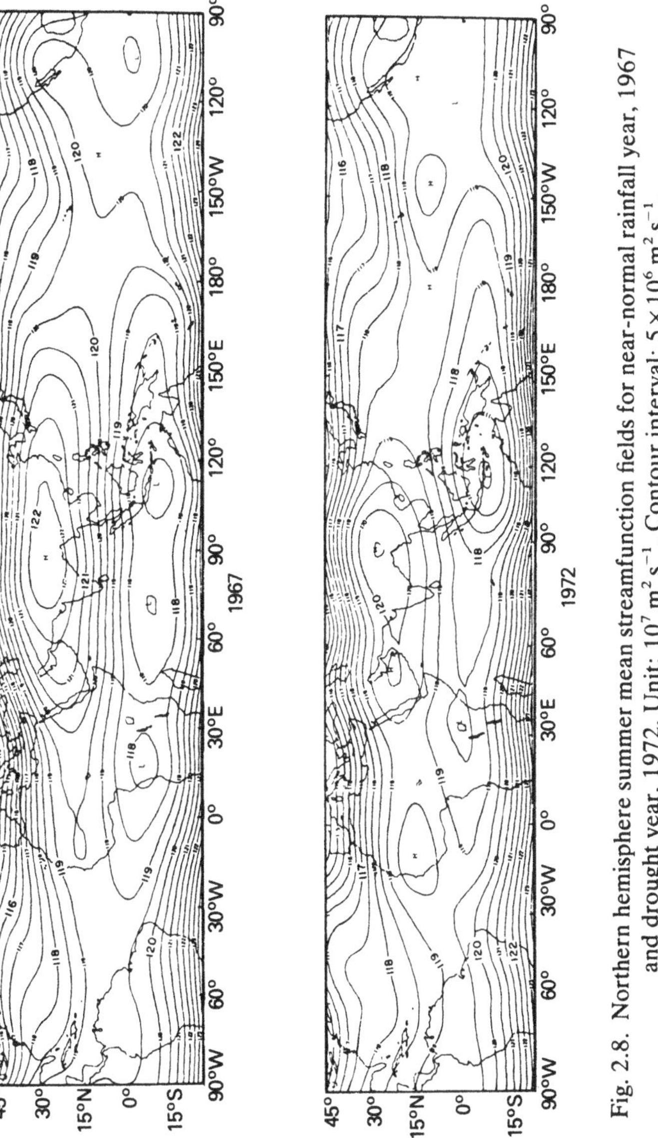

Fig. 2.8. Northern hemisphere summer mean streamfunction fields for near-normal rainfall year, 1967 and drought year, 1972. Unit: $10^7\ m^2\ s^{-1}$. Contour interval: $5 \times 10^6\ m^2\ s^{-1}$.

(ii) The easterlies associated with the tropical easterly jet are much weaker during 1972. This is quite clear from the spacing of the stream-function isopleths. This is evidently related to the much weaker Tibetan anticyclone during 1972.

Total horizontal motion field. The total motion field is divided into seasonal mean streamlines and isotachs and is shown in Figs. 2.9 and 2.10. The streamline geometry depicts the eastward shift of the Tibetan High during 1972 and also reflects many other features already noted. The speed field (Fig. 2.10, in m s^{-1}) is also similar to that of the zonal velocity. The major difference is the presence of a number of smaller-scale features in 1972 and a lack of large-scale organization in the broad-scale easterly belt along the tropical easterly jet. The diagrams of the zonal velocity, as well as those of the speed field, indicate clearly that the long waves (primarily wavenumber 1) near 10° N were very much more pronounced in 1967 (normal) than in 1972 (drought).

2.2.1 Differences in the scales of motion during the two summers

The scale resolution is depicted by estimates of percentage variances plotted as a function of the zonal wavenumber, and is based on zonal harmonic analysis of the u and v components for each day at each latitude during the two summer seasons. In addition, the scale resolution is also carried out for the summer mean-motion field to determine the character of the stationary disturbances.

Fig. 2.11 shows the results of a spectral analysis of the zonal wind for the two years. The dominant scale is found to be wavenumber 1 in the tropics south of 20° N. A very large proportion, between 50 and 90%, of the total variance for zonal wavenumber 1 is accounted for by the stationary part (darkened area) of the variance. As shown in several earlier studies (e.g., Krishnamurti, 1971*a*) the dominant zonal wavenumber 1 is associated with the broad belt of upper easterlies that extend from the Date Line to the western Atlantic ocean. As one proceeds northward towards 30° N, the dominant scale shifts to wavenumber 2, primarily due to the two relatively-strong belts of west-erlies ahead of the mid-oceanic troughs. Here again the stationary components account for roughly 50% of the total variance for this wavenumber. It is important to note that there does not seem to be much difference in the spectra of the zonal wind during the two summers, although it was noted earlier that the strength of the zonal flow (i.e.,

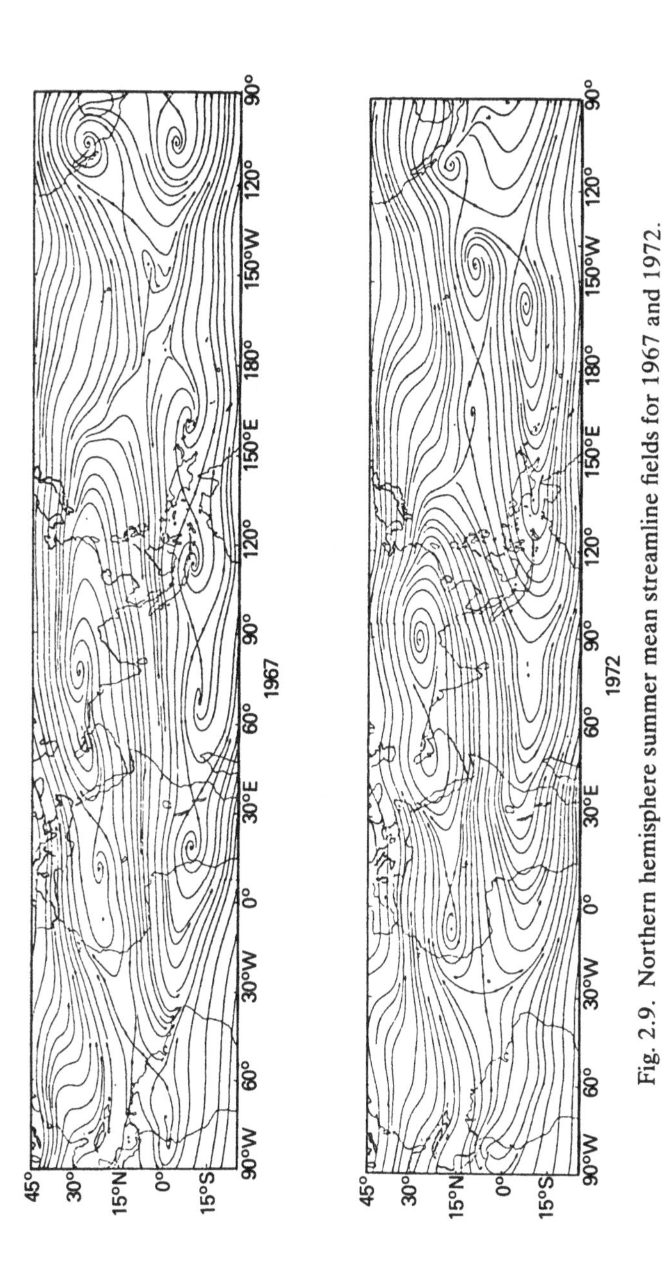

Fig. 2.9. Northern hemisphere summer mean streamline fields for 1967 and 1972.

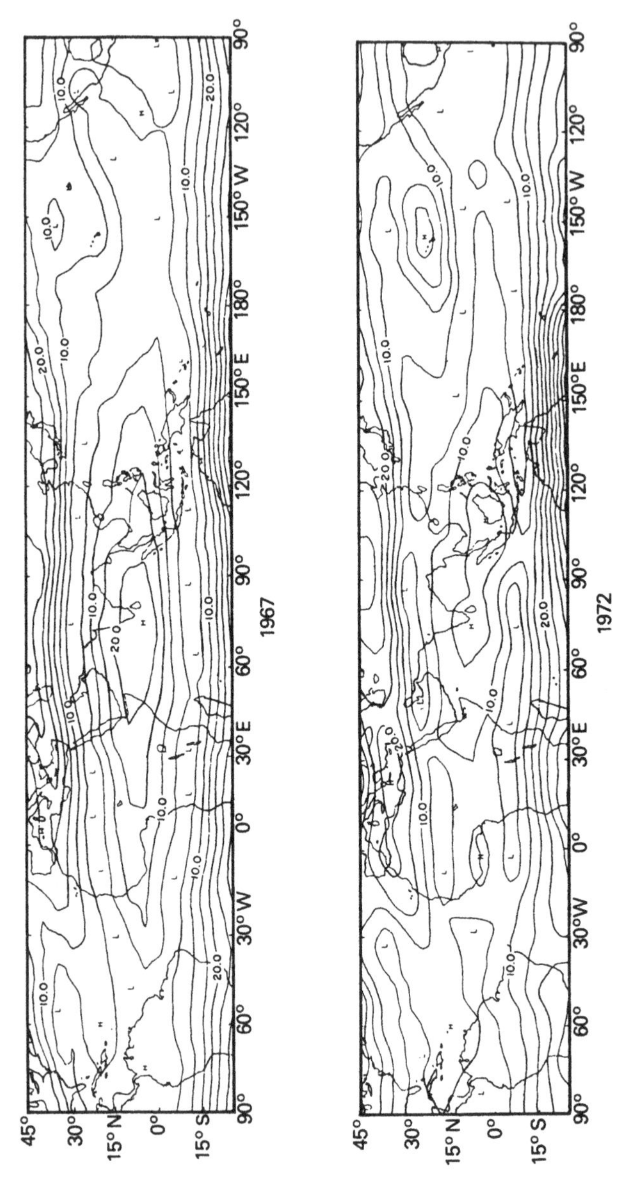

Fig. 2.10. Northern hemisphere summer mean isotach fields for 1967 and 1972. Unit: m s^{-1}. Contour interval: 5 m s^{-1}.

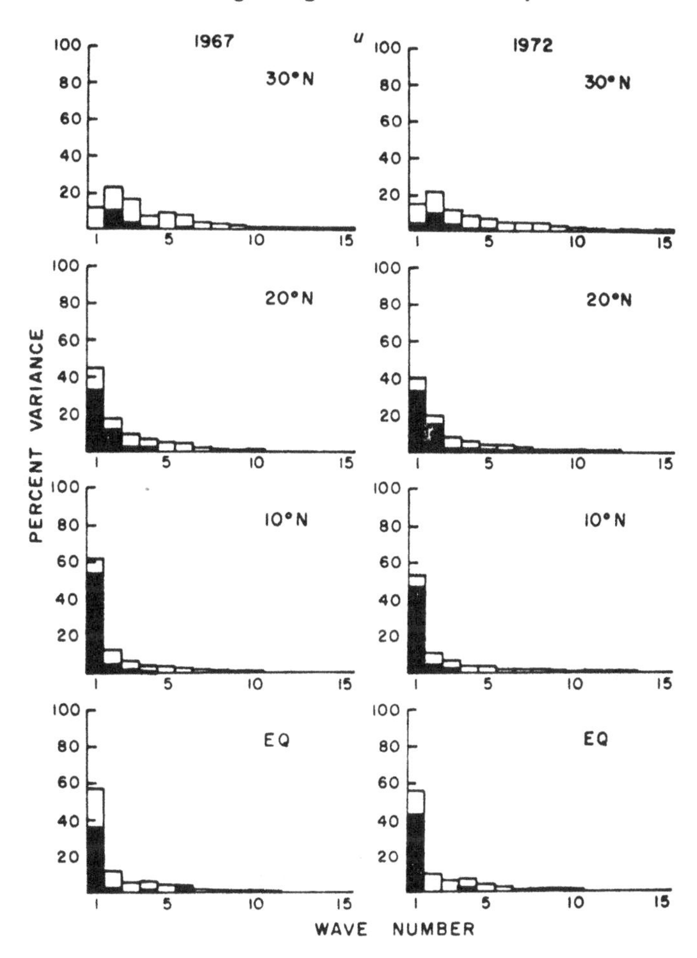

Fig. 2.11. Percentage variance of the zonal harmonics of the zonal wind spectra at selected latitude circles. The variance of the stationary part of the flows is darkened.

wavenumber 0) was stronger in 1967 (normal) than in 1972 (drought) in the tropical latitudes.

Fig. 2.12 shows the corresponding spectra for the v component. Here the spectra of the meridional wind are not peaked around the long waves as for the zonal flows. The overall spectra for the two years appear somewhat different. The primary difference seems to be in the relative role of the stationary components. At 30° N during 1967 (normal), a substantial proportion of the total variance is accounted for by the

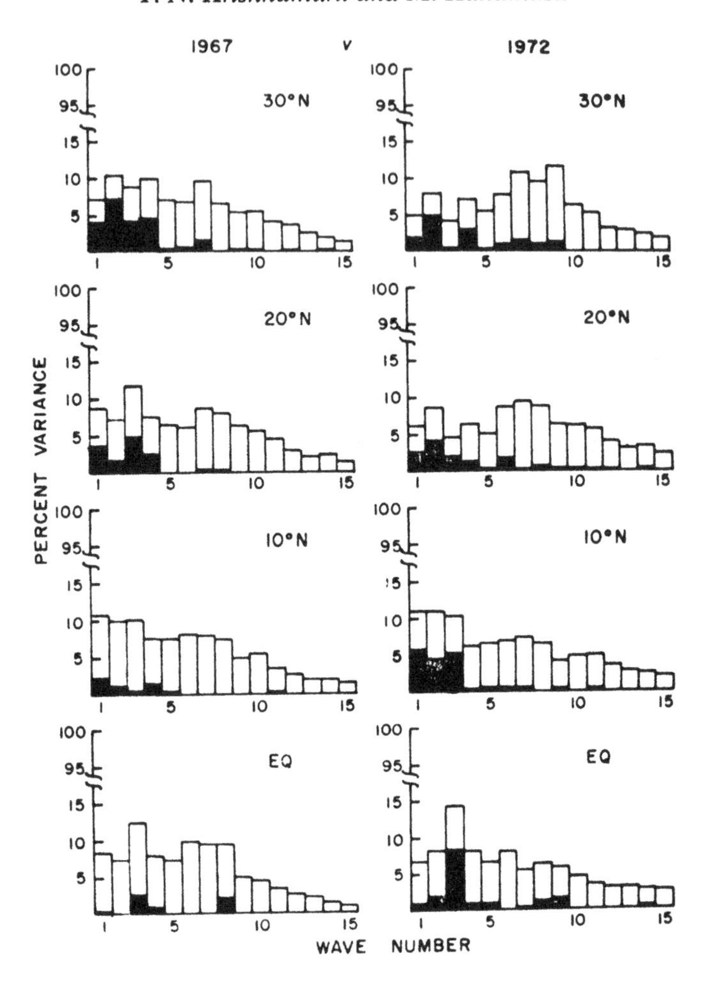

Fig. 2.12. Percentage variance of the zonal harmonics of the meridional wind spectra on selected latitude circles. The variance of the stationary part of the flows is darkened.

stationary component. The relative strength of the stationary component appears to decrease equatorward in 1967. The spectra of the meridional wind during 1972 appear quite different. A large proportion of the total variance of the long waves arises from the stationary component. The major difference clearly seems to be the dominance in the v -spectra of stationary long waves at 20° and 30° N during 1967 (normal) and at the Equator and 10° N during 1972 (drought).

2.2.2 *Cross-equatorial flows during the two summers*

An examination of the daily streamline charts from atlases (Krishnamurti and Rodgers, 1970; Krishnamurti *et al.*, 1975) reveals that the nature of the cross-equatorial flows was in fact different during the normal rainfall and the drought years. In particular, during 1972 a very striking quasi-stationary oscillation in the northerly flow near the Equator northeast of Australia was noted. This flow was associated with the circulations around a large anticyclone located near 140° E and 10° S. This strong northerly flow was conspicuously absent during 1967. In order to show the major differences in the nature of these cross-equatorial flows during the entire summer, time–zonal cross-sections of the meridional wind were prepared. These are shown in Figs. 2.13*a* and *b*. It should be noted that there is a net flux of mass directed (at 200 mb) from the northern to the southern hemisphere (since $\bar{v} < 0$). During 1967 there were three well-marked regions of northerly flows that appeared with some regularity at 40° W, 60° E, and 145° W. The intensity of the local meridional wind attained values as large as 20 ms^{-1}. The spectra of the meridional wind (shown in Fig. 2.12) at the Equator show that during 1967 the transient motions were a dominant part of the total motion field in most scales. The field of the cross-equatorial flows during the drought year presents another story. Fig. 2.13 shows the major quasi-stationary and persistent oscillations of the meridional wind at 100° E. Two other regions of relatively-weaker northerly flows appeared frequently around 20° W and 125° W. These three features provide a large variance for the stationary part of zonal wavenumber 3 of the meridional wind during 1972. The nature of the meridional flows near 100° E during 1972 and the dominant quasi-stationarity of wavenumber 3 appear to be striking features of the drought year. The question this raises is why there was a recurrent formation of upper anticyclones over the northeast of Australia during 1972. This may be related to the warm sea-surface temperatures.

2.2.3 *Some dynamical calculations contrasting the two summer regimes*

Zonally averaged quantities. Here the time average (overbar) of a zonally-averaged (square bracket) value of a parameter Q is denoted by the conventional notation $[\bar{Q}]$. Departure from the zonal average is denoted by an asterisk, and from the time average by a prime. The

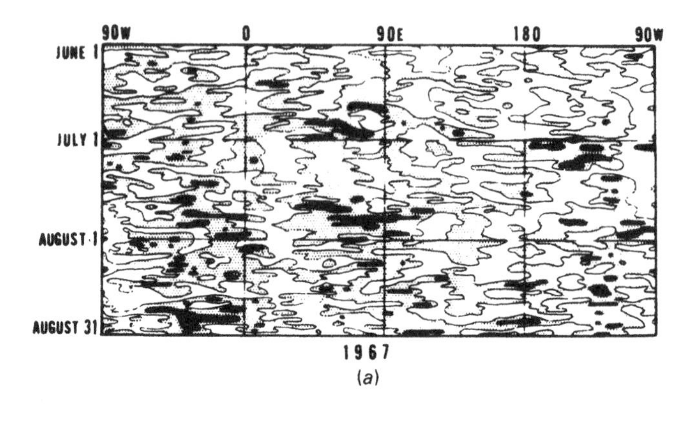

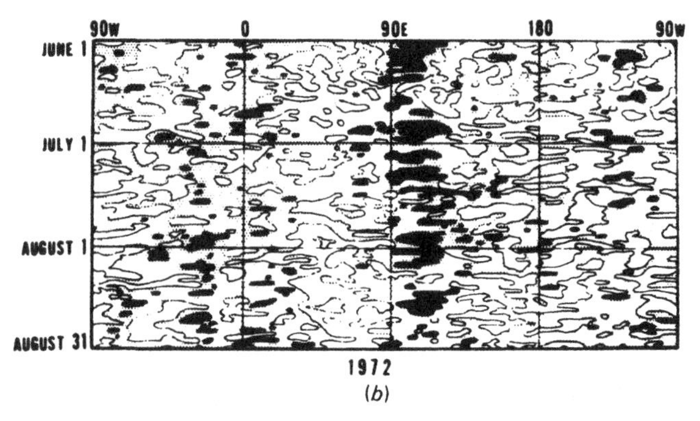

Fig. 2.13. Time–zonal cross-section of the meridional wind at the Equator during June, July and August of (*a*) 1967, and (*b*) 1972. Northerly winds are shaded and regions with $v < -10\ \mathrm{m\,s^{-1}}$ are stippled dark. The unshaded regions denote southerly flows. Note the persistent pulses of cross-equatorial northerly flows near 100° E in the cross-section for 1972.

meridional flux of momentum is examined following Lorenz (1967):

$$[\overline{uv}] = [\bar{u}][\bar{v}] + [\bar{u}^*\bar{v}^*] + [\overline{u'v'}]. \tag{2.3}$$

The three terms on the right-hand side of (2.3) are fluxes associated respectively with mean meridional circulation, standing eddies and transient eddies. Fig. 2.14 shows these terms for the years 1967 (normal) and 1972 (drought). The solid and dashed lines denote the results for the respective years. There are no significant differences of the fluxes of momentum by the standing and by the transient eddies between the two years.

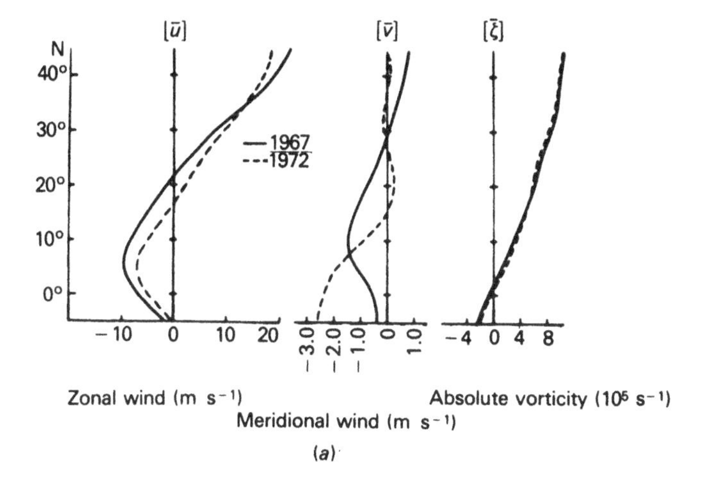

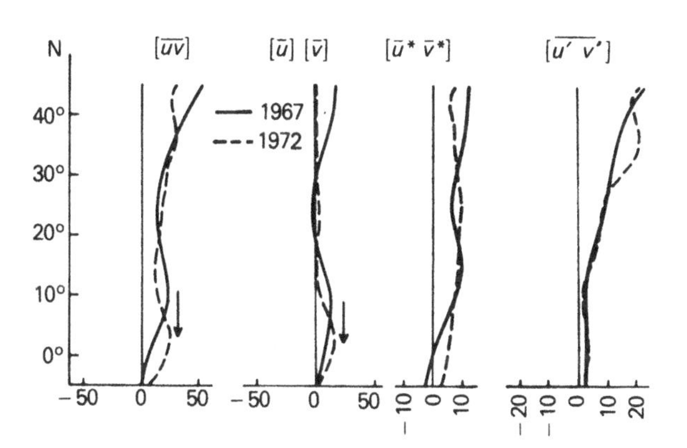

Fig. 2.14. (*a*) Latitudinal variations of zonally- and seasonally-averaged zonal wind, meridional wind and absolute vorticity for the near-normal rainfall year, 1967, and the drought year, 1972. (*b*) Latitudinal variation of meridional flux of momentum for the near-normal rainfall year, 1967, and the drought year, 1972. $[\overline{uv}]$ is total flux, $[\overline{u}][\overline{v}]$ is flux by meridional circulation, $[\overline{u^*}\ \overline{v^*}]$ is flux by standing eddies and $[\overline{u'v'}]$ is flux by transient eddies. Southward shifts of the profiles of the curves in the drought year, 1972, are indicated by arrows.

The major differences in the two curves for the total transport seem to arise due to an equatorward shift of the flow patterns by some 7° latitude. If the dashed line is moved bodily northward by 7°, then the two curves coincide. This equatorward shift is almost entirely due to the shift of maximum poleward flux of momentum by the mean meridional circulation as indicated. The meridional shift is also reflected in the resolution of these eddy fluxes as a function of zonal wavenumber; this will be discussed below.

Meridional flux of momentum as a function of scale. Here, the decomposition of the fluxes into their Fourier components is presented in Fig. 2.15.

At 10° N in 1967 (normal) the largest poleward fluxes of westerly momentum are carried by the Hadley cell (wavenumber 0) and by long

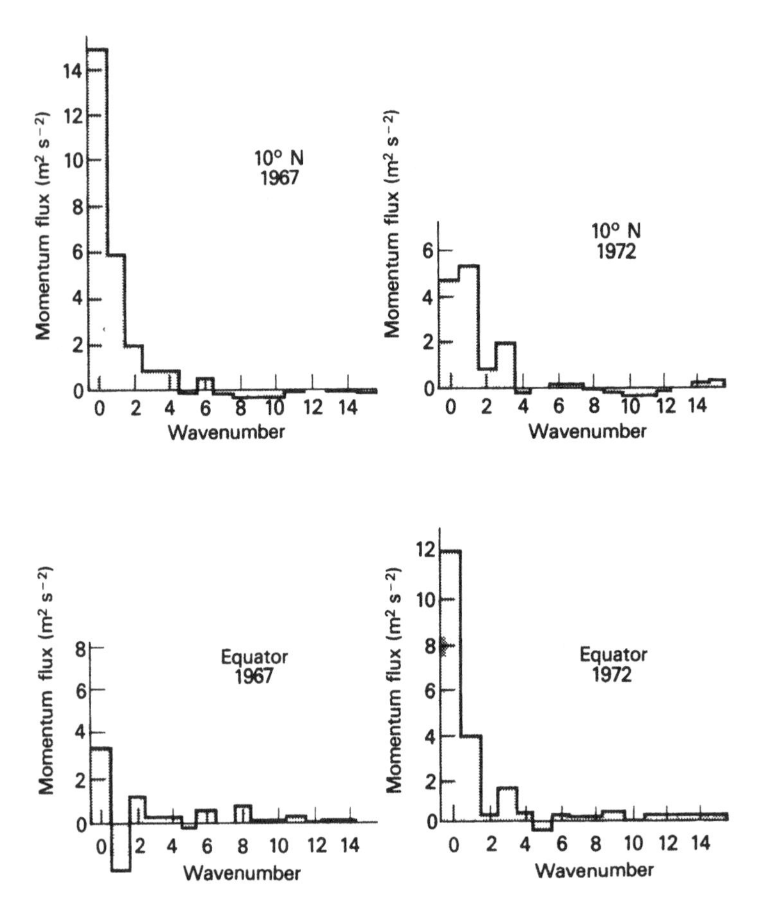

Fig. 2.15. Meridional flux of westerly momentum by various scales at 10° N and the Equator for near-normal rainfall year, 1967, and drought year, 1972. Wavenumber 0 denotes momentum flux by the mean meridional circulation.

waves (wavenumbers 1 and 2). The equatorward shift of this feature is quite apparent in this figure. In 1972, the large fluxes were at the Equator and were executed by the Hadley cell (wavenumber 0) and the long wave (wavenumber 1). Furthermore, the belt between the Equator and 10° N experienced a divergence of flux of westerly momentum during 1967 and a convergence of flux of westerly momentum during 1972. Thus the following argument can be offered for the weaker mean zonal easterlies in the belt between the Equator and 10° N during 1972. The equatorward shift of the major flow regimes results in a shift of the latitude of the largest poleward flux from 10° N to the Equator between these two years. Thus the belt of easterlies seems to encounter different configurations of convergence of momentum flux in these two years. The explanation for the weaker mean zonal easterlies during 1972 thus requires an understanding of this major shift of the circulation patterns. This question is examined in other sections.

Energy transformations. In this paper the nonlinear wave-zonal and wave–wave barotropic energetics following Saltzman (1957, 1970) and Kanamitsu *et al.* (1972) will be examined. The energy equations are not presented here.

The motivation for this exercise is that the interest is not in seeing whether there are any differences in the primary energy exchanges in the drought versus the normal rainfall year; some differences are to be expected since the flow patterns, as shown in this paper, were somewhat dissimilar. Here, the internal wave–wave and wave–zonal exchanges for the months June, July and August for 1967 (normal) and 1972 (drought) will be compared. Fig. 2.16*a* shows these energy exchange diagrams for 1967 in the tropical belt 15° S to 15° N. The planetary-scale monsoons will be compared in terms of the broad-scale zonal asymmetries of a number of parameters for a near-normal rainfall year. The major southeastward shift in the circulation patterns during 1972 and their anomalous character lead one to believe that major differences in energetics must exist during the drought year.

Fig. 2.16*b* illustrates the corresponding energy exchanges for the drought year, 1972. Here the most striking result is that the zonal flow loses energy to all other scales. The primary mechanism within the wave–zonal exchanges is described by the term

$$\langle K_Z K_E \rangle = \overline{[u]\partial[u^*v^*]_n/\partial y}, \qquad (2.4)$$

where n refers to a zonal harmonic, in this case 1 and 2, $[u]$ is the

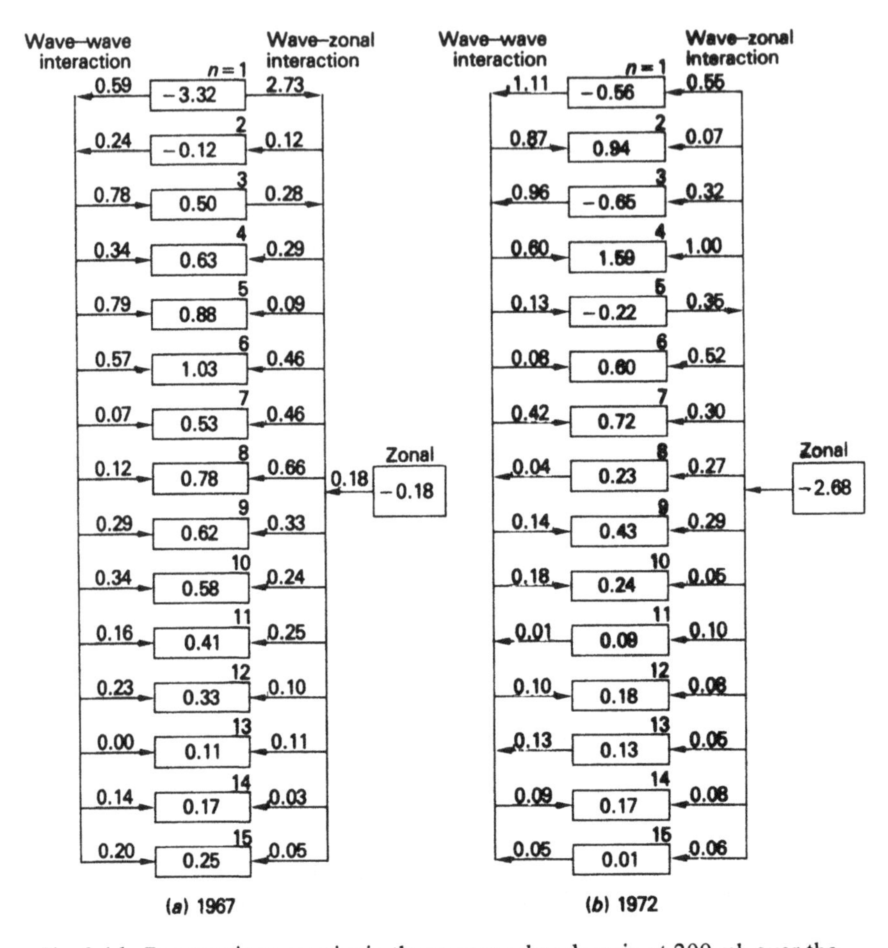

Fig. 2.16. Barotropic energetics in the wavenumber domain at 200 mb over the tropical global belt (15° S to 15° N) for two summer seasons (June, July and August): (*a*) 1967; and (*b*) 1972. The numbers in the boxes denote the gain or loss of energy for a wave (wavenumbers 1–15) due to interaction with other waves (denoted wave–wave interaction) and with zonal flow (denoted wave–zonal interaction). Units in 10^{-5} m^2 s^{-3}.

zonally-averaged zonal flow, and $\partial[u^*v^*]_n/\partial y$ is the divergence of flux of westerly momentum by a harmonic component n. Earlier, it was remarked that during 1972 $[\bar{u}]$ was more westerly than in 1967. Pronounced tilts of long waves were noted in both years in the belt 15° S to 15° N, together with a divergence of flux of westerly momentum (Fig. 2.14) for both years. Thus, the reversal in sign of the barotropic energy exchange in the belt 15° S to 15° N is attributable primarily to the weakening of the mean zonal easterlies during 1972. This feature of the barotropic energetics is somewhat similar to tropical winter-season exchanges noted by Krishnamurti, Kanamitsu *et al.* (1973) Thus, one might speculate that the 1972 summertime circulation did not develop fully into a strong easterly regime as in normal years. The wave–wave exchanges were largest from wavenumbers 1 to all other waves. Some irregular features seem to be present in these mutual wave–wave exchanges among the various other individual harmonic components. In order to synthesize this picture, a composite of results averaged over several harmonics is examined. Here it was decided to isolate zonal wavenumber 3 and group the other waves in the following manner: long waves *L*, wavenumbers 1 and 2; short waves S, wavenumbers 4 through 15. Wavenumber 3 was used as a frame of reference in the composite diagram in order to identify types of exchanges which might have contributed towards its maintenance and quasi-stationarity during 1972. As stated earlier, the spectrum of the meridional wind at the Equator had a pronounced peak around zonal wavenumber 3 and a large proportion of the total variance was accounted for by the standing wave (Fig. 2.12). This feature was absent during 1967 when there were no large amplitudes in the standing components. Fig. 2.17 presents three-component energy interactions for 1967 and 1972 using wavenumber 3 as a centre of reference.

In 1972 wavenumber 3 received energy from the zonal flows; however, it lost more energy to all other waves via wave–wave interactions. Hence, the maintenance of zonal wavenumber 3 must invoke boundary forcing or baroclinic effects. In view of the large number of typhoons in the vicinity of the region of strong meridional flows around 100° E, it seems that the baroclinic effects must have been important.

During 1967 wavenumber 3 lost energy to the zonal flow and received energy from all other waves via wave–wave interactions. Thus the maintenance of the transient wavenumber 3 may be attributed to wave–wave interactions.

2.3. Vector motion films

Vector motion format films, utilizing real data, have been one of the frequently used tools for studying tropical climate at Florida State University. The films use vector arrows to show Lagrangian motions during three-month periods. The horizontal motion field is bilinearly interpolated in space and time from analysed fields, and the arrows move along the horizontal motion fields at 200 mb, their direction always

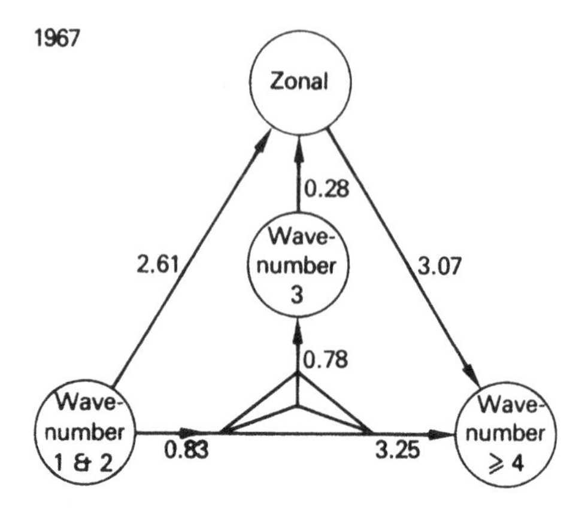

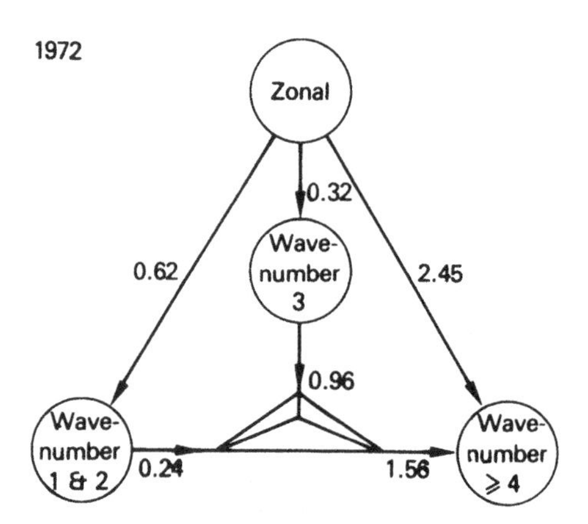

Fig. 2.17. Barotropic energetics between 4 different scales, i.e. zonal, long waves (wavenumbers 1 and 2), wavenumber 3 and shorter waves (other waves) obtained by averaging several wavenumbers. Units: $10^{-5}\ \mathrm{m^2\ s^{-3}}$.

matching that of the local velocity, and their length being proportional to the speed. The number of arrows is kept constant by assigning and reassigning their location to newer vectors as old vectors leave the domain of the film display. Special provision is made also to avoid excessive clustering or arrow-void regions. The films are designed to display the analysed streamline charts correctly at the viewing of the analysis periods. The films are more than just a curiosity, since they contain a history of the time evolution of phenomena that cannot be described in a few words. This format has been used to study many sequences of weather maps on a variety of phenomena from timescales of a few days to a decade. A summary of circulation features for 1967 and 1972 based on the vector motion films is as follows:

(i) The films clearly show that the easterly jet was better defined in 1967 than in 1972. During 1972 the flows along the jet exhibited many smaller scales.

(ii) The flows over west Africa during 1967 clearly showed a west African upper anticyclone (Krishnamurti, 1971a) while in 1972 the flows were broken into several vigorous eddies. Westerly flows over west Africa were closer to the Equator during 1972 and the easterly jet was much weaker.

(iii) Flows along the tropical easterly jet frequently turn towards the winter hemisphere and cross the Equator. During 1967 this feature was dominant over the eastern Atlantic Ocean, while in 1972 it was very persistent over the eastern Indian Ocean. The cross-equatorial flows show an intermittancy with frequent surges and weak periods.

(iv) A striking feature downstream of the easterly jet in the eastern Atlantic Ocean is the formation of near-equatorial large eddies whose scale is of the order of a few thousand kilometers with a lifetime of 2 to 3 days. This feature has been mapped by the authors for several years over the eastern Atlantic Ocean. This phenomenon was very conspicuous in 1967 and was totally absent during 1972 when the jet was weak. Recent studies of The Global Atmospheric Research Programme Atlantic Tropical Experiment (GATE) observations suggest that they form due to barotropic instability of the easterly jet. The counterclockwise eddies form on the cyclonic shear side of the easterly jet in the northern hemisphere and slowly drift towards the southern hemisphere subtropics. When they form in the northern hemisphere, they possess a low pressure at the centre; however, as they drift towards the subtropics of the southern hemisphere a pressure reversal to a high pressure must occur. The pressure adjustment to the flows presents an interesting problem.

(v) The films clearly show that the flows during 1972 were very much influenced by the upper anticyclone over the western Pacific Ocean. Its role was somewhat similar to that of blocking high-pressure cells that influence the circulations downstream from it. The southeastward shift of the large-scale semipermanent features during 1972 was apparent.

In conclusion, it may be stated that the results of dynamical calculations summarized here and also published elsewhere were, to a large extent, motivated by the events seen from the data analyses for the two years.

2.4 Concluding remarks

The anomalous tropical circulation during the summer of 1972 may be related to a sequence of events starting from the warm near-equatorial sea-surface temperatures over the eastern and central Pacific Ocean. The preponderance of warm-core ITCZ wave disturbances over the western Pacific and a larger than normal frequency of typhoons may be related to these warmer than normal sea-surface temperatures. A series of upper anticyclones are found over these typhoons. This large frequency of upper anticyclones over the western Pacific near 15° N, in effect, produces a shift in the anticyclone climatology of the tropical upper troposphere. The shift is towards the southeast. This quasi-permanent anticyclone is found to act as a blocking high at these upper levels. A consequence of this blocking is the manifestation of intermittent cross-equatorial pulses of meridional mass flux in the eastern Indian Ocean, a phenomenon which normally occurs in the eastern Atlantic Ocean. Furthermore, the blocking seems to weaken the high-level easterlies between the longitudes of the International Date Line and the Greenwich Meridian. The weaker easterlies seem to affect the storm tracks as well as the development of wave disturbances. The latter were examined from calculations of the necessary conditions for the existence of combined barotropic–baroclinic instability. This was not satisfied by the weaker zonal flows over west Africa.

Thus, this complex sequence of events might have been important for the anomalous circulation regimes and the widespread drought over the monsoonal regions during 1972.

This work was supported by NSF Grant No. ATM75-18945.

References

Berlage, H. P. (1966) The southern oscillation and world weather. *Mededelingen en Verhandelingen*, Koninkijk Nederlands Meteorologisch Institut, No. 88, 152 pp.

Chang, C. P. (1977) Some theoretical problems of the planetary scale monsoons. *Pure and Applied Geophysics*, **115**, 1089–110.

Chang, C. P. and Miller III, C. R. (1977) Comparison of easterly waves in the tropical Pacific during two contrasting periods of sea surface temperature anomalies. *J. Atmos. Sci.*, **34**, 615–28.

Crutcher, H. L. and Davis, O. M. (1969) *U.S. Navy Marine Climate Atlas of the World.* **VIII**, 'The World', NAVAIR 50-1C-54, Naval Weather Service Command, 1 March.

Depradine, C. (1978) An investigation of tropical large-scale wave motions during GATE, 1974. Ph.D. dissertation, Department of Meteorology, Florida State University, Tallahassee, Florida, 184 pp.

Frank, N. L. (1970) On the nature of upper tropospheric cold core cyclones over the tropical Atlantic. Ph.D. dissertation, Florida State University, Tallahassee, Florida, 248 pp.

Kanamitsu, M., Krishnamurti, T. N. and Depradine, C. (1972) On scale interactions in the tropics during northern summer. *J. Atmos. Sci.*, **29**, 698–706.

Kanamitsu, M., Krishnamurti, T. N. and Depradine, C. (1977) Monsoonal quasi-stationary ultralong waves of the tropical troposphere predicted by a real data prediction over a global tropical belt. *Pure and Applied Geophysics*, **115**, 1187–208.

Kanamitsu, M. and Krishnamurti, T. N. (1978) Northern summer tropical circulations during drought and normal rainfall months. *Mon. Wea. Rev.*, **106**, 331–47.

Krishnamurti, T. N. (1971*a*) Observational study of the tropical upper tropospheric motion field during the northern hemisphere summer. *J. Appl. Meteor.*, **10**, 1066–96.

Krishnamurti, T. N. (1971*b*) Tropical east–west circulations during northern summer. *J. Atmos. Sci.*, **28**, 1342–7.

Krishnamurti, T. N. and Rodgers, E. D. (1970) 200 mb wind field during June, July, August 1967. Technical report no. 70–2, Florida State University, Tallahassee, Florida.

Krishnamurti, T. N., Daggapaty, S. M., Fein, J., Kanamitsu, M. and Lee, J. D. (1973) Tibetan high and upper tropospheric tropical circulations during northern summer. *Bull. Amer. Meteor. Soc.*, **54**, 1234–49.

Krishnamurti, T. N., Kanamitsu, M., Koss, W. J. and Lee, J. D. (1973) Tropical east–west circulations during the northern winter. *J. Atmos. Sci.*, **30**, 780–7.

Krishnamurti, T. N., Astling, E. and Kanamitsu, M. (1975) 200 mb wind, June, July, and August, 1972. Department of Meteorology, Florida State University, 111 pp.

Lorenz, E. N. (1967) *The Nature and Theory of the General Circulation of the Atmosphere.* WMO, 161 pp.

Namias, J. (1976) Some statistical and synoptic characteristics associated with El Niño. *J. Phys. Oceanogr.*, **6**, 130–8.

Newell, R. E., Kidson, J. W., Vincent, D. G. and Boer, G. J. (1972) *The General Circulation of the Tropical Atmosphere and Interactions with Extra-Tropical Latitudes.* Vol. I, MIT Press, Cambridge, Mass., 258 pp.

Pelissier, J. M. (1972) A numerical model of tropical upper tropospheric cold core cyclones including parameterized infrared cooling. Proc. of conference on atmospheric radiation. American Meteorological Society, Boston, Mass., pp. 260–65.

Reed, R. J. and Recker, E. E. (1971) Structure and properties of synoptic scale wave disturbances in the equatorial western Pacific. *J. Atmos. Sci.*, **28**, 1117–33.

Saltzman, B. (1957) Equations governing the energetics of the large scales of atmospheric turbulence in the domain of wavenumber. *J. Meteor.*, **14**, 425–31.

Saltzman, B. (1970) Large-scale atmospheric energetics in the wavenumber domain. *Rev. Geophys.*, **8**, 289–302.

Tsay, C. (1978) Growth and decay of large-scale waves in the tropical upper troposphere during the northern summer. *J. Atmos. Sci.*, **36**, 24–31.

Walker, G. T. (1923) Correlation in seasonal variation of weather VIII: A preliminary study of world weather. *Memoirs of the India Meteorological Department*, Vol. 24, part 4, Calcutta, pp. 75–131.

Walker, G. T. (1924) Correlation in seasonal variation of weather IX: A further study of world weather. *Memoirs of the India Meteorological Department*, Vol. 24, part 9, Calcutta, pp. 275–332.

Wyrtki, K. (1975) El Niño – The dynamic response of the equatorial Pacific Ocean to atmospheric forcing. *J. Phys. Oceanogr.*, **5**, 572–84.

3

The annual oscillation of the tropospheric temperature in the northern hemisphere

R. K. VERMA AND D. R. SIKKA

Mean monthly temperature data at 850, 700, 500, 300, 200 and 100 mb from about 120 radiosonde stations over the northern hemisphere are subjected to harmonic analysis to study the annual oscillation of tropospheric temperature. Two main features stand out in this analysis. Firstly, over the tropics at individual stations double maxima occur – one at the lowest level, and the other at about 300 mb. Secondly, the amplitude of the annual oscillation is large over the tropics and subtropics with the maximum value appearing over the Asiatic monsoon region. There is a phase lag varying from 30 to 45 days between the two tropospheric maxima – the upper-tropospheric one occurring later and towards the end of July. The lower-tropospheric maximum is linked with the sensible heating. The physical forcing mechanisms of the upper-tropospheric maximum are discussed. The contribution of latent heating to the high amplitude of the upper-tropospheric maximum over the monsoon region is emphasized. A schematic model for the tropospheric heating and its link with the dynamics of the Asiatic summer monsoon is suggested.

3.1 Introduction

A study of Smagorinsky (1953) showed that atmospheric motions are affected not only by the zonally-asymmetric distribution of heating and cooling, but also by the major orographic influences. The large-scale circulation associated with the summer monsoon is described in the classical literature as the interaction of the annual cycle of solar radiation and the differential effective heat capacities of land masses and oceans. The most striking feature of the Asian summer monsoon is its annual periodicity. The annual cycle in the mean temperature accounts for a very large percentage of the variance in monthly mean temperatures, amply justifying the study of its structural features. Asnani and Mishra (1975) carried out a theoretical study of the annual temperature cycle in a diabatic heating model of the monsoon over the Asiatic region. The circulation in their model was forced by two heat sources in the vertical – one at the ground and the other in the midtroposphere. Webster (1972) carried out a theoretical analysis of the response of the tropical atmosphere to local, steady forcing considering latent heating as the dominating source, and a further such study is described in this volume (Chapter 10). Dickinson (1971) also showed the importance of the thermal source in maintaining the zonal wind distribution in the tropics, the heat source, having a relatively sharp maximum at the equator, being associated with latent heat release in the tropical rain belt.

Observational studies of the large-scale annual oscillation have been reported by many authors. For the zonally-averaged temperature field, Oort and Rasmusson (1971) showed the distribution of the amplitudes and phases of the annual oscillation at different pressure levels. Annual thickness oscillations of the 1000 to 850 mb and 1000 to 300 mb layers were described by Asnani and Verma (1975). The present study attempts to bring out the observational characteristics of the annual temperature oscillation in the troposphere. Amplitudes and phases of this oscillation at individual stations are first examined. Zonal asymmetry in these parameters is also described.

3.2 Data sources and analysis procedures

About 120 stations over the northern hemisphere, fairly evenly distributed over latitudinal belts between the Equator and 70° N were selected. Mean monthly temperatures at 850, 700, 500, 300, 200 and 100 mb over these stations are taken from the table published by the World Meteorological Organization (1965). The data are subjected to harmonic analysis to obtain the amplitude and phase of the first harmonic at different levels for each station.

Amplitudes and phases are plotted for all stations at each of the six levels and hand-analysed to ascertain horizontal variations. The zero of the phase angle is counted from mid-January since the data consist of normals commencing in January.

3.3 Results and discussion

3.3.1 Vertical variation of amplitude of the annual oscillation in temperature at individual stations over the northern hemisphere

As the main purpose of the study is to describe the annual cycle of the large-scale heating and investigate its link with the Asiatic summer monsoon as well as its implications for the large-scale atmospheric behaviour in other regions, the individual stations were grouped into the following three belts:

(i) eastern tropical belt (Equator to 30° N, 0° to 180° E);
(ii) western tropical belt (Equator to 30° N, 0° to 180° W);
(iii) stations outside the tropical belt.

Vertical variations in amplitude of the annual oscillation in temperature at selected stations, in these belts, are shown in Figs. 3.1, 3.2 and 3.3.

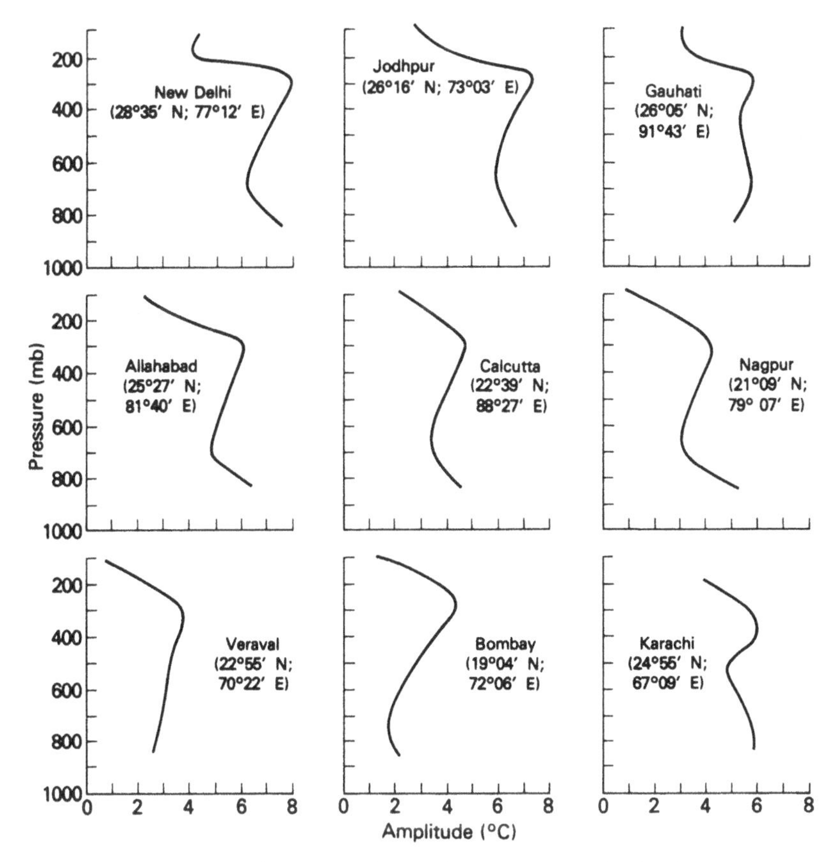

Fig. 3.1. Vertical variation of amplitude of the annual oscillation in temperature at selected stations over the east tropical belt.

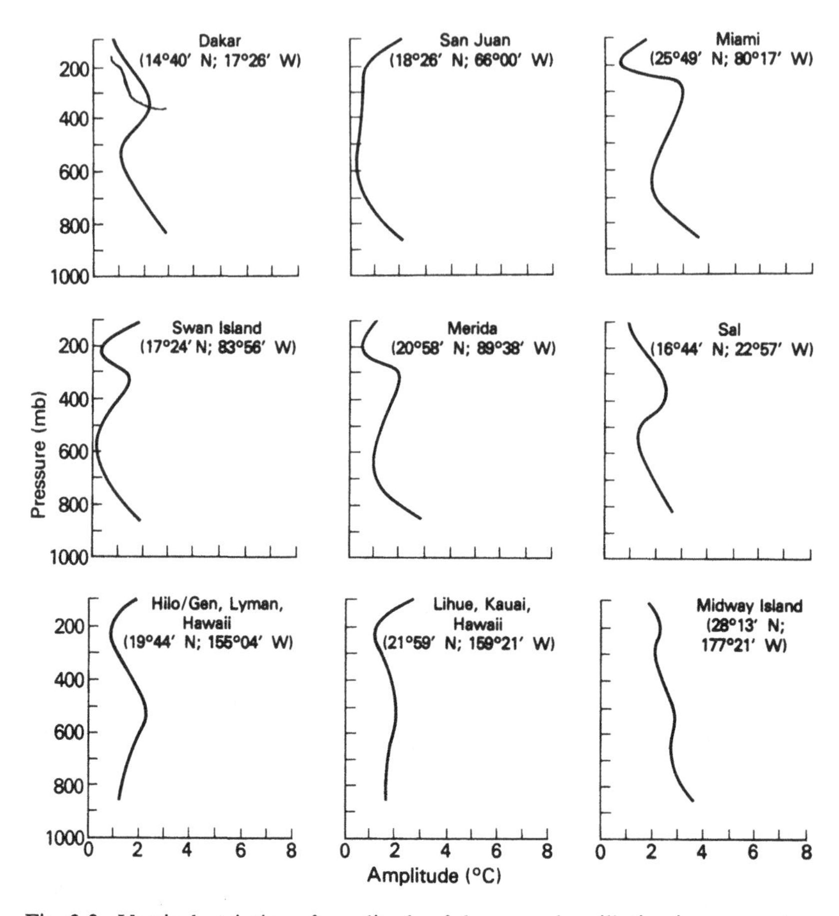

Fig. 3.2. Vertical variation of amplitude of the annual oscillation in temperature at selected stations over the west tropical belt.

(i) Fig. 3.1 refers mainly to the data for Indian stations. These stations show characteristic double maxima in the vertical, one in the lower troposphere near the surface and the other between 400 and 300 mb. Banerjee and Sharma (1967) also showed the presence of double maxima for Indian stations.

The amplitudes of the upper maxima are between 6 and 8 °C. There is a phase lag of about 30 to 40 days between the two maxima – the upper one occurring later and towards the end of July or the beginning of August. The lower-tropospheric maximum is obviously linked with the sensible heating. The primary physical factor leading to the second maximum between 400 and 300 mb is not so apparent. It is possible that this may

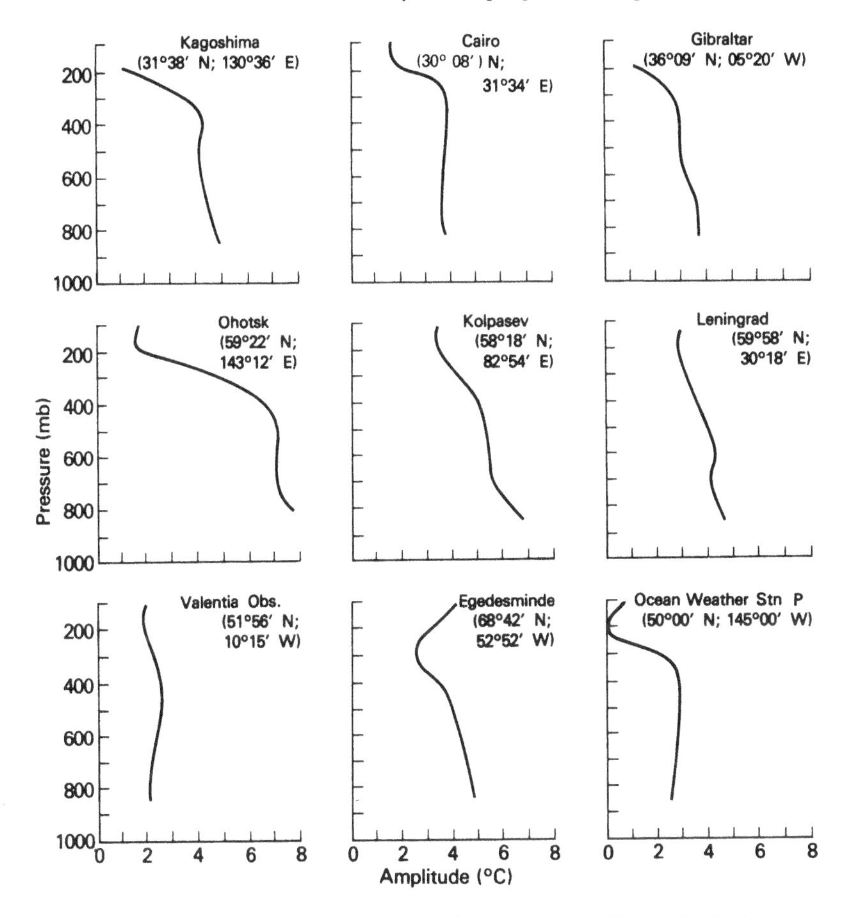

Fig. 3.3. Vertical variation of amplitude of the annual oscillation in the temperature at selected stations over the extra tropical belt.

be the abundant latent heat release in deep cumulus convection in midsummer over the monsoon trough region.

(ii) Fig. 3.2 refers to the stations from the western tropical belt. The amplitudes of the lower- and upper-tropospheric maxima are significantly less than those shown in Fig. 3.1.

(iii) Fig. 3.3 refers to stations outside the tropics. The first three are in the subtropics and show almost constant amplitudes of about 8 °C from 850 to 300 mb and a sharp fall from 300 to 200 mb. The other six stations are in higher latitudes (between 50° and 70° N), three in each of the eastern and western hemispheres. The stations in the eastern hemisphere are characterized by large amplitudes (between 12 °C and 18 °C) in the lower troposphere, decreasing to a minimum of about 4 °C at 200 mb. Such large annual variations occur in land areas affected by both polar and tropical air masses. The particular stations selected in the western hemisphere have somewhat smaller amplitudes and vertical profiles which are more or less similar to those of subtropical stations.

3.3.2 Horizontal distribution of the amplitudes of the annual oscillation in temperature over the northern hemisphere

These are shown in Figs. 3.4 and 3.5 for 850, 700, 500, 300, 200 and 100 mb. The following are the main features:

(i) At all levels, amplitudes near the Equator are very small (less than 1 °C).

(ii) In the lower troposphere, the amplitudes are large over Eurasia and North America between 50 and 60° N, reflecting large temperature

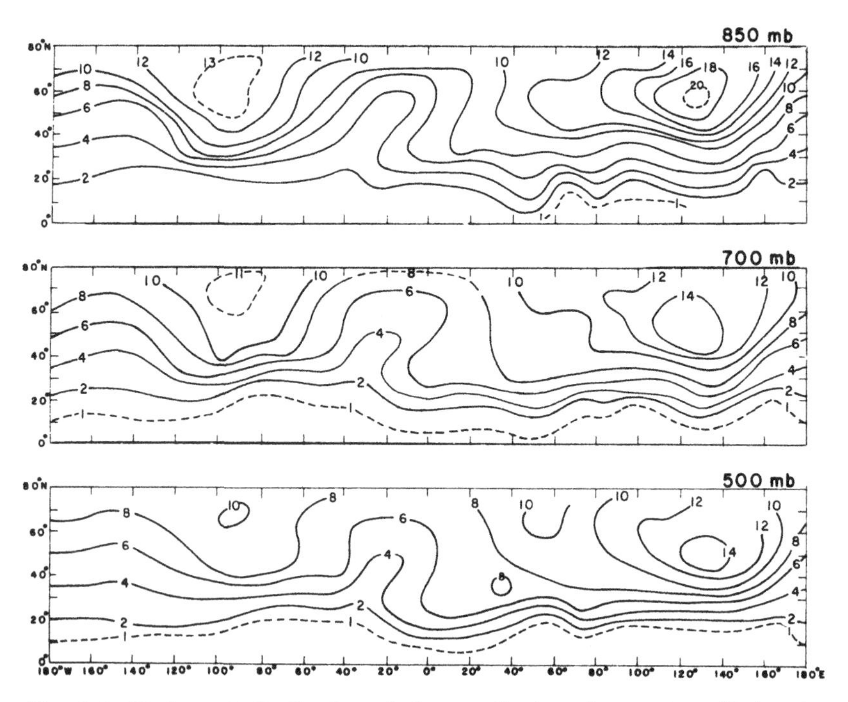

Fig. 3.4. Horizontal distribution of the amplitudes of annual oscillation in temperature over the northern hemisphere at the 850, 700 and 500 mb levels. Contours are labelled in °C.

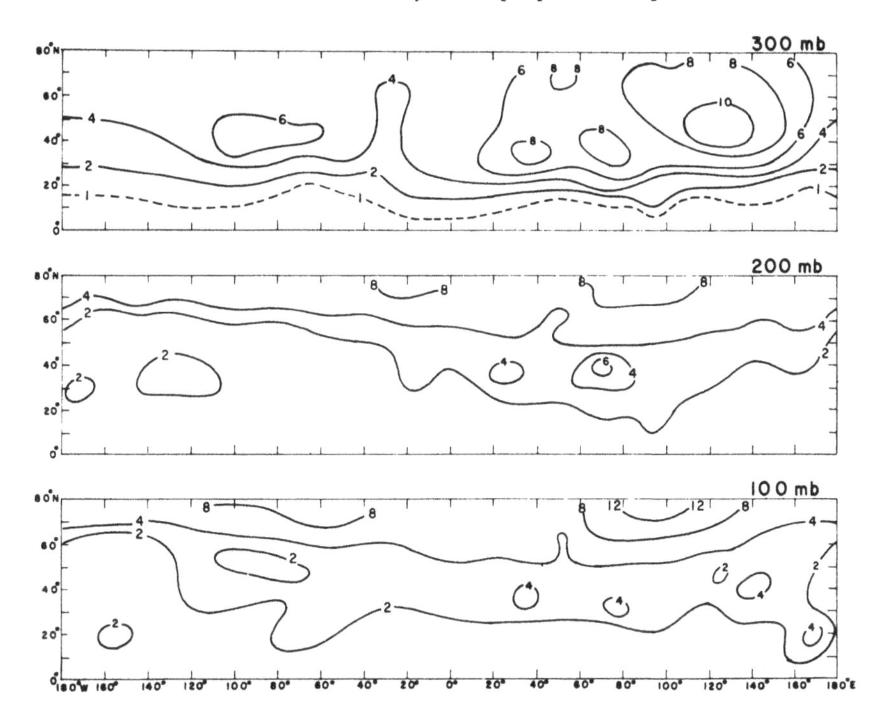

Fig. 3.5. Horizontal distribution of the amplitudes of annual oscillation in temperature over the northern hemisphere at the 300, 200 and 100 mb levels. Contours are labelled in °C.

contrasts between summer and winter. The isopleths clearly bring out the contrasting response of the air in the lower troposphere over the large land masses of Eurasia and North America on the one hand, and the Atlantic and Pacific oceans on the other.

(iii) Over the subtropics and midlatitudes, amplitudes generally decrease from 500 to 100 mb. However, the behaviour is somewhat different in the region 30 to 45° N and 40 to 120° E, where the amplitudes are rather large; between 500 and 300 mb (~8 °C). There is also a relative maximum at 200 mb. The large-scale horizontal distribution of amplitudes indicates the prevalence of two maxima (over the continents) and two minima (over the oceans) in the lower- and midtroposphere. At 200 mb the zonal asymmetry in the amplitude is characterized by a maximum over the region of the Tibetan anticyclone and a minimum in the mid-Pacific ocean.

3.3.3 *Spatial distribution of phases*

The phase indicates the time of occurrence of maximum temperature during the annual cycle. The horizontal and vertical distributions of the phases of the annual oscillation in temperature over the northern hemisphere are shown in Fig. 3.6.

(i) At 850 mb and 700 mb the distributions of phases are almost identical, hence only the 850 mb phases are shown. Over the land masses of Eurasia and North America, outside the tropics, the temperature maxima of the annual oscillation occur towards the end of July. The tropical land masses show somewhat different behaviour. Over the tropical belt (10 to 30° N) of North America and northern Africa, the phases lie between 10 and 20 July. The maxima for the Indian and south Asian regions occur between 10 and 20 June. The early occurrence (about 10 June) of the maximum in the annual temperature wave over peninsular India and the south Asian region is associated with the onset of the moist summer monsoon and its cloudiness and rainfall. In comparison

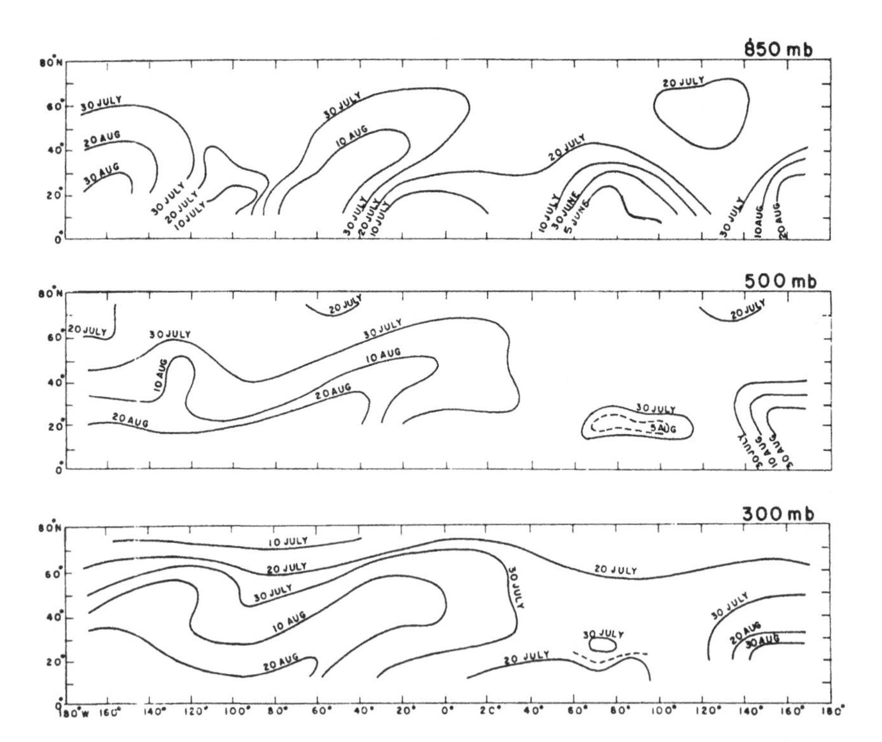

Fig. 3.6. Horizontal distribution of the phases of annual oscillation in temperature over the northern hemisphere at the 850, 500 and 300 mb levels.

with the continental stations, the oceanic parts of the tropical Atlantic and the Pacific show maxima in the temperature much later (10 to 20 August). The large difference (40 to 45 days) in the time of occurrence of the maximum temperature between continental and the Atlantic and Pacific oceanic regions to some extent reflects the slower response of the ocean to the solar heating than that of continental land masses. Monsoonal influences over the Indian region are found to result in a time of maximum temperature even earlier than that over other continental regions at the same latitude.

(ii) At 500 mb, there is again a distinct difference in the same latitudinal belt between the phases in the south Asian monsoon region and the North American region on the one hand and the oceanic regions on the other. The maximum in the temperature over the south Asian monsoon region occurs at the end of July or beginning of August (30 July to 5 August) and is retarded by about 45 to 50 days relative to the lower-tropospheric maximum. Over the North American tropical belt the 500 mb maximum temperature occurs by the middle of August and is retarded by about 30 days with respect to the lower-tropospheric maximum. Over the oceanic belt there is no appreciable lag between the lower-tropospheric and midtropospheric temperature.

(iii) At 300 and 200 mb, the phases over the Asiatic monsoon region indicate a reversal in the time of occurrence of maximum heating in the sense that these occur about 10 days earlier than at the layers immediately below. In the western counterpart there is hardly any phase reversal. Oort and Rasmusson's (1971) phase diagram of the annual oscillation in the zonal mean temperature indicates a similar phase reversal between 300 and 200 mb over the tropical belt. However, the horizontal distributions presented here bring out the longitudinal asymmetry more clearly.

(iv) At 100 mb, the phases, which relate to the lower stratosphere, are complicated and are not presented here.

(v) In equatorial regions, the annual temperature variation is of very small amplitude at all levels. The phases show large variations and are not reproduced here.

3.3.4 Asymmetries between the eastern and western hemispheres

Amplitudes are picked up at 10-degree longitude–latitude grid-points from the analyses, and zonal averages are calculated at 10, 20, 30, 40, 50, 60 and 70 degree latitudes for each of the eastern and western

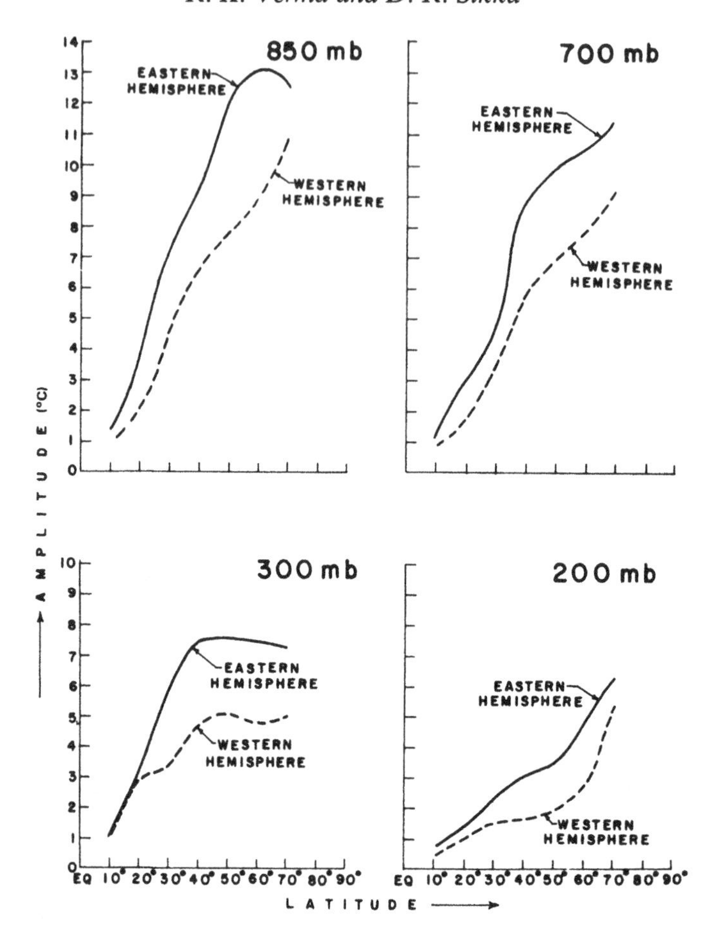

Fig. 3.7. Latitudinal variation of the amplitude of annual oscillation in temperature over the eastern and western hemisphere.

hemispheres. Latitudinal cross-sections of these amplitudes are drawn at various levels. Out of these, cross-sections for 850, 700, 300 and 200 mb are shown in Fig. 3.7.

At all levels from 850 to 100 mb and at all latitudes from 10° to 70° N, the amplitudes over the eastern hemisphere are larger than those over the western hemisphere. The largest differences occur in the latitudinal belt 40 to 50° N. Larger temperature amplitudes in the eastern hemisphere are related to the Eurasian continent and they may be assumed to reflect the planetary-scale asymmetry existing in the tropospheric heating pattern.

3.3.5 Discussion of the results

The chief features which emerge from the results discussed above are:

(i) Horizontal variations in the amplitudes and the phases of the annual temperature at different levels occur on a very large scale and are connected with the distribution of continents and oceans. This is particularly apparent in the tropical and subtropical belt (20° N to 45° N). The amplitudes are larger over the continents than over the oceans. These differences are obviously linked to the large-scale differential response of the land and the ocean surfaces to the solar heating. The amplitudes are larger for the land-dominated eastern hemisphere than for the ocean-dominated western hemisphere.

(ii) The amplitudes fall slowly from the surface to 500 mb and then increase again at 300 mb. The double maxima are more marked in the eastern hemisphere, particularly over the Indian monsoon region.

(iii) The vertical distribution of the phases of the annual wave also show a pattern over the Indian monsoon region different from that found elsewhere. The maximum temperature in the lower troposphere (850 mb) occurs by about the second or third week of June and in the upper troposphere (300 and 200 mb), it occurs by the end of July or beginning of August.

The maximum temperature over the oceans occurs by the middle of August in the lower troposphere, and there is hardly any phase shift in the vertical. The maximum temperature over the continental parts of the tropics between 20 and 30° N outside the Indian monsoon region occurs in the lower troposphere by the middle of July and in the upper troposphere by about the second week of August.

The amplitude and phase differences in the temperature between the Indian monsoon region and the oceanic regions on the one hand and the continents on the other are very conspicuous and require explanation. These are presumably linked with the influence of the southwest monsoon which brings about enormous cloudiness and large-scale precipitation during the summer season over India. As the monsoon sets in over the extreme south of peninsular India by the beginning of June and then steadily advances northwards, the resulting cloudiness cuts off the solar heating and so the phases in the lower-tropospheric levels over India lag by about 30 to 45 days those over the corresponding latitudes in the remaining parts of the northern hemisphere.

The conspicuous second maximum at 300 mb, which occurs over the tropical latitudes of the eastern hemisphere, is more difficult to explain. It is suggested that it is linked with the release of latent heat and an attempt is made in the following paragraphs to support this suggestion.

Heating of the tropical atmosphere with particular reference to monsoonal upper tropospheric maximum heating. Neglecting horizontal advection, the total heating in a unit column of the atmosphere may be represented as

$$Q_T = Q_R + Q_S + Q_L, \tag{3.1}$$

where

Q_R = radiational heating, including absorption of solar short-wave radiation and long-wave emission in the column,

Q_S = heating or cooling by the turbulent transfer of sensible heat between the atmosphere and its lower boundary, and

Q_L = Heating or cooling due to condensation or evaporation of water vapour.

Asakura and Katayama (1964) have estimated the mean values of the various terms of (3.1) for January and July over the northern hemisphere. Their results for the zonal average and rms value of each term at 0°, 10° N, 20° N and 30° N show that the longitudinal variation of Q_R and Q_S is appreciably smaller than that of Q_L. In fact, latent heat release accounts for nearly all of the total variance. Smagorinsky (1953) showed that large-scale asymmetries of non-adiabatic heating or cooling have a strong influence on the planetary-scale stationary circulation. The annual distribution of precipitation in the tropics reveals a large longitudinal variation. For example, each of the three tropical continents possesses relative precipitation maxima. Most estimates of the distribution of the release of latent heat utilize an observed distribution of precipitation. Webster (1972) assumed a direct relationship between cloudiness, precipitation and latent heat release and noted, for the period June to August, the dominance of the extremely strong heat source in the vicinity of the Indian subcontinent.

It is now generally recognized that the primary thermal source in the tropics is the release of latent heat in the convective clouds along the intertropical convergence zone (ITCZ). According to the hypothesis of Riehl and Malkus (1958), the release of latent heat occurs primarily in hot cumulus towers penetrating from the surface boundary layer through the stable middle layers into the upper-tropospheric layers. Kuo's (1965, 1974) model of cumulus parameterization indicates that the maximum heating should occur at 300 mb. Dickinson (1971), in his analytic model for the zonal winds in the tropics, also used a vertical distribution of diabatic heating due to latent heat release in the tropics which is zero at the surface and the tropopause and peaks at 300 mb. Reed and Recker (1971) in their observational study of the composite easterly wave in the equatorial western Pacific Ocean computed a maximum in the latent heating at 400 mb (11 °C per day) in the trough region.

Most of the precipitation and hence latent heat release in the summer monsoon over India occurs on the synoptic scale. The quasi-stationary nature of the rainfall belt in the monsoon season justifies the interpretation of the latent heat release as a time average for the summer season. The peak of the summer monsoon rainfall over central and northern India is reached by the end of July or beginning of August. The rainfall over the monsoon trough region (20 to 30° N over India) and farther north over Tibet, as well as over southeast Asia, occurs in association with convective clouds. Fig. 3.8 shows the distribution of thunderstorm days (with total frequency greater than 10 days) during the months of July and August over India and its neighbours given in the publications of the India Meteorological Department (1943) and the World Meteorological Organization (1953). Combining cloud observations from satellites with the aerological data and some qualitative considerations of the heat budget during the summer, Flohn (1968) has shown that the Tibetan Highlands act as a heat engine with an enormous convective activity in the southeastern sector where giant cumulonimbus cells play a major role in continuously carrying heat upwards into the upper troposphere. The warmest area in the 300 to 500 mb layer is situated between latitudes 27° N and 33° N and longitudes 76° E and 94° E, centred near 30° N, 85° E. This is shown by dotted lines as 3900 and 3920 contours in Fig. 3.8. Flohn has also shown that convective clouds form a very large percentage (greater than 80%) of the total cloudiness over the Tibetan and adjoining Himalayan region. According to him, the release of latent heat of condensation, with a marked degree of persistency over the Himalayan and adjacent parts of India, contributes towards the occurrence of a quasi-stationary heat centre during the summer monsoon season in the middle and upper troposphere.

It is therefore suggested that the second maximum at 400 to 300 mb layer noticed in the annual oscillation in the temperature over the monsoon region is linked to the latent heat release in deep cumulus clouds.

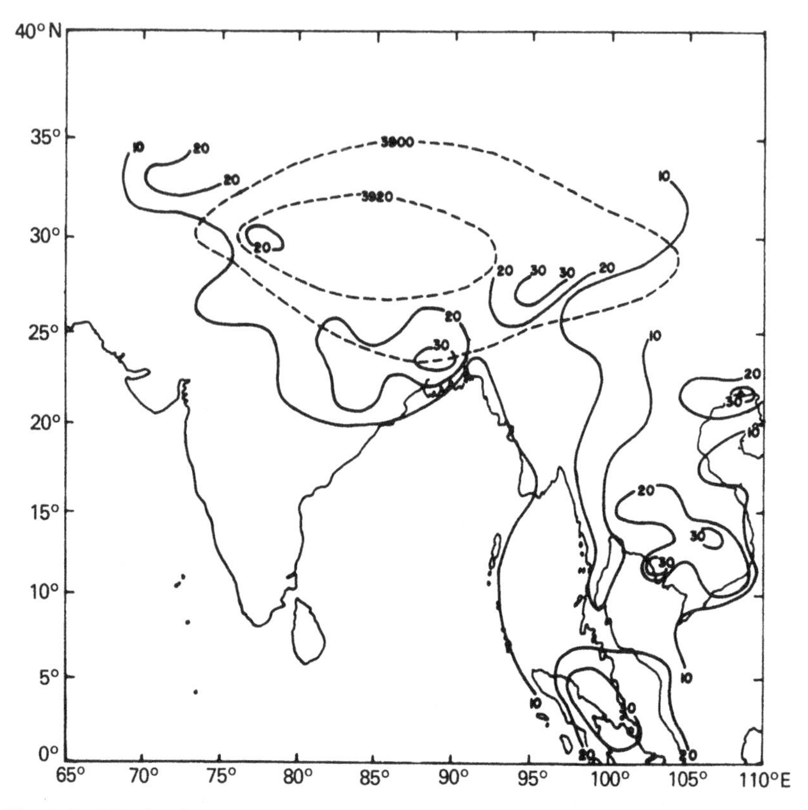

Fig. 3.8. Distribution of thunderstorm days (>10 days) during the months of July and August over India and neighbouring southeast Asia. 300 to 500 mb thicknesses (>3900 geopotential meters) are shown as dashed lines.

3.4 A suggested heating model for the Asiatic summer monsoon region

The vertical variations of amplitude and phase of the annual temperature oscillation over the Asiatic summer monsoon region, as discussed in sections 3.3.1 to 3.3.5, suggest the heating model shown in the schematic diagram in Fig. 3.9 with the following characteristics:

(i) Double heating maxima occur in the vertical, one at 850 mb (the lowest level considered) and the other at about 400 mb.

(ii) The sensible heating maximum occurs at 850 mb in the middle of June and decreases up to 200 mb. A time lag of about 45 days exists due to sensible heating from the surface to 200 mb.

(iii) Superimposed on this heating is another heating maximum, centred at about 400 mb and occurring towards the end of July and early August, which may be ascribed to latent heat release.

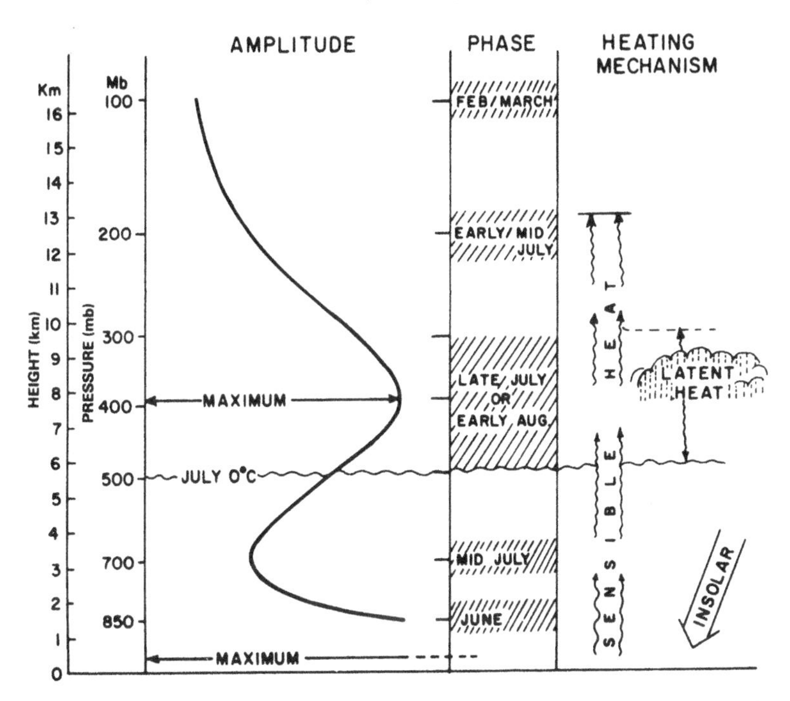

Fig. 3.9. Suggested heating model over the Asiatic summer monsoon region.

Numerical models of the Asiatic summer monsoon must be capable of reproducing these features if they are to be fully realistic.

References

Asakura, T. and Katayama, A. (1964) Normal distribution of heat sources and sinks in the lower troposphere over the northern hemisphere. *J. Meteor. Soc. Japan*. **42**, 209–44.

Asnani, G. C. and Mishra, S. K. (1975) Diabatic heating model of the Indian monsoon. *Mon. Wea. Rev.*, **103**, 115–30.

Asnani, G. C. and Verma, R. K. (1975) Annual and semi-annual thickness oscillations in the northern hemisphere. RR-020, Indian Institute of Tropical Meteorology, Poona.

Banerjee, A. K. and Sharma, K. K. (1967) A study of the seasonal oscillations in the upper air temperature over India. *Ind. J. Meteor. Geophys.*, **18**, 69–74.

Dickinson, R. E. (1971) Analytic model for zonal winds in the tropics. *Mon. Wea. Rev.*, **99**, 501–23.

Flohn, H. (1968) Contributions to a meteorology of the Tibetan Highlands. Atmospheric Science Paper No. 130, Dept. of Atmospheric Science, Colorado State University, Colorado.

India Meteorological Department (1943) *Climatological Atlas for Airmen.*

Kuo, H. L. (1965) On the formation and intensification of tropical cyclones through latent heat release by cumulus convection. *J. Atmos. Sci.*, **22**, 40–63.

Kuo, H. L. (1974) Further studies of the parameterization of the influence of cumulus convection on the large-scale flow. *J. Atmos. Sci.*, **31**, 1232–40.

Oort, A. H. and Rasmusson, E. M. (1971) Atmospheric circulation statistics. NOAA professional paper No. 5.

Reed, R. J. and Recker, E. E. (1971) Structure and properties of synoptic scale wave disturbances in the equatorial western Pacific. *J. Atmos. Sci.*, **28**, 1117–33.

Riehl, H. and Malkus, J. S. (1958) On the heat balance of the equatorial trough zone. *Geophysica*, **6**, 503–38.

Smagorinsky, J. (1953) The dynamical influence of large-scale heat sources and sinks on the quasi-stationary mean motions of the atmosphere. *Quart. J. Roy. Meteor. Soc.*, **79**, 342–66.

Webster, P. J. (1972) Response of the tropical atmosphere to local, steady forcing. *Mon. Wea. Rev.*, **100**, 518–41.

World Meteorological Organization (1953) *World distribution of thunderstorm days.* Part I, No. 21, TP 6.

World Meteorological Organization (1965) *Short period averages for 1951–1960 and provisional average values for climat temp and climat temp ship stations.* No. 170, TP 84.

4

Summer mean energetics for standing and transient eddies in the wavenumber domain†

T. MURAKAMI

Energy equations, similar to those proposed by Saltzman, were applied to four specific latitudinal belts to investigate the atmospheric energetics in the wavenumber domain during three summers in 1970–2. The selected latitudinal belts are: region 1 (30.8° to 44.6° N); region 2 (14.8° to 30.8° N); region 3 (0° to 14.8° N); and region 4 (14.8° S to 0°). Energy exchanges due to wave–wave and wave–zonal mean flow interactions are important for the maintenance of eddy kinetic energy, over all regions. Almost all waves furnish their kinetic energy to zonal mean flows via wave–zonal flow interaction.

A computational model to partition kinetic energy exchanges into the standing- (summer mean) and transient-wave motions has been proposed. Planetary-scale standing waves 1 to 3, over regions 2 and 3, lose large amounts of kinetic energy to transient eddies via 'standing to transient' wave interactions. The majority of transient waves, over regions 2 and 4, act as a kinetic energy drain for standing-wave motions.

4.1 Introduction

Scale interactions are usually defined in terms of zonal wavenumber. In such a scale resolution, the monsoon circulation can be identified as a

† Contribution No. 77-12, Department of Meteorology, University of Hawaii.

distinct low wavenumber mode. The Fourier representation of atmospheric motion has shed much light upon the processes of generation, dissipation, and transfer of kinetic energy in and among scales of atmospheric motions in extratropical and tropical regions. Saltzman (1970) summarized the results of several studies of energy interactions in the Fourier domain, in extratropical regions where synoptic-scale baroclinic disturbances are dominant. Kanamitsu *et al.* (1972) measured the wave–wave and wave–zonal mean flow interactions at 200 mb over a tropical belt (15° S to 15° N) for the 1967 summer. Similar computations were made by Unninayar and Murakami over a tropical zone (5° S to 19.6° N) for the 1970 summer.

Using a semispectral representation of the one-level nonlinear barotropic vorticity equation, Colton (1973) showed that synoptic disturbances remove energy from planetary-scale waves during their growing stages and feed energy back to the planetary-scale waves when decaying. Colton then speculated that by including some damping mechanisms for transient synoptic waves, they would extract more energy than they would receive from planetary waves; namely, that in the long-term average, transient synoptic waves would gain energy via barotropic instability processes from the standing planetary-scale circulation. Clearly much more effort is needed to investigate the mechanisms through which transient eddies are generated and maintained via barotropic interaction with steady planetary-scale motions during summer. This is elaborated in §§ 4.5 and 4.6.

4.2 Data and computational procedures

In this study, 200 mb wind data as determined by operational analyses at the National Meteorological Center (NMC) during three summer (June to August) seasons in 1970–2, were utilized. Murakami's (1978) study confirmed that NMC data was adequate, at least qualitatively, in describing some of the characteristic features of the 200 mb monsoon circulations during 1970–2 summers. The reader is directed to this paper for detailed information on the expected errors in NMC analyses and their possible effects on the study of the summer monsoon circulation energetics. (For brevity, Murakami's (1978) paper will hereafter be referred to as M8.)

Daily u and v data at 200 mb were separated into zonal mean and eddy components. Perturbations u' and v' were then expanded into Fourier series in real form. For example, u' was expressed as:†

$$u' = \sum_{n=1}^{10} (u_{n,1} \cos n\lambda + u_{n,2} \sin n\lambda). \tag{4.1}$$

§ 4.3 details the computational procedures used to measure wave–wave and wave–zonal mean flow interactions in the wavenumber domain. In § 4.5, a computational model is proposed to investigate the role of barotropic processes in maintaining the time-averaged motions and transient waves during summer.

4.3 Kinetic energy equation in the wavenumber domain

Using the same symbols as Saltzman (1970), the energy equation for eddy kinetic energy can be written as

$$\frac{\partial K(n)}{\partial t} = M(n) + L(n) + C(n) - D(n), \tag{4.2}$$

where,

$$K(n) = \tfrac{1}{4} \sum_{i=1}^{2} [u_{n,i}^2 + v_{n,i}^2],$$

$$\begin{aligned} M(n) = & -\left[\overline{u'v'}^n \frac{\partial}{\partial y}\left(\frac{\bar{u}}{\cos\phi}\right) + \overline{v'v'}^n \frac{\partial \bar{v}}{\cos\phi\, \partial y} - \frac{\tan\phi}{a}\overline{u'u'}^n\right] \\ & -\left[\overline{u'\omega'}^n \frac{\partial \bar{u}}{\partial p} + \overline{v'\omega'}^n \frac{\partial \bar{v}}{\partial p}\right], \end{aligned} \tag{4.3}$$

$$L(n) = -[\overline{u'N_u(v', u')'}^n + \overline{v'N_v(v', v')'}^n]. \tag{4.4}$$

The terms $M(n)$ and $L(n)$ were expressed in real form, rather than in the complex form of Saltzman (1970). Similar expressions were used in Murakami (1965). In (4.2) to (4.4), we define

$$N_u(v', u') = u' \frac{\partial u'}{\cos\phi\, \partial x} + v' \frac{\partial (u'\cos\phi)}{\cos^2\phi\, \partial y} + \omega' \frac{\partial u'}{\partial p}, \tag{4.5}$$

$$N_v(v', v') = u' \frac{\partial v'}{\cos\phi\, \partial x} + v' \frac{\partial v'}{\cos\phi\, \partial y} + \frac{\tan\phi}{a} u'^2 + \omega' \frac{\partial v'}{\partial p} \tag{4.6}$$

$$\overline{A'B'}^n = \tfrac{1}{2}(A_{n,1}B_{n,1} + A_{n,2}B_{n,2}), \tag{4.7}$$

where A and B are any of the variables u, v, ω, N_u, and N_v.

† A list of symbols is given at the end of this chapter.

$M(n)$ indicates the rate of transfer of kinetic energy between the zonally-averaged flows and eddies of wavenumber n. The rate of kinetic energy transfer to eddies of wavenumber n from eddies of all wavenumbers is represented by $L(n)$.

The $C(n)$ term represents the conversion of eddy available potential energy to eddy kinetic energy for wavenumber n. Since data for temperature and vertical velocity was not available, no attempt was made to measure $C(n)$. Likewise, the terms related to vertical motion in (4.3) to (4.6) were omitted when measuring $M(n)$ and $L(n)$. The frictional term $D(n)$ was estimated as a residual in (4.2). Since the $D(n)$ term is not based on direct measurement, one must be cautious in drawing inferences about the contribution of this term to the maintenance of eddy kinetic energy $K(n)$.

In view of the substantially differing results of energetic computations for midlatitude regions (Saltzman, 1970) as compared to tropical regions (Kanamitsu *et al.*, 1972), care must be exercised that the energy equations not be applied to too large a latitudinal span, lest entirely different circulation regimes be compounded together. In the present study, we have chosen four different latitudinal belts as follows:

Region 1 (30.8° to 44.6° N): strong three-year summer mean zonal (westerly) winds at 200 mb, reaching 23 m s^{-1} at around 41° N (Fig. 4.1*a*).

Region 2 (14.8° to 30.8° N): weak westerly 200 mb zonal mean winds with anticyclonic lateral shear.

Region 3 (0° to 14.8° N): easterly zonal mean winds with a maximum magnitude of 7 m s^{-1} at 5° N, and westerly momentum flux divergence due to both standing and transient waves.

Region 4 (14.8° S to 0°): large lateral shear in 200 mb zonal mean winds between the Equator (−5 m s^{-1}) and 14.8° S (10 m s^{-1}).

In § 4.4, energetic differences between different latitudinal belts are investigated. The interaction terms $M(n)$ and $L(n)$, calculated daily at 200 mb, were averaged to obtain three-year summer mean values.

4.4 Wave–wave and wave–zonal mean flow interactions during summer

In Table 4.1, the three-year summer mean $M(n)$ is negative and substantial for all wavenumbers over regions 1, 2, and 3. This implies that all waves over the entire northern hemisphere zone (44.6° N to 0°) behave as a source of kinetic energy for zonal mean flows. Over region 1,

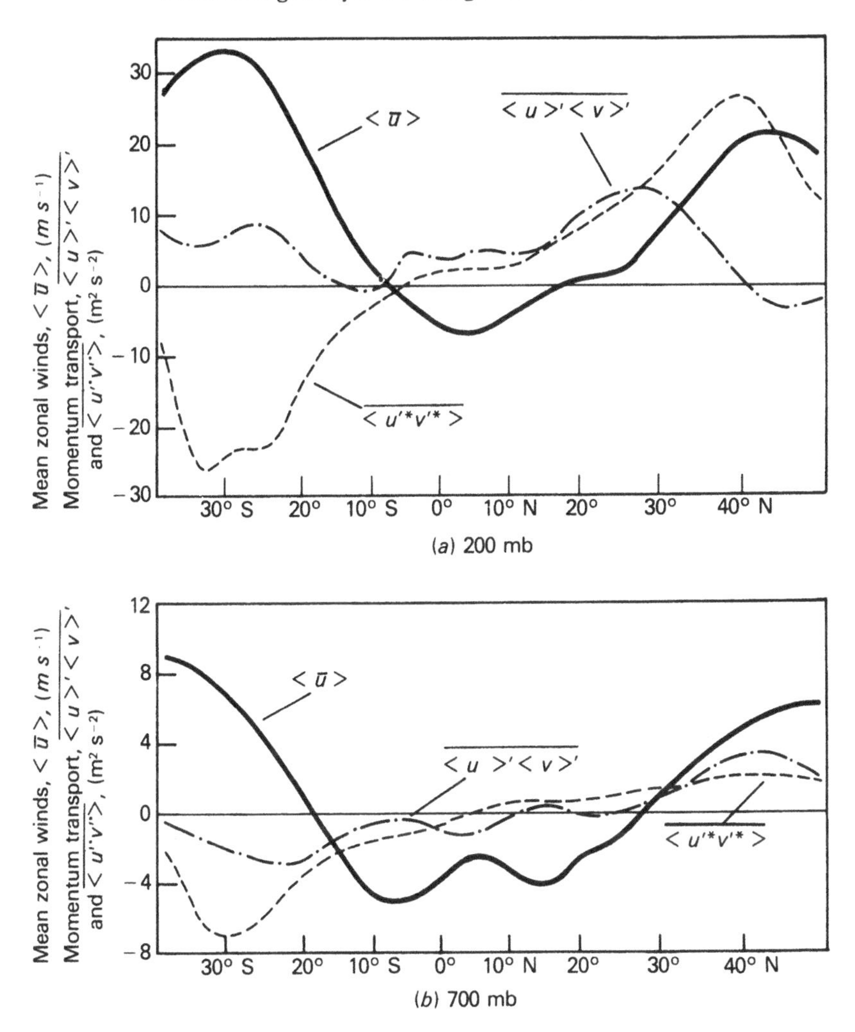

Fig. 4.1. Three-year (1970–2) summer (June to August) mean zonal winds $\langle\bar{u}\rangle$, the momentum transport due to standing eddies, $\overline{\langle u\rangle'\langle v\rangle'}$, and transient eddies, $\overline{\langle u'^* v'^*\rangle}$ at (*a*) 200 mb and (*b*) 700 mb. (After Murakami, 1978.)

wavenumbers 2 to 3 and 5 to 10 contribute most. In regions 2 and 3, the contribution due to $n = 1$ to 3 is prominent. As shown in Fig. 4.1*a*, a rather strong anticyclonic shear exists in 200 mb zonal mean winds between the Equator (-5 m s^{-1}), and 41° N ($+23$ m s^{-1}). In the intervening latitudes, the wave axis possesses a distinct northeast–southwest tilt, indicating a northward westerly momentum flux against the $\langle\bar{u}\rangle$ gradient. This results in kinetic energy transfers from waves to zonal mean flows.

TABLE 4.1. *Three-year (1970–2) summer mean* $K(n)$, *in units of* $10^{-1}\ m^2\ s^{-2}$, *and* $M(n)$ *and* $L(n)$, *in units of* $10^{-7}\ m^2\ s^{-3}$, *measured from (4.3) and (4.4) at 200 mb over regions 1 to 4.*

	Region 1 (30.8° to 44.6° N)			Region 2 (14.8° to 30.8° N)		
	K	*M*	*L*	*K*	*M*	*L*
n						
1	258	−81	−568	237	−272	19
2	169	−502	−68	122	−162	−459
3	149	−412	73	86	−182	186
4	133	−60	−16	55	−70	44
5	149	−220	206	41	−34	55
6	121	−184	264	44	−76	−6
7	135	−515	21	44	−81	−5
8	91	−215	171	33	−49	37
9	90	−273	34	34	−64	28
10	57	−220	−7	26	−54	56
	Region 3 (0° to 14.8° N)			**Region 4 (0° to 14.8° S)**		
	K	*M*	*L*	*K*	*M*	*L*
n						
1	210	−50	−21	187	67	45
2	64	−42	−129	66	63	−20
3	49	−62	−101	50	−54	65
4	32	−8	−10	47	−22	74
5	28	−18	−11	45	−61	16
6	24	−32	−17	33	−45	72
7	26	−30	8	32	−6	44
8	19	−7	−17	25	2	13
9	16	−11	12	18	−8	22
10	15	−14	19	14	−2	5

Fig. 4.1*b* shows corresponding data at 700 mb where the anticyclonic shear is much weaker.

Table 4.1 also illustrates the differences in $L(n)$ processes among regions 1 to 4. A large negative $L(1)$ over region 1 indicates that the wavenumber 1 loses substantial amounts of kinetic energy to other waves via wave–wave interactions. Over region 2, an energy loss for wavenumber 2 is pronounced. Conversely, waves of wavenumbers 3 and 5 to 9 over region 1 (3 to 5 and 8 to 10 over region 2) receive kinetic energy through barotropic nonlinear $L(n)$ processes. The above results are in marked contrast to those of Fig. 2 of Saltzman (1970), representing summer mean $L(n)$ terms over a latitudinal belt between about 15° and 80° N. In his analysis, the largest exchanges of kinetic energy between waves took place at higher latitudes not considered in the present study. Saltzman's results indicate significant transfers of kinetic energy from cyclone-scale waves of $n=6$ to 10, which derive their energies convectively through baroclinic instability, to both smaller ($n>10$) and larger (1 and 3) scales. Perhaps, this discrepancy between the two studies reflects the large difference in the nature of barotropic nonlinear $L(n)$ interactions between different latitudinal zones.

In Table 4.1, the majority of waves in region 3 are seen to lose kinetic energy to other waves via the $L(n)$ effect. Region 4, on the other hand, is characterized by the gain of kinetic energy for most waves except wavenumber 2. The sum for all wavenumbers, $\sum_{n=1}^{10} L(n)$, computed over each region, is as follows (in units of $10^{-7}\ \mathrm{m}^2\ \mathrm{s}^{-3}$): +110 over region 1; −45 for region 2; −267 in region 3; +336 over region 4. The net gain of kinetic energy over region 4 is approximately balanced by the energy loss in regions 2 and 3. This means that the wave activity over the south tropical belt (0° to 14.8° S) is perhaps supported, in some way, by receiving energy from the north tropical belt (0° to 30.8° N). This point will be discussed further in § 4.6.

When considering all regions 1 to 4, the $\sum_{n=1}^{10} L(n)$ sum amounts to only +134 units. This total sum of $L(n)$, when partitioned into each wavenumber and each region, is extremely small, which is gratifying in view of the complexity of the calculations of the $L(n)$ term as defined in (4.4). From this smallness, it may be said that NMC operational data adequately describes the characteristic nature of nonlinear coupling $L(n)$ processes during summer.

It is likely that a record of three summer seasons in 1970–2 is not sufficient to establish significant statistics. Computed results indicate that an individual year can be highly anomalous with respect to long-term average conditions. For example, $L(4)$ and $L(5)$ computed over region 1 in each summer are as follows (again, in $10^{-7}\ \mathrm{m}^2\ \mathrm{s}^{-3}$) 164 and −335 in 1970; −37 and 156 for 1971; −175 and 797 in 1972.

4.5 Energy equations for standing- and transient-wave components

Daily u and v data were separated into time-running mean and transient components. For example, u was expressed as:

$$u = \langle u \rangle + u^*. \qquad (4.8)$$

Thus, u^* is defined as the transient component of u. The term $\langle u\rangle$ changes slowly in time through interaction with transient eddies.

We then define transient eddy kinetic energy, $K_1(n)$, and standing eddy kinetic energy, $K_2(n)$, as follows:

$$K_1(n)=\tfrac{1}{4}\sum_{i=1}^{2}\left[\langle u^{*2}_{n,i}\rangle+\langle v^{*2}_{n,i}\rangle\right], \tag{4.9}$$

$$K_2(n)=\tfrac{1}{4}\sum_{i=1}^{2}\left[\langle u\rangle^2_{n,i}+\langle v\rangle^2_{n,i}\right]. \tag{4.10}$$

The energy equations for $K_1(n)$ and $K_2(n)$ may be written approximately as follows:

$$\frac{\partial K_1(n)}{\partial t}=M_1(n)+M_2(n)+L_1(n)+L_2(n)+C_1(n), \tag{4.11}$$

$$\frac{\partial K_2(n)}{\partial t}=M_3(n)+M_4(n)+L_3(n)+L_4(n)+C_2(n), \tag{4.12}$$

in which the term related to friction is omitted. The energy interaction terms are defined by

$$M_1(n)=-\left[\langle\overline{u'^*v'^*}^n\rangle\frac{\partial}{\partial y}\left(\frac{\langle\bar{u}\rangle}{\cos\phi}\right)+\langle\overline{v'^*v'^*}^n\rangle\frac{\partial\langle\bar{v}\rangle}{\cos\phi\,\partial y}\right.$$
$$\left.-\frac{\tan\phi}{a}\langle\overline{u'^*u'^*}^n\rangle\langle\bar{v}\rangle\right], \tag{4.13}$$

$$M_2(n)=-\left[\left\langle\overline{u'^*v'}^n\frac{\partial}{\partial y}\left(\frac{\overline{u^*}}{\cos\phi}\right)+\overline{v'^*v'}^n\frac{\partial\overline{v^*}}{\cos\phi\,\partial y}-\frac{\tan\phi}{a}\overline{u'^*u'}^n\overline{v^*}\right\rangle\right], \tag{4.14}$$

$$M_3(n)=-\left[\overline{\langle u\rangle'\langle v\rangle'}^n\frac{\partial}{\partial y}\left(\frac{\langle\bar{u}\rangle}{\cos\phi}\right)+\overline{\langle v\rangle'\langle v\rangle'}^n\frac{\partial\langle\bar{v}\rangle}{\cos\phi\,\partial y}\right.$$
$$\left.-\frac{\tan\phi}{a}\overline{\langle u\rangle'\langle u\rangle'}^n\langle\bar{v}\rangle\right], \tag{4.15}$$

$$M_4(n)=-\left[\left\langle\overline{\langle u\rangle'v'^*}^n\frac{\partial}{\partial y}\left(\frac{\overline{u^*}}{\cos\phi}\right)+\overline{\langle v\rangle'v'^*}^n\frac{\partial\overline{v^*}}{\cos\phi\,\partial y}\right.\right.$$
$$\left.\left.-\frac{\tan\phi}{a}\overline{\langle u\rangle'u'^*}^n\overline{v^*}\right\rangle\right], \tag{4.16}$$

$$L_1(n) = -[\langle \overline{u'^* N_u(\langle v\rangle', u'^*)'}^n + \overline{v'^* N_v(\langle v\rangle', v'^*)'}^n \rangle]$$
$$-[\langle \overline{u'^* N_u(v^*, \langle u\rangle')'}^n + \overline{v'^* N_v(v^*, \langle v\rangle')'}^n \rangle], \quad (4.17)$$

$$L_2(n) = -[\langle \overline{u'^* N_u(v'^*, u'^*)'^*}^n + \overline{v'^* N_v(v'^*, v'^*)'^*}^n \rangle], \quad (4.18)$$

$$L_3(n) = -[\langle \overline{\langle u\rangle' N_u(v^*, u'^*)'}^n + \overline{\langle v\rangle' N_v(v^*, v'^*)'}^n \rangle], \quad (4.19)$$

$$L_4(n) = -[\overline{\langle u\rangle' N_u(\langle v\rangle', \langle u\rangle')'}^n + \overline{\langle v\rangle' N_v(\langle v\rangle', \langle v\rangle')'}^n]. \quad (4.20)$$

In (4.13) to (4.20), the terms related to vertical motion are omitted. For the convenience of the reader, the physical meanings of the terms $M_i(n)$ and $L_i(n)$, ($i = 1$ to 4), are summarized as follows:

Wave–zonal mean flow interaction

$M_1(n)$ Interaction between the time-averaged zonal mean flows and transient wave n.

$M_2(n)$ Interaction between the transient zonal mean flows, and transient and standing eddies of wavenumber n.†

$M_3(n)$ Interaction between the time-averaged zonal mean flows and standing wave n.

$M_4(n)$ Interaction between the transient zonal mean flows, and standing and transient eddies of wavenumber n.

Wave–wave interaction

$L_1(n)$ Interaction between transient eddies and standing waves.

$L_2(n)$ Kinetic energy transfer to transient wave n from transient waves of all wavenumbers.

$L_3(n)$ Kinetic energy transfer to standing wave n from transient waves of all wavenumbers.

$L_4(n)$ Kinetic energy transfer to standing wave n from standing waves of all wavenumbers.

In general, $L_3(n)$ is not equal to $-L_1(n)$ for a particular wavenumber n. When considering all wavenumbers, we find the following relationship:

$$\sum_{n=1}^{\infty} (L_1(n) + L_3(n)) = 0, \quad (4.21)$$

where the effects due to energy fluxes across the boundaries are omitted.

It is also possible to derive the following important relationships:

$$\langle M(n)\rangle = \sum_{i=1}^{4} M_i(n), \qquad \langle L(n)\rangle = \sum_{i=1}^{4} L_i(n). \quad (4.22)$$

† In (4.14), note that $v' = \langle v\rangle' + v'^*$, and $u' = \langle u\rangle' + u'^*$. In (4.17), $v^* = \bar{v}^* + v'^*$.

Thus, (4.11) and (4.12) are consistent with the original kinetic energy equation (4.2) for any wavenumber n.

4.6 Summer mean kinetic energy exchanges for transient and standing waves

Table 4.2 illustrates the processes through which the kinetic energy of transient- and standing-wave motions is maintained over *region 2*. The kinetic energy for standing waves, $K_2(n)$, is a maximum at $n = 1$ and decreases sharply with increasing wavenumber. In fact, $K_2(1)$ is two orders of magnitude larger than $K_2(10)$. The predominance of standing

TABLE 4.2. *Three-year summer mean energy exchange terms for transient- and standing-wave motions at 200 mb over region 2. Units for $K_i(n)$, where $i=1$ to 2: $10^{-1}\ m^2\ s^{-2}$. Units for $M_i(n)$, $L_i(n)$, where $i=1$ to 4: $10^{-7}\ m^2\ s^{-3}$.*

	Region 2 (14.8° to 30.8° N)				
	K_1	M_1	M_2	L_1	L_2
n	Transient waves				
1	58	−26	−52	−6	34
2	48	−24	23	3	−2
3	53	−55	−13	85	−47
4	41	−39	0	43	14
5	38	−40	2	39	11
6	39	−56	0	25	−9
7	41	−74	−8	49	−29
8	30	−29	−7	27	36
9	31	−51	−10	29	0
10	25	−49	−1	46	14
	K_2	M_3	M_4	L_3	L_4
n	Standing waves				
1	180	−201	9	−50	42
2	74	−161	1	−47	−410
3	33	−116	3	−118	269
4	14	−37	7	−71	59
5	3	5	0	13	−7
6	5	−19	0	−37	16
7	3	0	2	−15	−8
8	2	−9	0	−12	−12
9	3	−3	2	−14	13
10	1	−2	0	−3	0

waves 1 and 2 is depicted in Fig. 4.2*a*; namely at 200 mb the streamfunction field is dominated by a pronounced (monsoonal) anticyclone near India and distinct mid-oceanic troughs over the north Pacific and Atlantic Oceans. In comparison, transient eddy kinetic energy, $K_1(n)$, is substantial even for high wavenumbers with a secondary maximum (41 units) at $n=7$. Note that $K_1(n) > K_2(n)$ for $n \geq 3$. Here, the point of emphasis is the importance of transient eddies as an integral part of the summer circulation system. The activity of transient eddies at 200 mb is very pronounced in and around the upper oceanic trough over the tropical north Pacific (Fig. 4.2*b*).

Interestingly, all transient waves over region 2 furnish kinetic energy to summer mean zonal flows via $M_1(n)$ type interactions with a maximum contribution from wavenumber 7. As for $M_3(n)$ type interactions, planetary-scale standing waves of $n=1$ to 3 largely contribute to the maintenance of summer mean zonal flows. This is evidently related to a very pronounced north–southwest tilt of summer mean streamfunction fields (Fig. 4.2*a*).

Table 4.2 shows that all standing waves except wavenumber 5 lose their kinetic energy to transient waves due to $L_3(n)$ interactions. Wavenumbers 1 to 4 and 6 contribute most. Conversely, all transient waves except wavenumber 1 gain kinetic energy from standing waves via $L_1(n)$. Transient waves of $n=8$ and 10 also receive appreciable amounts of kinetic energy from other transient waves (perhaps $n=3$, 6 and 7) through barotropic nonlinear $L_2(n)$ interactions. Therefore, both $L_1(n)$ and $L_2(n)$ energy exchange processes are important for the development of transient short ($n=8$ to 10) waves in a subtropical belt between 15° and 30.8° N during summer.

In Table 4.3 for *region 3*, large and negative values of $L_3(n)$ for planetary-scale standing waves 1 to 3 imply that these waves lose kinetic energy due to 'standing to transient' wave interactions. The net loss for all standing waves, $\sum_{n=1}^{10} L_3(n)$, amounts to -105 units. In contrast, $L_1(n)$ is small and even negative for certain wavenumbers, and contradicts expectations of large gain of transient eddy kinetic energy from standing-wave motions. In fact, the net sum for all transient waves, $\sum_{n=1}^{10} L_1(n)$, is -4 units. This results in a large imbalance between $\sum_{n=1}^{10} L_1(n)$ and $\sum_{n=1}^{10} L_3(n)$ over region 3, which is puzzling. Perhaps a part of the answer lies in the kinetic energy fluxes across the boundaries of region 3. It is possible that the surplus energy furnished by 'standing to transient' wave interactions over region 3 is transported to other regions, eventually facilitating the development of transient eddies there.

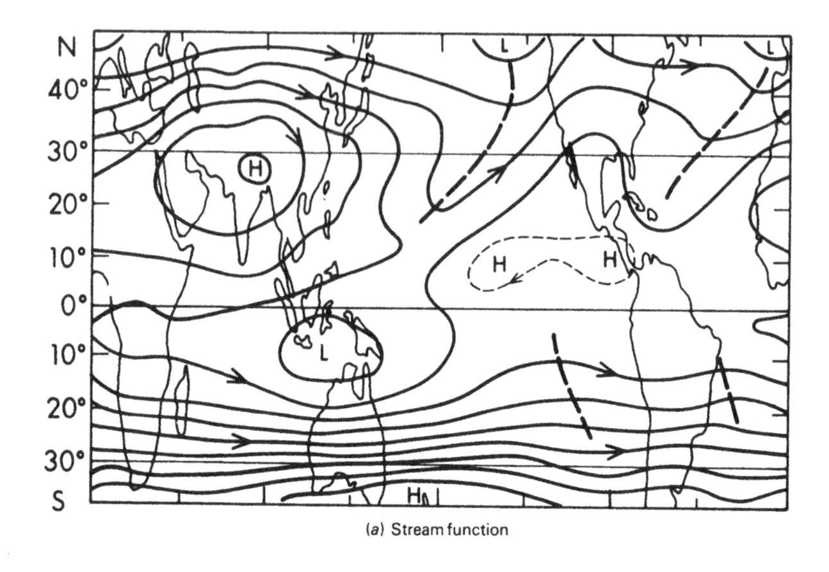

(a) Stream function

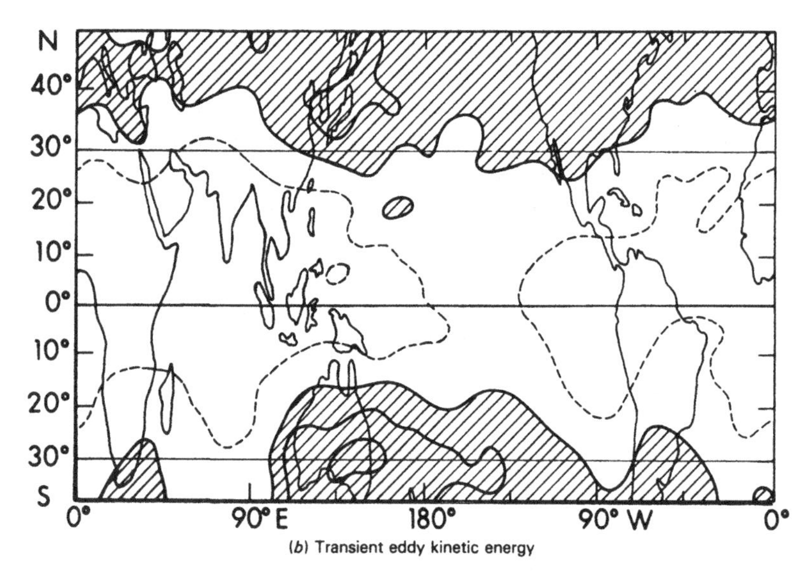

(b) Transient eddy kinetic energy

Fig. 4.2. (a) Three-year (1970–2) summer (June to August) mean 200 mb streamfunction. The contour interval is $10^7\ m^2\ s^{-3}$. Major troughs are shown by heavy dashed lines. (b) Three-year summer mean transient kinetic energy, K^*, at 200 mb. The contour interval is $10^2\ m^2\ s^{-2}$. Shading indicates regions where K^* exceeds $10^2\ m^2\ s^{-2}$. The dashed contour is for $50\ m^2\ s^{-2}$. (After Murakami, 1978.)

TABLE 4.3. *Three-year summer mean energy exchange terms for transient- and standing-wave motions at 200 mb over region 3. Units for $K_i(n)$, where i=1 to 2: $10^{-1}\ m^2\ s^{-2}$. Units for $M_i(n)$, $L_i(n)$, where i=1 to 4: $10^{-7}\ m^2\ s^{-3}$.*

	Region 3 (0° to 14.8° N)				
	K_1	M_1	M_2	L_1	L_2
n	Transient waves				
1	37	−5	−3	13	−7
2	34	−13	0	−1	−18
3	29	−11	3	15	4
4	27	−18	0	−11	9
5	24	−20	0	−16	−4
6	22	−23	−1	−1	−12
7	24	−26	0	−5	6
8	17	−13	3	−8	−4
9	15	−13	1	10	0
10	14	−14	0	5	16
	K_2	M_3	M_4	L_3	L_4
n	Standing waves				
1	173	−52	12	−15	−11
2	30	−24	−1	−38	−69
3	20	−51	0	−36	−84
4	4	9	0	−7	1
5	4	2	0	0	11
6	2	−7	1	−6	4
7	2	−1	0	3	5
8	2	2	1	−3	0
9	1	0	0	1	0
10	1	0	0	1	−1

In Table 4.3, $L_4(n)$ is somewhat large and negative for planetary-scale standing waves $n = 1$ to 3 over region 3, indicating that these waves lose large amounts of kinetic energy to other standing waves. In comparison, $L_4(n)$ is positive but small for $n = 4$ to 9. Thus, there is a large imbalance between the loss of standing eddy kinetic energy for $n = 1$ to 3 and its gain for $n = 4$ to 9. Perhaps, region 3 planetary-scale standing waves 1 to 3 contribute to maintaining standing waves elsewhere.

TABLE 4.4. *Three-year summer mean energy exchange for transient- and standing-wave motions at 200 mb over region 4. Units for $K_i(n)$, where i = 1 to 2: 10^{-1} m^2 s^{-2}. Units for $M_i(n)$, $L_i(n)$, where i = 1 to 4: 10^{-7} m^2 s^{-3}.*

	Region 4 (0° to 14.8° S)				
	K_1	M_1	M_2	L_1	L_2
n	Transient waves				
1	64	−3	−4	23	−21
2	52	4	−6	3	6
3	43	−42	4	29	36
4	41	−32	2	37	−7
5	41	−50	−5	3	26
6	31	−43	0	56	19
7	30	−17	4	27	21
8	21	−17	3	15	0
9	16	−15	1	3	21
10	14	−4	0	4	−1
	K_2	M_3	M_4	L_3	L_4
n	Standing waves				
1	123	67	10	−35	81
2	14	67	0	18	−46
3	7	−16	1	0	2
4	6	10	−1	35	10
5	4	−4	0	−12	0
6	2	0	0	0	−1
7	2	7	0	0	−3
8	3	16	0	−2	2
9	2	6	0	−2	0
10	1	3	0	4	0

Table 4.4 describes the three-year mean energetics for transient and standing waves over region 4. The main results are summarized as follows:

(i) $L_3(n)$ is positive for $n = 2$, 4 and 10, but negative for all other wavenumbers. The sum, $\sum_{n=1}^{10} L_3(n)$, amounts to only −1 unit.

(ii) For all wavenumbers, $L_1(n)$ is positive and somewhat large, implying energy flow from the standing- to transient-wave motions. The gain for all transient waves, $\sum_{n=1}^{10} L_1(n)$, totals 199 units, which is two orders of magnitude greater than the loss for all standing waves within region 4 (−1 unit). Thus, it is evident that standing waves in other latitudinal belts should contribute to the development of transient eddies over region 4. This interpretation can be justified as follows: the net sum $\sum_{n=1}^{10} L_1(n)$ over regions 2 to 4, amounting to 534, approximately balances the net kinetic energy loss for standing waves $\sum_{n=1}^{10} L_3(n)$ over the same regions, totaling −469 (in units of 10^{-7} m^2 s^{-3}). As predicted in (4.22), such approximate balance can exist only for an extensive latitudinal zone with relatively small boundary effects.

4.7 Concluding remarks

This study shows that the character of energetic processes in the wavenumber domain varies considerably with latitudinal region, during the northern summer. To summarize the main results of this study:

(i) Planetary-scale standing waves 1 to 3, over regions 2 and 3, behave as a source of kinetic energy for summer mean zonally-averaged flows. These standing eddies also lose their kinetic energy to transient eddies via 'standing to transient' wave interactions.

(ii) Almost all transient waves, in regions 2 and 4, receive large amounts of kinetic energy from standing waves.

(iii) All transient waves furnish their kinetic energy to summer mean zonally-averaged flows with large contributions from wavenumbers ≥3.

The foregoing should be considered preliminary until verified with data which is more complete (both in accuracy and length of record) than NMC operational data.

The author is indebted to Mrs M. Frydrych for programming assistance and Mrs S. Arita for typing the manuscript.

This research has been supported by the Global Atmospheric Research Program, Climate Dynamics Research Section, National Science Foundation, under Grant No. ATM 76-02502.

List of symbols

a	radius of earth
λ	longitude
ϕ	latitude
x, y	eastward and northward Mercator coordinate axes
p	pressure
t	time
u, v	zonal and meridional winds
ω	vertical p-velocity

$\boldsymbol{v}$	vector (u, v, ω)
$(\bar{\ })$	globally-averaged zonal mean
$(\ )'$	departure from zonal mean
$[\ ]$	meridional average over the domain
$\langle\ \rangle$	time average
$(\ )^*$	departure from time average
$(\ )_{n,1}$	Fourier cosine component for wavenumber n
$(\ )_{n,2}$	Fourier sine component for wavenumber n

References

Colton, D. E. (1973) Barotropic scale interactions in the tropical troposphere during the northern summer. *J. Atmos. Sci.*, **30**, 1287–1302.

Kanamitsu, N., Krishnamurti, T. N. and Depradine, C. (1972) On scale interactions in the tropics during northern summer. *J. Atmos. Sci.*, **29**, 698–706.

Murakami, T. (1965) Energy cycle of the stratospheric warming in early 1958. *J. Meteor. Soc., Japan*, **43**, 262–83.

Murakami, T. (1978) Regional energetics of the 200 mb summer circulations. *Mon. Wea. Rev.*, **106**, 614–28.

Saltzman, B. (1970) Large-scale atmospheric energetics in the wave-number domain. *Rev. Geophys. Space Phys.*, **8**, 289–302.

5

Monitoring the monsoon outflow from geosynchronous satellite data†

J. C. SADLER AND J. T. LIM

Vectors derived from cloud displacements observed from geosynchronous satellites are now a major source of wind data over a large portion of the western hemisphere under surveillance of instruments aboard the United States satellites – *Goes East* and *Goes West*. The planned distribution of geostationary satellites during the Global Weather Experiment (FGGE) will provide such data for the entire globe between approximately 50° N and 50° S, and for the first time the wind data base over the tropical belt for levels near the surface and in the upper troposphere will rival that of higher latitudes.

This paper evaluates the use of these data in producing monthly mean circulations of the upper troposphere and discusses some preliminary comparisons with radio–wind finding (rawin) observations.

5.1 Data procedures

The daily operational wind vectors determined by the National Environmental Satellite Service comprise the data base. Fig. 5.1, showing a typical daily distribution of wind vectors, illustrates that the data are not uniformly distributed. Also, the data are not randomly distributed because of their dependence on the presence of cirrus clouds which in turn depend on the atmospheric circulation systems. Some of the circulation systems biases have been discussed by Gruber *et al.* (1971). Table 5.1

† Contribution No. 78-1 of the Department of Meteorology, University of Hawaii.

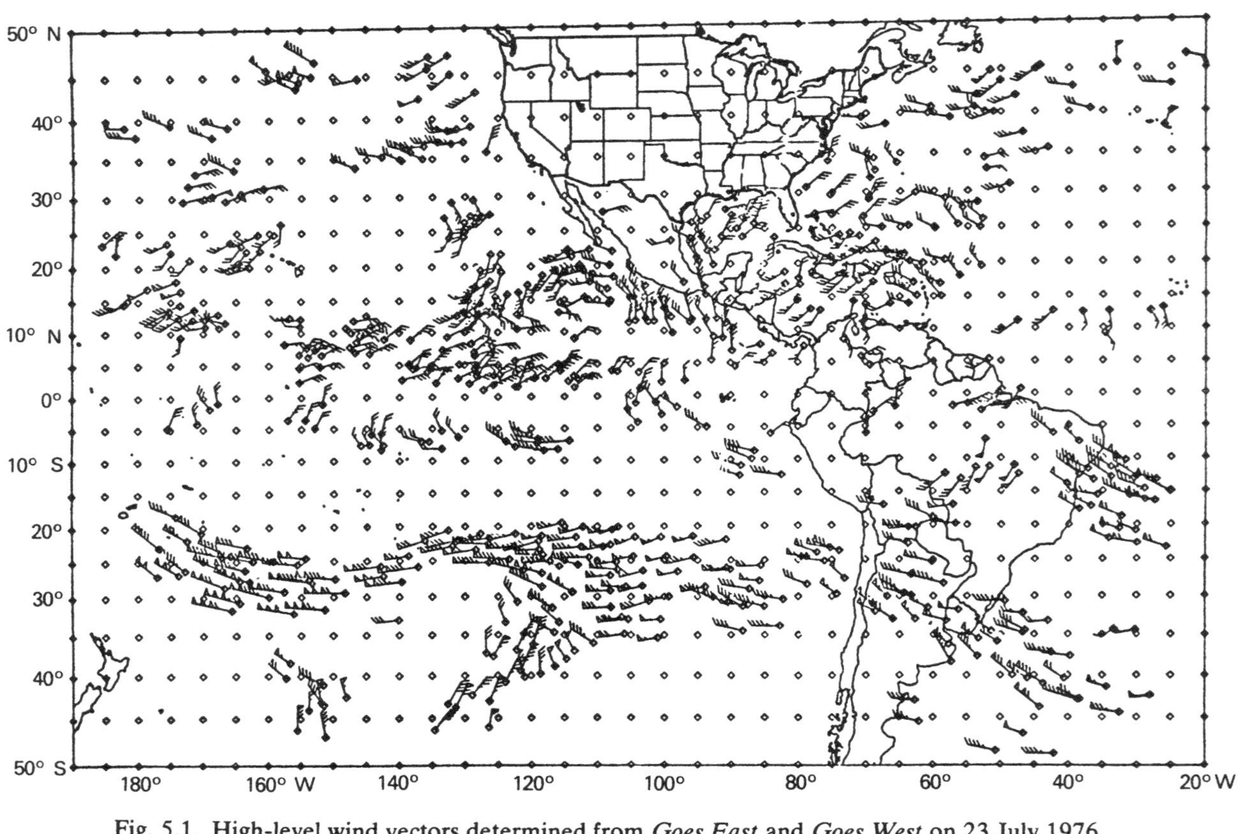

Fig. 5.1. High-level wind vectors determined from *Goes East* and *Goes West* on 23 July 1976.

TABLE 5.1. *The number of days in August 1976 with at least one high-level wind in a 5-degree square.*

	175°W	170°	165°	160°	155°	150°	145°	140°	135°	130°	125°	120°	115°	110°	105°	100°	95°	90°	85°	80°	75°	70°	65°	60°	55°	50°	45°	40°	35°	30°W
50° N – 45°	10	12	17	13	13	17	14	16	13	15	10	0	0	0	0	0	0	0	0	0	0	0	0	2	7	15	15	15	11	5
45° – 40°	10	18	16	18	17	15	16	13	10	7	11	0	0	0	0	0	0	0	0	0	0	0	17	20	26	25	20	17	11	10
40° – 35°	14	9	11	16	16	17	15	9	13	13	10	8	0	0	0	0	0	0	0	0	0	18	19	18	19	15	14	16	14	12
35° – 30°	14	14	17	14	14	16	14	11	3	4	8	5	0	1	0	0	0	0	1	0	16	24	24	23	23	20	11	13	13	10
30° – 25°	15	20	20	15	11	9	10	9	5	4	3	5	5	5	6	5	8	23	22	9	26	24	27	29	25	19	10	12	12	7
25° – 20°	21	22	17	17	6	8	9	16	6	4	4	6	17	18	13	13	14	24	26	21	25	27	26	24	23	15	13	12	9	5
20° – 15°	19	21	12	13	15	15	16	21	16	11	14	21	23	26	19	24	15	16	24	26	25	27	24	22	20	19	8	6	9	4
15° – 10°	17	22	23	16	20	15	19	13	23	20	24	21	19	18	23	25	22	20	16	27	29	23	23	19	19	16	22	12	13	13
10° – 5°	10	21	16	16	13	14	18	14	13	18	18	21	21	22	22	22	23	21	22	12	9	8	15	6	13	10	6	4	5	4
5° – 0°	12	17	18	21	20	18	18	8	13	11	14	18	23	19	21	21	24	27	25	27	12	10	10	14	11	12	5	3	2	2
0° – 5°	14	17	16	22	18	14	12	10	5	8	7	9	11	10	8	13	15	18	16	13	5	6	11	13	16	11	8	5	1	1
5° – 10°	13	9	11	14	12	10	7	4	6	8	6	8	6	3	6	5	11	13	13	7	5	12	12	15	10	9	5	2	1	0
10° – 15°	9	9	9	8	4	4	3	4	2	2	3	8	5	1	1	3	5	5	8	6	2	12	19	11	8	7	6	4	2	0
15° – 20°	7	10	10	7	11	10	4	21	5	6	8	7	6	5	8	7	3	3	2	4	5	7	14	17	15	19	16	10	8	0
20° – 25°	11	14	15	15	12	15	19	20	20	24	22	17	15	13	8	12	11	7	3	5	5	6	6	12	16	17	14	15	19	15
25° – 30°	14	12	15	18	19	18	20	15	21	25	26	25	22	20	23	15	17	11	14	9	12	5	7	13	17	16	17	21	17	19
30° – 35°	12	14	17	14	18	17	18	10	14	28	24	23	24	21	17	19	19	17	15	13	16	16	18	18	18	19	17	20	19	16
35° – 40°	5	11	11	15	11	14	12	9	14	16	19	18	18	17	15	15	15	13	18	18	15	14	17	19	15	23	18	19	13	14
40° – 45°	0	4	6	8	8	10	6	1	8	8	13	10	15	8	11	15	9	14	12	10	12	7	11	11	18	14	10	9	11	9
45° – 50° S	0	0	0	0	1	0	3	1	5	3	3	3	3	5	5	2	4	6	5	3	1	1	5	5	7	6	6	6	3	2

is a typical monthly distribution of data during the northern summer. Data are most seriously deficient just south of the Equator in the eastern Pacific where the cross-equatorial outflow from the northern summer monsoon converges with the southern winter westerlies and sinks (Sadler, 1977).

From the daily data, monthly resultant wind vectors were computed for 2° latitude–10° longitude rectangles, then plotted and analysed subjectively.

5.2 Some examples of analysed resultant wind fields

The wind fields have been analysed beginning October 1974. For illustration here, the mean monthly circulations during the evolution of the eastern Pacific summer monsoon in 1976 are shown together with those for three Julys to illustrate the interannual variability during midsummer. Comparisons are made with the long-term mean 200 mb circulation determined from aircraft and rawin observations (Sadler, 1975).

(i) April 1976 (Fig. 5.2). April is a transition month. The South American anticyclone has weakened since midsummer and moved northeastward to near 5° S. The summer tropical upper-tropospheric trough (TUTT) of the central south Pacific remains in evidence. Westerlies dominate the equatorial region over the Pacific and Atlantic. The direction and speed patterns resemble the long-term mean.

(ii) May 1976 (Fig. 5.3). May is the month of normal monsoon onset in the eastern north Pacific (Sadler, 1963). The upper circulation in 1976 indicates a near-normal year. The outflow from the monsoon is well established from 125° W to the coast of South America but cross-equatorial speeds of 10 m s^{-1} are much stronger than the long-term mean. The circulation and isotach patterns over the remainder of the area are quite similar to the long-term means.

(iii) June 1976 (Fig. 5.4). Since May, the subtropical ridge has undergone normal changes – shifting northward from 8° N to 16° N over central America, extending westward to 140° W and linking over northern South America with the westward extending ridge of Africa. The cross-equatorial outflow branch of the monsoon in the eastern Pacific is stronger than the long-term mean at 200 mb, resulting in the buffer systems being some 5° south of normal in this region. In addition the tropical westerlies near 20° N in the central Atlantic and the Hawaiian region of the Pacific are considerably stronger than the long-term mean.

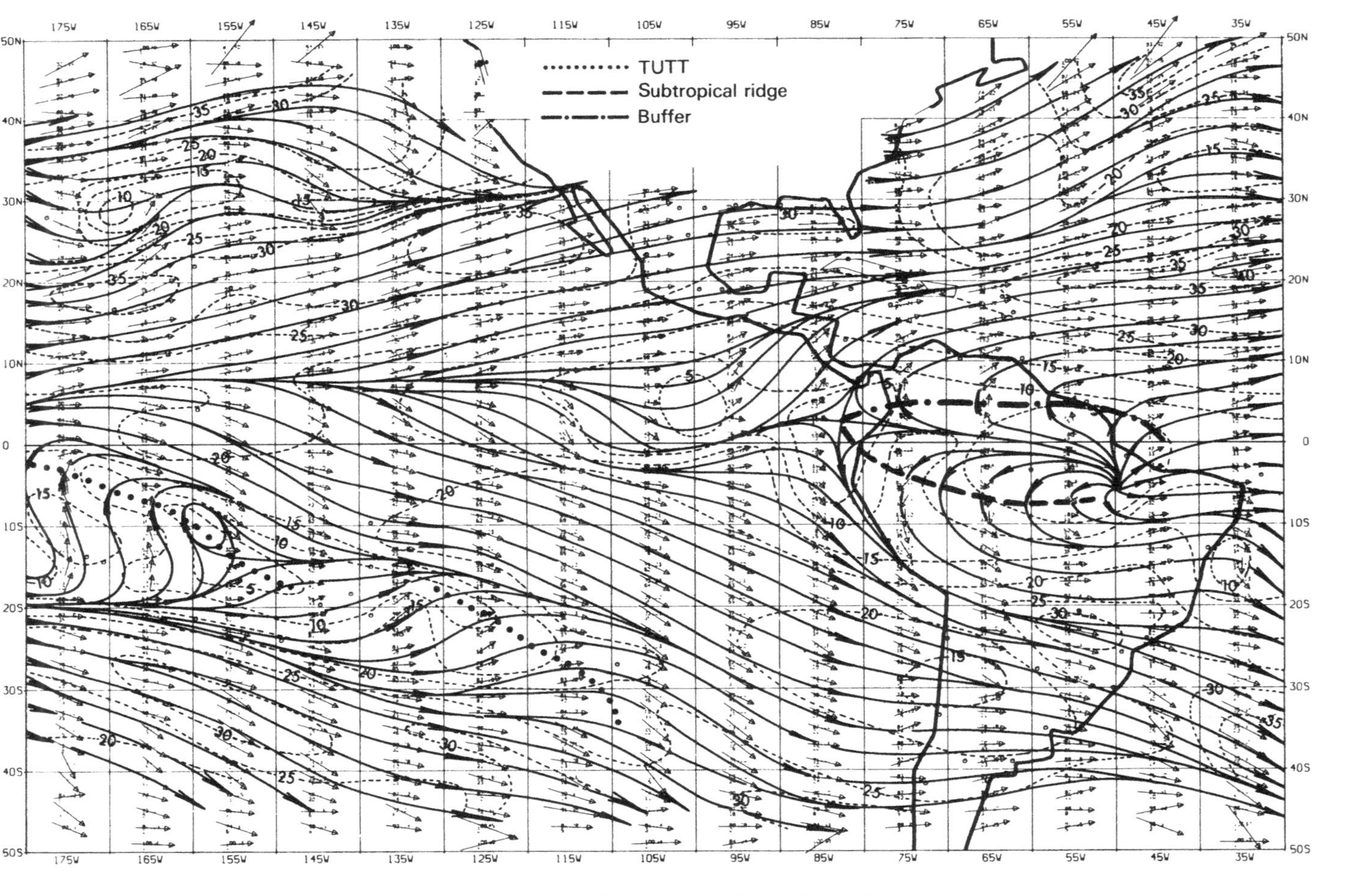

Fig. 5.2. The mean upper-tropospheric circulation in April 1976 from satellite data.

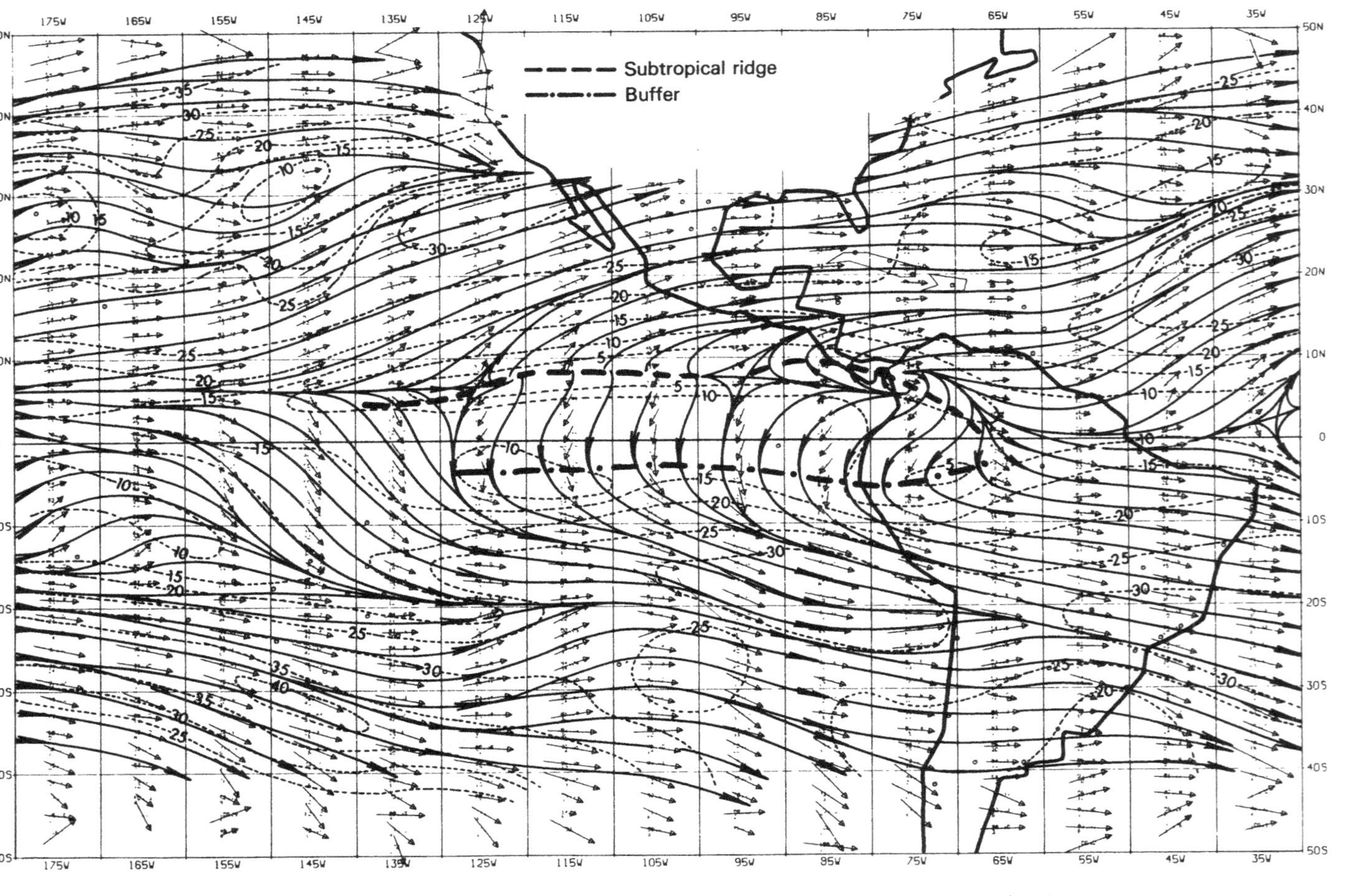

Fig. 5.3. The mean upper-tropospheric circulation in May 1976 from satellite data.

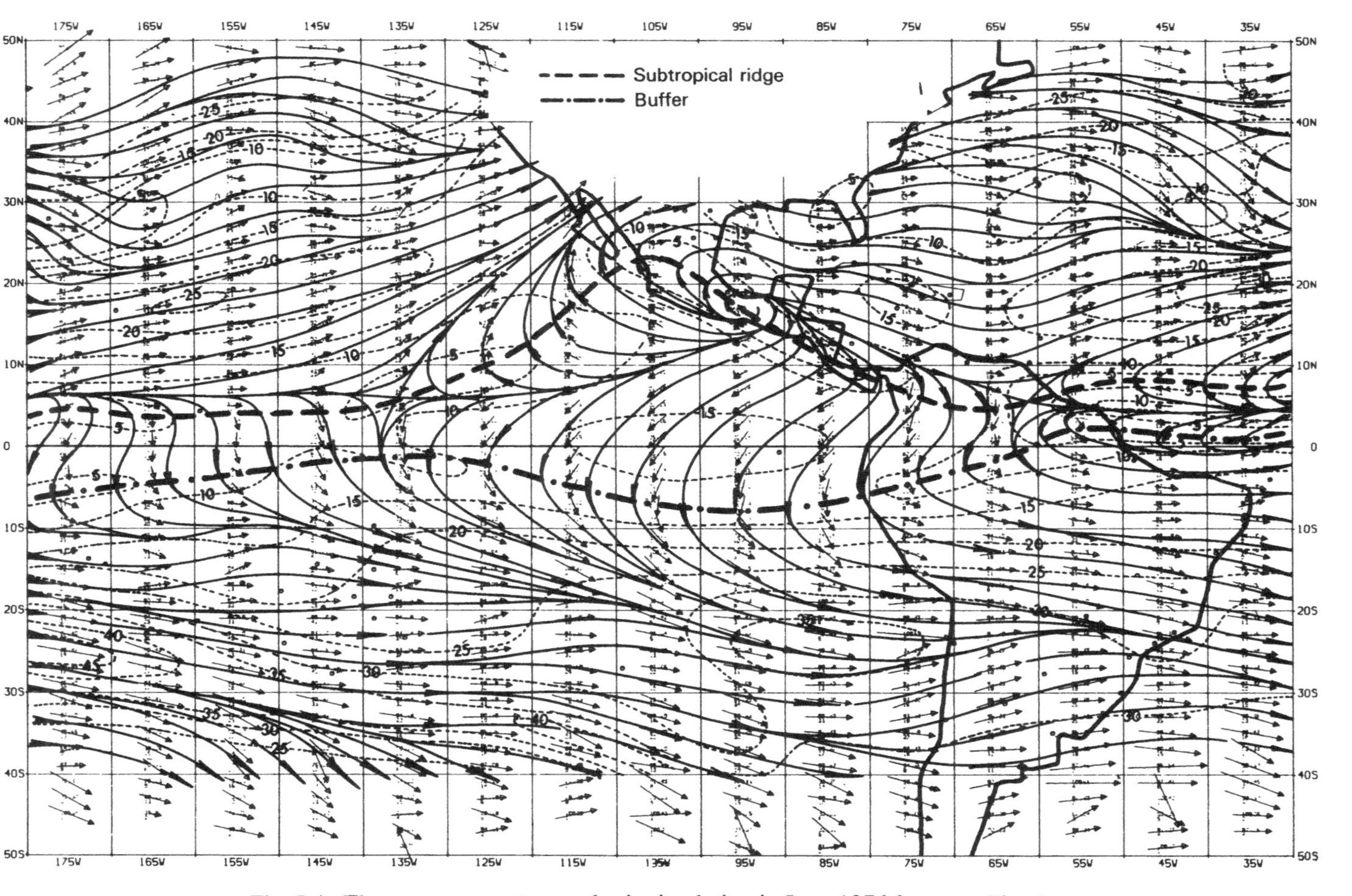

Fig. 5.4. The mean upper-tropospheric circulation in June 1976 from satellite data.

(iv) July 1976 (Fig. 5.5). There are major deviations from the long-term July mean circulation. The anticyclone over Baja California lies 10° west and 5° south of normal while the normally continuous ridge across the southern United States is broken by a trough centred near 100 °W. The subequatorial ridge is not continuous across northern South America and the central Atlantic segment is some 8° north of normal and oriented west-northwest instead of west-southwest. Thus the TUTT is discontinuous in the central Atlantic and much weaker than normal in the western Atlantic. The cross-equatorial flow in the eastern pacific continues to be stronger than the long-term 200 mb mean and the buffer is some 5° south of normal.

(v) Interannual variability. The circulations during the three Julys of 1975, 1976 and 1977, of Figs. 5.6, 5.5 and 5.7, respectively, show significant variability in the position and intensity of the major systems and currents. Some further specific points of note are the following

In contrast to the ill-defined and segmented systems of 1976 discussed earlier, those of 1977 are sharp, strong and well defined. In contrast to the other years the TUTTs over both the Atlantic and Pacific link across central America. A continuous well-defined subequatorial ridge is associated with the strong TUTTs, and a strong zonal subtropical ridge lies across the southern United States, with strong tropical westerly currents in the Caribbean and east-central north Pacific and a strong subtropical easterly current across southern Florida.

The circulations in 1975 are intermediate between the extremes of 1976 and 1977; however, there are two features of note. The tropical westerlies in the central Atlantic near 25° N and the cross-equatorial flow in the eastern Pacific are very strong compared with both the other two years and with the long-term mean.

The cross-equatorial outflow from the eastern north pacific monsoon differs among the three years, but for all three (and for all of the summer months analysed) it is stronger than the long-term mean. The long-term mean in this area is based on aircraft observations which were all made below 13.0 km (40 000 ft) and were concentrated near 11.5 km (35 000 ft). Cirrus outflow from this deep convective region probably occurs at a higher level with greater speeds. if this region is analogous to the Indian monsoon region the maximum speed near 15 km (48 000 ft) is double that at 11.5 km (Ramage and Raman, 1972).

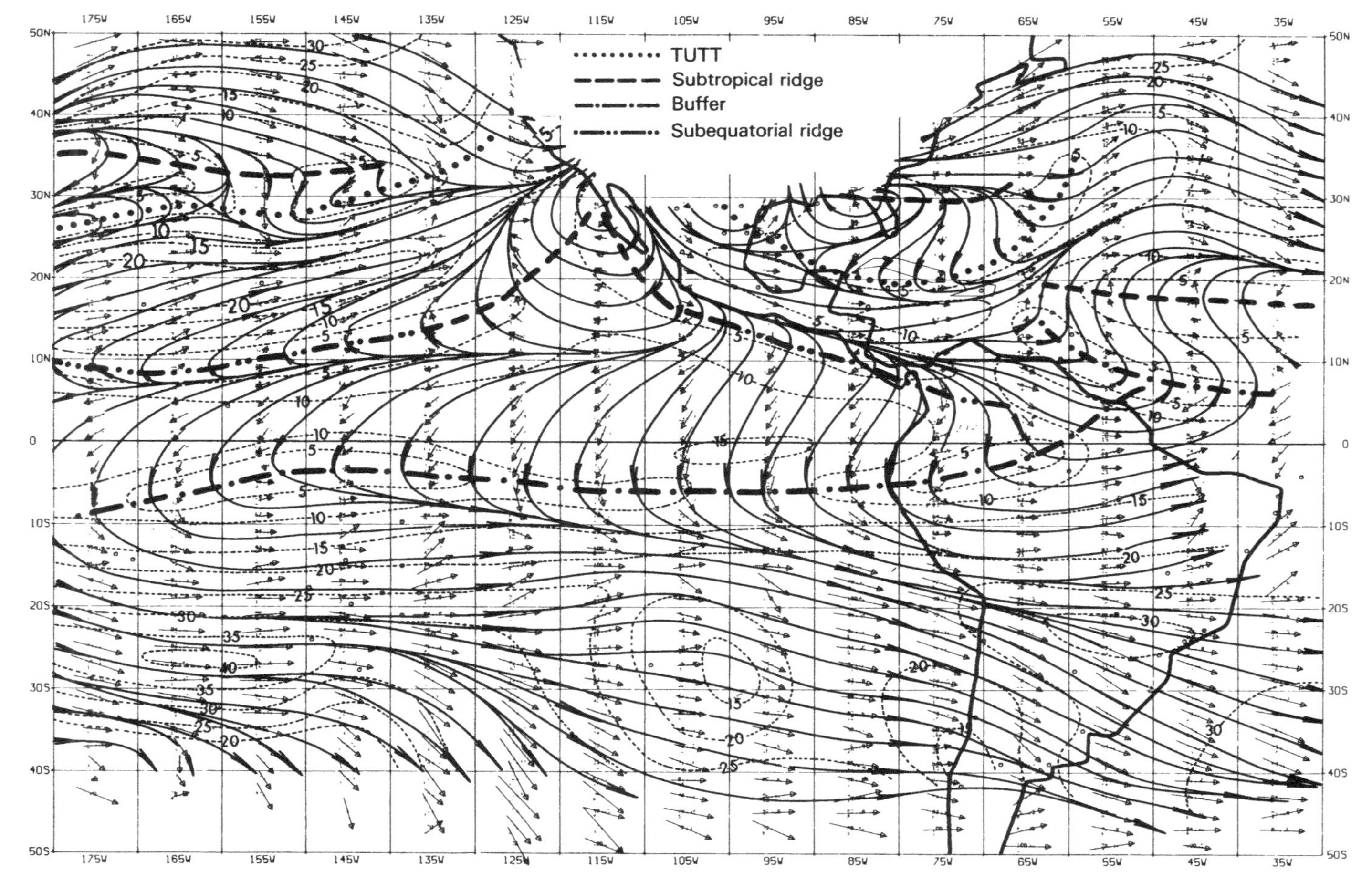

Fig. 5.5. The mean upper-tropospheric circulation in July 1976 from satellite data.

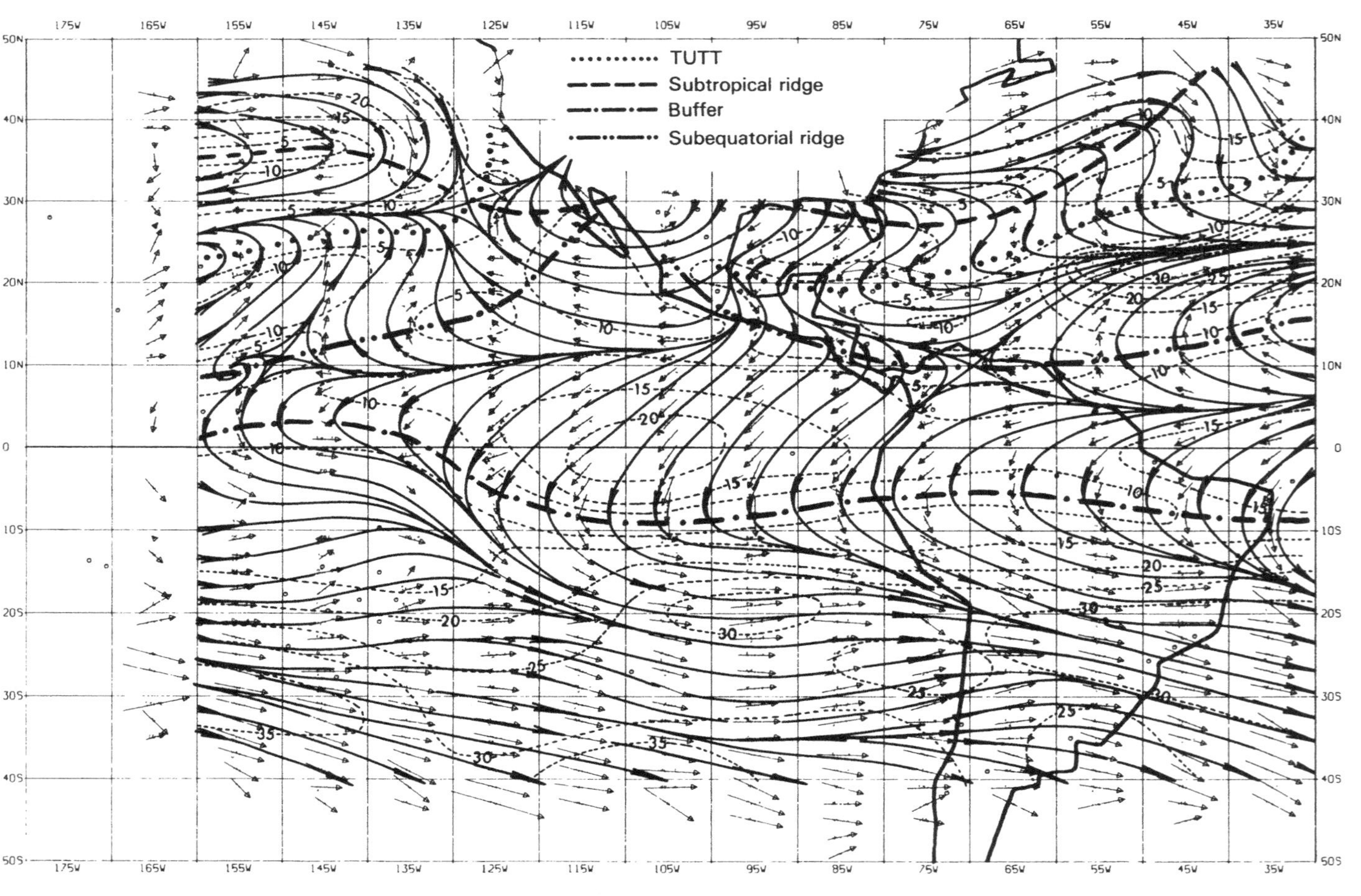

Fig. 5.6. The mean upper-tropospheric circulation in July 1975 from satellite data.

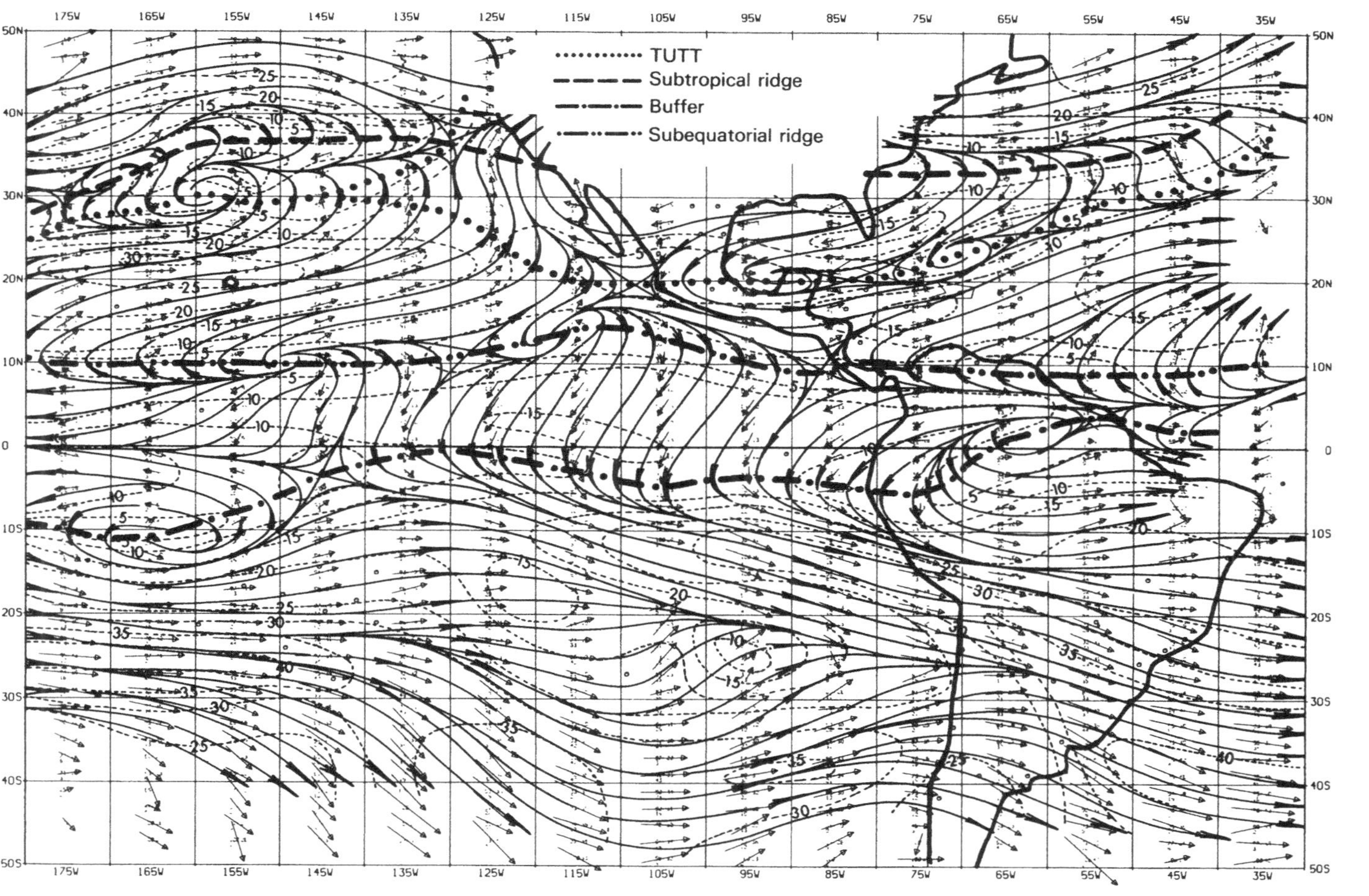

Fig. 5.7. The mean upper-tropospheric circulation in July 1977 from satellite data.

5.3 Comparison between satellite-derived winds and rawins

The preceding analyses demonstrate that the winds derived from geosynchronous satellite data provide mean monthly wind charts which are useful within themselves for monitoring annual and interannual changes in the upper-tropospheric circulation. However, to prevent a discontinuity in the climatic data base between the pre-satellite and satellite eras, a project has been started to integrate the satellite, rawin and aircraft wind observations. We report here some preliminary comparisons with rawin observations. The errors in determining satellite winds and their comparison with rawins have been previously discussed by Hubert and Whitney (1971) and some subsequent comparisons were made by Poteat (1973). These were semicontrolled experiments in that the observations for comparison were within a specified time and distance of each other. We compare the reported monthly mean resultant rawins with the monthly mean resultant satellite winds determined from the analyses. This adds the problem of data bias noted earlier.

(i) Comparison data. Sixty-six stations within the area of the analyses report monthly mean resultant winds regularly in the *Monthly Climatic Data for the World.* They are distributed as shown in Fig. 5.8. Note that there are no stations in the tropical eastern Pacific monsoon region. The satellite-based resultant winds were measured subjectively from the

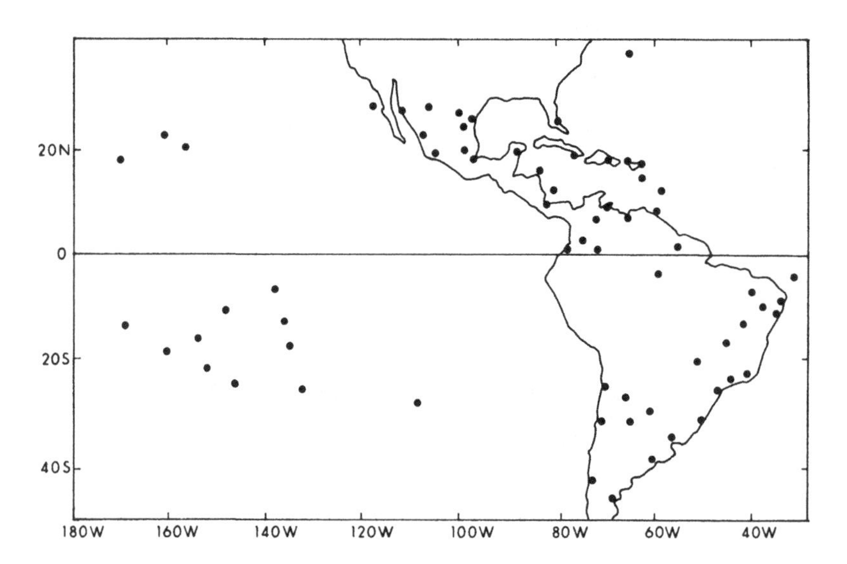

Fig. 5.8. Distribution of rawinsonde stations over the area of analysis.

analyses at each rawin station for comparison with the rawin data at the standard reporting levels of 300, 200, 150 and 100 mb. Seven months of data, from April to October 1976, were included in this preliminary test in order to include the transition and summer monsoon months of the eastern north Pacific.

(ii) Level of best fit. The cumulative frequency of the differences in speed and direction between the satellite-derived winds and the rawins are shown in Figs. 5.9 and 5.10, respectively. Both indicate that the level of best fit is 200 mb but that there is little to choose between 200 and 150 mb, suggesting that the resultant satellite winds represent a layer-average wind from about 11.5 to 15.5 km. The worst fit is at 100 mb near the tropopause. The deviations at 300 mb are about midway between those at the levels of best and worst fit.

Hubert and Whitney (1971) determined the level of best fit to be 300 mb, and it is interesting that at 300 mb our cumulative differences in direction are the same as theirs; however, our data show significant improvement at 200 and 150 mb in comparison to 300 mb. Our cumulative frequency results at 200 mb are similar to those of Poteat (1973). We

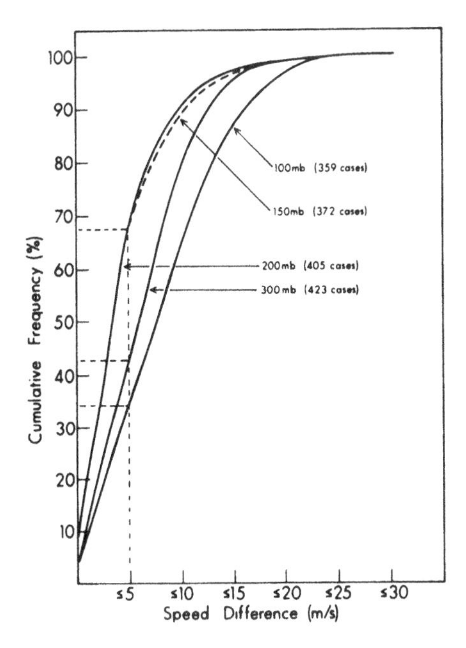

Fig. 5.9. Cumulative frequency distribution of differences in speed between satellite-derived winds and rawins.

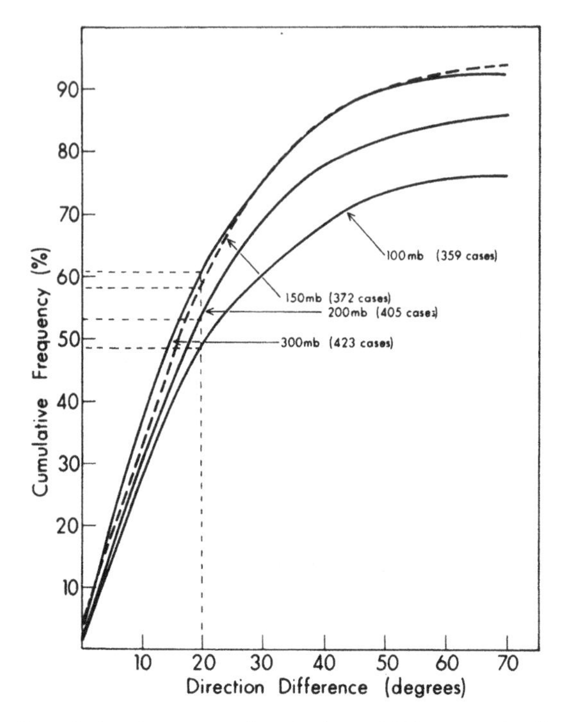

Fig. 5.10. Cumulative frequency distribution of differences in direction between satellite-derived winds and rawins.

have less deviation in speed (82% versus 78%, less than 8 m s^{-1}) but more deviation in direction (61% versus 73%, less than 20°).

Although they were not determined from this small data sample, we feel there are some systematic plus or minus differences between the mean monthly satellite-derived winds and rawins which can be applied to improve the relationship. These will be determined from a much larger data sample.

(iii) Interseasonal variation. The monthly average differences in the speed and direction between the satellite-derived winds and rawins are presented in Table 5.2. The levels of best fit, at 200 and 150 mb, are consistent through the seven-month period. There is an interseasonal change in the differences, particularly in the direction, where the July and August values are about double those of April. This is due in large part to the interseasonal change in the complexity of the flow patterns. The simple pattern of April (Fig. 5.2) gives way in July (Figs. 5.5, 5.6 and 5.7) and August to a complex northern hemisphere pattern of troughs, ridges, cyclones, and anticyclones with accompanying large direction changes in

TABLE 5.2. *The average differences in speed and direction between satellite-derived winds and rawins. Number of cases in brackets.*

1976	300 mb		200 mb		150 mb		100 mb	
	Direction	Speed	Direction	Speed	Direction	Speed	Direction	Speed
April	22 (62)	7	22 (60)	4	20 (56)	5	36 (54)	10
May	22 (60)	8	19 (57)	5	19 (52)	5	20 (52)	10
June	29 (64)	8	19 (60)	5	22 (56)	5	37 (54)	10
July	45 (60)	6	37 (58)	5	36 (51)	5	67 (51)	6
Aug.	57 (61)	6	41 (58)	6	39 (53)	6	87 (49)	9
Sept.	33 (60)	7	22 (57)	5	28 (52)	5	49 (51)	9
Oct.	26 (57)	6	16 (55)	4	18 (52)	4	24 (49)	9
7 month	33 (424)	7	25 (405)	5	26 (372)	5	43 (360)	9

light wind fields. Also, as shown in the next section, the wind speed and deviation in direction are negatively correlated.

(iv) Variations with wind speed. The satellite winds were stratified by wind-speed ranges and the average deviation in speed and direction between the satellite winds and the 200 mb rawins were determined for each range. Results for the combined seven months are shown in Table 5.3. The speed deviations increase and the directional deviations decrease with increasing current speed.

(v) Geographical variations. The geographical variation in the deviations between the satellite-derived winds and the rawins are complex over this large area. As noted in (iii) and (iv) above, the deviations vary with seasons and with wind speed, both of which vary geographically, especially with latitude. In addition, the meteorological system bias varies geographically. We have done little on this last factor except to use a small subsample to verify the findings of Gruber *et al.* (1971), that the satellite

TABLE 5.3. *Average differences in speed and direction between the satellite-derived winds and the 200 mb rawins for various speed ranges during April–October 1976.*

	Satellite wind speed range (m s^{-1})						
	0–5	6–10	11–15	16–20	21–25	26–30	≥31
Sample	43	61	58	68	74	64	40
Speed (m s^{-1})	2	4	6	5	5	6	7
Direction	63°	36°	29°	16°	16°	16°	13°

winds are biased toward more poleward flow in the midlatitude westerlies. Even this bias, which depends on the position and intensity of midlatitude troughs, varies geographically. We plan a geographical stratification from a much larger data sample.

5.4 Comments

A major purpose of this project is to develop circulation indices in the lower latitudes for long-range forecasting and climatic purposes similar to and in support of those available in higher latitudes. The demonstrated capability of the satellite winds to furnish the data base for determining the upper-level circulation, with a fineness of detail never before possible, promotes optimism. The type and areal extent of the indices are yet to be determined. Some obvious candidates are: (*a*) a cross-equatorial meridional wind index for selected longitudinal bands to monitor an outflow branch of summer monsoon areas; (*b*) a zonal-averaged meridional wind index at the Equator to monitor interhemispheric exchange; (*c*) a zonal-averaged meridional wind index at some intermediate latitude to monitor the interaction between the tropics and higher latitudes. This index should also be maintained for subzones to monitor the preferred areas of strong interaction; (*d*) zonal wind indices at the Equator for selected longitude bands to monitor the east–west or Walker circulations.

This research is supported by the National Science Foundation under Grant GX42007 of the North Pacific Experiment (NORPAX).

Mr Louis Oda drafted the figures and Mrs S. Arita typed the manuscript.

References

Gruber, A., Herman, L. and Krueger, A. (1971) The use of satellite cloud motions for estimating the circulation over the tropics. *Mon. Wea. Rev.*, **99**, 739–43.

Hubert, L. F. and Whitney, L. F. (1971) Wind estimation from geostationary satellites. *Mon. Wea. Rev.*, **99**, 665–72.

Poteat, K. O. (1973) A comparison of satellite-derived, low-level and cirrus-level winds with conventional wind observations. *J. Appl. Meteor.*, **12**, 1416–19.

Ramage, C. S. and Raman, C. R. V. (1972) *Meteorological Atlas of the International Indian Ocean Expedition*, Vol. 2, Upper Air. National Science Foundation, Washington, DC.

Sadler, J. C. (1963) TIROS observations of the summer circulation and weather patterns of the eastern North Pacific. *Proceedings of the Symposium on Tropical Meteorology*, New Zealand Meteorology Service, 553–71.

Sadler, J. C. (1975) The upper tropospheric circulation over the global tropics. Report UHMET 75-05, Department of Meteorology, University of Hawaii, 35 pp.
Sadler, J. C. (1977) The upper tropospheric circulation over the global tropics, Part II–Statistics. Report UHMET 77-02, Department of Meteorology, University of Hawaii.

Part Ib

Modelling and theoretical studies

6

Predictability of monsoons

J. G. CHARNEY AND J. SHUKLA

It is shown by numerical simulation that the variability of average pressure and rainfall for July due to short-period flow instabilities occurring in the absence of boundary anomalies can account for most of the observed variability at midlatitudes but not at low latitudes. On the basis of the available evidence it is suggested that a large part of the low-latitude variability is due to boundary anomalies in such quantities as sea-surface temperature, albedo and soil moisture, which, having longer time constants, are more predictable than the flow instabilities. Additional variability due to long-period natural fluctuations would likewise be more predictable.

The degree of predictability of monsoons is a matter of considerable social and economic importance. Large agrarian populations exist in monsoon areas, and monsoon rains have a critical influence on food production and human welfare. The long-range prediction of average rainfall could be of immense value for water management and agricultural planning. In this chapter, evidence that mean flow conditions and precipitation patterns at low latitudes are in principle more predictable than those at high latitudes is presented. Among the low-latitude circulations are the African and Asian monsoons and perhaps also the southeasterly monsoon flow east of the North American Cordillera. In particular we wish to show that the natural flow instabilities on synoptic scales account for most of the interannual variability of monthly mean quantities at midlatitudes, but cannot explain the observed variability at low latitudes. The latter, we suggest, is due partly to fluctuations in such boundary parameters as sea-surface temperature, albedo, ground moisture and vegetation and partly to flow fluctuations of large planetary scales. Since the boundary parameters and the planetary-scale flows vary on much larger time-scales than the synoptic-scale flow instabilities, they should be predictable for longer periods of time. They have not yet been so predicted, but we believe that they can be.

The possibility of longer-range prediction at low latitudes occurred to one of the authors of this chapter during the course of an investigation, by Charney *et al.* (1975, 1977), of the effects of albedo and ground moisture changes on mean July rainfall in, or adjacent to, monsoonal regions. Numerical simulation experiments showed that large changes in rainfall could easily be produced by changes of albedo. To assess the significance of these changes, three randomized numerical experiments were performed with the general circulation model of the Goddard Institute for Space Studies (GISS), in each of which a random alteration to a previously calculated control integration was made by introducing at each grid-point over the globe a random perturbation of temperature, wind and sea-level pressure in the middle of June and continuing the integration through July. The spatial variability of the random perturbations corresponded to Gaussian distributions with zero means, and standard deviations of 1 °C in temperature, 3 m s^{-1} in horizontal wind components and 1 mb in surface pressure. Owing to flow instabilities, the individual perturbed flows began immediately to differ from one another and from the control flow until by early July the daily weather fluctuations had become so different that it seemed possible to regard the variability of the July averages as approximating the variability of the model flow from one July to another. The experiments showed remarkable persistence in mean July rainfall despite the random fluctuations in the daily weather patterns. On closer examination it was found that while the variances among the model-generated July averages approximated the variances among the observed July averages in midlatitudes, they fell far short in low latitudes. To account for the low-latitude discrepancy, we were led to consider the factors responsible for the actual variability at low and at middle latitudes. Two of the most important are the natural fluctuations due to flow instabilities that would exist under prescribed boundary conditions, and the fluctuations due to anomalous variations in the boundary conditions themselves. The two are not independent because there is close coupling between the flow instabilities and the boundary anomalies. However, the boundary anomalies are not observed to be so large as to change the statistical properties of the flow instabilities significantly, and therefore the latter may be studied independently. The reverse is not necessarily true because the boundary anomalies manifest themselves largely through shifts in position and intensity of the flow instabilities.

Let us consider first the natural fluctuations. Lorenz (1969) has shown that deterministic predictability is limited fundamentally by the growth of uncertainty due to flow instabilities. If the uncertainty is initially confined to observations of the very small-scale components of the flow, it will spread to larger and larger scales by a combination of flow instability and nonlinear interaction, until eventually it will extend to the principal weather fluctuations and cause them to become random and unpredictable. Since the rate of growth of uncertainty at small scales is extremely rapid, it is of little avail to increase the accuracy and spatial resolution of the observations beyond what is required to specify the main, energy-bearing, synoptic scales of motion. Uncertainty grows to synoptic scales within a day or two, after which it is the basic instabilities at these scales, and their nonlinear interactions with still larger scales, that determine the further growth of error. When the error enters the synoptic scales, its doubling time is about two days, and as the error amplitude increases and uncertainty is extended to still larger scales, the growth rates diminish. Numerical simulations of error growth set an upper limit to the predictability of the transient synoptic-scale motions of the atmosphere of about two weeks (Charney *et al.*, 1966; Smagorinsky, 1967). However, the larger-scale components of the flow remain predictable for longer periods.

It was in the framework of these ideas that the numerical studies of variability in mean July conditions were performed. Insertion of a rather large random perturbation (1 °C in temperature, 3 m s^{-1} in horizontal wind components and 1 mb in surface pressure) at each grid-point of the numerical model on June 18 caused individual synoptic systems to become effectively random by the beginning of July. It was then assumed that each such July simulation represented a random element of an ensemble whose statistical properties would approximate those of an actual time series of Julys if the seasonally varying Earth's surface boundary conditions were replaced by their seasonally varying climatological averages and if the variances due to timescales of the order of a month or longer could be ignored. We were, of course, aware that the longer fluctuating timescales could also contribute to the total variance, but were unable to carry out longer-period integrations to test their effects because of unavailability of additional computer time. Four

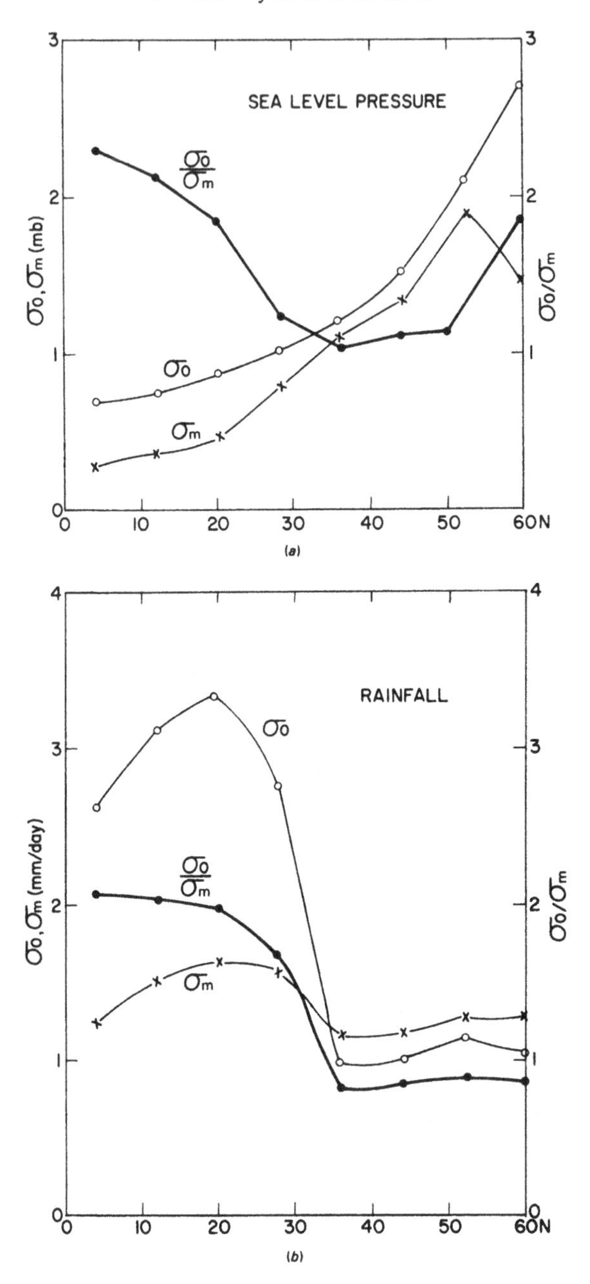

Fig. 6.1. Model and observed zonally-averaged standard deviations as functions of latitude, and their ratio, for: (*a*) mean July sea-level pressure; and (*b*) rainfall. Observed values are for land stations and model values are for grid-points over land.

six-week integrations with four independent sets of randomly perturbed initial conditions were available for comparison. Figs. 6.1*a* and 6.1*b* show the zonally-averaged standard deviations, $\sigma_m(p)$ and $\sigma_m(r)$, of the mean July sea-level pressures and rainfall among the four model runs as functions of latitude. For comparison, the observed standard deviations, $\sigma_o(p)$ and $\sigma_o(r)$, compiled from climatological data are also shown, together with the ratios of the observed standard deviations to the model standard deviations. The observed standard deviations were calculated for 380 stations in the northern hemisphere from mean July sea-level pressure and rainfall data for the 10 years, 1966–75.† It is seen that the ratios σ_o/σ_m are less than 1.5 for pressure at latitudes between 25° N and 55° N and for rainfall at latitudes greater than 30° N; at other latitudes the ratio is greater than 1.5 and for low latitudes it is greater than 2. Since the observations are for land stations, the model values are also taken for the grid-points over land. It is likely that, at all latitudes, the ratio σ_o/σ_m for both sea-level pressure and rainfall may be larger over the oceans compared to the land. Table 6.1 shows areal averages of σ_o, σ_m and σ_o/σ_m for both sea-level pressure and rainfall over three monsoonal and three midlatitude regions for which observational data were available. Again, it is seen that the calculated variability falls far short of the observed variability in the monsoonal areas, whereas it accounts for most of the observed variability in midlatitudes, the contrast being particularly marked for rainfall. Monsoon rainfall, and in general most low-latitude rainfall, is convective in character and is not as well simulated as other physical processes in numerical general-circulation models. However, since the GISS model simulates travelling synoptic disturbances, and in fact tends to overpredict rainfall over land, we have had no reason to assume that the low-latitude variability due to natural fluctuations on timescales of less than a month is underpredicted, and we conclude that synoptic fluctuations are capable of explaining most of the July variability at the higher latitudes, but not at the lower latitudes.

† The standard deviation $\sigma(a)$ at a grid-point or a station was defined as

$$\left(\frac{1}{n-1}\sum_1^n (a_i-\bar{a})^2\right)^{\frac{1}{2}}$$

where

$$\bar{a}=\frac{1}{n}\sum_1^n a_i.$$

For the model runs $n=4$, for the observations $n=10$.

TABLE 6.1. *Areally-averaged modelled and observed standard deviations of July mean sea-level pressure and rainfall for monsoonal and midlatitude regions. $\bar{\sigma}_o$ is the average value of standard deviation for the observations; $\bar{\sigma}_m$ is the average value of standard deviation for the model.*

		Sea-level pressure (mb)			Rainfall (mm per day)		
		$\bar{\sigma}_o$	$\bar{\sigma}_m$	$\bar{\sigma}_o/\bar{\sigma}_m$	σ_o	$\bar{\sigma}_m$	$\bar{\sigma}_o/\bar{\sigma}_m$
African monsoon:	0° –20° N 10 °W–50 °E	0.713	0.354	2.01	1.900	1.328	1.43
Indian monsoon:	0° –30° N 70 °E–90 °E	0.702	0.455	1.54	4.390	1.433	3.06
N. American monsoon:	20 °N–40 °N 100 °W–70 °W	1.384	0.845	1.64	1.817	0.945	1.92
Europe midlatitude:	40 °N–60 °N 10 °W–50 °E	2.528	1.98	1.28	1.057	1.14	0.927
Asia midlatitude:	40 °N–60 °N 110 °E–170 °E	1.738	1.37	1.27	1.506	1.27	1.19
N. America midlatitude:	40 °N–60 °N 90 °W–60 °W	1.339	1.28	1.05	1.442	1.183	1.22

Chervin *et al.* (1976) have used the general-circulation model of the National Center for Atmospheric Research to calculate model standard deviations of the mean January 850 mb temperature from five random January simulations. Comparing with observed standard deviations determined by Crutcher and Meserve (1970) from 14 mean January analyses, they find the model standard deviations to be systematically too small over the oceans, and they attribute this to the absence of ocean temperature anomalies in the model simulations. It can be seen from their charts that ratios of observed to model sigmas are especially large at low latitudes over both land and ocean, in agreement with our results for July.

Madden (1976) has attempted to infer the natural variability of monthly mean sea-level pressure for the months of January, April, July and October directly from a 74-year record of observations, combining successive 96-day sequences centred around each of these months, subtracting suitably smoothed climatological means, filtering out all non-random spectral components of frequency less than $\frac{1}{96}$ cycles per day, and assuming that the smaller-frequency spectral components are constant in amplitude and random in phase, i.e., by assuming white noise at frequencies less than $\frac{1}{96}$ cycles per day. The residual or "natural" variability is then due to the random daily synoptic noise fluctuations, and it is assumed that these may be modelled by a first-order Markov process which gives white noise at sufficiently low frequencies. The deviation of the natural from the observed variability is then considered to be due to a potentially predictable signal. Madden's definition of natural variability differs from ours by including fluctuations of period up to 96 days and boundary-induced fluctuations of periods less than 96 days, but it resembles ours by excluding systematic fluctuations or 'signals' of longer period. Madden was unable to compare his results with observation much below 30° N, but in regions where comparison with observation were possible, his results agree qualitatively with ours, and his conclusion, that the natural variability can explain the actual interannual variability from about 40° N to 60° N, whereas it falls far short south of 40° N, is likewise in agreement with ours.

We now cite evidence for our hypothesis that fluctuations in boundary conditions can account for a large part of the additional variability at low latitudes, whereas they are usually too small to produce a signal strong enough to be discerned above the noise of the natural variability in midlatitudes.

Let us first consider the midlatitude signal to noise ratio. Here, investigation has concentrated on the possible influence of sea-surface temperature anomalies on downstream mean flow conditions. A number of numerical simulation experiments have been performed to evaluate the effects of such anomalies in the Pacific and Atlantic oceans on the downstream flow. We may cite those of Spar and Atlas (1975), Spar (1973*a*, *b*) and Chervin *et al.* (1976). In none of these did significant changes appear downstream when the anomalies were comparable in amplitude to those normally observed; in some cases small systematic changes were observed but only over the boundary anomalies themselves as might have been expected. However, Kutzbach *et al.* (1977) re-examined the results of Chervin *et al.* (1976) and claimed to have found systematic changes in the storm tracks and intensities. While we cannot rule out the possibility that significant downstream changes may be induced by unusually large or widespread sea-surface temperature anomalies (Davis, 1978; Namias, 1971, 1978), we may infer from the existing numerical simulations that large signal to noise ratios are not usually to be expected.

The case for a strong signal to noise ratio at low latitudes stands in sharp contrast. Numerical simulation experiments of Rowntree (1972, 1976) gave support to Bjerknes' (1966, 1969, 1972) hypothesis based on observation, that equatorial sea-surface temperature anomalies produce large changes in the mean tropical Hadley circulation. Another such experiment by Shukla (1975) showed that a decrease in sea-surface temperature of between 1 to 3 °C over an area in the Arabian Sea of approximately 1000 km × 1000 km produced a statistically significant reduction of more than 40% in the monsoon rainfall over India. The reality of this effect is further supported by the observational evidence found by Ellis (1952) and by Shukla and Misra (1977) for the existence of a significant relationship between sea-surface temperature anomalies in the Indian Ocean at about 10° N (along ship lanes) and rainfall over India. With respect to albedo, observational evidence for large changes accompanying the recent drought in sub-Saharan Africa has been presented by Otterman *et al.* (1976) and by Norton and Mosher (1977) from analyses of satellite observations. The effects of such changes on rainfall in monsoonal and semi-arid border regions have been shown by Charney *et al.* (1977) also to be large and to be easily discernible above the random noise.

The low-latitude variability in flow pattern and rainfall due to fluctuations in ocean surface temperatures tends to be large because of the highly nonlinear relation between saturation vapour pressure and temperature, because of the instability of the tropical atmosphere for moist adiabatic processes, and because small temperature changes induce large thermal winds.

To account theoretically for the smallness of the ratio of 'natural' synoptic-scale variance to observed variance at low latitudes, one must show that the flow instabilities at low latitudes are intrinsically weak and that the more intense instabilities of midlatitudes cannot influence the low-latitude circulation by mechanical propagation of energy. The theories of baroclinic instability and wave-energy propagation seem to provide a partial answer. Charney (1947), Eady (1949), Kuo (1952) and others have shown that the flow instabilities of midlatitudes are due primarily to the strong meridional temperature gradients that exist there. Such gradients are not to be found at low latitudes. Also, it may be shown that the midlatitude instabilities cannot propagate their energies into the low latitude easterlies. A simple extension by Charney (1969) of a theory of Charney and Drazin (1961) shows that eastward-propagating Rossby waves originating in the midlatitude westerlies are trapped by zonal easterlies, and since the tropical circulations south of 30° N are dominated by easterlies, they are protected from the disturbances in the westerlies. The GISS general circulation model used for calculating the natural variances has been found to represent the large midlatitude temperature gradients and the low-latitude easterlies quite well (Stone *et al.*, 1977). Although intrusions of large-amplitude midlatitude disturbances into the Indian monsoon sometimes occur (Ramaswamy, 1962), according to Ramamurthy (1969) such cases are infrequent and confined mainly to upper levels. At these levels the mean zonal winds do sometimes become westerly and thus permit the southward propagation of midlatitude disturbances. In the limited six-week integration period of our numerical experiment such events did not occur. However, a subsequent long-period extension of one of the runs did produce an intrusion of a midlatitude trough over India. Events of this kind, though infrequent, could obviously increase the model variance.

We have presented evidence from numerical experiments and from observation that natural fluctuations of short period (less than 30 days in our numerical experiments, and less than 96 days in Madden's analysis of observations) are adequate to account for most of the observed variability at midlatitudes, but are not capable of explaining the variability at low latitudes, say below 30° N. In addition, we have cited evidence based on numerical simulation experiments and statistical analyses of observations that anomalies in sea-surface temperature and in ground albedo are capable of producing large variances. Since these anomalies are usually of long duration, the possibility arises that mean monthly conditions at low latitudes, such as monsoon rainfall, may be predictable with some accuracy. It remains, of course, to be seen whether, and to what extent, the combination of the boundary anomalies with the flow instabilities will account for the actual variances. Natural variations of long period, such as fluctuations in the tropical branch of the Hadley circulation driven by fluctuations in the eddy transports of heat and momentum at midlatitudes, may very well contribute. But if such fluctuations exist, it seems likely that they, too, being of longer period, will be predictable for longer periods of time. Thus we suggest that the synoptic-scale flow instabilities which limit prediction so drastically at midlatitudes have less influence at low latitudes and therefore leave room for the longer-period and more predictable signals to make themselves felt.

Clearly, more observational, numerical and theoretical analyses must be performed to assess the relative importance of the various factors producing tropical variability. Above all, we consider that attempts at longer-range prediction in low latitudes should concentrate first on the observation and prediction of fluctuations in such variables as sea-surface temperature, vegetative cover, albedo and ground moisture. Such observations can in principle be made from polar orbiting or geostationary satellites.

References

Bjerknes, J. (1966) A possible response of the atmospheric Hadley Circulation to equatorial anomalies of ocean temperature. *Tellus*, **18**, 820.

Bjerknes, J. (1969) Atmospheric teleconnections from equatorial Pacific. *Mon. Wea. Rev.*, **97**, 163.

Bjerknes, J. (1972) Large-scale atmospheric response to the 1964–65 Pacific equatorial warming. *J. Phys. Oceanogr.*, **2**, 212.

Charney, J. G. (1947) The dynamics of long waves in a baroclinic westerly current. *J. Meteor.*, **4**, 135.

Charney, J. G. (1969) A further note on large-scale motions in the tropics. *J. Atmos. Sci.*, **26**, 182.

Charney, J. G. and Drazin, P. G. (1961) Propagation of planetary scale disturbances from the lower into the upper atmosphere. *J. Geophys. Res.*, **66**, 83.

Charney, J. G. *et al.* (Panel on Int'l. Meteor. Cooperation to the Comm. on Atmos. Sci., NAS/NRC) (1966) The feasibility of a global observation and analysis experiment. *Bull. Amer. Met. Soc.*, **47**, 200.

Charney, J. G., Stone, P. H. and Quirk, W. J. (1975) Drought in the Sahara: A biogeophysical feedback mechanism. *Science*, **187**, 434.

Charney, J. G., Quirk, W. J., Chow, S. and Kornfield, J. (1977) A comparative study of the effects of albedo change on drought in semi-arid regions. *J. Atmos. Sci.*, **34**, 1366.

Chervin, R. M., Washington, W. M. and Schneider, S. H. (1976) Testing the statistical significance of the response of the NCAR general circulation model to North Pacific Ocean surface temperature anomalies. *J. Atmos. Sci.*, **33**, 413.

Crutcher, H. L. and Meserve, J. M. (1970) Selected level heights, temperatures and dew points for the Northern Hemisphere. NAVAIR 50-1C-52 Rev., Naval Weather Service Command, Washington, DC.

Davis, R. E. (1978) Predictability of sea level pressure anomalies over the North Pacific Ocean. *J. Phys. Oceanogr.*, **8**, 233.

Eady, E. T. (1949) Long waves and cyclone waves. *Tellus*, **1**, 33.

Ellis, R. S. (1952) A preliminary study of a relation between surface temperature of the North Indian Ocean and precipitation over India. M.S. thesis, Department of Meteorology, Florida State University.

Kuo, H. L. (1952) Three-dimensional disturbances in a baroclinic zonal current. *J. Meteor.*, **9**, 260.

Kutzbach, J. E., Chervin, R. M. and Houghton, D. D. (1977) Response of the NCAR general circulation model to prescribed changes in ocean surface temperature. *J. Atmos. Sci.*, **34**, 1200.

Lorenz, E. N. (1969) Three approaches to atmospheric predictability. *Bull. Amer. Met. Soc.*, **50**, 345.

Madden, R. A. (1976) Estimates of the natural variability of time-averaged sea level pressure. *Mon. Wea. Rev.*, **104**, 942.

Namias, J. (1971) The 1968–69 winter as an outgrowth of sea and air coupling during antecedent seasons. *J. Phys. Oceanogr.*, **1**, 65.

Namias, J. (1978) Multiple causes of the North America abnormal winter 1976–77. *Mon. Wea. Rev.*, **106**, 279.

Norton, C. C. and Mosher, F. R. (1977) An investigation of surface albedo variations during the recent Sahel drought. Report on NASA Grant NGR 50-002-215, Space Science and Engineering Center, University of Wisconsin.
Otterman, J., Walter, L. S. and Schmugge, T. J. (1976) Observations from ERTS of overgrazing and cultivation impact on the earth's surface. *Space Research*, **16**, 15.
Ramamurthy, K. (1969) Forecasting Manual, Part IV, India Meteorological Department.
Ramaswamy, C. (1962) Break in the Indian summer monsoon as a phenomenon of interaction between the easterly and the sub-tropical westerly jet streams. *Tellus*, **14**, 337.
Rowntree, P. R. (1972) The influence of tropical east Pacific Ocean temperatures on the atmosphere. *Quart. J. Roy. Meteor. Soc.*, **98**, 290.
Rowntree, P. R. (1976) Response of the atmosphere to a tropical Atlantic Ocean temperature anomaly. *Quart. J. Roy. Meteor. Soc.*, **102**, 607.
Shukla, J. (1975) Effect of Arabian sea surface temperature anomaly on Indian summer monsoon: A numerical experiment with the GFDL model. *J. Atmos. Sci.*, **32**, 503.
Shukla, J. and Misra, B. M. (1977) Relationships between sea surface temperature and wind speed over the central Arabian Sea and monsoon rainfall over India. *Mon. Wea. Rev.*, **105**, 998.
Smagorinsky, J. (1967) Problems and promises of deterministic extended-range forecasting. *Bull. Amer. Met. Soc.*, **50**, 286.
Spar, J. (1973*a*) Transequatorial effects of sea surface temperature anomalies in a global general circulation model. *Mon. Wea. Rev.*, **101**, 554.
Spar, J. (1973*b*) Supplementary notes on sea surface temperature anomalies and model-generated meteorological histories. *Mon. Wea. Rev.*, **101**, 767.
Spar, J. and Atlas, R. (1975) Atmospheric response to variations in sea surface temperature. *J. Appl. Meteor.*, **14**, 1235.
Stone, P. H., Chow, S. and Quirk, W. J. (1977) The July climate and a comparison of the January and July climates simulated by the GISS general circulation model. *Mon. Wea. Rev.*, **105**, 170.

7

A review of general-circulation model experiments on the Indian monsoon

W. M. WASHINGTON

After simulating the global aspects of climate for a number of years, general circulation models (GCMs) have recently been applied to studies of the monsoon. Despite differences in models, simulated large-scale features agree quite well with observations. The Somali Jet, tropical jet stream, and pressure and wind patterns are well simulated, but many small-scale features are not. Recent experiments give insight into the role of mountains and ocean-surface temperature anomalies in the monsoon circulation. Previous work with such models is reviewed, and several speculations about the use of GCMs for monsoon simulation are made.

7.1 Introduction

General circulation models (GCMs) have recently been applied to the study of the summer and winter monsoon patterns over eastern Africa, southern Asia, and the nearby oceans. For convenience, we refer to the entire region as the Indian monsoon region. Although GCMs have led to a better understanding of the large-scale features of the monsoon, they have not properly simulated all the small-scale features. This review discusses the 'state of the art' of Indian monsoon simulations and points out problems in, and prospects for, improving our understanding of this interesting and important meteorological phenomenon.

Washington (1970) experimented with a 5-degree, latitude–longitude grid version of the National Center for Atmospheric Research (NCAR) GCM showing the basic features of the monsoon. Even in this early experiment, the strong cross-equatorial jet near Somalia, the formation of a tropical easterly jet, and the low-level westerly flow in the vicinity of India were apparent. However, the precipitation pattern was incorrect, especially over land. This was partly because the model at that time assumed the entire Earth's surface to be saturated, in contrast to later versions that included a simplified ground hydrology parameterization for variable soil moisture (Washington, 1974; Washington and Williamson, 1977). Another reason for errors in precipitation patterns over the Indian subcontinent was the failure of the model to effectively handle small-scale precipitation, a problem still plaguing present-day GCMs.

Murakami *et al.* (1970) and Godbole (1973) carried out several experiments with a zonally-symmetric GCM by applying mean conditions at 80 °E. They were able to simulate the principal areas of zonal westerlies and easterlies in their cross-sections. Alyea (1972) used a quasi-geostrophic model to simulate the July flow patterns over the northern hemisphere, but he failed to generate the proper monsoon structure. The reasons for failure are not known, but Abbott (1973), with a geostrophic model similar to Alyea's, generated a monsoon circulation structure much closer to reality. It is conjectured that Alyea's model failed because of the heating and cooling distributions used – not because of the simplified quasi-geostrophic assumption. Manabe *et al.* (1974) carried out a seasonal calculation with the Geophysical Fluid Dynamics Laboratory (GFDL) GCM, examining the tropical regions. They simulated the global circulation patterns quite well and produced a remarkably realistic simulation of the large-scale features of the summer and winter monsoons. Gilchrist's (1974, 1976) technical notes from the British Meteorological Office GCM also showed good simulation of the summer circulation patterns. Washington and Daggupaty (1975), and Hahn and Manabe (1975) published results of the NCAR and GFDL GCMs, respectively, giving detailed analyses of the summer monsoon circulation.

Other July global simulations have been discussed, but without a great deal of attention to the monsoon circulation itself (Mintz *et al.*, 1972, for the model developed at the University of California at Los Angeles (UCLA); Stone *et al.*, 1977, for the Goddard Institute for Space Studies (GISS) model; and more recently, Gates and Schlesinger, 1977, using the Rand GCM).

Gilchrist (1977) examined the horizontal distributions of winds in the lower and upper troposphere, precipitation amounts, and sea-level pressure in the GFDL, Rand, NCAR, and British Meteorological Office models. All models simulated qualitatively the upper-tropospheric tropical easterly jet, the Somali Jet, and the low-level westerly flow in the vicinity of India. However, the precipitation amounts in the models varied greatly from each other and from those observed. Model simulations are usually averaged for a monthly mean, and precipitation statistics outside areas of large rainfall are highly variable in the actual atmosphere as well as in GCMs on a monthly timescale.

The basic GCM equations are presented in § 7.2 and the basic features simulated by most current GCMs are discussed in § 7.3. In § 7.4, GCM sensitivity experiments are summarized and the factors influencing the African–Asian summer monsoon are examined. § 7.5 reviews what is needed to improve GCMs and speculates on what types of experiments will further our understanding of the monsoon circulation.

7.2 Basic equations

Following Kasahara (1974), we present the basic equations in Cartesian coordinates where x, y, z are directed eastward, northward, and upward, respectively.

The equation of horizontal motion may be expressed as

$$\frac{\mathrm{d}\boldsymbol{V}}{\mathrm{d}t} + f\boldsymbol{k} \times \boldsymbol{V} = -\frac{1}{\rho}\nabla p + \boldsymbol{F}, \tag{7.1}$$

$$\left.\begin{aligned} &\boldsymbol{V} = u\boldsymbol{i} + v\boldsymbol{j}, \qquad \nabla = \boldsymbol{i}\frac{\partial}{\partial x} + \boldsymbol{j}\frac{\partial}{\partial y}, \\ &\frac{\mathrm{d}}{\mathrm{d}t} = \frac{\partial}{\partial t} + \boldsymbol{V} \cdot \nabla + w\frac{\partial}{\partial z}, \end{aligned}\right\} \tag{7.2}$$

in which $\boldsymbol{i}$, $\boldsymbol{j}$, and $\boldsymbol{k}$ denote unit vectors in x, y, and z coordinates, respectively; ∇ is the horizontal del operator, $\boldsymbol{V}$ horizontal velocity, u and v the x and y components of $\boldsymbol{V}$, w vertical velocity, $\mathrm{d}/\mathrm{d}t$ total derivative, f Coriolis parameter ($\equiv 2\Omega \sin \phi$), Ω angular velocity of the earth's rotation, ϕ geographical latitude, ρ density, p pressure, and $\boldsymbol{F}$ frictional force per unit mass.

For large-scale motions, the hydrostatic equation

$$\frac{\partial p}{\partial z} = -\rho g, \tag{7.3}$$

where g denotes the Earth's gravity, is a good approximation to the vertical equation of motion.

The mass continuity equation can be written as

$$\frac{\mathrm{d}}{\mathrm{d}t}\ln \rho + \nabla \cdot \boldsymbol{V} + \frac{\partial w}{\partial z} = 0, \tag{7.4}$$

or

$$\frac{\partial \rho}{\partial t} + \nabla \cdot (\rho \boldsymbol{V}) + \frac{\partial (\rho w)}{\partial z} = 0. \tag{7.5}$$

The first law of thermodynamics may be expressed by

$$\frac{\mathrm{d}}{\mathrm{d}t}\ln \theta = \frac{Q}{c_p T}, \tag{7.6}$$

where θ is the potential temperature defined by

$$\theta \equiv T(p_0/p)^\kappa, \tag{7.7}$$

with $\kappa = R/c_p$ and $p_0 = 1013$ mb; and T is temperature given by the ideal gas law

$$p = \rho R T, \tag{7.8}$$

where R represents the specific gas constant. In (7.6), c_p signifies the specific heat at constant pressure and Q the rate of heating/cooling per unit mass per unit time. Here R and c_p are related through

$$R = c_p - c_v, \tag{7.9}$$

where c_v denotes the specific heat at constant volume. Both c_p and c_v are assumed constant.

Another form of the first law of thermodynamics is derived from (7.6), (7.7) and (7.8) as

$$c_p \frac{\mathrm{d}T}{\mathrm{d}t} - \frac{1}{\rho}\frac{\mathrm{d}p}{\mathrm{d}t} = Q. \tag{7.10}$$

Equations (7.1) to (7.10) constitute the basic dynamical principles of GCMs (see Kasahara, 1974, for details on alternate vertical coordinate systems). For large-scale predictions, it is important to take into account water vapour in the atmosphere and, thus, prognostic equations similar to (7.1) must be included with sources and sinks for moisture. If the frictional term $\boldsymbol{F}$ and the heating term Q can be expressed by the dependent variables $\boldsymbol{V}$, w, p, and ρ (or T), the system, together with the proper boundary conditions, is complete.

The physical processes incorporated into the GCMs discussed here are solar and infrared radiation, cloudiness (specified or computed), latent heat release due to small- and large-scale precipitation, and subgrid-scale transports of momentum, moisture, and sensible heat. Over the oceans, surface temperature is specified from climatological data, while over land, sea ice, and snow-covered regions it is computed from a surface energy balance.

7.3 GCM simulations of large-scale features of the Indian monsoon

Because of the high specific heat of water, continents tend to be warmer than oceans in summer and cooler than oceans in winter. This differential heating results in pressure and temperature changes that cause much of the circulation difference between the two seasons. In particular, in summer, relatively low pressure exists over continents, while in winter, relatively high pressure exists, suggesting that without the Coriolis effect there would be low-level inflow into the land areas in summer and outflow in winter. At higher levels, the pattern reverses. The entire system produces a direct circulation, with upward flow mainly over the region with heating and with sinking mainly over cooling regions.

Specifically, in the Indian monsoon region, the summer heating of the Indian subcontinent and the Arabian–African regions causes low pressure, whereas high pressure exists over the Indian Ocean. Because of the Coriolis effect, wind does not flow directly onto the land but instead low-level westerlies form parallel to the strong pressure gradient along the boundary between the Asian–Indian continent and the ocean areas to the south. Reversal of this heating pattern in winter can be seen in Hahn and Manabe (1975) and by comparing Washington (1976) with Washington and Daggupaty (1975).

7.3.1 Wind patterns

One of the best methods of visualizing the patterns in GCM data simulating the Indian monsoon is to examine the wind fields. In Fig. 7.1 the winds in the lower troposphere are from the GFDL model by Hahn and Manabe (1975), the NCAR GCM (Washington and Daggupaty, 1975), and the British Meteorological Office model (Gilchrist, 1977). The observed data are from Atkinson and Sadler (1970). Each GCM differs from the observed and other models in detail, but the large-scale overall features, the Somali Jet, and the westerlies near India appear in all model

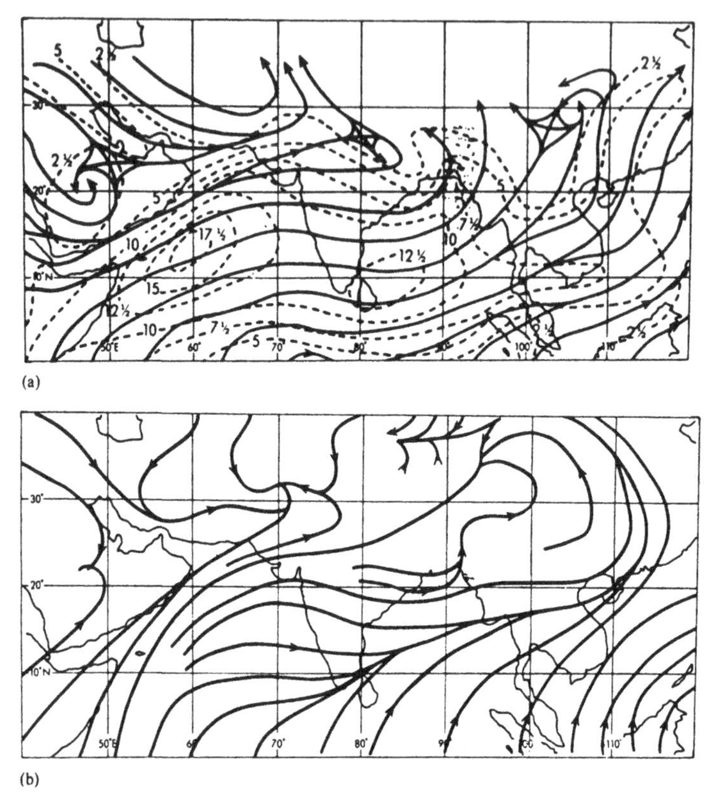

Fig. 7.1. Low-level wind distribution (streamlines and isotachs) over the monsoon area (from Gilchrist, 1977). Isotachs are labelled in m s^{-1}. (*a*) Observed climatological gradient level winds (Atkinson and Sadler, 1970). (*b*) Near-surface winds from the GFDL model with isotachs omitted. The streamlines are estimated from published streak diagrams (Hahn and Manabe, 1975). (*c*) Average winds at 1.5 km from days 91 to 120 of a July integration of the NCAR model (Washington and Daggupaty, 1975). (*d*) Average winds at the lowest level for the Meteorological Office model. (From Gilchrist, 1977.)

simulations. The winds differ from model to model partly because the streamlines and isotachs are shown at different levels in each model. The models also show convergence to a monsoon trough centred north of the Equator.

The Somali Jet transports a great deal of moisture from the southern Indian Ocean across the Equator to the Arabian Sea. This large flux of moisture is an important component of the Indian monsoon (Hahn and Manabe, 1975; Shukla, 1975). Washington (1976) and Manabe *et al.* (1974) depict the reversal of the Somali Jet and thus a southward transport in winter. The upper-tropospheric flow for the three models

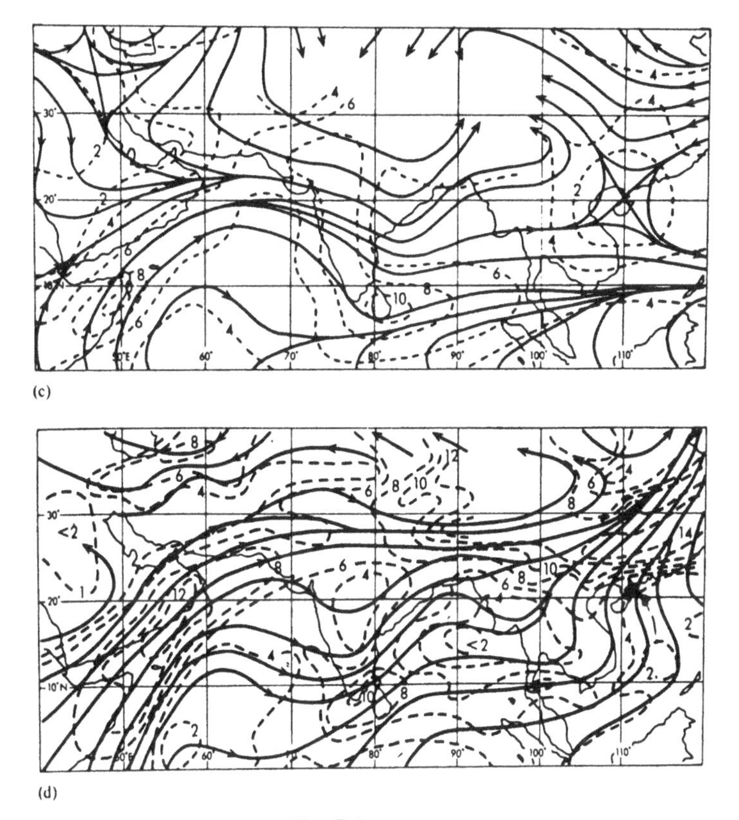

Fig. 7.1.—*cont.*

appears in Gilchrist (1977). Basic features are simulated in all circulations, but quantitative differences do appear in the various models. For example, an anticyclone over the Himalayan region, the northern hemisphere jet stream, and an easterly jet appear at different locations in each simulation. Although not detailed here, Ramage and Raman (1972) show a very large interannual variability in the observed data of the monsoon circulation. It is not clear how much the model difference from reality is due to model variability or to data variability. Gates and Schlesinger (1977) do not show upper-level winds, but from the temperature structure it is possible to infer the existence of a large anticyclone in the Himalayas. In simplest terms, the temperature structure is probably caused by sensible heat from the Earth's surface and is augmented by latent heat from precipitation over the continent so that a reversed temperature gradient is generated from the land region in the vicinity of the Himalayas to the equatorial region. This reversed temperature gradient implies an easterly jet from thermal wind considerations.

7.3.2 *Precipitation patterns*

Gilchrist (1977) also shows precipitation patterns for the three models. All models have precipitation with a maximum over the central Indian Ocean with increasing amounts toward Burma. This compares with the observed data of Möller (1951) which show considerable precipitation near Burma and in the surrounding regions and a secondary maximum along the Western Ghats. The latter maximum is not seen in the GCM results because the mountains are smoothed. Precipitation amounts exceed 2 cm per day in the NCAR and GFDL results which is excessive if compared with Möller's (1951) data. Although the monthly mean precipitation in the models is highly variable from one year to the next, it appears that the NCAR and GFDL models generate too much precipitation over the tropical oceans. No large amounts of precipitation occur over the northern parts of India, in particular the Ganges Valley. This valley does not appear in the mountain distributions in the NCAR and GFDL GCMs because such features are smoothed. Without a Ganges Valley to allow proper surface values of moisture in the boundary layer of the model, the moist static energy is underestimated. The generation of large amounts of precipitation in the Ganges Valley comes from a weak low-pressure system or monsoon depressions forming in the Bay of Bengal. Presumably, much of the moisture driving these disturbances comes from the lower troposphere. The disturbances move up the Ganges Valley towards the northwest releasing large amounts of precipitation, but their scale is of the order of a few degrees of latitude and longitude and is not resolvable by existing GCMs. Because of their size and intermittence, such monsoon depressions will probably have to be parameterized in future GCMs. The same problems with precipitation fields appear in the Rand model (Gates and Schlesinger, 1977).

7.4 Numerical sensitivity experiments

7.4.1 *Effect of mountains*

Experiments on the effect of mountains on the Asian monsoon circulation were performed by Hahn and Manabe (1975). By removing the mountains and repeating a seasonal simulation from 25 March to 31 July, they found that the mountains strongly affect the formation of low pressure over the Indian and southeast Asian subcontinents. In experiments without mountains, the low-pressure system moved far to the north and east of the Himalayas (Fig. 7.2) and the moist air penetrated

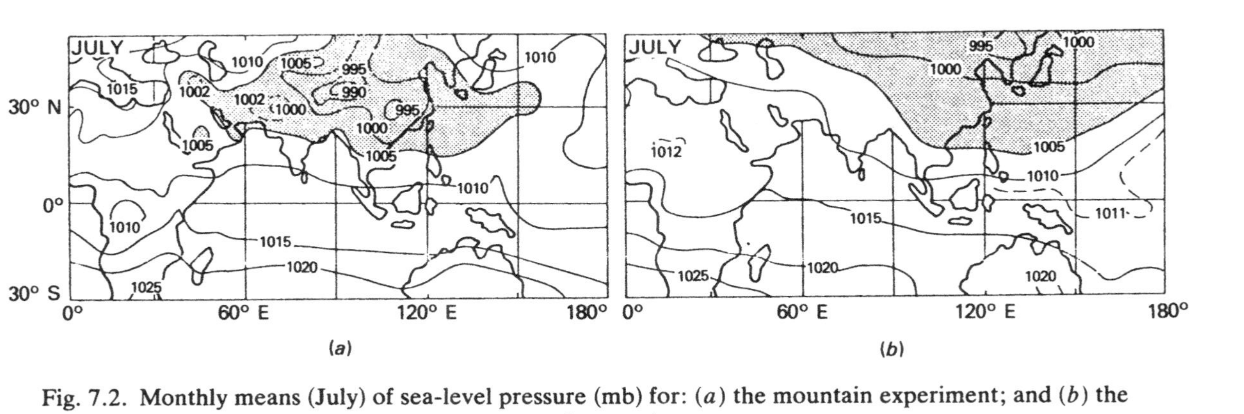

Fig. 7.2. Monthly means (July) of sea-level pressure (mb) for: (*a*) the mountain experiment; and (*b*) the no-mountain experiment.

farther north into India. They concluded that the no-mountain experiment extended the monsoon circulation into the Asian continent more than did the control experiment with mountains. In the mountain simulation, the onset of the monsoon occurred rather abruptly so that the jet stream quickly moved north of the Himalayas into its summertime position. In the simulation without mountains, the transition from spring to summer was gradual. They attributed these differences to the mechanical and thermodynamical effects of the Tibetan Plateau. However, the basic monsoon patterns are quite similar in the figures for the two experiments. The thermal influence is more significant than the topographical influence in the creation of such features as the tropical easterlies, the Somali Jet, and other large-scale features of the monsoon. Mountains, however, block the flow, preventing the colder air from penetrating farther south. It is possible to conclude from the Hahn and Manabe (1975) experiments that mountains modulate the placement of dominant features.

7.4.2 Effects of ocean temperature anomaly patterns

Two papers have been published on the effects of ocean-surface temperature anomalies on the summer monsoon (Shukla, 1975; Washington *et al.*, 1977). Shukla introduced into a seasonal experiment with the GFDL model a cold sea-surface temperature anomaly on 15 June in the vicinity of the Somali coast and western Arabian Sea. The anomaly pattern had a maximum of −3 °C near the coast and decreased linearly to 1500 km east of the coast (Fig. 7.3). Shukla found that the decrease in ocean temperature led to a decrease in precipitation over India and the surrounding region. His hypothesis was that cold temperatures result in a decrease in evaporation and thus a decrease in the moisture available for downstream precipitation. He plotted (Fig. 7.4) the rate of precipitation averaged over the verification area for the standard and anomaly experiments showing the expected result of decreased precipitation in the anomaly experiment. However, after approximately 34 days, the difference between the anomaly and the standard run was very small. He calculated from the standard model statistics a measure of the noise level. It appears from his analysis that the effect of the anomaly is significant, but it is difficult to determine if this is a transient response of the model or a difference that would be maintained, because the difference between anomaly and control experiments is small at the end of the experiment (Fig. 7.4).

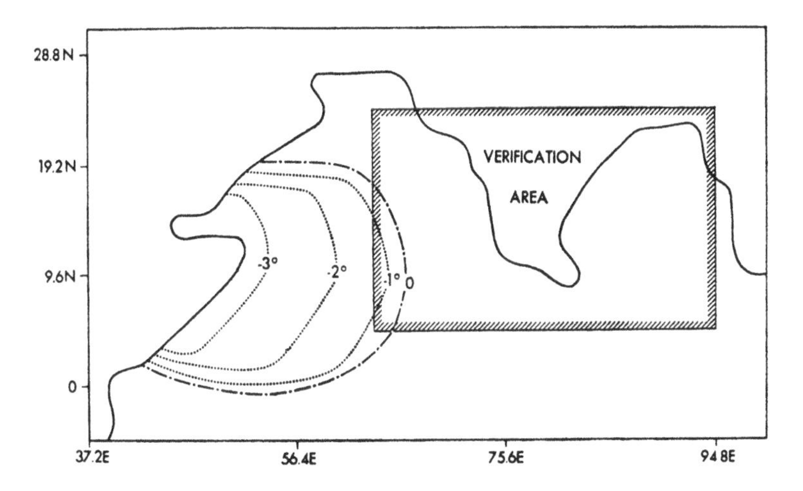

Fig. 7.3. Sea-surface temperature (°C) over the western Arabian Sea and location of verification area.

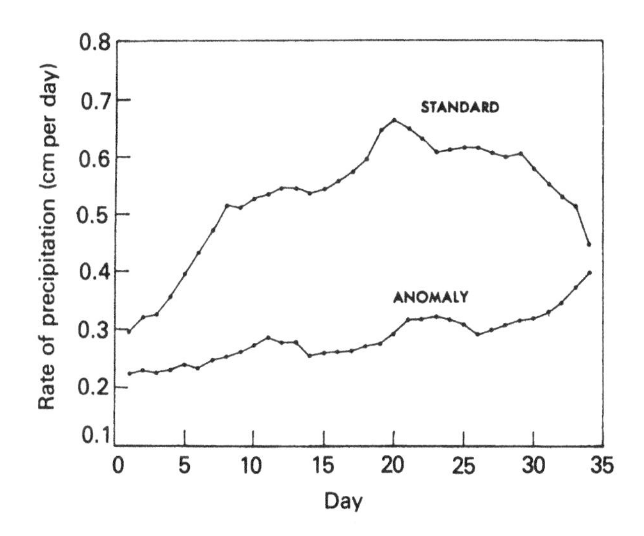

Fig. 7.4. Rate of precipitation (cm per day) averaged over the verification area for the standard and anomaly runs.

In a second set of experiments, Washington *et al.* (1977) imposed prescribed temperature changes, first in the western Arabian Sea, then in the eastern Arabian Sea, and finally in the central Indian Ocean (Fig. 7.5). The pattern of changes in the western Arabian Sea was identical to Shukla's so that comparison of the response of the two models could be

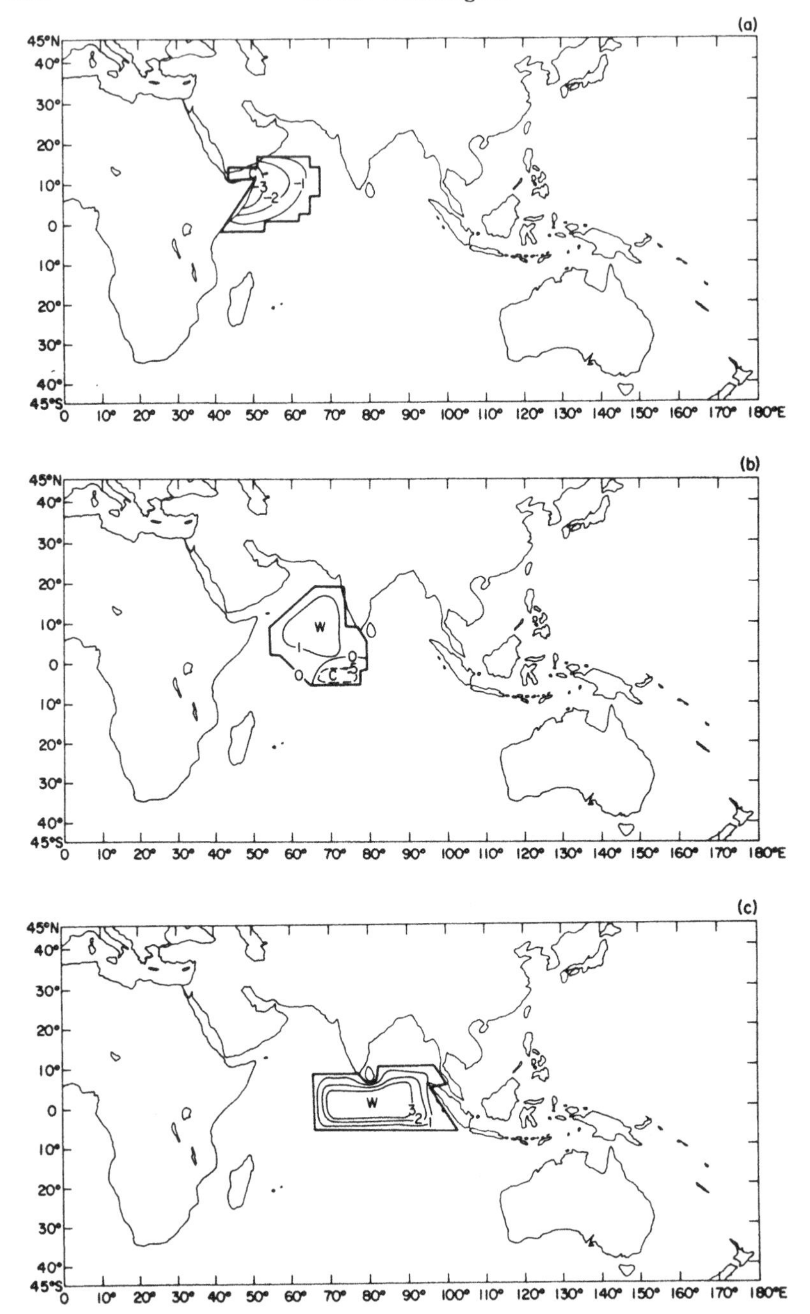

Fig. 7.5. Ocean-surface temperature anomaly distributions: (*a*) western Arabian Sea; (*b*) eastern Arabian Sea; and (*c*) central Indian Ocean. (The contours are labelled in °C.)

made. The method of measuring statistical significance with the NCAR model is much closer to the traditional methods of evaluating the noise level in prescribed change experiments. All experiments were for a perpetual July in which the Sun's declination remains constant. The control experiment started from an atmosphere at rest and ran 120 days. The random experiments used to obtain the noise-level statistics were started on day 60 of the control run. A small alteration of the initial condition was introduced and allowed to grow with time. A standard deviation was estimated for the mean climate from day 91 to 120 - 30 days after introduction of the random perturbation. The use of three control experiments in the climate estimation allowed two degrees of freedom. The anomalies were introduced on day 60 of each experiment, and the difference between the anomaly and the control experiments was divided by the standard deviation to yield a normalized response which can be related to the Student's *t* statistic. This allows calculation of the significance level for the response. Regions where the normalized response is greater than 5, implying a significance level of approximately 5%, are shaded in subsequent figures. The colder ocean temperatures in Fig. 7.6 show a statistically significant local response (shaded region) in the model, in that there is sinking motion on the eastern edge of the anomaly pattern and the precipitation is less. Also, consistent with Shukla's results, less precipitation is found over the Indian subcontinent. However, the statistical significance is not high and fails our criterion in not being sufficiently above the noise level to ensure that the response is real, not random.

In an NCAR experiment with an anomaly pattern in the eastern Arabian Sea, the ocean temperature increase has a maximum of 1 °C west of India and a minimum of −0.5 °C near the Equator (Fig. 7.5*b*). This particular pattern was chosen because it appeared similar to the observations of Ramage and Raman (1972) during the International Indian Ocean Expedition in July 1963. The effect of this pattern on the model simulation (Fig. 7.7) was rising motion and increasing precipitation over the area with a warm anomaly in ocean temperature, and sinking motion and lessened precipitation over the southern portion of the anomaly pattern where the temperature decreased. The statistical significance in the region of upward motion satisfied our criterion. There appeared to be no systematic change in precipitation over the Indian subcontinent. The model's response suggests that such an ocean temperature anomaly pattern in this region does not appreciably affect precipitation over India itself (Fig. 7.7).

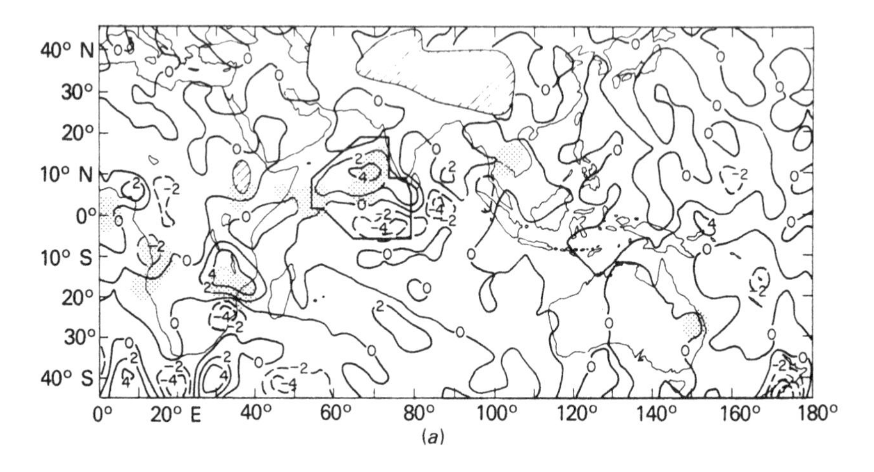

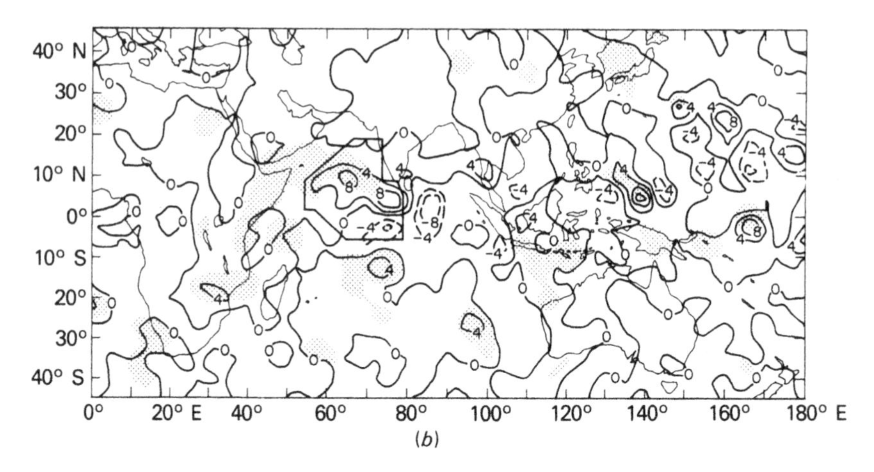

Fig. 7.6. (*a*) Difference of vertical velocity at 3 km and (*b*) precipitation for the western Arabian Sea experiment minus average of ensemble time averaged over days 91 to 120. Contour intervals are: 2 mm s^{-1} for vertical velocity and 4 mm per day for precipitation. The anomaly region is outlined with a heavy line. Stippled areas correspond to normalized responses greater than or equal to five.

The third NCAR experiment introduced a temperature change of up to 3 °C in the central Indian Ocean and between longitude 65° E and 100° E (Fig. 7.5*c*). The purpose of this experiment was to study modification of the southern equatorial trough (SET) and its effects on the overall monsoon circulation. The response in the model (Fig. 7.8) to this warm ocean temperature anomaly was to increase precipitation and vertical velocity directly over the warm anomaly pattern. Both increases

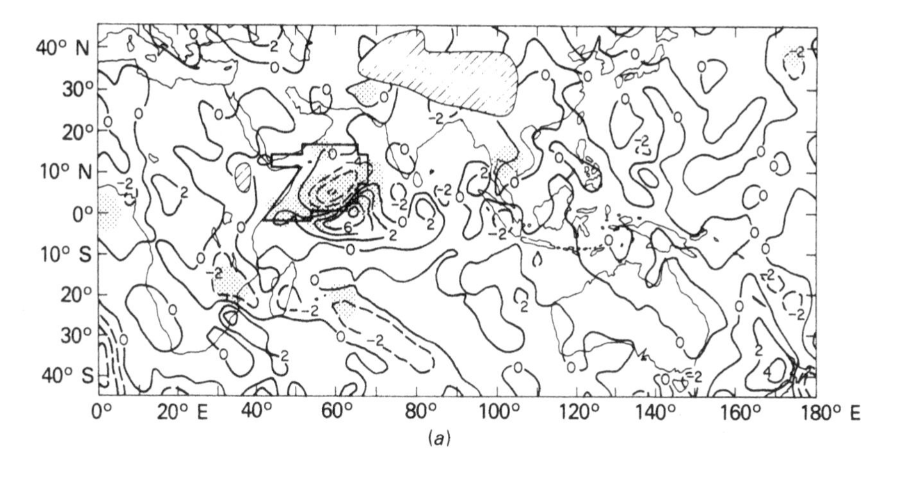

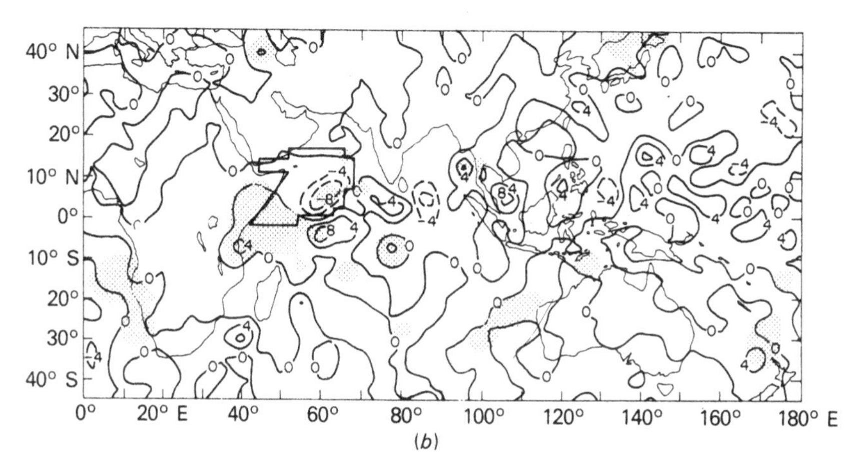

Fig. 7.7. (*a*) Difference of vertical velocity at 3 km and (*b*) precipitation for the western Arabian Sea experiment minus average of ensemble time averaged over days 91 to 120. Contour intervals are: 2 mm s^{-1} for vertical velocity and 4 mm per day for precipitation. The anomaly region is outlined with a heavy line. Stippled areas correspond to normalized responses greater than or equal to five.

appeared to be statistically significant. An interesting feature of the results is that there seems to be an equatorial teleconnection between the upward motion over the central Indian Ocean and that over Malaysia and the western Pacific. This pattern shows more upward motion over the anomaly and less over other regions. The strength of the Somali Jet increases in the anomaly experiment. Also, more inflow to the region across the Bay of Bengal occurs in the experiment (Washington *et al.*,

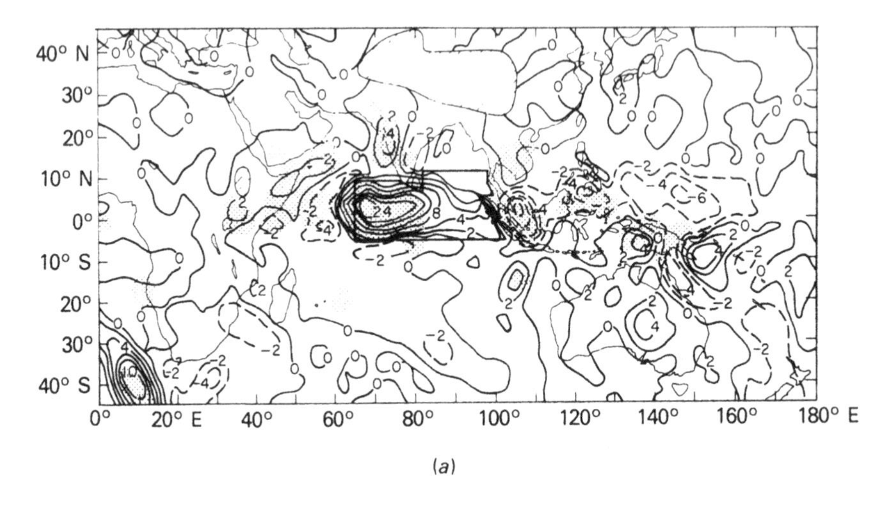

(*a*)

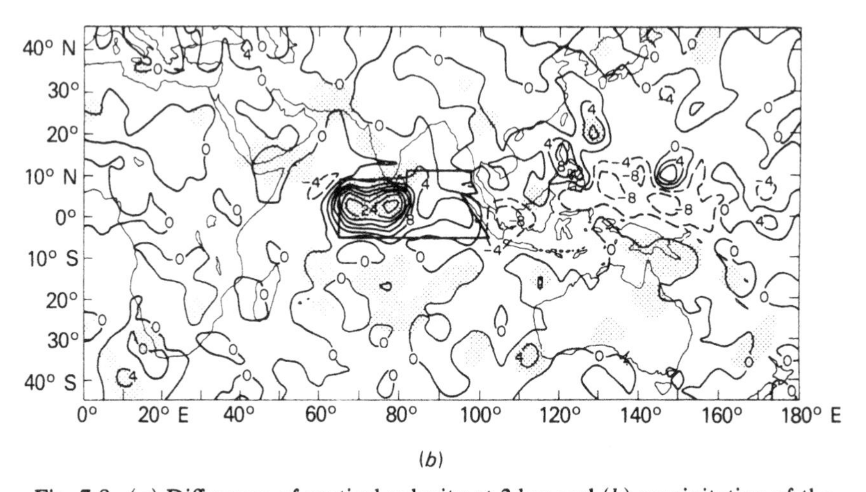

(*b*)

Fig. 7.8. (*a*) Difference of vertical velocity at 3 km and (*b*) precipitation of the central Indian Ocean experiment minus average of ensemble. Contour intervals are 2 mm s^{-1} for vertical velocity and 4 mm per day for precipitation. The anomaly region is outlined with a heavy line. Stippled areas correspond to normalized responses greater than or equal to five.

1977). The tropical easterly jet stream has slowed over the anomaly region and shows a large diffluence immediately over the anomaly.

The three NCAR experiments, in addition to Shukla's earlier experiment, suggest that a local response between ocean temperatures and precipitation exists. However, the extent to which the Indian Ocean surface-temperature anomalies correlate with precipitation anomalies over the Indian subcontinent remains to be established. It is hoped that the Monsoon Experiment (MONEX) will shed some light on this issue.

7.4.3 Ice age simulations

July simulations with GCMs of the hemispheric or global circulation have been carried out for the last ice age – approximately 18 000 BP – by Alyea (1972), Saltzman and Vernekar (1972), Williams *et al.* (1974), Gates (1976), and Manabe and Hahn (1977). Williams *et al.* and Manabe and Hahn found a weaker summer monsoon circulation in the ice age simulation than in the present-day simulations, with the southwesterlies in particular not as strong. Manabe and Hahn's analysis further indicates that this is primarily caused by increased albedo in south Asia rather than by colder sea-surface temperatures during the ice age. They speculate that this may be caused by the negative correlation between the extent of snow cover over Asia and the amount of monsoon precipitation during the following summer.

7.4.4 Simulation of the winter monsoon

In a seasonal simulation with the GFDL model, Manabe *et al.* (1974) showed that over the Arabian Sea the winter flow is from the northeast and is a reversal of the southwest flow in summer. A more detailed discussion of the winter monsoon simulation with the NCAR model is given in Washington (1976) where the low- and high-level wind patterns are compared with observed data. The flow pattern over the Arabian Sea has a maximum of 7.5 m s^{-1} which compares favourably with the observed data of Sadler and Harris (1970) and Findlater (1971). Also, the cross-section along the Equator from 35° E to 75° E shows that the model generates a shallow jet close to the African Highlands, but in the model the jet is much wider than that observed by Findlater. In the pressure field, the pattern was opposite to that in summer: over northern India, the Arabian area, and the Sahara there was relatively high pressure and over the Arabian Sea and near Kenya relatively low pressure. This pressure gradient drives the low-level jet near the Kenya coast toward the south. Comparisons of cloudiness and precipitation showed patterns consistent with observations in that there was a maximum precipitation zone south of the Equator near Madagascar and relatively little precipitation over the Sahara, Arabia, and India regions. The maxima in the observed cloudiness agreed well with the precipitation maxima.

7.5 Future development

It is believed that nearly all aspects of GCMs need improvement. Kasahara (1977) outlined some problems in orography associated with present-day GCMs. The differences in z, pressure, and sigma coordinates are not as sizeable in the vicinity of mountains as once believed, because in any system the hydrostatic component must be subtracted from the pressure gradient along the coordinate surface to obtain the horizontal pressure gradient. The subtraction of two large terms with opposite sign invariably leads to very large truncation errors. Removing this type of error from the Indian monsoon region calls for high horizontal and vertical resolutions, perhaps simply a nested grid within a low-resolution global domain, so that features such as very steep mountain ranges (Himalayas or Western Ghats) can be more accurately incorporated. This poses special numerical problems for flows between the coarse and nested grids, problems that in principle cannot be totally eliminated. For instance, because of phase error truncations, a simple linear wave moving into a region of both coarse and fine grids will slow down more in the coarse grid, so that the wave exciting the fine grid will interfere with the slower wave from the coarse grid. Several numerical techniques are being developed to minimize this computational error, but it will exist any time there is a discontinuity in grid size, whether abruptly or smoothly varying. Perhaps the solution rests with computer speed. With fast computers, high-resolution GCMs will reduce truncation errors from orography (H. Sundqvist discusses the problems of incorporation of orography into numerical models of the atmosphere in Chapter 39). An added benefit would be that small-scale dynamic features could be simulated more accurately.

Other improvements required in GCMs are the treatment of solar and infrared radiation, convection (particularly moist processes), cloudiness, precipitation, and boundary-layer fluxes of momentum, moisture, and sensible heat. The inclusion of vegetation and improved ground hydrology, as proposed by Deardorff (1978), should have a sizeable impact on monsoon features because the energy and momentum exchanges at the interface between the atmosphere and the Earth's surface are important.

The future for GCM simulation is optimistic if one views the problems in present models as surmountable. Given the brief history of modelling of the Indian monsoon, the prospects are promising.

The author thanks Robert Chervin and Claire Parkinson for many editorial suggestions. This research was carried out at the National Center for Atmospheric

Research, Boulder, Colorado, which is sponsored by the National Science Foundation.

References

Abbott, D. A. (1973) Scale interactions of forced quasistationary planetary waves at low latitudes. Technical Report No. 73-2, Department of Meteorology, Florida State University, 190 pp.

Alyea, F. N. (1972) Numerical simulation of an ice age paleoclimate. Atmos. Sci. Paper No. 193, Department of Atmospheric Science, Colorado State University, 120 pp.

Atkinson, G. C. and Sadler, J. C. (1970) Mean cloudiness and gradient level wind charts over the tropics. Vol. 2, Charts. Technical Report 215, US Air Weather Service, 48 pp.

Deardorff, J. (1978) Efficient prediction of ground temperatures and moisture with inclusion of a layer of vegetation. *J. Geophys. Res.*, **83**, 1889–1903.

Findlater, J. (1971) Mean monthly airflow at low levels over the western Indian Ocean. *Geophys. Mem.* (HMSO, London), **16**, 115, 1–53.

Gates, W. L. (1976) Modeling the ice-age climate. *Science* **191**, 1138–44.

Gates, W. L. and Schlesinger, M. E. (1977) Numerical simulation of the January and July global climate with a two-level atmospheric model. *J. Atmos. Sci.*, **34**, 36–76.

Gilchrist, A. (1974) A general circulation model of the atmosphere incorporating an explicit boundary layer top. Technical Note II/29, Meteorological Office, Bracknell, 35 pp. (Available from Dynamical Climatology Branch, Meteorological Office, London Road, Bracknell, UK.)

Gilchrist, A. (1976) Tropical results from a July integration of a general circulation model incorporating an explicit boundary layer top. Technical Note II/46, Meteorological Office, Bracknell.

Gilchrist, A. (1977) The simulation of the Asian summer monsoon by general circulation models. *Pure and Applied Geophysics*, **115**, 1431–48.

Godbole, R. V. (1973) Numerical simulation of the Indian summer monsoon. *Ind. J. Meteor. Geophys.*, **24**, 1–14.

Hahn, D. G. and Manabe, S. (1975) The role of mountains in the south Asian monsoon circulation. *J. Atmos. Sci.*, **32**, 1515–41.

Kasahara, A. (1974) Various vertical coordinate systems used for numerical weather prediction. *Mon. Wea. Rev.*, **102**, 509–22.

Kasahara, A. (1977) The effect of mountains on synoptic-scale flows. Presented at First Planning Meeting for the GARP Sub-programme on Air Flow Around Mountains, Venice, Italy, 24–28 October 1977. (NCAR Ms. 0901/77-1, National Center for Atmospheric Research, Boulder, Colorado.)

Manabe, S., Hahn, D. G. and Holloway, J. L., Jr. (1974) The seasonal variation of the tropical circulation as simulated by a global model of the atmosphere. *J. Atmos. Sci.*, **31**, 43–83.

Manabe, S. and Hahn, D. G. (1977) Simulation of the tropical climate of an ice age. *J. Geophys. Res.*, **82**, 3880–911.

Mintz, Y., Katayama, A. and Arakawa, A. (1972) Numerical simulation of the seasonally and interannually varying tropospheric circulation. Climatic Impact Assessment Program. *Proc. Survey Conference* Feb. 1972, ed. A. E. Barrington, US Department of Transportation, 194–216.

Möller, F. (1951) Vierteljahrskarten des Niederschlags für die ganze Erde. *Petermanns Geogr. Mitt.*, **95**, 1–7.

Murakami, T., Godbole, R. V. and Kelkar, R. R. (1970) Numerical simulation of the monsoon along 80° E. *Proc. Conf. Summer Monsoon of Southeast Asia*, Navy Weather Research Facility, Norfolk, Virginia, 39–51.

Ramage, C. S. and Raman, C. R. V. (1972) *Meteorological Atlas of the International Indian Ocean Expedition*, Vol. 2, Upper Air. National Science Foundation, Washington, DC, 121 pp.

Sadler, J. C. and Harris, B. E. (1970) The mean tropospheric circulation and cloudiness over southwest Asia and neighbouring areas. Science Report No. 1, AFCRL-70-0489.

Saltzman, B. and Vernekar, A. D. (1972) Global equilibrium solutions for the zonally averaged macroclimate. *J. Geophys. Res.*, **77**, 3936–45.

Shukla, J. (1975) Effect of Arabian Sea surface temperature anomaly on Indian summer monsoon: A numerical experiment with the GFDL model. *J. Atmos. Sci.*, **32**, 503–11.

Stone, P. H., Chow, S. and Quirk, W. J. (1977) The July climate and a comparison of the January and July climates simulated by the GISS general circulation model. *Mon. Wea. Rev.*, **105**, 170–94.

Washington, W. M. (1970) On the simulation of the Indian monsoon and tropical easterly jet stream with the NCAR general circulation model. *Proceedings of the Symposium on Tropical Meteorology*, 2–11 June 1970, University of Hawaii, Honolulu, Hawaii, sponsored by the American Meteorological Society, WMO JVI-1-6.

Washington, W. M. (1974) Brief description of NCAR global circulation model. In *Modeling for the First GARP Global Experiment*, Report No. 14 of GARP Publication Series. WMO, Geneva, 61–78.

Washington, W. M. (1976) Numerical simulation of the Asian–African winter monsoon. *Mon. Wea. Rev.*, **104**, 1023–28.

Washington, W. M. and Daggupaty, S. M. (1975) Numerical simulation with the NCAR global circulation model of the mean conditions during the Asian–African summer monsoon. *Mon. Wea. Rev.*, **103**, 105–14.

Washington, W. M., Chervin, R. M. and Rao, G. V. (1977) Effects of a variety of Indian Ocean surface temperature anomaly patterns on the summer monsoon circulation: Experiments with the NCAR general circulation model. *Pure and Applied Geophysics*, **115**, 1335–56.

Washington, W. M. and Williamson, D. L. (1977) A description of the NCAR global circulation models. In *Methods in Computational Physics*, ed. J. Chang, Vol. 17, General circulation models of the atmosphere, pp. 111–72, Academic Press, New York.

Williams, J., Barry, R. G. and Washington, W. M. (1974) Simulation of the atmospheric circulation using the NCAR global circulation model with ice age boundary conditions. *J. Appl. Meteor.*, **13**, 305–17.

8

Simulation of the Asian summer monsoon by an 11-layer general-circulation model

A. GILCHRIST

Gilchrist (1977) has described the simulation of the Asian summer monsoon by a 5-layer general-circulation model and has compared its results with those of other models that have been published. The diabatic heat sources and sinks for the model in the monsoon region were considered, since both on general theoretical grounds and as a matter of observation they are expected to be important in creating the monsoon circulation. From the simulation, it appeared that the surface-pressure trough and many features of the overall flow were related particularly to the low-level sensible heat input to the boundary layer.

Attention was drawn to certain shortcomings in the 5-layer model simulation. Primarily they were:

(i) The monsoon surface-pressure trough was weaker than indicated by climatology as was the near-surface flow, and the position of the surface trough was in error in being over the Indian continent rather than over the Bay of Bengal.

(ii) The rainfall distribution was poor in a number of respects, but particularly in failing to give adequate amounts of rain over the Ganges basin and the southern slopes of the Himalayas.

These errors were shared in substantial measure by the other general-circulation models considered.

In this chapter, the southwest monsoon simulation of the 5-layer model is compared with that of an 11-layer general-circulation model, which, like the simpler model, was developed in the Meteorological Office.

8.1 Brief description of the models

The formulation of the 5-layer model is given in Corby *et al.* (1977). It has an irregular grid-mesh on the sphere designed to achieve a quasi-constant mesh-length of approximately 330 km. It uses 'sigma' coordinates in the vertical, the layer boundaries being at sigma levels $\sigma = 1.0, 0.8, 0.6, 0.4, 0.2$ and 0, where sigma is the ratio of pressure to the pressure at the Earth's surface. Levels representing the layers are at $\sigma = 0.9, 0.7, 0.5, 0.3$ and 0.1. The 11-layer model is a higher-resolution version with a grid-length of 220 km approximately, and layers and levels as shown in Table 8.1. A detailed description of the 11-layer model can be found in Saker (1975). Forecast results from it are described by Carson and Cullen (1977).

The parameterization of radiative heating and cooling in the two models was basically similar in the two simulations reported here. It followed the method described by Corby *et al.* (1977). It is such that cloud is taken into account implicitly through empirical emissivities. Thus, the infrared

TABLE 8.1. *Layers and levels for the 11-layer model.*

Level	Layer boundaries (σ value)	Levels (σ value)
	0.0	
1		0.022
	0.06	
2		0.089
	0.125	
3		0.157
	0.195	
4		0.230
	0.270	
5		0.317
	0.370	
6		0.436
	0.510	
7		0.577
	0.650	
8		0.717
	0.790	
9		0.844
	0.900	
10		0.937
	0.975	
11		0.987
	1.000	

cooling rate is expressible as

$$dT/dt = aT^4 + bT_*^4$$

where a and b are the empirical emissivities and are functions of pressure and latitude only. The effects of radiative interchange with the Earth's surface are represented by bT_*^4, which is taken as zero except when dealing with the lowest layer. The solar heating rates are functions of latitude and pressure only and have been derived from the data of Rodgers (1967).

The convective exchange parameterization used in the 11-layer model is similar to the original scheme described by Corby *et al.* (1972). It is based on the concept of mixing between contiguous layers, the mass of air involved in the mixing process being dependent on the degree of vertical instability. However, the scheme as described incorporated an empirically estimated factor ε, which is probably most readily thought of as the amount by which air at the lower level taking part in convection is warmer than its environment. ε was given a value of about 2 K. The scheme in the 11-layer model did not incorporate this empirical factor, and therefore, where the temperature structure is determined largely by the convective scheme (and this is true over most of the tropical oceans) the lapse rate of temperature tends to be too large and the upper atmosphere consequently too cold. The 5-layer model used an extension of the original scheme which allows for mixing between non-contiguous layers, thus simulating more satisfactorily the effects of deep convection. This variant of the scheme has been described in Corby *et al.* (1977). It is to be noted that air initiating convection was again assumed to be warmer than its environment by about 2 K, and therefore this model tends to produce a tropical upper troposphere that is warmer than in the 11-layer model simulation. Near the Equator over the Indian Ocean at about 300 mb, for example, the difference in temperatures between the two simulations is around 4–5 K. One of the shortcomings of the original convection scheme and one of the motivations for amending it was that it tended to produce localized areas of very intense rainfall, particularly in the tropics; thus we may note the very large values from the 11-layer model near the Himalayas, demonstrating that this effect can manifest itself even in averages over a 30-day period.

Taking the boundary layer to be that part of the troposphere nearest the Earth's surface in which there are significant vertical eddy fluxes of momentum, heat, and water vapour due to interaction between the atmosphere and the underlying surface, it constitutes the lowest layer of the 5-layer model with a depth $\Delta\sigma$ equal to 0.2. For the higher-resolution model however, there are three layers within the boundary layer, the total depth $\Delta\sigma$ being 0.21. Because of these differences, the parameterizations of surface exchanges and boundary-layer fluxes differ more substantially than the radiative or convective parameterizations. Broadly, the higher-resolution model follows the proposals of Clarke (1970), using his method 1, which requires several levels within the boundary layer. In deriving the appropriate drag and exchange coefficients which are functions of the bulk Richardson number, the roughness lengths assumed are 10^{-1} m for land and 10^{-4} m for sea points.

The formulation for the 5-layer model may be regarded as a form of Clarke's method 2 appropriate to a bulk treatment of the boundary layer. However, there are significant differences. Conditions at the surface of the Earth are allocated to one of only four categories, viz. land/sea and stable/unstable, so that the exchange coefficients are available without iterative procedures. Appropriate values for the turning of the surface wind with respect to the geostrophic for each category are also assumed. Additionally, a parameter representing the depth of the boundary layer is carried as an explicit variable, and values of the temperature and humidity mixing ratio at its top are obtained by interpolation from the values at the model's levels. The formulae for the vertical turbulent fluxes at the surface then depend upon differences between the surface and the top of the boundary layer. However, it is unrealistic to assume that the atmosphere could not be heated until a dry adiabatic lapse rate has been established through the whole boundary layer which is typically 800–900 m deep over the tropical oceans, since on many occasions a shallower adiabatic layer surmounted by a cloudy saturated layer results from boundary-layer mixing. To allow for these circumstances, the potential temperature difference is taken as

$$(\theta_* - \theta_{\mathrm{T}})$$

when a parcel of air rising from the surface would not condense, but when condensation would occur, θ_* is incremented by a term that allows for the increase of potential temperature due to the release of latent heat in the upper part of the boundary layer, i.e.

$$\theta'_* = \theta_* + (L/c_{\mathrm{p}})(q_* - q_{\mathrm{T}}^{\mathrm{sat}}), \quad q_* > q_{\mathrm{T}}^{\mathrm{sat}}.$$

Here θ denotes potential temperature calculated with respect to a reference pressure of p_* (so that $\theta_* = T_*$); suffixes $*$ and T denote values at the Earth's surface and at the top of the boundary layer respectively, and q_T^{sat} is the saturation value of the mixing ratio at temperature T_T. In deriving the mixing ratio q_*, it is assumed that the relative humidity of air over the sea is 80% and over land 50%.

There is an important difference between the models which particularly affects their calculation of evaporation over land. In the 5-layer model the relative humidity of air near the surface is fixed, irrespective of the values of rainfall or evaporation, and this determines the surface mixing ratio value for the calculation of the evaporation. In contrast, the 11-layer model includes an interactive ground hydrology. A running total of soil moisture is maintained at each grid-point. It may increase as a result of rainfall, downward water vapour flux (dew) or by melting snow; it is decreased by evaporation. The maximum value of the soil moisture content is taken as 15 cm, any additions above this being assumed to constitute ground-water run-off. In calculating the evaporation, the difference in mixing ratio between the surface and the lowest model level is taken as

$$\alpha(q_*^{sat} - q_l)$$

where l denotes the lowest level of the model and * a surface value; q_*^{sat} is the saturation mixing ratio at temperature T_*. This expression is used both explicitly in the formula for the evaporation and implicitly in determining the relevant transfer coefficient dependent on the bulk Richardson number; α is given the value of 1 (i.e. the evaporation is the same as the potential evaporation) unless the soil moisture content is less than or equal to 5 cm and there is a hydrolapse. In the latter event α takes the value of a fifth of the soil moisture content in centimetres.

In the 5-layer model the albedo of land surfaces was taken as a zonal mean value derived by linear interpolation from the following:

Latitude	0	10	20	30
Albedo	0.146	0.188	0.225	0.212

In the higher-resolution model all land had an albedo of 0.2.

8.2 The simulations

The simulations have been derived from global integrations for July, with the solar declination set at the value appropriate to the middle of the month and the sea-surface temperatures fixed at their climatological values. The 5-layer model simulation is taken from the last 30 days of a 100-day integration, which started from cold, isothermal stationary conditions, while the 11-layer model simulation is taken from the last 30 days of a 50-day integration from real atmospheric initial conditions. In each case, the integrations are considered to be long enough for the influence of the initial conditions in the period chosen for study to be negligible. Both integrations were aimed at studying the general circulation of the global atmosphere, and were not specifically intended for examination of the southwest monsoon.

Following the approach of Gilchrist (1977) we consider first the sensible heat inputs into the boundary layer. These are shown for the two models in Figs. 8.1*a* and *b*. The values are the differences between the

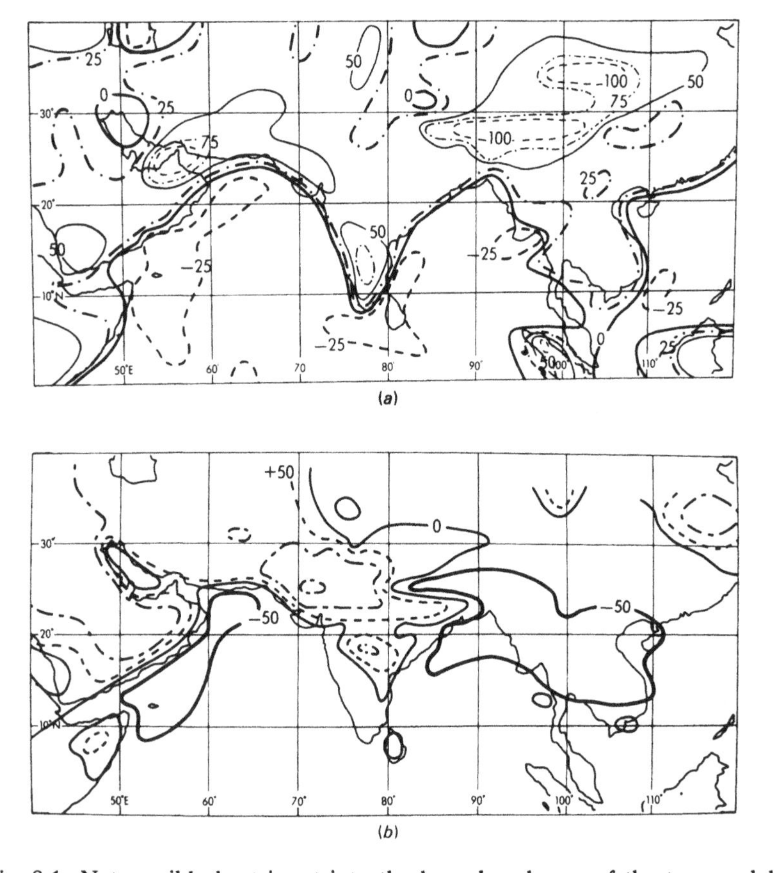

Fig. 8.1. Net sensible heat input into the boundary layers of the two models averaged over 30 days from integrations for July (units: W m^{-2}). (*a*) 5-layer model: the boundary layer is between 0.8 p_* and p_* where p_* is the pressure at the surface of the Earth (from Gilchrist, 1977). (*b*) 11-layer model: the boundary layer is between 0.79 p_* and p_*.

sensible heat input at the Earth's surface and the radiational cooling within the boundary layer (i.e. one and three model layers for the lower- and higher-resolution models respectively). An approximation has been made in ignoring the heat input resulting from condensation; this could be important in the boundary layer of the 11-layer model, since it may be large if there is shallow convection. It could be a significant contribution to the resultant heating in areas of high rainfall, but must be small or zero over such low-rainfall regions as the western Arabian Sea. There are large differences in the two fields. Thus in the 5-layer model, values vary from about $-25\ \mathrm{W\,m^{-2}}$ over western parts of the Arabian Sea to about $50\ \mathrm{W\,m^{-2}}$ over the Indian subcontinent. For the 11-layer model, the equivalent values are approximately $-75\ \mathrm{W\,m^{-2}}$ and 75–$100\ \mathrm{W\,m^{-2}}$ respectively. The gradient in low-level heating applied in the 11-layer model is therefore about double that in the 5-layer.

The difference is probably attributable primarily to two causes:

(i) the different formulation of the surface exchanges;

(ii) the different partitioning of solar radiation reaching the ground because only one of the models includes a variable ground hydrology.

In the 11-layer model the surface exchanges are functions of the differences of potential temperature and humidity between the surface and the lowest level, at about 100 m. This is shallower than in the 5-layer model allowing greater variability both in space and time. Over the western half of the Arabian Sea, the high-resolution model sensible heat exchanges are negative and typically around $-20\ \mathrm{W\,m^{-2}}$. At the same positions, the evaporation is substantial and about 5–6 mm per day, indicating that the lowest layer is generally unstable due to the effect of humidity on the density, but the potential temperature nevertheless increases with height. These sensible heat values are considerably more negative than calculated by Ramage *et al.* (1972). The latter used similar formulae applied to surface observations with a constant transfer coefficient of about 1.35×10^{-3}, a value generally considered to be near the optimum for tropical oceans (see for example the report of the US GATE Central Program Workshop, 1977). The 5-layer model values of sensible heat exchange are close to those of Ramage *et al.*; mainly they are small and positive, the only negative areas being over the cold water close to the African and Arabian coasts. Because there is horizontal advection of colder air within the boundary layer at most places over the Arabian Sea and there is little heating due to condensation, at least in the western part of the Sea, a steady temperature can only be achieved as a result of subsidence. Vertical velocities at the top of the boundary layer are such as to produce an adiabatic warming of about 1 K per day. The requirement for strong subsidence is consistent with the vigorous monsoon circulation simulated by the 11-layer model, with pressure gradients and winds that tend to be stronger than observed. For example, Fig. 8.3 shows the mean 30-day winds at the lowest model level. Maximum winds off the Arabian coast are in excess of 17.5 m s^{-1}, and seem too strong when compared with the surface winds of about 10 m s^{-1} in the same area shown by Ramage *et al.* for July and August of 1963 and 1964. Strengthening of the simulated low-level winds tends to increase the surface exchange values and therefore there may be an element of positive feedback in the interaction of these two quantities.

Over land, the large differences in the heating of the boundary layer result mainly from the variable ground hydrology in the high-resolution model. In the 5-layer model the areas of largest sensible heat input tend to coincide with the areas of highest moisture content, since it is there that the smallest proportion of the heat available at the surface is used up in evaporation. This leads, for example, to the large sensible heat values over mountains in Fig. 8.1*b*. With a variable ground hydrology, a model is capable of giving the greatest heat input where the soil and probably also the air is driest. Thus, the 11-layer model heating within the boundary layer shows maximum values over desert regions such as Arabia and northwest India and over the drier areas within the Indian subcontinent. The distribution of heating over land in Fig. 8.1*b* appears much more realistic than that in Fig. 8.1*a* and this has undoubtedly led to improvements in the monsoon simulation particularly with regard to the rainfall distribution (see Fig. 8.5*b*).

The surface-pressure distributions for the two models are shown in Figs. 8.2*a* and *b*, and they may be compared with the long period climatological chart by Ramage *et al.* (1972), in Fig. 8.2*c*. It is evident that the models have very different simulations; in the 5-layer model the north–south pressure gradient is weak, the difference in pressure between 1.5° N and 22.5° N over the Arabian Sea being about 5 mb while the comparable value for the high-resolution model is about 14 mb. An approximate value taken from climatological charts is about 10 mb. Thus the lower-resolution model tends to underestimate and the higher-resolution model tends to overestimate the surface-pressure gradients. This is consistent with the differences in heating of the boundary layer already discussed. It is probable, however, that an additional factor is the upper-air temperatures in the two models. As already noted, the upper atmosphere over the southern Indian Ocean tends to be notably colder in the 11- than in the

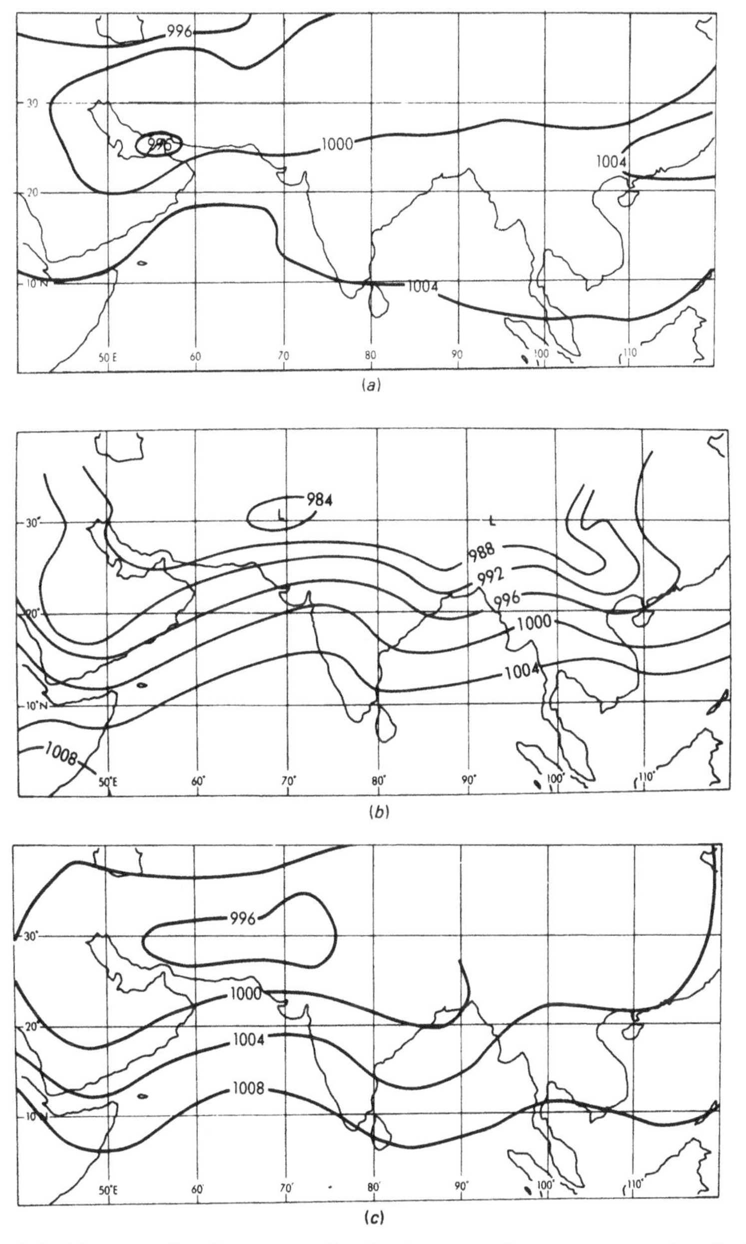

Fig. 8.2. Mean sea-level pressure distribution over the monsoon region (units: mb). (*a*) 30-day average values for the 5-layer model from a July integration (from Gilchrist, 1977). (*b*) 30-day average values for the 11-layer model from a July integration. (*c*) Observed climatological values for July (from Gilchrist, 1977).

5-layer model. Over northern India, particularly over the mountainous areas, the difference in temperatures is less, partly perhaps because the convection is a less dominant influence in determining the lapse rates, but probably also because the mountains form an elevated heat source, tending to maintain high temperatures in both models. Thus at upper levels, relative to the 5-layer model the 11-layer model has a thickness gradient tending to strengthen the direct overturning of air rising in the heated continental areas and sinking over the cooler oceanic regions. Such an effect would act in the same direction as the surface heating contrast to produce a relatively vigorous monsoon circulation in the 11-layer model simulation.

Although the pressure gradient in the 11-layer model is too strong, it otherwise provides a very good simulation of the surface-pressure distribution. It may be noted particularly that the trough is correctly placed over the Bay of Bengal, thus correcting one of the outstanding faults of the 5-layer model. It seems most likely that this improvement is caused by the introduction of a variable ground hydrology.

The winds at low levels of the 11-layer model are shown in Fig. 8.3*a* (for $\sigma = 0.987$) and Fig. 8.3*b* (for $\sigma = 0.844$). Low-level winds in the 5-layer model and in other models, as well as the climatological winds at the top of the boundary layer are shown in Gilchrist (1977) and reproduced by Washington in Chapter 7 of this book (Fig. 7.1*d*). The relative vigour of the high-resolution simulation is evident in the strength of the winds. Whereas the 5-layer model gives underestimates of the climatological values, the 11-layer model winds are too strong. The general pattern of the latter, however, follows the observed winds well. In particular, the direction of the winds over the Arabian Sea are backed with respect to those in the lower-resolution simulation, and closer to the observed direction. Also the strongest winds at $\sigma = 0.844$ extend further eastwards towards India and this too seems to be more in accordance with observation. At the lower level, $\sigma = 0.987$, the direction of the winds in the Bay of Bengal with a direct current on to the southern slopes of the Himalayas is a consequence of the more realistic simulation of the surface trough in this area.

Figs. 8.4*a*, *b* and *c* show the upper winds near the 200 mb level for the 5-layer model, the 11-layer model and for the observed climatology respectively. The salient feature is the contrast in wind strength between the two models. Again, the high-resolution simulation is stronger than indicated by observations. In general, the 11-layer model simulation is good, though, in common with other simulations (Gilchrist, 1977) the

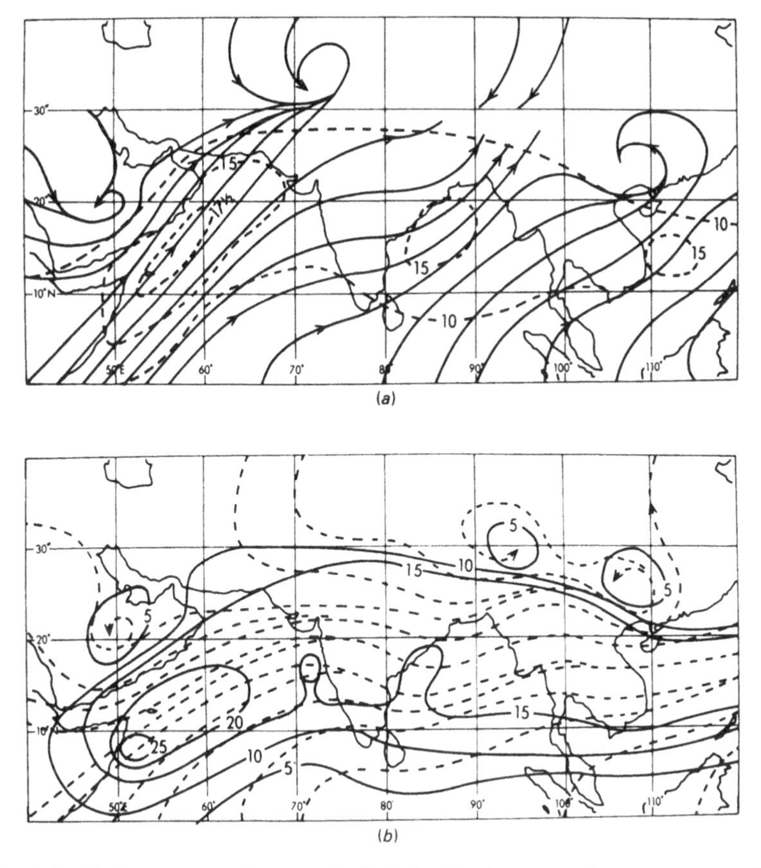

Fig. 8.3. 30-day average low-level wind distribution over the monsoon region (streamlines and isotachs) from a July integration of the 11-layer model (isotach units: m s^{-1}). (*a*) 30-day average values at the level $\sigma = 0.987\ p_*$ where p_* is the pressure at the Earth's surface. (*b*) Values at the level $\sigma = 0.844\ p_*$.

winds lack the significantly northerly component which is a feature of the climatological chart, Fig. 8.4*c*.

Figs. 8.5*a* and *b* illustrate the rainfall simulations in the two models. Both show patterns that are correct in a number of important respects. For example, both models obtain a dry area over the western Arabian Sea with rainfall increasing eastwards and over the Western Ghats of India they both have maximum rainfall due to air rising over the mountains. However, there is a marked deficiency of rain in the simulations over the Ganges Valley and northern India. The model simulations appear to be more characteristic of 'break-monsoon' conditions than those of the

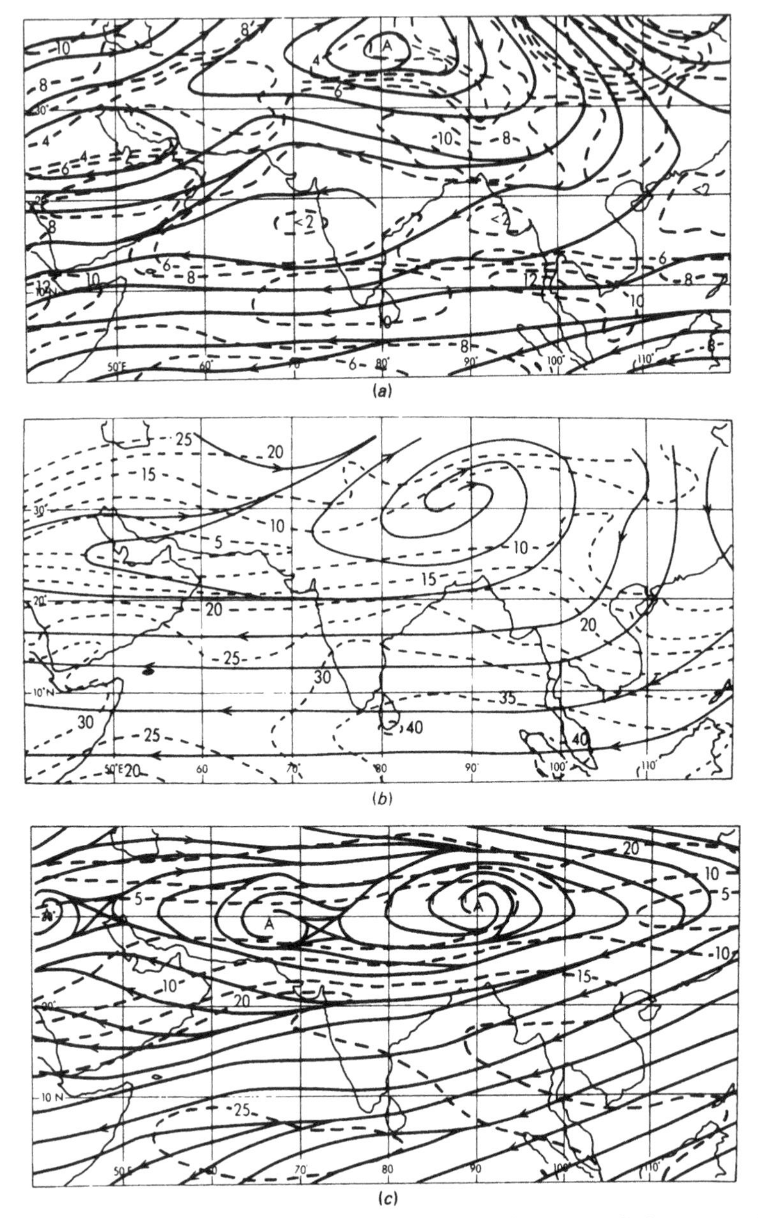

Fig. 8.4. Wind distribution (streamlines and isotachs) at about 200 mb over the monsoon region (isotach units: m s^{-1}). (*a*) 30-day average values from a July integration of the 5-layer model. They are averages of values at the levels $\sigma = 0.1$ and $\sigma = 0.3$ to give an estimate near the level of the climatological winds (from Gilchrist, 1977). (*b*) 30-day average values from a July integration of the 11-layer model. (*c*) Observed climatological winds at 200 mb for July (from Gilchrist, 1977, after Ramage and Raman, 1972).

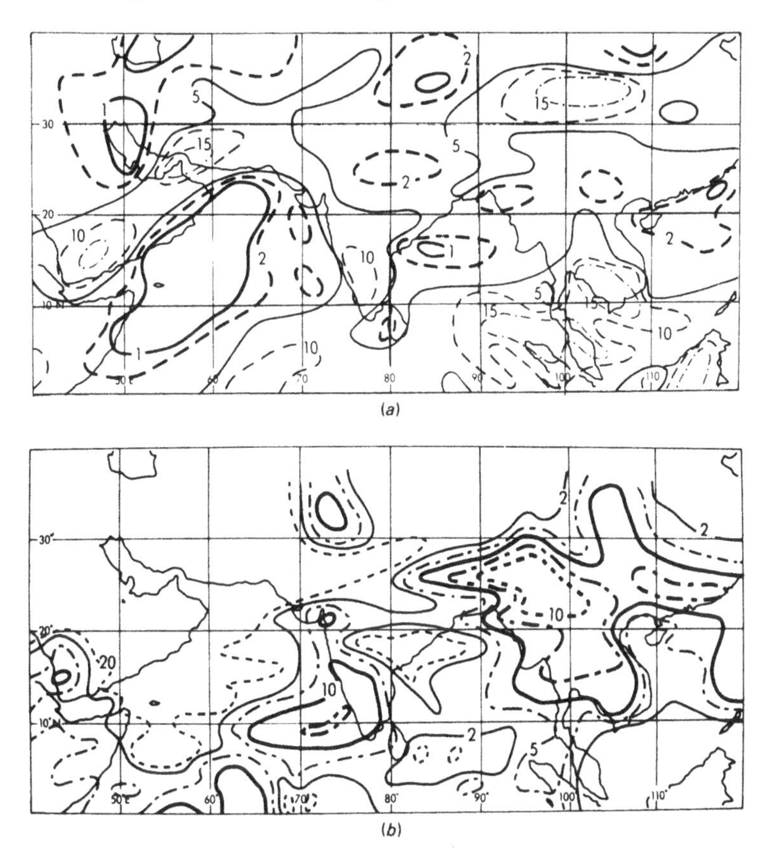

Fig. 8.5. Rainfall distribution over the monsoon region (units: mm per day). (*a*) 30-day average values from a July integration of the 5-layer model (from Gilchrist, 1977). (*b*) 30-day average from a July integration of the 11-layer model.

monsoon in general. This may be associated with their failure to create monsoon depressions in the Bay of Bengal since it is these that are generally responsible for heavy rains over northern India. The reason for this failure is not known, the fundamental mechanisms for the development of monsoon depressions being as yet unclear. In the Himalayas region the 11-layer simulation is to be preferred; it obtains heavy rain there as a result of the low-level flow from the Bay of Bengal rising over the southern slopes and initiating vertical instability. The maximum rainfall obtained, however, is excessive and caused by the convective parameterization which is sensitive to minor variations of heat input at or near the surface and this may create small-scale stationary features characterized by heavy rainfall. In this sense, the variation of the convective scheme adopted for the 5-layer model is less sensitive. The incorporation of a variable ground hydrology in the high-resolution model has led to removal of marked rainfall maxima over Arabia and the desert regions of Pakistan, which mar the 5-layer model simulations.

8.3 Conclusions

The simulations of the southwest monsoon by two different models developed in the Meteorological Office have been compared. Both simulated the overall large-scale features with considerable realism, though there were also significant errors. The models differed in resolution, the lower-resolution simulation being characterized by a lack of strength while the higher-resolution model obtained an over-vigorous circulation. This difference is unlikely to be a consequence of the resolution itself, though detailed variations within the simulations may well be. The resolution of even the low-resolution model was adequate to describe the primary monsoon circulation, while neither model reproduced important smaller-scale features such as Bay of Bengal depressions or Arabian Sea cyclones. Improvement of the general standard of monsoon simulations will undoubtedly require the simulation of such features, but at this time it is not clear what physical mechanisms need to be reproduced.

It seems more likely that the differences in the monsoon simulations in the two models were mainly the result of differences in their physical parameterizations. The diabatic heating of the atmosphere in the monsoon region was significantly different in the two cases, and undoubtedly this had a strong influence on the resultant circulations. While it is probable that the interaction of the monsoon with other large-scale features of the global circulation influences its development and intensity, it seems necessary to try to understand first of all how far it is a consequence of essentially local effects. The simulations tend to lend support to the thesis that the monsoon circulation is sensitive to the diabatic heating in the region over and around the Indian subcontinent. It might be conjectured that this sensitivity will show itself in significant variations in the strength of the monsoon as a result of variations in the heat input such as might be caused by extensive sea-surface temperature anomalies, changes in surface albedo, moisture availability at the surface or cloudiness. To the extent that such features can be predicted, it may be possible to forsee certain characteristics in particular years.

References

Carson, D. J. and Cullen, M. J. P. (1977) Intercomparison of short-range numerical forecasts using finite difference and finite element models from the U.K. Meteorological Office. *Beit. Phys. Atmos.*, **50**, 1–15.

Clarke, R. H. (1970) Recommended methods for the treatment of the boundary layer in numerical models. *Aust. Meteor. Mag.*, **18**, 51–73.

Corby, G. A., Gilchrist, A. and Newson, R. L. (1972) A general circulation model of the atmosphere suitable for long period integrations. *Quart. J. Roy. Met. Soc.*, **98**, 809–33.

Corby, G. A., Gilchrist, A. and Rowntree, P. R. (1977) The U.K. Meteorological Office 5-level general circulation model. *Methods in Computational Physics.* Vol. 17, pp. 67–110, Academic Press, New York.

Gilchrist, A. (1977). The simulation of the Asian Summer Monsoon, *Pure and Applied Geophysics,* **115**, 1431–48.

Ramage, C. and Raman, C. V. R. (1972) *Meteorological Atlas of the International Indian Ocean Expedition.* Vol. 2. National Science Foundation, Washington, DC.

Ramage, C. S., Miller, F. R. and Jeffries, C. (1972) *Meteorological Atlas of the International Indian Ocean Expedition.* Vol. 1. National Science Foundation, Washington, DC.

Rodgers, C. D. (1967) The radiative heat budget of the troposphere and lower stratosphere. Planetary Circulation Project, Report No. A2, Department of Meteorology, MIT, Cambridge, Mass.

Saker, N. J. (1975) An 11-layer general circulation model. Meteorological Office, London, Met. O. 20 Technical Report II/30.

Report of the US GATE Central Program Workshop (1977) NSF and NOAA, held at NCAR, July–August, 1977.

9

Analysis of monsoonal quasi-stationary systems as revealed in a real-data prediction experiment

M. KANAMITSU

Using the results from a real-data medium-range (6-day) prediction experiment over the global tropics, the monsoonal subtropical quasi-stationary large-scale motion system is examined. Analyses of the structure and the barotropic and the baroclinic energetics in the wavenumber domain are performed and are contrasted with climatological observational studies. The mechanisms of the maintenance of the waves are discussed and the differing roles of each wave are stressed.

9.1 Introduction

A large number of observational and numerical studies have been carried out in recent years to investigate the general circulation of the tropical atmosphere. The observational studies were greatly affected by the recent development of a wide variety of global observational systems (satellite, aircraft, etc.). Most of these studies have dealt with regional or global monthly/seasonal mean analyses of the tropical flow fields (e.g., Aspliden *et al.*, 1965; Flohn, 1971 and Newell *et al.*, 1972) and provided characteristic mean features of the tropospheric motion fields. However, not many long sequences of daily analyses required for a quantitative investigation have ever been performed. Only two such studies are available (Krishnamurti and Rogers, 1970 and Krishnamurti *et al.*, 1975). These studies provided daily analyses of the upper-tropospheric motion fields during northern summer seasons. More recently, it has been recognized that the analyses of the tropical motion fields prepared at the US National Meteorological Center (NMC) have been improved as a result of extensive data-gathering efforts and have become useful data sources.

These daily analyses have yielded a number of quantitative studies of the tropical general circulation. Krishnamurti (1971*a*, *b*), Kanamitsu *et al.* (1972), Kanamitsu and Krishnamurti (1978) and Zangvil (1975) have studied motion fields and energetics of the circulations during the northern summer season. Holton and Colton (1972) and Fein (1977) have studied the vorticity budget of the ultra-long waves during the same period. Krishnamurti *et al.* (1974) and Murakami and Unninayar (1977) have focused on winter tropical circulations. Several other recent observational studies are included in this book (see Chapters 1 to 5).

The observational studies are, however, still severely limited by the lack of adequate observational systems. For example, temperature and moisture analyses over the entire tropical troposphere are still not satisfactory because of the lack of accurate observing systems. Accordingly, several essential features such as thermal structures and the related baroclinic energetics of the tropical troposphere are still not well investigated. In this respect, the First GARP Global Experiment (FGGE) and the Monsoon Experiment (MONEX) are expected to fill such gaps.

Numerical studies of the tropical general circulation have been performed mostly by general-circulation model groups. These studies have been very successful in simulating the gross features of the tropical circulation (Manabe *et al.*, 1974; Washington and Daggupaty, 1975; Gilchrist, 1974; Stone *et al.*, 1977). However in all these studies, discrepancies between the simulated and the real atmosphere seem to have caused difficulties when the results were applied to the real atmosphere. Despite this, simulation experiments have contributed significantly to the understanding of the tropical general circulation.

These observational and numerical studies have shown that the tropical tropospheric motion field is dominated by the quasi-stationary very-large-scale cyclonic and anticyclonic circulation systems. Furthermore, these systems have been shown to be maintained by the east–west land–ocean heating contrast. The importance of this zonal asymmetry in the tropics was the major finding of these studies. The reader should refer to Krishnamurti *et al.* (1973) for a more comprehensive account of this subject.

The main purpose of this chapter is to present the analysis of the tropical circulation revealed in a real-data prediction experiment. In the following sections the merit and the validity of the use of predictions in the analysis of the tropical general circulation is discussed. The experiment and the analysis of the motion field are then described. By separating the scales, detailed three-dimensional structures of the quasi-stationary very-large-scale systems will be presented and compared with the observational and numerical studies mentioned earlier in this section.

9.2 Limitations of observational and numerical studies

Quantitative observational studies are generally limited by the accuracy of the observational data. In the case of the large-scale motion fields, for example, it is widely believed that the wind observations are not accurate enough to give a reasonably accurate vertical motion field unless it is computed as a mean over a very long period. In middle and high latitudes, however, pressure analysis alone can provide the vertical motion field with a fairly good accuracy using the geostrophic approximation. Since this approximation is not a good assumption in very low latitudes, the vertical motion field must be determined from the wind field. This is considered to create large errors in the quantitative investigation of the tropical circulation.

Another problem of observational studies is the evaluation of the effects of various diabatic heating processes. Some of these can be evaluated from the time tendencies of the variables, but the poor accuracy of the observations again results in serious errors. Good physical models are generally required to isolate the effects of various diabatic heating processes, but they are not easy to formulate since many of the details of such processes in the real atmosphere are still not well understood.

These problems suggest that observational and numerical studies must be combined in some manner. One of the solutions to this problem is to use observational data to define an initial state and make numerical predictions. This kind of study has two advantages. One is the mutual adjustment of the mass and velocity fields which takes place in the course of the prediction. This adjustment establishes a quasi-balanced state in which all the observational and analysis errors are considered to be removed from the original data sets. The second advantage is that all the contributions of the various physical processes can be evaluated independently and are tested against the observations by forecast verifications. One disadvantage which may arise in this type of study is that the prediction cannot be made for too long a period, otherwise it departs from the real atmosphere and approaches the model's climatology. Because of this, long-term mean motion fields may not be studied using a prediction experiment. However, for tropical latitudes, since the most dominant systems are quasi-stationary and are always identifiable on daily weather maps, it may be possible to use short-term prediction experiments for the study of long-term mean characteristics. As will be shown later, this turns out to be the case in the present experiment.

9.3 The data and the model

0000 GMT 6 August 1972 was selected as an initial state for this experiment. Extensive efforts were made to collect the conventional upper-air observations, commercial aircraft reports and satellite-derived winds. Subjective analyses were performed for the winds, temperature and moisture fields with the help of climatological maps and satellite pictures over the global tropics. The analysed fields were manually tabulated at the five standard pressure levels up to 200 mb.

The main features of the prediction model are as follows:

(i) Primitive equation formulation with pressure as vertical coordinate.

(ii) 4-layers in the vertical with a resolution of 200 mb.

(iii) A domain covering the global tropical belt extending from 25° S to 45° N with a horizontal resolution of 2.5/2.5 degree latitude–longitude grid.

(iv) Dependent variables; east–west and north–south component of the wind (u and v respectively), vertical velocity in the pressure coordinate system (ω), height of the constant pressure surface (z), potential temperature (θ) and specific humidity (q).

(v) Specified variables; sea-surface temperature, smoothed terrain height and surface albedo over land areas.

(vi) Solid wall conditions with a large diffusion coefficient at the north and south boundaries.

The physical processes included in the model are as follows:

(i) shortwave radiation with diurnal variation;

(ii) longwave radiation;

(iii) large scale condensation;

(iv) moist convection;

(v) dry convective adjustment;

(vi) surface boundary layer;
(vii) topography;
(viii) subgrid-scale diffusion processes.

A more detailed description of the model is given in Kanamitsu (1975).

The initial data were processed through dynamic iterative initialization to remove some of the imbalances between the mass and the velocity fields resulting from the subjective analyses. The prediction was made for six days. During the first 24 hours, the entire fields passed through additional adjustment processes and, after 36 hours, a proper quasi-balanced state was established.

Figs 9.1, 9.2 and 9.3 show predicted streamline fields at 200 mb with observed winds plotted. The predictions are quite good up to at least 72 hours for transient disturbance scales but gradually depart from the observations. However, the quasi-stationary circulation systems (e.g., the Tibetan High, mid-oceanic troughs and the Mexican High) are very well maintained during the 6-day prediction.

For the purpose of separating various scales of motion, zonal Fourier analysis is applied to the predicted fields. Fig. 9.4 shows the time variation of the kinetic energy of the very long waves (wavenumbers 1, 2 and 3) during the prediction. It is seen that the kinetic energies of wavenumbers 1 and 2 stay nearly the same, but that of wavenumber 3 changes significantly, particularly after 120 hours. The time variations of the phase of the waves (of the meridional wind field and the vorticity field) show the quasi-stationary nature of the wavenumbers 1 and 2, and the transient nature of the wavenumber 3.

The present study will concentrate on the analysis of the quasi-stationary wavenumbers 1 and 2. These waves represent the most characteristic zonally-asymmetric features of the tropical/subtropical circulation.

9.4 The structures of the ultra-long waves

Horizontal structure of the waves in the upper troposphere. Fig. 9.5 shows the potential temperature distribution for wavenumber 1 at the 300 mb level. The maximum amplitude of the wave is located at around 35° N where the difference between the minimum and the maximum reaches 10 °C. This difference agrees with the climatological potential-temperature difference between the land and the ocean at this latitude (Newell *et al.*, 1972). The wave has a very weak horizontal tilt north of 15° N but has

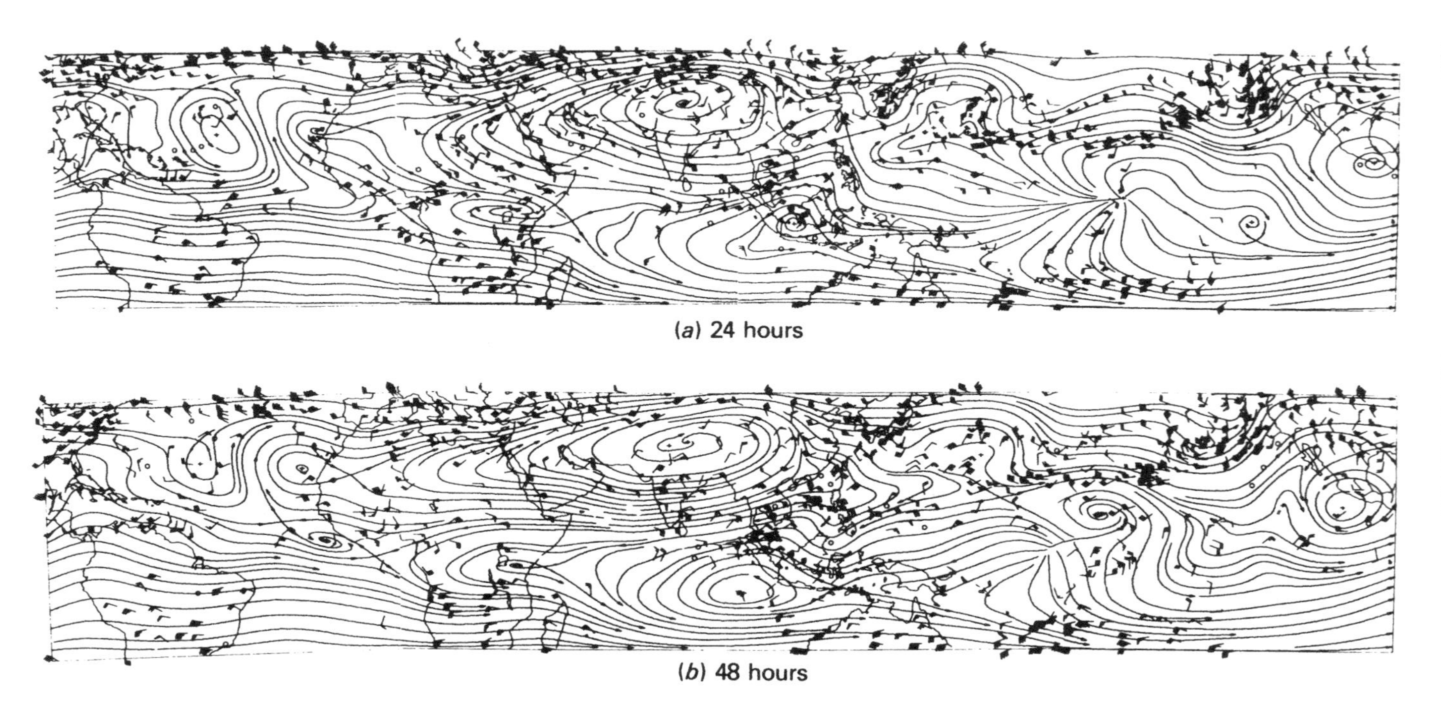

Fig. 9.1. Predicted streamlines at 200 mb for: (*a*) 24 hours (7 August 1972, 912 observations); and (*b*) 48 hours (8 August 1972, 859 observations). Wind speed and direction is represented by feathers, each full filament of which represents 10 knots (approximately 5 m s^{-1}).

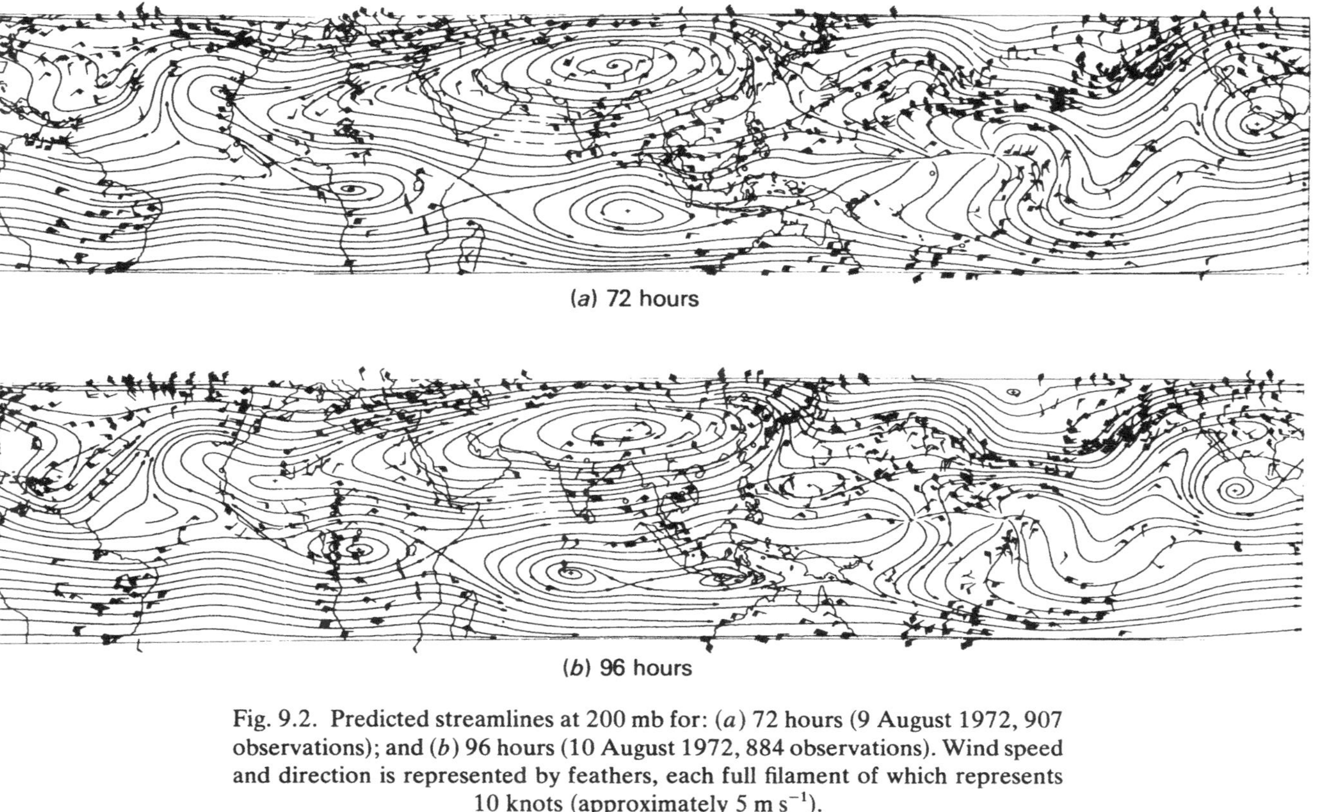

Fig. 9.2. Predicted streamlines at 200 mb for: (*a*) 72 hours (9 August 1972, 907 observations); and (*b*) 96 hours (10 August 1972, 884 observations). Wind speed and direction is represented by feathers, each full filament of which represents 10 knots (approximately 5 m s^{-1}).

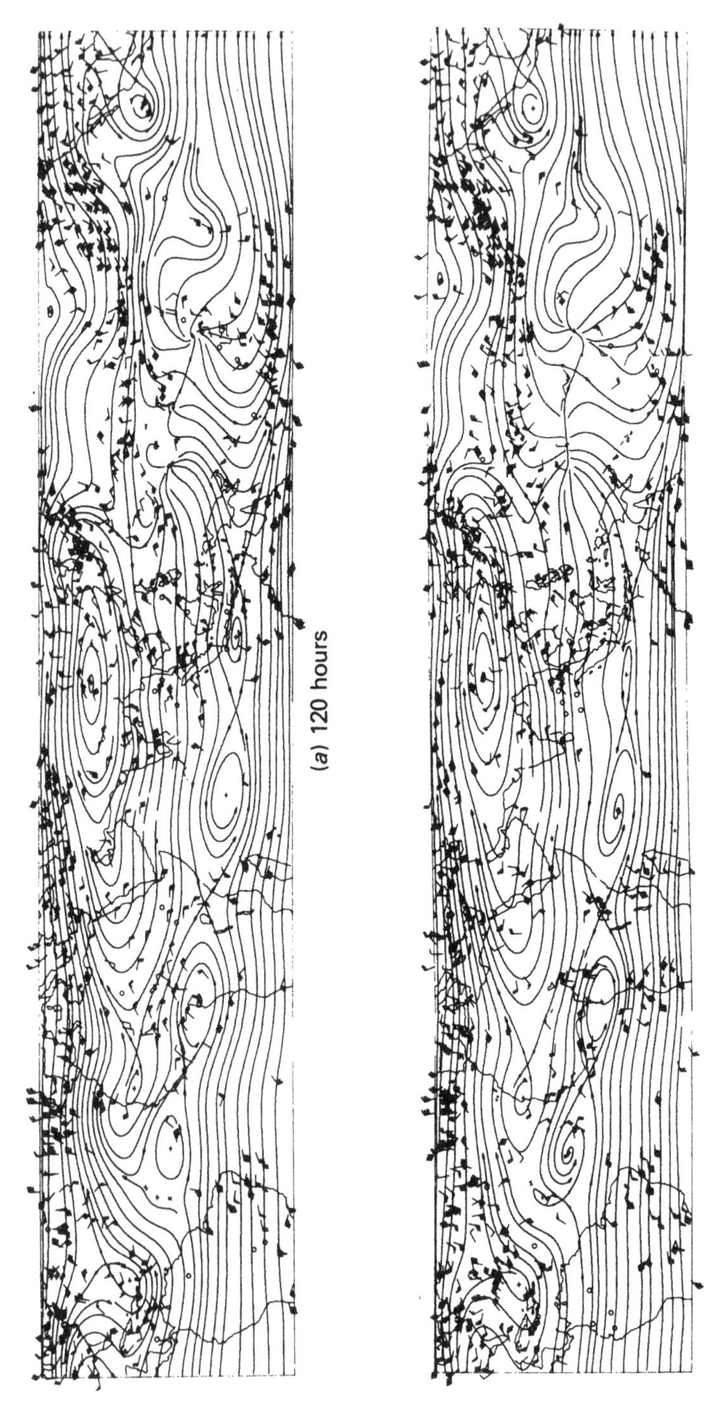

Fig. 9.3. Predicted streamlines at 200 mb for: (*a*) 120 hours (11 August 1972, 823 observations); and (*b*) 144 hours (12 August 1972, 818 observations). Wind speed and direction is represented by feathers, each full filament of which represents 10 knots (approximately 5 m s^{-1}).

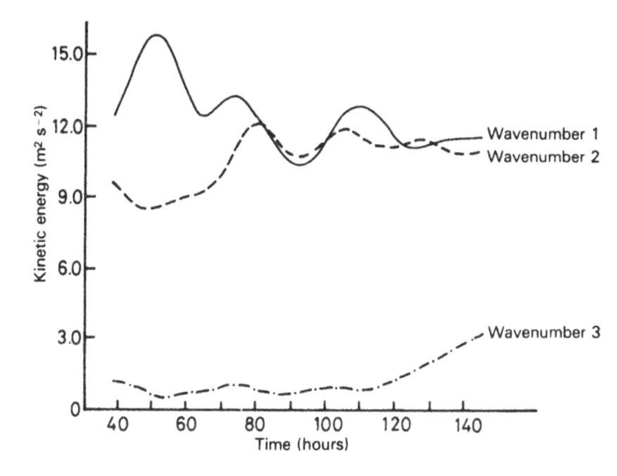

Fig. 9.4. Time variations of the kinetic energy of wavenumbers 1, 2 and 3 at 300 mb between 15° N and 30° N.

strong tilt to the south. The meridional wind field of wavenumber 1 (Fig. 9.6) shows its maximum amplitude at 30° N. The ridge is located at around 90° E where the temperature is a maximum. This system apparently exhibits a characteristic feature of the Tibetan High. The wave has very small horizontal tilt north of 20° N. Close examination of Figs. 9.5 and 9.6 indicates that sensible heat is transported southward by the wave over the area north of 10° N.

The wave south of 15° N apparently has a structure quite different from that of the wave further north. In the following discussion, attention will be focused on the waves north of 15° N, which are manifestations of the typical tropical quasi-stationary circulation systems.

The meridional and the zonal wind components of wavenumber 2 are shown in Figs. 9.7 and 9.8. It is very clear that the wave has very strong southwest to northeast tilt and evidently transports westerly momentum to the north over the area between 15° N and 30° N, thus removing westerly momentum from the equatorial easterlies. This wave corresponds to the mid-Atlantic and mid-Pacific trough systems observed in climatological studies.

Vertical structure of the wave at 30° N. As was shown above, the Tibetan High, Mexican High and the mid-oceanic troughs have their maximum intensities at around 30° N. Fig. 9.9 is the vertical cross-section of the meridional wind for wavenumber 1. The maximum amplitude is located at 200 mb and the minimum at 400 mb. The wave has a very strong tilt with height, the waves at 200 mb and 1000 mb being almost exactly out of phase. Therefore, the upper anticyclonic circulation (Tibetan High) at 90° E is associated with the lower cyclonic circulation (monsoon trough) and vice versa at around 90° W. It should be noted that 90° E and 90° W correspond roughly to the land and ocean, respectively, in terms of the zonal Fourier analysis of the land–ocean distributions at this latitude.

Fig. 9.10 shows the cross-section of the temperature for wavenumber 1. The wave has a nearly vertical axis with amplitude increasing with height. It should be emphasized here that, although the thermal structure of the troposphere during the summer season, i.e., warm over land and cold over ocean, is widely known, the present experiment indicates that the temperature contrast extends through the entire troposphere with its maximum in the upper troposphere.

The vertical motion field for wavenumber 1 is shown in Fig. 9.11. The upward motion at around 90° E and sinking motion at around 90° W correspond to the east–west circulation observed in the climatology. This circulation is apparently thermally direct and the potential energy is evidently converted to kinetic energy in this scale. The maximum amplitude of the vertical motion is located at 800 mb. This rather low altitude of the maximum vertical motion seems to be due to the dynamical effect of the topography, but further studies are required to make more comments on this matter.

In Fig. 9.12, moist convective heating on the scale of wavenumber 1 is presented. The maximum heating is located roughly at the temperature maximum showing generation of available potential energy of this scale. Other diabatic heating terms are at least an order of magnitude smaller than the convective heating at this level.

In conclusion, the three-dimensional structure of the ultra-long waves presented here is strikingly similar to the observed long-term mean ultra-long waves. This indicates the exceptionally pronounced stationary nature of the very-large-scale systems in subtropical latitudes.

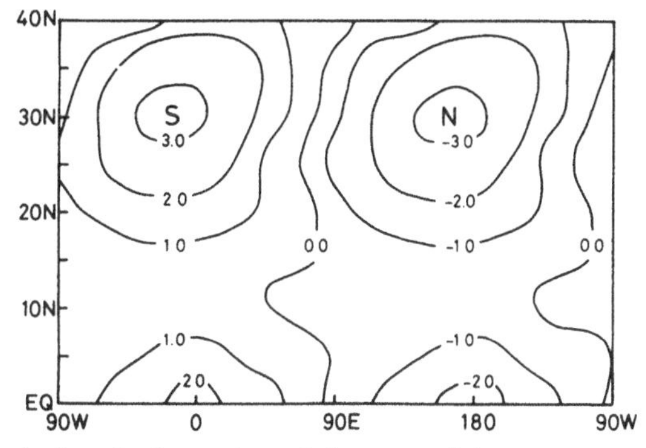

Fig. 9.5. Latitude–longitude section of the potential temperature at 300 mb for wavenumber 1 (contour interval: 1 °C).

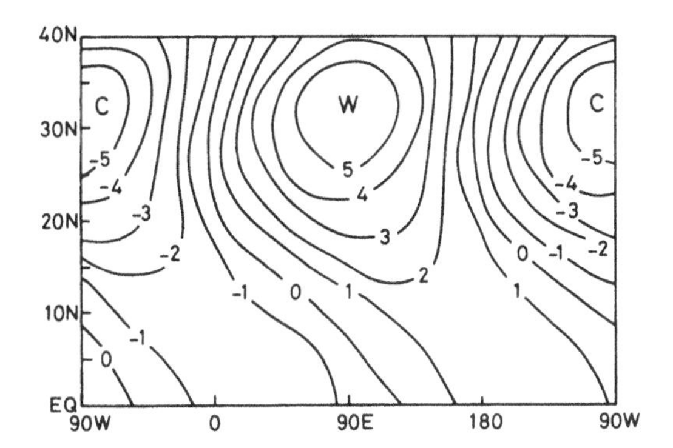

Fig. 9.6. Latitude–longitude section of the meridional component of the wind at 200 mb for wavenumber 1 (contour interval: 1 m s^{-1}).

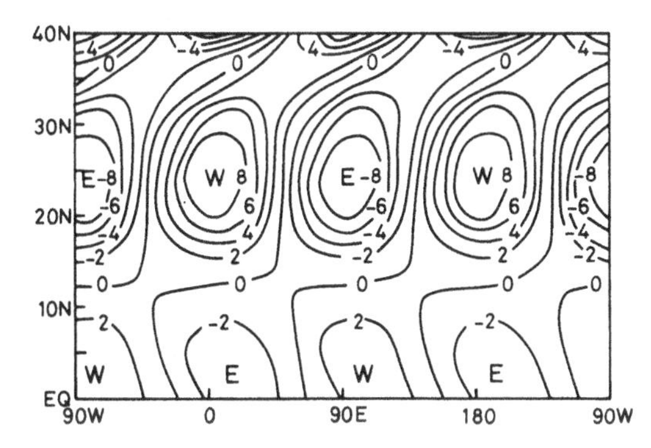

Fig. 9.7. Latitude–longitude section of the zonal component of the wind at 200 mb for wavenumber 2 (contour interval: 2 m s^{-1}).

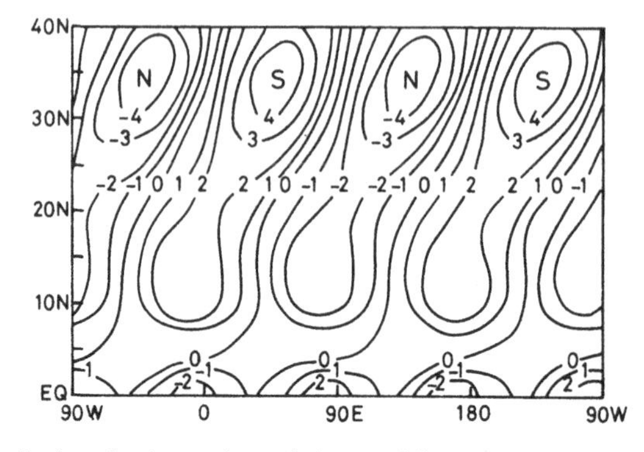

Fig. 9.8. Latitude–longitude section of the meridional component of the wind at 200 mb for wavenumber 2 (contour interval: 1 m s^{-1}).

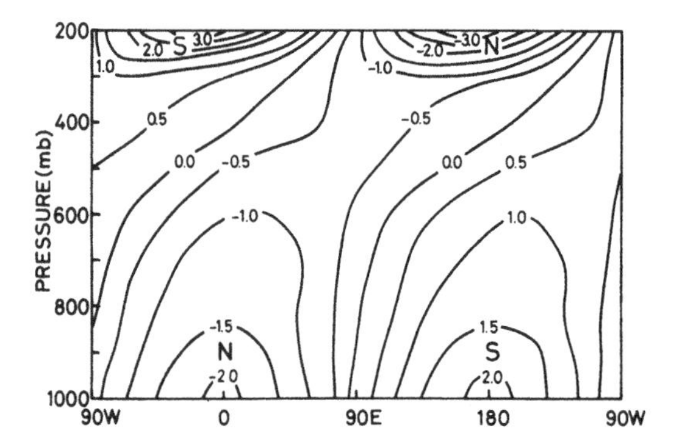

Fig. 9.9. Longitude–pressure cross-section of the meridional wind component for wavenumber 1 at 30° N (contour interval: 0.5 m s^{-1}).

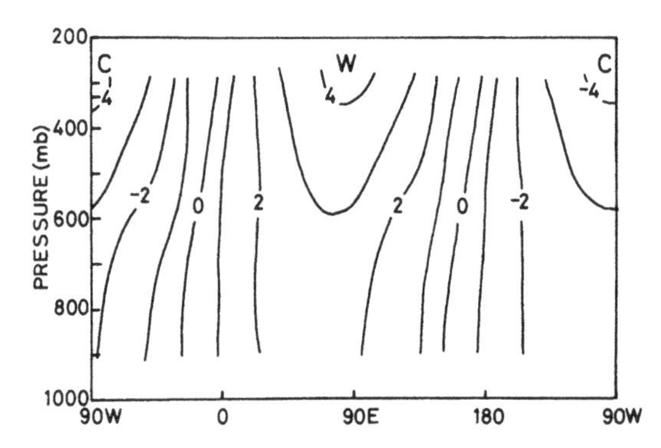

Fig. 9.10. Longitude–pressure cross-section of the temperature for wavenumber 1 at 30° N (contour interval: 1 °C).

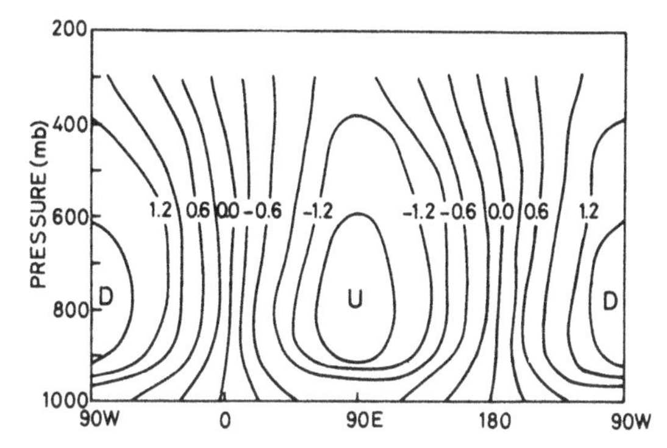

Fig. 9.11. Longitude–pressure cross-section of the vertical motion for wavenumber 1 at 30° N (contour interval 0.3×10^{-4} mb s^{-1}).

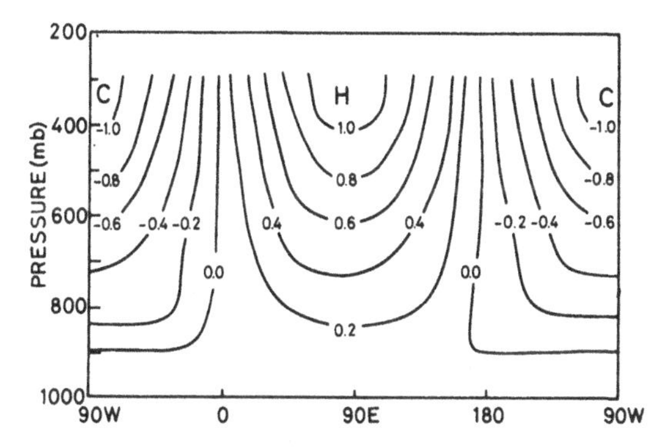

Fig. 9.12. Longitude–pressure cross-section of the moist convective heating for wavenumber 1 at 30° N (contour interval 0.2×10^{-5} K s^{-1}).

9.5 Energetics of the ultra-long waves

Energy generation, conversion and dissipation processes are evaluated by computing corresponding terms of the energy equation in the wavenumber domain. The complete expressions and their physical meanings for the equation used in this study are given in Saltzman (1970). Fig. 9.13 shows the time-mean energy components and conversions during the prediction period computed at 300 mb over the global subtropical belt extending from 15° N to 30° N. The selection of this latitudinal band is determined from the horizontal structure of the stationary waves discussed earlier. In this figure, numbers in the boxes show the available potential energy (P) and the kinetic energy (K) of the wavenumber n ($n = 0$ is the zonal mean state), and are expressed in units of m^2 s^{-2}.

Fig. 9.13 indicates that the energy in wavenumbers 1 and 2 accounts for 70 to 80% of the total energy. The energy of the zonal state is considerably smaller. The dominance of the ultra-long waves in the domain has been noted in observational studies (Krishnamurti, 1971*a*).

The available potential energy of wavenumber 1 is generated mostly by the diabatic heating (moist convective heating), but it also receives a fairly

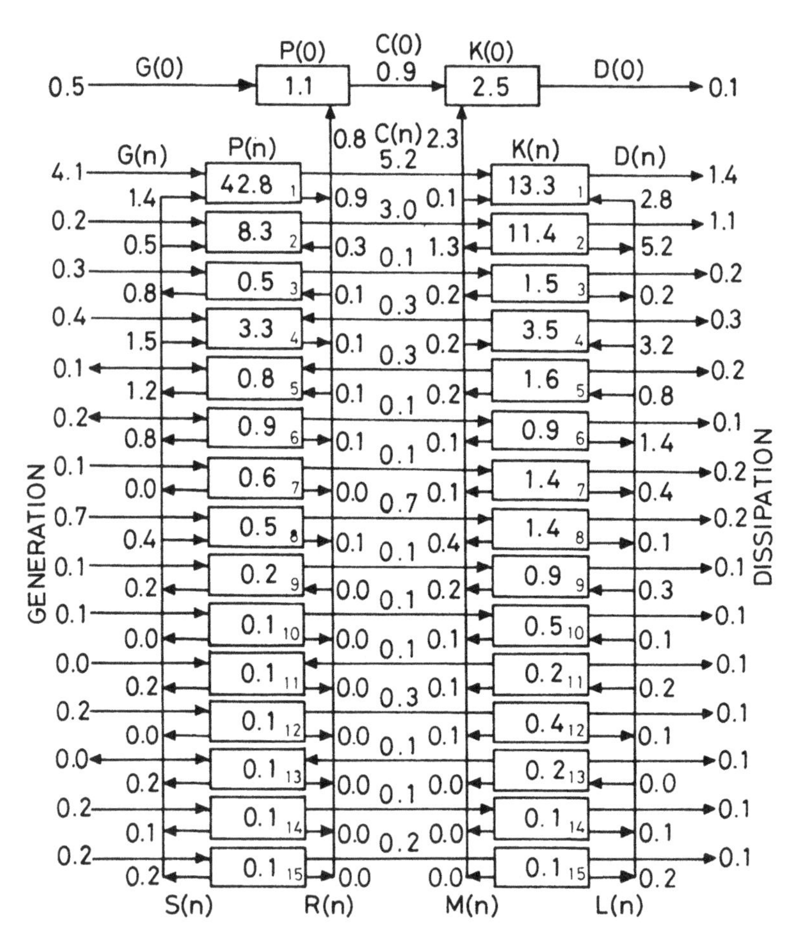

Fig. 9.13. Barotropic and baroclinic energetics in the wavenumber domain computed at 300 mb over the subtropical belt between 15° N and 30° N from the prediction experiment. *P* and *K* denote available potential energy and kinetic energy respectively. Arrows going into and out of the boxes indicate energy generation (*G*), conversion (*C*), exchange between different scales (*S, L*), exchange between the waves and the zonal state (*R, M*) and dissipation (*D*). Small numbers in the boxes are wavenumbers (units: $m^2\,s^{-2}$ for *P* and *K*; $10^{-5}\,m^2\,s^{-3}$ for conversions).

large amount from other smaller scales. A large part of the available potential energy of wavenumber 1 is converted to kinetic energy of the same scale by thermally direct east–west overturnings shown in Figs. 9.10 and 9.11. The wave also supplies available potential energy to the zonal mean state through heat transport up the temperature gradient as shown in Figs. 9.5 and 9.6 (the temperature maximum of the zonal mean state is located at 5° N at 300 mb). There are no observational studies of the scale interactions of the available potential energy for comparison with the present study. The barotropic energetic conversions show that wavenumber 1 receives a large amount of kinetic energy from smaller scales but interacts very weakly with the zonal motion. These barotropic conversions are somewhat different from those observed by Kanamitsu *et al.* (1972), but since the domain in their study is different (15° S to 15° N) from that of the present study, direct comparison is not meaningful.

The available potential energy of wavenumber 2 is also generated by diabatic heating, but the magnitude is considerably smaller. The wave receives available potential energy from other scales as well as from the zonal mean state. The conversion from potential to kinetic energy also takes place in wavenumber 2. In the barotropic conversion processes, the wave acts as a strong kinetic energy source for other scales of motion as well as for the zonal motion. This is to be expected from the strong horizontal tilt of the wave (Fig. 9.7).

It is an important finding of this study that the barotropic characteristics of the quasi-stationary wavenumbers 1 and 2 are quite different, i.e., the former is barotropically passive while the latter is very active.

The energy diagram shows, furthermore, that other smaller scales (up to wavenumber 8) also play important roles on the energetics of the subtropics. In particular the secondary maximum of the kinetic energy at wavenumber 8 indicates that the transient disturbances are another predominant phenomenon. It is also noted that the scale interactions of wavenumbers 1 and 2 favour the existence of more intense smaller-scale motion systems.

Some imbalances of the energy budget apparent in some of the wavenumbers are considered to arise from several factors, namely the omission of boundary terms, truncation errors and the real transient nature of the waves (especially of the smaller-scale waves) during the prediction.

Similar energy diagrams have been computed for the entire troposphere. The results may be briefly summarized as follows. Most of the generation and conversion processes are the same as those at 300 mb.

The largest difference is the conversion of available potential energy between the ultra-long waves and the zonal mean state. The waves receive available potential energy from the zonal mean state for the entire troposphere through heat transport down the temperature gradient which is evidently large in the lower troposphere. The interaction of the ultra-long waves with the smaller scales also reverses sign. These differences in the horizontal heat transport are due to the vertical variation of the correlation between the wind and the temperature, which is evident from the vertical structure of the waves (Figs. 9.9 and 9.10). More complete discussions of the energetics of this experiment may be found in Kanamitsu (1977).

9.6 Summary and conclusions

The real-data medium-range prediction experiment was performed to examine the monsoonal quasi-stationary ultra-long waves. The prediction is very successful in maintaining the large-scale quasi-stationary systems during the 6-day forecast. Zonal Fourier analysis was applied to the predicted fields to determine the structures of these systems.

The systems are composed mainly of zonal wavenumbers 1 and 2. These waves have maximum amplitudes at around 30° N and extend to 15° N. Wavenumber 1 has a small horizontal tilt while wavenumber 2 has a strong southwest to northeast tilt.

At 30° N, the wave clearly shows warm rising motion over the land area with a cyclonic circulation at low levels and an anticyclonic circulation at upper levels. These features are exactly the reverse over the ocean. The largest land–ocean temperature contrast is found at 300 mb where the heating (moist convective) contrast is also a maximum.

The baroclinic and the barotropic energetics of the waves are computed. The energy of wavenumbers 1 and 2 account for almost 80% of the total energy (including the energy of the zonal mean state). The available potential energies of these waves are generated by the moist convective heating. They are also provided with a fairly large amount of available potential energy from other scales. Wavenumber 1 transports sensible heat up the gradient of the temperature and increases the zonal available potential energy.

Both waves convert available potential energy to kinetic energy in thermally direct east–west circulations.

Major differences arise between wavenumbers 1 and 2 in their barotropic energy conversions. Wavenumber 1 operates as a kinetic energy sink for the smaller scales and interacts weakly with the zonal motion. On the other hand, wavenumber 2 acts as a kinetic energy source for other waves and also for the zonal motion.

In short, the quasi-stationary ultra-long waves described in the present study are driven by the moist convective heating over the continental monsoon area. These scales of motion drive other smaller-scale motions through barotropic kinetic energy conversion processes.

These features are very similar to those of the observed summer mean state. This fact indicates the degree of stationarity of the very-large-scale systems in the tropics and subtropics.

It should be noted that very small changes in the quasi-stationary ultra-long waves can cause drastic changes in the weather as well as in the climate in low latitudes. In this respect, it is necessary to investigate the time-varying properties of the ultra-long waves, especially in relation to the changes in heating and the nonlinear interactions among the waves.

References

Aspliden, C. I., Dean, G. A. and Landers, H. (1965) Satellite study, tropical North Atlantic, 1963. Technical Report, Florida State University, Tallahassee, Florida, 32306, USA.

Fein, J. (1977) Global vorticity budget over the tropics and subtropics at 200 mb during northern hemisphere summer. *Pure and Applied Geophysics*, **115**.

Flohn, H. (1971) Tropical circulation pattern. *Bonner Meteor. Abhand.* **15**, 55 pp.

Gilchrist, A. (1974) A general circulation model of the atmosphere incorporating an explicit boundary layer top. Technical Note NO. II/29, Meteorological Office, Bracknell, UK, 35 pp.

Holton, J. R. and Colton, D. E. (1972) A diagnostic study of the vorticity balance at 200 mb in the tropics during the northern summer. *J. Atmos. Sci.*, **29**, 1124–8.

Kanamitsu, M. (1975) On numerical prediction over a global tropical belt. Report No. 75-1, Department of Meteorology, Florida State University, Tallahassee, Florida, 32306, USA.

Kanamitsu, M. (1977) Monsoonal quasi-stationary ultralong waves of the tropical troposphere predicted by a real data prediction over a global tropical belt. *Pure and Applied Geophysics*, **115**, 1187–1208.

Kanamitsu, M., Krishnamurti, T. N. and Depradine, C. (1972) On scale interactions in the tropics during northern summer. *J. Atmos. Sci.*, **29**, 698–706.

Kanamitsu, M. and Krishnamurti, T. N. (1978) Northern summer tropical circulations during drought and normal rainfall years. *Mon. Wea. Rev.*, **106**, 331–47.

Krishnamurti, T. N. (1971*a*) Observational study of the tropical upper tropospheric motion field during the northern hemisphere summer. *J. Appl. Meteor.*, **10**, 1066–96.

Krishnamurti, T. N. (1971*b*) Tropical east–west circulations during the northern summer. *J. Atmos. Sci.*, **28**, 1342–7.
Krishnamurti, T. N. and Rogers, E. B. (1970) 200 mb wind field June, July and August, 1967. Report No. 70–2, Department of Meteorology, Florida State University, Tallahassee, Florida, 32306, USA.
Krishnamurti, T. N., Daggupaty, S. M., Fein, J., Kanamitsu, M. and Lee, J. D. (1973) Tibetan high and upper troposperic tropical circulations during northern summer. *Bull. Amer. Meteor. Soc.*, **54**, 1234–49.
Krishnamurti, T. N., Kanamitsu, M., Koss, W. J. and Lee, J. D. (1974) Tropical east–west circulations during the northern winter. *J. Atmos. Sci.*, **30**, 780–7.
Krishnamurti, T. N., Astling, E. G. and Kanamitsu, M. (1975) 200 mb wind field June, July and August, 1972. Report No. 75–1, Department of Meteorology, Florida State University, Tallahassee, Florida, 32306, USA.
Manabe, S., Hahn, D. G. and Holloway, J. L. Jr. (1974) The seasonal variation of the tropical circulation as simulated by a global model of the atmosphere. *J. Atmos. Sci.*, **31**, 43–83.
Murakami, T. and Unninayar, M. S. (1977) Atmospheric circulation during December 1970 through February 1971. *Mon. Wea. Rev.*, **105**, 1024–38.
Newell, R. E., Kidson, J. W., Vincent, D. G. and Boer, G. J. (1972) *The General Circulation of the Tropical Atmosphere and Interactions with Extratropical Latitudes*. Vol. 1. MIT Press, Cambridge, Mass., 258 pp.
Saltzman, B. (1970) Large-scale atmospheric energetics in the wavenumber domain. *Rev. Geophys.*, **8**, 289–302.
Stone, P. H., Chow, S. and Quirk, W. J. (1977) The July climate and a comparison of the January and July climates simulated by the GISS general circulation model. *Mon. Wea. Rev.*, **105**, 170–94.
Washington, W. M. and Daggupaty, S. M. (1975) Numerical simulation with the NCAR global circulation model of the mean conditions during the Asian–African summer monsoon. *Mon. Wea. Rev.*, **103**, 105–14.
Zangvil, A. (1975) Temporal and spatial behavior of large scale disturbances in tropical cloudiness derived from satellite brightness data. *Mon. Wea. Rev.*, **103**, 904–20.

10

A model of the seasonally-varying planetary-scale monsoon

P. J. WEBSTER

A simple ocean–atmosphere climate model is used to test the hypothesis that the mean and transient structure of the monsoon system and the phasing of the response of the system relative to the solar declination are primarily functions of the differential heating of the land and ocean regions and the interaction of their individual responses. The model used is a dry version of the primitive equation domain-averaged model of Webster and Lau (1977*a*) which is the simplest model capable of incorporating a continental region and an interactive ocean system.

Both mean and seasonal transient features of the monsoon are well simulated by the model. The influences of the oceans to the south and east of the continents are found to be of equal importance in determining the phase of the atmospheric response. Zonal convergences of heat and momentum are found to be much greater than the meridional fluxes at low latitudes. Magnitudes of the heat convergence into the ocean domain from the continent in summer are at least as large as the radiative and boundary sensible heating in the ocean regions.

The results of the model are critically evaluated and the fields compared with the limited observations that exist in the monsoon region. The degree of complication that the model would have to assume to account for sub-seasonal variations is considered.

10.1 Introduction

Based on a criterion of seasonal surface-pressure persistence and wind reversal, the climatological monsoon region is generally defined to be contained within the subtropics and tropics of the eastern hemisphere

(Ramage, 1971). Meteorologically, such geographical restrictions have little meaning. Diagnostic and model studies (e.g. Krishnamurti, 1971*a*, *b*; Webster, 1972) suggest that the monsoon circulation extends eastward to at least the central Atlantic Ocean and westward to the eastern Pacific Ocean. The latitudinal extent is of a similar scale extending from the southern Indian Ocean (Krishnamurti and Bhalme, 1976) to northern Eurasia (Blackmon *et al.*, 1977). The spatial extent of the monsoon is not restricted to summer (northern hemisphere chronology) as the winter monsoon circulation is at least as large as the summer counterpart.

The latitudinal structure of the monsoon system has received considerable attention despite the data limitations in the ocean regions to the south of Asia. A picture has emerged which appears consistent from both observation and theoretical viewpoints. Based on Koteswaram (1958), the summer monsoon may be thought of as consisting of two meridional cells; a Hadley cell located between 20° N and the Equator, and a monsoon cell located from 20° N to 40° N. The two cells possess a common ascending branch at 20 °N, which coincides with the hot summer continent. Associated with the cell structure are the strong upper-tropospheric easterly winds emanating from the outflow of the ascending leg which owes much of its vigour to the enormous release of latent heat associated with the monsoon rains. In winter, the maximum disturbed and precipitating region is in the southern hemisphere. At this time, the northern hemisphere upper troposphere is dominated by strong westerlies.

The observed summer structure is simulated by the zonally-averaged model of Murakami *et al.* (1970) which was developed to study the mean seasonal monsoon along 80° E. The multi-dimensional general-circulation models (Washington and Daggupaty, 1975; Hahn and Manabe, 1975) exhibit similar features in the vicinity of the Asian continent. The large models add one further perspective. They show that the latitudinal structure discussed above is not representative far away from 80° E and that the monsoon system exhibits considerable meridional structure.

The variation in longitude of the monsoon system has received scant attention. Beyond the observational works of Krishnamurti *et al.* (1973), few studies exist, principally because the sparse data in the Pacific region render diagnostic investigation to be a dubious proposition, with little chance of obtaining meaningful results.

Hopefully, the observational aspirations of the First GARP Global Experiment (FGGE) and the Monsoon Experiment (MONEX) will provide a useful data set.†

The simulation of the longitudinal variation of the monsoon is achievable by the large-scale multi-dimensional general-circulation models (e.g. Hahn and Manabe, 1975). Such models include many important physical mechanisms, feedback processes and detailed spatial resolution which unfortunately, render experimentation expensive and difficult, especially when processes are evolving over seasonal timescales. Normally, the logistical problems of the general-circulation models are overcome by the development of simpler 'phenomonological' models aimed modestly at elucidating one or a few basic physical mechanisms rather than producing detailed simulations of the atmospheric structure. In the case of the longitudinal variation of monsoons, the development of a framework of even the simplest phenomenological model is difficult. This is because the lateral structure of the monsoon demands the use of a model with more degrees of freedom than a zonally-averaged model (e.g. as used by Murakami *et al.*, 1970), as in the monsoon regions ocean–continent contrast occurs in longitude as well as latitude.

Webster and Lau (1977*a*, *b*) have shown that in order to obtain a meaningful simulation of the mean seasonal state of a monsoon system, it is important to include an interactive ocean model adjunct to the atmospheric model. It would, therefore, seem imperative that this should be included if one wishes to describe the seasonal and annual variation of the monsoon. This is because the ocean also possesses a strong seasonal variation in its state which is out-of-phase with solar heating and thus out-of-phase with continental heating. Furthermore the variation is a strong function of latitude.

In summary, it appears that the study of the time-varying monsoon requires the use of a general-circulation model which possesses an interactive ocean model, or the development of a phenomonological model which allows variations in three dimensions, is variable in time and also contains an interactive ocean. The model we chose for this study is the phenomonological 'domain-average' model of Webster and Lau (1977*a*) which incorporates the three requirements outlined above with

† Both FGGE and MONEX are sub-programmes of the Global Atmospheric Research Programme–GARP.

the additional attribute that it is sufficiently simple to allow extended integration.

The aim of this chapter is to re-establish the basic premise of Halley (1686), that the differential heating of the atmosphere by the ocean and the continental regions is the *primum mobile* of the monsoon circulation, and to understand the mechanisms which control its seasonal and annual variation. We are particularly interested in the manner adjacent ocean regions influence the monsoon circulation and, in turn, are influenced by the monsoon circulation itself. Lastly, we are interested in achieving a first-order estimate of the relative magnitudes of the zonal fluxes of heat and momentum, which allow a two-way communication between the continental region and the adjacent oceans, and the meridional fluxes of heat and momentum.

In essence, we are testing the hypothesis that the gross features of the monsoon and its seasonal variation are primarily due to the interaction of a variable solar heating and the differential and dynamic response of the continents, atmosphere and oceans. In the experiment, we purposely neglect such features as the hydrology cycle even though we expect it to be responsible for considerable sub-seasonal variations and for much of the vigour of the mean seasonal fields. We leave the discussion of the role of the hydrology cycle and the timescales it imposes on the monsoon system to Webster and Chou (1980*a*, *b*).

10.2 The model

The earth–atmosphere–ocean system is hypothesized to consist of a large continental region occupying roughly 40% of the northern hemisphere with the remainder of the earth covered by ocean. The land mass is constrained to occupy the eastern hemisphere north of 18° N. The model geography is shown in Fig. 10.1*a*.

The equations of motion for the atmosphere are written in primitive form and averaged in longitude over each domain separately to form two coupled sets of mean equations. After domain averaging, the equations relate functions of latitude, height and time. The dynamic coupling between the two domains is accomplished by zonal fluxes of heat and momentum and by zonal pressure–work terms. The coupling terms originate from domain averaging the zonal derivatives and are indicated by the broad arrows in Fig. 10.1.*b*.

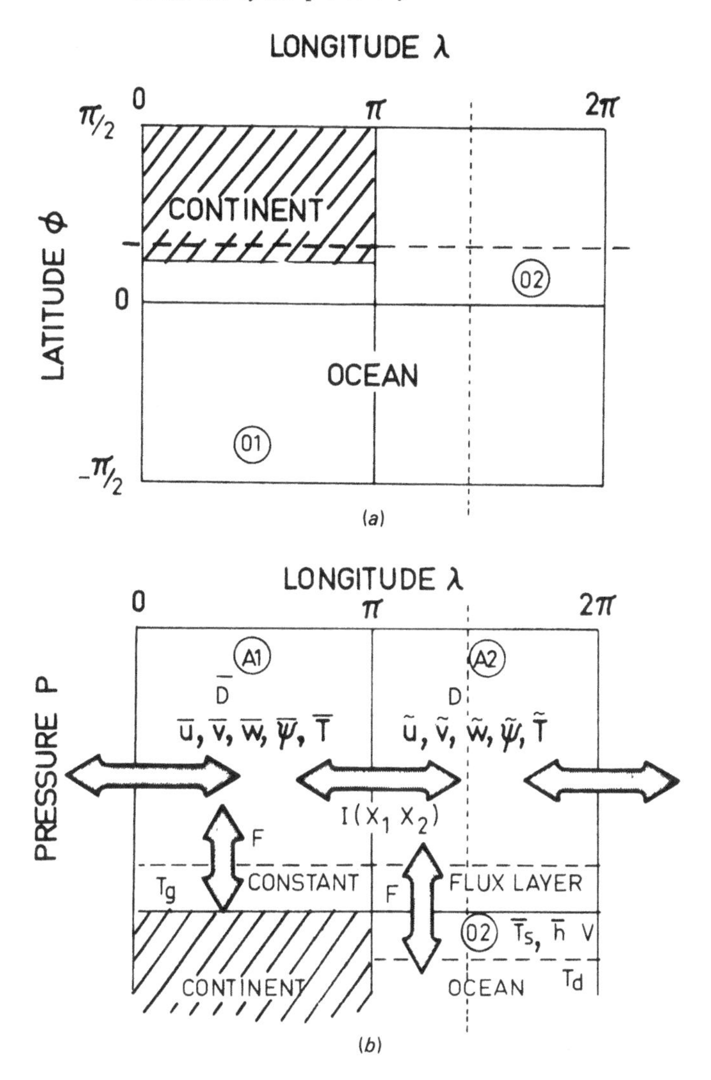

Fig. 10.1. (*a*) Latitude–longitude and (*b*) latitude–pressure sections of the two-atmosphere domain-averaged model. Panel (*b*) refers to the dashed line cut through the model northern hemisphere in (*a*). The meridional dotted line denotes a further domain subdivision for later reference. In this figure T_g is the ground temperature, $\bar{T}_s$ is the sea-surface temperature, h is the depth of the mixed layer, T_d is the temperature of the sub-mixed layer and the I functions indicate the interdomain fluxes. Atmosphere domains 1 and 2 are labelled (A1) and (A2) respectively. Ocean domains 1 and 2 are labelled (01) and (02) respectively.

The basic atmospheric set for one domain (e.g. eastern hemisphere) is

$$\left.\begin{aligned}
\bar{u}_t+M(\bar{u})-f\bar{v}&=D(u')+I(\bar{u})+F(\bar{u}),\\
\bar{v}_t+M(\bar{v})-\frac{1}{a}\bar{\psi}_\phi+f\bar{u}&=D(v')+I(\bar{v})+F(\bar{v}),\\
\bar{T}_t+M(\bar{T})&=D(T')+I(\bar{T})+F(\bar{T})+\bar{\dot{Q}},\\
\bar{q}_t+M(\bar{q})&=D(q')+I(\bar{q})+F(\bar{q})+\bar{S},\\
\bar{\nabla}\cdot(\bar{V})&=I(m),\\
\bar{\psi}_p&=-R\bar{T}/p.
\end{aligned}\right\}\qquad(10.1)$$

A similar set exists for the western hemisphere. A two-layer vertical structure is adopted. In (10.1), $\bar{u}$ and $\bar{v}$ represent the horizontal components of the velocity vector $\bar{V}$; $\bar{T}$, $\bar{q}$ and $\bar{\psi}$ represent the mean domain temperature, specific humidity and geopotential. Non-adiabatic effects are represented by $\bar{\dot{Q}}$ and moisture sources and sinks by $\bar{S}$.

The primed quantities denote variables of scales smaller than the domain which are not resolvable by the model. The D and M operators represent vertical and meridional fluxes of momentum, heat and moisture by perturbation (i.e. space-scales less than one domain) and mean domain-averaged processes, respectively. The I functions indicate the lateral or interdomain fluxes. The M operators are functions of the mean domain quantities and are thus explicitly determined.

Webster and Lau (1977*a*) present a detailed analysis of the domain-averaging procedure and the parameterization of the D and I operators. It suffices to indicate that the parameterization upon which the meridional flux is based depends primarily upon baroclinic instability theory (Stone, 1974; Green, 1970), whilst the zonal flux parameterization depends upon the theory of slowly varying or quasi-stationary waves.

The F functions of (10.1) indicate vertical fluxes of momentum, heat and moisture between the lower boundary and the atmosphere. The determination of F (also $\bar{\dot{Q}}$ and $\bar{S}$) requires a detailed knowledge of the atmosphere (provided by (10.1)) and its underlying surface, which is provided by the use of a simple ocean model described in detail by Webster and Lau (1977*a*). In essence the model allows for a variable surface temperatures of the ocean and continent at a particular latitude to sub-mixed-layer ocean and ice extent as a function of longitude, latitude and time.

Via the processes described above, changes in surface conditions resulting from a varying solar declination are communicated both meridionally and zonally on various space and time scales. As we expect the surface temperatures of the ocean and continent at a particular latitude to be at different phases relative to the solar declination, we anticipate that the I operators will be important terms since through them lateral interdomain communication occurs.

10.3 Results

Results are presented for two experiments. The first experiment attempts to simulate the mean seasonal structure of the monsoon. To accomplish this the model was forced by solar heating with declination held at both extreme solstice positions and integration continued until an equilibrium solution was approached. In the second experiment the solar declination was allowed to vary through an annual cycle.

10.3.1 Basic fields

Figs. 10.2 and 10.3 show the resultant mean surface temperature and tropospheric zonal wind distributions for the summer and winter seasons. The solid lines in all diagrams correspond to the continent–ocean domain (i.e. the eastern hemisphere of Fig. 10.1) and the dashed lines refer to the ocean domain (the western hemisphere).

Fig. 10.2 illustrates the considerable differences between the two extreme seasons and between the adjacent domains during each season. During winter the mean eastward temperature gradient is reversed with oceans being considerably warmer than the continents. We will see in § 10.3.2 that the reversal of the zonal temperature gradient is responsible for the zonal heat flux reversals from summer to winter.

During winter, the surface temperatures decrease northward rapidly from maxima to the south of the Equator to minima in the arctic regions. In summer, the ocean domain maintains a gradient similar to its winter structure, whereas the continent–ocean domain exhibits a maximum temperature near 20° N thus forming a reversed latitudinal temperature gradient between the continent and the Equator. The winter temperature structure is responsible for the intense westerlies in the northern hemisphere (Fig. 10.3). The reversed temperature gradient of summer forces strong upper-level easterlies in the tropical northern hemisphere. Weaker westerlies occur aloft at all seasons over the oceans and are related to the weaker latitudinal temperature gradients. In summary, the

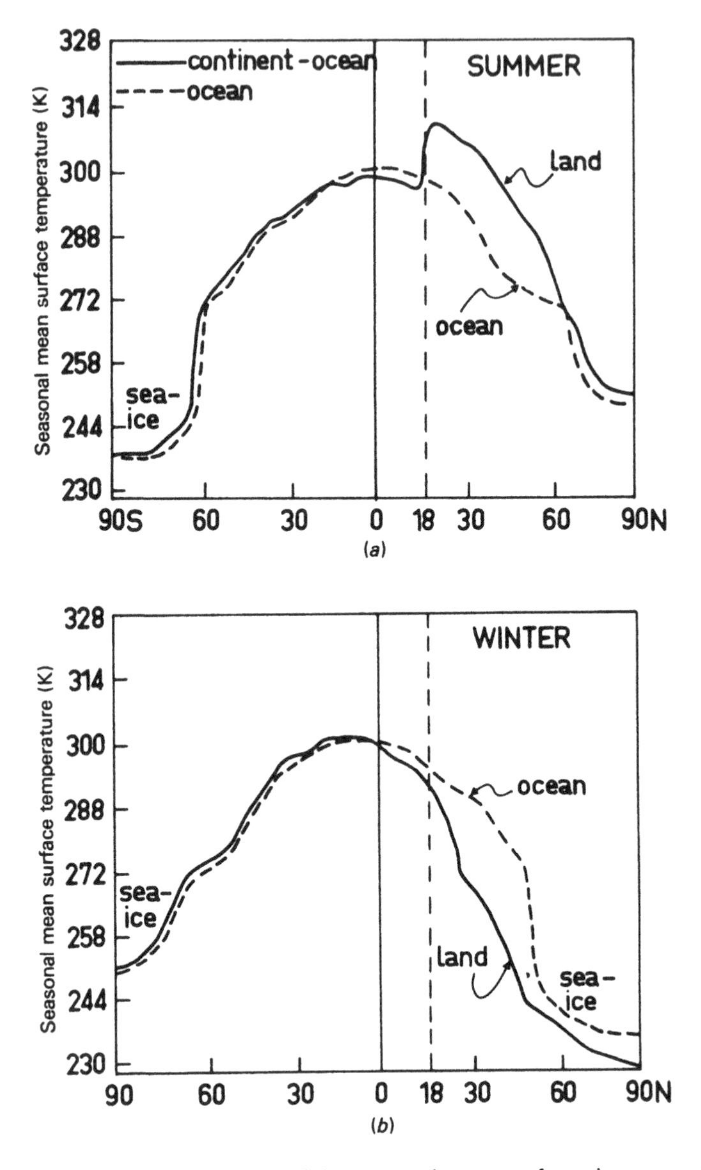

Fig. 10.2. Latitudinal distribution of the seasonal mean surface air temperature for the atmospheric domain over the continent–ocean domain and the ocean domain for: (*a*) the summer solstice; and (*b*) the winter solstice. The vertical dashed line represents the land margin.

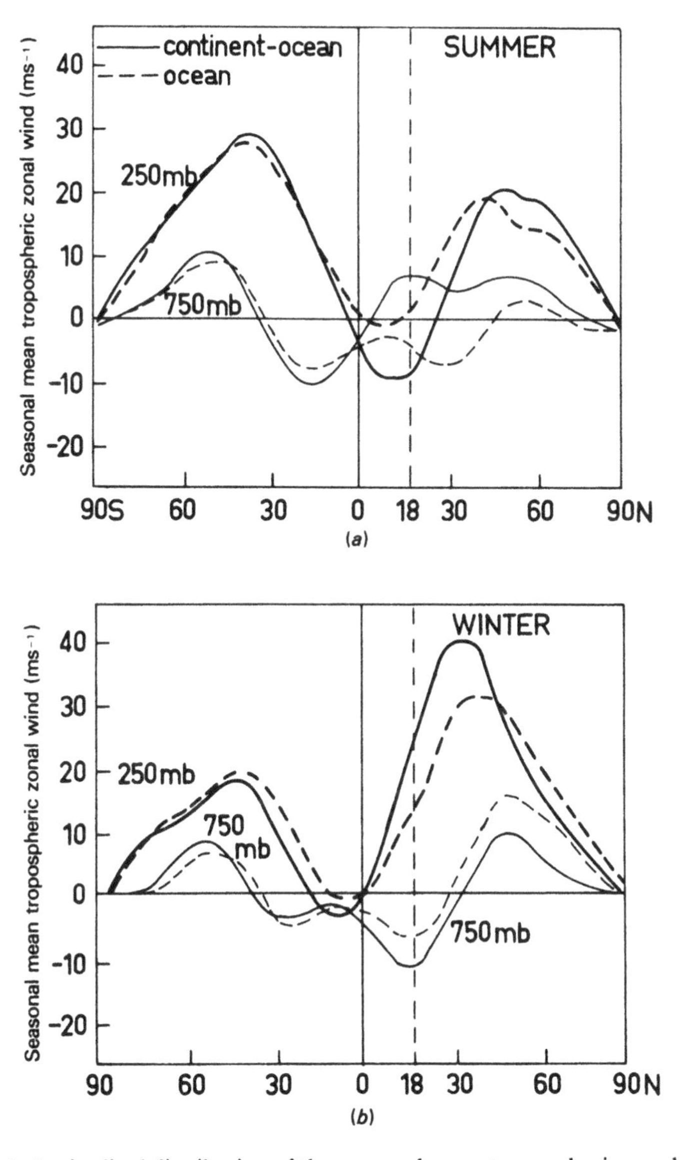

Fig. 10.3. Latitudinal distribution of the seasonal mean tropospheric zonal wind for the atmospheric domain over the continent–ocean domain and the ocean domain at 250 mb and 750 mb for: (*a*) the summer solstice; and (*b*) the winter solstice. The vertical dashed line represents the land margin.

domain-averaged model simulates many aspects of the mean seasonal structure of the monsoon. Furthermore, the features of Figs. 10.2 and 10.3 appear explainable in terms of the mechanisms proposed by Halley (1686).

The seasonal variation of the domain mean quantities is shown in Figs. 10.4 to 10.8. Due to the differing characteristic timescales of the ocean, land and atmosphere, the model requires extended integration before a reproducible annual cycle is produced. The results refer to the fourth year of integration.

Fig. 10.4 shows a smooth transition of surface temperature as a function of time of year. The maximum surface temperature over land occurs only a few weeks after the summer solstice. However, the variation of the sea-surface temperature in the adjacent ocean lags some two months behind the maximum declination. This is accounted for by the manner in which the ocean changes its temperature structure at the surface, which is accomplished by wind mixing and advection in addition to solar heating. This is in contrast to the continent which is closer to an equilibrium with the radiative forcing.

As mentioned in the introduction to this chapter, the smoothness of the seasonal transition is disturbed considerably by the addition of hydrological effects. However, as we are attempting merely to elucidate the seasonal timescale macro-monsoon, we will ignore sub-seasonal effects and refer the reader to Webster and Chou (1980*a*, *b*).

Variations of temperature at 750 and 250 mb through the year are shown in Figs. 10.5*a* and *b* for both domains. The largest variations occur in the lower troposphere over the continent and the smallest in the upper troposphere over the ocean. Between Figs. 10.4 and 10.5 we find examples of the influence adjacent domains have upon each other. In Fig. 10.4 a six to eight week phase lag between the surface temperatures over the ocean and the adjacent ocean domain is found. The 750 mb temperatures of each domain are more nearly in phase with each other, their maximum temperature in summer falling between the peaks of the surface temperatures. Over land the lower-tropospheric temperature lags the land surface temperature by nearly one month, but leads the ocean temperature by only two weeks. The phase variations are illustrated in Fig. 10.6 where the surface and 750 mb temperatures are plotted as functions of time. The reductions in phase difference of the variation of the free atmospheric temperature will be shown to be manifestations of zonal fluxes of heat.

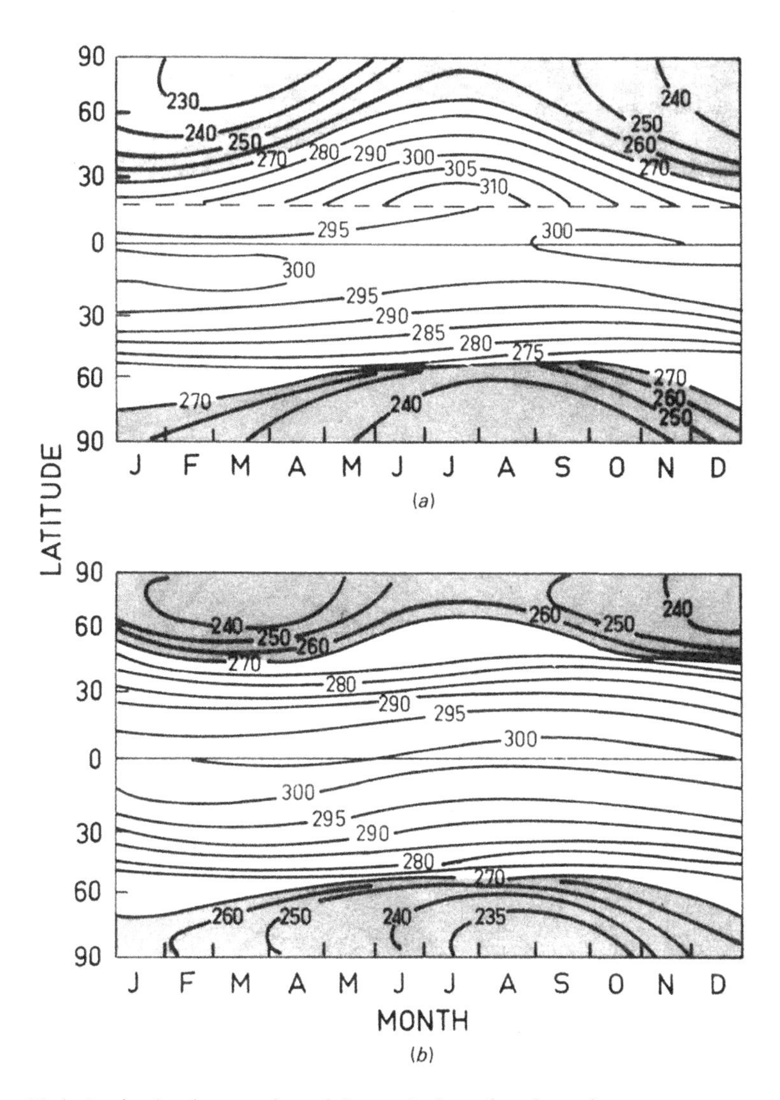

Fig. 10.4. Latitude–time section of the evolution of surface air temperature over a one-year period for the atmospheric domain over (*a*) the continent–ocean domain and (*b*) the ocean domain. The letters on the abscissa denote months. Shaded areas indicate temperatures less than 270 K which correspond to sea ice in the oceans. The dashed line represents the land margin. Contours are labelled in kelvin.

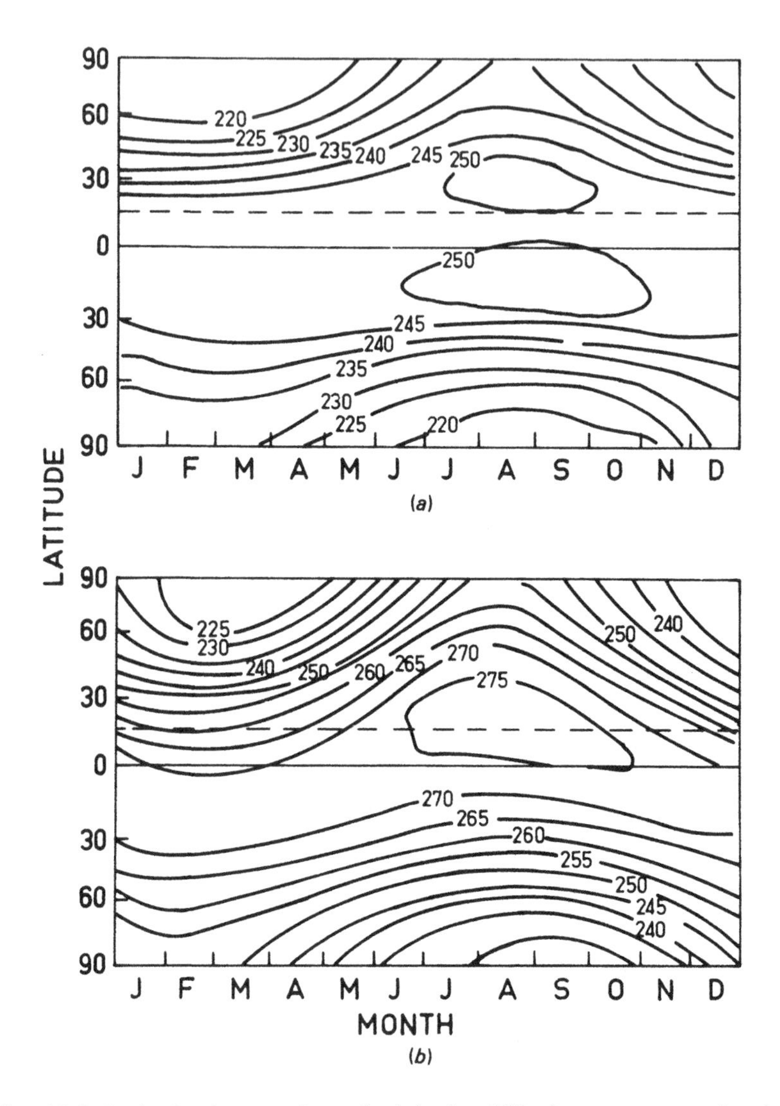

Fig. 10.5. Latitude–time section of: (*a*) the 250 mb temperature in the continent–ocean domain; (*b*) the 750 mb temperature in the continent–ocean domain; (*c*) the 250 mb temperature in the ocean domain; and (*d*) the 750 mb temperature in the ocean domain. The dashed line represents the land margin. Contours are labelled in kelvin.

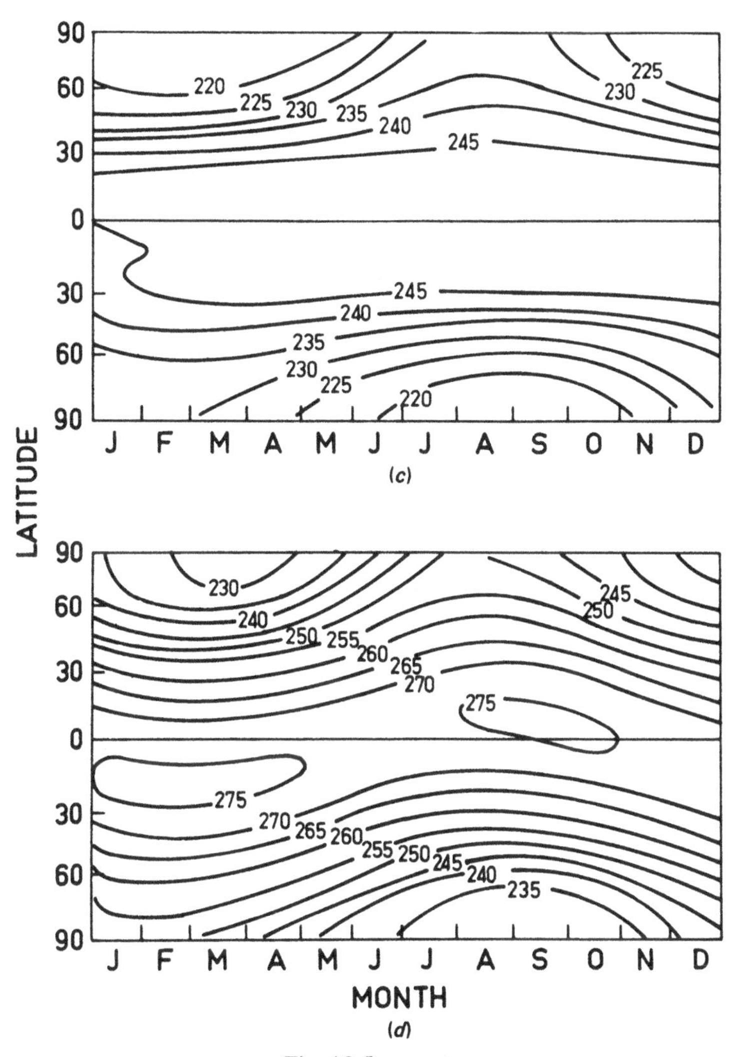
90
60
30
0
30
60
90
220
225
230
235
240
245
225
230
245
240
235
230
225
220
J F M A M J J A S O N D
(c)
LATITUDE
230
240
250
255
260
265
270
275
245
250
275
270
265
260
255
250
245
240
235
MONTH
(d)

Fig. 10.5.—*cont.*

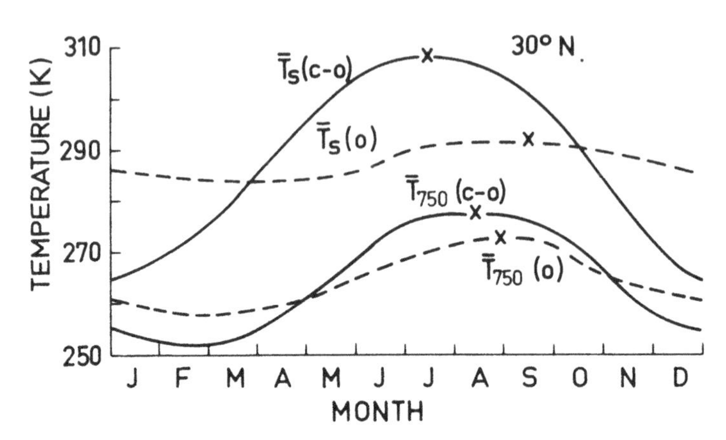

Fig. 10.6. Variation of the surface air temperature ($\bar{T}_s$) and 750 mb temperature ($\bar{T}_{750}$) as a function of time at 30° N for the continent–ocean domain (c-o) and the ocean domain (o). Crosses denote times of maximum temperature.

The seasonal variations are of the zonal velocity component are shown in Figs. 10.7*a* and 10.7*b* for the continent–ocean and ocean domain respectively. A comparison of the upper-tropospheric wind fields shows substantial differences in the form of the evolving zonal wind field. Over the continent, the strong westerlies weaken rapidly and are replaced by moderate easterlies between the Equator and the continental margin. At the same time, moderate westerlies develop in the lower troposphere and the upper-level meridional wind turns to the south (Fig. 10.8). Considered together, the evolving velocity field corresponds to the development of the low-level southwesterly monsoon with the northeasterly outflow aloft. Over the ocean the easterlies attain only weak intensity, which is a result of zonal interaction between domains at low latitudes.

Fig. 10.9 presents a series of observed mean latitude–time fields of zonal velocity at 200 mb along various meridians. Sections were based on the data of Sadler (1975) and the shaded regions indicate easterlies. Two basic patterns emerge from the analyses; one indicative of the 'ocean' sections (150° E, 180° and 150° W) and the other of the 'continental' sections (80° E, 100° E and 115° W). The continental sections show a rapid build-up and poleward extension of the upper-level easterlies during earlier summer. In contrast, the ocean sections possess extremely weak easterlies or even weak westerlies.

The model zonal velocity fields shown in Fig. 10.7 synthesize the essence of the observed fields. However, the addition of a third domain (indicated in Fig. 10.1 as the dotted line through the ocean domain) allows lateral wind-driven ocean heat transports and permits a closer approximation of the observations. Fig. 10.10 shows the three-domain model zonal velocity distributions, and in particular the rapid diminution of upper-level easterly winds noted in Fig. 10.9. Details of experiments using the multi-domain version of the model are described by Lau (1978).

10.3.2 Derived fields

Four derived fields are discussed. These are the meridional convergences of momentum and heat by the baroclinic eddies (the D functions of (10.1)) and the zonal convergences of heat and momentum (the I functions of (10.1)). The seasonal variations are shown in Figs. 10.11 to 10.14.

The seasonal variation of the convergence of meridional eddy momentum ($D(\bar{u})$, $D(\bar{v})$), at 250 mb in each domain is shown in Fig. 10.11. The maximum convergence, which is strongest in the upper troposphere, alternates between the two winter hemispheres near 45° latitude. Weaker divergent areas occur near 20° from the Equator. In this manner the eddies are acting as extremely effective transporters of momentum upward and poleward in order to maintain the positive zonal wind regime against the frictional momentum loss at the surface. In the continent–ocean domain, the momentum convergence is stronger than in the adjacent ocean which is a manifestation of the stronger baroclinicity over the continental region as may be observed in Figs. 10.2, 10.4 and 10.5.

The convergence of heat by the horizontal eddies at 250 mb ($D(\bar{T})$) is shown in Fig. 10.12. The strong poleward transport of heat by the baroclinic eddies is apparent in all domains with a strong divergence from the subtropics converging into the winter midlatitudes. Strong heat fluxes are again more associated with the continent–ocean domain than with the ocean domain. As the eddy heat flux is closely related to the baroclinicity of the atmosphere, as are the eddy momentum fluxes (Green, 1970; Stone, 1974), it is not surprising to find similar patterns of response.

The major effect of the meridional eddy heat flux is to relax the latitudinal temperature gradient. Coupled with the eddy momentum fluxes, they produce a thermally indirect circulation near the latitude of diminishing heat flux convergence. Weak indirect cells in the winter hemisphere may be discerned from Fig. 10.8. During the northern summer, the eddy heat divergence region has moved considerably north following the region of maximum baroclinicity poleward. At this time, the removal of accumulated heat from the heated continent in the subtropics

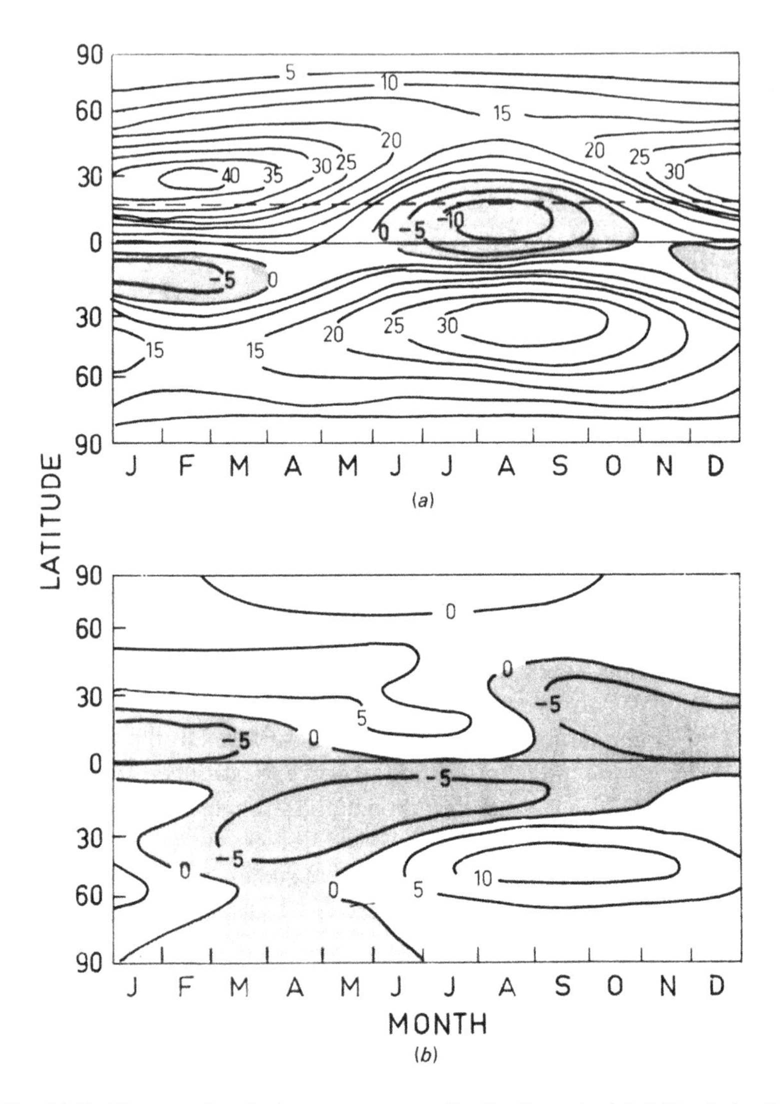

Fig. 10.7. The zonal velocity component distribution at: (a) 250 mb in the continent–ocean domain; (*b*) 750 mb in the continent–ocean domain; (*c*) 250 mb in the ocean domain; and (*d*) 750 mb in the ocean domain. The dashed line represents the land margin. Contours are labelled in m s^{-1}.

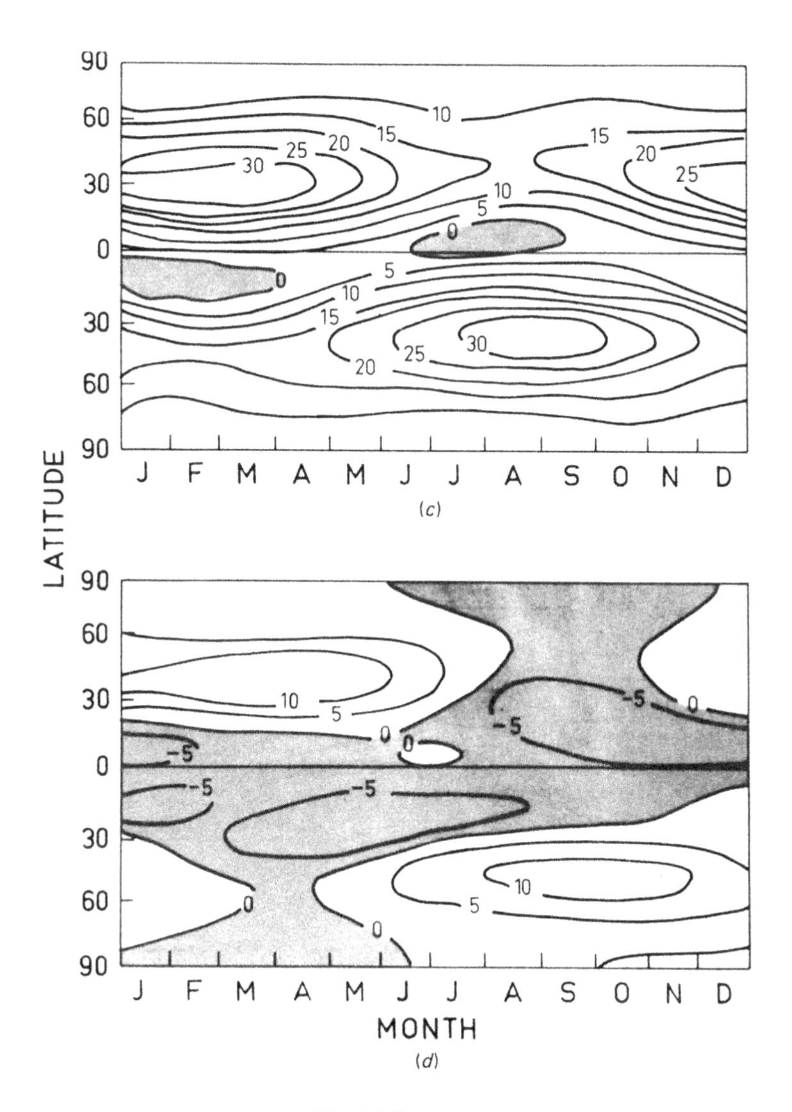
LATITUDE
90
60
30
0
30
60
90
10
15
20
25
30
5
0
J F M A M J J A S O N D
(c)
-5
MONTH
(d)

Fig. 10.7.—*cont.*

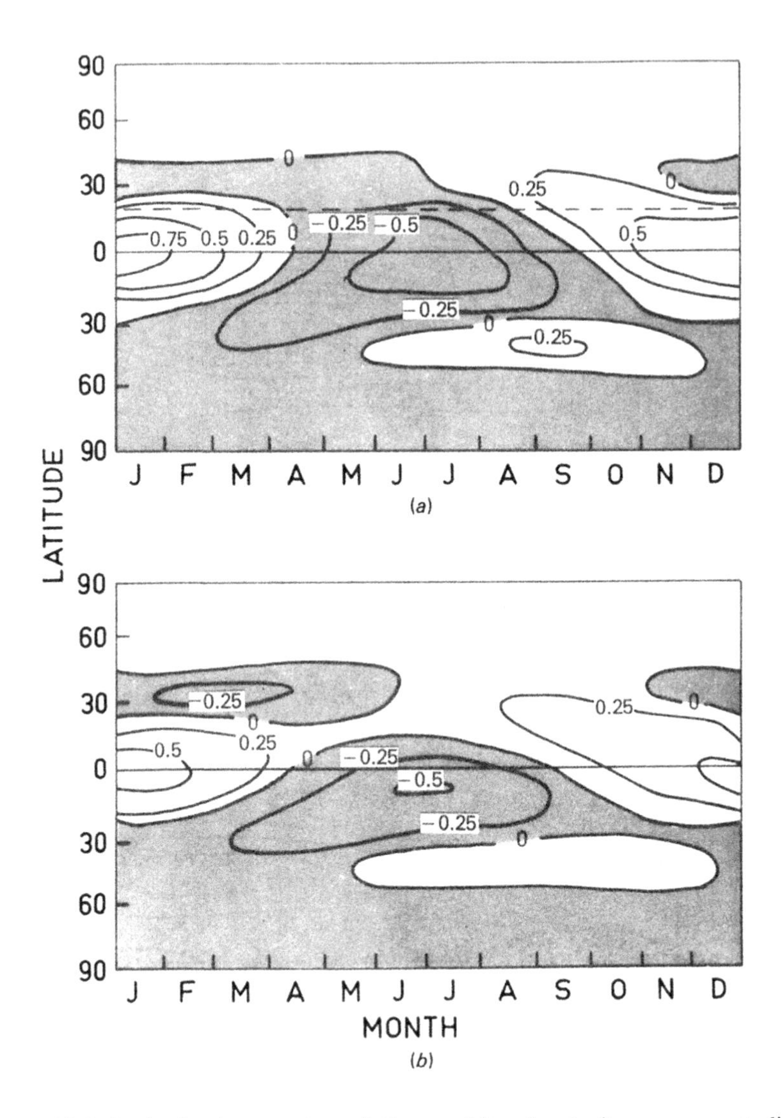

Fig. 10.8. Latitude–time section of the meridional velocity component distribution at 250 mb for: (*a*) the continent–ocean domain; and (*b*) the ocean domain. The dashed line represents the land margin. Contours are labelled in m s^{-1}.

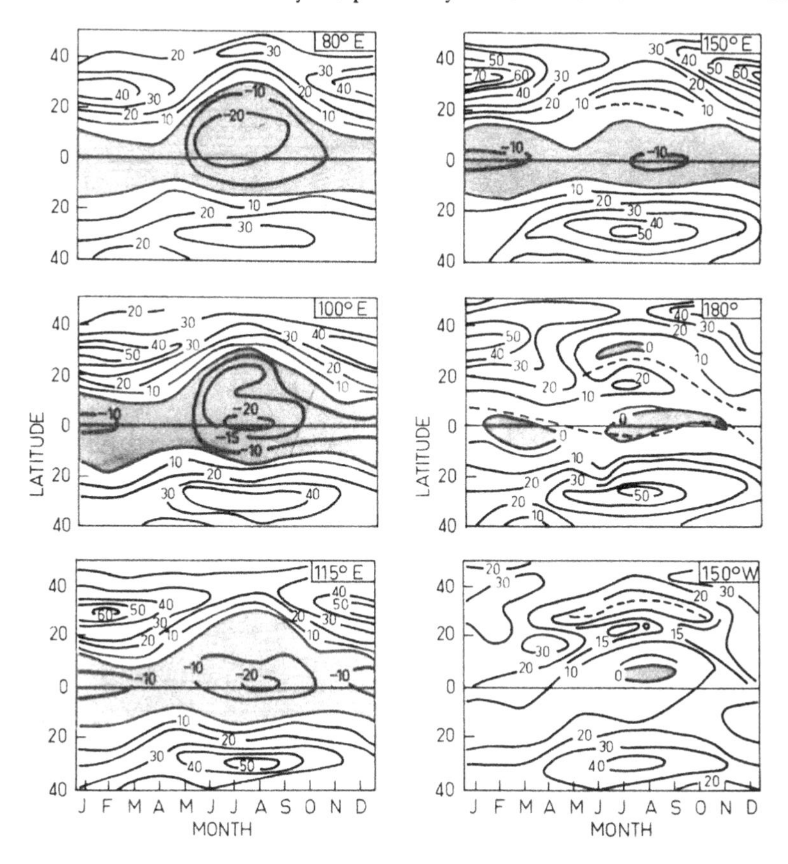

Fig. 10.9. Latitude–time section of the observed zonal velocity component at 200 mb for selected meridions. Dashed lines denote troughs and heavy lines denote ridges. Contours are labelled in m s^{-1}. (Adapted from Sadler, 1975.)

is shared by lateral effects. At low latitudes in summer, zonal fluxes of heat (and momentum) become important components of the general circulation at low latitudes.

The seasonal variations of the zonal fluxes of momentum and heat are shown in Figs. 10.13 and 10.14. In the northern hemisphere at both levels in the vertical, the maximum convergence of momentum into the ocean domain is centred in the subtropics and midlatitudes. There, the zonal momentum fluxes are strongly related to the difference in zonal wind speed between the two domains (Fig. 10.7). For example, in the northern hemisphere winter and spring, westward momentum is carried from the continent–ocean domain into the ocean domain causing a westerly

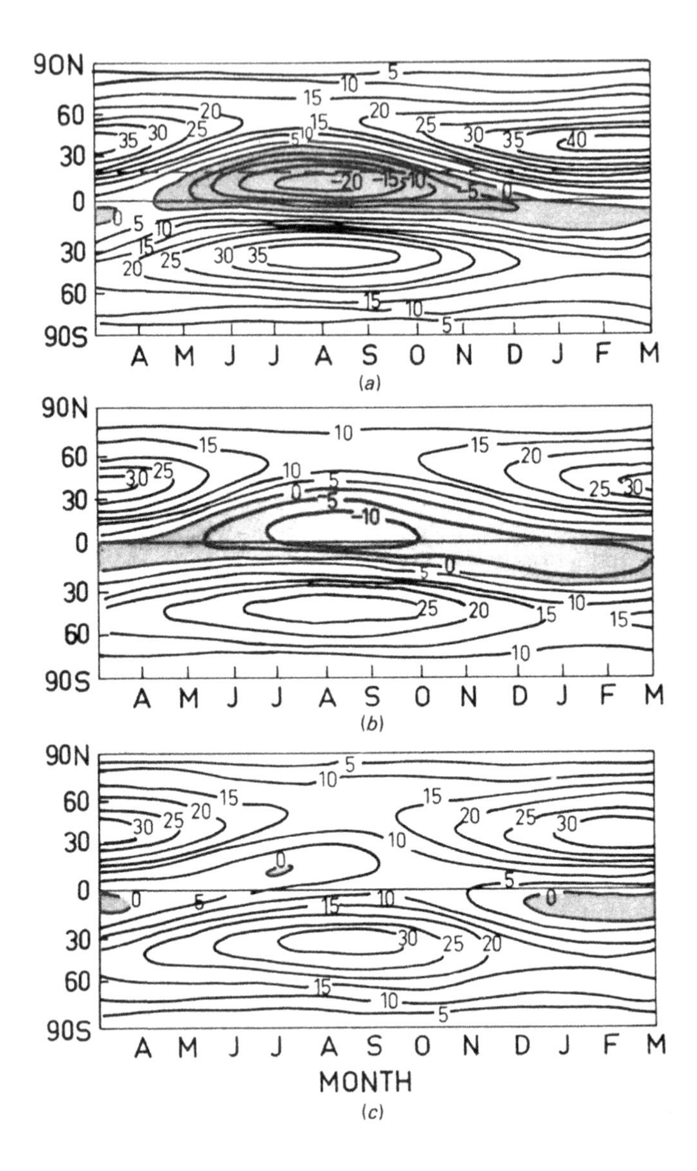

Fig. 10.10. Zonal velocity field evolution at 250 mb for a three-domain version of the model as shown in Fig. 10.1. (*a*) Continent–ocean domain, (*b*) western ocean domain; and (*c*) eastern ocean domain. Contours are labelled in m s^{-1}. Shaded regions denote easterlies. (From Lau, 1978.)

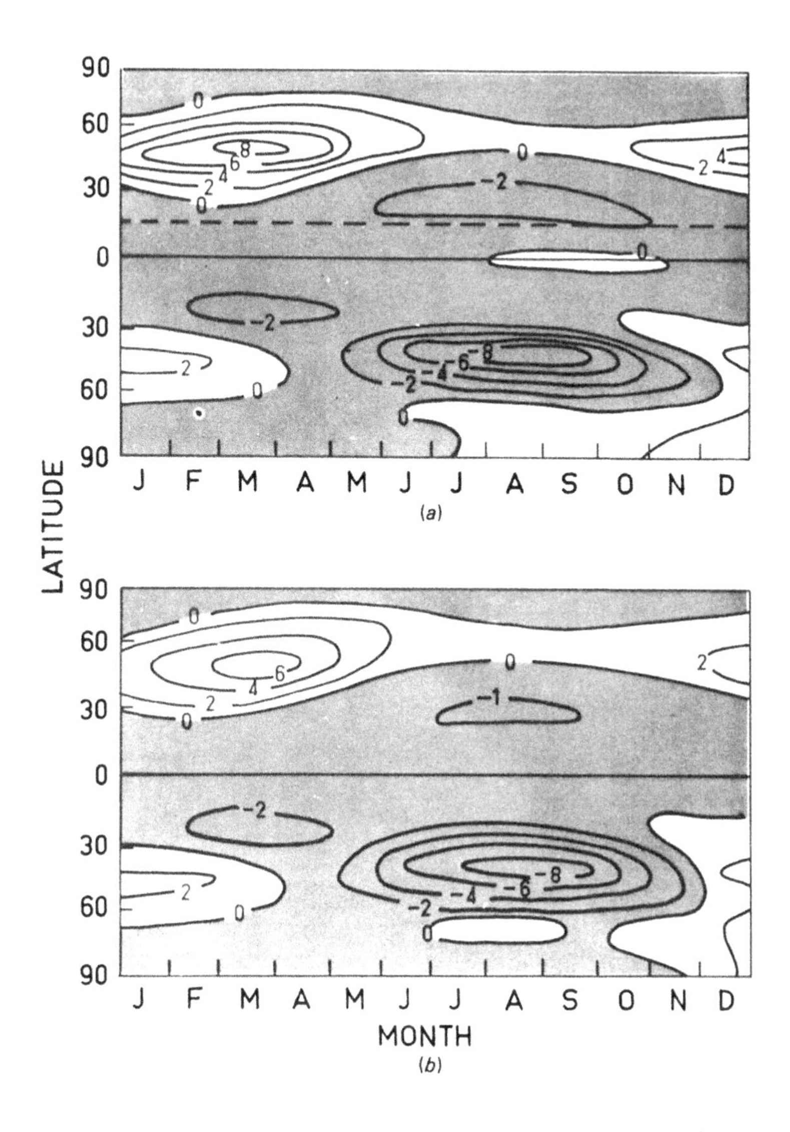

Fig. 10.11. Latitude–time section of the meridional convergence of westerly momentum at 250 mb by the transient eddies for: (*a*) the continent–ocean domain; and (*b*) the ocean domain. Negative (i.e. divergent) regions are shaded. Contours are labelled in 10^{-5} m s^{-2}.

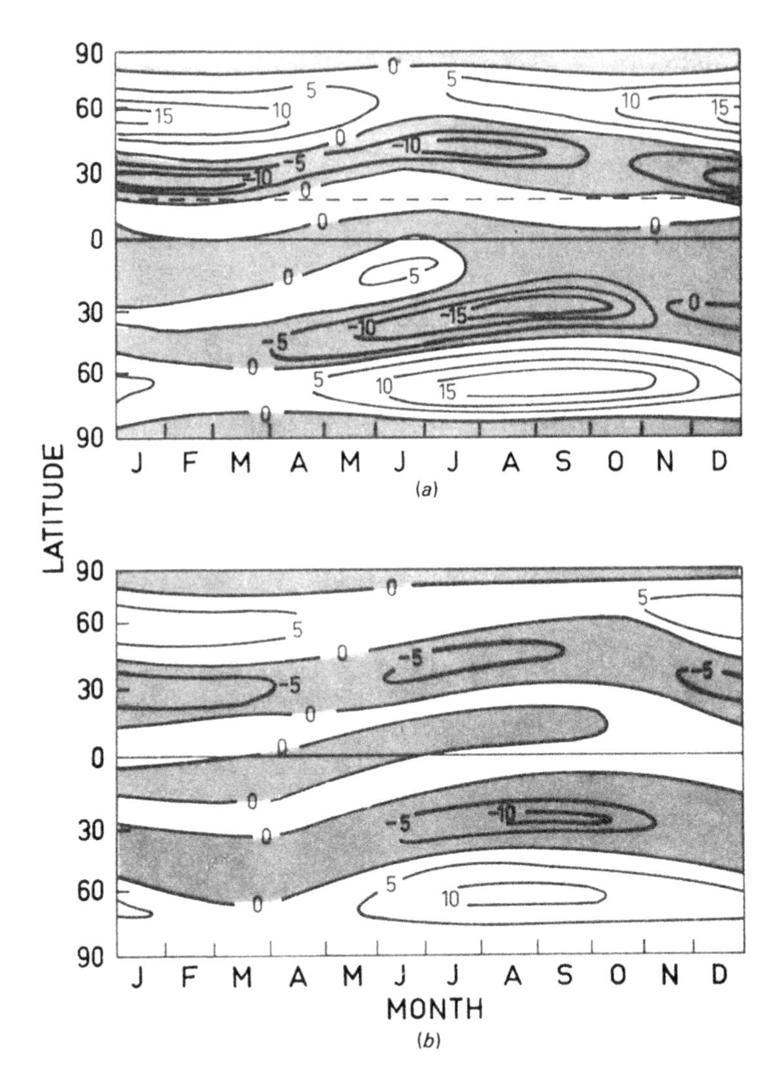

Fig. 10.12. Latitude–time section of the meridional convergence of northward heat flux at 250 mb by the transient eddies for: (*a*) the continent–ocean domain; and (*b*) the ocean domain. Negative (i.e. divergent) regions are shaded. Contours are labelled in 10^{-5} K s^{-1}.

acceleration in the ocean domain. In summary, the sense of the zonal momentum flux is from the domain of higher westerly momentum (e.g. the northern hemisphere continental region in winter) to regions of lower westerly momentum. If the adjacent domains possess similar magnitudes of westerly momentum, the convergence is small as can be seen in the southern hemisphere during all seasons.

Between the midlatitudes and the Equator, the character of the zonal momentum flux has changed considerably. The fluxes have become larger relative to the meridional fluxes and approach parity with them near the Equator. The sense of the flux has also changed. In midlatitudes, the seasonal variation of the flux convergence is in phase in the vertical. In the tropics the flux convergence is completely out of phase. The total effect in summer, for example, is to enhance the tropical upper-level easterly winds to the south of the continent and favour general westerlies or weak easterlies in the ocean domain. At lower levels the westerlies are enhanced at low latitudes of the continent–ocean domain and the easterlies accelerated over the adjacent ocean region.

Fig. 10.14 shows the annual distribution of the zonal convergence of heat into the ocean domain ($I(\bar{T})$) at 250 and 750 mb. Such fluxes are effected via planetary-scale quasi-stationary waves forced by the thermal contrast between the continent and ocean. The convergences are strongest in the northern hemisphere during summer and occur in the lower troposphere where the thermal contrasts are largest. The resultant heating rate within the ocean domain is nearly 1 K per day which is of the same order as the radiative and the boundary sensible heating. During winter, the continental region is cooler than the ocean and the heat convergence is into the continental regions. In the southern hemisphere at both tropospheric levels and at 250 mb in the northern hemisphere, the zonal heat flux divergence is insignificant, which is a manifestation of a small zonal temperature gradient. This is in sharp contrast to the low latitudes of the northern hemisphere, where in summer the zonal fluxes of heat completely dominate the meridional heat flux.

In summary, the zonal fluxes of heat are generally down the temperature gradient and transport heat to the cooler domain in an attempt to reduce the zonal temperature gradient. To accomplish this the parameterized quasi-stationary waves change polarity between summer and winter, in a similar manner to the quasi-stationary waves of the atmosphere, in order to adjust to the reversing zonal temperature gradient caused by the seasonally varying differential heating of the land and ocean regions.

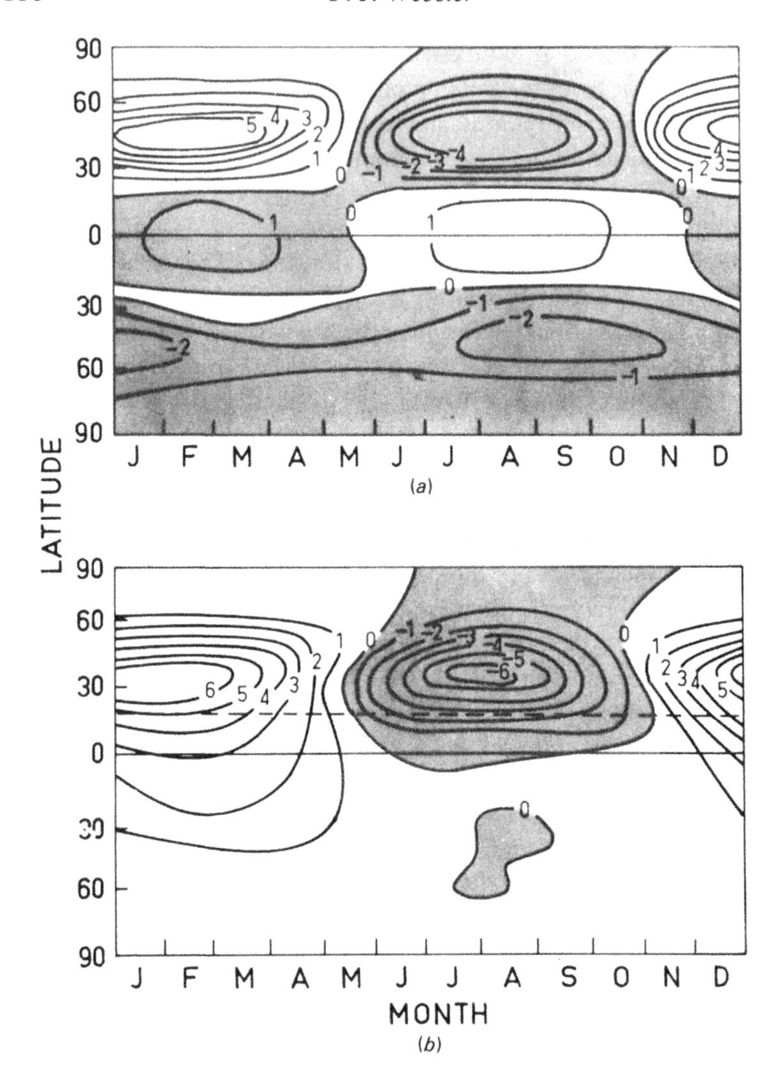

Fig. 10.13. Latitude–time section of the zonal convergence of westerly momentum into the domain at: (*a*) 250 mb; and (*b*) 750 mb. Negative regions are shaded. Contours are labelled in 10^{-6} m s^{-2}.

Distributions of the three-dimensional variation of the meridional fluxes of heat and momentum have been calculated from observations by Newell *et al.* (1972). In spite of the simplicity of the model geometry, the simulated fluxes agree well in magnitude, position and seasonal variation with the fluxes derived from observations. Similar distributions of the zonal heat flux are not evident in the literature so that no observational

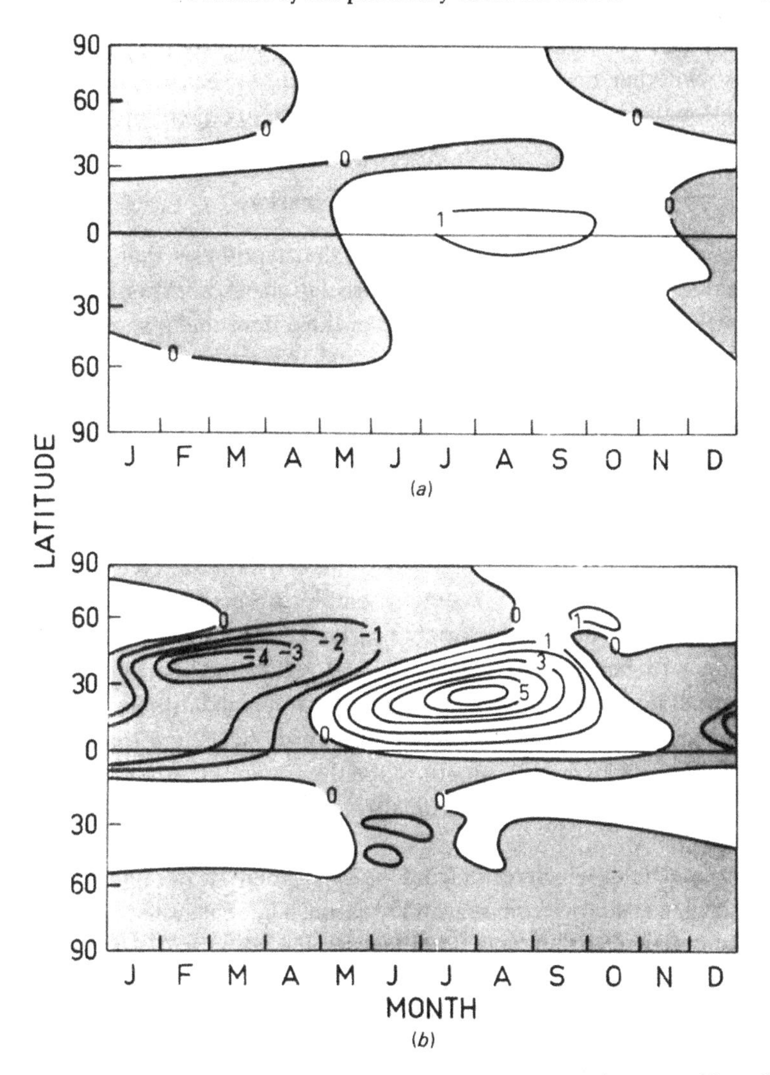

Fig. 10.14. Latitude–time section of the zonal convergence of westward heat flux $(u'T')_x$ into the ocean domain at: (*a*) 250 mb; and (*b*) 750 mb. Negative regions are shaded. Contours are labelled in 10^{-6} K s^{-1}.

comparison exists for the model results. Only Oort and Chan (1977) appear to have considered zonal interactions between 'monsoon' and 'extra-monsoon' regions for the angular momentum and kinetic energy balance. Unfortunately, comparison is difficult as Oort and Chan's boundary fluxes are averaged between 0° and 30° N and for the complete vertical column. Over the same spatial bounds the model shows considerable detail in the various fluxes. Consequently it is difficult to establish whether or not the agreement found between the model and observational estimates emanate from the same flux distributions.

10.4 Concluding remarks

The basic aim of the study was to test the hypothesis that the gross features of the large-scale monsoon circulation, in both its mean and evolving states, are primarily functions of the differential response of the oceans and the continents to solar heating. In particular we have been interested in the mutual zonal and meridional influence of adjacent ocean and continental regions and the manner in which they alter the monsoon sequence.

Although hydrological effects were omitted from the model, many features of the large-scale monsoon were well simulated. In particular, the relative phasing of the response of the low-latitude continent and ocean-surface temperatures keenly resembles observed variations. The tropospheric response to the variable surface temperature was followed and shown to be strongly related to the zonal and meridional convergences of heat and momentum. In middle and high latitudes the zonal fluxes are considerably smaller than their meridional counterparts. Only at low latitudes, between about 30° N and the Equator do they dominate. There, the zonal fluxes play a critical role in lateral communication of monsoon effects.

The model fields which result from the experiment show relatively slow and smooth transitions from season to season. The real monsoon circulation contains more abrupt changes and sub-seasonal timescale phenomena. The features of the monsoon which the model best depicts are similar to fields which would be obtained after running a long-period time filter through atmospheric data in monsoon regions. In that sense, one may claim that the model contains sufficient physical ingredients to account for the slowly evolving seasonal character of the monsoon. However, as the model fails to reproduce more rapid sub-seasonal variations, it would seem that additional effects must be aggregated in order that they may be simulated. Such an effect is the hydrology cycle. In a specific study of the sub-seasonal transition of the monsoon system, Webster and Chou (1980*a*, *b*) found hydrology and interactive clouds necessary in obtaining scales of motion which were super-synoptic but sub-seasonal.

Perhaps the most important result of the study is that it points out the significance of zonal interactions of the monsoon regions with adjacent domains. This has ramifications for both future modelling and observational studies. In order to allow for the proper atmospheric transitions of the monsoon, a model of the monsoon must contain more degrees of freedom than a zonally-averaged model. At the same time, the results of the study indicate that if objectives are relatively modest, one can use a model which is significantly simpler than a general-circulation model. This allows considerable flexibility in experimentation and avoids the logistical problems and cost involved in resolving very small scales of motion. However the burden is then transferred to the development of parameterization schemes which allow for potential rectification of the unresolved small-scale features upon motions of the scale of the domain.

Finally, it is important that observational studies be made to allow direct calculations of the magnitudes of the zonal fluxes of heat and momentum and testing of the inferences drawn from the model results. The combined MONEX and FGGE data sets will be useful in this regard.

References

Blackmon, M., Wallace, J. M., Lau, N. G. and Mullen, S. (1977) An observational study of the northern hemisphere wintertime circulation. *J. Atmos. Sci.*, **34**, 1040–53.

Green, J. S. A. (1970) Transfer properties of the large scale eddies and the general circulation of the atmosphere. *Quart. J. Roy. Meteor. Soc.*, **96**, 157–85.

Hahn, D. G. and Manabe, S. (1975) The role of mountains in the South Asian monsoon circulation. *J. Atmos. Sci.*, **32**, 1515–41.

Halley, E. (1686) An historical account of the trade winds and monsoons observable in the seas between and near the tropics with an attempt to assign a physical cause of the said winds. *Phil. Trans. Roy. Soc. London*, **16**, 153–68.

Koteswaram P. (1958) *Monsoons of the World*. India Meteorological Department, p. 105.

Krishnamurti, T. N. (1971*a*) Tropical east–west circulations during the northern summer. *J. Atmos. Sci.*, **28**, 1342–7.

Krishnamurti, T. N. (1971*b*) Observational study of the tropical upper troposphere motion field during the northern hemisphere summer. *J. Appl. Meteor.*, **10**, 1066–96.

Krishnamurti, T. N., Kanamitsu, N., Koss, W. J. and Lee, J. D. (1973) Tropical east–west circulations during the northern winter. *J. Atmos. Sci.*, **30**, 780–7.

Krishnamurti, T. N. and Bhalme, H. N. (1976) Oscillations of a monsoon system, Part I. Observational Aspects. *J. Atmos. Sci.*, **33**, 1937–54.

Lau, K. M. W. (1978) A large-scale ocean atmosphere interaction model. Ph.D. Dissertation, Department of Meteorology, University of Washington, Seattle, 220 pp.

Murakami, T., Godbole, R. V. and Kelkar, R. R. (1970) Numerical simulation of the monsoon along 80° E, In *Proceedings Conf. Summer Monsoon South-east Asia*, ed. C. S. Ramage, pp. 39–51, Navy Weather Research Facility, Norfolk, Virginia.

Newell, R. E., Kidson, J. W., Vincent, D. G., and Boer, G. J. (1972) *The General Circulation of the Tropical Atmosphere and Interaction with Extratropical Latitudes.* Vol. 1, MIT Press, Cambridge, Mass. 258 pp.

Oort, A. H. and Chan, P. H. (1977) On the role of the Asian Monsoon in the angular momentum and kinetic energy balances of the tropics. *Pure and Applied Geophysics*, **115**, 1167–86.

Ramage, C. (1971) *Monsoon Meteorology.* Academic Press, New York, 296 pp.

Sadler, J. C. (1975) The upper tropospheric circulation over the global tropics, Report UHMET 75–05, Department of Meteorology, University of Hawaii, 35 pp.

Stone, P. H. (1974) The meridional variation of the eddy fluxes by baroclinic waves and their parameterization. *J. Atmos. Sci.*, **31**, 444–56.

Washington, W. M. and Daggupaty, S. M. (1975) Numerical simulation with the NCAR global circulation model of the mean conditions during the Asian–African summer monsoon. *Mon. Wea. Rev.*, **103**, 105–14.

Webster, P. J. (1972) Response of the tropical atmosphere to local steady forcing. *Mon. Wea. Rev.*, **100**, 518–41.

Webster, P. J. and Lau, K. M. W. (1977*a*) A simple ocean–atmosphere climate model: Basic model and a simple experiment. *J. Atmos. Sci.*, **34**, 1063–84.

Webster, P. J. and Lau, K. M. W. (1977*b*) Simulation of the global monsoon sequence by a simple ocean–atmospheric interaction model. *Proceedings of the International Symposium on Monsoons, New Delhi, March, 1977* (in press).

Webster, P. J. and Chou, L. (1980*a*) Seasonal structure of a simple monsoon system. *J. Atmos. Sci.* (in press).

Webster, P. J. and Chou, L. (1980*b*) Low frequency transients of a simple monsoon system. *J. Atmos. Sci.* (in press).

11

Wave interactions in the equatorial atmosphere – an analytical study

P. K. DAS, M. P. SINGH AND R. C. RAGHAVA

The effect of zonal motion on wave interactions is considered in an equatorial atmosphere. Extending the formulation of Matsuno (1966), Duffy (1974), and Domaracki and Loesch (1977), perturbations are imposed on a zonal current which varies slowly with time. Considering the conditions for resonance between wave triads, coupling coefficients are computed between the zonal motion and the equatorial waves. It is shown that these coupling coefficients do not involve an energy exchange between the zonal current and the perturbed motion, but that the zonal current does lead to phase changes in the complex amplitude of resonant waves. A stability analysis is made to indicate how the zonal current could alter the growth of interacting waves.

11.1 Introduction

When the governing equations of a system are nonlinear, one of the analytical methods for obtaining the solution is the perturbation technique. In such a situation the system is reduced to a non-dimensional form by referring different physical variables to characteristic scales. The dependent variables are expanded asymptotically in terms of small parameters which occur in a natural way depending on the physics of the problem. It is assumed that each term of the perturbation series is smaller in magnitude than its preceding term throughout the region of interest. If this condition breaks down in any part of the region, the asymptotic expansion is no longer valid. An important reason why perturbation solutions are often not uniformly valid is concerned with the 'large integrated effect'. Thus, the disturbance imparted to a system may be small, but its cumulative effect may become quite large. This generates 'secular terms' which lead to the break-down of the asymptotic solution. For periodic motion, it can be demonstrated that secular terms are produced by resonance. In such a situation, the physics of the problem admits more than one characteristic scale, and the different scales are disparate. Let us, for example, consider two disparate scales, called the 'short' and 'long' scales which represent, respectively, the fast and slow changes of the system. The secular terms can be avoided if we introduce both 'slow' and 'fast' variables when the system is made non-dimensional. If there are more than two disparate scales, an extension of this technique to cover such cases is fairly straightforward. This method, known as the 'method of multiple scales', has been applied to several problems in atmospheric science.

It is recognized, for example, that the atmosphere interacts with the Earth's surface through a boundary layer. One can design several scales of length and time for motion within the boundary layer. Thus the time taken for the decay of vorticity of an atmospheric vortex to $1/e$ of its initial value by friction is about four days. This is its spin-down time. But, the spin-down time for molecular diffusion is about 100 days. A multiple-scale analysis for an Ekman boundary layer by St-Maurice and Veronis (1975) brought out the interesting result that, for a stratified fluid, the structure of the final steady state is only resolved when the motion with longest timescale is evaluated.

Multiple scales were used by Blumen (1972) to study geostrophic adjustment. This raises important problems for the integration of primitive equation models. It seeks to determine how pressure and wind, which are initially unbalanced, ultimately reach a balanced state of equilibrium. In this problem the removal of secular terms leads to a conservation theorem for an important meteorological variable, namely, the potential vorticity of the system. More recently, multiple scales have been used to study resonant interactions between atmospheric waves. Thus, Newell (1969) found resonant interactions between Rossby waves, while Stone (1969) studied a similar problem with baroclinic waves. Side-band resonance was considered by Shabbar (1971) for Rossby waves, and Lindzen (1971) examined the vertical propagation of waves near the Equator. These investigations have been extended to beta-plane models by Duffy (1974) and by Domaracki and Loesch (1977). In this chapter multiple-scale analysis is applied to study interactions between a slowly changing zonal flow and atmospheric waves near the Equator.

11.2 Scaling for meteorological problems

A simple meteorological model utilizes the well-known shallow water equations. These describe the dynamics of an incompressible fluid with a free surface and assume hydrostatic balance. Near the Equator the Coriolis parameter (f) is assumed to be proportional to the latitude, i.e.

$$f = \beta y, \tag{11.1}$$

where

$$\beta = \mathrm{d}f/\mathrm{d}y.$$

Using Cartesian coordinates (x, y, z) in which the x axis is directed eastward, y northward and z is upward, the basic equations are

$$u_t + (uu_x + vu_y) - \beta yv + \phi_x = 0, \tag{11.2a}$$

$$v_t + (uv_x + vv_y) + \beta yu + \phi_y = 0, \tag{11.2b}$$

$$\phi_t + [(u\phi)_x + (v\phi)_y] + gH(u_x + v_y) = 0, \tag{11.2c}$$

where u, v are the eastward and northward components of motion, ϕ is the geopotential, H stands for the undisturbed depth of the fluid and g is the acceleration due to gravity. The partial derivatives are indicated by subscripts.

To render (11.2 a, b and c) non-dimensional it is necessary to define a characteristic length and time for the system. The velocity of shallow water gravity waves is $(gH)^{1/2}(=c)$, consequently, as we wish to consider slower moving systems it is necessary to define another characteristic velocity (U), which is smaller than c, for scaling the governing equations. It is necessary to emphasize that g, the gravitational acceleration, should be weighted by $\Delta\theta/\theta$ where θ is the mean potential temperature of the atmosphere and $\Delta\theta$ is the difference in θ between the top and bottom of the atmosphere. But, if $\Delta\theta/\theta \sim 0.1$, c would still be, approximately, 100 m s^{-1}. On the other hand, internal gravity waves of small vertical wavelength (2 km) have a horizontal phase speed of only 3 m s^{-1}, the characteristic velocity c does not, therefore, correspond to internal gravity waves. For U, a typical speed corresponding to the phase speed of westward moving Rossby waves (5–10 m s^{-1}) may be chosen.

It is now possible to scale the governing equations in terms of β, c and U. Define a characteristic length (L) and time (T) by

$$L=(c/\beta)^{\frac{1}{2}}, \qquad T=(1/\beta c)^{\frac{1}{2}}. \tag{11.3}$$

Putting $\beta = 2.3\times 10^{-11}$ m^{-1} s^{-1} and $c = 100$ m s^{-1} gives $L = 2000$ km and $T = 6$ hours.

A small parameter for expansion of the dependent and independent variables may now be taken as

$$\varepsilon = U/c. \tag{11.4}$$

It is interesting to note that if U were used instead of c in (11.3), and L and T were taken as

$$L=(U/\beta)^{\frac{1}{2}}, \qquad T=(1/\beta U)^{\frac{1}{2}}, \tag{11.5}$$

then, for $U = 10$ m s^{-1}, the numerical values of L and T would be 650 km and 18 hours. But, this scaling is not permissible, because the nonlinear advective terms in (11.2) are scaled by UT/L. This is equal to unity according to (11.5), but is equal to ε if the scaling is by (11.3).

Putting

$$(x, y)=L(x^*, y^*), \qquad (u, v)=U(u^*, v^*), \qquad \phi=cU\phi^*, \tag{11.6}$$

the non-dimensional form of (11.2) is

$$u_t+\varepsilon(uu_x+vu_y)-yv+\phi_x=0, \tag{11.7a}$$

$$v_t+\varepsilon(uv_x+vv_y)+yu+\phi_y=0, \tag{11.7b}$$

$$\phi_t+\varepsilon[(u\phi)_x+(v\phi)_y]+u_x+v_y=0. \tag{11.7c}$$

where, for convenience, the asterisks have been dropped for non-dimensional variables.

11.3 Interactions with zonal flow

Domaracki and Loesch (1977) started with an atmosphere at rest. As a variation of this problem, an atmosphere which is initially in zonal motion is considered. The initial state is taken to satisfy

$$y\bar{u}+\bar{\phi}_y=0, \tag{11.8}$$

where $\bar{u}(y)$ represents a specified zonal flow which is balanced through (11.8) by a mean geopotential $\bar{\phi}$. This is the equivalent of geostrophic balance near the Equator for a beta-plane model.

A slow time and space variable (T, X) may be defined by

$$T=\varepsilon t, \qquad X=\varepsilon x, \tag{11.9}$$

whence

$$\frac{\partial}{\partial t}\equiv\frac{\partial}{\partial t}+\varepsilon\frac{\partial}{\partial T}+\cdots, \tag{11.10a}$$

$$\frac{\partial}{\partial x}\equiv\frac{\partial}{\partial x}+\varepsilon\frac{\partial}{\partial X}+\cdots \tag{11.10b}$$

The dependent variables may be expanded as

$$u=\bar{u}(y, T)+u^{(0)}+\varepsilon u^{(1)}+\cdots, \tag{11.11a}$$

$$v=v^{(0)}+\varepsilon v^{(1)}+\cdots, \tag{11.11b}$$

$$\phi=\bar{\phi}(y, T)+\phi^{(0)}+\varepsilon\phi^{(1)}\cdots \tag{11.11c}$$

Putting (11.11) into (11.7), the equation for the meridional velocity (v) is

$$\begin{aligned}&\mathcal{L}(v^{(0)})+\varepsilon\mathcal{L}(v^{(1)})+\varepsilon(3v^{(0)}_{ttT}+y^2v^{(0)}_T-v^{(0)}_{yyT}-v^{(0)}_{xxT}\\&-v^{(0)}_X-2v^{(0)}_{xtX})+\varepsilon(\text{nonlinear terms})+0(\varepsilon^2, \varepsilon^3, \ldots)=0,\end{aligned} \tag{11.12}$$

where

$$\mathcal{L}\equiv\frac{\partial^3}{\partial t^3}+y^2\frac{\partial}{\partial t}-\frac{\partial^3}{\partial y^2\,\partial t}-\frac{\partial^3}{\partial x^2\,\partial t}-\frac{\partial}{\partial x}.$$

The fourth term in (11.12) contains 12 nonlinear terms which have been listed by Domaracki and Loesch (1977). They will not be reproduced here. But, as this model starts with zonal motion, additional nonlinear terms emerge. They are

$$\begin{aligned}&\bar{u}(v^{(0)}_{xtt}-v^{(0)}_{xxx}-yu^{(0)}_{xt}-\phi^{(0)}_{xyt}+u^{(0)}_{yxx}+y\phi^{(0)}_{xx})\\&+\bar{u}_y(u^{(0)}_{xx}+v^{(0)}_{xy}-yv^{(0)}_t-\phi^{(0)}_{xt})+\bar{u}_{yy}v^{(0)}_x\\&+\bar{\phi}(yu^{(0)}_{xx}+yv^{(0)}_{yx}-u^{(0)}_{xyt}-v^{(0)}_{yyt})\\&+\bar{\phi}_y(yv^{(0)}_x-u^{(0)}_{xt}-2v^{(0)}_{yt})-\bar{\phi}_{yy}v^{(0)}_t.\end{aligned} \tag{11.13}$$

These terms do not appear when the initial state is an atmosphere at rest. It is interesting to note that they depend not only on $\bar{u}$ and $\bar{\phi}$, but also on their latitudinal gradients.

11.4 Zero-order solution

The zero-order solutions of (11.12) were given by Matsuno (1966). They are

$$u^{(0)} = \sum_{n=0}^{\infty} \sum_{j=1}^{3} [-\tfrac{1}{2}(\sigma_j + k_j)\psi_{n+1} - n(\sigma_j - k_j)\psi_{n-1} + cc], \quad (11.14)$$

$$v^{(0)} = \sum_{n=0}^{\infty} \sum_{j=1}^{3} [-i(\sigma_j^2 - k_j^2)\psi_n + cc],$$

$$\phi^{(0)} = \sum_{n=0}^{\infty} \sum_{j=1}^{3} [-\tfrac{1}{2}(\sigma_j + k_j)\psi_{n+1} + n(\sigma_j - k_j)\psi_{n-1} + cc], \quad (11.15)$$

where

$$\psi_n = A_j^{(0)} \exp(i\theta_j) \exp(-\tfrac{1}{2}y^2) H_n(y),$$

$$\theta_j = k_j x - \sigma_j t.$$

The eigenfunctions for y dependence are Hermite polynomials ($H_n(y)$). The subscript n represents a wavenumber in the y direction. For each n, there are three roots denoted by the subscript j. Zonal wavenumbers are denoted by k_j and frequencies by σ_j; cc denotes the complex conjugate of the preceeding terms. The amplitude of the zero order wave is given by $A_j^{(0)}$. The dispersion relation is

$$\sigma_j^2 - k_j^2 - k_j/\sigma_j = 2n + 1, \qquad n = 0, 1, 2, \ldots \quad (11.16)$$

Matsuno pointed out that a Kelvin wave could be represented by putting $n = -1$ and $\sigma_j = k_j$.

For $n = 0$, the roots of (11.16) indicate a gravity wave propagating towards the east, and a mixed Rossby-gravity wave moving westwards.

For the other integral values of $n (n = 1, 2, 3, \ldots)$, both eastward and westward propagating gravity waves are obtained, together with a westward propagating Rossby wave.

11.5 Wave–zonal interactions

When first-order solutions are considered, the wave solutions of the zero-order equations generate forced modes. Resonance occurs when the frequency of a forced mode coincides with the natural frequency of the first-order system. As there are three roots ($j = 1, 2, 3$) for each meridional wave number (n), the conditions for resonant interactions are

$$k_1 + k_2 + k_3 = 0, \quad (11.17a)$$

$$\sigma_1 + \sigma_2 + \sigma_3 = 0, \quad (11.17b)$$

$$n_1+n_2+n_3 \text{ is odd}, \tag{11.17c}$$

$$\sigma_j^2-k_j^2-k_j/\sigma_j=2n+1. \tag{11.17d}$$

The third condition ensures that resonance north of the Equator is not cancelled by similar resonance to the south of the Equator. It is a less restrictive condition than making the sum of meridional wavenumbers vanish.

From (11.12), the first-order equation is

$$\mathscr{L}(v^{(1)})=-(3v_{ttT}^{(0)}+y^2v_T^{(0)}-v_{xxT}^{(0)}-v_{yyT}^{(0)}-v_X^{(0)}-2v_{xtX}^{(0)})$$
$$+(\text{Nonlinear terms containing products of } \bar{u},\, u^{(0)},\, v^{(0)},\, \phi^{(0)}). \tag{11.18}$$

The secular terms in (11.18) should be removed if the expansions (11.10) and (11.11) are to be uniformly valid. This is achieved by making the non-homogeneous part of (11.9) orthogonal to solutions of the zero-order equations (Nayfeh, 1973). The algebra involved is lengthy, but it provides an equation for the growth of the zero-order wave ($A_j^{(0)}$). In order to examine how the growth is influenced by the zonal current, consider a sinusoidal zonal flow changing slowly with time, i.e.

$$\bar{u}=B(T)\cos(2\pi y/\lambda). \tag{11.19}$$

By applying the resonance conditions (11.17), and removing the secular terms in (11.18), we find

$$\left(\frac{\partial}{\partial T}+\alpha_1\frac{\partial}{\partial x}\right)A_1^{(0)}+\mathrm{i}\gamma_1A_1^{(0)}B+cc=\mathrm{i}\rho_1A_2^*A_3^*+cc, \tag{11.20}$$

where α_1 is the group velocity of the growing wave and ρ_1 is a coupling coefficient between the two other waves which make up a triad with $A_1^{(0)}$. Numerical values of the coupling coefficients (ρ) have been tabulated by Domaracki and Loesch (1977). The difference between (11.20) and the work of Domaracki and Loesch (1977) lies in the addition of an interaction between $A_1^{(0)}$ and B, the amplitude of the zonal current. This is represented by the coefficient (γ_1).

The computation of γ involves integrals of the form

$$\int_{-\infty}^{\infty}\exp(-y^2)\sin(2\pi y/\lambda)H_n(y)H_{n+2m+1}(y)\,\mathrm{d}y$$

and

$$\int_{-\infty}^{\infty}\exp(-y^2)\cos(2\pi y/\lambda)H_n(y)H_{n+2m}(y)\,\mathrm{d}y,$$

which are available in tables of integrals of Hermite polynomials. Longuet-Higgins and Gill (1967) suggest that the additional interaction term does not alter the exchange of energy between waves, but it changes the phase of $A_1^{(0)}$. Let

$$A_1^{(0)} = R(T)\exp(\mathrm{i}\phi(T)), \tag{11.21}$$

where $R(T)$ and $\phi(T)$ represent the modulus and phase of $A_1^{(0)}$. As α_1 is small, we may neglect the advective term in (11.20). Putting (11.21) in (11.20) and equating imaginary parts we have

$$\partial\phi/\partial T = -\gamma_1 B(T),$$

whence

$$\phi = -\gamma_1 \int B(T)\,\mathrm{d}T. \tag{11.22}$$

The changes in ϕ, which are brought about by γ_1 and the zonal current (B), represent a turning of the complex amplitude $A_1^{(0)}$ without altering its magnitude. How does this depend on the wavelength of the zonal current? To examine this question, γ was computed for three typical wavenumbers. The numerical values of γ are shown in Table 11.1.

The wavenumbers in Table 11.1 represent mixed Rossby waves. Table 11.1 does not indicate any preferred wavelengths (λ) which impart positive or negative vorticity to $A_1^{(0)}$ but, as we can see, there are waves, such as k_2, for which the interaction coefficient (γ) is always negative. This implies a sharp rotation of $A_1^{(0)}$ by nonlinear interaction with the zonal current. The formation of small-scale vortices near the Equator could be attributed to such an interaction.

The variation of γ with λ is shown in Fig. 11.1 for $n = 1, 2, 3$. Fig. 11.2 shows its variation for coupling with the Kelvin wave ($n = -1$). From Fig. 11.2 we see that negative values of γ are obtained when $\lambda \leq 16.2 \times 10^3$ km. In this range, the zonal current imparts positive rotation to the Kelvin wave.

TABLE 11.1. *Wave–zonal flow interaction coefficients* (γ).

Wavelength 10^3 km	γ_1 ($k_1 = -1.0432$, $\sigma_1 = 0.2594$ and $n_1 = 1$)	γ_2 ($k_2 = -1.5194$, $\sigma_2 = 0.2091$ and $n_2 = 2$)	γ_3 ($k_3 = 1.1631$, $\sigma_3 = -0.1395$ and $n_3 = 3$)
1.8	0.0007	−0.0064	−0.0327
3.0	0.4514	−1.0607	−0.6932
15.0	−1.3232	−0.8634	0.4642
30.0	−0.1371	−0.5758	0.3413

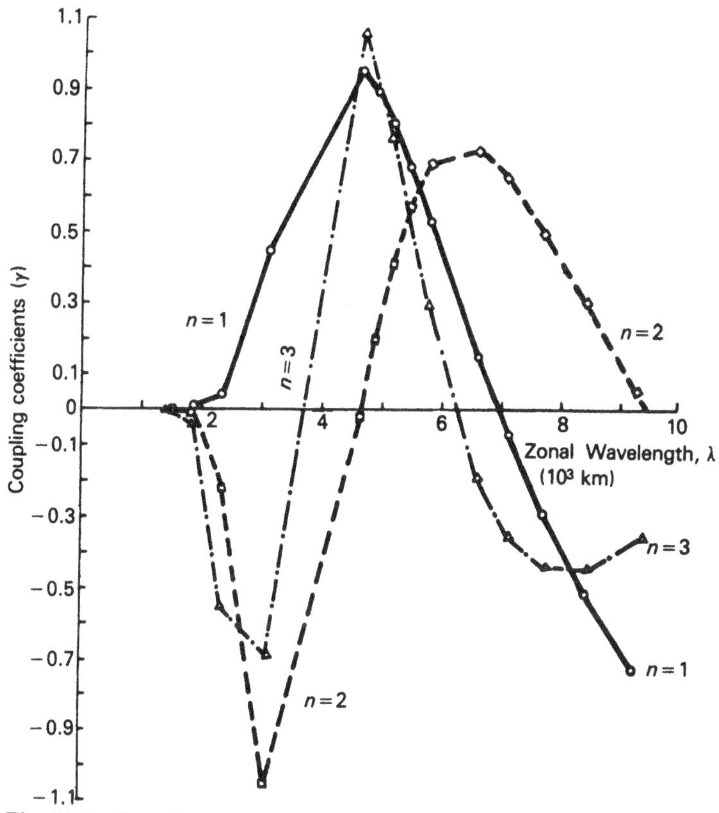

Fig. 11.1. Coupling coefficients for zonal wavelengths (λ in 10^3 km).

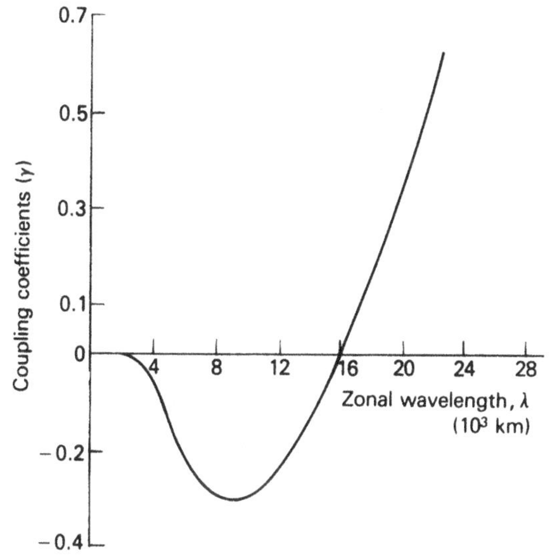

Fig. 11.2. Coupling coefficients for Kelvin wave interactions (λ in 10^3 km).

11.6 Stability of a stationary wave

Following Lorenz (1972) and Duffy (1974) we will examine the stability of a stationary wave. Let

$$A_1^{(0)} = a_1^{(0)},$$

$$A_2^{(0)} = A_3^{(0)} = 0.$$

Inserting a perturbation, we put

$$A_1^{(0)} = a_1^{(0)} + A_1',$$

$$A_2^{(0)} = A_2',$$

$$A_3^{(0)} = A_3'.$$

On neglecting variations in the zonal direction and substituting the above in (11.20), we find

$$\frac{\partial A_1'}{\partial T} = 0,$$

$$\frac{\partial A_2'}{\partial T} + \mathrm{i}\gamma_2 B A_2' - \mathrm{i}\rho_2 a_1^{(0)} A_3' = 0,$$

$$\frac{\partial A_3'}{\partial T} + \mathrm{i}\gamma_3 B A_3' - \mathrm{i}\rho_3 a_1^{(0)} A_2' = 0.$$

Putting

$$A_2' \approx \exp \mathrm{i}(kx - \omega T)$$

and ignoring x dependency, the following inequality is obtained for ω to become imaginary

$$B^2 < -\frac{4\rho_2\rho_3}{[\gamma_2 - \gamma_3]^2} |a_1^{(0)}|^2.$$

As B^2 and the denominator are positive, we infer that ρ_2 and ρ_3 must be of opposite signs for instability. We observe from Table 11.1 that this condition is satisfied for a few values of λ.

11.7 Summary and conclusions

Using the shallow water equations, coupling coefficients have been computed between a slowly varying sinusoidal zonal flow and wave triads composed of (*a*) mixed Rossby–gravity and (*b*) Kelvin waves. The main conclusions of the study are:

(i) The zonal flow is observed to impart large rotation to the equatorial waves.

(ii) There appears to be a range of wavelengths (λ), for which the zonal flow rotates Kelvin waves; for other equatorial waves this effect is not so well marked.

(iii) The stability of a wave depends on the coupling coefficients (ρ) being of opposite sign.

References

Blumen, W. (1972) Geostrophic adjustment. *Rev. Geophys. and Space Physics*, **10**, 2, 485–528.

Domaracki, A. and Loesch, A. (1977) Non-linear interactions among equatorial waves. *J. Atmos. Sci.*, **34**, 486–98.

Duffy, D. G. (1974) Resonant interactions of inertial-gravity and Rossby waves. *J. Atmos. Sci.*, **31**, 1218–31.

Lindzen, R. S. (1971) Equatorial planetary waves in shear. *J. Atmos. Sci.*, **28**, 609–22.

Longuet-Higgins, M. S. and Gill, A. E. (1967) Resonant interactions between planetary waves. *Proc. Roy. Soc. London*, **A299**, 120–40.

Lorenz, E. N. (1972) Barotropic instability of Rossby wave motion, *J. Atmos. Sci.*, **29**, 258–64.

Matsuno, T. (1966) Quasi-geostrophic motions in the equatorial area. *J. Met. Soc. Japan*, **44**, 25–43.

Nayfeh, A. (1973) *Perturbation Methods*. Wiley and Sons, New York, 425 pp.

Newell, A. C. (1969) Rossby wave packet interactions. *J. Fluid Mech.*, **35**, 255–71.

Shabbar, M. (1971) Side-band resonance mechanism in the atmosphere supporting Rossby waves. *J. Atmos. Sci.*, **28**, 345–9.

Stone, P. H. (1969) The meridional structure of baroclinic waves. *J. Atmos. Sci.*, **26**, 376–89.

St-Maurice, J. P. and Veronis, G. (1975) A multi-scaling analysis of the spin-up problem. *J. Fluid Mech.*, **68**, 3, 417–45.

Part II

The summer monsoon over the Indian subcontinent and east Africa

Introduction

R. P. PEARCE

Many studies have been carried out of the rainfall characteristics of most of the regions affected by the monsoons, and extensive records exist in many countries. In more recent years, however, the emphasis in regional observational studies has tended to be placed more on the analysis of weather patterns at various levels in the atmosphere in order to identify the structures of particular weather systems responsible for rainfall. Research has also been carried out which attempts to evaluate the various terms in the moisture budget in particular regions, for instance the net water vapour transport into the region and the accompanying evaporation and rainfall. Chapters 12 to 28 describe such regional studies.

Chapter 12 concentrates on the role of the upper-tropospheric anticyclone during the summer monsoon onset, while in Chapters 13 and 14 aspects of the structures of weather systems responsible for particular monsoon rain events are described. Such events are shown, in Chapter 15, to be closely related to the pattern of sea-surface temperature in the Arabian Sea and Bay of Bengal. The rainfall distribution in the Himalayas is discussed in Chapter 16, and the rainfall over India as a whole is discussed in the full water vapour budget context in Chapter 17.

One of the basic problems of tropical meteorology is to determine the precise way in which the atmosphere organizes the upward transports of heat, moisture and momentum on a variety of different scales. At the smaller-scale end of the spectrum there are convective clouds, some of which produce rain and others which do not, while at the larger-scale end there are the large rain areas associated with disturbances such as monsoon depressions. Recent advances have been made in the theory of scale interaction processes and the results of one such theory have been used to determine from observational data the partitioning of the scales at the convective end of the range. These studies are described in Chapter 18. As part of the joint Indo-Soviet experiment in the Indian Ocean in 1977, constant pressure balloons were released and tracked over the Arabian Sea at various times during the monsoon. The computed trajectories provide insight into the variation of the low-level winds over this data-sparse area during one monsoon season, and these are described in Chapter 19. One of the more spectacular manifestations of the monsoon circulation is the jet-like flow at low levels just off the coast of east Africa which extends from south of the equator up to the Horn of Africa where it turns sharply east. This jet, which is a regular feature of the monsoon, transports large amounts of water vapour from the Indian Ocean onto the Indian subcontinent, and it is obviously of importance to attempt to relate the rainfall over India to the intensity of this jet. This is the topic of Chapter 20, and Chapter 21 describes the structure of the jet as observed from aircraft. In the last chapter, the rainfall regimes over Lake Victoria are described in relation to the monsoon, a topic which illustrates that the area affected by this phenomenon extends well into continental Africa.

Theoretical studies of regional aspects of the monsoon have been pursued for many years. It is now possible to construct regional numerical models which integrate finite-difference analogues of the full set of dynamical equations. They can be used to study regional phenomena in somewhat idealized but fairly realistic numerical experiments. Examples of such models are described in Chapters 23 and 24 with, in Chapter 24, a comparison of the model output with the observed structure of a midtropospheric depression off the west coast of India. This study provides a good example of the use of a numerical model to identify the basic dynamical processes responsible for the development of this particular system. The studies in Chapters 25 to 28 are all somewhat more idealized but, again, are directed towards understanding the dynamical processes which result in the occurrence of monsoonal disturbances. Such studies are a necessary complement to the integration of sophisticated numerical models described in Part I which incorporate the full complexity of these physical processes. It is only by carrying out experiments with more idealized models that it becomes possible to determine the relative importance of the various processes involved. Thus, in Chapter 25, it is suggested that the phenomenon of downstream development could be a process involved in the amplification of monsoon depressions. In Chapters 25 and 26, the relative roles of barotropic and baroclinic instability are discussed in the context of the monsoon circulation, and in Chapter 28 the role of orography in maintaining the low-level trough over the Bay of Bengal is investigated using an analytical technique.

Part IIa

Observational studies

12

The monsoon as reflected in the behaviour of the tropical high-pressure belt

Y. P. RAO

The southwest monsoon develops in the domain of the subtropical high-pressure belt. Certain characteristics of this feature are discussed in relation to the development of the monsoon.

12.1 Introduction

The area of the southwest monsoon is substantially in the domain of the subtropical high-pressure belt. In the lower troposphere, monsoon 'lows' replace the 'high', but in the upper troposphere the 'high' persists. Fig. 12.1 shows the mean monthly displacement of the ridge line from January to June over India as traced by Banerjee *et al.* (1976). The shift is largest near 150 mb. Studies have been in progress on the relationship between the monthly displacement of the ridge and the development of the monsoon in different years.

12.2 Monsoon rainfall

Banerjee *et al.* (1976) found monsoon rains over India to be substantial when the ridge at 500 mb in April is well displaced to the north, but poor when it lags to the south. For the purpose of this study India is divided into

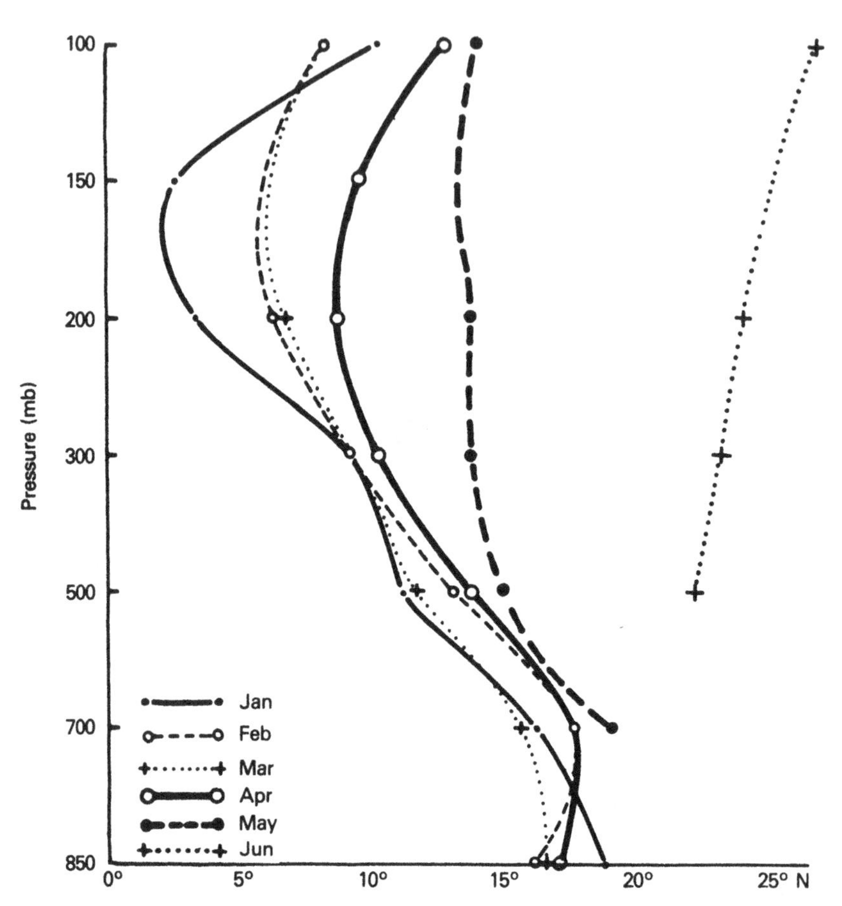

Fig. 12.1. The location of the subtropical ridge at 75° E from January to June (from Banerjee *et al.*, 1976).

31 subdivisions. The percentage of subdivisions receiving rainfall of not less than 20% of the normal from June to September, is designated as R. The latitude of the subtropical ridge over the Indian peninsula in April is denoted L. These parameters are related by

$$(R)^{\frac{1}{3}} = 0.124\,L + 2.5.$$

When L is 17.5, $R = 100$, i.e. no subdivision is expected to be deficient in rainfall by 20% or more. The expectation is the same when the ridge is at a still higher latitude. Fig. 12.2 shows the relationship between the number of rainfall-deficient subdivisions and the position of the subtropical ridge for about three decades. This formula seems to describe, with

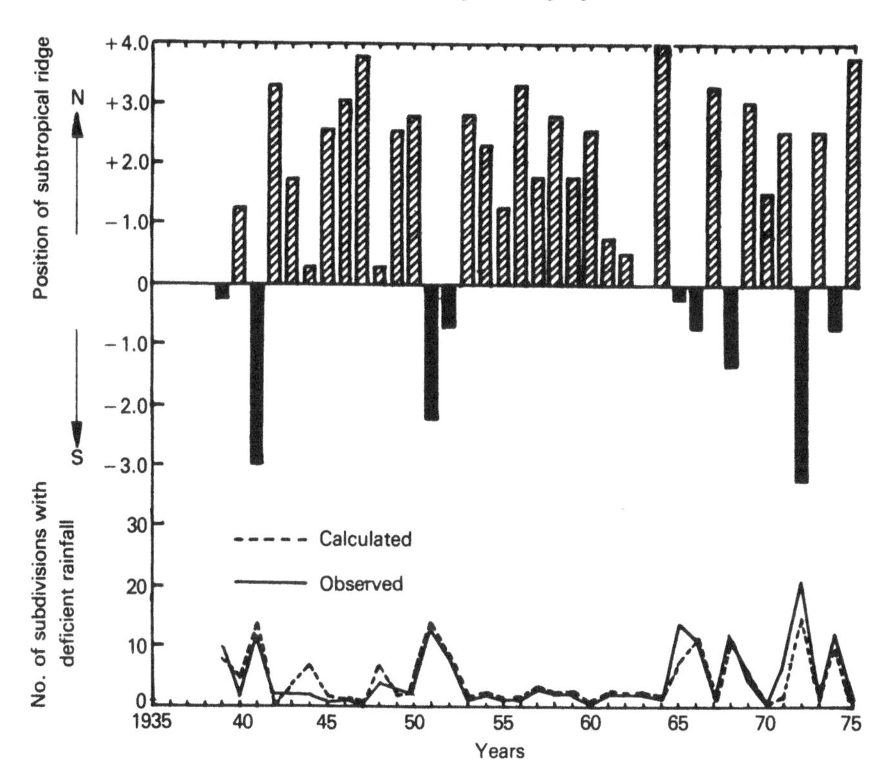

Fig. 12.2. The relationship between the position of the subtropical ridge and the number of rainfall-deficient subdivisions for the years 1935 to 1975. The zero position of the subtropical ridge is taken as 13.5° N.

reasonable accuracy, years of widespread deficit in monsoon rainfall in terms of the position of the subtropical ridge in April.

12.3 The onset of the monsoon in Kerala

Kerala is the southernmost state on the west coast of India. The onset of the monsoon in Kerala signals the beginning of the advance into the rest of the mainland. V. Thapliyal and R. P. Sarker have developed a regression equation, based on the 200 mb winds at Trivandrum and Madras in February, and the January wind at the same level at Forrest (30°50′ S, 128°06′ E) in Australia. The mean wind direction is 225° at Madras and 160° at Trivandrum, indicating the ridge line in between. The difference in the wind direction between Madras and Trivandrum has a correlation of 0.74 with the date of onset of the monsoon over Kerala.

The mean wind direction at Forrest, in January, is 262° and the correlation between the wind direction and Kerala onset is −0.70. The multiple correlation using both the factors is 0.93. The regression equation shows a capability of predicting a late onset, which is very important.

12.4 The eastern 'high'

Recently the India Meteorological Department has prepared mean upper wind charts for ten-day periods. The development of the 'high' in the upper troposphere over the Bay of Bengal and its northward shift is a very prominent feature from about the middle of April. From 21 to 30 April, the axis of the 'high' is at 15° N at 100 mb over the Bay and the 'high' is less distinct at 150 mb. In the first ten days of May this 'high' descends to 200 mb. Now the ridge line is at 19° N at 100 mb and 14° N at 200 mb. In the next ten days the ridge lines are at 21° N and 17° N at 100 and 200 mb respectively. From 21 to 31 May, the ridge lines are at 23° N and 21° N at these levels. Progress of this 'high' has to be monitored to establish the time of monsoon onset. This may be another way of projecting the build-up of the upper-tropospheric easterlies. The extension of the eastern 'high' in the premonsoon months is consistent with the intensification of the Walker circulation during the monsoon build-up and relates the monsoon rains in India with preceding conditions in the Pacific.

Reference

Banerjee, A. K., Sen, P. N. and Raman, C. R. V. (1976) On foreshadowing southwest monsoon rainfall over India with mid-tropospheric circulation anomaly of April. Prepublished Scientific Report No. 76/8, pp. 1–9, India Meteorological Department.

13

On the onset of the Indian southwest monsoon and the monsoon general circulation

I. SUBBARAMAYYA AND R. RAMANADHAM

The problem of the onset of the southwest monsoon over India is critically reviewed and the need for a procedure to objectively determine the northern limit of the monsoon (NLM) during the period of its advance is brought out. A procedure suggested by Subbaramayya and Bhanukumar (1977) has been used including the satellite cloud pictures for demarcating the NLM during the 1977 period of the advance of the monsoon.

The nature of the three-dimensional circulation of the Asian summer monsoon constructed by the authors of this chapter is reappraised and the important branches of the circulation are pointed out. It is suggested that this model of the general circulation is to be further examined using extensive aerological data.

13.1 Introduction

The southwest monsoon season is the rainy season for the Indian sub-continent and this monsoon accounts for more than 80% of the annual rainfall. The total rainfall in the period June to September amounts to about 3.8×10^{15} kg. Its value at the rate of 1 paisa a pot of water comes to 13 000 billion rupees, and therefore it is a very important recurring natural resource to the country. Similar amounts of rainfall are given by this monsoon to several countries in south and southeast Asia and therefore it is an important meteorological phenomenon of great economic significance. Meteorologically, it is equally important because the heat liberated by the monsoon rains in the free atmosphere over the Indian region is about 8.4×10^{21} J in the monsoon season. It is almost equal to the total shortwave radiation received by the same area.

Not only the total rainfall during the monsoon season but, more than that, the rainfall distribution is very important from the agricultural point of view. Among several important factors, one is the starting date of the monsoon rains or the onset of the monsoon. A delay in the onset by 10 to 15 days would adversely affect the crop output, while an early onset might not be utilized to advantage without an advance forecast.

Two aspects of the southwest monsoon are presented in this chapter: (i) the onset of the monsoon; and (ii) the monsoon general circulation.

13.2 Onset of the monsoon

Though the word monsoon strictly refers to the wind, it is customary to use it for the associated rains. Accordingly, the word 'onset' of the monsoon is used for the starting of the characteristic monsoon rains. To identify the onset, the India Meteorological Department (1943) considered the 'characteristic rise' in the trend of the cumulative rainfall curve at any station and prepared a chart of the normal dates of onset. By a similar procedure the normal dates of withdrawal were also determined.

There are difficulties in determining the normal date of onset of the monsoon in some localities where the characteristic rise is not clear, as pointed out by Ananthakrishnan and Rajagopalachari (1964). It is all the more difficult to fix the date of onset of the monsoon accurately in individual years, and lack of a specific criterion for demarcating the monsoon boundary has led to conflicting results. This has been adequately discussed by Ananthakrishnan *et al.* (1967). In view of these problems Subbaramayya and Bhanukumar (1978) have critically examined the onset of the monsoon at a number of stations and studied the synoptic events associated with its advance over the Indian subcontinent during several years. It was found that the advance of the monsoon is the result of a series of synoptic and sub-synoptic rain-bearing systems progressing northwards and westwards in contrast to the eastward-moving disturbances of the winter and premonsoon seasons. The same concept was used and the progress of the monsoon in the period May to June 1977 has been obtained from the spreading of the rains associated with westward-moving rain-bearing systems. The results are presented in this chapter.

13.3 Advance of the monsoon in 1977

The synoptic and sub-synoptic systems and their movement over India and neighbouring seas during the period May to June 1977 have been studied from the surface weather charts given in the Indian Daily Weather Reports. Satellite cloud pictures for the month of June 77, available at Visakhapatnam, have also been consulted to further confirm the locations of the disturbances and their movement.

The tracks of the disturbances are presented in Fig. 13.1. During the period 1 to 10 May, seven westerly disturbances passed over north India. In the extreme south, three weak easterly waves were noticed. One of them developed into a cyclonic storm in the Bay of Bengal and moved in a north-northeasterly direction and crossed the Bangaladesh coast on 13 May. In the period 11 to 20 May, four westerly disturbances were observed. In the south, two minor troughs formed and persisted each for a couple of days in the southeast Arabian Sea off the Kerala coast. Two troughs moved westward in south Bay and two more minor troughs formed and persisted near the south Coromandel coast corresponding to rains in the Bay Islands, Sri Lanka and southern parts of Kerala and Tamilnadu. Simultaneously, strong, persistent west-southwesterly winds occurred over south Bay. From 26 to 31 May, two westerly disturbances appeared over north India. Two troughs over the Sri Lanka area, one large trough further north, two troughs over the north Andaman Sea and one trough over upper Burma formed and slowly moved northwestward. A minor trough persisted off the Kerela–Karnataka coast in the Arabian Sea. Under the influence of these systems the monsoon advanced slightly northwards over the south peninsula, the north Andaman Sea and lower Burma. The trough system off the Andhra coast and that over upper Burma did not give much rain or contribute substantially to the monsoon progress.

During the period 1 to 10 June, only one westerly disturbance occurred. This persisted for some time over the Punjab and the adjoining area and subsequently dissipated. Three low-pressure systems appeared over upper Burma. One rapidly moved westward to east Uttar Pradesh and the other two moved southwestward to the northeast of the Bay of Bengal. An earlier trough over the north Andaman Sea moved northwestward and persisted over east Madhya Pradesh and the adjoining area. Two minor troughs in the southwest Bay of Bengal moved northwestward and dissolved over the south Madras coast. The minor trough that persisted off the Kerala–Karnataka coast earlier, now developed into a depression

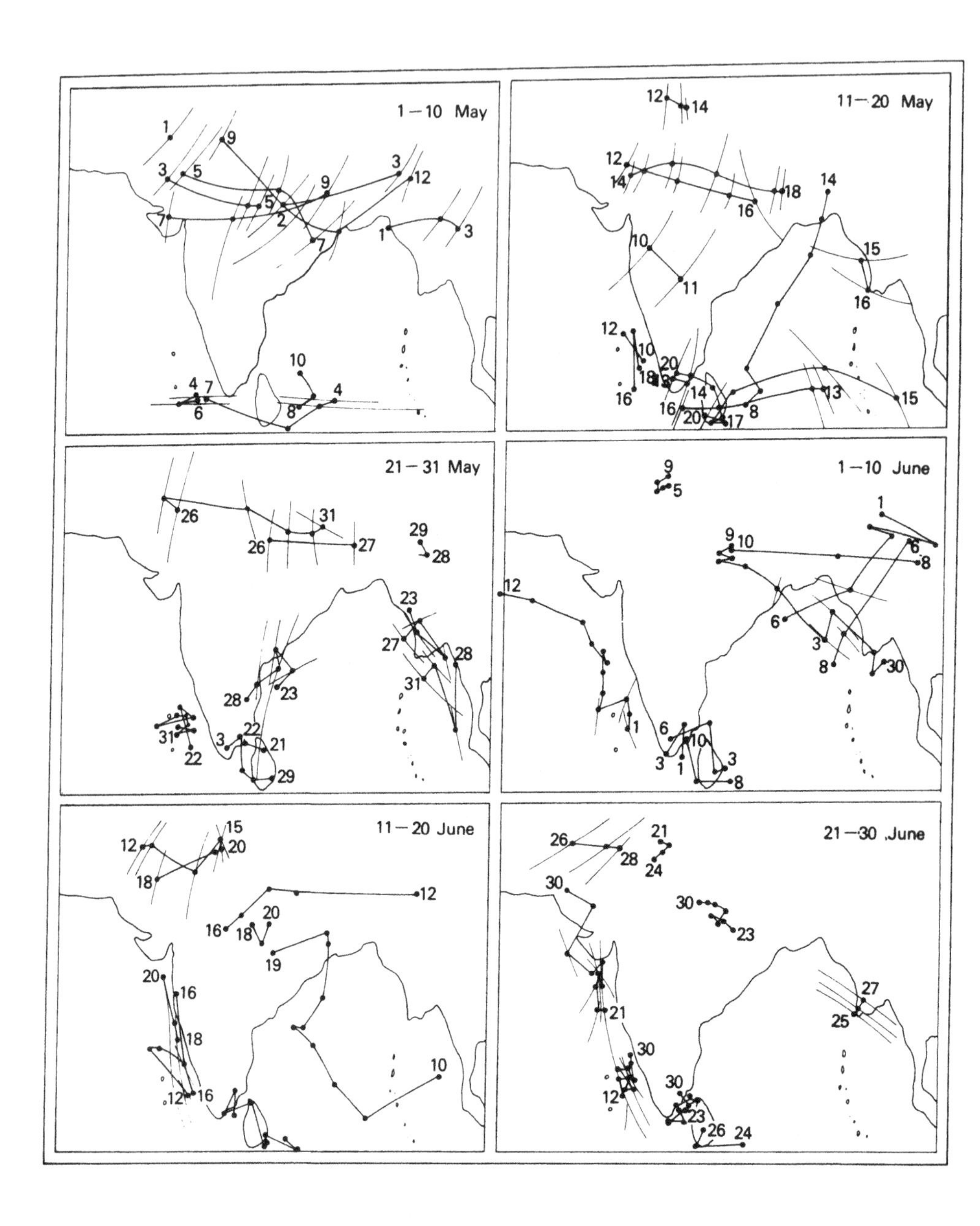

Fig. 13.1. The location and movement of low-level troughs and lows during May and June 1977. The numbers at the beginning and end of the lines of movement are the days of month (the core dates are given in each frame), and the dots indicate the positions on consecutive days.

and moved in a north-northwesterly direction. Subsequently, it developed into a cyclonic storm and moved west-northwestward. As a result of these synoptic systems the monsoon rains spread westward near the head of the Bay and northward over the south peninsula. In the period 11 to 20 June two westerly disturbances moved eastward in the extreme northwest of the country. A low-pressure area moved from the Andaman Sea to the southeast of the Bay and later moved west-northwestward for sometime, and then westward. One trough was active off the Karnataka coast from 12 to 16 June and another trough off Konkan coast intensified and extended to the north. Under the influence of the Konkan trough and the Bay system, the monsoon advanced over the north peninsula. In the period 21–30 June, a weak but extended trough occurred near the north Andaman Sea and the activity of minor troughs continued near Sri Lanka and off the Karnataka coast. The trough off the Konkan coast deepened and extended into Sourastra and Kutch, and the monsoon advanced in that region. Two westerly disturbances appeared over Pakistan and Punjab and Haryana. An induced low developed north of Kutch on the extended Konkan trough, which moved westward, spreading the monsoon rains further north. A low-pressure system developed over east Madhya Pradesh and slowly moved to west Uttar Pradesh. As a result of this the monsoon advanced northwestward covering parts of north India.

The foregoing observations clearly show that the advance of the monsoon is mainly affected by northwestward moving synoptic systems. The day-to-day advance of the monsoon determined from the spreading of the rains due to northwestward moving systems is presented in Fig. 13.2*a*. The normal dates of onset are given in Fig. 13.2*b*. A comparison of the two charts show that the monsoon set in early by about 10 days in the south Bay of Bengal, Sri Lanka, Kerala and Laccadev islands, was normal in northeast India and was delayed by 3 to 7 days over the rest of the country.

13.4 Monsoon general circulation

The general circulation associated with the Asian summer monsoon was constructed by Subbaramayya and Ramanadham (1966) using the results of wind studies by different workers in different areas in the region from the west Pacific to the Atlantic (Fassig, 1933; Stone, 1942; Hess, 1948; Hubert, 1949; David and Sanson, 1952; Hay, 1953; Riehl, 1954; Koteswaram, 1958; Rao, 1960; Subbaramayya, 1961). This is reproduced in Fig. 13.3.

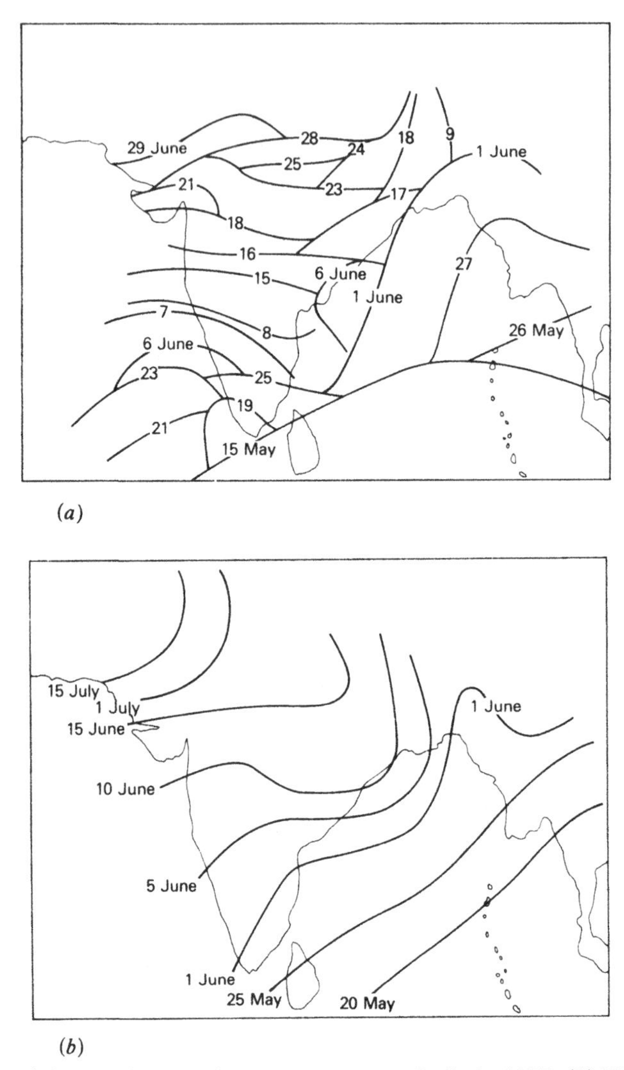

Fig. 13.2 (*a*) The advance of the monsoon over India in 1977. (*b*) The normal dates of onset of the monsoon.

The essential features of the circulation are: (i) the lower-tropospheric monsoon air comes from the southern hemisphere; (ii) the monsoon air-stream, i.e. the flow with a westerly component, has greater thickness and northward spread in the eastern region than in the west; (iii) there is general ascending motion north of the Equator in the lower troposphere; (iv) the easterlies in the upper troposphere are primarily subsiding and are associated with a northerly component; (v) most of the upper-

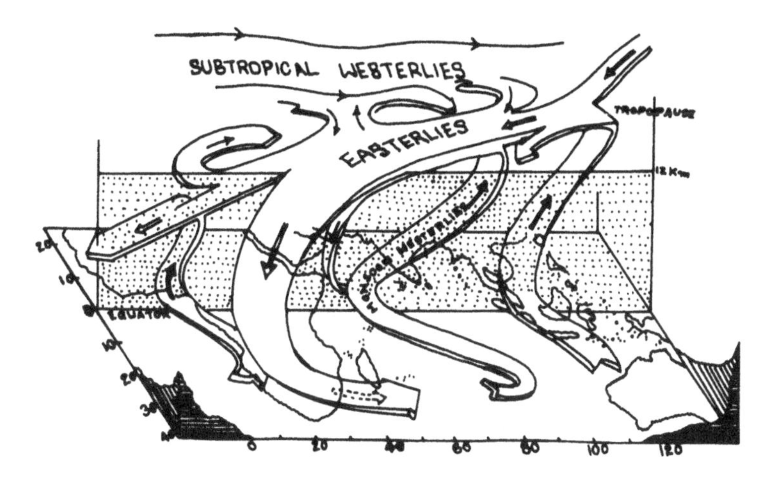

Fig. 13.3. A schematic picture of the three-dimensional circulation over the monsoon area.

tropospheric air crosses the Equator and subsides to low levels in the southern hemisphere; and (vi) there is exchange of air between the upper easterlies and the subtropical westerlies in the eastern and western flanks of the Tibetan and Middle East anticyclones.

The above model of the circulation basically shows that the Hadley cell of the southern hemisphere extends into the northern hemisphere.

There are several branches of the flow in the above model and the transports of air need to be evaluated for a proper understanding of the circulation. The authors of this chapter have estimated the air flow in the monsoon westerlies across peninsular India as 7×10^{15} kg per day. This estimate is strikingly similar to the mass flow across the Equator over the Arabian Sea estimated by Findlater (1969). An attempt was made by Saha (1970) to make a more accurate estimate of the cross-equatorial air and moisture fluxes over the Arabian Sea, and he found that they constitute 60 to 80% of the respective eastward fluxes near the west coast of India between 26° N and the Equator. The average transport of air by the upper easterlies over India was estimated as 12×10^{15} kg per day. The authors of this chapter have further indicated that this could be the rate of mass transport in the total monsoon circulation. The mass circulation round the globe in the Hadley cell of the winter hemisphere was estimated as 19×10^{15} kg per day by Kidson *et al.* (1969). This shows that the monsoon circulation is comparable in size to the Hadley cell on the global scale.

Estimates of mass transport in the different other branches of the circulation are yet to be made. Unfortunately, the present net-work of upper-air stations is not adequate, but it is hoped that accurate estimates of air transports in the different branches of the monsoon circulation will be possible following MONEX-79.

References

Ananthakrishnan, R., Acharya, V. R. and Ramakrishnan, A. R. (1967) India Meteorological Department Forecasting Manual, Part 11, 18: Monsoons of India. p. 9.

Ananthakrishnan, R. and Rajagopalachari, P. J. (1964) Pattern of monsoon rainfall distribution over India and neighbourhood. Proceedings of the Symposium on Tropical Meteorology, New Zealand Meteorology Service, p. 192.

David, D. A. and Sanson, H. W. (1952) Easterly jet over east Africa. *Weather*, **7**, 343–4.

Fassig, O. L. (1933) The trade winds of the eastern Caribbean. *Trans. Amer. Geophys. Union*, **14**, 69.

Findlater, J. (1969) Interhemispheric transport of air in the lower troposphere over western Indian Ocean. *Quart. J. Roy. Meteor. Soc.*, **95**, 400–3.

Hay, R. F. M. (1953) High level strong easterlies over Singapore and Hongkong. *Weather*, **8**, 206–8.

Hess, S. L. (1948) Some new meridional cross-sections through the atmosphere. *J. Meteor.*, **5**, 293–300.

Hubert, L. F. (1949) High tropospheric westerlies of the equatorial west Pacific Ocean. *J. Meteor.*, **6**, 216.

India Meteorological Department (1943) *Climatological Atlas for Airmen.*

Kidson, J. W., Vincent, D. G. and Newell, R. E. (1969) Observational studies of the general circulation of the tropics: Long-term mean values. *Quart. J. Roy. Meteor. Soc.*, **95**, 258–87.

Koteswaram, P. (1958) The easterly jet stream in the tropics. *Tellus*, **10**, 43–57.

Rao, Y. P. (1960) Interhemispherical features of the general circulation of the atmosphere. *Quart. J. Roy. Meteor. Soc.*, **86**, 156–66.

Riehl, H. (1954) *Tropical Meteorology*, McGraw-Hill, New York.

Saha, K. R. (1970) Air and water vapour transport across the Equator in western Indian Ocean during northern summer. *Tellus*, **22**, 681–7.

Stone, R. G. (1942) On the mean circulation of the atmosphere over the Caribbean. *Bull. Amer. Meteor. Soc.*, **23**, 4.

Subbaramayya, I. (1961) Studies on the Indian southwest monsoon. Ph.D. Thesis, Andhra University, Waltair, India.

Subbaramayya, I. and Bhanukumar, O.S.R.U. (1978) The onset and the northern limit of the southwest monsoon over India. *Meteor. Magazine*, **107**, 37.

Subbaramayya, I. and Ramanadham, R. (1966) The Asian summer monsoon circulation. *J. Meteor. Soc. Japan*, **44**, 167–72.

14

Medium-range forecasting of monsoon rains

P. S. PANT

For the planning of agricultural operations as well as the operation of large dams and reservoirs, medium-range forecasts of precipitation covering a period of 5 to 10 days are of great potential value.

Medium-range forecasting of monsoon activity requires first the identification of slowly evolving changes in the monsoon circulation. A study of periodicities in the monsoon circulation by power spectrum analysis of the zonal and meridional components of winds shows the existence of periodicities of 6 to 8 days' duration. Synoptic experience also indicates that each spell of active monsoon conditions is of about one week's duration.

A recent study of the monsoon circulation during Phase II of MONSOON-77 (25 June to 2 August 1977) revealed that during a weak phase of the monsoon a second trough formed in the south Bay of Bengal in the lower troposphere, while the trough in the north persisted. Low-pressure systems formed in this low-latitude trough and caused rainfall over the southern peninsula. This circulation pattern resulted in two distinct belts of excess rainfall, one in the north and another in the south, with relatively less rainfall over central parts. During the course of a week, the southern trough shifted northwards and resulted in the initiation of another active spell of monsoon rains over central parts.

Some of the observed oscillations of the monsoon circulation seem to be linked with features of the hemispheric circulation as revealed by variations of the zonal index. This opens up the possibility of developing new methods of medium-range forecasting of monsoon activity.

14.1 Introduction

Over the years, the demand has been growing for forecasts of precipitation and other meteorological conditions for periods of a week for use in planning agricultural operations and in the field of water resources management. In order to meet this demand, attempts were made in the past to identify slowly evolving changes in the monsoon circulation pattern to enable the meteorologist to foresee the evolution of the precipitation pattern. Rahmatullah (1952) identified different types of flow pattern in the monsoon season which lasted 4 to 8 days. Investigations were also carried out using as a basis 5-day mean 700 mb charts both for winter and monsoon periods (Pant, 1964*a* and *b*). These investigations resulted in the identification of a few interesting slowly evolving circulation changes. An objective method based on linear regression was also developed to predict the 5-day mean 700 mb charts (Pant and Natarajan, 1964).

The three significant features of the monsoon which attract the attention of meteorologists are: (i) its onset; (ii) strong monsoon conditions; and (iii) break-monsoon conditions. Rao (1976) presented an excellent summary of the available knowledge of these features. The initial studies of the onset of the monsoon over India were based on rainfall data. Yin (1949) identified certain interesting changes in the upper-tropospheric circulation features associated with the onset. Pant (1964*b*) studied the significant features of the 700 mb circulation associated with monsoon onset and progression. Ananthakrishnan *et al.* (1968) studied the upper-tropospheric westerlies over north India in relation to monsoon onset.

Strong monsoon conditions usually commence with the formation of a depression or a circulation in the lower troposphere over the head of the Bay of Bengal and last as long as the life of the system during its passage west-northwestwards. This phase is characterized by the east–west trough at 700 mb around 20° to 25° N.

Break-monsoon conditions have been extensively studied by Koteswaram (1950), Raman (1955) and Ramaswamy (1962, 1973). During the period of a break, the monsoon trough is along the foothills of the Himalayas, and westerlies extend right up to the hills in the lower troposphere. The westerlies over the southern peninsula are either weak or develop waviness.

MONSOON-77 data have revealed yet another interesting circulation pattern in which a trough in the northern plains occurs in conjunction with a second trough extending east–west from the southern peninsula to the Bay of Bengal. This circulation resulted in two east–west bands of excess precipitation: one over the northern provinces and the second extending from Lakshadweep across the southern peninsula to the Andaman and Nicobar Islands. The purpose of this chapter is to review the important circulation patterns associated with the three significant phases with special reference to MONSOON-77 and deal in greater detail with the evolution that was revealed by the data of the MONSOON-77 experiment.

The changes in the zonal index at the time of the observed slowly evolving changes were also studied. The association between the hemispheric circulation and the observed monsoon circulation is discussed with the help of the zonal index. For this purpose Kats' (1960) zonal index for the belt 45° to 90° E has been utilized.

14.2 The onset and progress of the monsoon

Earlier studies have shown that onset of the monsoon over the extreme south peninsula is associated with the formation of the monsoon trough at 700 mb extending from the south Arabian Sea to the south Bay, and the establishment of steady westerly winds over Kerala (Pant, 1964*b*; Pant and Vernekar, 1963). In 1977 the monsoon set in over Kerala towards the end of May and covered the entire country by the end of June. The progress of the monsoon in 1977, as indicated by the onset of rain, is shown in Fig. 14.1. It will be seen that its initial progress in 1977 was rather slow until about the middle of June. At 700 mb a well-marked east-west trough formed over the southern parts of the peninsula and the adjoining seas at 700 mb in association with the onset of the monsoon over Kerala as can be seen from Fig. 14.2*b*. This trough was located slightly further south before the onset (Fig. 14.2*a*). Fig. 14.2*c* illustrates the circulation pattern at 700 mb on 8 June, when the monsoon rains have extended up to about 15° N.

The monsoon extended over the entire country in the last week of June and was quite active. The circulation at 700 mb associated with strong monsoon conditions on 28 June 1977 is illustrated in Fig. 14.2*d*. The rainfall distribution over the country corresponding to the above-mentioned four locations of the monsoon trough at 700 mb is shown in Fig. 14.3.

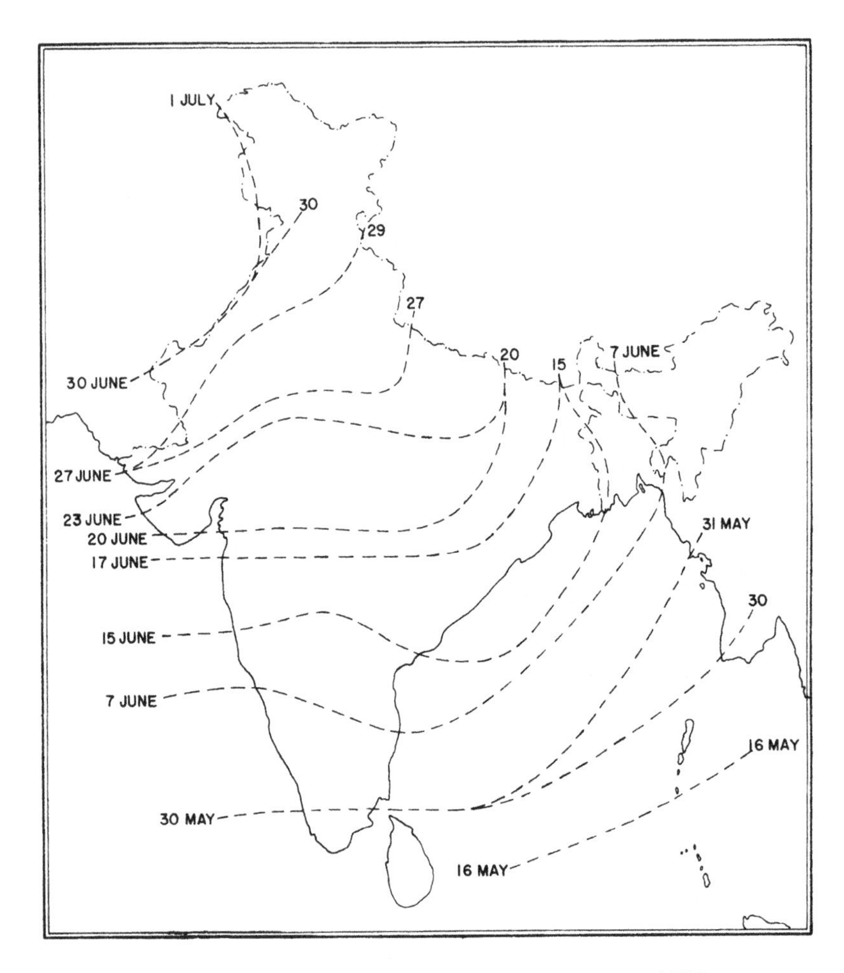

Fig. 14.1. The progress of the southwest monsoon, 1977.

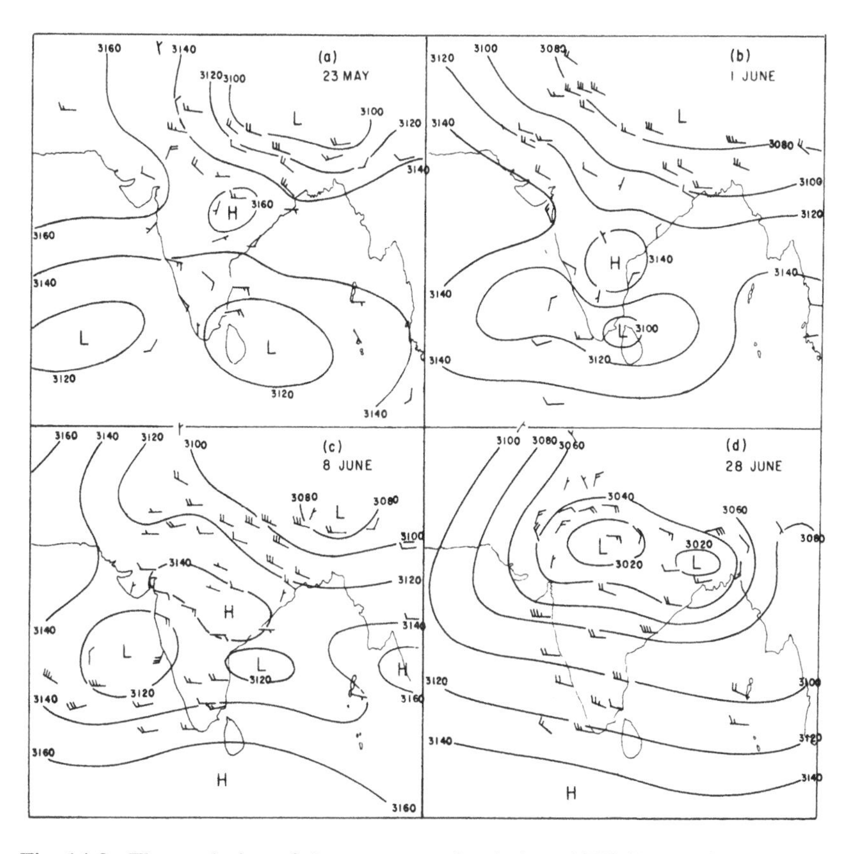

Fig. 14.2. The evolution of the monsoon circulation, 1977 (700 mb). Contours are labelled in gpm

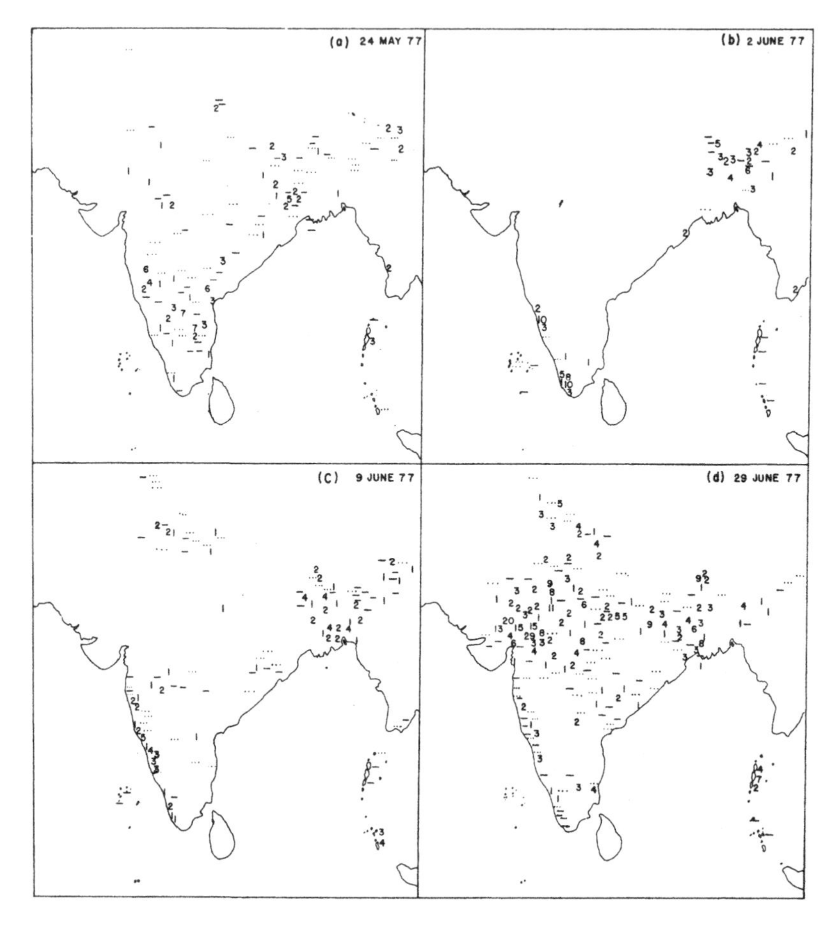

Fig. 14.3. Distribution of 24-hour rainfall (mm) (ending 0300 GMT) during the evolution of the 1977 monsoon.

Fig. 14.5. The evolution of the monsoon circulation from 15 to 23 July 1977 (700 mb). Contours are labelled in g m^{-1}.

14.3 The weak phase of the monsoon

An interesting evolution of the monsoon circulation took place from about 12 July until 22 July. During this period, the 700 mb circulation revealed the simultaneous existence of one trough extending over the northern parts of the country and a second east–west trough over the south Bay of Bengal and the adjoining southern peninsula. This circulation pattern was associated with two belts of excessive precipitation – one over the northern area and the second over the peninsula, Andaman and Nicobar Islands and Lakshadweep. During the same period the central parts of the country had deficient rainfall. This distribution of precipitation is illustrated by Fig. 14.4 which shows the percentage departure of rainfall from normal for the week ending 20 July 1977.

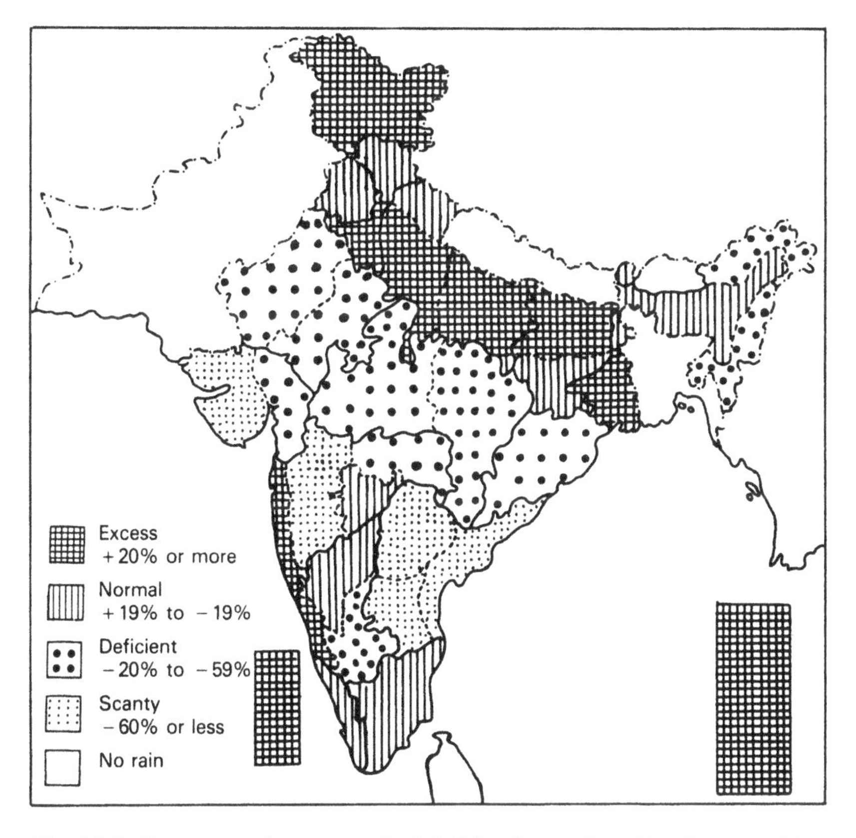

Fig. 14.4. Percentage departure of rainfall for the week ending 20 July 1977.

The corresponding circulation pattern is shown in Fig. 14.5*a*. This differs from a typical break condition in that there are two distinct troughs, one in the north and a second in the south; whereas during a break, westerlies extend right up to the foothills without any trough in the north, though there may be a trough in the south. Thus, this weak monsoon condition is distinct from a typical break in two respects, (i) the simultaneous existence of two troughs, and (ii) a belt of excess precipitation over the plains of north India; while the common feature between weak and break-monsoon conditions is the deficient rainfall over the central parts of the country.

From 12 to about 22 July the trough in the south Bay of Bengal gradually shifted northwards culminating in strong monsoon conditions around 22 July. Fig. 14.5 illustrates the evolution of the monsoon

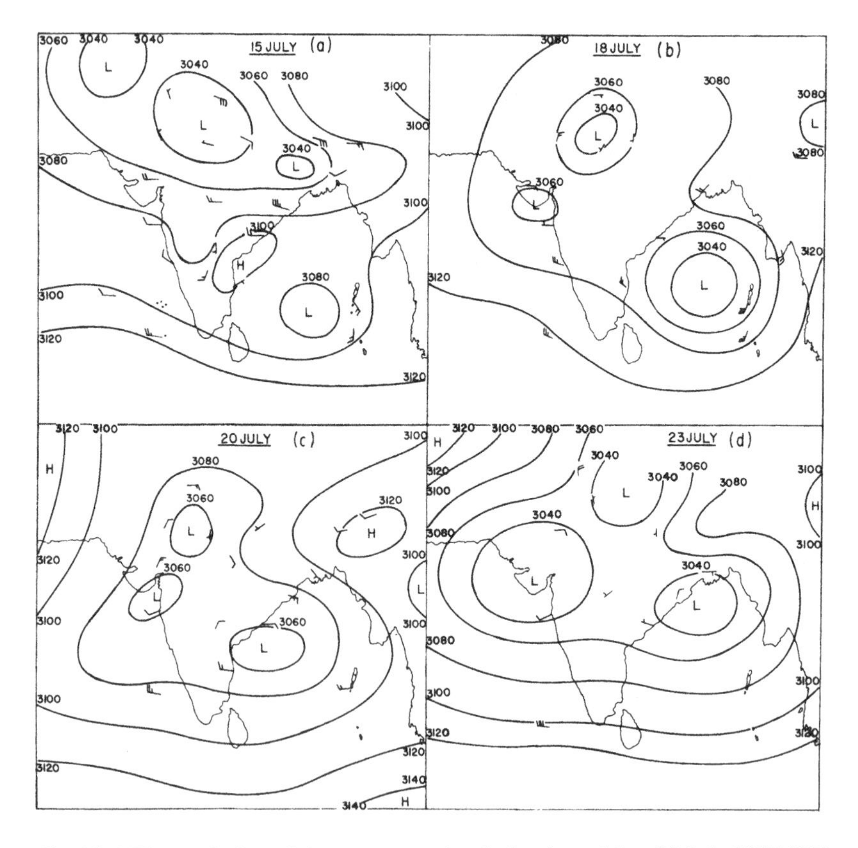

Fig. 14.5. The evolution of the monsoon circulation from 15 to 23 July 1977 (700 mb). Contours are labelled in gpm.

circulation at 700 mb during this period from a relatively weak situation to strong monsoon conditions over a period of about 10 days.

14.4 Break-monsoon conditions

The monsoon was weak during the second and third weeks in July but there was no typical monsoon break. It was only during the second and third weeks of August that a regular break in the monsoon occurred with excess or normal rainfall being confined to the peninsula and to the foothills of the Himalayas. This is very well illustrated by the rainfall departures for the week ending 17 August 1977 – see Fig. 14.6. The 700 mb circulation pattern for a typical day (15 August) during the period is shown in Fig. 14.7 and the rainfall distribution over the country on 16 August in Fig. 14.8.

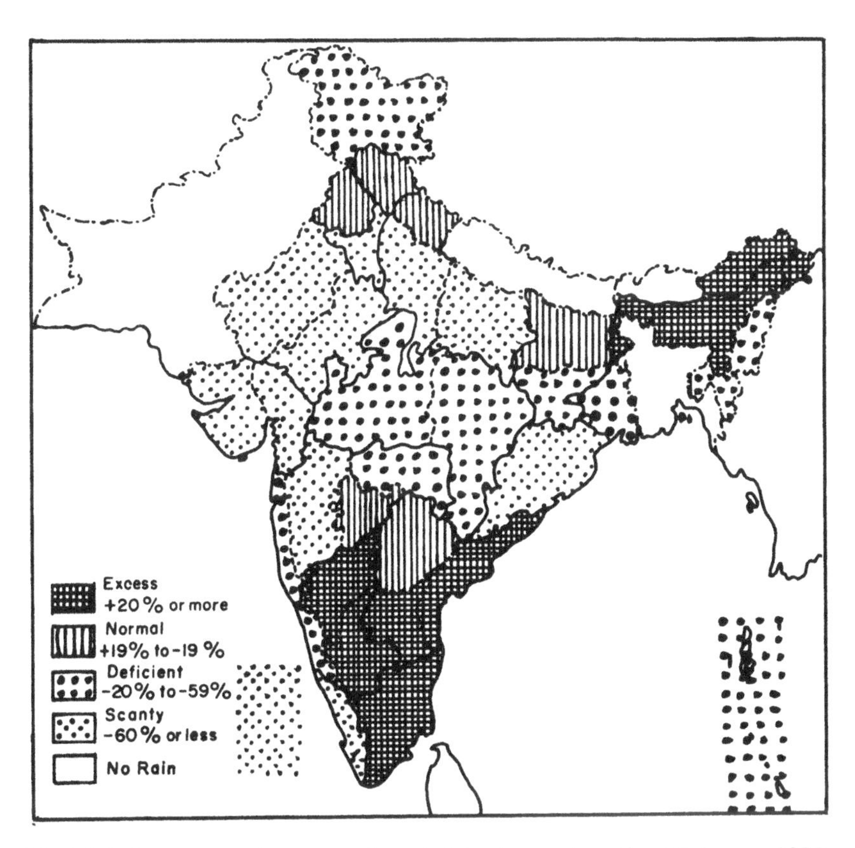

Fig. 14.6. Percentage departure of rainfall for the week ending 17 August 1977.

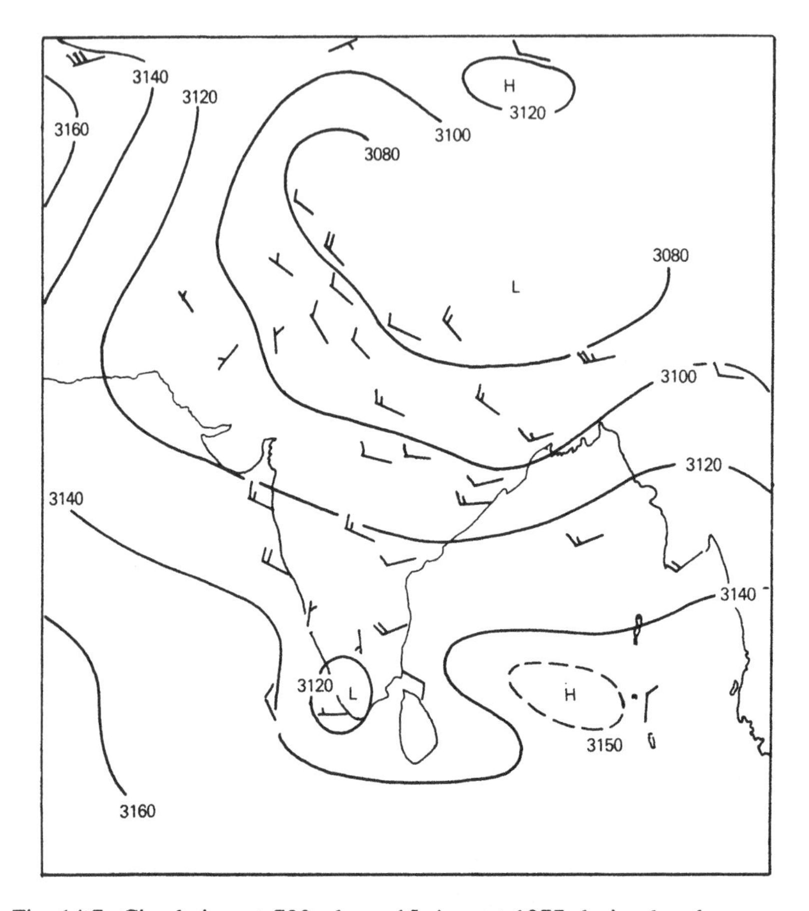

Fig. 14.7. Circulation at 700 mb on 15 August 1977 during break monsoon (1200 GMT). Contours are labelled in gpm.

Koteswaram (1950) connects breaks with westward-moving lows at low latitudes (10° N) in the Bay, prominent at 700 mb. These low-pressure systems form in the trough in the south Bay, which persists for several days under weak monsoon conditions. They move towards the southern peninsula and result in a second belt of excess rainfall in the Andaman and Nicobar Islands, southern peninsula and Lakshadweep. Sometimes the changeover from a weak or break-monsoon condition to an active monsoon condition is gradual as happened in the third week of July, 1977.

On the basis of an exhaustive study of satellite cloud pictures and the aerological data from Gan, Hamilton (1971) finds that a 'break' is often

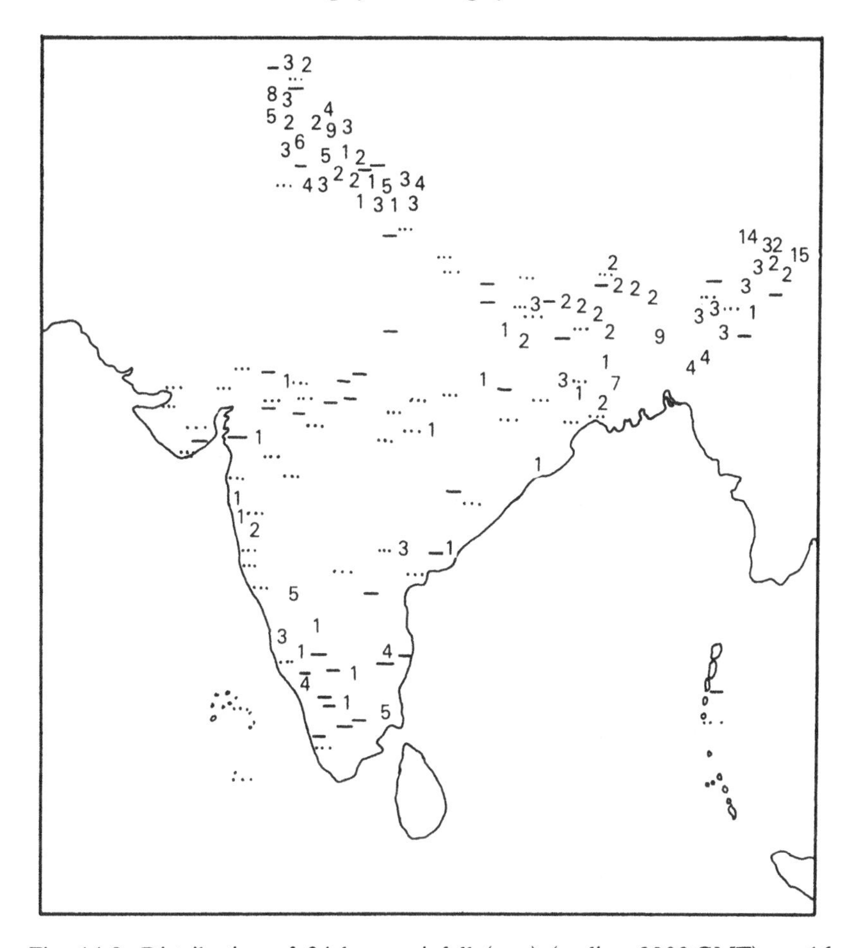

Fig. 14.8. Distribution of 24-hour rainfall (mm) (ending 0300 GMT) on 16 August 1977 (break monsoon).

accompanied by the development of a trough in the lower troposphere northeast of Gan. This result also fits in with the formation of an east–west trough over the south Bay in the lower troposphere referred to in this study.

14.5 Zonal index and the monsoon activity

According to Ramaswamy (1962, 1965), during break conditions a pronounced low-index circulation prevails in the midlatitude westerlies north of the Himalayas, with a large amplitude trough protruding into the Indo-Pakistan region at 500 mb and aloft. Thus, the influence of the zonal index on the monsoon circulation has been indicated in earlier studies. However, a complete appreciation of the linkage between the two is still to be achieved.

During the MONSOON-77 experiment the zonal indices for the eastern hemisphere from 45° W to 90° E were regularly received from Moscow on board the Soviet flagship *Akad Shirshov* and an attempt was made to study the relation between the variation in the zonal index and the changes in the monsoon activity. It was observed that during the initial period of the monsoon, i.e. the first fortnight of June, the zonal index remained low and progress of the monsoon was sluggish. The monsoon began advancing when the zonal index began rising after the middle of June – see Fig. 14.9. It has not been possible to recognize any close relation between the changes in position of the monsoon trough and the zonal index during the other phases. However, an examination of the monthly mean zonal indices for the years 1972 to 1975 has shown that, in general, zonal indices for the two years 1972 and 1974, when the monsoon activity was much below normal, were lower than for the years 1973 and 1975, when monsoon activity was much stronger. This is illustrated in Fig. 14.10. Particularly, the mean zonal index for the months of July, August and September during the excellent monsoon season of 1975 was much higher than for the other years considered in the study.

While complete data for 1977 are not available, the mean July values for this year are higher than for the years 1972 to 1974. The monsoon rainfall for 1977 was also generally above normal over the country.

One reasonable inference that can be drawn from this data is that while the zonal index is closely related to monsoon activity taking the season as a whole, shorter-period variations within the season seem to be related to other factors as well as to the zonal index. Hence, it has not been possible

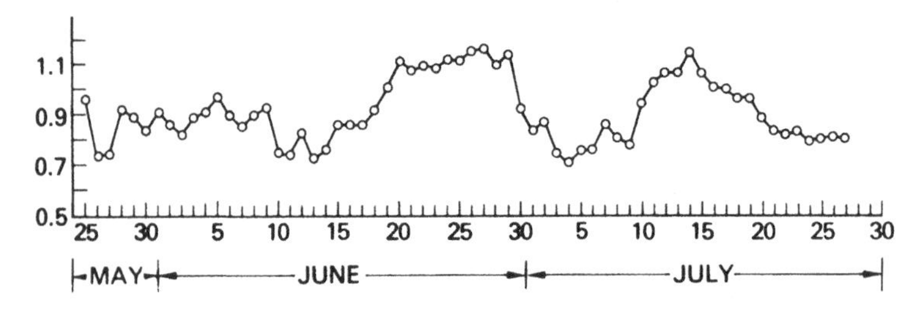

Fig. 14.9. 500 mb (Kat's (1960)) zonal index during June and July 1977 (45° W to 90° E).

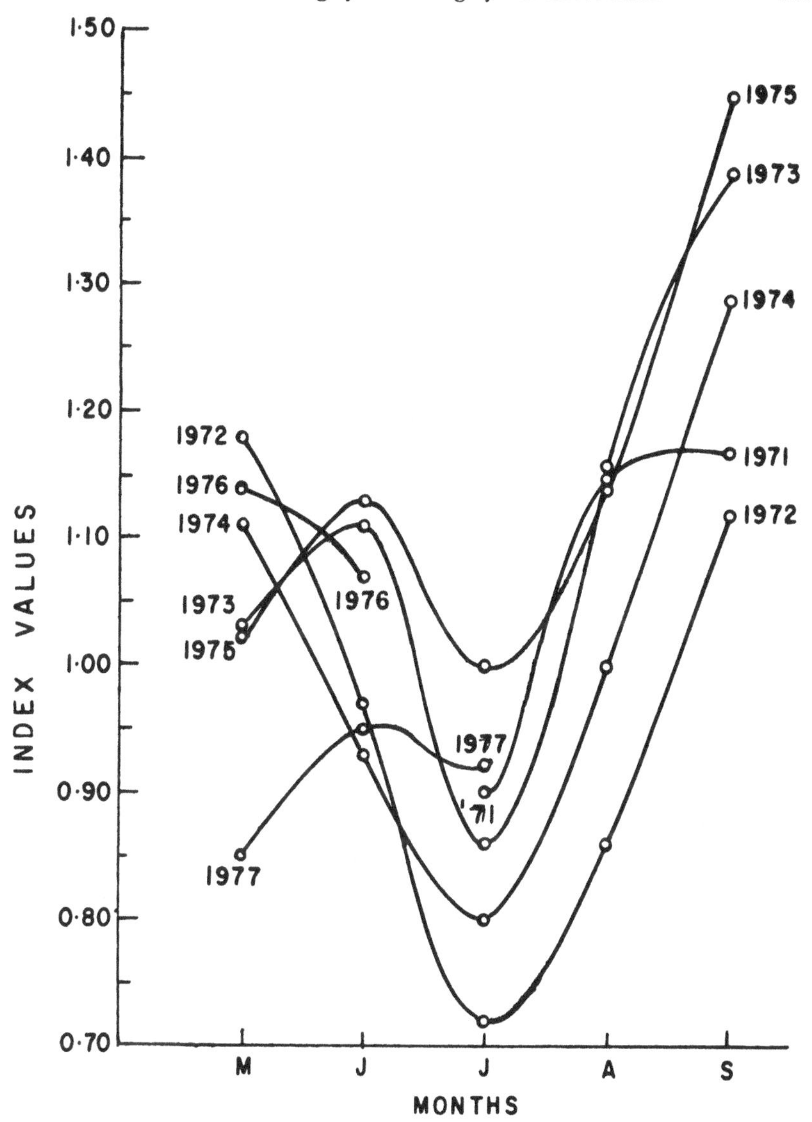

Fig. 14.10. Monthly mean zonal index (Kats, 1960).

to establish a close relation between the different phases of the monsoon and the zonal index.

14.6 Forecasting of monsoon precipitation

The evolution of the monsoon circulations, particularly the slowly evolving ones, do help synoptic meteorologists to indicate the areas of activity as well as the trend in weather over different parts. But, the problem of forecasting circulations over a period of a week remains to be solved. Experiments with the quasi-geostrophic model at the India Meteorological Department, even for short-range forecasting, have not been too encouraging for the monsoon season.

Until suitable numerical models for the monsoon region are developed, it will be necessary to explore further the relation between monsoon activity and parameters such as the zonal index and anomalies in sea-surface temperatures.

14.7 Summary and conclusions

The formation of the trough in the south Bay which persists for about a week is an important feature of a weak monsoon condition, and low-pressure systems that affect the southern peninsula originate or move along this region. The zonal index seems to be closely related to the monsoon activity over periods of a month or season. However, over shorter periods of a few days, synoptic and sub-synoptic systems seem to influence monsoon activity.

I am grateful to Dr B. S. Chuchukalov, leader of the Soviet scientists participating in MONSOON-77, who arranged to receive Kats' zonal index on the vessel *Akad Shirshov*. I am also grateful to my colleagues Messrs V. R. Seshadri and R. Jambunathan for their assistance in the preparation of charts and diagrams.

References

Ananthakrishnan, R., Srinivasan, V., Ramakrishnan, A. R. and Jambunathan, R. (1968) Synoptic features associated with the onset of the southwest monsoon over Kerala. India Meteorological Department Forecasting Manual Report IV–18–2.

Hamilton, M. G. (1971) Some aspects of break and active monsoon over southern Asia during summer. *Tellus*, **29**, 335–44.

Kats, A. L. (1960) Seasonal variations of the general circulation of the atmosphere and long-range prognosis. *Gidrometeoizdat*, Leningrad.

Koteswaram, P. (1950) Upper level lows in low latitudes in the Indian area during the southwest monsoon season and breaks in the monsoon. *Ind. J. Meteor. Geophys.*, **1**, 162–4.

Pant, P. S. (1964*a*) Forecasting winter precipitation over north India 3–7 days ahead – the synoptic approach. *Ind. J. Meteor. Geophys.*, **15**, 3, 347–58.

Pant, P. S. (1964*b*) Onset of monsoon over India. *Ind. J. Meteor. Geophys.*, **15**, 3, 375–80.

Pant, P. S. and Natarajan, T. R. (1964) Some characteristics of troughs observed on 5-day mean 700 mb charts during the winter season. *Ind. J. Meteor. Geophys.*, **15**, 4.

Pant, P. S. and Vernekar, A. D. (1963) A study of the onset of monsoon. Proceedings of IGY Symposium, Vol. II, February 1961, pp. 20–41, Council of Scientific and Industrial Research, New Delhi.

Rahmatullah, M. (1952) Synoptic aspects of the monsoon circulation and rainfall over Indo-Pakistan. *J. Meteor.*, **9**, 176–9.

Raman, C. R. V. (1955) Breaks in the Indian southwest monsoon and typhoons in the southwest Pacific. *Current Science*, **24**, 219–20.

Ramaswamy, C. (1962) Breaks in the Indian summer monsoon as a phenomenon of interaction between the easterly and the subtropical westerly jet-streams. *Tellus*, **14**, 337–49.

Ramaswamy, C. (1965) On synoptic methods of forecasting the vagaries of the southwest monsoon over India and the neighbouring countries. Proc. Symp. Meteor. Results IIOE-Bombay, 317–27.

Ramaswamy, C. (1973) A normal period of large-scale 'break' in the southwest monsoon over India. *Current Science*, **42**, 15, 517–23.

Rao, Y. P. (1976) Southwest Monsoon – Meteorological Monograph, Synoptic Meteorology 1/76. India Meteorological Department.

Yin, M. T. (1949) A synoptic aerological study of the onset of the summer monsoon over India and Burma. *J. Meteor*, **6**, 393–400.

15

Sea-surface temperature and the monsoon

P. R. PISHAROTY

The first part of the chapter presents a brief review of the main large-scale ocean–atmosphere interaction mechanisms determining the sea-surface temperature (SST) in the tropics: the second part considers the use of SST anomalies in the Indian Ocean for predicting monsoon rainfall.

The International Indian Ocean Expedition provided evidence for the first time that variations of the SST over the Arabian Sea are far greater than are usually given in the normal charts. During the height of the summer monsoon, the SST in the Somali–Socotra area is sometimes as low as 14 °C; similarly, off the west coast of India between 15° N and 20° N the SST is occasionally as high as 31 °C.

Using monthly mean SST charts and weekly synoptic SST charts, a few occasions can be found when the SST showed marked anomalies. These were used for forecasting some of the features of the monsoon weather one to two weeks ahead, with significant success. If the data were available in real time (and not a week later), the same forecasts could be made for two to three weeks ahead.

15.1 Introduction

It is well known that the ultimate source of energy for all atmospheric motions is solar radiation incident on the Earth. The atmosphere receives most of the thermal energy for the maintenance of its large-scale circulation in the form of sensible heat and latent heat of water vapour. The atmosphere feeds back part of its kinetic energy through wind stress into the energy budget of the ocean circulation. The preferred locations for the ocean-to-atmosphere sensible heat and latent heat supply are primarily within the tropical belt, and secondarily in the limited areas of ocean currents of warm water found in the middle and high latitudes. It is, therefore, obvious that oceans play a large role in weather developments, through seasonal and year-to-year variations of these means of thermal energy supply to the atmosphere.

The average annual latitudinal distribution of the components of the poleward energy transfer in the Earth's atmosphere system is shown in Fig. 15.1. It is based on the assumption that there is no storage of heat in the atmosphere and in the oceans over one year, although one cannot be sure that such a balance is attained, particularly in the oceans, every year. It shows that the meridional transport of thermal energy due to ocean currents and of latent heat (derived through evaporation of water) are significant fractions of the total poleward flux of thermal energy.

While the role of the oceans in weather developments was realized in principle, it was simplified to that of producing maritime air masses of more or less uniform characteristics; the direct use of sea-surface temperature values was very limited. Palmen (1948) has shown that the lowest water temperature for which warm-core tropical cyclones are likely to form is about 26 to 27 °C. The UK Meteorological Service has

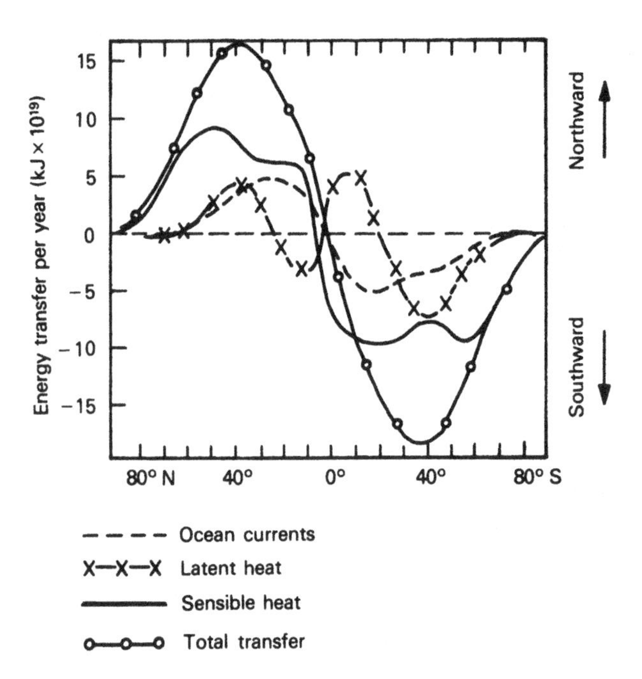

Fig. 15.1. Poleward transport of heat. (After Barry and Chorley, 1968.)

been making use of anomalies of sea-surface temperatures over the Atlantic in issuing their monthly weather forecasts.

15.2 Contribution of J. Bjerknes

It was J. Bjerknes who uncovered for us a class of long-period and medium-period ocean–atmosphere interactions which promises to supply a physical basis to account for seasonal and extra-seasonal fluctuations of the weather averaged over periods of a month, a season, or a year. Modelling experiments will help us to decipher these seasonal fluctuations. These experiments may use the sea-surface temperatures as one of the parameters, first as an *initial* input and later as a *derived* input. It is likely that through such modelling experiments we will be able to foreshadow the abnormalities in the medium-period (2–4 weeks) averages and the long-period (1–3 years) averages of rainfall and temperature.

The methodology of Bjerknes is based on past observations. He uses synoptic charts and anomaly charts of atmospheric pressure and of sea-surface temperatures, and employs cogent physical reasoning involving geostrophic currents, Ekman drifts, upwelling and downwelling, radiative and advective heat fluxes, sensible heat and latent heat exchanges at the sea surface, changes in the depth of the thermocline, etc.

Bjerknes, in his paper: 'The recent warming of the North Atlantic' (1959) has explained the observed warming south of 50° N as having been caused in part by increasing wind drag which has speeded up the Antilles Current and the Gulf Stream, and in part by an increase in the thickness of the warm surface layer in areas of anticyclonic wind drag. Such an increase is caused by the Ekman drift convergence in the water. Between 50° and 57° N the corresponding surface cooling of water is connected with a thinning out of the surface layer by an increasing cyclonic wind drag, this time by the Ekman drift divergence away from the cyclonic area. The same physical reasoning is followed in his study of the 'El Niño' (Bjerknes, 1961), as well as in his article: 'Atlantic air–sea interaction' (Bjerknes, 1964).

The reasoning behind changes in the sea-surface temperatures associated with Ekman drift is repeated in the following paragraphs, as it is felt that this kind of influence plays a large role in the Indian seas during the monsoon period.

An anticyclonic vortex in the ocean which is decreasing in intensity with depth must be of the warm-core type, in other words, the warm surface layer must have maximum thickness near the centre. This will be brought about by the Ekman drift, normal to and towards the right of the wind stress at the surface of the ocean. (This surface wind will be about 20° to the left of, and significantly lower than, the geostrophic wind.) The magnitude of the Ekman drift E_D is given by:

$$E_D = \frac{\tau}{\rho 2\omega \sin\phi} = \frac{\rho_a C_D W^2}{\rho 2\omega \sin\phi}$$

where

τ	is the wind stress,
ρ	is the density of the sea water,
$2\omega \sin\phi$	is the Coriolis parameter,
W	is the wind speed at the anemometer level,
ρ_a	is the density of air at the surface,
C_D	is an empirical constant, which may be taken as 0.001 for wind velocities less than 5 m s^{-1}, and as 0.0025 for wind velocities of 10 m s^{-1} or more (Roll, 1965).

The creation of a downward bulge in the thermocline, the lower limit of surface warm water, under such an anticyclonic circulation is illustrated in Fig. 15.2*a*. If the variation of the Coriolis force with latitude is taken into consideration, there is convergence in the northward flow and divergence in the southward flow; consequently the axis of the oceanic vortex is somewhat displaced towards the west with respect to the atmospheric vortex.

Conversely, a cyclonic current system decreasing with depth is characterized by a minimum depth of warm water at the centre, and therefore, an upward bulge in the thermocline. This is illustrated in Fig. 15.2*b*. It is brought about by an Ekman drift away from the cyclonic centre.

Similar conditions apply to the thickening and thinning of the layer above the thermocline, whenever anticyclonic and cyclonic vorticity are induced in the surface currents through appropriate wind systems.

When the warm water layer thickens, net direct solar heating produces a positive anomaly in the sea-surface temperature; this layer of water is a little *less* affected by the turbulent heat transfer into the colder water below the thermocline. Similarly when the layer becomes thinner than normal, a negative temperature anomaly is created, since the surface water there is *more* affected by the turbulent heat transfer downwards across the thermocline. Namias (1969, 1970, 1972 and 1973) has also given us a series of remarkable papers explaining how atmospheric

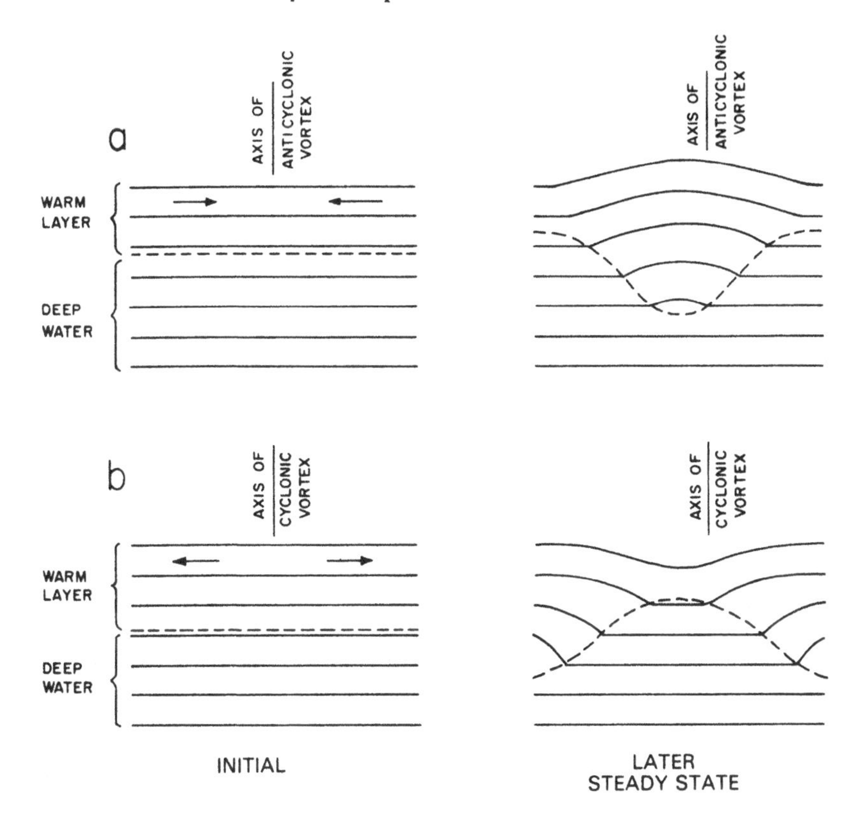

Fig. 15.2. Schematics of zonal, vertical profiles showing: (*a*) the thickening of the warm oceanic surface layer under the influence of anticyclonic stress; and (*b*) the thinning of the same layer under cyclonic stress. Continuous lines: sea-surface and interior isobaric surfaces. Dashed lines: density discontinuity surfaces. (After Bjerknes, 1964.)

circulations bring about seasonal anomalies in the sea-surface temperatures over the Pacific, and how these anomalies modify the locations of atmospheric highs and lows and thereby the tracks of extratropical cyclones.

15.3 Anomalies of sea-surface temperature over the tropics

Bjerknes has introduced two new concepts: (i) of changes in the intensity of the meridional Hadley circulation, and (ii) of the east–west displacements of an equatorial zonal circulation which he called the Walker circulation, both consequent on *anomalies* in the equatorial sea-surface temperatures. Some of the relevant papers are: 'A possible response of the atmospheric Hadley circulation to equatorial anomalies of ocean temperature' (Bjerknes, 1966), 'Atmospheric tele-connections from the equatorial Pacific' (Bjerknes, 1969). Of these, the first concept of anomalies in the meridional Hadley circulation has applications to the extratropical weather systems. These effects have been demonstrated by the modelling experiments of Rowntree (1972) and Spar (1973). The second concept, regarding the zonal circulation with the ascending and descending limbs along the thermal equator, with its location and intensity determined by the anomalies in the sea-surface temperatures over the tropical oceanic belt of the Pacific, is of great relevance to the Indian monsoon.

Apparently, there are two reasons why the equatorial belt of the Indian Ocean and the western Pacific Ocean can have a large influence on the summer monsoon of India: (i) the atmosphere over this belt is in its average state close to the threshold of latent vertical instability, so that even a small positive anomaly of heating and evaporation (resulting from an anomaly of 1 or 2 °C of the sea–air interface) can release vast amounts of atmospheric potential energy; (ii) the mass of air near the Equator has a vast amount of absolute angular momentum about the axis of the earth, so that its exchange with air at higher latitudes leads to a large transport of westerly angular momentum essential for the maintenance of the extra-tropical westerlies.

It was the data from the International Indian Ocean Expedition (IIOE), 1963–5 which first indicated that the Arabian Sea played a significant role in the summer monsoon. It was found that *in situ* evaporation from the Arabian Sea gave rise to an appreciable amount of water vapour flux across the west coast of India (Pisharoty, 1965; Saha and Bavadekar, 1973). A computation with the data of the Indo-Soviet Monsoon Experiment of 1973 showed the preponderance of a moisture flux divergence from the Arabian Sea during an active monsoon period. (Ramamurty *et al.*, 1976). An analysis of the ocean-surface temperatures recorded by ships of the expedition during 1963 and 1964 indicated a warm tongue of water off the west coast of India during the summer monsoon (Fig. 15.3). Based on this, it was hypothesized that this zone of warm water serves to increase the depth of the monsoon current to 4 or 5 km, just before it strikes the west coast of India. The occasional formation of a pressure trough on the windward side of the orographic

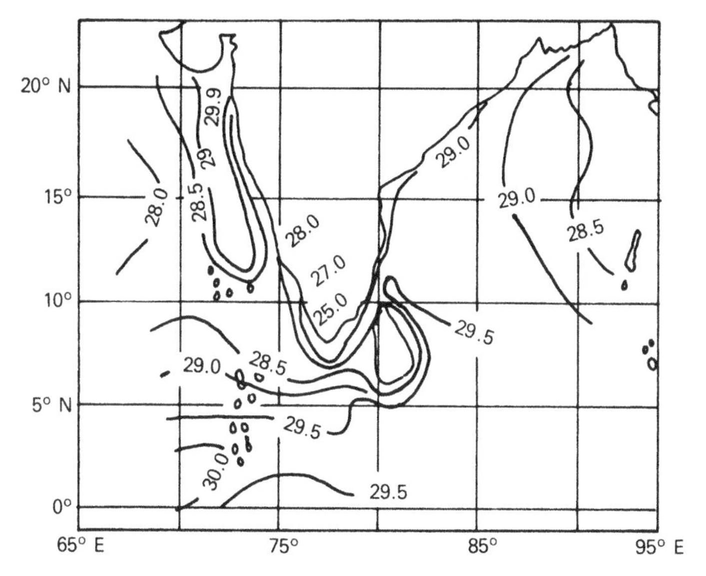

Fig. 15.3. Sea-surface temperatures recorded by the *Ins Kristna* cruise over the Arabian Sea (June to August 1963), and over the Bay of Bengal (After Anand *et al.*, 1968.)

barrier of the western Ghats is caused by a zone of unusually warm water off the coast (Pisharoty, 1968).

Infrared thermal scanner flights carried by the Space Applications Centre in May and June 1975 verified the appearance of this warm tongue, at first about 100 km off the coast of Cochin, and later off the Bombay coast, in each case about two days before the onset of the monsoon. Analyses of the sea-surface temperatures (SSTs) over the Arabian Sea have shown positive anomalies with strong monsoon conditions and negative anomalies with weak monsoon conditions over India (Jambunathan and Ramamurty, 1975). Computer modelling of monsoon rainfall with a negative anomaly of the SST over the west Arabian Sea has shown a decrease of rainfall over western India (Shukla, 1975). SSTs over the south central Arabian Sea have shown positive correlations with rainfall over the western and central parts of India (Shukla and Misra, 1977).

It is not at present possible to explain SST anomalies, except in the general terms described above. The rest of the chapter treats the SST anomalies as known over the north Indian Ocean, and consideration is given to their feedback on the monsoon circulation.

15.4 An example of the use of SST anomalies in forecasting the onset of the monsoon

While it may be difficult to forecast the ocean-surface temperatures, it may be possible to make use of actual SST values for weather forecasting. Oceanic areas, particularly those having an anticyclonic vorticity, have a large thermal inertia. For, as explained earlier, in these areas the surface layer of water is comparatively thick. Hence, one can feel that the feedback mechanisms from the atmosphere may require a period of three to four weeks for modifying the existing anomalies in SST. On this reasoning one may forecast anomalies in the future weather based on the *observed* anomalies in the SST. The operation of the NOAA (GOS) satellites have been providing some near-real-time composite sea-surface temperature values.

Such near-real-time data have shown significant SST anomalies over the Arabian Sea, beginning from the middle of April 1977.

Fig. 15.4 represents the synoptic chart of satellite-derived SSTs for 12 April 1977. The normal temperatures, as given by Alexander and Mobley (1974), are shown by the herring-bone lines. It will be seen that the thermal equator is around latitude 16° N. Normally on 12 April the thermal equator is expected to be south of 8.5° N—the latitude of the sun

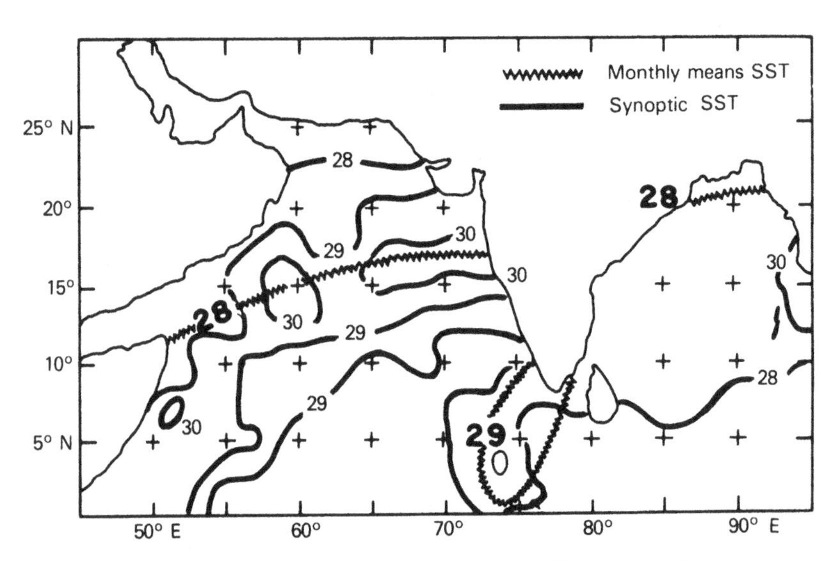

Fig. 15.4. Satellite-derived sea-surface temperatures (°C) over the Indian seas on 12 April 1977, along with monthly normals for April. (Courtesy A. Strong.)

for this date. The normal dates of onset of the monsoon over the different parts of India are given in Fig. 15.5. The standard deviation of the dates of onset is of the order of seven days for the different parts of the country. The situation had not changed significantly on the chart for 19 April.

Using all these 'pieces' of information, on 30 April a forecast was made of an early onset of the monsoon on those parts of the west coast of India north of 16° N, i.e. the Bombay coast and the Gujarat. An early onset of the monsoon over these parts is usually associated with a cyclonic storm off the Bombay coast. Hence a forecast of a cyclonic storm off the Bombay coast during the first week of June was also made.

A cyclonic storm did develop off the Bombay coast on 9 June. It may be recalled that the climatological frequency of the occurrence of a cyclonic storm in this area in the month of June is six in the eighty years, 1891–1970.

However, the monsoon came to Bombay on the normal date, and only five days earlier over Gujarat. There was a spell of heavy rains over

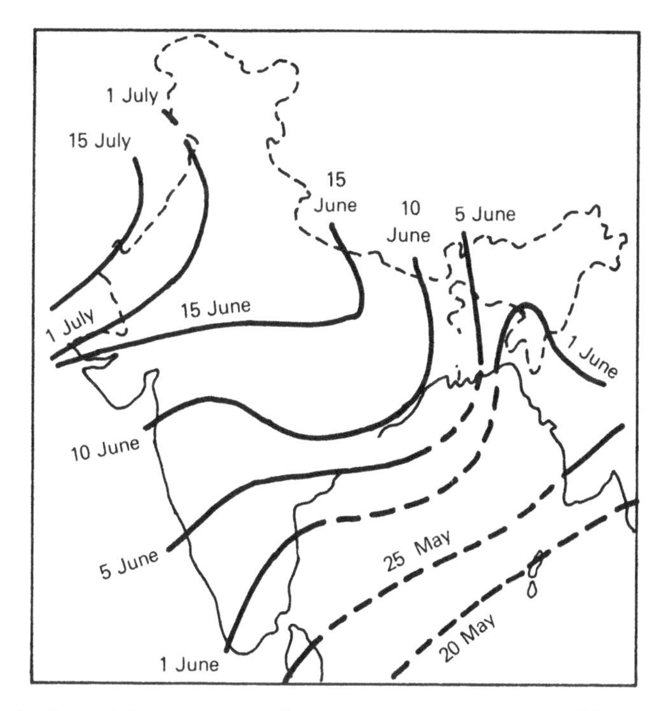

Fig. 15.5. Normal dates of onset of the summer monsoon over different parts of India. (India Meteorological Department.)

Kerala during the first week of May. But it has not been customary to call it a temporary advance of the monsoon when it occurs so early.

15.5 An example of the use of SST in forecasting a 'break' in the monsoon rainfall

Fig. 15.6 gives the SST over the Arabian Sea and the Bay of Bengal for 26 July 1977, along with the average values. There is a distinct anomaly of cooler temperatures over the east central Arabian Sea and the north Bay of Bengal.

Encouraged by the partial success with the forecast issued on 30 April, another forecast was issued on 7 August, based purely on the SST anomaly on 26 July. The forecast was that there would be a lull in the rains over most parts of the country (with an increase along, and near the eastern Himalayas) commencing between 10 and 12 August. This is technically known as a 'break' in the monsoon. Such a break did occur and lasted about two weeks: climatologically such breaks do occur in August, although generally they occur around the third week of August. Figs. 15.7 and 15.8 are the weekly rainfall charts provided by the India Meteorological Department for the weeks ending 17 August and 24

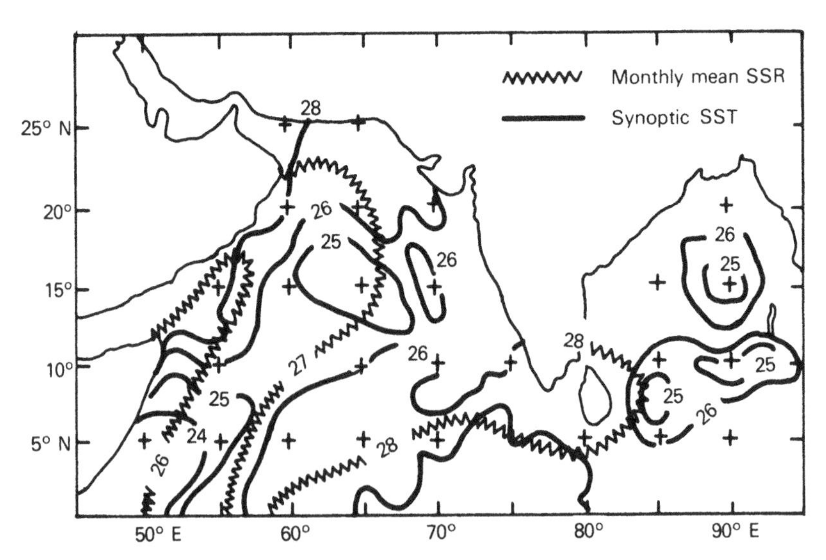

Fig. 15.6. Satellite-derived sea-surface temperatures (°C) over the Indian seas on 26 July 1977 along with the monthly normals for July. (Courtesy A. Strong.)

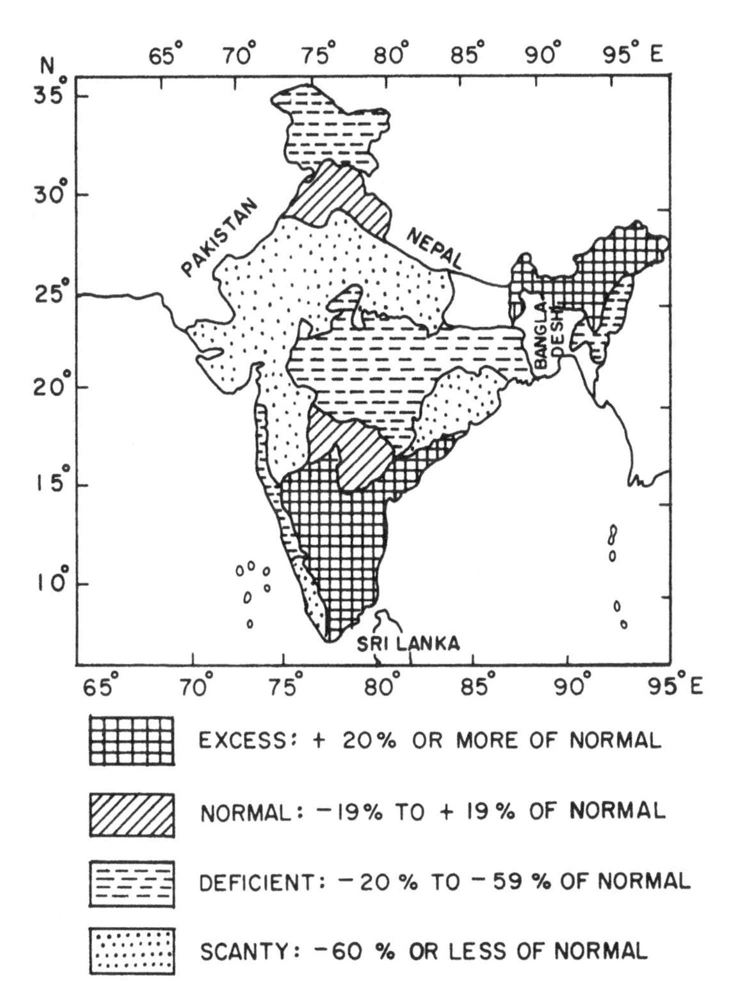

Fig. 15.7. Abnormality of the rainfall over India during the week 11 to 17 August 1977.

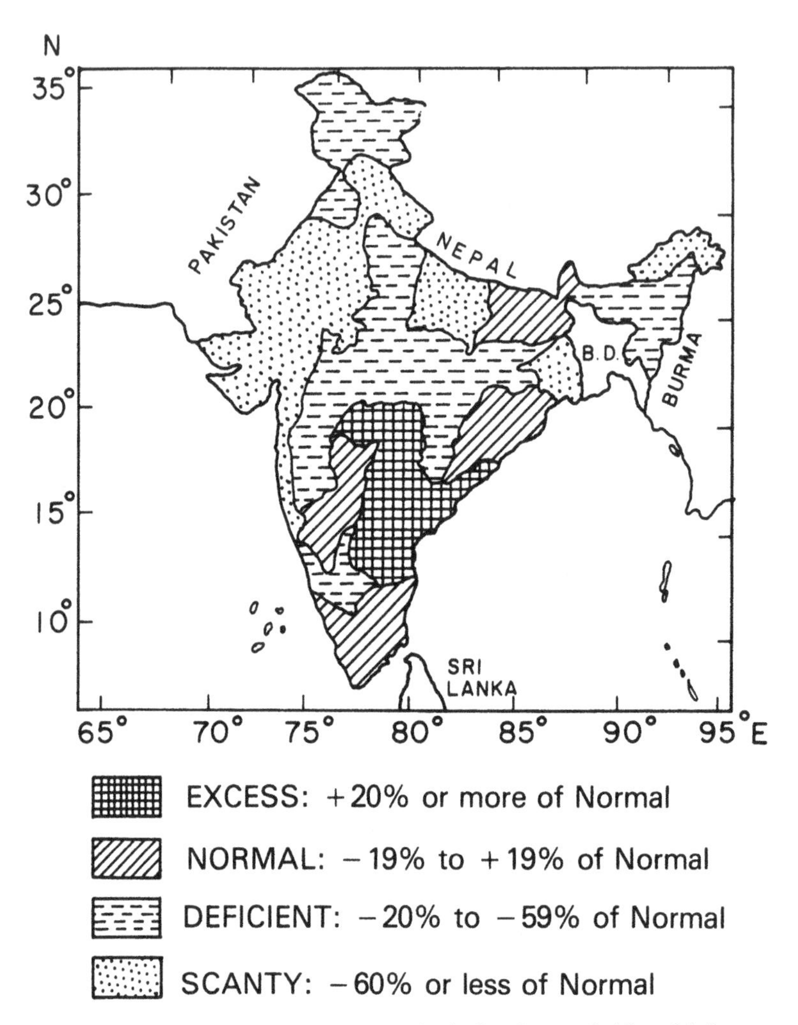

Fig. 15.8. Abnormality of the rainfall over India for the week 18 to 24 August 1977.

August. They show the deficiencies of rainfall during these two weeks.

Positive anomalies in SST persisted over the east central and northeast Arabian Sea throughout May, June, and the first week of July. Fig. 15.9 is the rainfall chart for the period 1 June to 27 July; it shows that the rainfall was above normal all over the western part of India. The hypothesis that this above-normal rainfall is directly related to the positive anomaly of the SST mentioned above, needs further investigation.

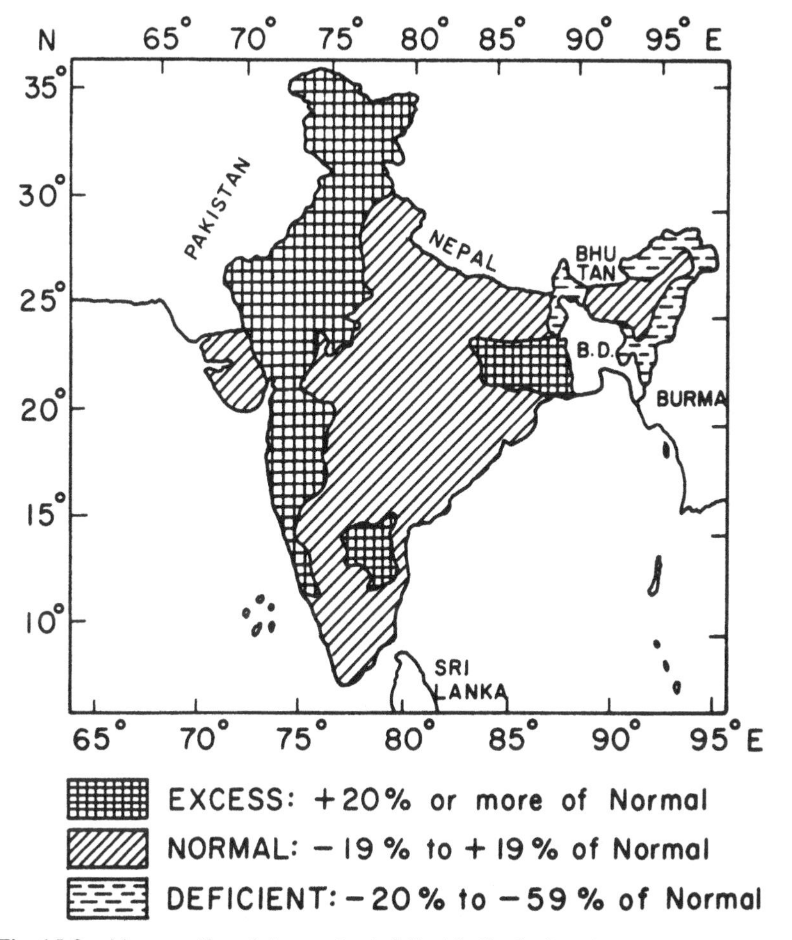

Fig. 15.9. Abnormality of the total rainfall of India during the two-month period 1 June to 27 July 1977.

15.6 Concluding remarks

Charney (see Chapter 6 of this book) has suggested that the ratio of the variance of the *observed* rainfall to the variance of the rainfall caused by flow instabilities consequent on temperature errors is much more over the tropics than over the middle latitudes. He has attributed this to the possibility of a large variance in the actual sea-surface temperatures over the tropics. Therefore, medium-range forecasting would appear to be feasible over the tropics if proper use were made of the SST values. The two examples of forecasting mentioned in this chapter lend support to this view.

A special acknowledgement is due to Dr Alan Strong, National Environmental Satellite Service, NOAA, who kindly provided the author with the synoptic GOSCOMP Sea Surface Temperatures obtained and processed by NOAA. The author is grateful to the India Meteorological Department for the rainfall charts reproduced as Figs. 15.7, 15.8 and 15.9.

References

Alexander, R. C. and Mobley, R. L. (1974) *Monthly Average Sea Surface Temperatures and Ice Pack limits on a 1° Global Grid.* R-1310-ARPA Rand Corporation, Santa Monica, California, USA.

Anand, S. P., Murty, C. B., Jayaraman, R. and Aggarwal, B. M. (1968) Temperature and oxygen in the Arabian Sea and the Bay of Bengal. *Symposium on the Indian Ocean. Bull. Nat. Inst. Sci. India,* Part-I, **38**, 1–24.

Barry, R. G. and Chorley, R. J. (1968) *Atmosphere, Weather and Climate,* Methuen, London, pp. 379.

Bjerknes, J. (1959) The recent warming of the North Atlantic. In *Rossby Memorial Volume,* ed. B. Bolin, Rockfeller, Inst. Press in Association with Oxford University Press, pp. 65–73.

Bjerknes, J. (1961) 'El Niño' study based on analysis of ocean surface temperatures 1935–57. *Inter-American Tropical Tuna Commission Bulletin,* **5**, No. 3.

Bjerknes, J. (1964) Atlantic air–sea interaction. *Advances in Geophysics,* Vol. 10, pp. 1–81, Academic Press, New York.

Bjerknes, J. (1966) A possible response of the atmospheric Hadley circulation to equatorial anomalies of ocean temperature. *Tellus,* **18**, 820–9.

Bjerknes, J. (1969) Atmospheric tele-connections from the equatorial Pacific. *Mon. Wea. Rev.,* **97**, 163–72.

Jambunathan, R. and Ramamurty, K. (1975) Sea and air temperature distribution over the Arabian Sea. *Ind. J. Meteor. Geophys.,* **26**, 465–78.

Namias, J. (1969) Seasonal interactions between the north Pacific Ocean and the atmosphere during the 1960s. *Mon. Wea. Rev.,* **97**, 173–92.

Namias, J. (1970) Macroscale variations in sea surface temperatures over the north Pacific. *J. Geophys. Res.,* **75**, 565–82.

Namias, J. (1972) Experiments in objectively predicting some atmospheric and oceanic variables for the winter of 1971–72. *J. Appl. Meteor.,* **11**, 1164–74.

Namias, J. (1973) Thermal communication between the sea surface and the lower atmosphere. *J. Phys. Oceanogr.,* **3**, 373–8.

Palmen, E. (1948) On the formation and structure of tropical hurricanes. *Geophysica* (Helsinki), **3**, 26–38.

Pisharoty, P. R. (1965) Evaporation from the Arabian Sea and the Indian southwest monsoon. In *Proceedings of the International Indian Ocean Experiment,* ed. P. R. Pisharoty, pp. 43–54.

Pisharoty, P. R. (1968) Contribution of IIOE and weather satellites to monsoon meterology. *Space Research,* **8**, 1073–9.

Ramamurthy, K., Jambunathan, R. and Sikka, D. R. (1976) Moisture distribution and water vapour flux, southwest monsoon, 1973. *Ind. J. Meteor. Geophys,* **27**, 127–40.

Roll, H. U. (1965) *Physics of the Marine Atmosphere*. Academic Press, New York, pp. 156–60.

Rowntree, P. R. (1972) The influence of tropical east Pacific Ocean temperatures on the atmosphere. *Quart. J. Roy. Meteor. Soc.*, **98**, 290–321.

Saha, K. R. and Bavadekar, S. N. (1973) Water vapour budget and precipitation over the Arabian Sea during the northern summer. *Quart. J. Roy. Meteor. Soc.*, **99**, 273–8.

Shukla, J. (1975) Effect of Arabian Sea surface temperature anomaly on Indian summer monsoon. A numerical experiment with the GFDL model. *J. Atmos. Sci.*, **32**, 503–11.

Shukla, J. and Misra, B. M. (1977). Relationship between sea surface temperature and wind speed over central Arabian Sea, and monsoon rainfall over India. *Mon. Wea. Rev.*, **105**, 998–1002.

Spar, J. (1973) Transequatorial effects of sea surface temperature anomalies in a global general circulation model. *Mon. Wea. Rev.*, **101**, 554–63.

16

The effect of elevation on monsoon rainfall distribution in the central Himalayas†

O. N. DHAR AND P. R. RAKHECHA

In this study, an attempt is made to ascertain up to what elevation in the central Himalayas rainfall increases with height. In the central Himalayas there are about 50 rainfall stations which were installed from 1948 onwards in connection with the Kosi dam project and other flood control projects. The mean monsoon (June to October) rainfall data of these stations have been utilized in this study with a view to obtaining a suitable relationship between rainfall and elevation in this section of the Himalayas. This study has shown: (i) that there exist no linear relationships between elevation and monsoon rainfall; (ii) that the elevation and rainfall parameters can best be related by a polynomial of fourth degree; and (iii) rainfall-elevation profiles show that the zones of maximum rainfall occur near the foothills and at an elevation of 2.0 to 2.4 km. Beyond this elevation, rainfall decreases continuously as elevation increases until the great Himalayan range is reached.

16.1 Introduction

Hill (1881) made a detailed study of the distribution of rainfall in the northwest Himalayas and found that rainfall increases with elevation up to a height of about 1.2 km and thereafter it diminishes as the elevation

† Preliminary results of this study were published in *Indian Journal of Power and River Valley Development*, **26**, NO. 6, 179–86.

increases. In the Sierra Nevada mountains in the USA, the rainfall increases up to a height of 1.5 km (Linsley *et al.* 1949). Rumley (1965) investigated the distribution of rainfall with elevation in the Andes mountains in Ecuador and found two zones of maximum rainfall along the western and eastern slopes at elevations of 1.0 and 1.4 km respectively. In this study an attempt has been made to investigate the variation of rainfall with elevation in the central Himalayas (see Fig. 16.1) which are also called the Kosi Himalayas. In and near this section of the Himalayas there exist more than 50 rainfall stations which were installed some 15 to 20 years ago in connection with hydrological investigations of the proposed Kosi dam project at Barahkshetra in east Nepal.

In the central Himalayas, the southwest monsoon normally sets in by about the end of the first week of June and withdraws by about the first week of October (Dhar, 1960) and contributes more than 80% of the mean annual rainfall (Nayava, 1974). Dhar and Narayanan (1965) found that even at remote stations in the interior of the Himalayas, close to the great Himalayan range, rainfall has a pronounced monsoonal character. In this region, since the bulk of the rainfall occurs during the southwest monsoon season (i.e. June to October), this season was chosen for studying the variation of rainfall with elevation.

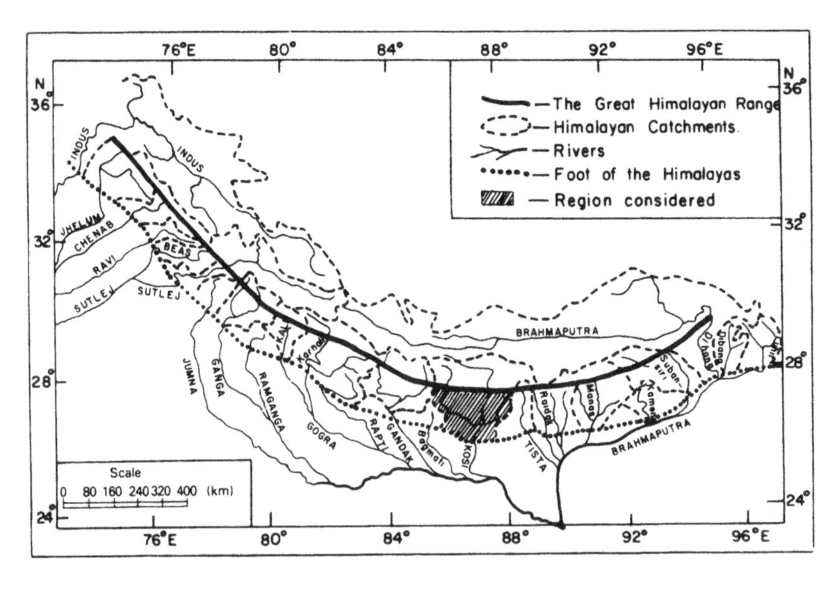

Fig. 16.1. The Great Himalayan range and catchments of the Himalayan rivers. (From Dhar and Bhattacharya, 1976.)

16.2 Brief description of the Himalayas

The northern mountain barrier that separates the Indian subcontinent from the Tibetan plateau is known as the Himalayas. Latitudinally this great barrier is divided into three almost parallel ranges, i.e. the great Himalayan range, the lesser Himalayan range and the Siwalik range, covering an area about 2400 km long and 250 to 400 km wide in the form of a convex arc with convexity towards the south. In between these mountain ranges there are a number of narrow valleys. The great Himalayan range has an average height of about 6.0 km and contains some of the highest peaks in the world e.g. Mt. Everest (8848 m) and Kanchenjunga (8598 m). Further south are the lesser Himalayas, which are also known as the middle Himalayas, whose mean height is about 2.0 to 3.3 km with an average width of about 60 to 80 km. South of the lesser Himalayas is the Siwalik range, or the outer Himalàyas, the average height of which varies from 0.9 to 1.2 km with an average width of 10 to 50 km.

16.3 Data used

In and near the Kosi basin in the central Himalayas, there are more than 44 rain-gauge stations for which rainfall data are available for a period of 15 to 20 years from 1948 onwards. The geographical locations of these stations are shown in Fig. 16.2. The names of these stations together with their respective approximate heights above mean sea level and the mean monsoon (June to October) rainfall are given in Table 16.1. These data were used as the basic data for this study.

16.4 Rainfall–elevation relationships

The relationships between rainfall and elevation in the central Himalayas were derived by using two different statistical approaches. The methods and results obtained are described in the following sections.

16.4.1 Orthogonal polynomial analysis using rainfall–elevation data

In order to ascertain whether there is any evidence of association between rainfall and elevation in the central Himalayas, the correlation coefficient between these two parameters was first determined. This was 0.188, which is very low, thereby indicating that rainfall does not increase

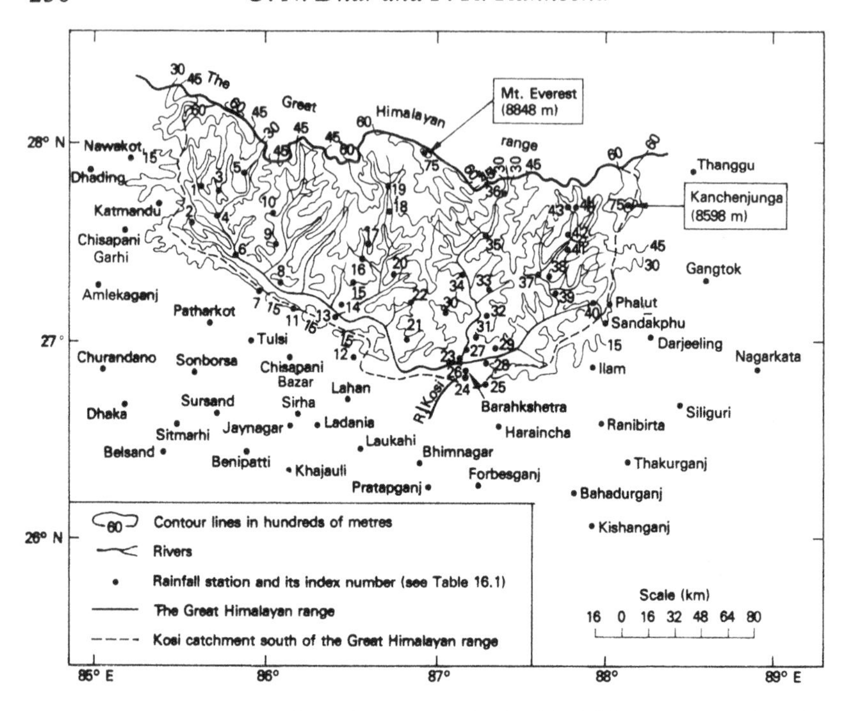

Fig. 16.2. The network of rainfall stations in the Kosi Himalayas south of the Great Himalayan range. (From Dhar and Bhattacharya, 1976.)

linearly with elevation. An attempt was, therefore, made to fit a curvilinear relation between rainfall and elevation by employing the technique of orthogonal polynomial regression analysis. Polynomials of different degrees were tested by the Student '*t*' test to derive the best-fit polynomial. It was found that these two parameters can best be related by a polynomial of the fourth degree at 95% level of significance. The relation derived between rainfall, Y (in decimetres) and mean elevation X (in km) is

$$Y = 21.81 - 40.58X + 40.12X^2 - 12.40X^3 + 1.06X^4. \quad (16.1)$$

The functional relation between rainfall and elevation obtained by (16.1) was plotted in Fig. 16.3 as curve No. I. This figure shows that the first zone of maximum rainfall occurs near the foothill region of the Himalayas. Rainfall starts decreasing from this zone and reaches a minimum at an elevation of 0.6 to 0.8 km. Thereafter, it again starts to increase and another zone of maximum rainfall is obtained at an elevation of 2.0 to 2.4 km. Beyond this elevation, it is observed that rainfall sharply decreases as the elevation increases.

TABLE 16.1. *Mean monsoon rainfall of stations in the central Himalayas.*

Station	Station No. as shown in Fig. 16.2	Approximate elevation above mean sea level (m)	Mean monsoon (June–October) rainfall (mm).
Nawalpur	1	1645	2091
Dhulikhel	2	1372	1299
Chautara	3	1676	1669
Dolalghat	4	792	844
Ghumthang	5	2134	3529
Nepalthok	6	579	873
Sindulgarhi	7	1463	2016
Ramechap	8	1219	755
Melung	9	1573	1557
Charikote	10	1981	1907
Kalimate	11	1417	1739
Udaipur Garhi	12	1390	1726
Kurleghat	13	610	766
Manebhanjyang	14	1615	915
Okhaldunga	15	2103	1502
Pekarnas	16	2134	1553
Paphlu	17	2316	1505
Chaurikharkha	18	2438	1955
Namchebazar	19	3200	792
Aisyalukharha	20	2450	2091
Khotang	21	1295	966
Dwarpa	22	1515	1277
Tribeni	23	143	1574
Chatra	24	115	1919
Dharan Bazar	25	500	2117
Barahkshetra	26	146	2185
Machuaghat	27	158	1205
Mulghat	28	341	759
Dhankuta	29	1524	694
Bhojpur	30	1524	954
Munga	31	1317	1050
Leguaghat	32	400	618
Chainpur	33	1329	1039
Dingla	34	1375	1662
Num	35	1676	2332
Chepua	36	2591	1805
Dumuhan	37	914	1398
Taplejung	38	1768	1536
Angbang	39	1219	979
Mameng Jagat	40	1829	1546
Taplethok	41	1372	1869
Lungthang	42	2438	1880
Walungchung Gola	43	3048	1291
Pangthangdoma	44	2818	1172

16.4.2 Relationship derived by the ratio method

Rainfall along the Himalayas decreases from east to west and as such, the central and western regions of the Himalayas receive less rainfall than the eastern region. In order to make the rainfall parameters constant in the region under study, ratios between the rainfall at each station in the central Himalayas and that of the nearest plain-area station were determined. These ratios were grouped together for different ranges of elevation and are given in Table 16.2.

TABLE 16.2. *Average monsoon rainfall ratios in the central Himalayas bounded by 85° E to 88° E and north of 26° N to 30° N (see Fig. 16.2).*

Height above mean sea level (m)	Mean height (m)	Ratio of average monsoon rainfall at hill-area to that at plain stations
0 to 500	233	1.05
500 to 1000	703	0.73
1000 to 1500	1321	1.00
1500 to 2000	1657	1.33
2000 to 2500	2224	1.65
2500 and above	2857	0.83

Employing the technique of orthogonal polynomial analysis described earlier, a curvilinear relation was derived between the rainfall ratios and the elevation. The relation between rainfall ratios, R, and the mean elevation, X (in km), is

$$R = 1.43 - 1.98X + 1.46X^2 - 0.03X^3 - 0.09X^4. \qquad (16.2)$$

The distribution of rainfall computed by (16.2) was plotted as curve No. II in Fig. 16.3. Curve No. II reveals characteristics similar to those of curve No. I of Fig. 16.3. It is thus seen that the rainfall curves, obtained by using the two different approaches, give almost similar results.

According to Hill (1881) the zone of maximum rainfall in the northwest Himalayas is at an elevation of about 1.2 km, but this study has shown that in the central Himalayas the first zone of maximum rainfall is near the foot of the Himalayas and the second maximum zone is at an elevation of 2.0 to 2.4 km.

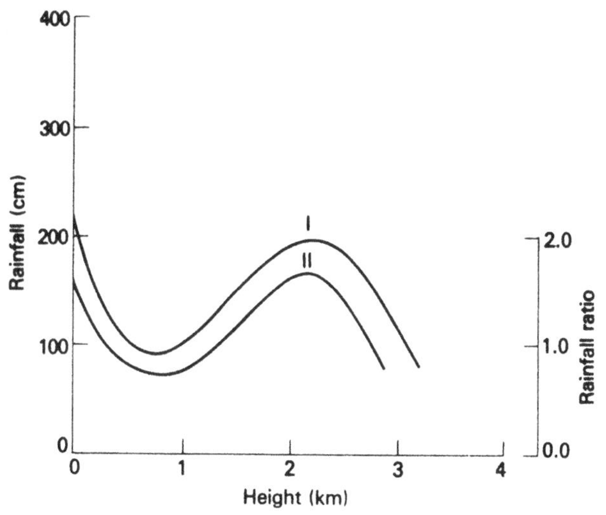

Fig. 16.3. Monsoon rainfall curves obtained by the orthogonal polynomial method. Curve I is based upon actual data; curve II is based upon ratios.

16.5 Schematic representation of rainfall in the central Himalayas

In order to examine the rainfall gradient from south to north in the central Himalayas, two cross-sections, i.e. one in the east and the other in the west, were considered. The mean elevation profiles for these two sections are shown in Fig. 16.4. This figure clearly brings out the three major parallel ranges of the Himalayas; which are the Siwalik, the lesser and the great Himalayas. The mean observed monsoon rain profiles of the stations falling in these two sections were then superposed upon their respective ground profiles. Fig. 16.4 shows the variation between observed monsoon rainfall and elevation in these two sections of the central Himalayas. The main features of the observed rainfall curves in Fig. 16.4 are consistent with the best-fit rainfall–elevation curves shown in Fig. 16.3.

16.6 Conclusions

From the foregoing, the following broad conclusions can be drawn:

(i) Based upon the available rainfall data of stations in the central Himalayas, the correlation coefficient between the elevation and the mean monsoon rainfall was found to be 0.188. This shows that no significant linear relationship exists between elevation and monsoon rainfall. However, it was found that these two parameters can best be related by a polynomial of fourth degree;

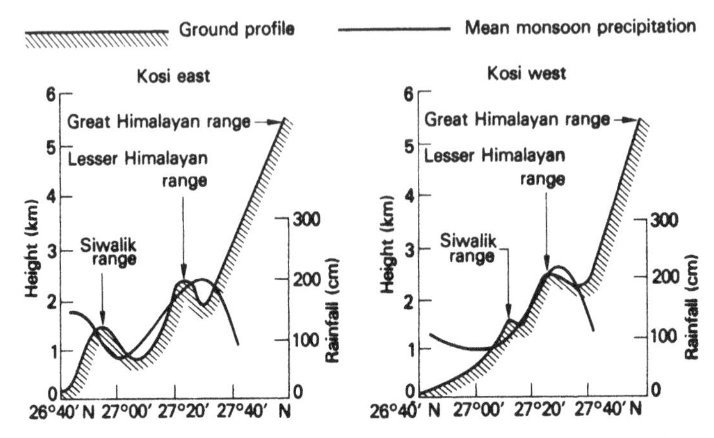

Fig. 16.4. A schematic representation of the mean monsoon precipitation across the Kosi Himalayas.

(ii) The rain profiles obtained from the polynomial equations indicate that one zone of maximum rainfall is located just near the foot of the Himalayas, and the other zone at an elevation of 2.0 to 2.4 km. The first zone of minimum rainfall is located at an elevation of about 0.6 to 0.8 km. Rainfall decreases sharply with the increase of elevation after the second zone of maximum rainfall is reached.

References

Dhar, O. N. (1960) The diurnal variation of rainfall at Barahkshetra and Kathmandu during monsoon months. *Ind. J. Meteor. Geophys.*, **11**, 2, 153–6.

Dhar, O. N. and Bhattacharya, B. K. (1976) *Ind. J. Power and River Valley Development*, **26**, 6, 179–86.

Dhar, O. N. and Narayanan, J. (1965) A study of precipitation distribution in the neighbourhood of Mt. Everest. *Ind. J. Meteor. Geophys.*, **16**, 2, 229–40.

Hill, S. A. (1881) Meteorology of northwest Himalayas. *Memoirs of the India Meteorological Department*, **1**, 377–426.

Linsley, R. K., Kholer, M. A. and Paulhus, J. R. H. (1949) *Applied Hydrology*, McGraw-Hill, New York, pp. 92–5.

Nayava, J. L. (1974) Heavy monsoon rainfall in Nepal. *Weather*, **29**, 12, 443–50.

Rumley, G. B. (1965) An investigation of the distribution of rainfall with elevation for selected stations in Ecuador. M.S. Thesis, Texas A & M University, USA.

17

Use of the equation of continuity of water vapour for computation of average precipitation over peninsular India during the summer monsoon

S. N. BAVADEKAR AND D. A. MOOLEY

Water vapour fluxes, computed across different walls of a triangular volume of peninsular India, bounded by Trivandrum, Bombay and Nagpur, were used to compute the net water vapour flux convergence on a monthly mean basis for the months June to September for the years 1967 to 1972. The precipitation rates over the region were computed by using the flux convergence values and the equation of continuity for water, and were then compared with the actual rainfall. The agreement between the computed precipitation and actual rainfall was found to be fairly close.

17.1 Introduction

Saha and Bavadekar (1973) computed the water vapour fluxes across different walls of a rectangular volume over the Arabian Sea during summer monsoon months of 1963 and 1964 and suggested that the major influx of water vapour into this rectangular volume was associated with the cross-equatorial flow (i.e. across the wall 42° E to 75° E) and the major outflow was across the section wall parallel to the west coast of India (0° to 26° N).

The onshore fluxes over the west coast of India were further studied by Saha and Bavadekar (1977) for the nine years 1964 to 1972 for the summer monsoon months and were correlated with the rainfall along the west coast of India and also over peninsular India bounded by Trivandrum, Bombay and Nagpur. The correlation coefficients were 0.87 for the west coast section (Trivandrum to Bombay) and 0.85 for the peninsular region, and these were statistically significant at the 1% level. In this chapter, precipitation values computed by the water vapour budget method for peninsular India are compared with the actual rainfall over the region.

17.2 Computation of water vapour fluxes

The water vapour fluxes are computed across the section walls of the triangular region (Fig. 17.1) of peninsular India. The expression for computing the net flux, F_n normal to a section wall is

$$F_n = \frac{1}{g}\left\{\sum_{p_t}^{p_b}\sum_{0}^{L}[(\bar{u}\bar{q}+\overline{u'q'})\cos\beta+(\bar{v}\bar{q}+\overline{v'q'})\sin\beta]\Delta L\,\Delta p\right\}, \quad (17.1)$$

where:

g	is the acceleration due to gravity;
p_b, p_t	are the pressures at the bottom and the top of the section wall respectively;
L	is the length of the section;
ΔL	is a length segment;
Δp	is a pressure interval;
u, v	are the components of the daily wind velocity, u positive towards the east and v positive towards the north;
β	is the angle of inclination of the section to the meridian.

The bar and prime denote time mean (i.e. mean for the month) and eddy components respectively.

17.3 Data and analysis

The positions of the radiosonde and pilot balloon stations used in the computations are shown in Fig. 17.1. For the particular section wall under consideration for the computation of the net flux by (17.1), the upper boundary was fixed at 450 mb. The pressure interval Δp was taken as 50 mb. The lower boundary was fixed at 1000 mb and the portion intercepted by orography on the wall was excluded from the computation of fluxes. The radiosonde and pilot balloon stations, the length segments

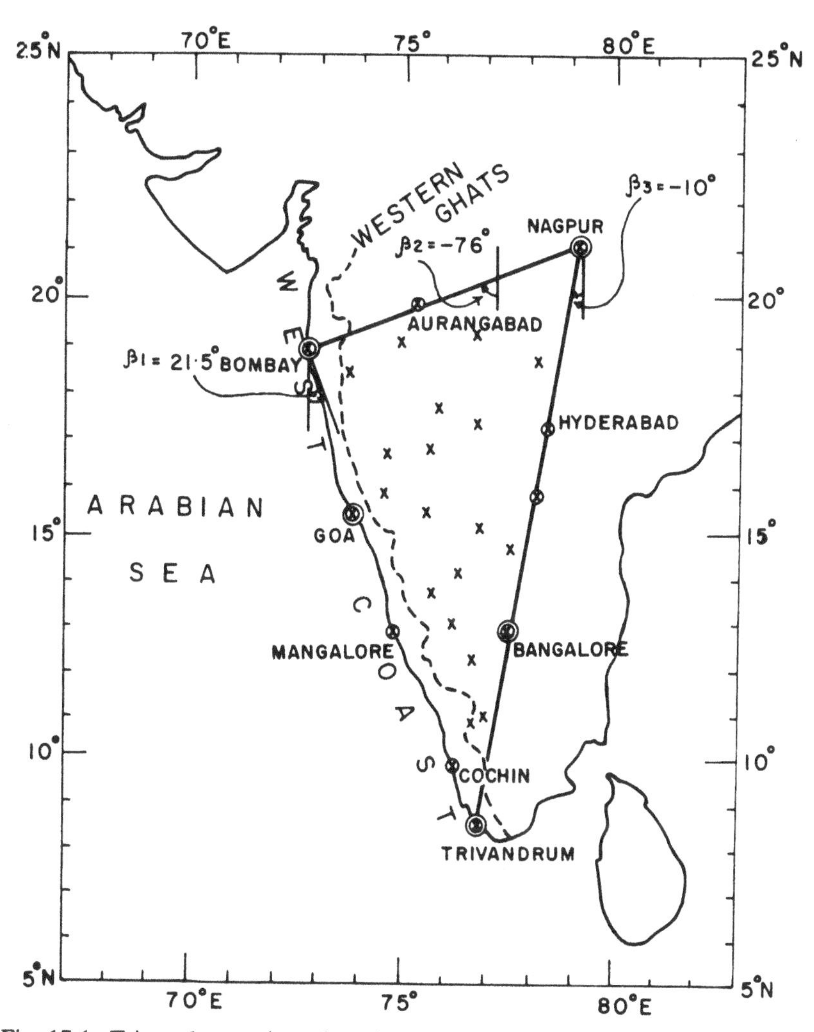

Fig. 17.1. Triangular portion of peninsular India showing the positions of the radiosonde, pilot balloon and rainfall stations (Bavadekar and Mooley, 1978).

used in the computation and the station heights etc. for the three section walls are tabulated in Table 17.1. The procedure for analysis is similar to that given by Saha and Bavadekar (1977).

17.4 Computation of the precipitation

The water vapour fluxes computed for the monsoon months of different years are tabulated in Table 17.2 for the three section walls. The average

TABLE 17.1. *Lengths of the segments used in three sections for the computation of water vapour fluxes.* RS *denotes a radiosonde station*; PB *denotes a pilot balloon station.*

Section	Stations	Station height above mean sea level (m)	RS/PB	Length of segment, ΔL (km)	
Trivandrum–Bombay	Trivandrum	64	RS		
	Cochin	03	PB	Trivandrum to 14° N	530
	Mangalore	22	PB		
	Goa	55	RS	14° N to Bombay	600
	Bombay	14	RS		
Bombay–Nagpur	Bombay	14	RS	Bombay to 76° E	330
	Aurangabad	600	PB	76° E to Nagpur	310
	Nagpur	300	RS		
Nagpur–Trivandrum	Nagpur	300	RS	Nagpur to Hyderabad	370
	Hyderabad	600	PB	Hyderabad to Bangalore	450
	Bangalore	950	RS	Bangalore to Trivandrum	450
	Trivandrum	64	RS		

precipitation P per day can be computed by using the equation of continuity for water vapour

$$P = -\frac{1}{g}\int_{p_t}^{p_b} \nabla \cdot (q\mathbf{V})\, dp + E, \tag{17.2}$$

where E is the average rate of evapotranspiration per day and the other symbols have their usual meanings.

The flux convergence represented by the first term on the right-hand side of the above equation is obtained on a monthly mean basis with the help of Table 17.2. If the value of E in (17.2) is known, then the precipitation rate P can be obtained.

17.5 Estimation of the evapotranspiration

The actual evapotranspiration E over the region was not readily available for the required period. The charts for the potential evapotranspiration E_p over the Indian region have been prepared on a monthly basis (Rao *et al.*, 1971). The average daily potential evapotranspiration was estimated

TABLE 17.2. *Water vapour fluxes across the boundary walls (inflow positive; outflow negative). Unit: 10^{10} tonnes per day.*

	West coast section					
	1967	68	69	70	71	72
June	2.57	2.44	2.72	3.47	3.97	1.91
July	4.36	5.14	4.82	5.04	3.88	4.16
August	4.20	3.61	2.22	3.76	2.78	3.32
September	1.40	1.18	1.35	1.66	1.46	0.54
	Bombay–76° E–Nagpur section					
	1967	68	69	70	71	72
June	0.41	0.60	0.50	0.79	0.51	0.49
July	0.50	0.52	0.84	0.43	0.56	0.59
August	*	0.85	0.42	0.70	0.42	0.74
September	0.66	0.55	0.77	0.62	0.26	0.64
	Trivandrum–Bangalore–Hyderabad–Nagpur section					
	1967	68	69	70	71	72
June	−2.74	−2.85	−3.07	−3.67	−4.16	−2.10
July	−4.33	−5.31	−5.35	−5.00	−4.22	−4.59
August	*	−4.37	−2.58	−4.00	−3.02	−3.80
September	−2.08	−1.57	*	−2.18	−1.44	−1.08

* Not computed due to inadequate data

from these charts for the monsoon months. The actual evapotranspiration E is obtained by using the following criteria,

$$E = E_p, \quad \text{if } P > E_p,$$

and

$$E = P - \text{storage rate} - \text{run off}, \quad \text{if } P < E_p.$$

In most of the cases P was found to be greater than E_p (see Table 17.3). Year to year variations of E are disregarded in this study.

17.6 Discussion of the results

From Table 17.2 it is clear that influx into the area takes place across the sections, Trivandrum–Bombay and Bombay–Nagpur. The major contribution is across the Trivandrum–Bombay section. The average daily fluxes in this section also show an increasing trend from June to July. The values drop in August and are lowest in September. This is consistent with the normal progress of the monsoon over the country and is found to be the case in almost every year except for 1971, when the maximum was

TABLE 17.3. *Moisture budget derived from fluxes and actual precipitation. Unit: 10^{10} tonnes per day. F_c is the flux convergence; E is the evapotranspiration; P_c is the computed precipitation; and P is the actual precipitation.*

Month		Year 1967	68	69	70	71	72
June	F_c	0.24	0.19	0.15	0.59	0.32	0.30
	E	0.16	0.16	0.16	0.16	0.16	0.16
	P_c	0.40	0.35	0.31	0.75	0.48	0.46
	P	0.44	0.44	0.41	0.55	0.52	0.41
July	F_c	0.53	0.35	0.31	0.47	0.22	0.16
	E	0.17	0.17	0.17	0.17	0.17	0.17
	P_c	0.70	0.52	0.48	0.64	0.39	0.33
	P	0.66	0.64	0.58	0.48	0.41	0.40
August	F_c	*	0.09	0.06	0.46	0.18	0.26
	E	0.16	0.16	0.16	0.16	0.16	0.16
	P_c	*	0.25	0.22	0.62	0.34	0.42
	P	0.27	0.25	0.27	0.62	0.34	0.20
September	F_c	−0.02	0.16	*	0.10	0.28	0.10
	E	0.16	0.16	0.16	0.16	0.16	0.16
	P_c	0.14	0.32	*	0.26	0.44	0.26
	P	0.24	0.30	0.37	0.27	0.26	0.14

* Not computed due to inadequate data

found in June. The section, Bombay–Nagpur does not show such a systematic behaviour. The outflow given by the negative values for the Trivandrum–Nagpur section also shows a systematic behaviour similar to that of the Trivandrum–Bombay section.

From the comparison of total influx and precipitation it is found that, on average, about 13% of the total influx over the triangular volume is converted into precipitation. The contribution expressed as a percentage of the net flux convergence is 8%, while 5% is evapotranspiration. These figures indicate that abundant moisture is available throughout the summer monsoon period but only a small fraction of it is converted into precipitation over peninsular India.

Fig. 17.2 shows how the water vapour fluxes are distributed in the vertical for the section. Trivandrum–Bombay. On an average, the maximum contribution is from the 900 to 850 mb layer (curve II). The cumulative contribution is shown by curve I. The layer up to 700 mb contributes 86%, with only 14% above. The latter contribution, though relatively insignificant, is comparable to the magnitude of the precipitation.

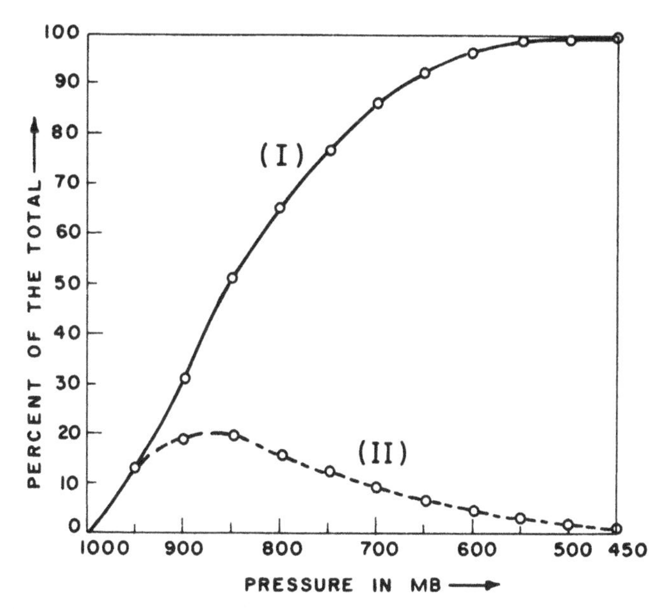

Fig. 17.2. The vertical distribution of water vapour flux for the Trivandrum–Bombay section. Curve I shows the cumulative contribution of water vapour fluxes in layers of 50 mb from 1000 mb to 450 mb. Curve II shows the contribution to the water vapour flux of individual 50 mb layers from 1000 to 450 mb.

17.7 Conclusion

The budget method of computing the precipitation has some limitations. The limitations arise from an inadequate network of radiosonde stations over the region under consideration and also from lack of knowledge of the actual evapotranspiration. The flux convergence term is the small difference of two large terms of total inflow and outflow and requires proper and careful analysis of wind and mixing ratios. The computed precipitation rates, in spite of these limitations, are fairly close to the estimated total rainfall.

The authors grateful thanks are due to Mrs M. S. Naik and Shri R. M. Khaladkar who assisted in the processing of data and Shri A. Girijavallabhan who typed the manuscript.

References

Bavadekar, S. N. and Mookey, D. A. (1978) *Tellus*, **30**, No. 6.

Rao, K. N., George, C. J. and Ramasastri, K. S. (1971) Potential evapotranspiration (PE) over India. Scientific Report No. 136, India Meteorological Department.

Saha, K. R. and Bavadekar, S. N. (1973) Water vapour budget and precipitation over the Arabian Sea during the northern summer. *Quart. J. Roy. Meteor. Soc.*, **99**, 273–8.

Saha, K. R. and Bavadekar, S. N. (1977) Moisture flux across the west coast of India and rainfall during the southwest monsoon. *Quart. J. Roy. Meteor. Soc.*, **103**, 370–4.

18

Determination of cloud cluster properties from MONSOON-77 data

Y. RAMANATHAN

The spectral diagnostic method based on the Arakawa–Schubert parameterization theory is applied to a case study of the active monsoon at the time of onset. The estimates of vertical mass flux, the vertical eddy transport of heat, the excess of temperature, the apparent heat source and the apparent moisture sink are found to be much higher than those for disturbed conditions in the equatorial Pacific. Conditional instability of the second kind (CISK) appears to be operative in the presence of a large vertical mass flux associated with shallow clouds.

18.1 Introduction

Precipitation during the Indian southwest monsoon is in the form of thundershowers, showers, or rain. The presence of cumulus and cumulonimbus clouds over the Arabian Sea during the active monsoon has been noted by several authors (Deshpande, 1964; Bunker and Chaffe, 1969). General-circulation studies (Manabe *et al.*, 1970) have verified the importance for the intensification of the monsoon circulation of the release of latent heat by convection in the ascending air. In recent years the interaction between cumulus-scale convection and large-scale motion has been the subject of several theoretical and diagnostic studies. A spectral diagnostic method has been developed on the basis of the parameterization theory formulated by Arakawa and Schubert (1974) and used by Cho and Ogura (1974) for the study of composite data of easterly waves, by Nitta (1975) for the Barbados Oceanographic and

Meteorological Experiment (BOMEX) phase III data, and by Yanai *et al.* (1976) for the intertropical convergence zone (ITCZ) Marshall Islands data. The above synoptic situations do not involve the same types of disturbances as an active monsoon. In this chapter, the spectral diagnostic method will be used in a case study over the Arabian Sea at the time of monsoon onset. § 18.2 gives a brief outline of the method and the assumptions involved, § 18.3 gives details of the computations, and the results are discussed in § 18.4.

Explanations of the symbols used are given in the list of symbols at the end of this chapter.

18.2 The computational method

The relevant equations are:

$$Q_1 \equiv \frac{\partial \bar{s}}{\partial t} + \overline{\nabla \cdot (sv)} + \frac{\partial}{\partial p}\bar{s}\bar{\omega} = Q_R + L(c-e) - \frac{\partial}{\partial p}\overline{s'\omega'}, \tag{18.1}$$

$$Q_2 \equiv -L\left(\frac{\partial \bar{q}}{\partial t} + \overline{\nabla \cdot (qv)} + \frac{\partial}{\partial p}\bar{q}\bar{\omega}\right) = L(c-e) + L\frac{\partial}{\partial p}\overline{q'\omega'}, \tag{18.2}$$

$$Q \equiv Q_1 - Q_2 - Q_R = -\frac{\partial}{\partial p}\overline{(s'+Lq')\omega'} = -\frac{\partial}{\partial p}\overline{h'\omega'}. \tag{18.3}$$

The flux terms on the right-hand sides represent the contributions of the clouds.

To evaluate these contributions the following assumptions are made:

(i) The cloud ensembles occupy a fraction $\sigma(\ll 1)$ of the total area of the clouds and environment.

(ii) $$\bar{M} = M_c + \tilde{M}. \tag{18.4}$$

(iii) $$\bar{s} \approx \tilde{s}, \quad \bar{q} \approx \tilde{q}, \quad \bar{h} \approx \tilde{h}. \tag{18.5}$$

(iv) $$-\overline{s'\omega'} = \sum M_c(s_c - \tilde{s}), \tag{18.6}$$

$$-\overline{q'\omega'} = \sum M_c(q_c - \tilde{q}), \tag{18.7}$$

where $\sum$ indicates summation over all cloud ensembles.

(v) The cloud ensemble average for any parameter ϕ is defined as

$$\hat{\phi} = \sum \phi_c M_c / \sum M_c. \tag{18.8}$$

(vi) Clouds are saturated and detrain only at the level at which they lose their buoyancy,

$$h_c = \tilde{h}^*, \qquad \tilde{q}_c = q^*, \qquad T_c = \tilde{T}^*. \tag{18.9}$$

(vii) No overshooting by cumulonimbus clouds is incorporated.

(viii) Downdrafts, multi-level detrainment and ice-phase transitions are ignored since quantitative knowledge of them is inadequate.

(ix) Evaporation of cloud water occurs only on detrainment, i.e.

$$e = \delta \hat{l}. \tag{18.10}$$

(x) Rainout is proportional to the liquid water content of the cloud at that level,

$$r = k\hat{l}. \tag{18.11}$$

(xi) Each cloud type is characterized by a constant fractional entrainment rate λ. The detrainment level $z_D(\lambda)$ is inversely proportional to λ (i.e. a cloud with a large entrainment parameter detrains at a lower level because the mixing with the drier environmental air causes the cloud to lose its buoyancy earlier).

(xii) Cloudbase for all clouds is assumed to be 950 mb.

(xiii) If $m_B(\lambda)$ is the cloudbase mass flux distribution function and $\eta(p, \lambda)$ is the normalized mass flux at various p-levels for λ-type clouds, the total cloud mass flux is given by

$$M_c(p) = \int_0^{\lambda_D(p)} m_B(\lambda)\eta(p, \lambda)\, d\lambda, \tag{18.12}$$

and the detraining mass flux of clouds having parameters $\lambda_D(p)$ and $\lambda_D(p) + (d\lambda_D/dp)\, dp$ is given by

$$\delta(p) = m_B[\lambda_D(p)]\eta[p, \lambda_D(p)]\frac{d\lambda_D(p)}{dp}, \tag{18.13}$$

$$\varepsilon(p) = \int_0^{\lambda_D(z)} \lambda m_B(\lambda)\eta(z, \lambda)\, d\lambda. \tag{18.14}$$

With the above assumptions for evaluating cloud-flux terms, (18.3) can be written (see Arakawa and Schubert, 1974)

$$Q \equiv Q_1 - Q_2 - Q_R = (\tilde{h}^* - \tilde{h})m_B[\lambda_D(p)]\eta[p, \lambda_D(p)]\frac{d\lambda_D(p)}{dp} - \frac{\partial \tilde{h}}{\partial p}\int_0^{\lambda_D(p)} m_B(\lambda)\eta(p, \lambda)\, d\lambda. \tag{18.15}$$

This is a Volterra integral equation of the second kind, and knowing λ, η, $\tilde{h}$ and $\tilde{h}^*$, $m_B(\lambda)$ is obtained by inverting (18.15) ($\tilde{h}$, $\tilde{h}^*$ can be evaluated from large-scale observations). The procedure for obtaining λ and η is to solve the following equations:

(i)
$$\frac{1}{\eta(p,\lambda)}\frac{\partial}{\partial p}\eta(p,\lambda)=-\frac{\lambda}{\rho g}, \tag{18.16}$$

giving the normalized sub-ensemble mass flux;

(ii)
$$h_c(p,\lambda)=\frac{1}{\eta(p,\lambda)}\left[h_B-\lambda\int_{p_B}^{p}\eta(p',\lambda)\tilde{h}(p')\frac{dp'}{\rho g}\right], \tag{18.17}$$

giving the moist static energy $h_c(p,\lambda)$ of the sub-ensemble;

(iii)
$$h_c(p,\lambda)=\tilde{h}^*(p_D). \tag{18.18}$$

expressing the condition of vanishing buoyancy.

Starting from a guessed value of $\lambda_D(p_D)$ for a detraining level, (18.16) to (18.18) are solved by Newton's iterative method to obtain explicit forms of $\eta(p,\lambda)$ for all possible p_D values.

18.3 The data used in the study

During the MONSOON-77 experiment four USSR research vessels were stationary in a polygon formation at a diagonal distance of 4° around the central position 66° E, 12.5° N (Fig. 18.1). The first phase of the experiment was from 7 June to 19 June 1977 immediately after the onset of the monsoon. The case study presented is for 7 June 1977 at 0000 hrs GMT which was an active monsoon situation. The low-level wind field is shown in Fig. 18.2. The easternmost ship reported complete cloud cover and the other three ships deep cumulus and nimbostratus. It was raining and the sky was overcast. The sea–air temperature difference was 5.1° C at the easternmost location and was about 1 °C for the other locations (Fig. 18.1). Pictures from the US satellite indicate complete cloud cover over the area. An average of four radiosonde/radiowind ascents was made each day, with a radiometersonde ascent from one ship.

The variables s, q, u and v were fitted by a plane-surface function of the form $D=ax+by+c$ by the least-squares method for each isobaric level. Data were interpolated at 18 vertical levels at 50 mb intervals from 950 mb to 100 mb. The horizontal gradients of s and q and the horizontal mass divergence were computed from the expression for the plane-surface function.

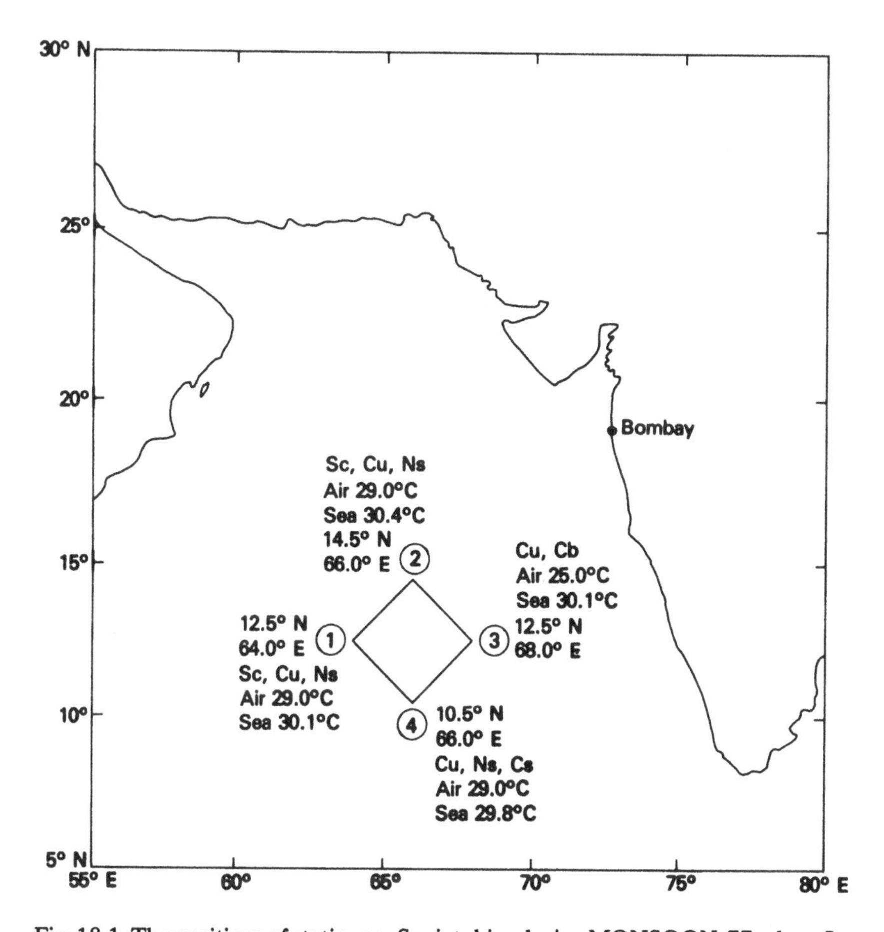

Fig. 18.1. The positions of stationary Soviet ships during MONSOON-77, phase I. The air and sea temperatures, and the nature of the cloud cover are given for 0000 GMT, 7 June 1977 at each location. The cloud types are denoted by Sc (stratocumulus), Cu (cumulus), Ns (nimbostratus), Cs (cirrostratus) and Cb (cumulonimbus).

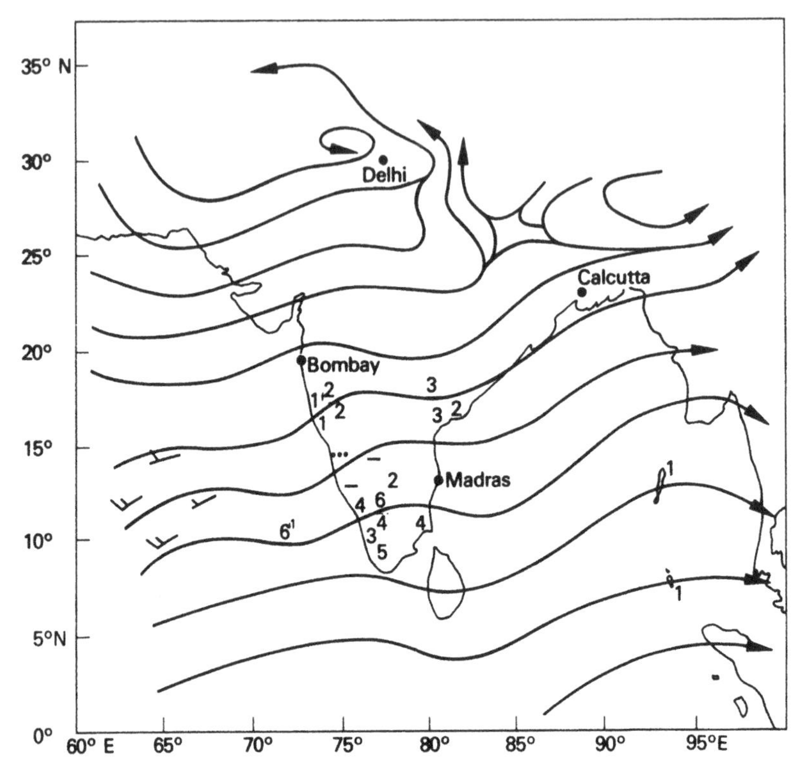

Fig. 18.2. The low-level wind field (0.9 km) at 0000 GMT, 7 June 1977. The numbers give rainfall (mm per day); – denotes no rain and · · · a trace.

18.4 Results and discussion

The computations are presented in Figs. 18.3 to 18.12:

(i) The divergence and ω values in Fig. 18.3 are very much higher than those computed for other situations. In the study by Yanai *et al.* (1976) on ITCZ data, which he categorized as a disturbed situation, the maximum divergence and vertical velocity were $-5\times10^{-6}\,\mathrm{s}^{-1}$ and $-1.5\,\mathrm{cm\,s}^{-1}$ respectively compared with $-2.5\times10^{-5}\,\mathrm{s}^{-1}$ and $-6.8\,\mathrm{cm\,s}^{-1}$ in the monsoon case. The level of non-divergence is at 400 mb, somewhat lower than in Yanai *et al.* (1976), where it was 300 mb.

(ii) Fig. 18.4 gives profiles of s, $\tilde{h}_0$ and $\tilde{h}^*$. The environment is almost saturated with a moist adiabatic lapse rate. The profiles indicate near

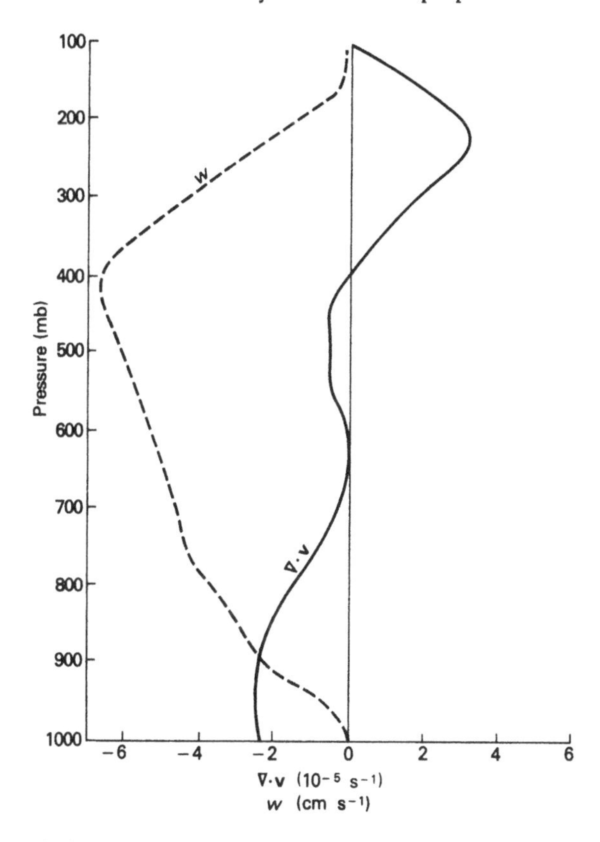

Fig. 18.3. Variation with pressure of divergence, $\nabla \cdot \mathbf{v}$, and vertical velocity, w.

neutral stability conditions up to 300 mb and are much different from BOMEX data (Nitta, 1975) where $\tilde{h}$ shows inversions. The minimum occurs at around 450 mb in contrast to a minimum at 625 mb in Yanai *et al.* (1976).

(iii) The main contribution for the apparent heat source (Q_1) (Fig. 18.5) is from the vertical advection which produced a warming throughout the troposphere with a maximum of about 45 °C per day at 400 mb due to the downward environmental motion compensating the upward mass flux in the clouds. The maximum heating in Yanai (1976) was 6.4 °C per day at 475 mb.

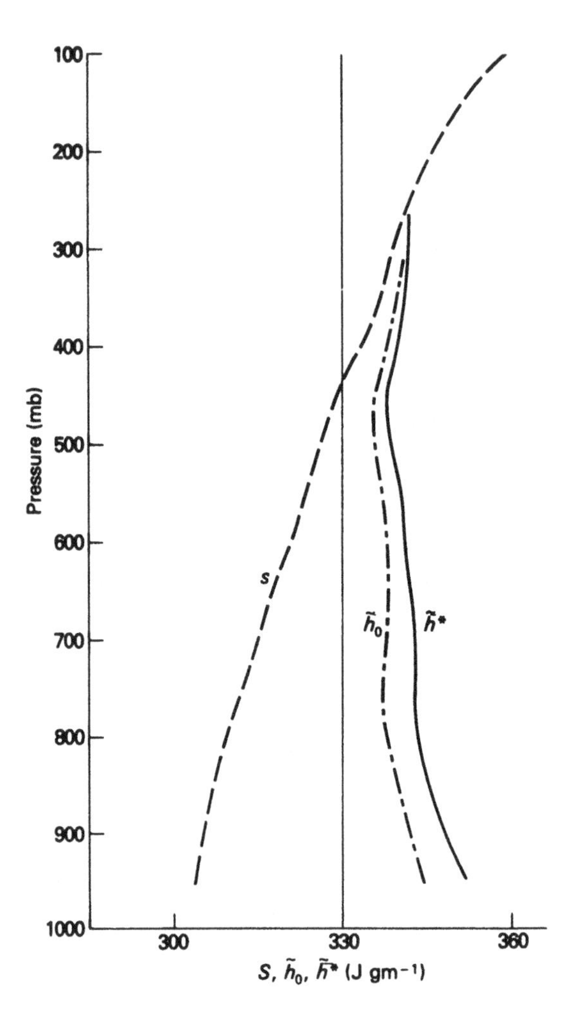

Fig. 18.4. Variation with pressure of s, h_0 and $\tilde{h}^*$.

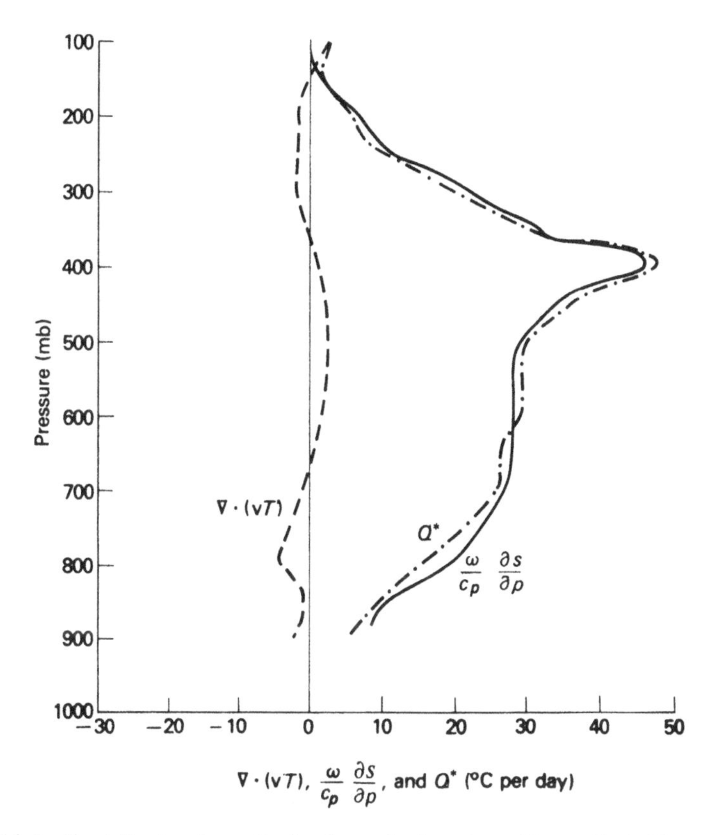

Fig. 18.5. Contribution from the horizontal advection of heat, $\nabla \cdot (\boldsymbol{v}T)$, and the vertical advection of heat, $(\omega/c_p)\,(\partial s/\partial p)$, to the temperature change at different levels. Q^* represents the sum of the two contributions.

(iv) There are two maxima in the Q_2 profile (Fig. 18.6), one around 800 mb (30 °C per day), and another around 500 mb (34 °C per day). The profiles in Yanai *et al.* (1976) show the same structure with a maximum sink of 5.8 °C per day around 775 mb and 4.8 °C per day around 500 mb. Horizontal advection produces a small drying effect but the vertical advection gives an enormous moistening effect at all levels.

(v) Fig. 18.7 gives the radiational heating rate (Q_R) from the radiometersonde ascent. This is much different from the climatological profiles used in Yanai *et al.* (1976) and others which have a uniform radiational cooling of 2° to 0 °C per day. There is a heating from 850 mb to 500 mb which is perhaps the top of cloud, but above this there is strong cooling of up to 3.5 °C per day; $|Q_1 - Q_2|$ is much larger than

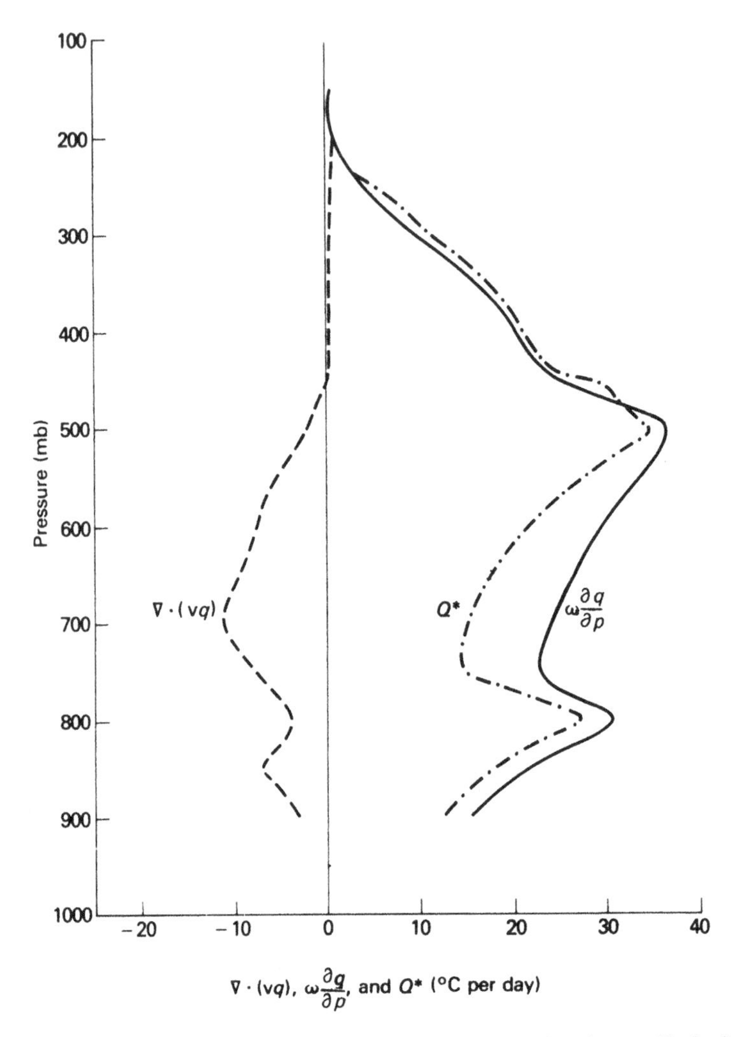

Fig. 18.6. Contribution from the horizontal advection of moisture, $\nabla \cdot (vq)$, and the vertical advection of moisture, $\omega \partial q/\partial p$, to the temperature change at different levels. Q^* represents the sum of the two contributions.

$|Q_R|$ in contrast to Yanai *et al.* (1976). Although convective heat release is at least an order of magnitude greater than radiational cooling, the cloud mass and energy fluxes are still extremely sensitive to $|Q_R|$ as shown by Yanai *et al.* (1976).

(vi) In all, eight types of clouds, detraining respectively in the thin layers centred around 900, 800, 700, 600, 500, 400, 300 and 200 mb, are

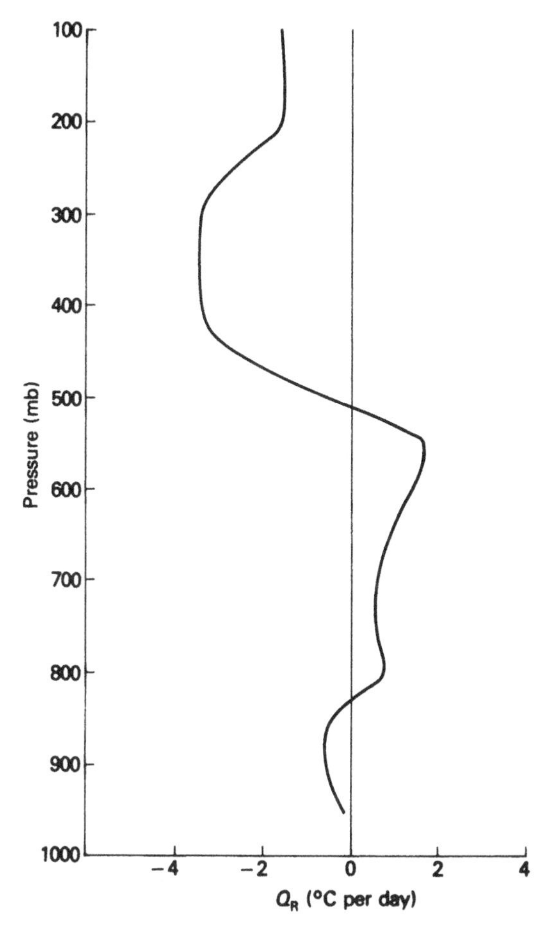

Fig. 18.7. Variations of the radiational heating, Q_R, with pressure.

considered. The λ values are from 28% km^{-1} for shallow clouds to 5% km^{-1} for deep clouds, if it is assumed, following Simpson and Wiggert (1969), that their radii vary from 1000 m to 4000 m (unlike the trade wind regime where radii vary from 70 m to 1000 m). As found in other regions, a large vertical mass flux is associated with shallow clouds (Fig. 18.8). The detrainment of moisture by shallow convection increases the moisture content of the lower troposphere. Also, cooling takes place by evaporation of detrained liquid water. These increase the buoyancy of the cumulus ensemble. There is also a significant contribution of mass flux at cloud base for clouds detraining at 500 mb and 300 mb (Fig. 18.8).

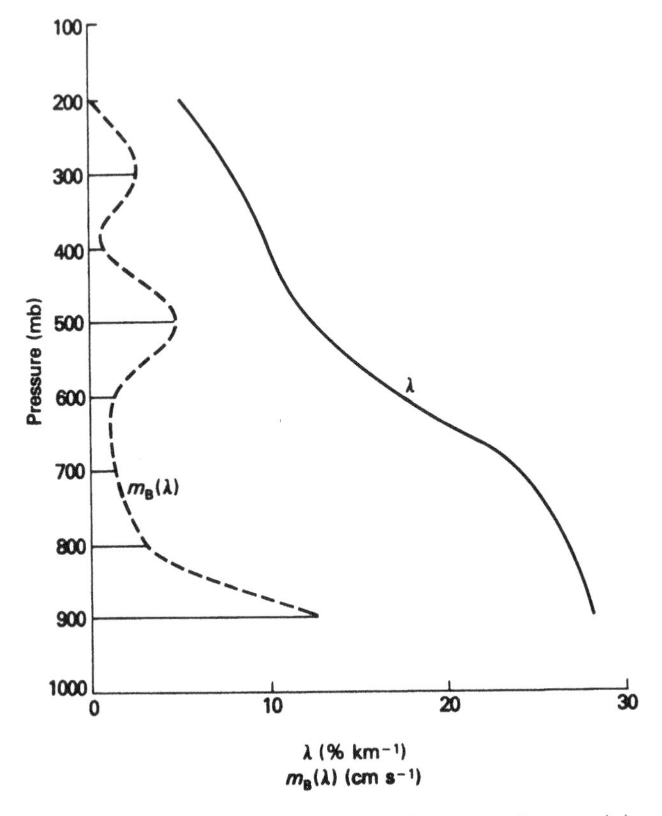

Fig. 18.8. Variation with pressure of the vertical mass flux, $m_B(\lambda)$, and the fractional entrainment rate, λ.

(vii) Fig. 18.9 shows the vertical eddy heat flux of total heat, which is a measure of cumulus activity. The values are several orders higher than those described by Yanai *et al.* (1976).

(viii) The vertical mass flux M_c is found to be very large near the cloudbase (Fig. 18.10). The cloudbase mass flux and vertical mass flux suggest a bimodel distribution observed in cloud clusters in Yanai *et al.* (1976) and Cho and Ogura (1974). However there are no deep clouds but only shallow and middle-level clouds with the second peak at about 600 mb.

(ix) Fig. 18.11 shows the heating of the environment due to cumulus convection. The maximum heating is around 500 mb as in the case described by Yanai *et al.* (1976). However the amount is 11.0 °C as against 3.1 °C per day in Yanai *et al.* (1976). The excess mixing ratio is

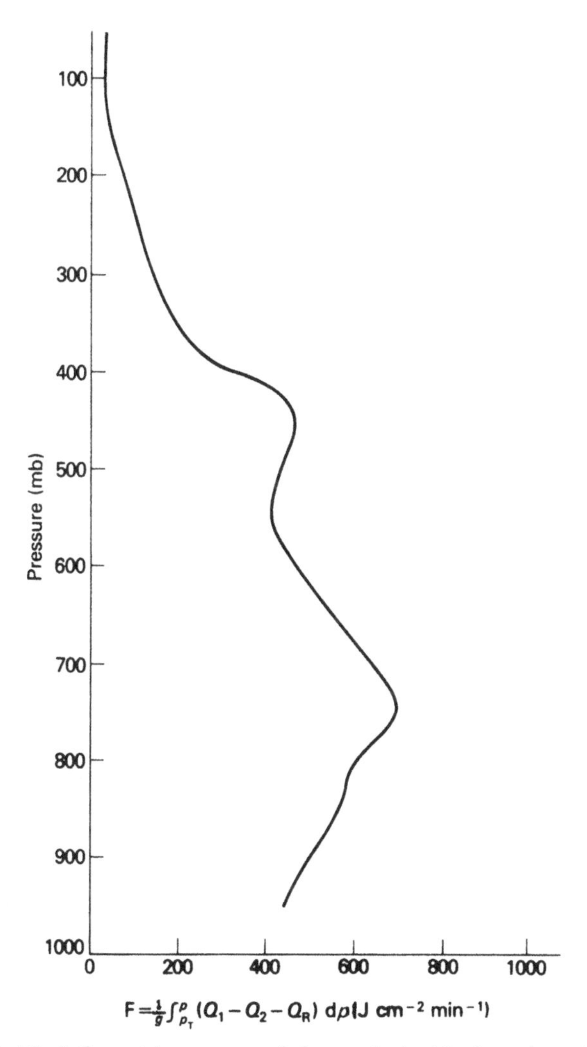

Fig. 18.9. Variation with pressure of the vertical eddy flux of total heat, *F*.

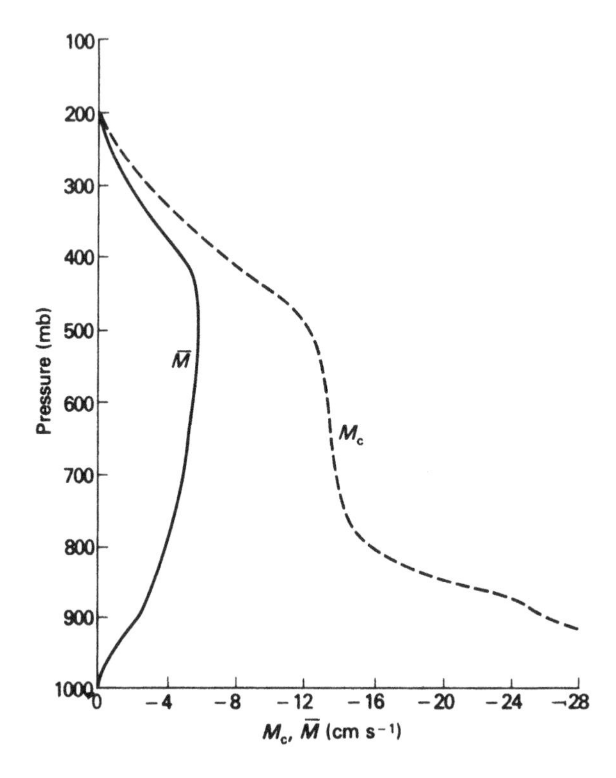

Fig. 18.10. Variation with pressure of the upwards mass flux, M_c, and the total mass flux, $\bar{M}$.

5 gm kg^{-1} and remains practically the same throughout the troposphere. This distribution resembles the disturbed situation during BOMEX (Nitta, 1975) and is different from the case described by Yanai *et al.* (1976) where a maximum of 5.4 gm kg^{-1} around 750 mb was calculated.

(x) The detrainment profile in Fig. 18.12 reveals the bimodal structure of shallow- and middle-level clouds. Entrainment values are much less than detrainment values.

18.5 Conclusion

The properties of cloud clusters in a case of disturbed conditions at the time of monsoon onset have been determined by the spectral diagnostic method. The vertical mass flux and the excess temperature were found to be much higher than the estimates obtained during disturbances in the

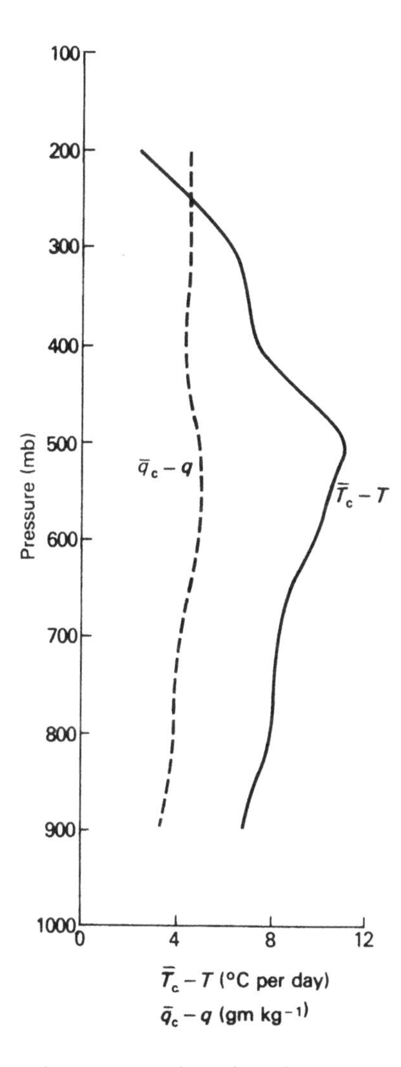

Fig. 18.11. Variation with pressure of heating of the environment, $\tilde{T}_c - T$, and the mixing ratio of the environment, $\tilde{q}_c - q$, due to cumulus convection.

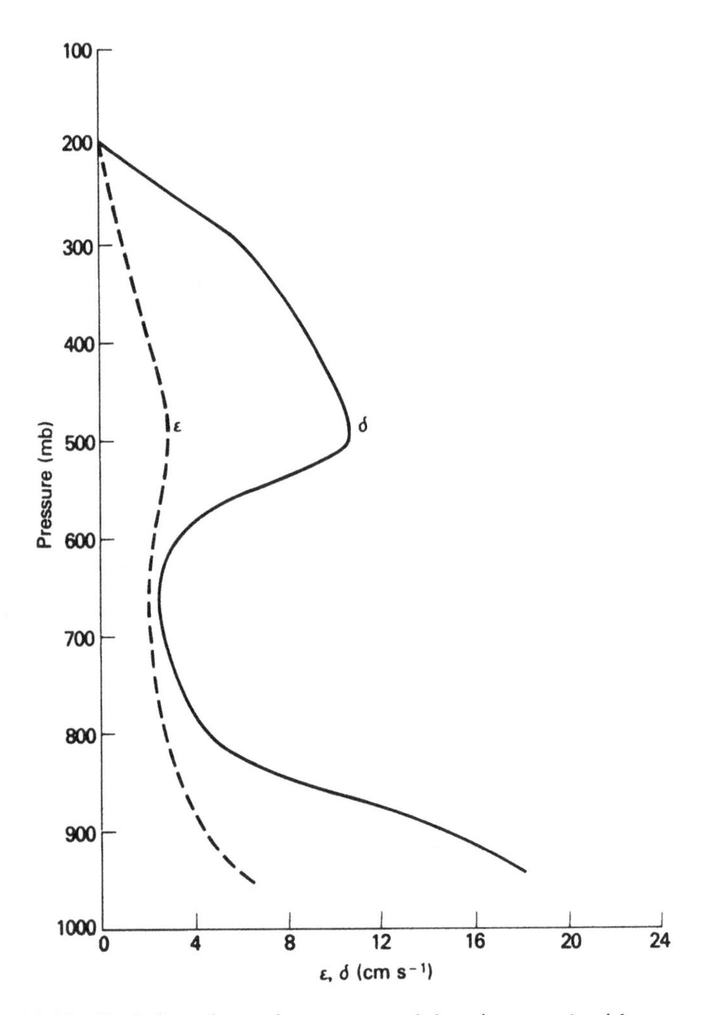

Fig. 18.12. Variation of entrainment, ε, and detrainment, δ, with pressure.

equatorial Pacific. CISK-type instability appears to be operative, a large vertical mass flux associated with shallow convection providing the triggering for deep convection.

The author thanks Dr P. K. Das for his interest and encouragement. Thanks are also due to Mr M. M. Srivastava for assistance in making the computations and to Mrs S. L. Arora for typing the manuscript.

List of symbols

c	rate of condensation of water vapour
e	amount of liquid water evaporated in air detrained from clouds
h	moist static energy $(c_pT + gz + Lq)$
h_c	moist static energy in clouds
$\tilde{h}^*$	saturation moist energy of the environment
k	ratio of rainout to liquid water content of cloud at a given level
$\hat{l}$	average mixing ratio of liquid water for clouds detraining at a particular level
L	latent heat of evaporation
$m_B(\lambda)$	vertical mass flux at cloudbase due to λ-type clouds
M_c	total vertical mass flux in clouds
M_i	vertical mass flux associated with the ith cloud
$\tilde{M}$	residual mass flux in the environment
p	pressure
q	mixing ratio of water vapour
q_c	mixing ratio in clouds
Q_1	large-scale apparent heat source
Q_2	large-scale apparent moisture sink
Q_R	radiational heating
r	rainout from condensed liquid water
s	dry static energy $(c_pT + gz)$
s_c	dry static energy due to clouds
T	temperature
u	eastward wind component
v	northward wind component
$\boldsymbol{v}$	horizontal wind vector ($\bar{}$ denotes an isobaric average and $'$ a departure)
$\nabla \cdot \boldsymbol{v}$	divergence
$z_D(\lambda)$	the detrainment level
δ	detrainment of cloud mass
ε	entrainment of cloud mass

λ	fractional entrainment rate
σ	areal fraction occupied by clouds
η	normalised mass flux distribution functions for p-levels
ω	vertical motion (dp/dt)
ρ	air density

References

Arakawa, A. and Schubert, W. H. (1974) Interaction of a cumulus cloud ensemble with the large-scale environment, Part I. *J. Atmos. Sci.*, **31**, 674–701.

Bunker, A. F. and Chaffe, M. (1969) Tropical Indian Ocean cloud. International Indian Ocean Experiment Meteorological Monograph 4, University of Hawaii, Honolulu.

Cho, H. R. and Ogura, Y. (1974) A relationship between cloud activity and the low-level convergence as observed in Reed–Recker's composite easterly waves. *J. Atmos. Sci.*, **31**, 2058–65.

Deshpande, D. V. (1964) Heights of cumulonimbus clouds over India during the south-west monsoon season. *Ind. J. Meteor. Geophys.* **7**, 225–40.

Manabe, S., Smagorinsky, J., Holloway, Jr. J. L. and Stone, H. M. (1970) Simulated climatology of a general circulation model with a hydrological cycle: III. Effects of increased horizontal computational resolution. *Mon. Wea. Rev.*, **98**, 175–212.

Nitta, T. (1975) Observational determination of cloud mass flux distributions. *J. Atmos. Sci.*, **32**, 73–91.

Simpson, J. and Wiggert, V. (1969) Models of precipitating cumulus towers. *Mon. Wea. Rev.*, **97**, 471–89.

Yanai, M., Chu, J. H., Stark, T. E. and Nitta, T. (1976) Response of deep and shallow tropical maritime cumuli to large-scale processes. *J. Atmos. Sci.*, **33**, 976–91.

19

Analysis of superpressure balloon trajectories and conventional observations over the Indian Ocean during different phases of the 1975 southwest monsoon

D. CADET AND P. OLORY-TOGBÉ

The trajectories of 45 superpressure balloons launched in the tropical boundary layer over the Indian Ocean during the 1975 summer monsoon are analysed in conjunction with conventional meteorological observations which consist mainly of ship reports. An assimilation of these randomly distributed data is performed to compute mean gridded fields (wind, pressure, air and sea-surface temperature) during four periods defined according to the level of monsoon activity over the Indian subcontinent. The main features of the southwest monsoon over the Indian Ocean are presented and discussed. More particularly, the differences in meteorological fields between the different periods are emphasized in order to determine the association between mean meteorological conditions over the Indian Ocean and monsoon activity over the subcontinent.

19.1 Introduction

Because of the lack of data over vast areas of the Indian Ocean, the details of the monsoonal airflow at low levels, where the development of the southwest monsoon takes place, still remains largely unknown.

All previous studies over the ocean have been related to monthly mean fields or local measurements, while transient phenomena of interest occurring with shorter periods over the Indian monsoon region as a whole are still largely uninvestigated. For example, one of the most important phenomena, which is the alternating of break and strong activity of the monsoon over the Indian subcontinent, takes place with a period of about two weeks. The study of the relationship between the variability of monsoon rainfall and that of the fields of other meteorological variables requires a knowledge of the variability of these fields averaged over a few days.

During the 1975 summer monsoon, 45 superpressure balloons were released from the Seychelles Islands (Cadet and Ovarlez, 1976) for flights within the tropical boundary layer. Quasi-Lagrangian low-level trajectories at the height of about 1000 m for periods of up to 10 days were obtained and gave valuable information about the mean monsoon circulation flow over the Indian Ocean as well as its transient features. At the same time, mean fields during different phases of the monsoon activity were derived from collected meteorological information (essentially ship reports). From the totality of these observations (Lagrangian and Eulerian), mean features of the monsoonal flow are deduced together with variations in mean fields during the different phases of the monsoon over India.

19.2 The data

The data consist of two sets: first, the trajectories obtained in 1975 (Fig. 19.1) which represent height-averaged quasi-Lagrangian paths of air parcels in the low-level southwesterlies over the Indian Ocean; Secondly, conventional Eulerian observations from meteorological stations and ship reports. Meteorological stations are scarce over the Indian Ocean. Ship reports are more frequently available along shipping routes but vary markedly from day to day. No data at all are available over a great part of the ocean. Thus features of the synoptic field may be undetectable using such conventional data. Nevertheless, on certain days we may get almost 300 surface winds over the ocean (Fig. 19.2), and these data give a general idea of the synoptic fields during the balloon experiment.

Each day, all the meteorological parameters, i.e. wind components, sea-surface and air temperatures and pressure, measured at different synoptic hours, are used to derive daily composite analyses. The grid adopted is a 2.5° latitude by 2.5° longitude mesh from 35° E to 90° E and from 25° S to 25° N. To derive these daily composite maps, a simple

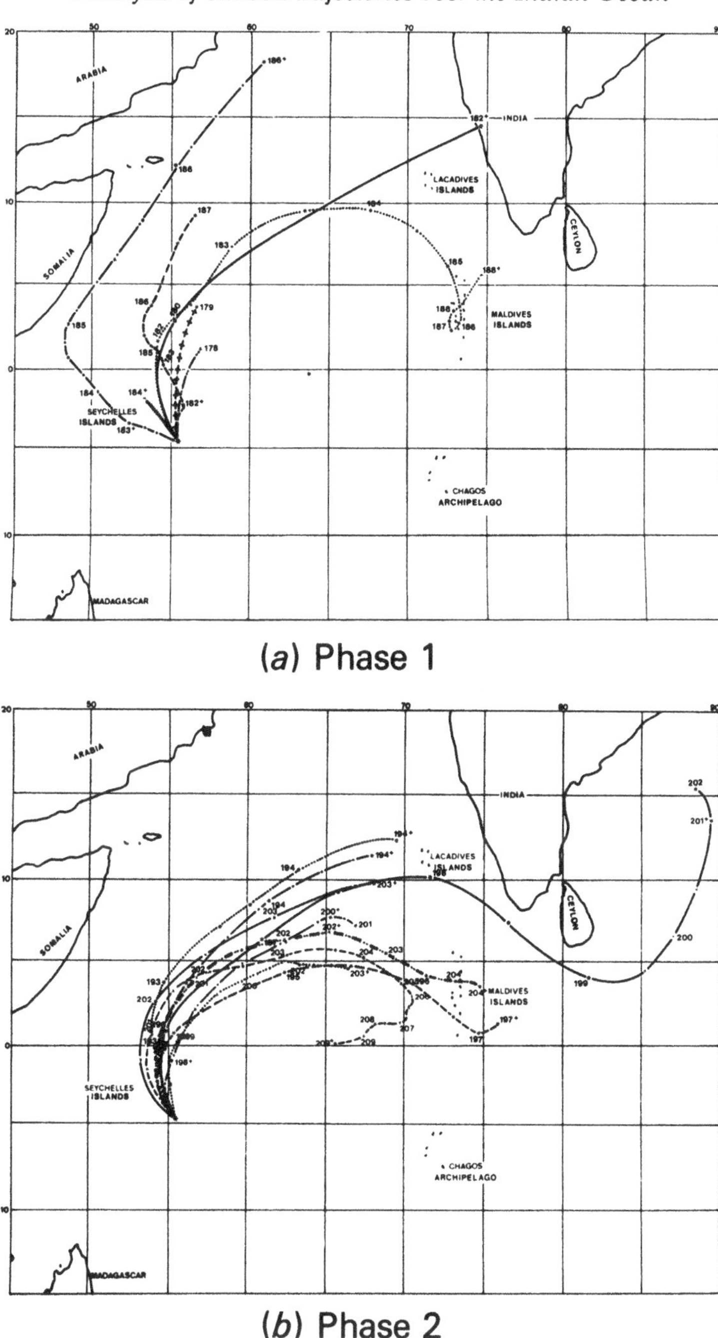

(a) Phase 1

(b) Phase 2

Fig. 19.1. Trajectories of the low-level superpressure ballons during summer 1975 for the four phases. The numbers along the trajectories refer to Julian days. (From Cadet and Ovarlez, 1976.)

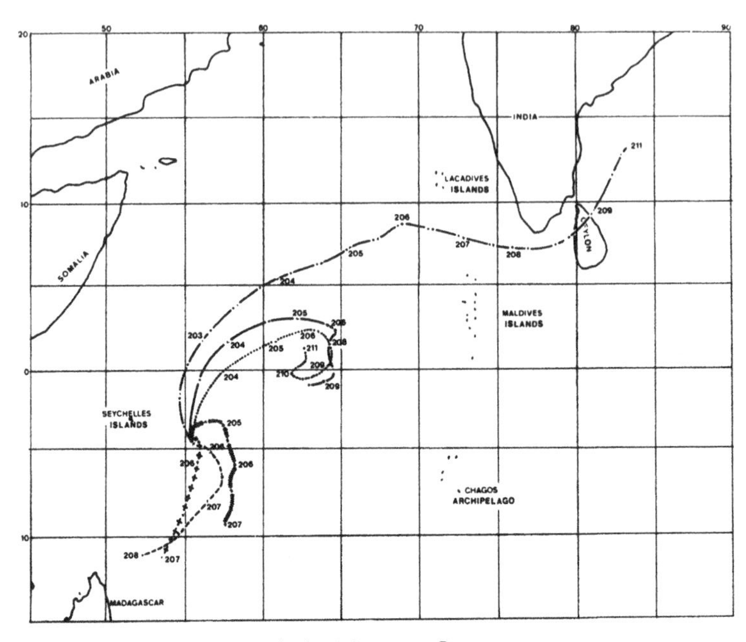

(*c*) Phase 3

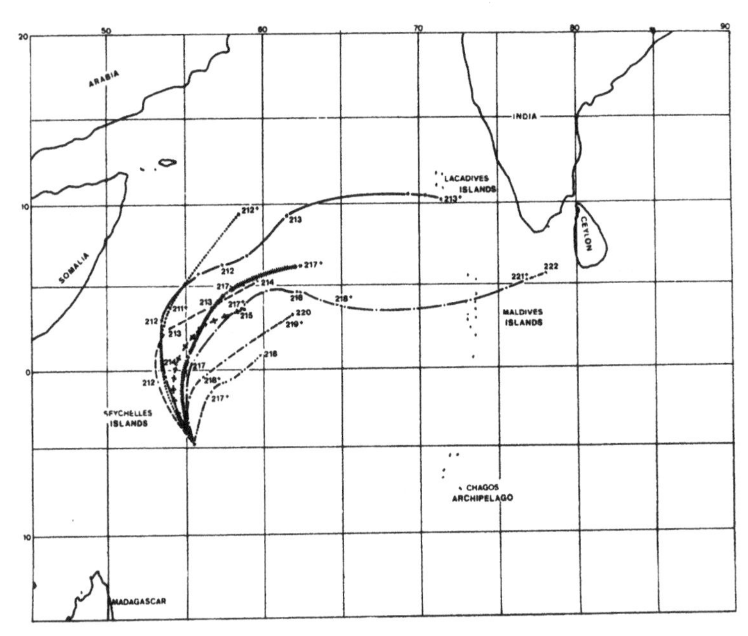

(*d*) Phase 4

Fig. 19.1—*cont.*

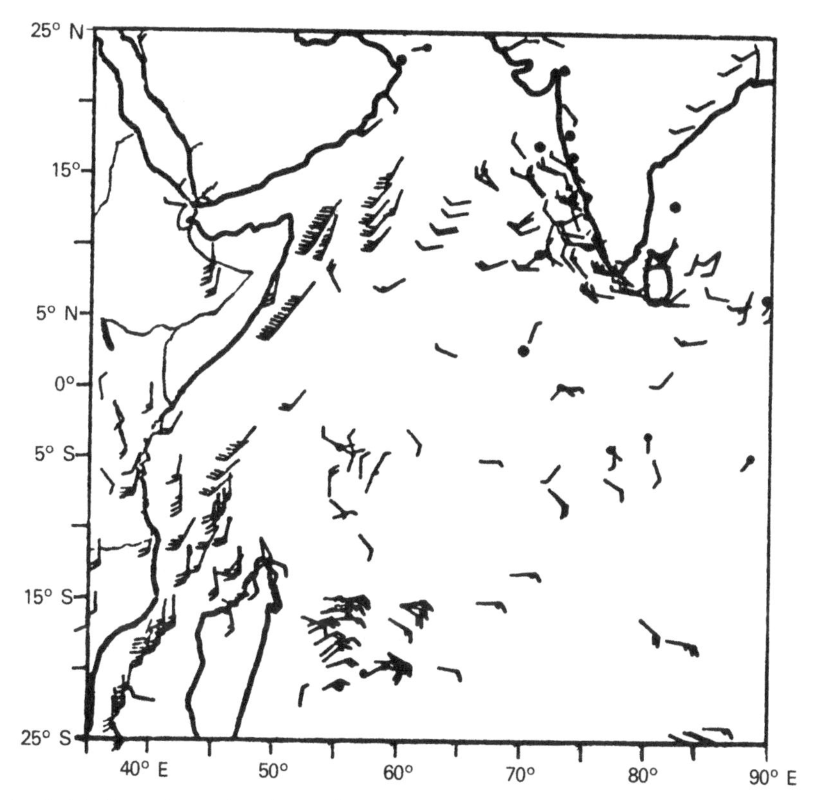

Fig. 19.2. Location of surface wind data on 25 July 1975. (From Cadet and Olory-Togbé, 1977*b*.)

interpolation scheme is used, since the derived charts are not used to perform an objective analysis. For a given day, D, a parameter u at each latitude–longitude point (l, L) is interpolated as follows

$$u(D, l, L) = \sum W_{ij} u(D, i, j) / \sum W_{ij},$$

where W_{ij} is a weighting function defined by

$$W_{ij} = (d_{ij}^2 - R^2)/(d_{ij}^2 + R^2), \quad \text{if } d \leq R,$$

$$W_{ij} = 0, \quad \text{if } d > R.$$

d_{ij} is the Cartesian distance of the observational point from the grid station; R is taken as 2000 km.

The balloons were released during three periods, and four different flow patterns were obtained from their trajectories (Cadet and Ovarlez, 1976). The monsoon activity was studied during the 1975 summer monsoon (Olory-Togbé, 1977) and different phases corresponding to variations in the activity of the monsoon over India were defined. These phases fit the different patterns of the flow field as reflected by the different trajectories. The mean field for each parameter during each phase is easily derived by calculating a mean value at each latitude–longitude point

$$\bar{u}(l, L) = \frac{1}{N} \sum_{D=1}^{N} u(D, l, L).$$

The composite mosaic of the US NOAA-4 satellite with images for the period of the experiment are used in order to perform a nephanalysis. The cloud amounts over the Indian Ocean are estimated in each 2.5° grid square. Each square is allocated a number between 0 and 9 according to the cloud amount, 0 corresponding to no cloud and 9 to complete cloud cover. This cloud amount analysis was made for each day from 25 June 1975 to 19 August 1975 and the period mean cloud coverage calculated for each of the four periods of the experiment.

19.3 The main features of the monsoonal flow over the Indian Ocean

The data taken as a whole, and more particularly the trajectories, show that the southwesterlies giving rise to monsoon activity over India are linked directly with the southeast trades of the southern hemisphere, the latter being deflected, largely as a result of the Coriolis effect, after crossing the Equator.

The wind fields (Figs. 19.4*a*, 19.5*a*, 19.6*a* and 19.7*a*) show the fundamental difference between the eastern and western parts of the equatorial Indian Ocean. East of 60° E, there is a confluence characterized by weak cross-equatorial flow (2.5 m s^{-1}) whereas in the western Arabian Sea, there is an area of strong winds with large cross-equatorial flow. The cross-equatorial flow which carries water vapour across the Arabian Sea and affects India, is very close to the east African coast and is concentrated into the low-level jet flowing off the Somali coast (Findlater, 1969*a*). This jet is certainly the most important dynamic component of the low-level circulation of the southwest monsoon. This explains discrepancies between the computations of the transport of air and water vapour northward across the Equator. A larger volume, closer to the truth, is found if the transport across the area of the western Arabian Sea is accurately estimated. Thus, Pisharoty's (1965) value for water vapour transport is half Saha's (1970), and the value of air transport by Findlater (1969*b*) is about one and a half Saha's (see Table 19.1).

TABLE 19.1. *Transport of water vapour and air across the Equator in the Indian Ocean. (After Desai et al.,* 1976)

Reference	Area considered	Transport of water vapour (10^{10} tonnes per day)	Transport of air 10^{12} tonnes per day)
Pisharoty (1965)	Between 42° and 75° E and layer surface to 450 mb	2.2 (July 1964)	
Saha (1970)	Same belt and depth as taken by Pisharoty	4.43 (July 1964)	5.02 (July 1964)
Findlater (1969*b*)	Belt between 35° and 75° E in the lower troposphere		7.63 (July mean)

The study of the wind fields and balloon trajectories gives some information about the origin of the westerlies. The wind measurements deduced from the successive balloon locations show that westerlies and even northwesterlies can be found up to 10° N. Westerlies can also be found (Fig. 19.6) down to 5° S. Westerlies do not necessarily originate in the northern hemisphere, as has been suggested by some meteorologists on the basis of studies using local data; they may be associated with southeast trades originating from the region of the southern hemisphere equatorial trough (SHET), crossing the Equator and being deflected by the Coriolis force. These westerlies enter the monsoon circulation only in the eastern Arabian Sea and southern Bay of Bengal and their contribution to the monsoon activity over India is certainly weak.

Another important feature of the monsoon circulation over the Indian Ocean is the occurrence of transient phenomena which could have an influence on the monsoon activity over India (Cadet and Olory-Togbé, 1977*a*).

19.4 The monsoon circulation during the different phases

Fig. 19.3 gives the average rainfall amount for six meteorological stations in central India. Four different phases of monsoon activity can be defined. The second period with a large rainfall amount is followed by a one-week period with a nearly zero rainfall amount. This break in the activity of the monsoon also appears on rainfall records of stations located along the western coast of India.

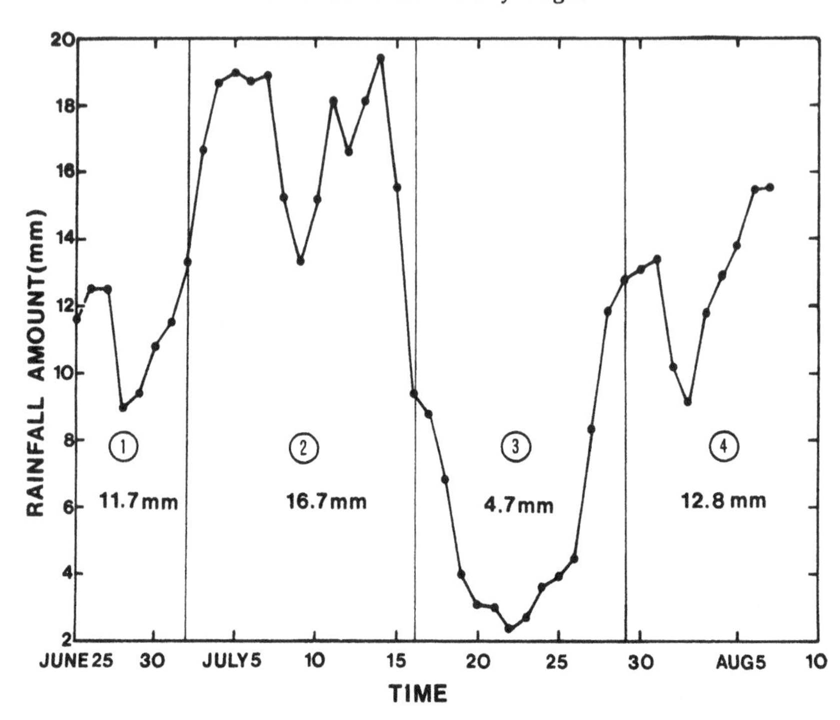

Fig. 19.3. The mean rainfall over central India. These values result from averaging the data of six meteorological stations in Madhya Pradesh state which are further smoothed by producing five-day running means. The four different phases of the monsoon activity are defined by changes in the mean rainfall amount. (From Cadet and Olory-Togbé, 1977*a*.)

Studies of break and active monsoons are often restricted to the analysis of satellite-determined cloudiness over the ocean and to the study of the relationship between this field and rainfall over India (Cadet and Olory-Togbé, 1977*b*; Ramaswamy and Kailasanathan, 1976; Hamilton, 1977; Srinivasan, 1968). The analysis reported in this chapter, though restricted in time, gives, in addition, the pressure and wind fields. It allows a comparison to be made between these fields during the different phases of the monsoon.

19.4.1 Phase 1: 25 June 1975 to 2 July 1975

During this phase, the trajectories show a change in direction of the airflow over the Seychelles Islands. Thus, at the beginning, air crossing the Equator near 55° E strikes the southwestern coast of India, whereas at the end it is embedded in the low-level Somali Jet and strikes the Indian coast further to the north.

The wind field (Fig. 19.4*a*) is typical of the monsoon circulation. The meridional pressure gradient is weaker in the eastern equatorial Indian Ocean than in its western part due to the occurrence of a trough centred around 75° E (Fig. 19.4*c*). The cloudiness field (Fig. 19.4*c*) shows cloudiness over central India, and a cloudy area appearing between 5° S and 15° S to the east of 60° E corresponds to the SHET.

19.4.2 Phase 2: 3 July to 16 July 1975

During this phase, rainfall amounts over India are large. The flow circulation is more stationary than during the previous phase but there is an important modification: after crossing the Equator, the northward-moving air reaches the monsoon current located over the Bay of Bengal. Thus, the air mass crossing the Equator to the east of 55° E and the water vapour it contains can influence the monsoon in the Bay of Bengal.

During this phase, the geographical extent of the Somali low-level Jet is increased. The intensity of the southeast trades in the southern hemisphere is also important as shown in Fig. 19.5*a*.

The previously noted trough in the southeast Indian Ocean is shifted to the east. Another important point is that between this phase and phase 1, the meridional pressure gradient is decreasing in the eastern equatorial Indian Ocean, whereas it is increasing along the African coast.

The large areas of cloudiness (Fig. 19.5*d*) over the Bay of Bengal and central India are associated with the occurrence of monsoon depressions producing large rainfall over central India. In the equatorial Indian Ocean, as suggested by the pressure field, the SHET is also in evidence.

19.4.3 Phase 3: 17 July to 19 July 1975

During this phase, which corresponds to a break in the activity of the monsoon, the balloon trajectories have a large southward rather than northward component. During the first days, after crossing northwards to the Equator, some balloons seem to go back southwards. The cyclonic circulation over the equatorial Indian Ocean around the SHET is well marked (Fig. 19.6*a*). Thus, westerlies are found near 5° S due to the occurrence of the SHET as shown on the pressure field (Fig. 19.6*c*) and cloudiness field (Fig. 19.6*d*). This is particularly typical of a break phase of the monsoon, i.e. it is associated with a strengthening of the SHET

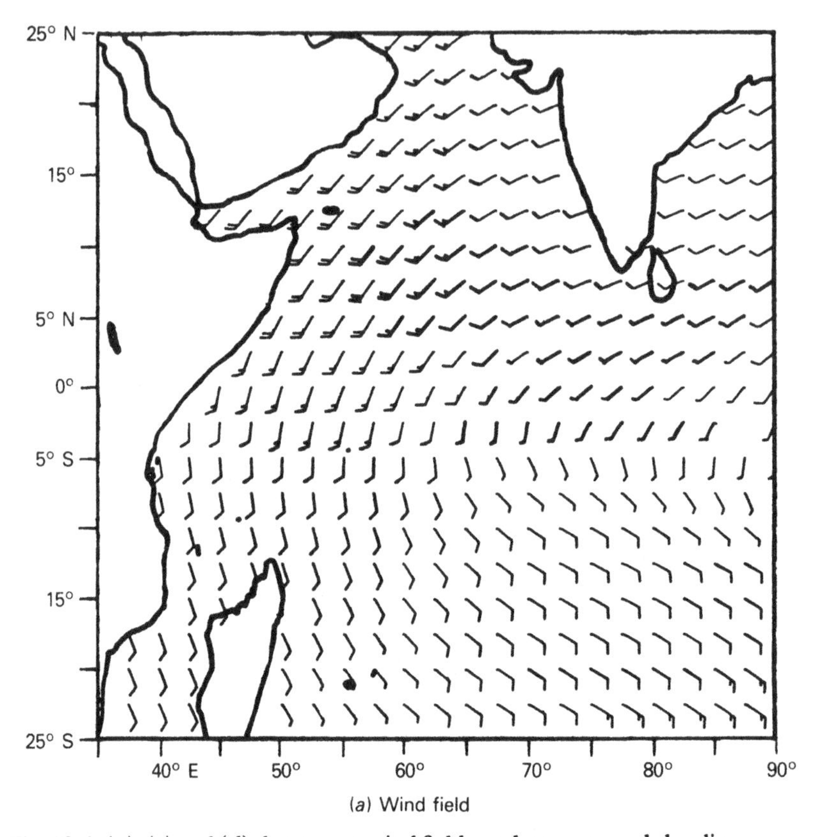

(a) Wind field

Fig. 19.4. (*a*), (*c*) and (*d*) show mean wind fields, msl pressure and cloudiness over the Indian Ocean (from Cadet and Olory-Togbé, 1977*b*), and (*b*) shows the wind intensity and direction from superpressure balloon trajectories (from Cadet and Ovarlez, 1976) for phase 1 (25 June to 2 July, 1975).

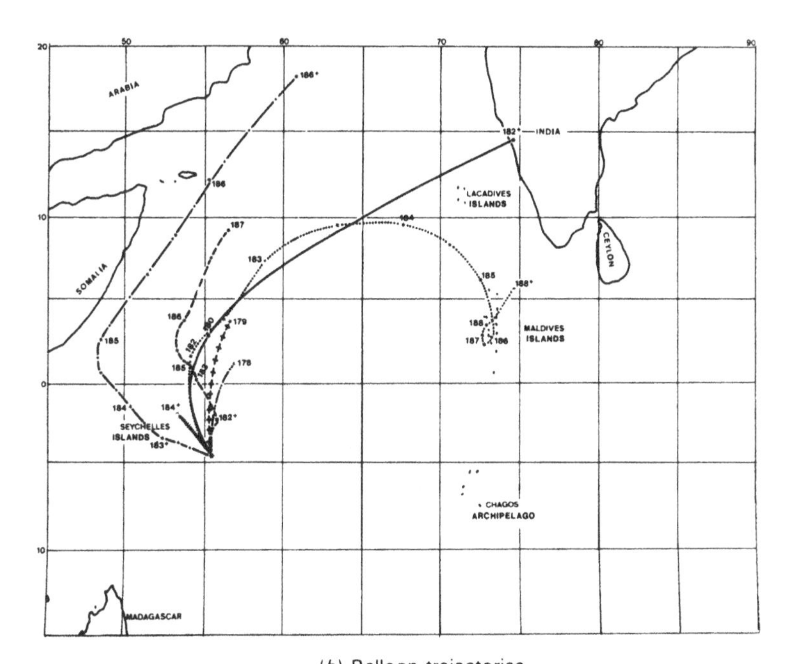

(*b*) Balloon trajectories

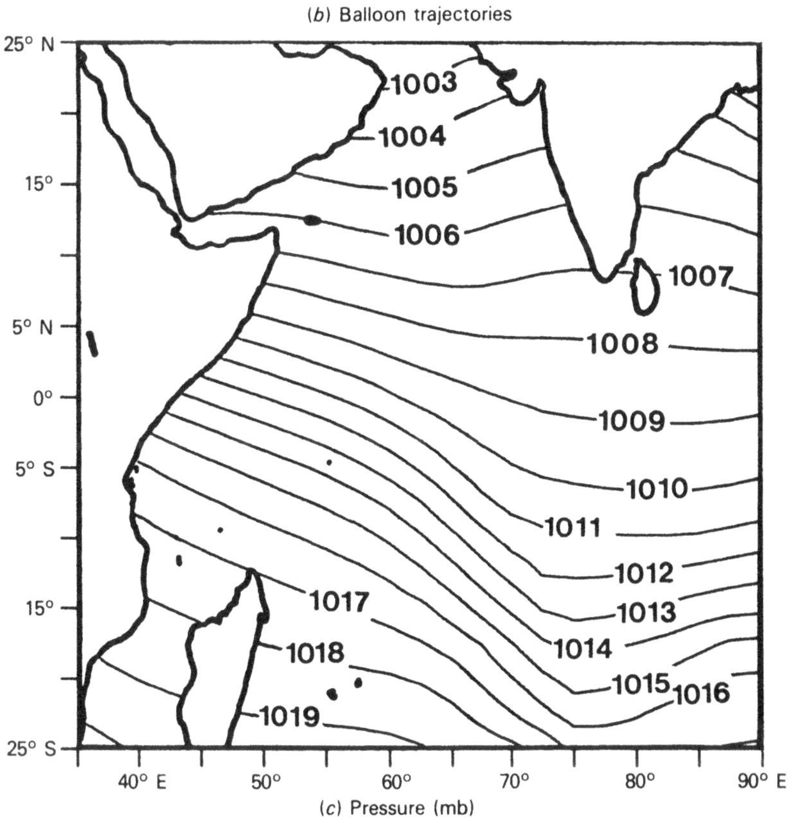

(*c*) Pressure (mb)

Fig. 19.4—*cont.*

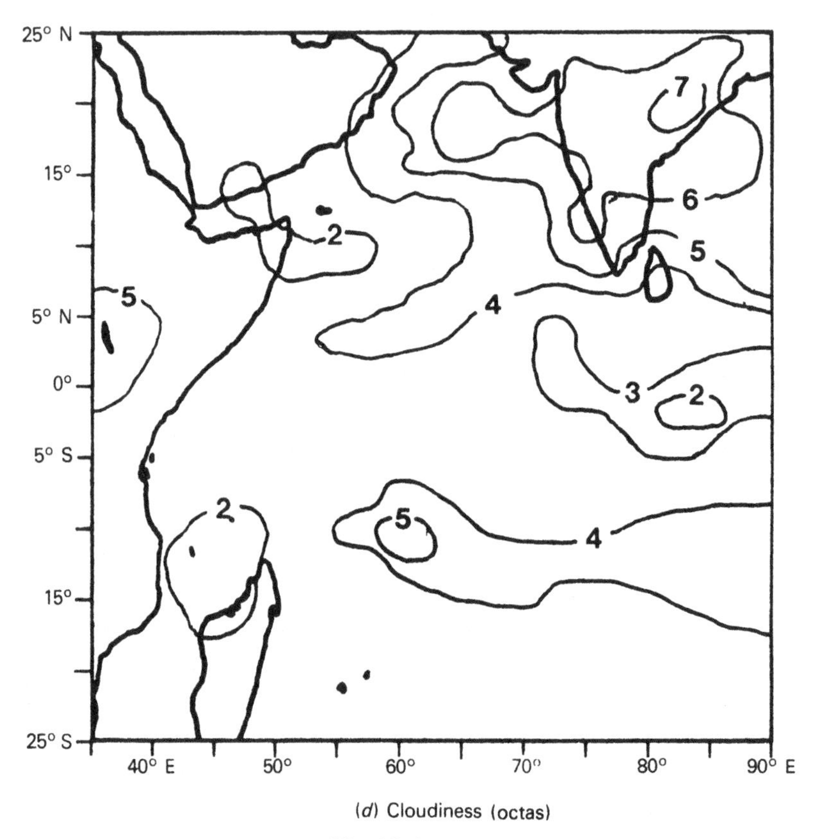

(*d*) Cloudiness (octas)

Fig. 19.4—*cont.*

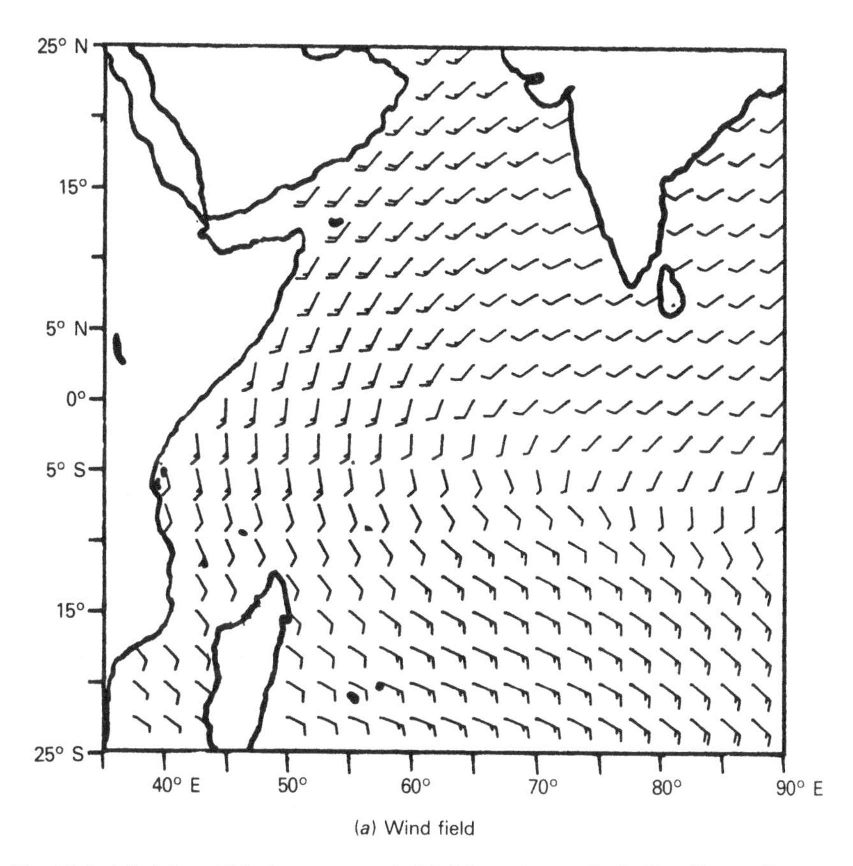

(*a*) Wind field

Fig. 19.5. (*a*), (*c*) and (*d*) show mean wind fields, msl over the Indian Ocean (from Cadet and Olory-Togbé, 1977*b*), and (*b*) shows the wind intensity and direction from superpressure balloon trajectories (from Cadet and Ovarlez, 1976) for phase 2 (3 July to 16 July 1975).

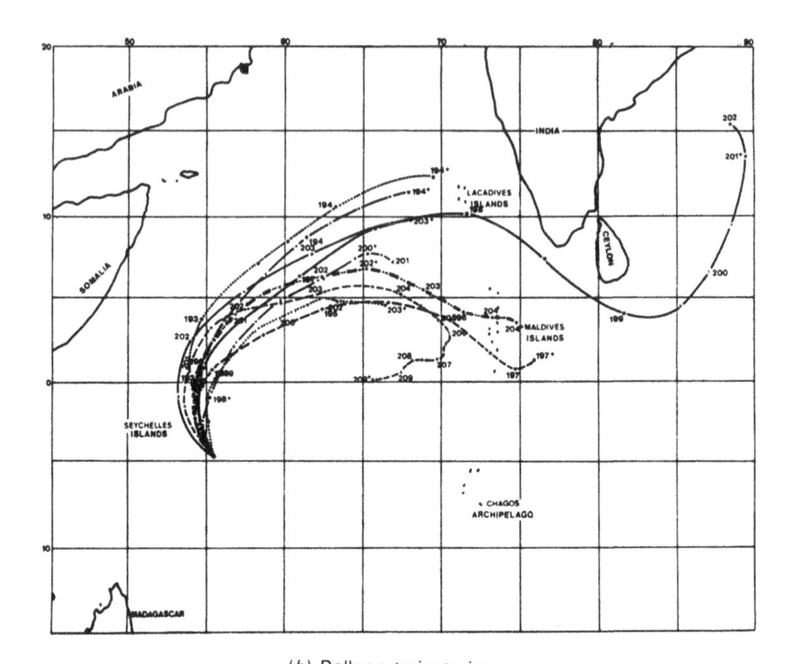

(*b*) Balloon trajectories

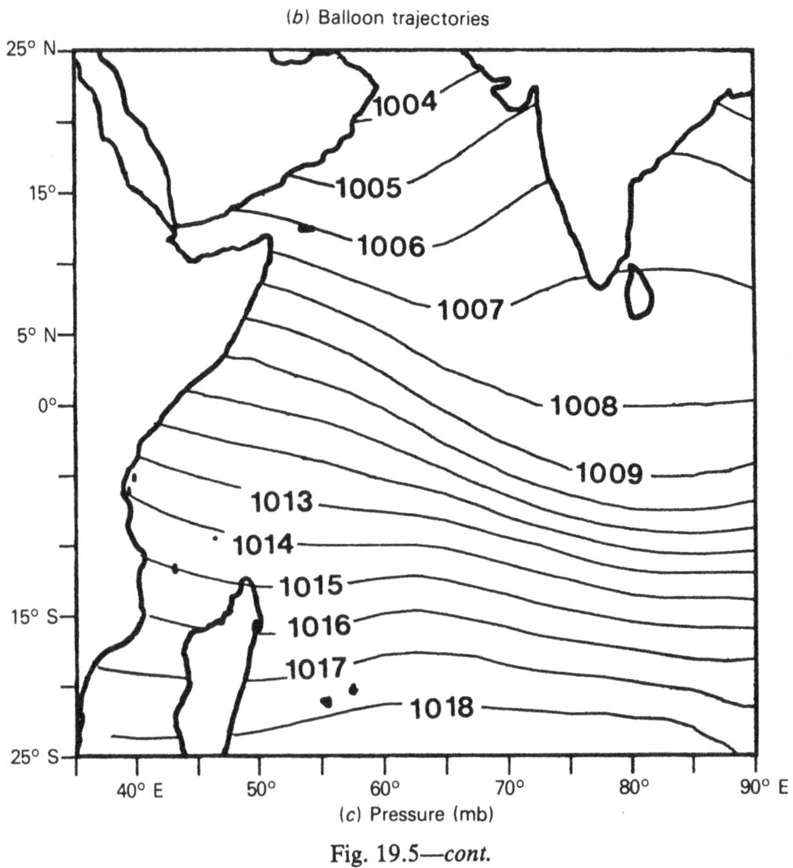

(*c*) Pressure (mb)

Fig. 19.5—*cont.*

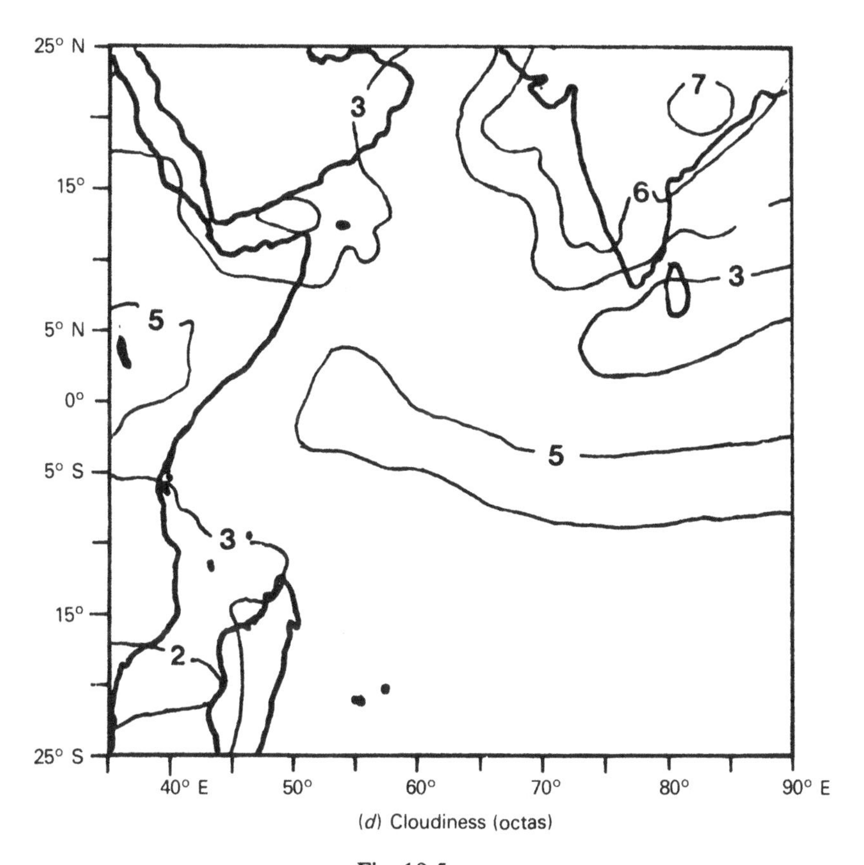

(*d*) Cloudiness (octas)

Fig. 19.5—*cont.*

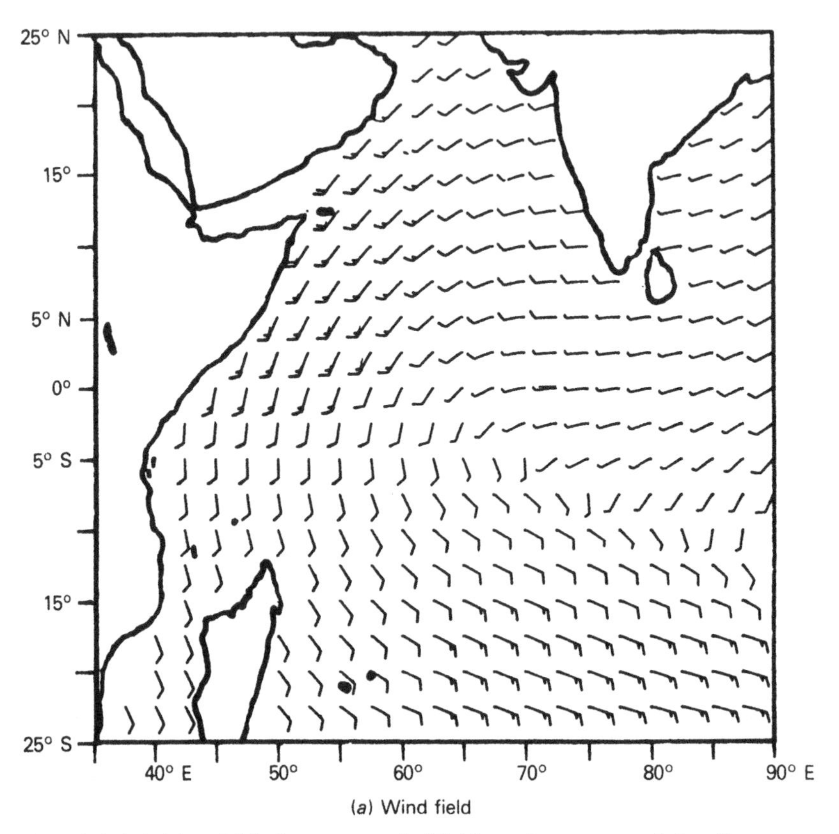

Fig. 19.6. (*a*), (*c*) and (*d*) show mean wind fields, msl pressure and cloudiness over the Indian Ocean (from Cadet and Olory-Togbé, 1977*b*), and (*b*) shows the wind intensity and direction from superpressure balloon trajectories (from Cadet and Ovarlez, 1976) for phase 3 (17 July to 29 July 1975).

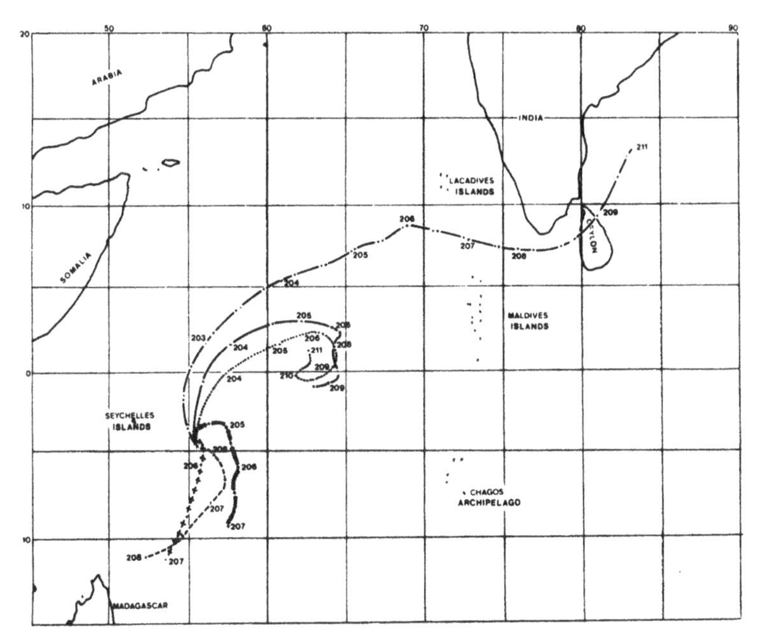

(*b*) Balloon trajectories

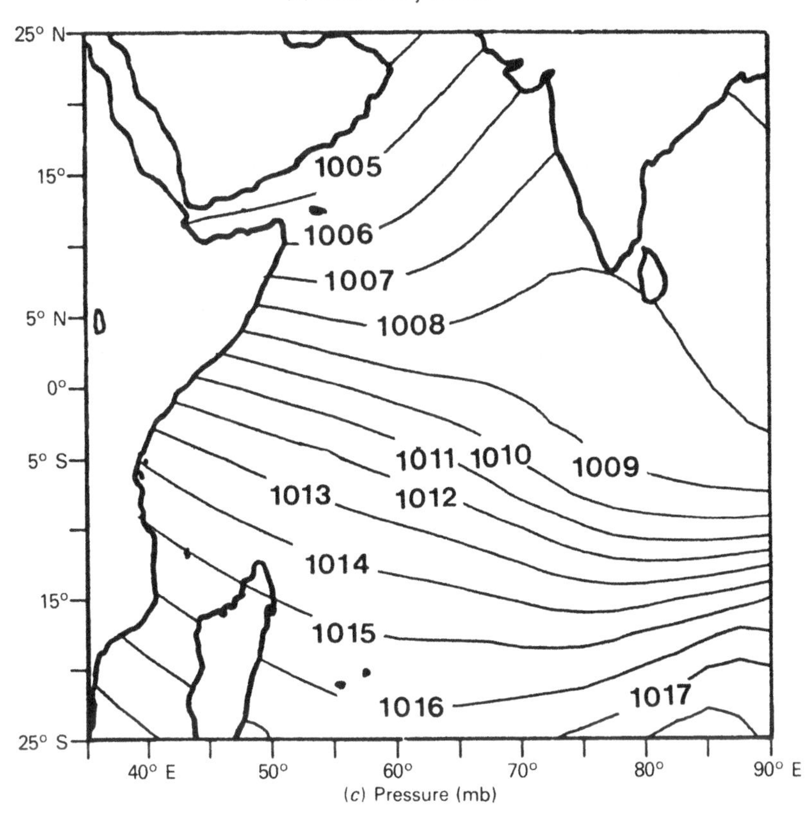

(*c*) Pressure (mb)

Fig. 19.6—*cont.*

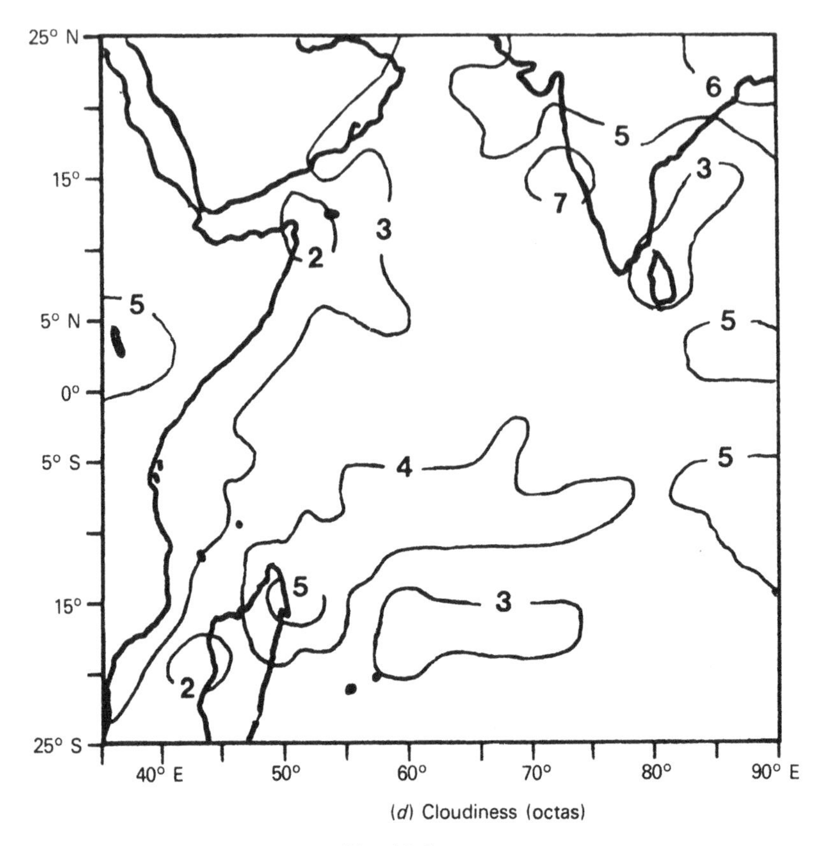

(*d*) Cloudiness (octas)

Fig. 19.6—*cont.*

(Maximum intensity) and a decrease of low-level cloud cover over central India.

At the end of this phase, the monsoon circulation is completely reversed over the Seychelles Islands and some balloons are flying towards Madagascar. This particular situation, which is associated with the propagation of a tropical disturbance (Cadet and Olory-Togbé, 1977*a*), is not apparent in the mean fields. This is because the phenomenon is quite local and very few data are available in the area; furthermore as the radius of influence used in the interpolating scheme is 20°, such features are smoothed out.

19.4.4 Phase 4: 30 July to 10 August 1975

The balloon trajectories and the mean wind field (Fig. 19.7*a*) show that the circulation is similar to that of phase 1, i.e. a phase of mean activity. This is also true for the pressure field (Fig. 19.7*c*) and cloudiness field (Fig. 19.7*d*) which shows a large cloud cover over central India, whereas cloudiness over the equatorial Indian Ocean is strongly reduced. These features are quite representative of this phase of activity.

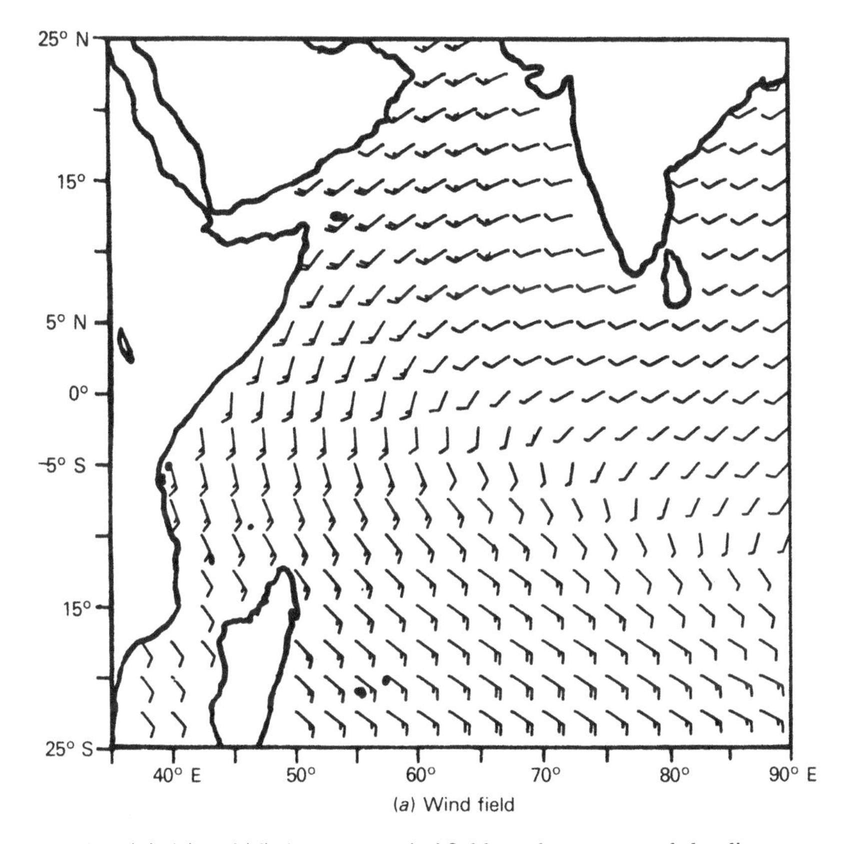

Fig. 19.7. (*a*), (*c*) and (*d*) show mean wind fields, msl pressure and cloudiness over the Indian Ocean (from Cadet and Olory-Togbé, 1977*b*), and (*b*) shows the wind intensity and direction from superpressure balloon trajectories (from Cadet and Ovarlez, 1976) for phase 4 (30 July to 10 August 1975).

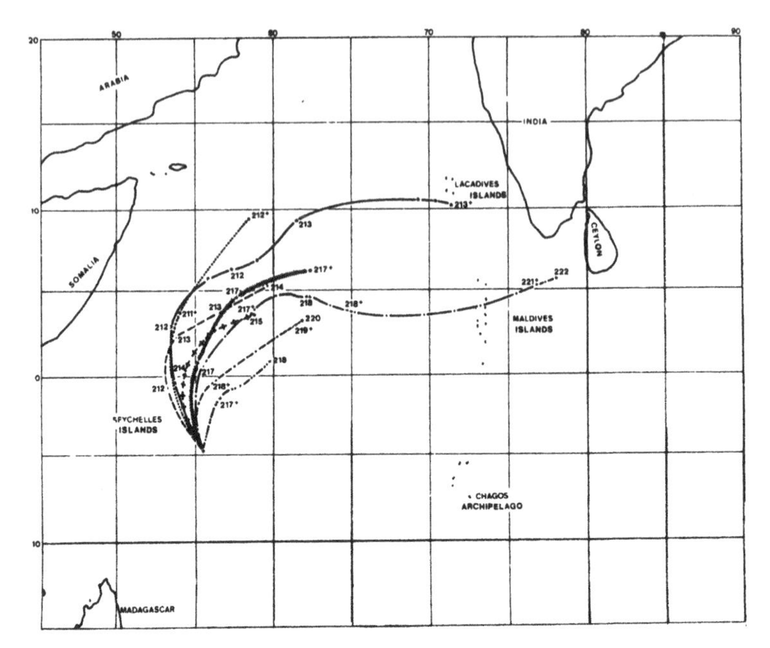

(*b*) Balloon trajectories

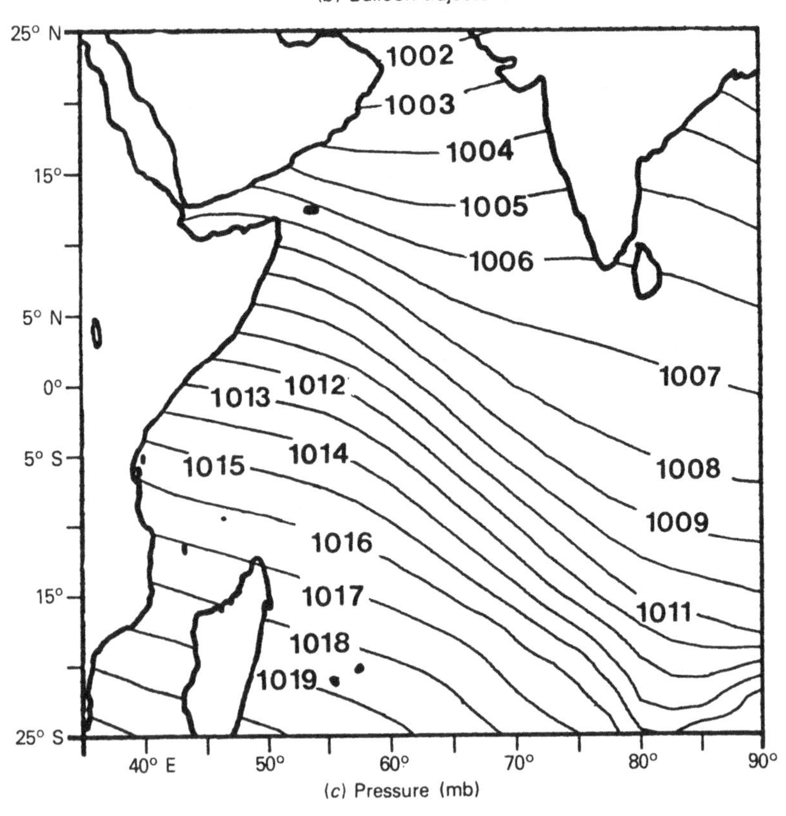

(*c*) Pressure (mb)

Fig. 19.7—*cont.*

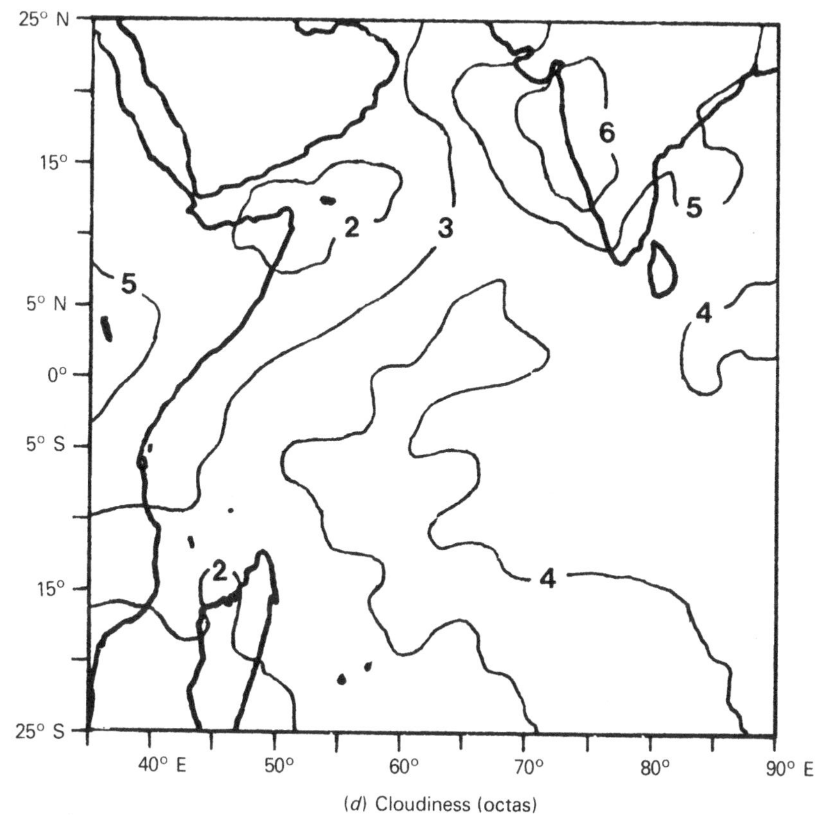

(*d*) Cloudiness (octas)

Fig. 19.7—*cont.*

19.5 Conclusions

An analysis of the low-level monsoonal flow over the Indian Ocean has been completed using Lagrangian trajectories and assimilated conventional data. The main features of the monsoonal flow over the Indian Ocean have been described. An analysis of the meteorological fields during different phases of activity of the monsoon reveals that the flow patterns are different between the phases. Thus, during the occurrence of a break, the typical features which can be noted are that the SHET is well marked over the equatorial Indian Ocean whereas cloudiness is reduced over central India (Murakami, 1976). The contrary is true during an average or intense phase; however the SHET can still be identified. It is also shown that a break is the end result of an evolution of the different fields (this is particularly noticable in pressure and cloudiness fields). This latter feature is consistent with recent findings about the pulsatory nature of the broad-scale monsoon system (Krishnamurti and Bhalme, 1976).

References

Cadet, D. and Olory-Togbé, P. (1977*a*) The propagation of tropical disturbances over the Indian Ocean during the summer monsoon. *Mon. Wea. Rev.*, **105**, 700–8.
Cadet, D. and Olory-Togbé, P. (1977*b*) Low-level air flow circulation over the Arabian Sea during the summer monsoon as deduced from satellite-tracked superpressure balloons. Part II – Analysis of the flow field. *Quart. J. Roy. Meteor. Soc.*, **104**, 971–7.
Cadet, D. and Ovarlez, H. (1976) Low-level air flow circulation over the Arabian Sea during the summer monsoon as deduced from satellite-tracked superpressure balloons. Part I – Balloon trajectories. *Quart. J. Roy. Meteor. Soc.*, **101**, 805–16.
Desai, B. N., Rangaghari, N., Subramanian, S. K. and Sambamoorthy, T. M. (1976) Conditions over the equatorial area during the period of monsoon experiment (MONEX) 1973. *Ind. J. Meteor. Hydrol. Geophys.*, **27**, 141–56.
Findlater, J. (1969*a*) A major low-level air current near the western Indian Ocean during the northern summer. *Quart. J. Roy. Meteor. Soc.*, **95**, 362–80.
Findlater, J. (1969*b*) Interhemispheric transport of air in the lower troposphere over the western Indian Ocean. *Quart. J. Roy. Meteor. Soc.*, **95**, 400–3.
Hamilton, M. G. (1977) Some aspects of break and active monsoons over southern Asia during summer. *Tellus*, **29**, 335–44.
Krishnamurti, T. N. and Bhalme, H. N. (1976) Oscillations of a monsoon system. Part I – Observational aspects. *J. Atmos. Sci.*, **33**, 1937–54.
Murakami, T. (1976) Cloudiness fluctuations during the summer monsoon. *J. Meteor. Soc. Japan*, **54**, 175–81.
Olory-Togbé, P. (1977) Contribution à l'étude de la mousson d'été au-dessus de l'océan Indien à l'aide de trajectoires de ballons plafonnants dans la couche limite. Thèse de spécialité. Université Pierre et Marie Curie, Paris.
Pisharoty, P. R. (1965) Evaporation from the Arabian Sea and the Indian Southwest monsoon. Proc. Symp. Meteor. Results IIOE-Bombay, pp. 22–6.
Ramaswamy C. and Kailasanathan, K. (1976) Weak monsoon over Keraba in relation to satellite-determined cloudiness and large-scale flow patterns. *Proc. Ind. Nat. Sci. Acad.*, **42**, 156–71.
Saha, K. R. (1970) Air and water vapor transport across the Equator in the western Indian Ocean during the northern summer. *Tellus*, **22**, 681–7.
Srinivasan, V. (1968) Some aspects of broad scale cloud distribution over the Indian Ocean during the Indian south-west monsoon season. *Ind. J. Meteor. Geophys.*, **23**, 39–54.

20

An experiment in monitoring cross-equatorial airflow at low level over Kenya and rainfall of western India during the northern summers

J. FINDLATER

The identification of the core of the major low-level air current of the northern summer monsoon at a topographically locked position over eastern Africa has led to experiments in monitoring the airflow and relating its pulsations to the rainfall of parts of western India.

Using five-day overlapping means, it is found that pulsations in the airflow across eastern Africa near the Equator are reflected in the rainfall of western Maharashtra Province, sometimes with a lag of a few days.

The mean airflow in July has also been compared with the mean July rainfall of western Maharashtra Province and, when two-year overlapping means are used, a pronounced lag of one year is evident. It is demonstrated that this lag might be usefully exploited in experimental work towards the development of long-range rainfall forecasting techniques.

20.1 Introduction

A special feature of the low-level airflow over eastern Africa and the western Indian Ocean during the northern summer is that it is organized into a relatively narrow high-speed transequatorial current in the western periphery of the monsoon regime. The flow is strongest where the current is blocked or guided by high ground, and is weakest in the vicinity of the oceanic Equator. The current is characterized by a system of daily low-level jet streams, sufficiently persistent to show up markedly in monthly-averaged wind data.

The characteristics of the major current and the daily low-level jet streams have been described in detail by Findlater (1966, 1967, 1969*a*, *b*, 1970, 1971*a*, *b*, 1972, 1974, 1977*b*), but the general form of the current at the 1 km level in July can be seen in Fig. 20.1.

The core of the current, although primarily an oceanic phenomenon, enters the African continent at about latitude 3° S and curves over the flat arid lands of eastern Kenya, eastern Ethiopia and northeastern Somalia

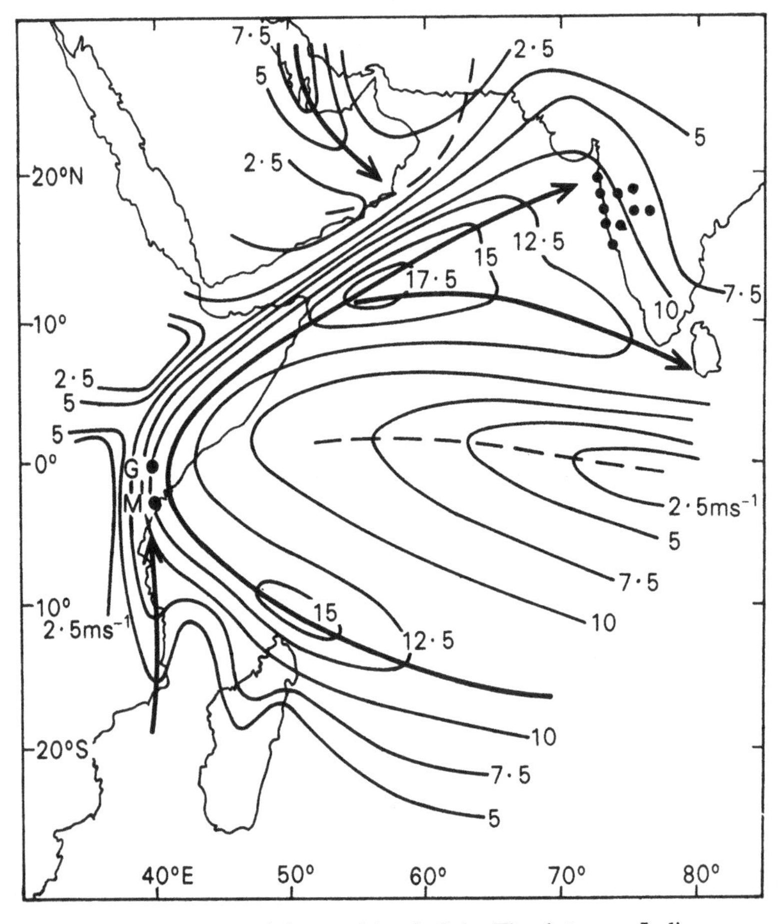

Fig. 20.1. Mean monthly airflow at 1 km in July. The dots over India represent the ten selected rainfall stations. Isotachs at intervals of 2.5 m s^{-1}. G denotes Garissa; M denotes Mombasa. (Crown copyright.)

before leaving the continent near latitude 9° N. The core stands off about 200 km from the edge of the highlands of eastern Africa. The height of the core is generally about 1 km over the ocean south of the Equator but it rises to 1.5 km over Kenya and to nearly 2 km over eastern Ethiopia. The height reduces rapidly over Somalia to become 1 km, or perhaps less, to the south of Socotra Island. The descent of the jet core over Somalia may be associated with the remarkable increase in mean daily surface wind speeds in July towards the northeast over that area, as shown in Fig. 20.2.

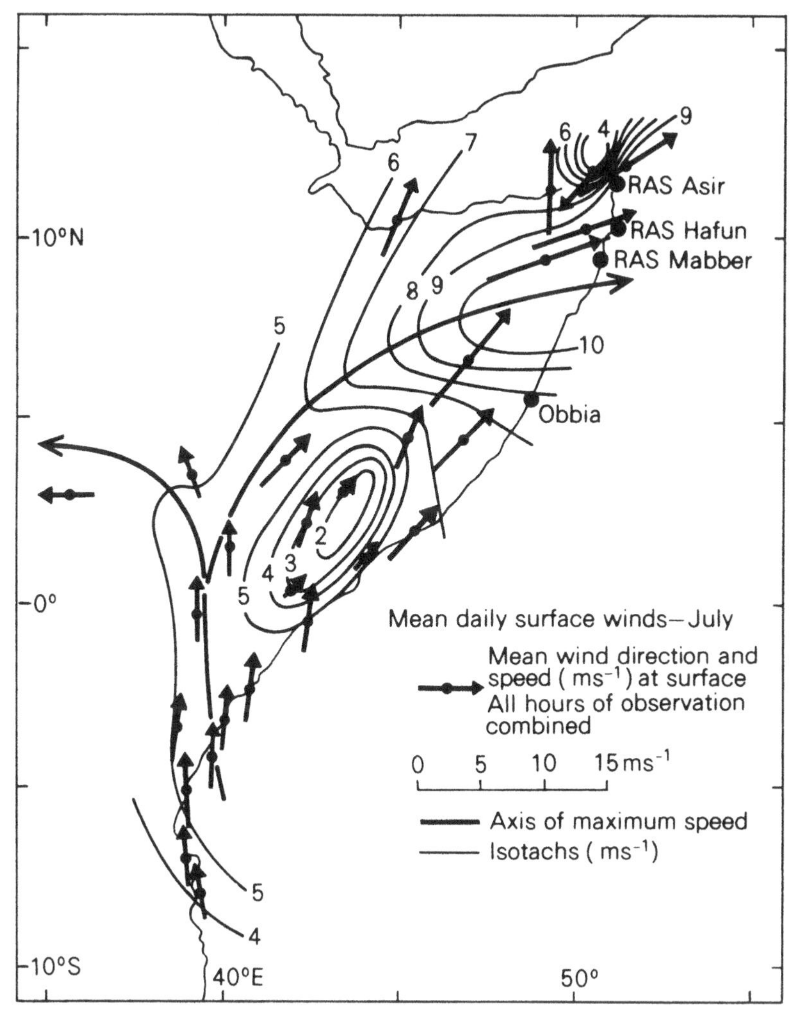

Fig. 20.2. Mean daily surface winds in Somalia for the years 1954–60 in the month of July. Isotachs at intervals of 1 m s^{-1}.

This figure illustrates the well-defined western edge of the strong surface winds which are known to blow off the coast of Somalia in this season. However, the area of strong surface winds, sometimes known as the Somali Jet, is only one manifestation of a much larger-scale phenomenon, the low-level jet stream system, which extends over some thirty degrees of latitude. In this system the winds at 1 to 2 km may accelerate to very high speeds on some occasions. Fig. 20.3 shows the

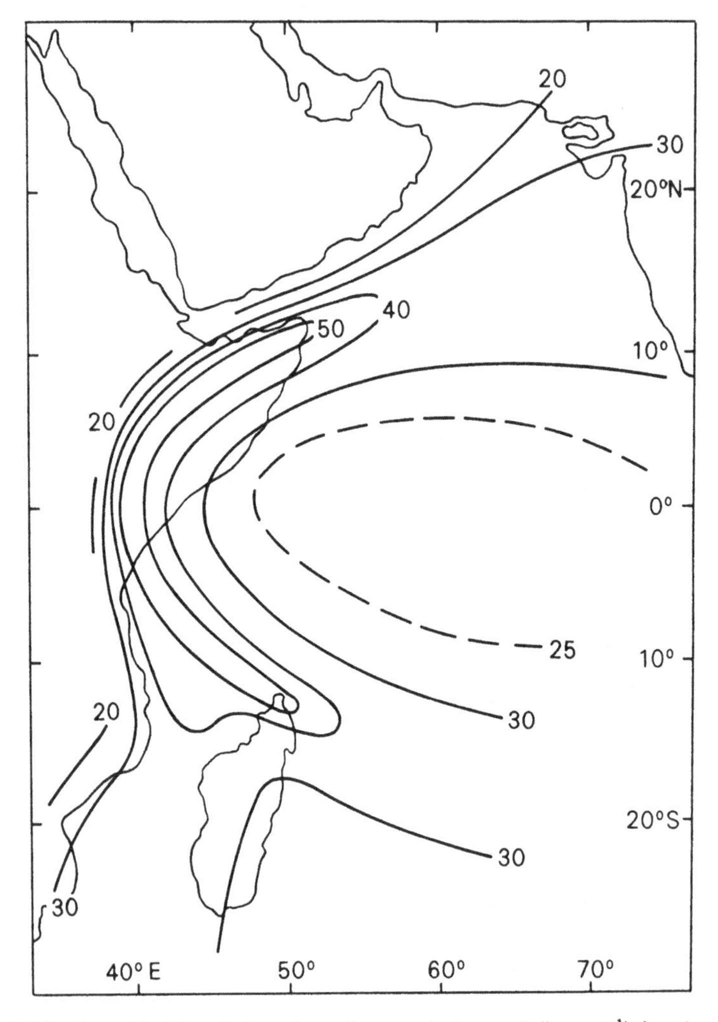

Fig. 20.3. Smoothed isopleths of maximum wind speed (in m s^{-1}) for the layer 600 to 2400 m above mean sea level during the period of the northern summer monsoon. (After Findlater, 1977*b*.)

maximum speeds known to have been recorded, and it is of interest that the highest speeds are limited to a narrow filament on the western edge of the monsoon circulation.

20.2 Monitoring the daily cross-equatorial flow

Vertical cross-sections along the Equator, as in Fig. 20.4, show the general form of the current. These studies used long-term mean monthly pilot balloon winds and illustrate the well-marked core of the current at longitude 40° E and at a height of 1.5 km. Research flights by meteorologically-instrumented aircraft, the *Turbo-Porter* in June 1970, and the *Electra* in June to July 1977 (Fig. 20.5), have confirmed the relative constancy of both the position and the height of the core near the Equator, although the detailed structure varies from day to day. An example of a strong jet explored by the *Electra* aircraft is shown in Fig. 20.6. This was an early morning flight and by the afternoon the core speed was reduced from 24.8 m s^{-1} to 17.3 m s^{-1} by convective mixing.

Pronounced diurnal variations in the strength of the jet can be illustrated by an analysis of pilot balloon data from Garissa, a station near the

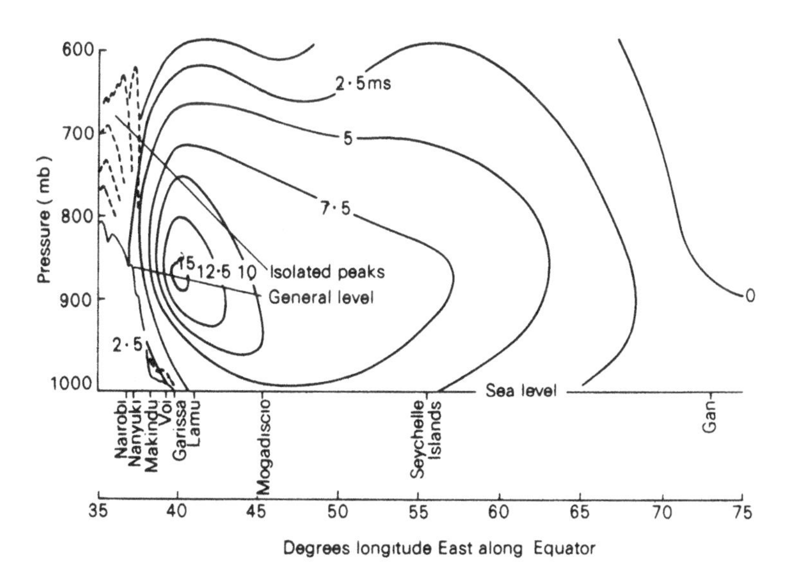

Fig. 20.4. Mean meridional airflow at the Equator in July. Positive values indicate flow from the south. Isotachs at intervals of 2.5 m s^{-1}. (After Findlater, 1969*b*.)

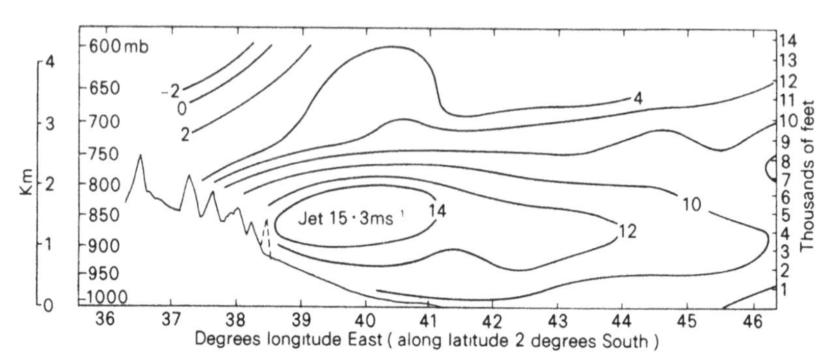

Fig. 20.5. Mean south components (in m s^{-1}) from ten flights by *Electra* aircraft in June and July 1977. All times of observation are combined.

core of the jet. Fig. 20.7 shows the south and east components of the upper winds for the month of July in the years 1962–71 for four times of day. The jet reaches its maximum speed in the early morning, whilst minimum speeds are recorded during the afternoon.

Because it has long been known by Indian meteorologists that strengthening of the surface southwesterly winds over the Arabian Sea heralded periods of heavier precipitation over the west coast of India, attempts have been made to monitor the airflow over eastern Africa in relation to the rainfall of parts of western India. Two upper wind stations in Kenya were selected as monitor stations because of their proximity to the core of the current. These were Mombasa (04°02′ S, 39°37′ E, 57 m) and Garissa

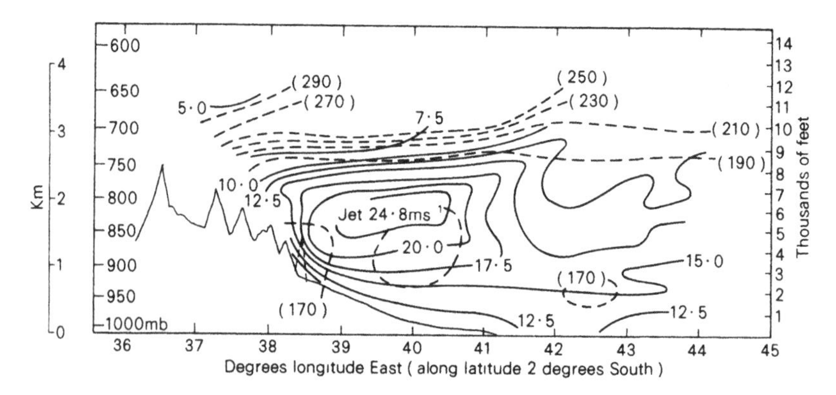

Fig. 20.6. The strong jet during the early morning of 4 July 1977 as located by *Electra* aircraft. Isotachs are solid lines, in m s^{-1}. Isogons are represented by dashed lines and labelled in degrees from the North.

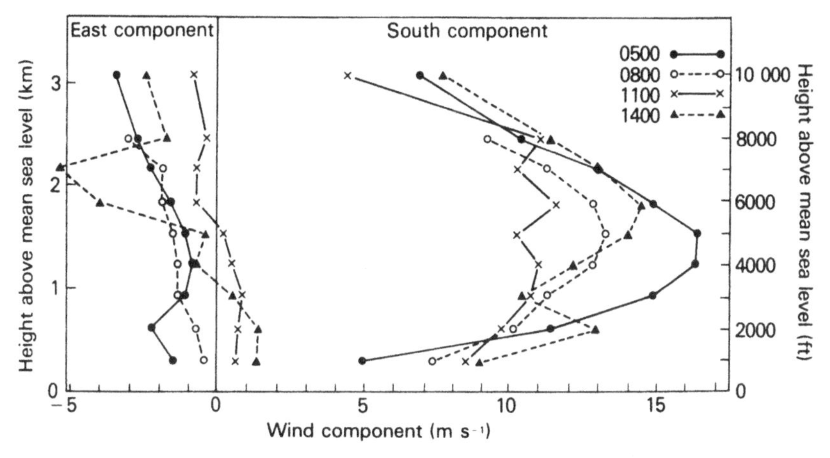

Fig. 20.7. Mean monthly wind components for July at Garissa, Kenya, in the years 1962–71, for four times of day (GMT). Local time in the area is three hours in advance of GMT. (After Findlater, 1977*b*.)

(00°29′ S, 39°38′ E, 128 m). The positions of these stations is shown in Fig. 20.1 and the mean profiles of the strong southerly winds at these stations in July are shown in Fig. 20.8.

When monitoring the flow, only the lower half of the jet-like profile was used because many more pilot balloons reach the 1.5 km level than the 3 km level. A monitor value, or wind index, was calculated for each day by taking the mean of the south components at 0.3, 0.6, 0.9, 1.2 and

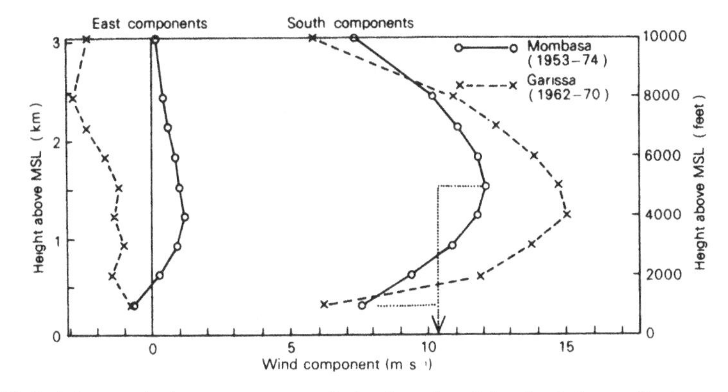

Fig. 20.8. Mean wind components of the low-level jet in July at Garissa and Mombasa. The index of cross-equatorial flow is the mean of the south components at 0.3, 0.6, 0.9, 1.2, and 1.5 km above mean sea level. For Mombasa the 22-year mean of the index is 10.3 m s^{-1}. (Crown copyright.)

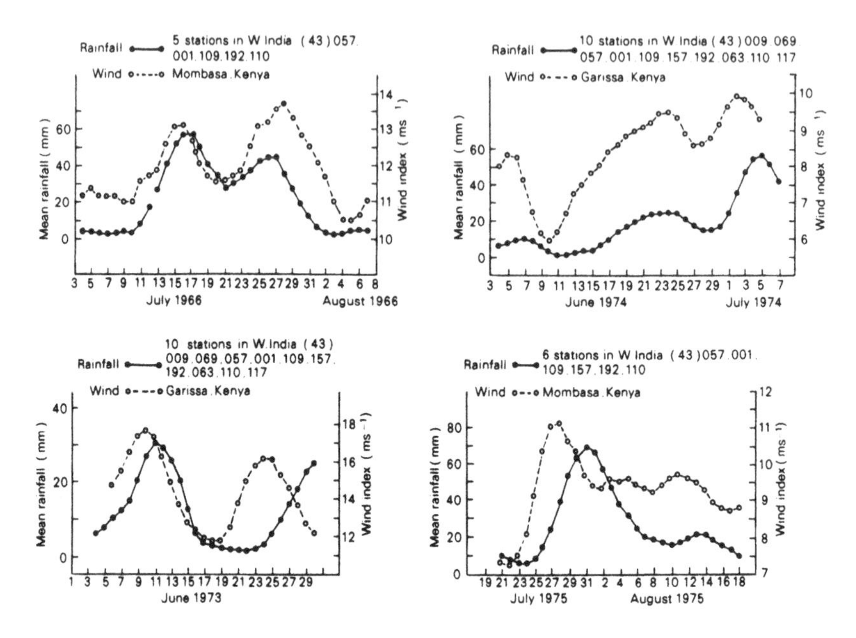

Fig. 20.9. Four examples of the relationship between cross-equatorial airflow over Kenya at Mombasa or Garissa and the mean rainfall at groups of stations in western Maharashtra province of India. The Indian stations are identified by station numbers. Values of the wind index and rainfall are 5-day overlapping means. (After Findlater, 1977*b*.)

1.5 km above mean sea level (corresponding to winds measured at 1000, 2000, 3000, 4000 and 5000 ft above mean sea level). Five-day overlapping means were formed for both the wind index and the mean rainfall at groups of stations in western India. Figs. 20.9 and 20.10 show some examples of the type of analyses produced, indicating a strong correspondence between the wind index and the rainfall of parts of western India.

The cross-equatorial wind index is not, of course, the only influence on the rainfall of western India, but these analyses, and many others not reproduced here, indicate that it can be a dominant influence for lengthy periods during the northern summer monsoon.

20.3 Monitoring the monthly cross-equatorial flow

Using the same method for calculating the cross-equatorial wind index, a mean value has been found for each month of July in the years 1953–76

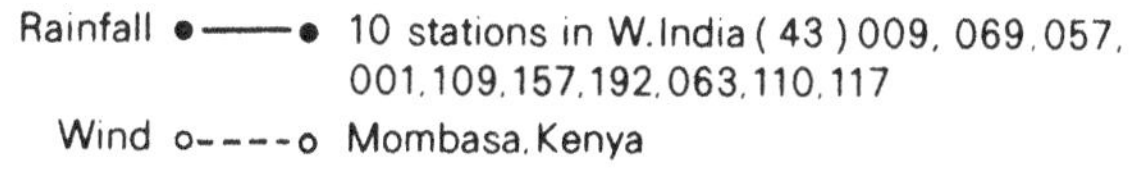

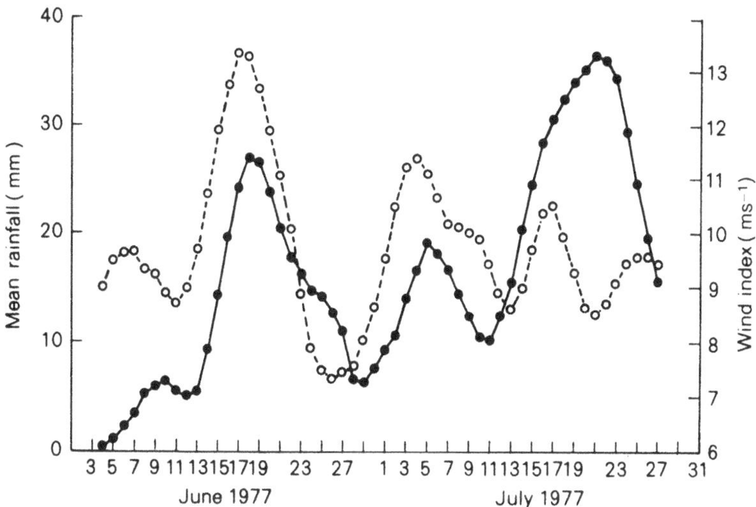

Fig. 20.10. Cross-equatorial airflow over Kenya in 1977 related to the rainfall at ten stations in western India. The Indian stations are identified by station numbers. Values of the wind index and rainfall are 5-day overlapping means.

inclusive. These values are shown in Fig. 20.11 together with the mean July rainfall of groups of stations in the western Maharashtra Province of India. Although the correspondence between the two parameters in individual years is not particularly good, a striking correspondence appears when two-year averages are used. The two-year averages are shown by dots joined by broken lines. Peaks, troughs and other changes in the two-year mean curve of the wind index are reflected in the rainfall curves one year later.

The change in the two-year mean wind index related to the change in the two-year mean rainfall one year later (for the group of ten stations) yields a correlation coefficient of 0.82. Using this relationship, the two-year mean rainfall may be predicted, and hence the mean July rainfall may be predicted one year ahead when that for the latest July is known. An experimental forecasting technique based upon these relationships, and using the dependent data, has yielded an average error in the prediction of 11%. The maximum error in the 22-year period, July only, was 29%. This experimental forecasting technique has been described by Findlater (1977*a*).

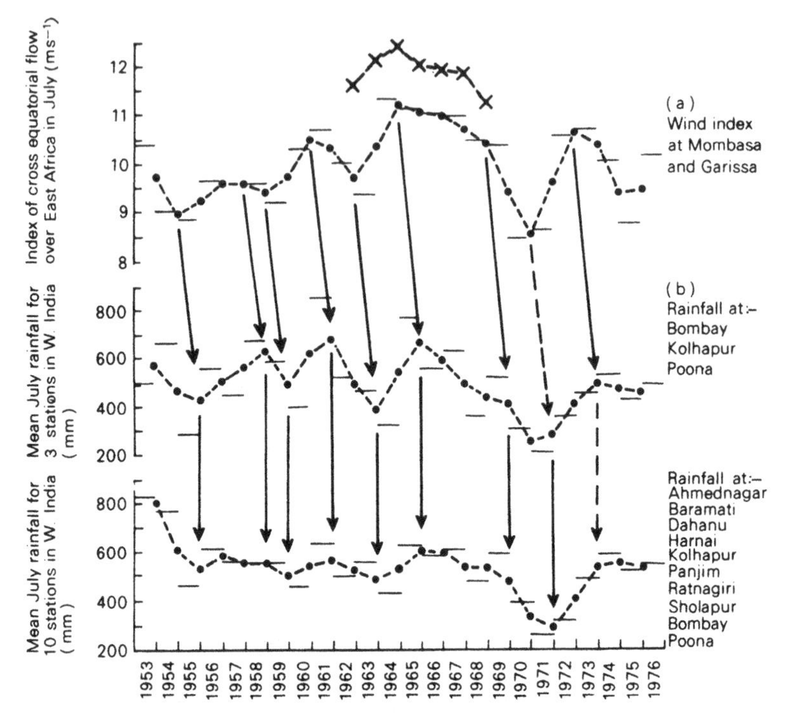

Fig. 20.11. Comparison of (*a*) the mean July index of cross-equatorial airflow over Kenya, (*b*) the mean July rainfall at three stations in western India, and (*c*) the mean July rainfall at ten stations in western India. (Crown copyright.)

20.4 Conclusion

The strong and persistent low-level jet stream which appears to be locked to the topography of eastern Africa during the northern summer can be monitored by upper wind stations near the Equator. Careful monitoring of the flow near the core of the current suggests some promise of advancing short- and long-range rainfall forecasting techniques for parts of western India.

References

Findlater, J. (1966) Cross-equatorial jet streams at low level over Kenya. *Meteor. Magazine*, **95**, 353–64.

Findlater, J. (1967) Some further evidence of cross-equatorial jet streams at low level over Kenya. *Meteor. Magazine*, **96**, 216–19.

Findlater, J. (1969*a*) A major low-level air current near the Indian Ocean during the northern summer. *Quart. J. Roy. Meteor. Soc.*, **95**, 362–80.

Findlater, J. (1969*b*) Interhemispheric transport of air in the lower troposphere over the western Indian Ocean. *Quart. J. Roy. Meteor. Soc.*, **95**, 400–3.
Findlater, J. (1970) Discussion of papers on a major low-level air current. *Quart. J. Roy. Meteor. Soc.* **96**, 551–4.
Findlater, J. (1971*a*) The strange winds of Ras Asir. *Meteor. Magazine*, **100**, 46–54.
Findlater, J. (1971*b*) Mean monthly airflow at low levels over the western Indian Ocean. *Geophys. Mem.* (HMSO, London), **16**, 115, 1–53.
Findlater, J. (1972) Aerial explorations of the low-level cross-equatorial current over eastern Africa. *Quart. J. Roy. Meteor. Soc.*, **98**, 274–89.
Findlater, J. (1974) An extreme wind speed in the low-level jet-stream system of the western Indian Ocean. *Meteor. Magazine*, **103**, 201–5.
Findlater, J. (1977*a*) A numerical index to monitor the Afro-Asian monsoon during the northern summers. *Meteor. Magazine*, **106**, 170–80.
Findlater, J. (1977*b*) Observational aspects of the low-level cross-equatorial jet stream. *Pure and Applied Geophys*, **115**, 1251–62.

21

Structure of the Somali Jet deduced from aerial observations taken during June–July, 1977

G. V. RAO AND H. M. E. VAN DE BOOGAARD

An aerial reconnaissance programme was sponsored by the US National Science Foundation in June and July 1977, under which NCAR's *Electra* was employed in thirteen flights to reconnoitre the Somali Jet by the east African coast. The base of the operations was Nairobi, Kenya.

Cross-sections along 2° S between 37° E and 44° E disclosing the kinematic structure of the jet on four different days were constructed from observations taken on these flights. The major points of interest of these cross-sections are the mesoscale variation of the jet, both in intensity and in its frequency of occurrence, and the existence of a secondary wind maximum over the Indian Ocean near the coast. The water vapour flux across 2° S for three different days (11 June, 15 June and 5 July) and one night (29 June) was computed for the section between 1000 mb and 700 mb and 37° E and 44° E. These fluxes revealed some day-to-day variations. The average flux was found to be 1.79×10^{10} tonnes per day. This shows the water vapour flux for August 1964 across the Equator between 1000 mb and 450 mb and between 420° E and 75° E previously to have been underestimated.

21.1 Introduction

Findlater (1971) described the climatology of the low-level monsoon flow in the south Indian Ocean. A major feature of this flow is a jet stream along the east African coast popularly called the Somali Jet. Findlater

(1972) was able to further describe the synoptic characteristics of this low-level jet as a result of aerial reconnaissance. His aerial studies disclosed that the structure of the jet is quite complex.

Significant amounts of water vapour and momentum are transported across the Equator by means of strong winds associated with this jet. This transport, from the southern hemisphere, takes place vigorously during the northern summer months. Thus, some relationship between the intensity of the jet and monsoon activity, particularly over western India, can be inferred to exist (Raghaven *et al.*, 1975). In view of this suspected coupling, and also because it is an interesting geophysical phenomenon, this jet is receiving considerable attention from the scientific community.

As part of the GARP Global Weather Experiment, 1979, a regional experiment, MONEX (Monsoon Experiment), has been conducted in the Indian Ocean, Arabian Sea and Bay of Bengal. As precursors to MONEX 1979 three field exercises took place in the summer of 1977. The USSR and India took upper-air observations in the eastern Arabian Sea during May, June and July. Krishnamurti of Florida State University, USA, and his collaborators gathered pilot balloon observations over Somalia during June and July 1977. Hart (University of Colorado) and other investigators carried out an aerial reconnaissance of the low-level winds over east Africa and the adjacent Indian Ocean using NCAR's *Electra*. A short description of this field exercise is given below.

21.2 The aerial reconnaissance programme

The base of the aerial operations was Nairobi, Kenya. The first mission took place on 9 June and the thirteenth, and last one, occurred on 4 July. The total flight time was 93 hours and each flight lasted an average of 6 hours. Eight of these missions were devoted principally to documenting the structure of the jet along 2° S between Nairobi and a point over the Indian Ocean at approximately 46° E. Two flights were made to document the characteristics of the flow between the northern tip of the Malagasy Republic and the Seychelles, and one to examine the flow between the Seychelles and Agalega. The remaining two flights were made from Nairobi eastwards to the coast and thence southwards to a point at about 7° S and 40° E. The cross-sections presented below are based on four of the eight cross-sectional flights.

21.3 Vertical cross-sections of the flow and humidity patterns

Fig. 21.1 shows the cross-sections of the zonal component of the wind, based on the aircraft data for 11 and 15 June. In both parts of the figure the saw-toothed path of the airplane is shown as a light solid line. The mesoscale variation of the zonal component from day to day and the existence of shallow westerlies near the coast are the principal features depicted in this figure. Obviously, these variations are related to the

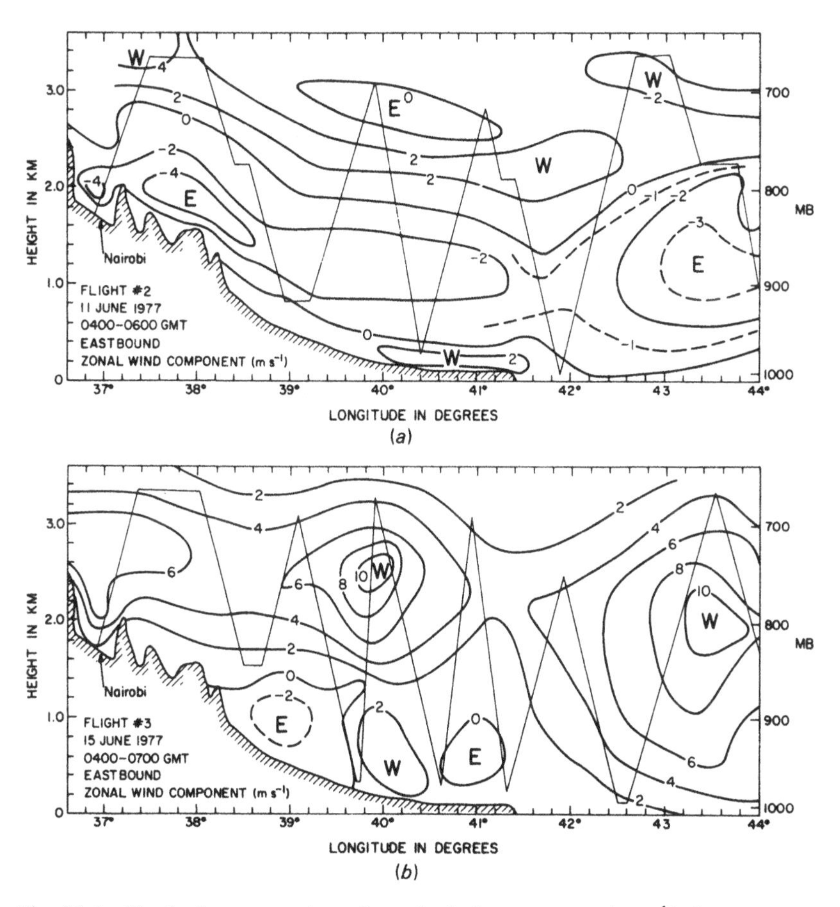

Fig. 21.1. Vertical cross-section of zonal wind component (m s^{-1}) along approximately 2° S, based on the eastbound flights from Nairobi for: (*a*) flight no. 2, 0400–0600 GMT, 11 June 1977; and (*b*) flight no. 3, 0400–0700 GMT, 15 June 1977. The beginning and conclusion of each of the eastbound missions are also shown on the figure. Notice the alternation of easterly and westerly layers of wind.

synoptic-scale changes, but the data on this scale needed to examine this problem in detail are at present lacking.

Fig. 21.2 shows the meridional component, which is stronger than its zonal counterpart. The major points of interest are the presence of a primary wind maximum situated over land at 1.4 and 1.0 km on 11 and 15 June, respectively, with pronounced vertical shears near the ground, and another wind maximum over the ocean. It is tempting to conclude from Fig. 21.2 that a stronger jet over land is likely to be depressed in elevation, but this conclusion does not hold true on other occasions.

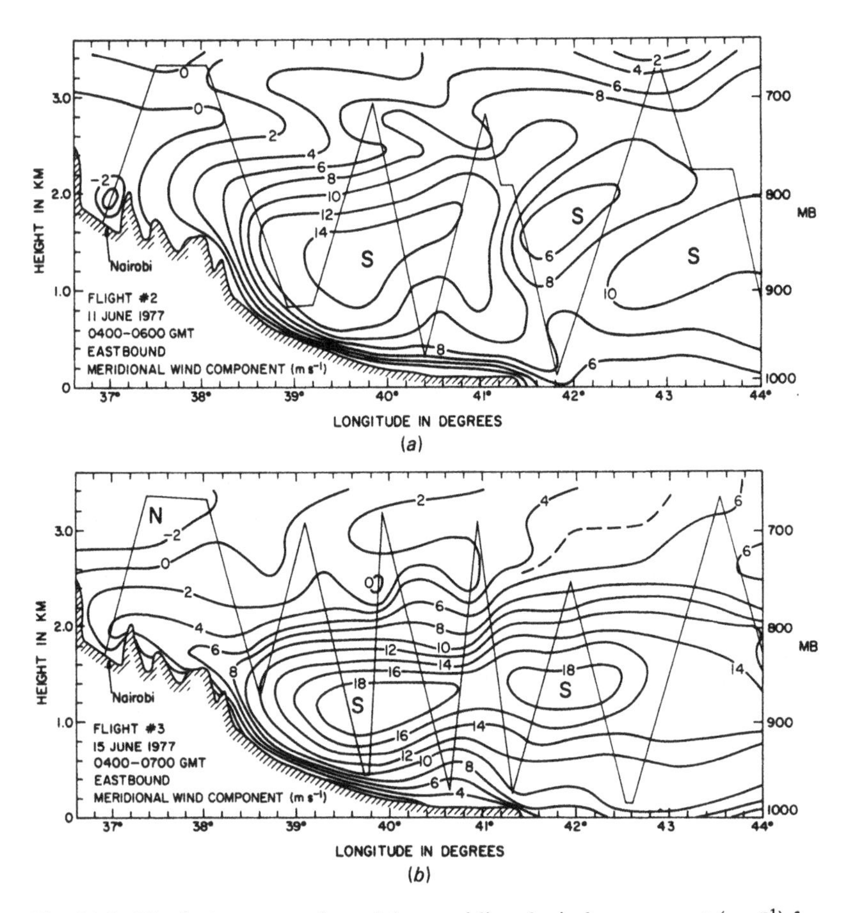

Fig. 21.2. Vertical cross-section of the meridional wind component (m s^{-1}) for: (*a*) flight no. 2, 0400–0600 GMT, 11 June 1977; and (*b*) flight no. 3, 0400–0700 GMT, 15 June 1977. Notice the oceanic wind maximum and pronounced vertical wind shears over land.

Fig. 21.3 shows the specific humidity distributions. At low levels, values as high as 17 g kg^{-1} over land and 18 g kg^{-1} over water are reached. On 11 June there was a convective disturbance immediately east of the coast (42°36′ E) which contributed to a rise of humidity values there. The convective clouds however, could not grow very high because of the presence of dry air at 3 km. There is some evidence that this dry air extended as far east as the Seychelles. For 15 June there was a comparatively rapid decrease of humidity.

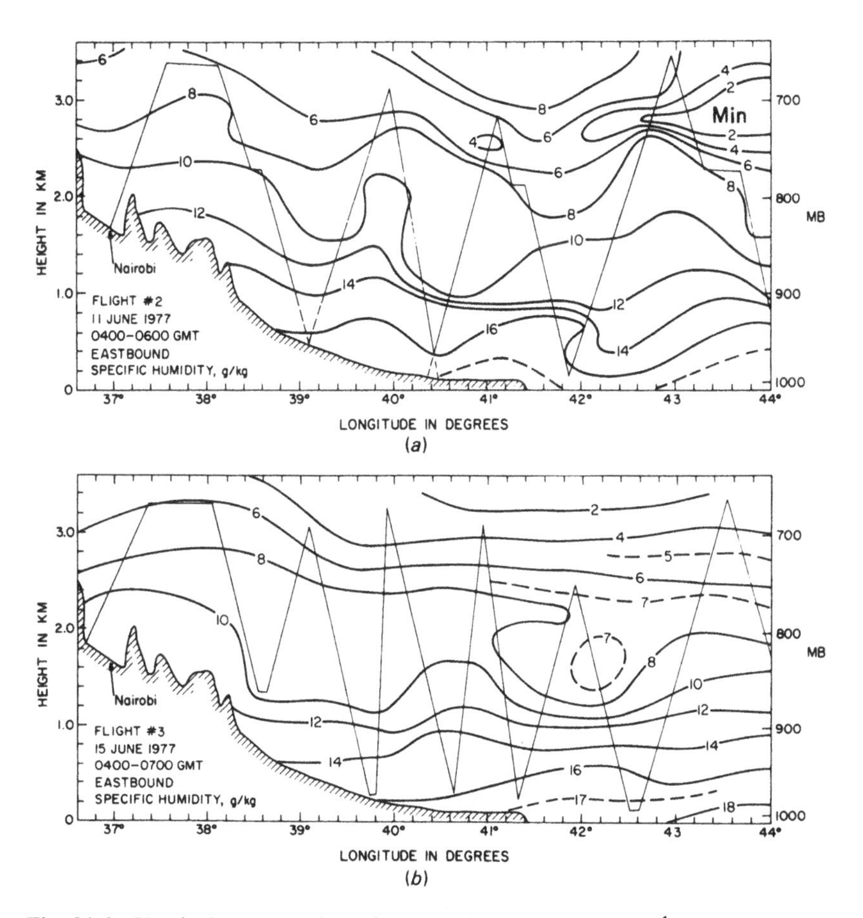

Fig. 21.3. Vertical cross-section of the specific humidity (g kg^{-1}) along approximately 2° S, based on the eastbound flights from Nairobi for: (*a*) flight no. 2, 0400–0600 GMT, 11 June 1977; and (*b*) flight no. 3, 0400–0700 GMT, 15 June 1977.

Figs. 21.4*a*, 21.5*a* and 21.6*a* show the corresponding patterns for 29 June and Figs. 21.4*b*, 21.5*b* and 21.6*b* the corresponding patterns for 4 July. The exploration on 29 June was unique because the outbound (eastbound) flight began at 1500 GMT, covered the early night and concluded by 1800 GMT. A significant finding of the night flight was an elevated southeasterly jet over the ocean. On 4 July the jet was found to be the strongest for the entire period of investigation. The intense meridional winds in Fig. 21.5*b* were elevated and are seen to occur near

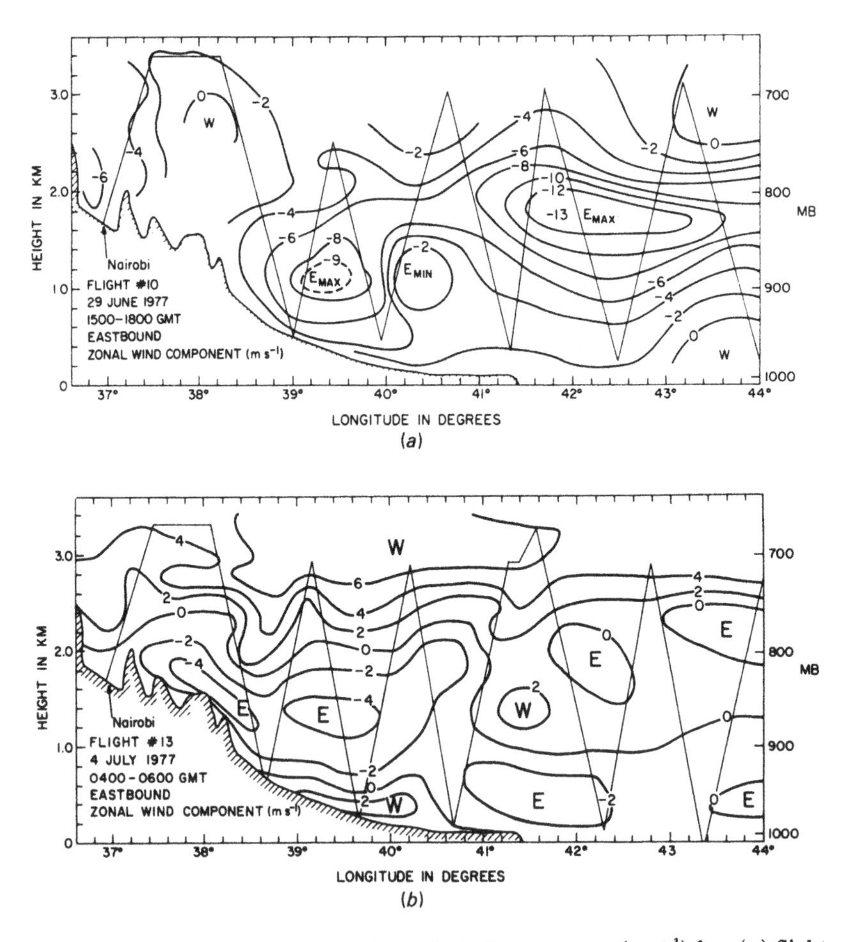

Fig. 21.4. Vertical cross-section of zonal wind component (m s^{-1}) for: (*a*) flight no. 10, 1500–1800 GMT, 29 June 1977; and (*b*) flight no. 13, 0400–0600 GMT, 4 July 1977.

2.0 km. The corresponding specific humidity distribution reveals low values at the altitudes where the winds are strong and the appearance of a dry tongue at about 2.6 km over the coast.

Figs. 21.7 and 21.8 show the northward fluxes of water vapour for the four days 11 June, 15 June, 29 June and 4 July. Important features are the two maxima of water vapour flux, one over the land and the other over the ocean. These flux maxima are lower than the corresponding wind maxima because of the rapid decrease of humidity with height.

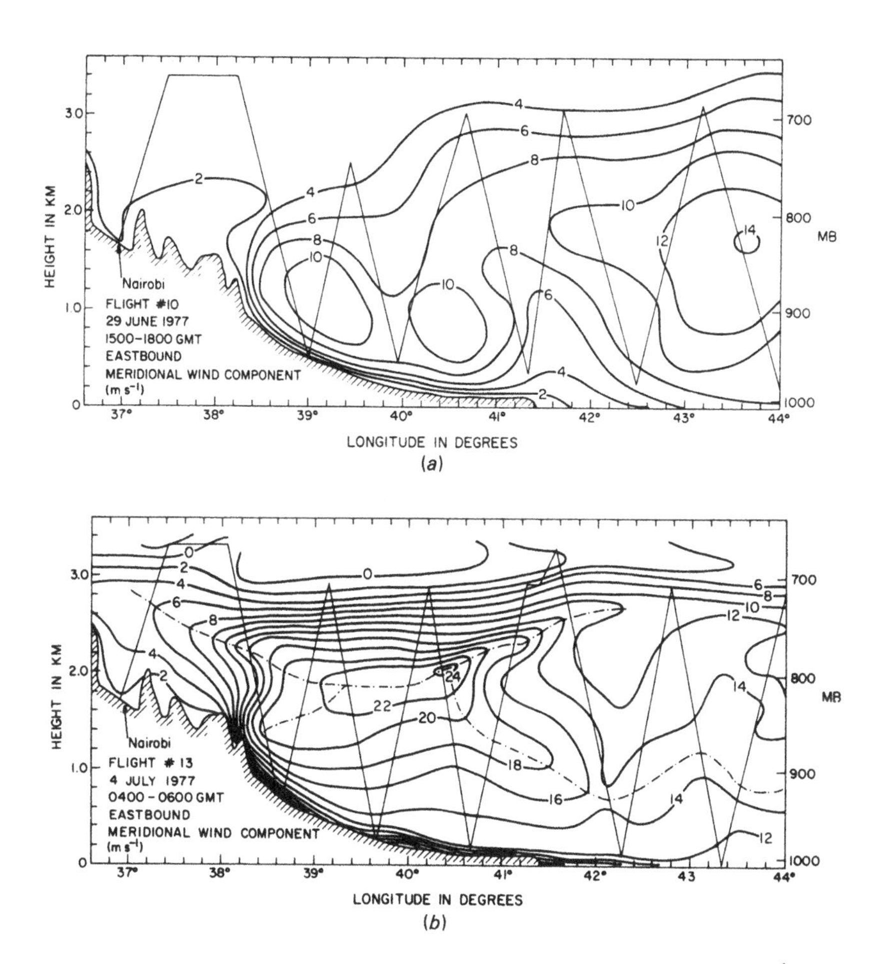

Fig. 21.5. Vertical cross-section of the meridional wind component (m s^{-1}) for: (*a*) flight no. 10, 1500–1800 GMT, 29 June 1977; and (*b*) flight no. 13, 0400–0600 GMT, 4 July 1977.

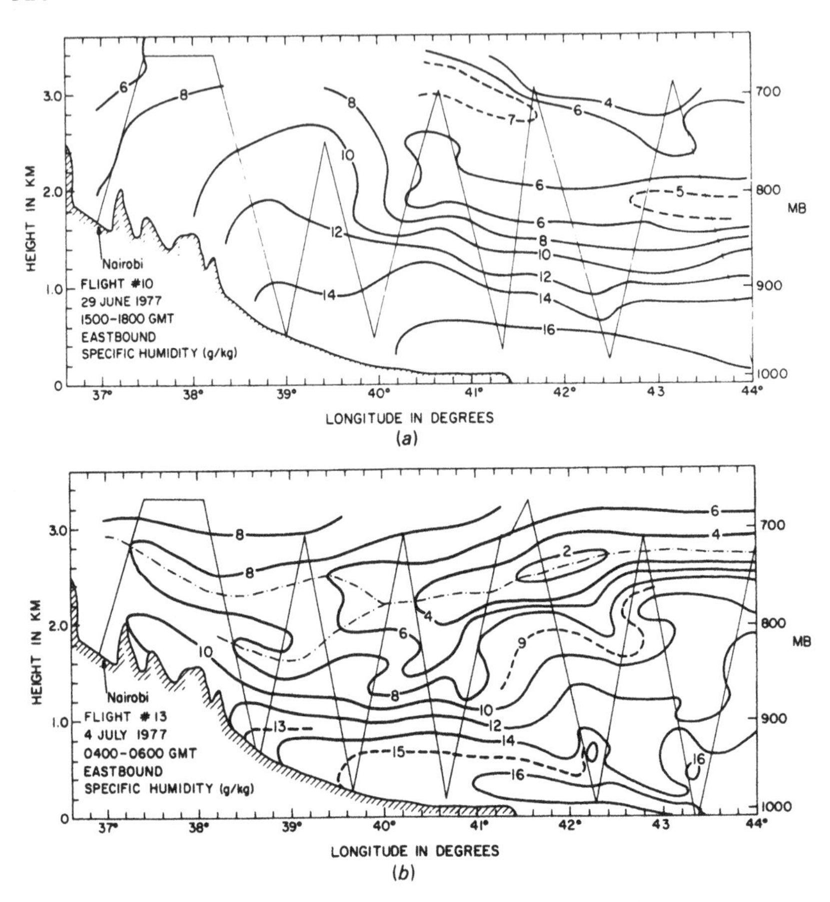

Fig. 21.6. Vertical cross-section of the specific humidity (g kg^{-1}) for: (*a*) flight no. 10, 1500–1800 GMT, 29 June 1977; and (*b*) flight no. 13, 0400–0600 GMT, 4 July 1977.

Based on Figs. 21.7 and 21.8, the integrated water vapour flux (F) can be computed. This is defined as

$$F = \frac{1}{g} \iint vq \, \mathrm{d}p \mathrm{d}L,$$

where g is the acceleration due to gravity, v the northward wind component, q the specific humidity, p the pressure and L eastwards distance. The integration is carried out for a cross-section extending from 1000 mb to 700 mb and from 37° E to 44° E. Table 21.1 shows F across 2° S for the four days 11 June, 15 June, 29 June and 4 July.

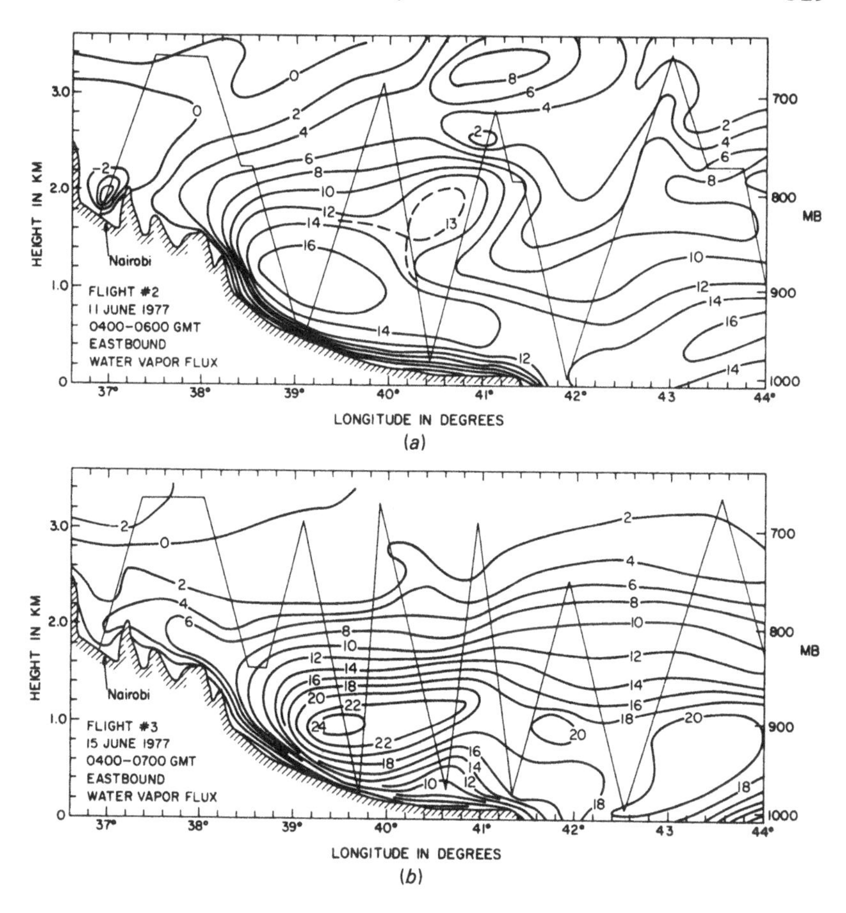

Fig. 21.7. Vertical cross-section of the water vapour flux (10^{-3} tonnes m^{-1} s^{-1} cb^{-1}) for: (*a*) flight no. 2, 0400–0600 GMT, 11 June 1977; and (*b*) flight no. 3, 0400–0700 GMT, 15 June 1977. Notice the primary maximum over land and secondary maximum over the ocean.

The average value of 1.79×10^{10} tonnes per day may be compared with the figure 3.38×10^{10} tonnes per day obtained by Saha and Bavadekar (1973). They evaluated the integral between the limits 1000 mb and 450 mb in altitude and 42° E and 75° E in longitude along the Equator, employing the upper-air data for August 1964, using special Seychelles and Gan data. Their cross-equatorial water vapour flux amount is obviously an underestimate because they omitted the contribution from 37° E to 42° E. Lack of data compelled them to exclude this portion of the contribution.

TABLE 2.1. *Integrated water vapour flux across the cross-sections shown in Figs. 21.7 and 21.8.*

Day	Water vapour flux (10^{10} tonnes per day)
11 June	1.50
15 June	1.97
29 June	1.43
4 July	2.26
Average	1.79

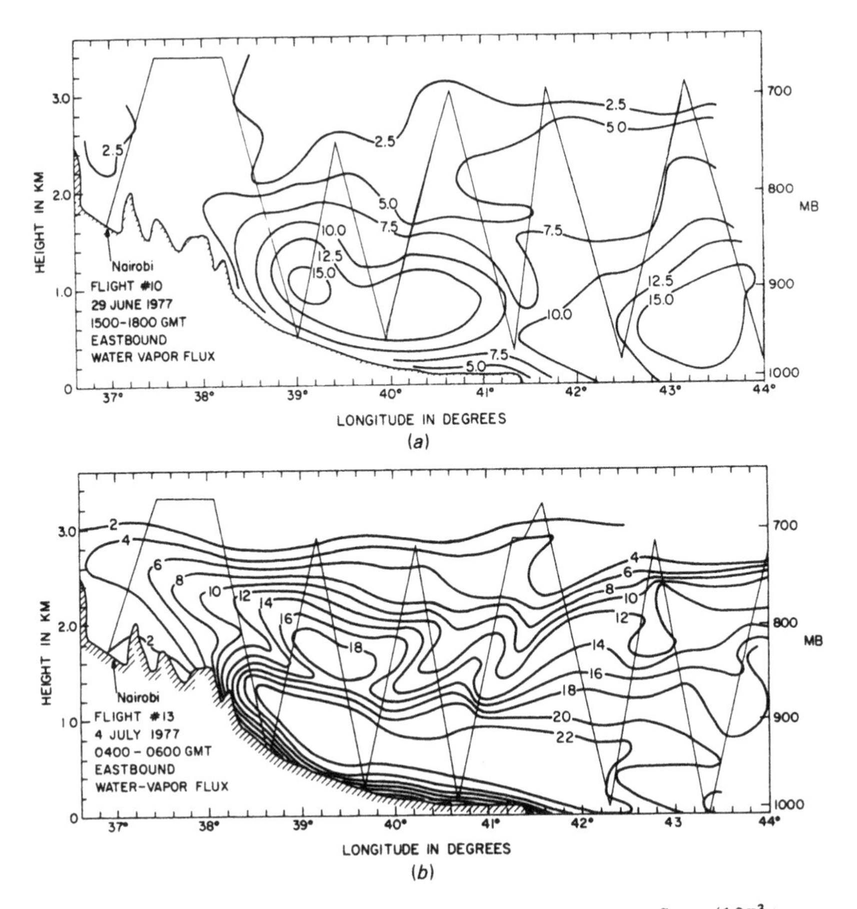

Fig. 21.8. Vertical cross-section of the water vapour flux (10^{-3} tonnes m^{-1} s^{-1} cb^{-1}) for: (*a*) flight no. 10, 1500–1800 GMT, 29 June 1977; and (*b*) flight no. 13, 0400–0600 GMT, 4 July 1977.

21.4 Summary and outlook

The kinematic structure of the low-level jet based on observations from four aerial missions from Nairobi has been presented. This structure reveals a mesoscale variation of the jet. The existence of a wind maximum over the ocean not far from the coast is also disclosed. The calculation further shows that significant amounts of water vapour are transported across 2° S between 37° E and 44° E.

Evidently, previous estimates of this quantity across the Equator were grossly underestimated because of lack of data near the African coast, where the winds are the strongest.

The aerial reconnaissance missions over east Africa, the analyses of the flight data, the presentation of the preliminary analyses at the IUTAM/IUGG meeting on monsoon dynamics at Delhi in December 1977, and the preparation of this article were made possible by different grants from the National Science Foundation. The last-mentioned effort was made possible by Grant No. ATM-7720631 of the GARP Division of Atmospheric Sciences, National Science Foundation, and by the NSF Institutional Grant (Beaumont Faculty Development Fund) to St Louis University. Thanks are due to the East African Meteorological Community for providing relevant data and facilities.

References

Findlater, J. (1971) Mean monthly airflow at low levels over the western Indian Ocean. *Geophys. Mem.* (HMSO, London), **16**, 115, 1–53.

Findlater, J. (1972) Aerial explorations of the low-level cross-equatorial current over Eastern Africa. *Quart. J. Roy. Meteor. Soc.*, **98**, 274–89.

Raghavan, K., Sikka, D. R. and Gujar, S. V. (1975) The influence of cross-equatorial flow over Kenya on the rainfall of Western India. *Quart. J. Roy. Meteor. Soc.*, **101**, 1003–5.

Saha, K. R. and Bavadekar, S. N. (1973) Water vapour budget and precipitation over the Arabian Sea during the northern summer. *Quart. J. Roy. Meteor. Soc.*, **99**, 273–8.

22

Certain aspects of monsoonal precipitation dynamics over Lake Victoria

R. K. DATTA

The rainfall over Lake Victoria is greatly influenced by the seasonal migration of the intertropical convergence zone (ITCZ), which is part of the monsoonal cycle over the Indian Ocean. Rainfall statistics during the rainy months April and November show that: (i) a maximum occurs on the western side of the lake in the early morning; and (ii) maxima occur on the eastern side in the late afternoon and around midnight. The factors determining these distributions are discussed.

22.1 Introduction

Lake Victoria is the biggest freshwater lake in the tropical regions of the world. It straddles the Equator and lies within the countries of Kenya, Tanzania and Uganda, having a surface area of approximately 70 000 km^2. Compared to the other freshwater lakes of the world, it is the second largest, being next to Lake Superior. The River Nile originates from this lake at Jinja in Uganda and, passing through Lakes Kyoga and Mobutu Sese-Seko, is joined by the Blue Nile from Ethiopia and then flows through the Sudan into Egypt. The catchment area of the Upper Nile up to Nimule at the Uganda–Sudan border is 41 000 km^2 of which 87% lies in Kenya, Uganda and Tanzania and the remaining 13% in Rwanda, Burundi and Zaire.

Since the River Nile has a direct bearing on the economic development of all these countries, study of the water balance of Lake Victoria, which regulates the flow of the Nile, is of immense importance. The unprecedented rise in the levels of the lakes in the early sixties led to a joint study by the World Meteorological Organisation (WMO) and the United Nations Food and Agricultural Organisation (FAO) in 1963, and subsequent initiation of the Hydromet Survey Project with the assistance of the United Nations Development Project (UNDP) and technical guidance from WMO in collaboration with five out of eight affected countries, namely Egypt, Kenya, Sudan, Tanzania and Uganda.

With the start of the project, existing hydrometeorological observatories in this region were upgraded and more were installed. About 17 island observatories for the measurement of precipitation over Lake Victoria were started or upgraded. But even today, the large open area of the lake remains without any observatory. Thus rainfall estimation over this vast area remains largely speculative. The estimates of various components of the water balance for the lake by various workers are summarized in Table 22.1.

TABLE 22.1. *Comparison of estimates of the annual water balance components for Lake Victoria by various workers (all parameters in mm).*

Parameters	Hurst (1952)	Meredieu (1961)	Baulny and Baker (1970)	WMO (1974)
Rainfall	1420	1145	1630	1582–690
Evaporation	1350	1130	1523	1423–96
Inflow	230	215–60	260	238
Outflow	305	305	306	426

Two interesting facts are brought out from Table 22.1:

(i) There is a large variation in the estimates of precipitation; between the lowest and highest figures there is a difference of about 50%.

(ii) A 20% error in the estimates of precipitation is equivalent to an error of over 100% in the inflow and outflow values. Thus, a small percentage error in precipitation can make a high precision of measurement of inflow and outflow pointless.

The magnitude of evaporation has been derived indirectly from the other three parameters; thus it has all the inherent limitations in accuracy. Apparently because of scanty observations over the lake surface, the estimation of precipitation has been based on the rainfall distribution assumed over the lake by different workers, e.g. Baulny and Baker (1970) considered 75 to 80% of the lake rainfall to be uniform with sharp gradients occurring in the vicinity of coastal stations and island stations, whereas in the Hydromet Project (WMO, 1974) isohyets were analysed giving a steady gradient between coastal stations on either side of the lake.

Thus, with the existing network, it is very necessary to understand the physics of formation of precipitation over the lake and thence to model the distribution if rainfall estimates are to be improved.

Various investigations have been carried out to identify the local circulation over Lake Victoria (Flohn and Fraedrich, 1966; Asnani, 1977). Fraedrich (1972) brought out a marked peak of precipitation with a climatological numerical model. This gave an excess of 500 mm per year over the lake compared with the hinterland, thereby suggesting a region of high precipitation over the lake as a result of nocturnal rainfall.

The movement of the intertropical convergence zone (ITCZ) greatly influences the weather-producing systems over the area. It moves as much as 20° north during July and August (summer of northern hemisphere) and approximately 15° S during January and February (summer of southern hemisphere)–see Fig. 22.1. Between these extreme positions, the ITCZ crosses Lake Victoria twice, once during April–May and again during November–December. This feature is well reflected in the rainfall figures derived by Faris (1977) for Lake Victoria (Fig. 22.2) which show two rainfall maxima, one around April and the other around November.

The seasonal changes in the large-scale flow pattern associated with the movement of the ITCZ are further modified on the mesoscale by the land–lake breeze effect (Cocheme, 1960).

The purpose of this chapter is: (i) to examine the physics of formation of precipitation over and around the lake and thence to produce a model spatial distribution of precipitation; and (ii) to provide details of available data and their interpretation.

For this purpose, it was found possible to obtain self-recording rain-gauge observations for various coastal and island stations, and to estimate from wind reports the time of occurrence of lake and land breezes.

Using as a basis the information on the time of day when there is a maximum frequency of rainstorm activity, and the time of occurrence of the land–lake breeze, an attempt is made to explain their association. Finally, a model of the spatial distribution of precipitation over the lake is constructed.

It is hoped that the results of this study will help provide a proper estimate of precipitation over the lake, thereby improving water balance estimates; also that they will provide insight into the processes associated with monsoon rainfall over this part of the world which could be useful for numerical simulations.

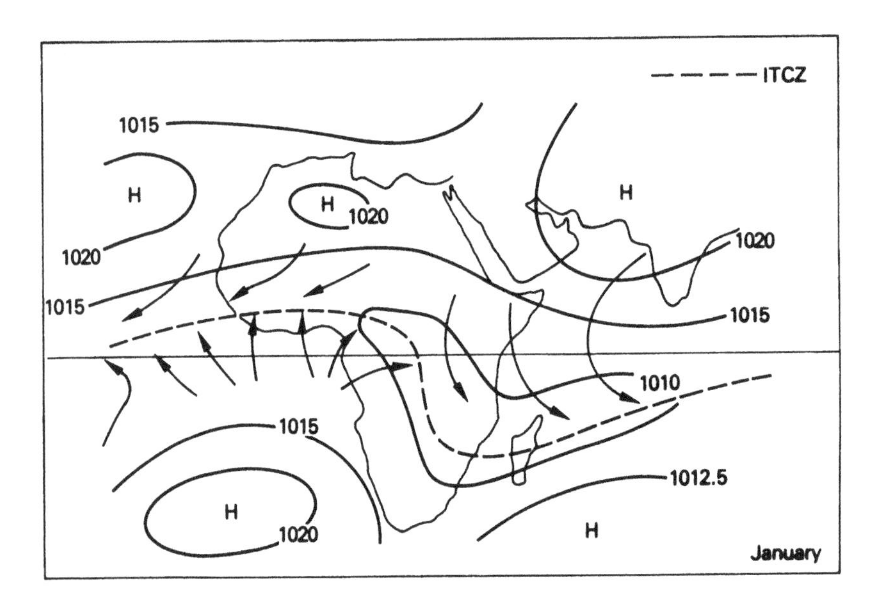

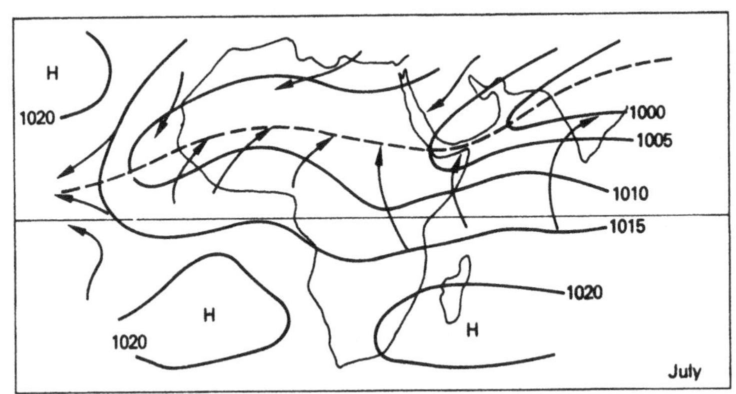

Fig. 22.1. Climatological mean sea-level isobars (mb) over Africa for January and July, and the surface position of the intertropical convergence zone (ITCZ).

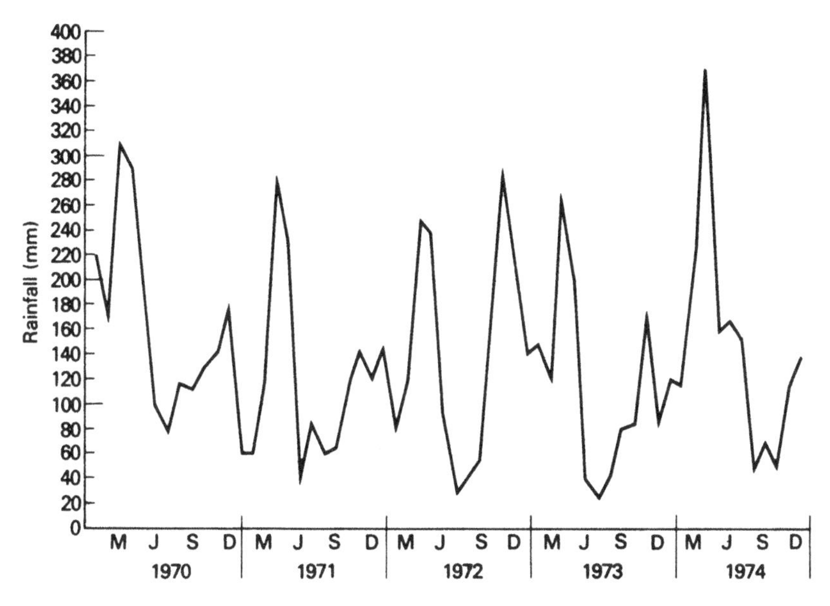

Fig. 22.2. Monthly rainfall over Lake Victoria (mm) 1970–5. (After Faris, 1977.)

22.2 Data sources

April and November are normally months of heavy rainfall, and therefore data have been selected for these two representative months for analysis. The list of stations used for this purpose is given in Table 22.2. They are also shown in Fig. 22.3.

TABLE 22.2. *Details of data analysed.*

Serial number	Name of station	Location with respect to lake	Period of data used
1	Entebbe	Northwest (west coast)	1970–6
2	Masaka	West coast	1972–6
3	Kishanda	West coast	1971–6
4	Kahunda	South/southwest (west coast)	1972, 1974–6
5	Koome	Island (west)	1969–70, 1975–6
6	Bukasa	Island (west)	1969, 1971–4
7	Ukerewe	Island (east)	1974–6
8	Rusinga	Island (east)	1973–6
9	Nabuyongo	Island (central)	1975–6
10	Muhuru Bay	East coast	1969, 1971–6
11	Kadenge	East coast	1971, 1973–6

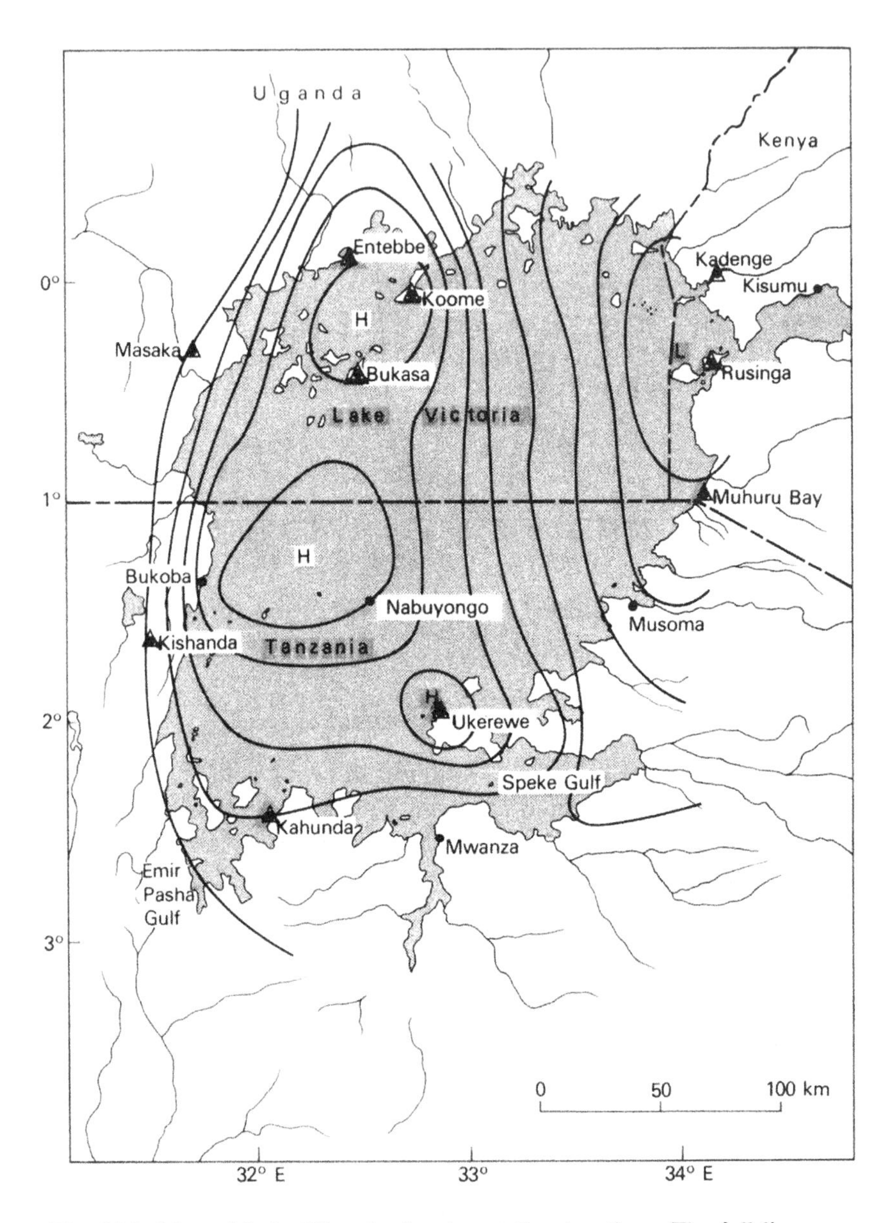

Fig. 22.3. Map of Lake Victoria showing station locations. The full lines are suggested isohyets for April and November.

The East African Meteorological publication (EAMD, 1968), giving diurnal variation of precipitation, is also referred to.

To investigate the time of occurrence of the land and lake breeze, autographic wind reports from some of the stations listed in Table 22.2 have been utilized, whereas more detailed information has been obtained from EAMD publication (1961) which summarizes the frequency of various wind directions at different hours of the day.

22.3 Method of analysis and results

22.3.1 Analysis of rainfall frequency and amounts

All available autographic rainfall records were used to summarize the total number of incidents of rain for each hour of the day in the months of April and November. The frequency graphs were plotted for all the stations listed in Table 22.2. It was evident that these were divided into three classes, i.e. stations located on the west coast of the lake, island stations, and stations located on the east coast.

Stations located on the west coast. Masaka, Kishanda, Entebbe and Kahunda, representing coastal stations on the west coast of the lake, exhibit the following features: (i) during April, at Kishanda, Masaka and Entebbe the frequency increases steadily from a minimum (zero) at midnight to a maximum at around 0800 hours and then falls comparatively rapidly to almost zero by 1600 hours, remaining the same until midnight; during November, the pattern of the plot is similar except that the maximum shifts to mid-day and is somewhat lower.

At Kahunda, which lies near the south tip of the lake, there is much the same probability of occurrence of rainfall between 0400 hours and 1400 hours. The frequency is zero or near zero after 1500 hours.

All these results are shown in Fig. 22.4.

The pattern during November at all the stations is similar to that at Kahunda, i.e. a maximum during the mornings and zero after 1500 hours (see Fig. 22.5).

The island stations. The island stations are Bukasa and Koome on the western portion of the lake, Nabuyongo in the centre and Ukerewe and Rusinga on the eastern side. Individual plots for the first three stations exhibit great similarity. Rusinga, however, is entirely different and

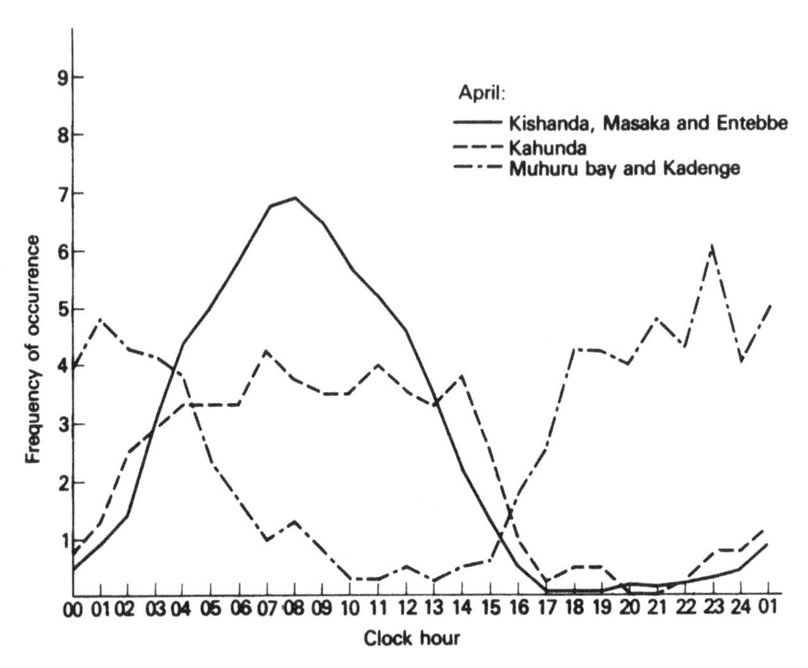

Fig. 22.4. Frequency of rainfall occurrence at various times of day during April on the west coast (Kishanda, Masaka and Entebbe), the south tip (Kahunda) and the east coast (Muhuru Bay and Kadenge). (After Faris, 1977.)

Ukerewe lies between the others. The salient features are:

(i) The plots for Koome and Bukasa are very similar in that most of the precipitation occurs during early morning and little or no precipitation during the afternoon, evening and night.

(ii) Nabuyongo, which is the most representative station available for the centre of the lake, is very similar to the above two island stations, i.e. most of the precipitation takes place during the early morning and little or no precipitation occurs during the evening and night. The maximum frequency is around 0600 hours. It is also significant to note from available data that the precipitation amounts at Nabuyongo are generally higher than those at the other island and coastal stations (see Fig. 22.6).

(iii) One would expect Ukerewe to exhibit a pattern similar to that for Rusinga on the eastern side of the lake (discussed below), but apparently it is similar to those of the three island stations on the western side with its lack of precipitation after 1500 hours. But this station does not exhibit a marked maximum during the early hours. The maximum rainfall amount is observed from 0600 to 0800.

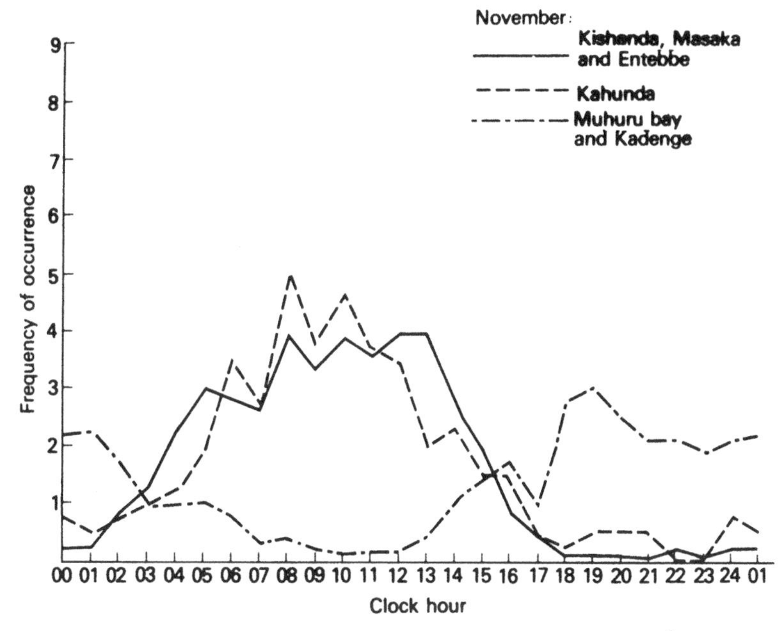

Fig. 22.5. Frequency of rainfall occurrence at various times of day during November on the west coast (Kishanda, Masaka and Entebbe), the south tip (Kahunda) and the east coast (Muhuru Bay and Kadenge).

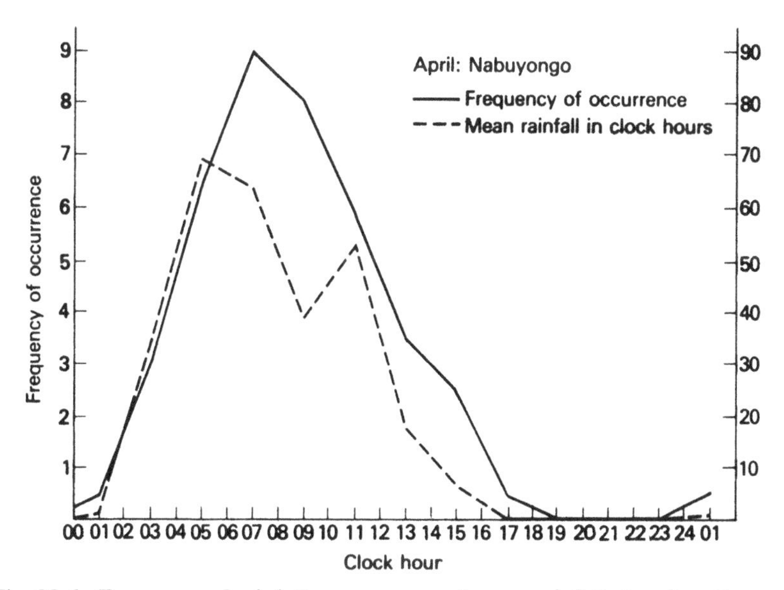

Fig. 22.6. Frequency of rainfall occurrence and mean rainfall at various times of the day during April in the central lake (Nabuyongo).

(iv) Rusinga, which lies at the northeastern end of the lake, presents an entirely different picture. It experiences a minimum frequency of rainfall during the early morning until noon, increasing gradually towards evening, with a maximum occurring towards midnight.

All these results are shown in Figs. 22.7 and 22.8.

Stations on the east coast of the lake. There are only two representative stations; these are Muhuru Bay and Kadenge. (Kisumu was not considered because of the gulf effect.) Both stations are similar to Rusinga, i.e. they exhibit (see Figs. 22.4 and 22.5):

(i) a sharp decrease in the rainfall frequency after midnight with a minimum around noon;

(ii) a sharp increase in the frequency after 1500 hours giving the first maximum around 1800 hours, the normal time for thunderstorm activity in the tropics associated with insolation. The second maximum is around midnight, possibly associated with radiative cloud-top cooling (Riehl, 1954).

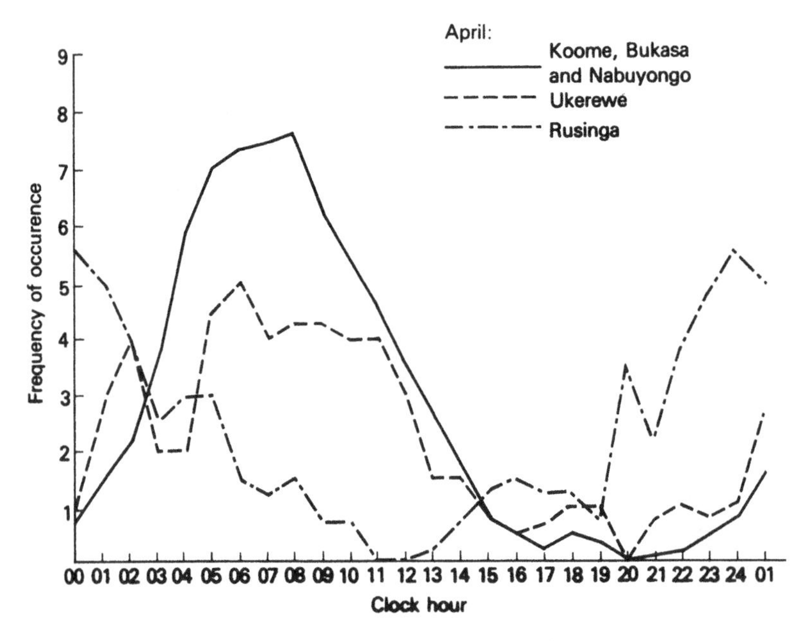

Fig. 22.7. Frequency of rainfall occurrence at various times of day during April at Ukerewe, Rusinga, and Koome, Bukasa and Nabuyongo. (After Faris, 1977)

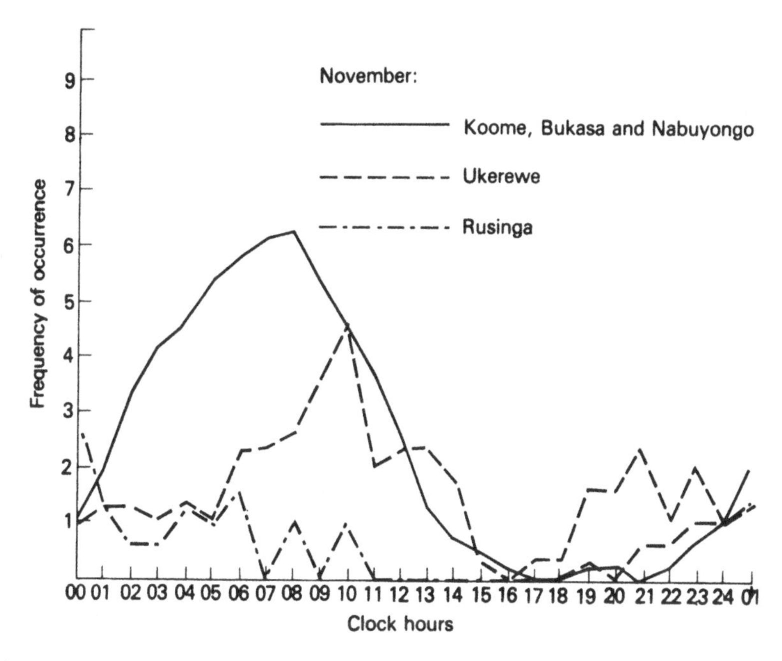

Fig. 22.8. Frequency of rainfall occurrence at various times of day during November at Ukerewe, Rusinga, and Koome, Bukasa and Nabuyongo.

22.4 Winds in the vicinity of the lake

The winds over the lake are mainly easterly with a slight southerly component north of the Equator during April and a slight northerly component during November. This flow pattern is modified by lake and land breezes giving rise to the well-known Lake Victoria low (Cocheme, 1960).

It is generally agreed that over and around Lake Victoria the rainfall distribution is, to a large extent, governed by the interaction between the large-scale flow and land and lake breezes. The precise mechanism has not, however, yet been explained. Using wind observations from station autographic records (see also EAMD publication, 1961) one can conclude that the lake breeze is a maximum around 1500 hours. This is the time when the easterly component reaches a maximum over stations on the west coast and the westerly component reaches its maximum over the east coast. The optimum intensity of the land breeze is encountered between 0600 and 0900 hours. Thus, over the west coast, the intensity of the lake breeze starts decreasing after 1500 hours as the land breeze component gradually builds up replacing the larger-scale easterlies until around 0600 when it reaches maximum intensity. On the east coast, however, it is the lake breeze which has to replace the large-scale easterlies. The only stations for which observations were available were Kisumu and Musoma. Kisumu, on the gulf, is not very representative and Musoma's observations are only for 0900 and 1500 hours. The 0900 observations for April show an easterly component on 140 out of 150 occasions, whereas the 1500 observations have a westerly component on 113 out of 150 occasions. Similar figures are available for the month of November. Thus the land breeze is at its maximum intensity from 0600 to 0900 hours and the effect of the lake breeze is a maximum around 1500 hours. The wind observations at the central island station, Nabuyongo, do not exhibit any set pattern, apparently because the station can be affected by land breezes from almost any direction.

22.5 Radiation data

Paramena (1977) has worked out the global mean monthly solar radiation for the years 1970–4. According to his study the significant features over Lake Victoria are similar during April and November. Figs. 22.9 and 22.10, taken from Paramena's study, depict a minimum over the central, western and northwestern parts of the lake. The maximum in the north-east part of the lake is also well marked.

22.6 A proposed mechanism leading to the rainfall distribution over the lake

From the summaries of observations of rainfall, wind and radiation, the following mechanism is proposed to explain the time and spatial differences in precipitation over and around the lake.

Over the eastern part of the lake, the region of high radiative input, the insolation together with the interaction of the westerly lake breeze with the prevailing easterlies, leads to thunderstorm activity during the afternoon and evening. Latent heat released in association with thunderstorm activity over this area is transported to the central and western portions of the lake by the prevailing winds, inhibiting thunderstorm activity there. Thus the central and western parts of the lake are characterized by the lack or absence of thunderstorm activity during the evening and night. However, interaction of the early morning westerly land breeze with the prevailing easterlies, with no transport of released latent heat from the

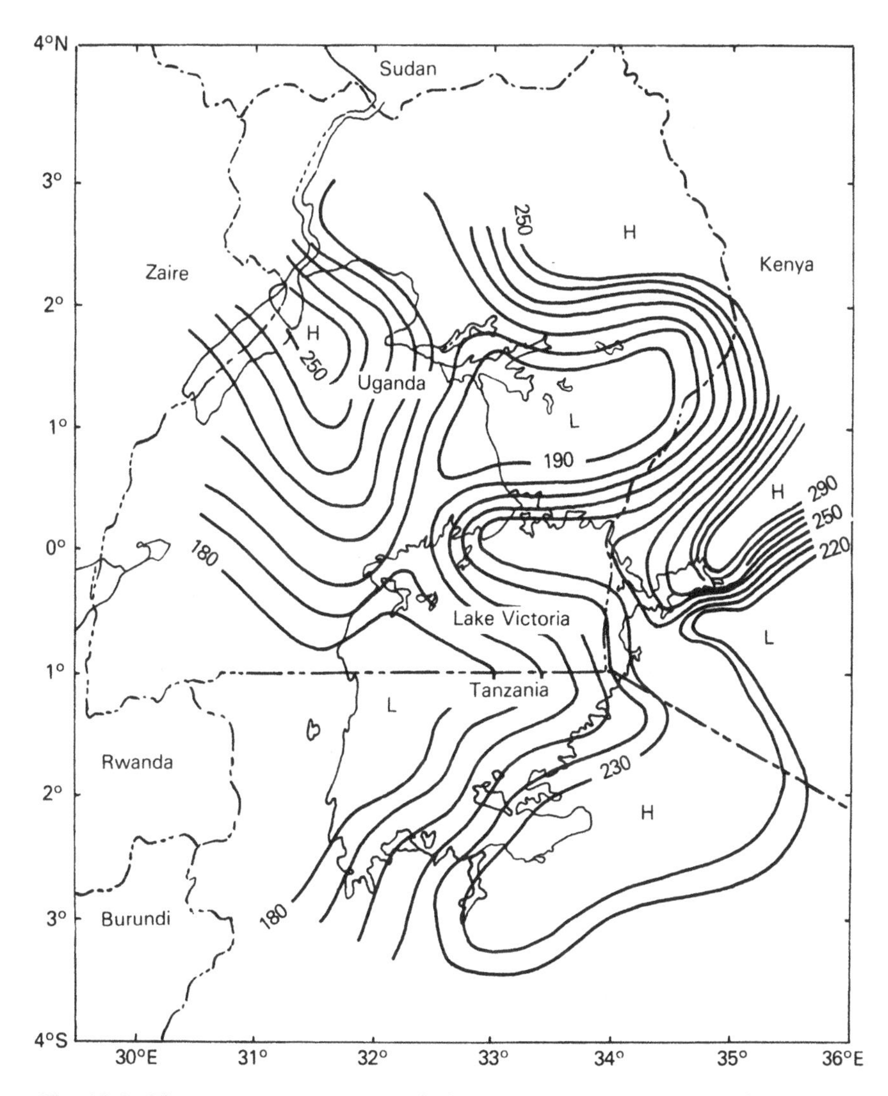

Fig. 22.9. Five-year mean global radiation over the project area during April. Contours are labelled in W m^{-2}. (After Paramena, 1977.)

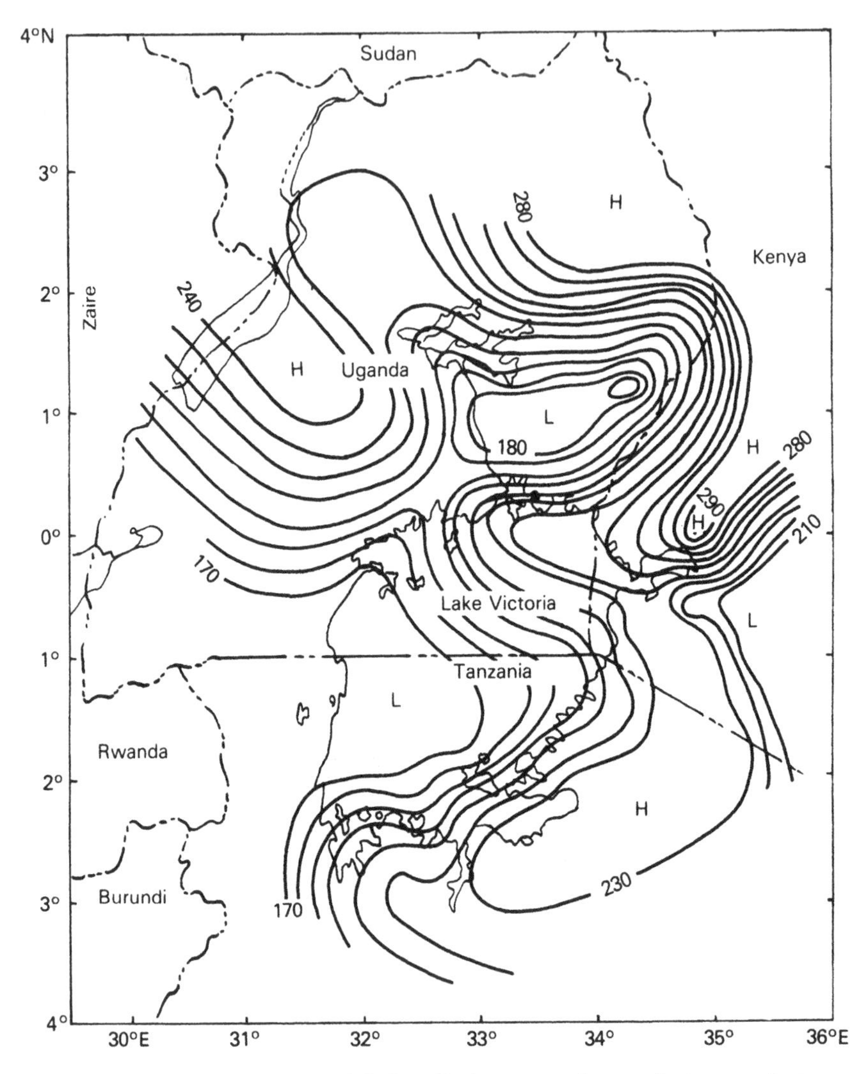

Fig. 22.10. Five-year mean global radiation over the project area during November. Contours are labelled in W m^{-2}. (After Paramena, 1977.)

east (since thunderstorm activity decreases rapidly after midnight over the eastern part of the lake), and nocturnal atmospheric cooling, provide the mechanism for vigorous activity in the early morning.

22.7 Conclusions

The rainfall data from self-recording rain gauges available for the present study covered only a short period, varying from 2 to 7 years. Wind observations were even more deficient. Within these limitations, the salient features of the study were:

(i) The main rainfall months for the region are April and November, associated with the movement of the ITCZ through the area as part of the monsoonal circulation change over the Indian Ocean.

(ii) The regions on the west coast, west lake and the central lake have similar rainfall patterns. The interaction of the land breeze with the prevailing easterlies plays a major role in determining the precipitation distribution over the region. The frequency of occurrences of precipitation, and its intensity, increase from the west coast up to the region where the land breeze can penetrate and provide a trigger action. The rainfall occurs mainly during the early morning. Little or no precipitation occurs during the evening and night.

(iii) The present observations of April rainfall, although scanty (see Table 22.3), support the conclusion of Fraedrich (1972) that there is a marked precipitation peak over the lake. However, this conclusion contradicts the assumption made by Baulny and Baker (1970) that there is a minimum of precipitation over the lake.

(iv) The regions on the east coast and over the eastern lake have similar rainfall patterns (excepting Ukerewe). The precipitation pattern can be explained by consideration of the fact that there is high land east of the lake and the region has a quasi-permanent high level of insolation leading to an interaction of the land/lake breeze with the prevailing easterlies. Little or no rainfall occurs during the easterly land breeze

TABLE 22.3. *April rainfall (mm).*

	Station						
	Masaka	Nabuyongo	Entebbe	Bukoba	Mwanza	Musoma	Kisumu
April 1975	158.0	394.9	171.3	332.3	89.1	73.1	164.5
April 1976	154.1	310.5	217.1	380.9	87.3	149.5	32.9

(nocturnal) period. But during the daytime, when the lake breeze occurs, there is a marked convergence between warm and almost dry easterlies and a cool moist lake breeze. Because of the location of the high land the precipitation normally increases eastward off the coast, thereby leading to a minimum in the rainfall over the northeastern tip of the lake.

(v) The interaction of the late-night/early-morning westerly land breeze with the prevailing easterlies provides the mechanism for the early morning precipitation over central and western portions of Lake Victoria.

(vi) Ukerewe presents an apparently anomalous case. The rainfall intensity and distribution are in contrast to those of Kahunda (on the east-southeast coast of Lake Victoria).

These conclusions are used to construct the idealized pattern of isohyets over the lake (Fig. 22.3) for both representative months.

I take great pleasure in recording my sincere thanks and gratitude to Professor J. Nemec, Dr G. W. Kite and the Director of the Hydromet Project for their comments and approval; to Messrs Okila, Otte and other staff of the Meteorological Data Centre for their help in providing the data used in this study; to Messrs A. A. M. Faris and J. A. Paramena for useful discussions and for the provision of Figs. 22.2, 22.4, 22.7, 22.9 and 22.10; to Dr P. K. Das and Professor R. P. Pearce for their review and useful suggestions; to the Organizing Secretary of the Symposium, Professor M. P. Singh, who kindly arranged funds for my travel; and to Miss Rachel Sematimba for typing the manuscript.

References

Asnani, G. C. (1977) Circulation over Lake Victoria. Lecture delivered at the second training course in Computer Application to Water Resources Management, Nairobi.

Baulny, H. L. de and Baker, D. (1970) The water balance of Lake Victoria. Technical Note, Water Development Department, Republic of Uganda.

Cocheme, J. (1960) Some streamlines and contours over the Equator. Proc. Symp. WMO and Munitalp, Nairobi.

EAMD (1961) Frequency of surface wind speeds and directions. EAMD, Nairobi.

EAMD (1968) Summary of the rainfall. EAMD, Nairobi.

Faris, A. A. M. (1977) Personal communication.

Flohn, H. and Fraedrich, K. (1966) Tegesperiodische Zirkulation und Niederschlagsverteilung am Victoria—See (Ostafrika). *Meteorologische Rundschau*, **6**, 157–65.

Fraedrich, K. (1972) A simple climatological model of the dynamics and energetics of the nocturnal circulation at Lake Victoria. *Quart. J. Roy. Meteor. Soc.*, **98**, 322–35.

Hurst, H. E. (1952) *The Nile Basin.* Constable, London.
Meredieu, D. E. (1961) FAO Report.
Paramena, J. A. (1977) A preliminary investigation of the spatial and temporal distributions of global and net global solar radiation over the catchments of Lakes Victoria, Kyoga and Mobutu Sese-Seko. Technical Report, UNDP/WMO Hydromet Project, Entebbe, June 1977.
Riehl, H. (1954) *Tropical Meteorology.* McGraw-Hill, New York.
World Meteorological Organization. (1974) RAF Technical Report No. 1, Hydromet Survey of the catchments of Lakes Victoria, Kyoga and Albert, Vol. 1, Part 1.

Part IIb

Modelling and theoretical studies

23

A numerical model of the monsoon trough

P. K. DAS AND H. S. BEDI

A regional primitive equation model is used to simulate the monsoon trough. The model has three layers and a boundary layer adjacent to the Earth's surface. It has smooth profiles for the Himalayas, Western Ghats, and the Burma mountains and includes radiative heating at the surface, but no precipitation.

The model has been integrated up to 8 days starting with an idealised wind and temperature field with meridional and vertical shear. It was observed that the monsoon trough could not be generated by topographic features alone, but the inclusion of radiative heating at the surface led to patterns resembling the monsoon trough. On decreasing the surface albedo near northwest India the monsoon trough was intensified. A Bowen ratio was used to estimate the flux of latent heat due to evaporation of soil moisture. Results suggest that a high rate of evaporation over northeast India leads to a southward extension of the monsoon trough.

23.1 Introduction

An elongated trough, running parallel to the southern periphery of the Himalayas, is an important feature of the monsoon. It is known as the monsoon trough, and its normal position in June is shown in Fig. 23.1. The line of symmetry, shown by a dashed line on the figure, is the axis of the trough. The trough is most marked at sea level, and rarely extends above 700 mb.

Short-period variations in monsoon rain are closely associated with the position of the trough. When the axis of the trough moves north and lies

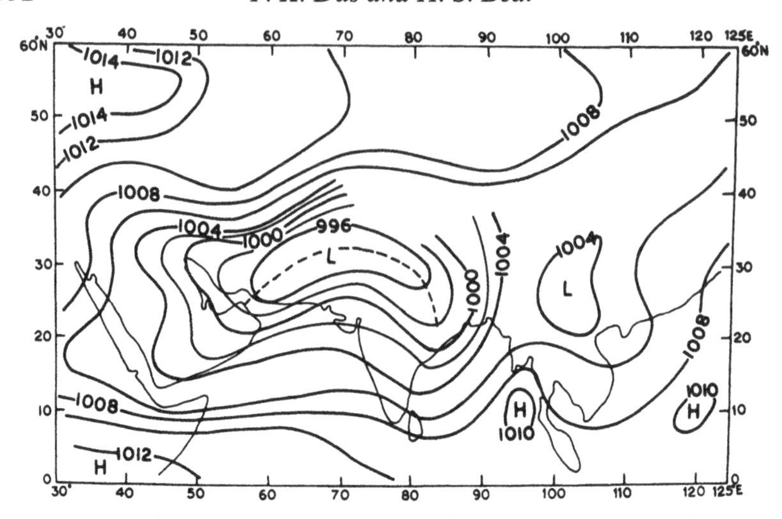

Fig. 23.1. The normal sea-level position of the monsoon trough (June). The line of symmetry shown by the dashed line is the axis of the trough.

close to the Himalayas, widespread and heavy rains occur in northeast India. In India this is known as a 'break' in the monsoon because the rainfall is entirely confined to northeast India. But, a southward movement of the trough leads to an abrupt cessation of rains in the northeast, and heavy rains then occur over the plains of north India and the central parts of the country.

It has been believed that the mountains are responsible for the existence of the monsoon trough. We wish to examine this question with a regional primitive equation model.

23.2 The design of the model

A modified version of the model constructed by Shuman and Hovermale (1968) was used. This particular model was chosen since it was readily adaptable to our computer facilities, and by the fact that it had been used at an operational centre (NMC) for several years (Baumhefner and Downey, 1976).

The model has its northern and southern boundaries at 60° N and the Equator. The western and eastern boundaries are along 0° E and 140° E. Cartesian coordinates are used with the x axis pointing east, and the y axis pointing north. A rigid upper lid is placed at 200 mb, while the lower boundary is made a coordinate surface coinciding with the Earth's topography. Its vertical structure is depicted in Fig. 23.2. There

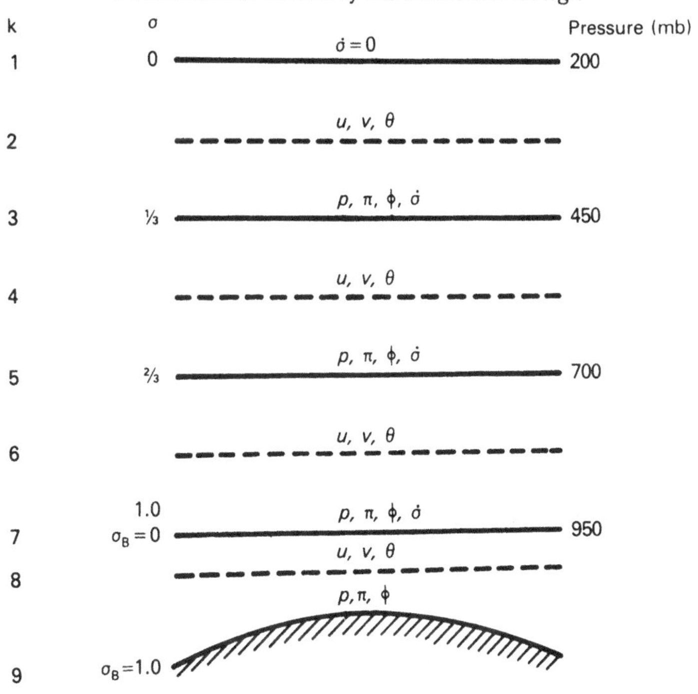

Fig. 23.2. The vertical structure of the model.

are three layers in the troposphere, with a boundary layer, 50 mb deep, adjacent to the Earth's surface.

A linear function of pressure is used as the vertical coordinate. This is

$$\sigma = (p - 200)/(p_S - 200), \tag{23.1}$$

where p is pressure (mb), and p_S stands for the surface pressure. Using σ as the vertical coordinate simplifies the upper and lower boundary conditions, which are

$$\dot{\sigma} = 0 \quad \text{for } \sigma = 0, \quad \text{and} \quad \sigma_B = 1, \tag{23.2}$$

where σ_B refers to the Earth's surface. The subscript B represents values of sigma within the boundary layer.

Despite this advantage, sigma coordinates do not handle truncation errors near steep mountains very well. This is discussed later.

The geopotentials of the 850, 700, 500, 300 and 200 mb surfaces along with sea-level pressure are used as input data. The horizontal resolution of the grid is 250 km, and variations in the map factor are neglected for simplicity.

The governing equations of the model consist of two equations for the change in horizontal momentum, an equation to express hydrostatic equilibrium, the first law of thermodynamics and the equation of continuity. A sixth equation is used to determine the vertical velocity $\dot{\sigma}$. The vertical velocities are then matched at the interface of each layer. These equations represent a closed system for six dependent variables. They are the three components of velocity u, v, $\dot{\sigma}$, the pressure (p), potential temperature (θ) and the geopotential (ϕ). Another function of pressure,

$$\pi = (p/1000)^{\kappa}, \quad \text{where } \kappa = R/c_p, \tag{23.3}$$

is used to determine the pressure gradient in the momentum equations. As shown in Fig. 23.2, p, π, ϕ, $\dot{\sigma}$ are evaluated at $\sigma = \frac{1}{3}, \frac{2}{3}$ and 1, while u, v and θ are computed at the mid-points of each layer.

Along the lateral walls, the time derivatives of all dependent variables are made to vanish, i.e.

$$u_t = v_t = 0, \qquad (p_\sigma)_t = \theta_t = 0, \tag{23.4}$$

where the subscripts denote partial derivatives.

The basic equations of the model have been described by Shuman and Hovermale (1968) and will not be repeated here, but the difficulties that were encountered in the model experiments will be discussed.

23.3 Inclusion of high mountains

The inclusion of high mountains in regional models is difficult because of large truncation errors in their vicinity. This difficulty is partly overcome by using smooth mountains. Accordingly the Himalayas are replaced by the elliptic profile

$$H(x, y) = H_0[1 - (x^2/a^2 + y^2/b^2)]. \tag{23.5}$$

The maximum height (H_0) is taken as 5 km, while the major and minor axes (a, b) are taken as 1750 and 1000 km along the zonal and meridional directions. The centre of the block is put at 35° N, 91.25° E. The other topographic features included are: (i) the Western Ghats; (ii) the Burma mountains; and (iii) the east African mountains. For these mountains smoothed profiles were used.

The gradients of geopotential are transformed to a surface of constant σ by

$$\nabla_p \phi = \nabla_\sigma \phi - \frac{\sigma}{p_S} \nabla p_S \frac{\partial \phi}{\partial \dot{\sigma}}, \tag{23.6}$$

where the subscripts denote surfaces of constant pressure and constant σ.

The pressure gradients are thus made up of two terms which are of nearly the same magnitude, but of opposite sign. The gradient of surface pressure is very large near steep mountains and small truncation errors are magnified. Phillips (1974) suggested that this source of error can be reduced if the hydrostatic component of the geopotential is removed, and the residual part taken as the new dependent variable. Following this suggestion new variables ϕ' and θ' are defined by

$$\phi' = \phi(x, y, p, t) - \bar{\phi}(p), \tag{23.7}$$

$$\theta' = \theta(x, y, p, t) - \bar{\theta}(p), \tag{23.8}$$

where $\bar{\phi}(p)$ is the mean geopotential on a surface of constant pressure. A mean potential temperature ($\bar{\theta}$) for the region is defined by

$$\bar{\theta}(p) = [\langle\phi_T\rangle - \langle\phi_S\rangle]/c_p[\langle p_S^\kappa\rangle - p_T^\kappa]. \tag{23.9}$$

The subscripts T and S refer to the top of the atmosphere (200 mb) and the Earth's surface, while $\langle\ \ \rangle$ denotes average values. Similarly, the mean geopotential is

$$\bar{\phi}(p) = [\langle\phi_S\rangle(p^\kappa - p_T^\kappa) - \langle\phi_T\rangle(\langle p_S^\kappa\rangle - p^\kappa)]/[\langle p_S^\kappa\rangle - p_T^\kappa]. \tag{23.10}$$

Over the mountains, the sea-level pressure is computed from

$$\phi_S - \phi_0 = R\bar{T}_S \ln(p_S/p_0), \tag{23.11}$$

where $p_0 = 1000$ mb, ϕ_0 is the geopotential for p_0 and $\bar{T}_S$ is a mean temperature for the column between p_S and p_0. Deviations from the mean values of ϕ and θ are computed by (23.7) and (23.8).

If the geopotential and potential temperatures are not reduced by this process, abnormal anticyclonic circulations are quickly generated over the Himalayas. It was found to be unrealistic to convert ϕ' from a pressure to a σ surface, or the output from a σ to a pressure surface, by using the relation for hydrostatic balance. The output is much improved using Lagrange's interpolation formula. This expresses ϕ' by the polynomial

$$\phi' = \sum L_k(p)\phi'_k, \tag{23.12}$$

where k is an index indicating the levels at which dependent variables are computed (Fig. 23.2), and

$$L_k(p) = [(p - p_S)(p - 850) \ldots (p - 200)]/ [(p_k - p_S)(p_k - 850) \ldots (p_k - 200)]. \tag{23.13}$$

Opinions differ at present as to the best ways of minimizing truncation errors near mountains, but it appears, from experience with this model, that working on deviations from a reference atmosphere, rather than using the total geopotential, improved the output.

23.4 Finite differences and smoothing

The finite-difference scheme of Shuman and Hovermale (1968) was used in this study. A limitation of the model is that it does not strictly conserve energy, but it is considered adequate for short-range predictions. Grotjahn (1976) found that the Shuman scheme was not as effective as a space-centred scheme, because of substantial group velocity errors in certain wavenumbers. But, Grotjahn was concerned with a barotropic atmosphere. In this context, it is to be noted that interest is mainly in wavelengths around 4000 km. A horizontal resolution of 250 km provides 16 grid-points per wave. But, admittedly, the scheme does not handle waves shorter than 1000 km very well. This leads to inadequate dispersion of inertia–gravity waves. Inertia–gravity waves are more dispersive and contain less energy than Rossby waves; it is not known how much of the energy of Rossby waves is dissipated into inertia–gravity waves. If this is large, it could lead to distortions in the results.

23.5 Initialization

A scheme of dynamic initialization was adopted. This consists of forward integration for one hour, followed by backward integration for another hour. One half the initial pressure and height fields were restored at the end of each cycle, the wind being allowed to freely adjust itself to the pressure field. Diffusion and diabatic heating were not included during initialization because they represent irreversible processes.

It was found necessary to repeat initialization after every 24 hours of integration to suppress the growth of inertia–gravity waves. This was a limitation of the experiment, but was probably caused by boundary conditions at the lateral walls which slowed down the outward dispersion of inertia–gravity waves. Experiments were conducted with high viscous damping near the boundaries, but no substantial improvement could be achieved by this means.

The pressure tendencies, $(p_\sigma)_t$, were monitored during the experiment. These are shown in Fig. 23.3. After each initialization, the effects of inertia–gravity waves were suppressed, and after 4 days of integration a

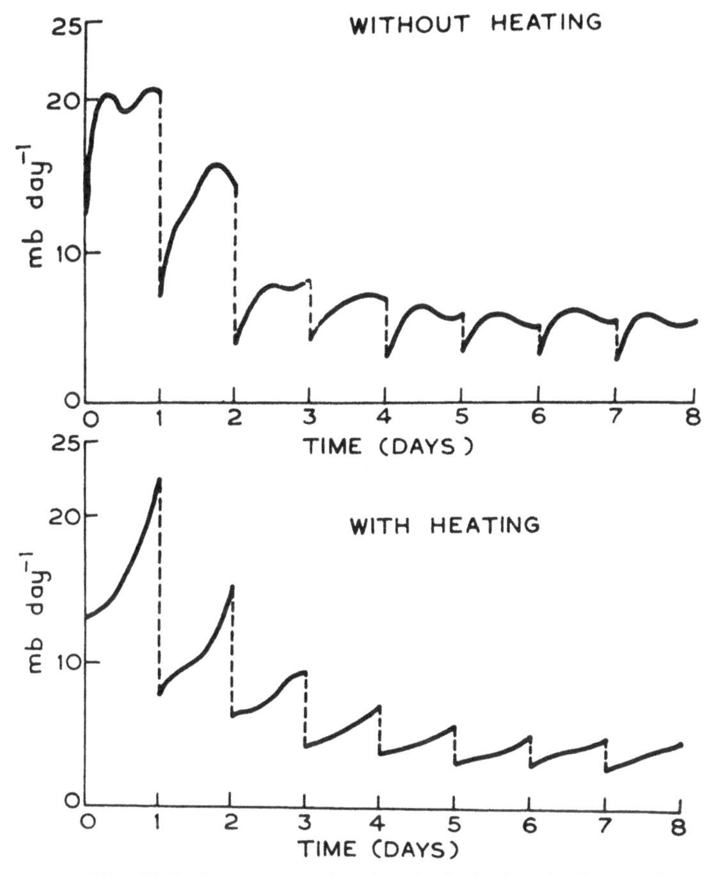

Fig. 23.3. Pressure tendencies, $(p_\sigma)_t$ during the integration.

steady state appeared to have been reached. Indeed, a subsequent experiment revealed that 24-hour initialization was not needed after 8 days. But, this result should be treated with caution because it is not known to what extent the energy of the longer waves, which contain a substantial part of the total energy, has been dissipated by this time.

23.6 Results

23.6.1 With mountains

The experiments were started with an initial profile of zonal winds. This is shown in Fig. 23.4. The initial state contained meridional as well as vertical shear. Thus, westerly winds at the surface were replaced by

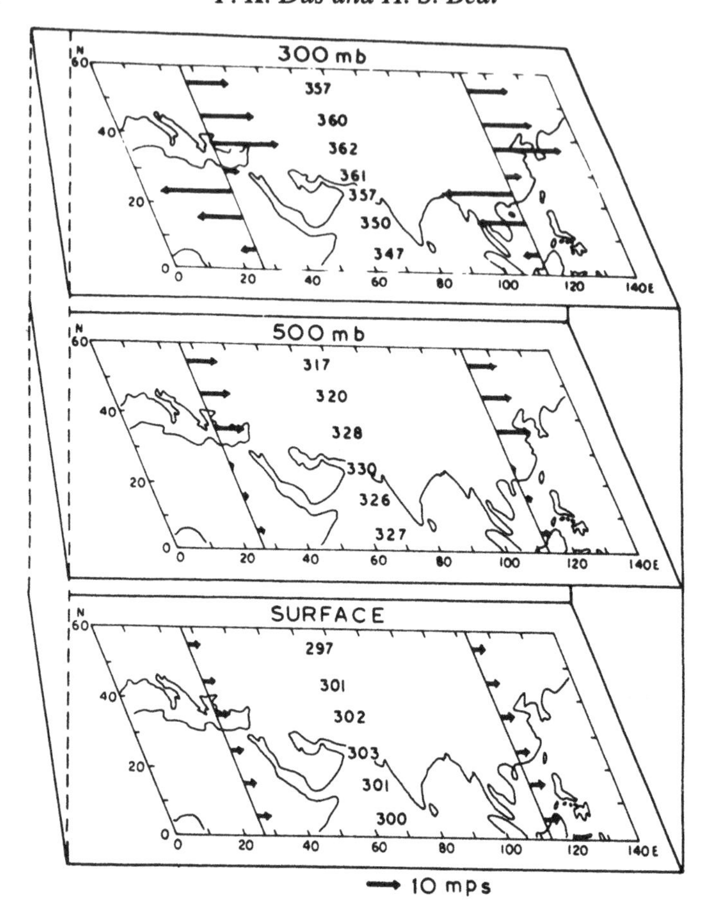

Fig. 23.4. Initial profiles at: (*a*) 300 mb; (*b*) 500 mb; and (*c*) Earth's surface. Zonal winds are indicated by arrows. The numbers in the figure represent temperature in Kelvin.

easterlies above 500 mb in the region south of 30° N. The strength of the upper easterly winds gradually increased with altitude up to 300 mb, and decreased thereafter. To the north of 30° N, the winds were westerly at all levels. This profile was chosen to simulate the zonal features of monsoon winds. The corresponding temperature profile was suitably adjusted to make it compatible with the winds. This was done using the thermal wind relationship.

In Fig. 23.5 the sea-level isobars after 8 days of model time are shown. The interesting features of this figure are: (i) the absence of the monsoon trough; (ii) a cyclonic rotation of isobars to the lee of the Western Ghats; and (iii) the formation of a low leeward of the Burma mountains.

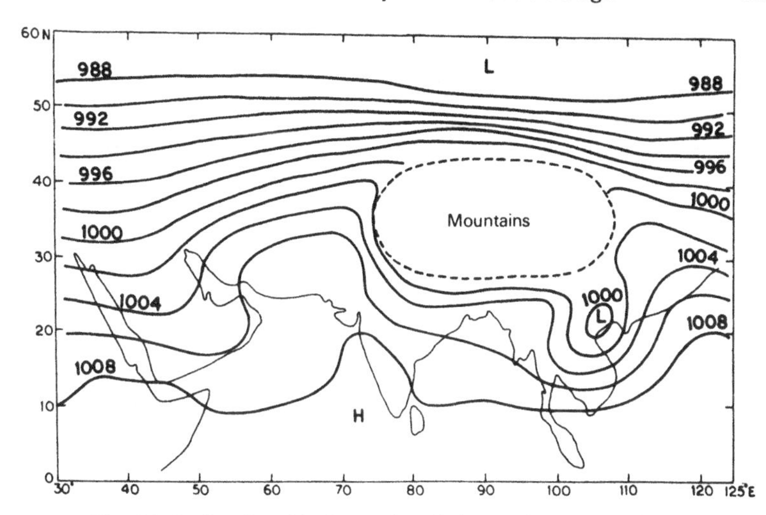

Fig. 23.5. Sea-level isobars after 8 days, with mountains.

23.6.2 With diabatic heating

The main features of the solar energy input were included in the second experiment (even though the model is of a dry atmosphere).

Let $\dot{Q}$ be the heating rate per unit mass. Integration over the depth of the boundary layer ($\sigma_B = 0, 1$) gives

$$\dot{Q} = \int [Q_0(1-\alpha) + \varepsilon\sigma_0 T_a^4 (a + b\sqrt{e}) - \varepsilon\sigma_0 T_E^4 + 4\varepsilon\sigma_0 T_A^3 (T_E - T_A)]\sigma \, d\sigma + K(\partial^2\theta/\partial z^2)_{\sigma_B=0.5}. \quad (23.14)$$

The first term in (23.14) represents the difference between incoming and outgoing radiation. Q_0 is the sum of the direct and diffuse solar radiation incident on the Earth's surface, α is the surface albedo, ε the emissivity and σ_0 denotes the Stefan–Boltzman constant. The temperatures of the Earth's surface, and of the atmosphere are T_E and T_A, respectively while e (17.3 mb) represents the water vapour pressure.

The various terms within brackets in (23.14) are: (i) net incoming solar radiation; (ii) an empirical expression for the long-wave counter radiation; (iii) the effective outgoing radiation; and (iv) a correction for the effect of air–ground temperature differences on outgoing radiation (Sellers, 1965). The numerical values of the constants were taken as

$$a = 0.605, \qquad b = 0.048.$$

In (23.14) the second term represents the flux of sensible heat. A constant value of K (10^4 cm^2 s^{-1}) was assumed.

For Q_0 values reported by Ramachandran and Kelkar (1977) for July were used. The average value of Q_0 over India was 260 W m^{-2}.

The numerical values of the albedo (α) adopted in the model are shown in Table 23.1. These values were taken as representative of the Himalayas and the deserts. Tibet was assumed to be a region of high

TABLE 23.1. *Numerical values of albedo (in percentage)*

Himalayas, including Tibet	50
Deserts of Rajasthan and adjoining Pakistan	30–40
Arabian desert	30–40
Land outside deserts and Himalayas	10
Oceanic regions	6

albedo, although there is uncertainty about its climatic features. Average minimum surface temperatures over Tibet in July are around 5° C, and satellite pictures over India suggest that the plateau experienced high cloud cover on most monsoon days. Measurements by Mani *et al.* (1977) at another high altitude station, Gulmarg (34° N, 74° E), in Kashmir, indicated an albedo as high as 44% on some days in June. The average albedo in June at Gulmarg is about 25%.

The isobars at sea level for the experiment with the inclusion of diabatic heating are shown in Fig. 23.6. The monsoon trough now appears as a

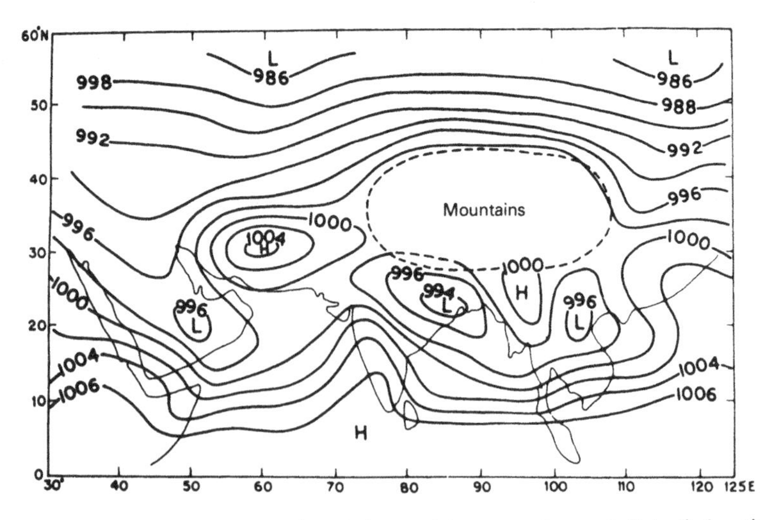

Fig. 23.6. Sea-level isobars after 8 days, with mountains and diabatic heating.

prominent feature, although it was not present earlier. This interesting result suggests that the formation of the trough is not entirely a mechanical effect of topographic barriers.

23.6.3 Albedo variations

For the third experiment, the albedo over the deserts of Rajasthan and adjoining areas of Pakistan was decreased from 30–40% to 10%. The sea-level isobars with this change are shown in Fig. 23.7. A comparison with Fig. 23.6 indicates an intensification of the monsoon trough by 2 mb. The central pressure of the trough is now reduced from 994 mb to 992 mb. This supports a strong feedback from albedo changes as suggested by Charney *et al.* (1977) for this region.

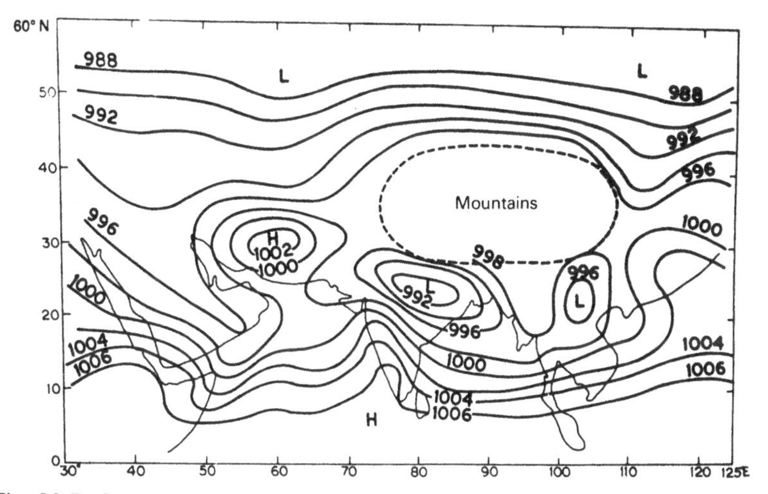

Fig. 23.7. Sea-level isobars after 8 days, with mountains, diabatic heating and changed albedo.

23.6.4 Soil mositure

Of the net radiation absorbed by the soil and plant surfaces, a part is dissipated as sensible heat and a part as latent heat for the evaporation of soil moisture. Let L be the latent heat of vaporization of water and E the amount of water available for evaporation. The ratio of the exchange of sensible heat (H) and latent heat (LE) between the Earth's surface and the atmosphere is termed the Bowen ratio (β).

When the surface is dry, β is high, but for a wet surface β is small and may even become negative. Simulating a hypothetical spell of heavy rain

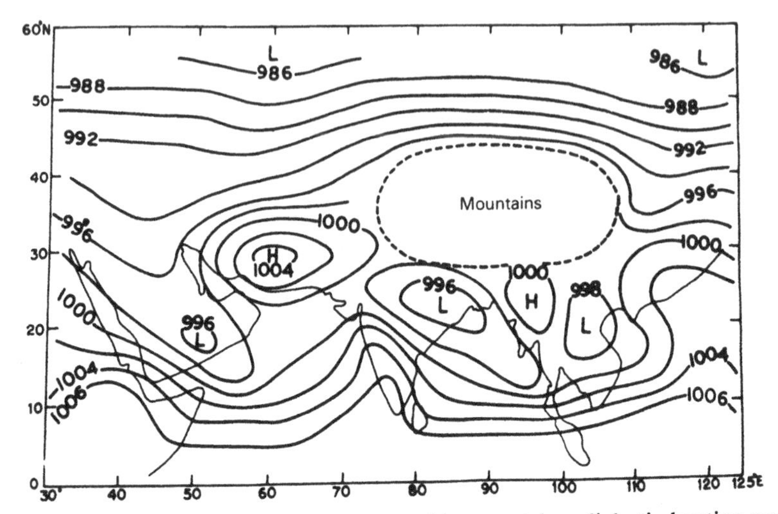

Fig. 23.8. Sea-level isobars after 8 days, with mountains, diabatic heating and flux of latent heat.

in the northeast, the values $\beta = 0.5$ in northeast India and $\beta = 10.0$ in the northwest, a dry region, were assigned. This increased the flux of sensible heat in (23.14) by a factor of 2 over northeast India, but there was only a marginal increase by 10% over the northwestern part of the country. The result of this change is shown in Fig. 23.8. A marked extension of the trough into the Bay of Bengal is found to occur as a consequence. It must be emphasized that this is a preliminary attempt to include the effect of soil moisture in the model.

23.7 Summary

The main results of this study may be summarized as follows:

(i) Starting with a zonal profile, it was found that the monsoon trough could not be generated at sea level by the mountains alone. A trough appeared if the main features of radiative heating were included.

(ii) A decrease in the albedo over northwest India led to an intensification of the trough by 2 mb.

(iii) Inclusion of the flux of latent heat by evaporation of soil moisture, led to a southward extension of the trough into the Bay of Bengal.

(iv) The regional model required initialization after every 24 hours, although a steady state appeared to develop after 4 days. This suggests slow dispersion of inertia–gravity waves by the model.

(v) Improvements occur in the output if deviations from a reference atmosphere are taken as dependent variables. A similar improvement was made as a result of using Lagrange's interpolation formula for conversion from pressure to σ coordinates and vice versa.

References

Baumhefner, D. and Downey, P. (1976) Forecast intercomparisons between large-scale numerical weather prediction models. *Proceedings of the International Conference on Simulation of Large-Scale Atmospheric Processes, Annalen der Meteorologie* (New Series), **11**, 205–8.

Charney, J. G., Quirk, W. J., Shu-Hsien Chow and Kornfield, J. (1977) A comparative study of the effects of albedo in semi-arid regions. *J. Atmos. Sci.*, **34**, 1366–85.

Grotjahn, R. (1976) Some consequences of finite differencing hyperbolic equations revealed by group velocity errors. *Proceedings of the International Conference on Simulation of Large-Scale Atmospheric Processes, Annalen der Meteorologie* (New Series), **11**, 220–3.

Mani, A., Chacko, O. and Desikan, V. (1977) Solar radiation measurement and studies of the atmospheric transmission at high altitude stations in India. *Ind. J. Meteor. Hydrol. Geophys.*, **28**, 51–62.

Phillips, N. M. (1974) Application of Arakawa's energy conserving layer model to the operational numerical weather prediction. Technical Note No. 104, NMC Washington, 40 pp.

Ramachandran, G. and Kelkar, R. R. (1977) Summer solar radiation and temperature in relation to monsoon rainfall. *Ind. J. Meteor. Hydrol. Geophys.*, **28**, 81–4.

Sellers, W. (1965) *Physical Climatology*, University of Chicago Press, 272 pp.

Shuman, F. G. and Hovermale, J. B. (1968) An operational six-layer primitive equation forecast model. *J. Appl. Meteor.*, **7**, 525–47.

24

On the monsoonal midtropospheric cyclogenesis over western India

R. W. BRODE AND M. K. MAK

The 1963 summer radiosonde data collected during the International Indian Ocean Expedition was used to determine the observed basic state prior to the development of an intense midtropospheric cyclone located over the west coast of India. This observed state was in turn used as the basis of a general quasi-geostrophic instability analysis. There is a dominant, most unstable mode, which in its overall structure and characteristics resembles the observed midtropospheric cyclone. This, therefore, lends further support to the basic hypothesis that midtropospheric cyclones originate as a result of the baroclinic instability of the broad southwest monsoonal flow, the direction and magnitude of which vary significantly with height in the lower troposphere. The results of this study also suggest that the initial development of the cyclone might occur most prominently at a lower-midtropospheric level and that in a later stage of development the level of maximum intensity might rise to the midtropospheric level due to the moist convection and accompanying vorticity transport by the clouds.

24.1 Introduction

This study seeks to determine the mechanisms of formation of an important type of summer monsoonal disturbance which often forms just off the west coast of India. Its structure is distinctly different both from the extratropical cyclones and tropical hurricanes. In particular, the cyclonic flow is significantly more intense at the midtropospheric level than at the surface and in the upper troposphere. For this reason, these systems are called midtropospheric cyclones (MTC). They have recently been reviewed by Carr (1977*a*). In our opinion, the initial formation of a MTC-like disturbance has yet to be demonstrated quantitatively to be a logical consequence of any particular proposed mechanism.

This study is a follow-up of Mak's (1975) analysis which will be referred to as M75. The conceptual basis of that study is that MTC might arise from an instability of the broad southwest monsoonal flow, the direction as well as the magnitude of which vary significantly with height. A class of idealized basic states was tested for qualitative compatibility with the meagre observational information available. The results were quite encouraging. Strong criticism has, however, been raised against Mak's hypothesis, mainly on the ground that the idealized basic flow used in M75 looks too idealistic. Is the agreement between the previous theoretical results and the observations merely fortuitous? In order to settle this controversy, we first determined from observed data the actual state over the northeastern Arabian Sea prior to the MTC formation. This data analysis is presented in § 24.2. Then we determined whether that basic state does have MTC-like unstable modes. The details and results are presented in §§ 24.3 and 24.4 respectively.

24.2 Determination of the observed basic state

The observed basic state required for the subsequent instability analysis consists of three elements: (i) the zonal velocity component $\bar{u}(p)$; (ii) the meridional velocity component $\bar{v}(p)$; and (iii) the static stability $\bar{\sigma}(p) = -R(\partial\bar{T}/\partial p - R\bar{T}/C_p p)/p$. These quantities were determined from a limited amount of radiosonde data collected in the summer of 1963 during the International Indian Ocean Expedition. The profiles of $\bar{u}$ and $\bar{v}$ over Bombay, averaged over two weeks immediately before the development of the MTC on June 26 1963, are shown in Fig. 24.1 (see Miller and Keshavamurty, 1968). The profiles of upper-tropospheric levels have been only slightly smoothed. This figure also shows the profile of $\bar{\sigma}$ over Bombay averaged over the month of June. The smoothing $\bar{u}$ and $\bar{v}$ at the upper levels is inconsequential because of the large values of $\bar{\sigma}$ at those levels.

There are certain significant differences as well as similarities between the observed profiles of $\bar{u}$, $\bar{v}$ and $\bar{\sigma}$, and the corresponding profiles used in M75. The observed $\bar{u}$ profile has a shear similar to that used in M75. However, it has a fairly deep layer of westerly flow and weak westerly shear in the lowest 100 mb. The observed $\bar{v}$ profile differs greatly from that used in M75. The magnitudes of $\bar{v}$ are by no means negligible,

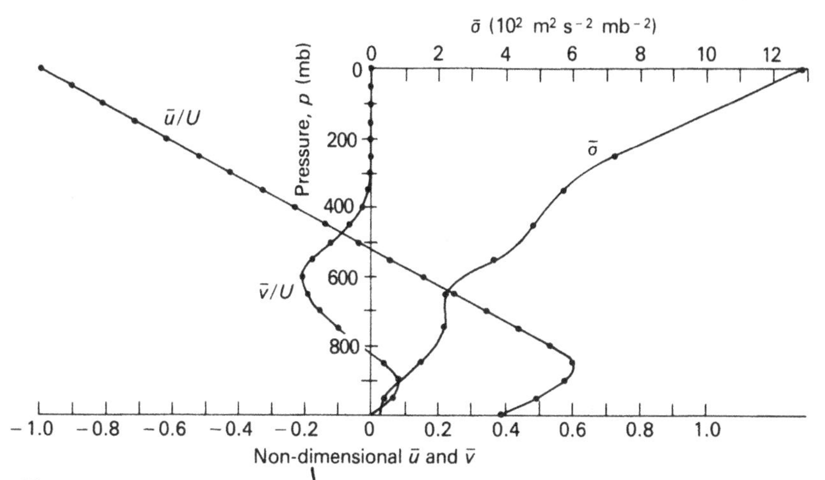

Fig. 24.1. The profiles of non-dimensional $\bar{u}$ and $\bar{v}$ and dimensional $\bar{\sigma}$ of the observed basic state used in the instability analysis. (From Brode and Mak, 1978.)

suggesting that the variations of the direction of the basic flow with height might play an important role. Furthermore, the vertical shear in $\bar{v}$ in the lower midtroposphere is large and is of the same sense as that used in M75. The observed $\bar{\sigma}$ profile shows large increases with height, whereas $\bar{\sigma}$ was treated as a constant in M75.

24.3 Formulation of the instability analysis

Following M75 we assume that, for the purpose of testing the relevance of the basic hypothesis of MTC cyclogenesis it is enough to consider quasi-geostrophic dynamics. The following are the dimensionless perturbation equations in p-coordinates (see M75 for details):

$$R_0\left(\frac{\partial u}{\partial t}+\bar{u}\frac{\partial u}{\partial x}+\bar{v}\frac{\partial u}{\partial y}+\omega\frac{\partial \bar{u}}{\partial p}\right)=-\frac{\partial \phi}{\partial x}+v, \tag{24.1a}$$

$$R_0\left(\frac{\partial v}{\partial t}+\bar{u}\frac{\partial v}{\partial x}+\bar{v}\frac{\partial v}{\partial y}+\omega\frac{\partial \bar{v}}{\partial p}\right)=-\frac{\partial \phi}{\partial y}-u, \tag{24.1b}$$

$$-\frac{\partial \phi}{\partial p}-\frac{T}{p}=0, \tag{24.1c}$$

$$\frac{\partial u}{\partial x}+\frac{\partial v}{\partial y}+\frac{\partial \omega}{\partial p}=0, \tag{24.1d}$$

$$R_0\left(\frac{\partial T}{\partial t}+\bar{u}\frac{\partial T}{\partial x}+\bar{v}\frac{\partial T}{\partial y}-up\frac{\partial \bar{v}}{\partial p}+vp\frac{\partial \bar{u}}{\partial p}\right)-\omega pS=0, \tag{24.1e}$$

where $S=(p_0/fL)^2\bar{\sigma}$ and $R_0=U/fL$, where U is the speed of the basic zonal flow at the top of the model, and L is the length of the perturbation in the x-direction. To examine the instability characteristics, a range of values of L, and consequently S, compatible with the quasi-geostrophic assumption were examined.

The usual power series expansion for each dependent variable, using R_0 as the small expansion parameter, would lead to a set of zeroth-order equations and a set of first-order equations. By combining these two sets of equations (see M75 for details of this procedure), one equation is finally obtained governing the zeroth-order component of the geopotential field $\phi_{(0)}$:

$$\left(\frac{\partial}{\partial t}+\bar{u}\frac{\partial}{\partial x}+\bar{v}\frac{\partial}{\partial y}\right)\left[\frac{\partial^2}{\partial x^2}+\frac{\partial^2}{\partial y^2}+\frac{\partial}{\partial p}\left(\frac{1}{S}\frac{\partial}{\partial p}\right)\right]\phi_{(0)}$$
$$-\left[\frac{\mathrm{d}}{\mathrm{d}p}\left(\frac{1}{S}\frac{\mathrm{d}\bar{u}}{\mathrm{d}p}\right)\frac{\partial\phi_{(0)}}{\partial x}+\frac{\mathrm{d}}{\mathrm{d}p}\left(\frac{1}{S}\frac{\mathrm{d}\bar{v}}{\mathrm{d}p}\right)\frac{\partial\phi_{(0)}}{\partial y}\right]=0, \quad (24.2)$$

where $\bar{u}$, $\bar{v}$, and S are prescribed functions of p. The first-order term of ω is related to $\phi_{(0)}$ by

$$\omega_{(1)}=-\frac{1}{S}\left(\frac{\partial}{\partial t}+\bar{u}\frac{\partial}{\partial x}+\bar{v}\frac{\partial}{\partial y}\right)\frac{\partial\phi_{(0)}}{\partial p}+\frac{1}{S}\left(\frac{\mathrm{d}\bar{u}}{\mathrm{d}p}\frac{\partial\phi_{(0)}}{\partial x}+\frac{\mathrm{d}\bar{v}}{\mathrm{d}p}\frac{\partial\phi_{(0)}}{\partial y}\right). \quad (24.3)$$

Using $\omega=0$ at $p=0$ and at $p=1$ as the boundary conditions then gives

$$\left(\frac{\partial}{\partial t}+\bar{u}\frac{\partial}{\partial x}+\bar{v}\frac{\partial}{\partial y}\right)\frac{\partial\phi_{(0)}}{\partial p}-\frac{\mathrm{d}\bar{u}}{\mathrm{d}p}\frac{\partial\phi_{(0)}}{\partial x}-\frac{\mathrm{d}\bar{v}}{\mathrm{d}p}\frac{\partial\phi_{(0)}}{\partial y}=0 \quad (24.4)$$

at $p=0$ and at $p=1$. The normal mode solution of (24.2) is

$$\phi_{(0)}=\mathrm{Re}\,\{A(p)\exp\left[\mathrm{i}(x+\gamma y-ct)\right]\}, \quad (24.5)$$

where γ is the non-dimensional wavenumber in the y-direction, c is the eigenvalue and $A(p)$ the eigenfunction of the problem. Substituting (24.5) into (24.2) we obtain

$$\left[\frac{\mathrm{d}^2}{\mathrm{d}p^2}-\frac{\mathrm{d}\ln S}{\mathrm{d}p}\frac{\mathrm{d}}{\mathrm{d}p}-(1+\gamma^2)S-M\right]A=0, \quad (24.6)$$

where

$$M=\frac{[\bar{u}''+\gamma\bar{v}''-(\bar{u}'+\gamma\bar{v}')(\ln S)']}{\bar{u}+\gamma\bar{v}-c}, \quad (24.7)$$

each prime indicating a derivative with respect to p. The boundary conditions (24.4) become

$$\frac{\mathrm{d}A}{\mathrm{d}p}=\frac{\bar{u}'+\gamma\bar{v}'}{\bar{u}+\gamma\bar{v}-c}A. \tag{24.8}$$

Equations (24.6), (24.7) and (24.8) are the generalized counterparts of equations (10), (11) and (12) in M75. They are solved numerically with the same iterative scheme as used in M75. One hundred grid-points were used in the computations to ensure adequate resolution. Both the eigenvalue, c, and eigenfunction, $A(p)$, are determined simultaneously by this method. The instability characteristics are embodied in c. To examine the structure of the most unstable wave, the zeroth-order fields, $\phi_{(0)}$, $T_{(0)}$, as well as the horizontal divergence, $\partial\omega_{(1)}/\partial p$, and the vertical velocity, $\omega_{(1)}$, fields are computed. All these quantities are related to $A(p)$ as follows:

$$\phi_{(0)}=A_r\cos x-A_i\sin x, \tag{24.9a}$$

$$T_{(0)}=-p\left(\frac{\mathrm{d}A_r}{\mathrm{d}p}\cos x-\frac{\mathrm{d}A_i}{\mathrm{d}p}\sin x\right), \tag{24.9b}$$

$$\begin{aligned}\frac{\partial\omega_{(1)}}{\partial p}&=(1+\gamma^2)[A_i(\bar{u}+\gamma\bar{v}-c_r)-A_rc_i]\cos x\\&\quad+(1+\gamma^2)[A_r(\bar{u}+\gamma\bar{v}-c_r)+A_ic_i]\sin x,\end{aligned} \tag{24.9c}$$

$$\omega_{(1)}=\int_0^p\frac{\partial\omega_{(1)}}{\partial p}\,\mathrm{d}p, \tag{24.9d}$$

where $A(p)=A_r(p)+\mathrm{i}A_i(p)$ and $c=c_r+\mathrm{i}c_i$. The subscripts r and i denote the real and imaginary parts of these quantities. The common exponentially amplifying factor $\exp(c_it)$ is not included in (24.9). The vertical structure of the wave is specifically examined on an (x, p) plane at latitude y_0 and at time t_0 such that $\gamma y_0-c_rt_0=2\pi$.

It is also instructive to examine the instability mechanism from the point of view of energy conversion processes. The energy equations can be written as

$$\frac{\partial K_E}{\partial t}=\tfrac{1}{2}\exp(2c_it)\left[-\frac{\mathrm{d}}{\mathrm{d}p}(\omega_r\Phi_r+\omega_i\Phi_i)+\left(\omega_r\frac{\mathrm{d}\Phi_r}{\mathrm{d}p}+\omega_i\frac{\mathrm{d}\Phi_i}{\mathrm{d}p}\right)\right], \tag{24.10}$$

$$\begin{aligned}\frac{\partial A_E}{\partial t}&=\frac{1}{2pS}\exp(2c_it)\left[(u_rT_r+u_iT_i)\frac{\mathrm{d}\bar{v}}{\mathrm{d}p}-(v_rT_r+v_iT_i)\frac{\mathrm{d}\bar{u}}{\mathrm{d}p}\right]\\&\quad-\tfrac{1}{2}\exp(2c_it)\left[\omega_r\frac{\mathrm{d}\Phi_r}{\mathrm{d}p}+\omega_i\frac{\mathrm{d}\Phi_i}{\mathrm{d}p}\right],\end{aligned} \tag{24.11}$$

where K_E and A_E respectively stand for the eddy kinetic energy and eddy available potential energy averaged over an (x, y) wavelength domain. The subscripts r and i again denote real and imaginary parts. The velocity components u and v are determined from Φ geostrophically. The first term on the right of (24.10) represents the convergence of wave-energy flux in the vertical, and the second term is the conversion from A_E to K_E. The first two terms on the right side of (24.11) stand for the two components of conversion of the available potential energy of the basic state to A_E, (i.e., $-\overline{u'T'}\, \partial\bar{T}/\partial x$, and $-\overline{v'T'}\, \partial\bar{T}/\partial y$, where perturbation variables are indicated by primes).

The computational procedure is first to determine the eigenvalue c for different combinations of γ and L within the range $-10 \leq \gamma \leq 10$ and $500 < L < 2500$ km. From the corresponding values of the growth rate, in terms of e-folding time $\tau (= L/c_i U)$, the most unstable mode is identified. The effective dimensional length scale of this mode is simply

$$\mathscr{L} = \frac{L}{\sqrt{(1+\gamma^2)}}. \tag{24.12}$$

The vertical structure of this mode is then examined in detail using (24.9). Its energetics are examined using (24.10) and (24.11).

24.4 Instability characteristics of the observed basic state

The observed basic state gives rise to two subsets of unstable modes in the (γ, L) plane. The most unstable mode was identified, in terms of e-folding time τ, from the corresponding values of the growth rate. They will be referred to as P modes and N modes respectively. The overlapping region on the (γ, L) plane of these two subsets of modes is more extensive than that in the case of the idealized basic state. Fig. 24.2 shows the variations with γ and L of the growth rate for the P modes. There is a well-defined maximum located at $L = 1080$ km and $\gamma = 0.7$. By (24.12), we found that the effective wavelength $\mathscr{L}$ is 900 km. The corresponding observed quantity of the MTC would be the radius of the cyclone, of the order of 750 km. Hence the agreement is good. The magnitude of the theoretical growth rate is however somewhat too slow compared with the observed overall rate from an incipient stage to the mature stage. This is not too surprising since moist convection has not been taken into account. A two-fold increase in the growth rate of baroclinic disturbances (including moist convection) over the dry model might be expected (Gall, 1976). The behaviour of the disturbance in a moist model might depend

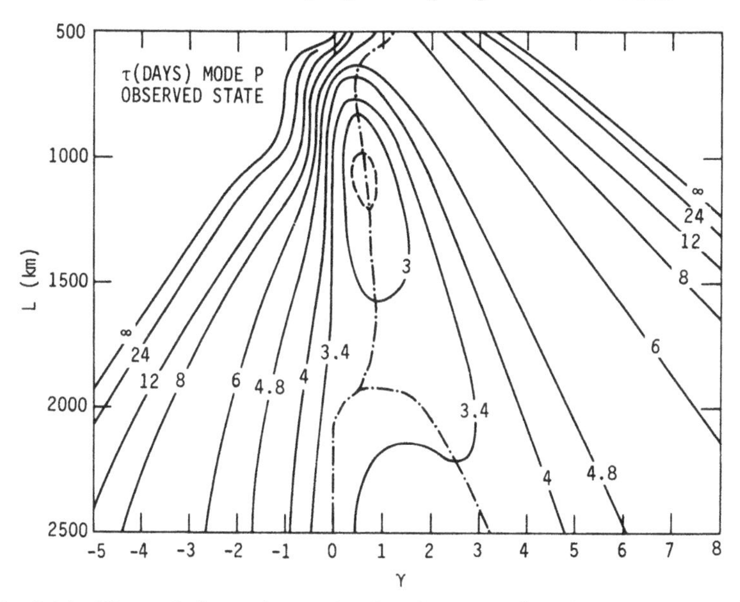

Fig. 24.2. The variations of the e-folding time $\tau = L/(c_i U)$ in days on the (γ, L) plane for the P modes associated with the observed basic state using $U = 24\ \mathrm{m\ s^{-1}}$. (From Brode and Mak, 1978.)

quite sensitively on the parameterization procedure (Carr, 1977*b*). Fig. 24.3 shows the phase speed, c_r, of the P modes. The dash–dot line indicates the positions of the maximum growth rate for each value of L. The nondimensional phase speed of the most unstable mode is about 0.23. This agrees much better with observation than that of the asymptotically most unstable mode for the idealized basic states. The P modes grow about twice as fast as the N modes in the region of the (γ, L) plane under consideration (compare Figs. 24.2 and 24.4). The most unstable N mode evidently has a small γ and a value of L much smaller than 500 km. It therefore has very little physical meaning in a quasi-geostrophic framework. No attempt was made to examine it in detail.

The results above indicate that the most unstable mode for the observed state is the one at $L = 1080$ km and $\gamma = 0.7$. We next present the cross-sectional structure of this mode in terms of $\phi_{(0)}$, $T_{(0)}$, $\partial\omega_{(1)}/\partial p$ and $\omega_{(1)}$ on (x, p) planes (Fig. 24.5). These results are derived from a normalized complex eigenfunction $A(p)$ whose maximum magnitude is set to unity. The prominent features of the structure may be summarized as follows. The geopotential field has a gross midtropospheric character. The vertical tilting is westward at the lower portion and eastward at the upper portion. The intensity of the perturbation is more concentrated in

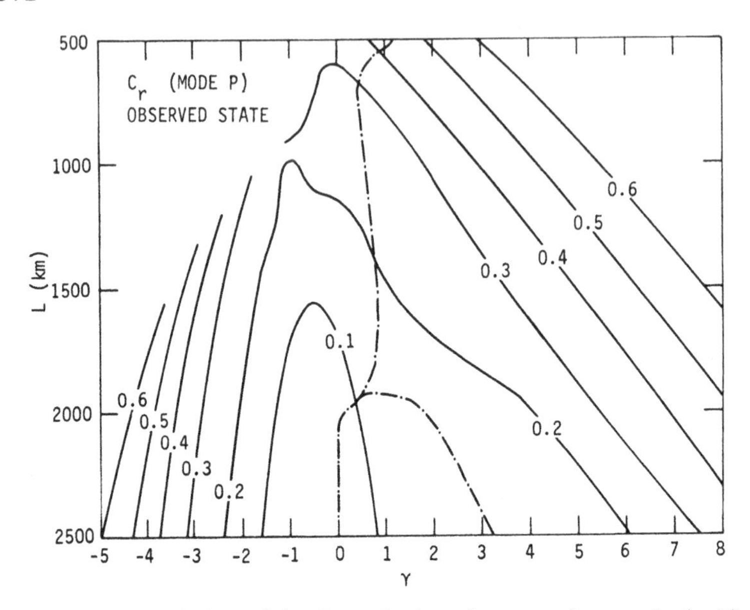

Fig. 24.3. The variations of the dimensionless phase speed, c_r, on the (γ, L) plane for the P modes associated with the observed basic state. (From Brode and Mak, 1978.)

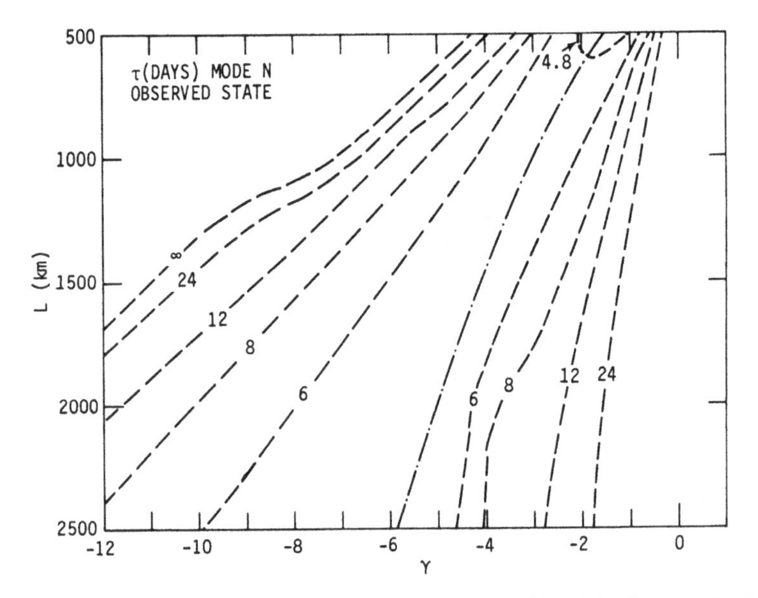

Fig. 24.4. The variations of the e-folding time $\tau = L/(c_i U)$ in days on the (γ, L) plane for the N modes associated with the observed basic state using $U = 24\ \mathrm{m\,s^{-1}}$. (From Brode and Mak, 1978.)

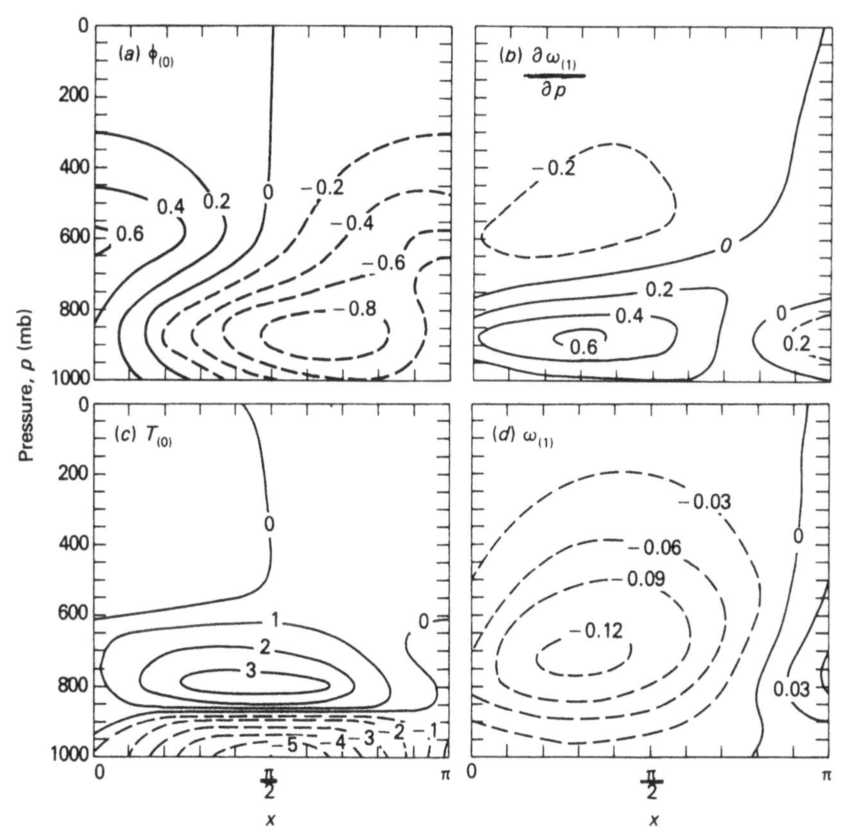

Fig. 24.5. The cross-sectional structure of the most unstable mode for the observed basic state in terms of: (a) the geopotential $\phi_{(0)}$; (b) the horizontal convergence $\partial\omega_{(1)}/\partial p$; ($c$) the temperature $T_{(0)}$; and (d) $\omega_{(1)}$. (From Brode and Mak, 1978.)

the midtroposphere than it is in the case of the idealized basic state. This is in better agreement with observation. The temperature field has a cold core below and a warm core above the level of maximum geopotential perturbation. The relative phase between $\phi_{(0)}$ and $T_{(0)}$ is such that the temperature structure described above is located at about one-eighth of a wavelength to the west of the trough. Fig. 24.5 also shows that at about three-eighths of a wavelength to the west of the trough, there is a maximum horizontal convergence, extending up to 700 mb with a weaker but deeper layer of divergence above. This divergence pattern results in a broad and deep layer of upward motion, with its maximum located somewhat to the west of the trough. The overall resemblance between this most unstable mode and the observed MTC is good. Unlike the most unstable mode for the idealized basic state, there is no single feature here that starkly contrasts with observation.

We do notice that the theoretical mode has its maximum intensity at a somewhat lower level than the observed MTC, 850 mb versus 650 mb. To assess the implication of this discrepancy, we should consider the inherent limitations of the model. Since moist convection is not taken into account, this model is only capable of revealing what might happen at the initial stage of MTC development. A close examination of the geopotential data reveals that the lower levels, say 850 and 900 mb, do appear to show development at an earlier time than the upper levels, say 600 and 700 mb (Figs. 24.6 and 24.7). On the basis of this observational informa-

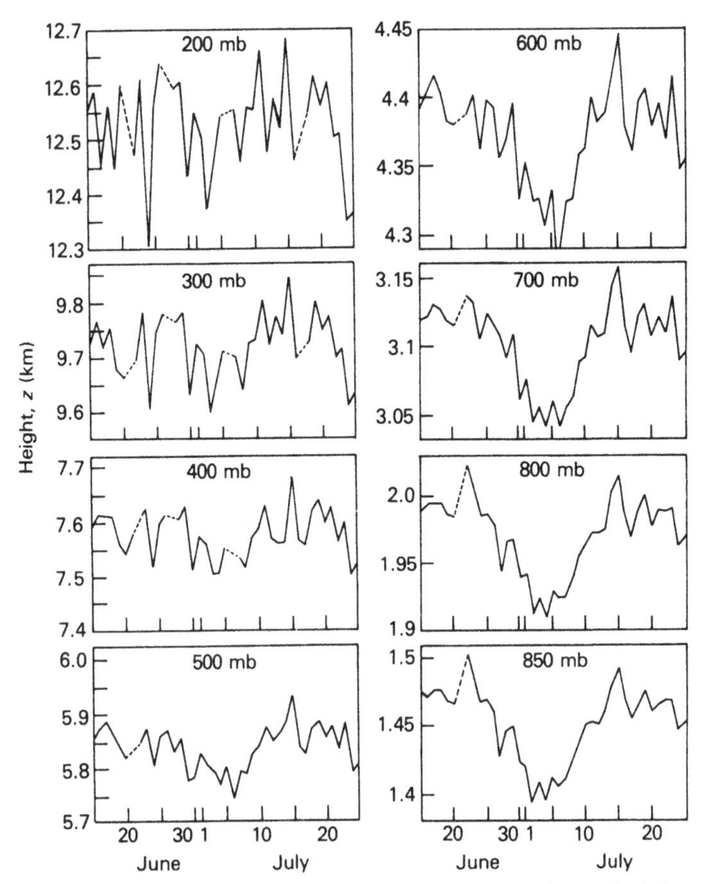

Fig. 24.6. Unsmoothed variations with time of the height of eight pressure surfaces over Bombay from 15 June to 25 July 1963. Dashed lines indicate missing data. (From Brode and Mak, 1978.)

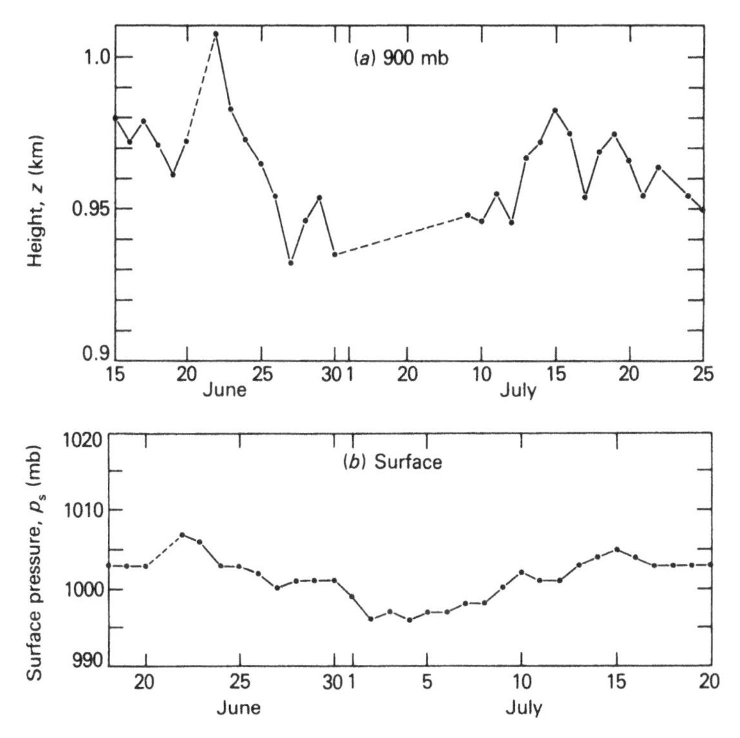

Fig. 24.7. Unsmoothed variations with time of (*a*) the height of the 900 mb pressure surface over Bombay, 15 June to 25 July 1963, and (*b*) the surface pressure over Bombay, 15 June to 25 July 1963. Dashed lines indicate missing data. (From Brode and Mak, 1978.)

tion in conjunction with our theoretical result, it seems justifiable to make the following suggestion. *The MTC develops first at a lower-midtropospheric level due to the generalized baroclinic instability at the initial stage. This dynamic mechanism gives rise to a large-scale ascent which is conducive to vigorous moist convection. The latter in turn significantly increases the growth rate and raises the level of maximum intensity to the midtroposphere.*

The energetics associated with the development of this most unstable mode, as computed using (24.10) and (24.11), are presented in Figs. 24.8 and 24.9. Fig. 24.8 shows the vertical distribution of the two terms on the right side of (24.10). It shows that eddy kinetic energy is generated from eddy available potential energy in the 875 to 575 mb layer. There is actually a destruction of K_E both below and above. The vertical wave-energy flux is such that there exists a vertical divergence in the layer

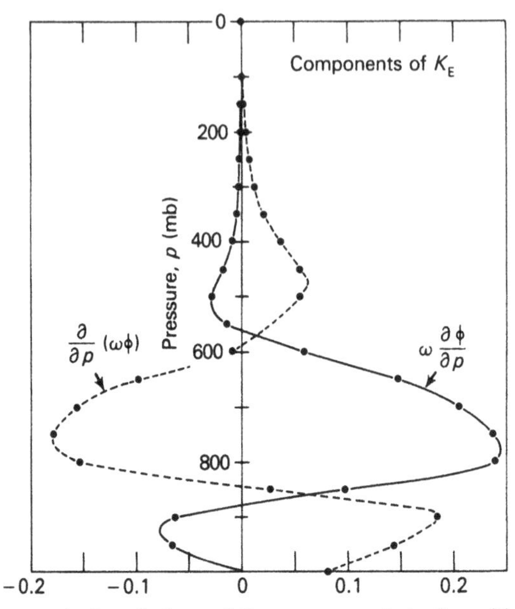

Fig. 24.8. The vertical variations of the components in the eddy kinetic energy equation (24.10) for mode P. (From Brode and Mak, 1978.)

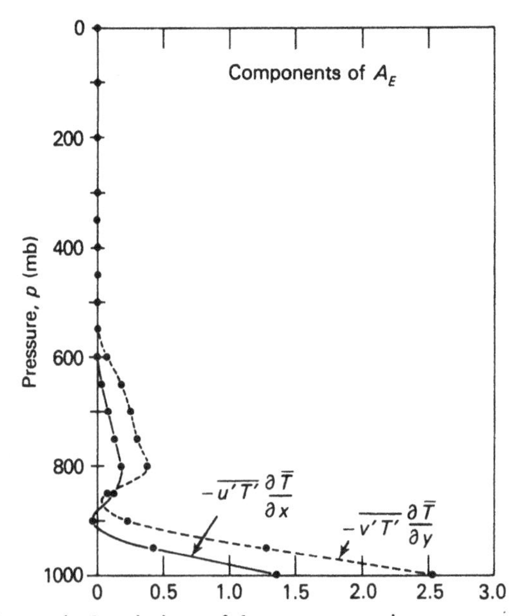

Fig. 24.9. The vertical variations of the two conversion processes from the zonal available potential energy to the eddy available potential energy in mode P, according to (24.11). (From Brode and Mak, 1978.)

850 to 600 mb, a strong convergence below 850 mb and a weaker convergence above 600 mb. Now we can explain why the unstable mode has a midtropospheric character. The intensity at the surface is relatively weak in spite of the strong convergence of wave-energy flux because there is a significant destruction of kinetic energy associated with the rising motion through the cold core. The intensity in the upper troposphere is weak for two reasons. Firstly, the convergence of wave-energy flux is weak due to the strong static stability. Secondly, there is also a small rate of destruction of kinetic energy associated with the upward motion through a secondary cold region. Fig. 24.9 shows that both processes of generating eddy available potential energy, $-\overline{u'T'}\,\partial\bar{T}/\partial x$ and $-\overline{v'T'}\,\partial\bar{T}/\partial y$, are comparable and reinforce one another. The second process is slightly larger than the first process. Fig. 24.9 indicates that the temperature perturbation should only have a large magnitude in the surface to 900 mb and the 825 to 600 mb layers with maxima occurring at the surface and at about 800 mb. In particular, there is cold advection in the trough region in the lowest 100 mb due to $-u'\bar{T}_x$ and $-v'\bar{T}_y$. From 850 mb to 600 mb, there is a warm advection because $\bar{T}_x$ and $\bar{T}_y$ reverse their signs. It has been proposed in the literature that the MTC thermal structure is due to evaporation of raindrops in the low levels and release of latent heat at higher levels (e.g., Miller and Keshavamurty, 1968). That is very likely to be an important factor at a later stage of development of an MTC, but these results indicate that it is not necessary to invoke the moist convective processes to explain the thermal structure at the initial stage.

One may be curious about the N mode even though it is not as important as the P mode. For the sake of completeness, its cross-sectional structure for $\gamma = -3.25$ and $L = 1100$ km is presented in Fig. 24.10 as a counterpart of Fig. 24.5. It is seen that this mode is a much shallower system with its maximum intensity at 950 mb. The vertical tilt is in the opposite sense to that of the P mode. The relative phase between the temperature field and the geopotential field is opposite to that in the P mode, as is the relative phase between the vertical velocity field and the geopotential field, leading, in this mode, to upward motion to the east of the trough. The energetics have the following characteristics. The kinetic energy budget is qualitatively similar to that of a P mode, except, of course, in that the processes are important only in a very shallow layer near the surface. On the other hand, unlike the P mode, the two processes $-\overline{u'T'}\,\partial\bar{T}/\partial x$ and $-\overline{v'T'}\,\partial\bar{T}/\partial y$ have opposite signs in the surface layer. The former is dominant by virtue of the large value of $|\gamma|$.

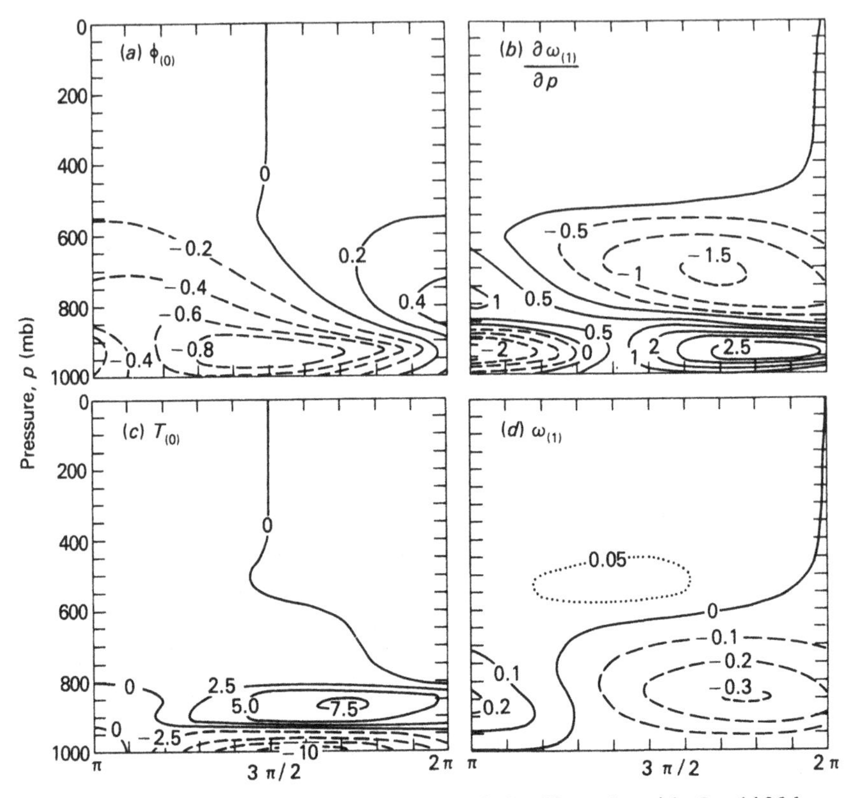

Fig. 24.10. The cross-sectional structure of the N mode with $L = 1100$ km, $\gamma = -3.25$, in terms of (*a*) the geopotential $\phi_{(0)}$, (*b*) the horizontal convergence $\partial\omega_{(1)}/\partial p$; (*c*) the temperature $T_{(0)}$; and (*d*) $\omega_{(1)}$.

24.5 Concluding remarks

This analysis may be taken as strong evidence for a simple self-contained explanation for the formation of a midtropospheric cyclone. By applying the general theory of quasi-geostrophic baroclinic instability to the specific conditions over the northeastern Arabian Sea, we have demonstrated that the formation of an MTC is a logical necessity. It should be emphasized that the findings of this investigation only imply that moist convection need not be a crucial factor in the initial development of the cyclone.

When the direction of a baroclinic flow varies with height, there are two distinct subsets of unstable modes. For the monsoonal basic state, the most unstable mode in one of these subsets closely resembles the MTC.

The other subset of modes have much shorter length scales and may be more readily influenced by non-geostrophic effects.

We thank Dr C. S. Ramage, University of Hawaii at Manoa, for providing us with the radiosonde data for the period June to August 1963 used in this study. This research was supported by the Global Atmospheric Research Program, Climatic Dynamics Research Section, National Science Foundation, under Grant ATM74-01188. Figs. 24.1–9 are reproduced from 'On the mechanism of the monsoonal mid-tropospheric cyclone formation' by R. W. Brode and M. K. Mak, published in the *Journal of the Atmospheric Sciences*, August 1978.

References

Brode, R. W. and Mak, M. K. (1978) On the mechanism of the monsoonal mid-tropospheric cyclone formation. *J. Atmos. Sci.*, **35**, 1473–84.

Carr, F. H. (1977*a*) Mid-tropospheric cyclones of the summer monsoon. *Pure and Applied Geophysics*, **115**, 1383–1412.

Carr, F. H. (1977*b*) Numerical simulation of a mid-tropospheric cyclone. Report No. FSU 77-1, Department of Meteorology. Florida State University.

Gall, R. L. (1976) The effects of released latent heat in growing baroclinic waves. *J. Atmos. Sci.*, **33**, 1686–1701.

Mak, M. K. (1975) The monsoonal mid-tropospheric cyclogenesis. *J. Atmos. Sci.*, **32**, 2246–53.

Miller, F. R. and Keshavarmurty, R. N. (1968) *Structure of an Arabian Sea Summer Monsoon System.* East–West Center Press, Honolulu, 94 pp.

25

Downstream development of baroclinic waves in the upper-tropospheric monsoon easterlies suggested by a simple model experiment

R. P. PEARCE

The analytical theories of barotropic and baroclinic development are first briefly reviewed. The results of a computer model investigation of the phenomenon of downstream development in midlatitudes are then presented and these are extended, using a highly simplified model, to developments in the monsoon upper easterlies. The results suggest that this could be a mechanism contributing to the development of monsoon depressions.

25.1 Introduction

The purpose of this contribution is two-fold. Firstly, to stress the importance and value of simple model experiments in the development of atmospheric dynamics. Modern computers present us with the opportunity of using numerical methods to extend dynamical theory to the study of much more realistic phenomena than was previously possible using purely analytical methods. Secondly, some results of a preliminary investigation of the phenomenon of downstream development are presented. This investigation has yielded interesting results in the study of midlatitude systems and, by using a simple model, it is possible to draw some preliminary conclusions concerning its possible importance as a primary mechanism in the development of some monsoon depressions. A suggestion that this might be the case was recently made by Krishnamurti *et al.* (1977).

The analytical theories of barotropic and baroclinic development are first briefly reviewed and a simple numerical model capable of simulating the latter process is described. This is validated using the results of exact analysis. The model is then used to investigate downstream development in midlatitudes and in the tropics, and the results of some preliminary experiments are presented.

25.2 Barotropic model studies

There are two main aspects of barotropic theory, both of crucial significance to the understanding of the atmospheric flow in low latitudes, which have been developed extensively using both analytic and numerical methods. The first is Rossby wave propagation, and the second is the stability of zonal flows $U(y)$ with a latitudinal (y) variation. Both aspects are studied starting from the barotropic vorticity equation, expressing conservation of absolute vorticity following fluid particles moving in the horizontal (xy) plane, i.e.

$$\left(\frac{\partial}{\partial t}+(U+u)\frac{\partial}{\partial x}+v\frac{\partial}{\partial y}\right)(v_x-u_y-U_y+f)=0. \tag{25.1}$$

If it is assumed that the flow is non-divergent, i.e. the perturbation velocity components u and v satisfy

$$u_x+v_y=0,$$

a perturbation stream function ψ may be introduced and (25.1) linearized to give

$$\left(\frac{\partial}{\partial t}+U\frac{\partial}{\partial x}\right)\nabla^2\psi+(\beta-U_{yy})\psi_x=0. \tag{25.2}$$

Rossby wave propagation may be studied in a simple way using this equation by setting U_{yy} to zero and seeking solutions of the form

$$\psi \,\alpha\, \exp\,[\mathrm{i}k(x-ct)+\mathrm{i}ly].$$

This immediately yields the well-known expressions for the wave perturbation relative phase speed,

$$c-U=-\frac{\beta}{k^2+l^2}$$

and the relative group velocity

$$c_g - U = \frac{\beta(k^2 - l^2)}{(k^2 + l^2)^2}.$$

Thus every wave component propagates westward relative to the flow. However, wavepackets disperse in such a way that the energy associated with wave components with a larger zonal than meridional wavelength propagates westward, and that associated with the shorter wavelengths propagates eastwards.

The main result of barotropic stability studies may be summarized as:

The necessary condition for barotropic instability is that, for a zonal flow $U(y)$, where y is a horizontal coordinate directed northwards, there must be some region in which the gradient of the absolute vorticity changes sign, i.e. it must have at least one zero in the region. It may be demonstrated as follows:

Writing $\beta = \partial f/\partial y$ where $f = 2\Omega \sin \phi$, Ω denoting the angular rotation rate of the earth and ϕ the latitude, the absolute vorticity ζ_a is

$$\zeta_a = f - U_y$$

and the gradient of the absolute vorticity is

$$\frac{\partial \zeta_a}{\partial y} = \beta - U_{yy}.$$

For barotropic instability to occur in a given region, this quantity must change sign.

The basic flow $U(y)$ in (25.2) is specified in an infinite channel of width $2Y$ and the boundary conditions are taken as

$$\psi = 0, \quad \text{on } y = Y \quad \text{and} \quad y = -Y, \qquad \text{for all } x, t. \tag{25.3}$$

If solutions are sought having the form

$$\psi(x, y, t) = F(y) \exp[\mathrm{i}k(x - ct)]$$

where $F(y)$ $(=F_r(y) + \mathrm{i}F_i(y))$ is a complex function of y only and c $(=c_r + \mathrm{i}c_i)$ is also in general complex, then (25.2) may be separated into its real and imaginary parts and the resulting equations combined to yield a single equation of the form

$$\frac{\mathrm{d}}{\mathrm{d}y}\left(F_i \frac{\mathrm{d}F_r}{\mathrm{d}y} - F_r \frac{\mathrm{d}F_i}{\mathrm{d}y}\right) = c_i(\beta - U_{yy})|F|^2/[(U - c_r)^2 + c_i^2].$$

Integration of this equation with respect to y over the interval $[-Y, Y]$ and use of the boundary conditions leads to the quoted necessary condition for instability. The June–August climatological mean meridional profile of U at 150 mb at 80° E, near the jet maximum, is shown in Fig. 25.1*a*, together with the corresponding distribution of $\partial\zeta_a/\partial y$. The necessary condition is seen to be satisfied near the jet maximum.

Extension of the theory, to determine growth rates and structures of waves which can amplify in particular initial flow configurations, has been carried out by several investigators, notably Kuo (1949, 1951, 1973), Lipps (1962, 1965, 1970), Yanai and Nitta (1968) and Tupaz (1977). The last is concerned with the behaviour of a jet with U varying in both the x- and y-directions.

Simmons (1977) has recently carried out a series of numerical studies to determine the growth rates and structures of barotropic waves on an easterly jet; he solved the matrix eigenvalue problem obtained by expanding the solution for ψ of the barotropic vorticity equation as a sum of spherical harmonics. This enabled the waves with complex phase speed to be identified for each zonal wavenumber and the one corresponding to the maximum growth rate for that wavenumber to be determined. Results are shown in Fig. 25.1*b* for a jet resembling the climatological mean at 150 mb at 80° E (Fig. 25.1*a*). There is a well-defined cut-off at a zonal wavelength of about 6000 km with a maximum growth rate at a

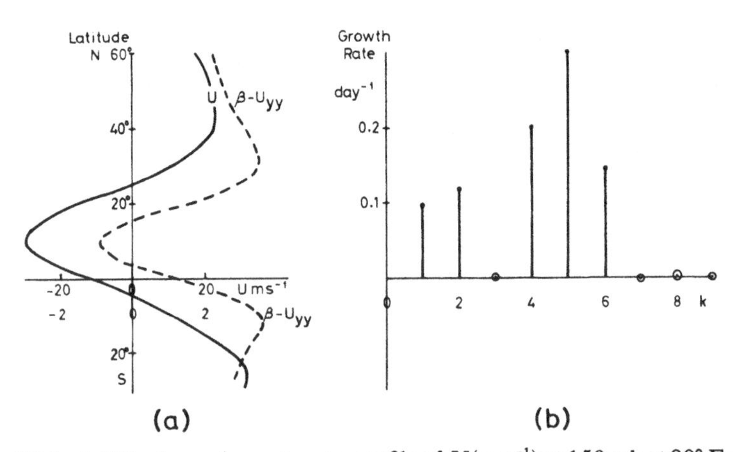

Fig. 25.1. (*a*) The June–August mean profile of $U(\mathrm{m\,s^{-1}})$ at 150 mb at 80° E with the corresponding distribution of $\beta - U_{yy} (= \partial\zeta_a/\partial y)$ in units of Ω/a where Ω is the Earth's angular rotation rate and a its radius. (*b*) Computed growth rates for barotropic waves of various zonal wavenumbers (k) on the jet shown in (*a*). The circles indicate that no amplifying waves exist for the corresponding wavenumber.

wavelength of about 7500 km. The most unstable waves have a phase speed towards the west of about 20° longitude per day.

The structure of the amplifying wave is such that the waves have troughs and ridges bowing eastwards, i.e. in a direction opposite to that of the bowing of the jet profile. Examination of the sign of $\overline{uv}$ shows that this configuration represents momentum transport away from the jet, i.e. it corresponds to a transfer of kinetic energy from the jet to the eddies.

The numerical studies by Colton (1973) and Tupaz (1977), of the barotropic instability of an easterly jet similar to that which is observed, both show a maximum disturbance amplitude downstream of the jet maximum, i.e. over the northwestern part of the Indian Ocean, rather than to the east over the Bay of Bengal.

25.3 Baroclinic model studies

25.3.1 Analytical theory

Global three-dimensional primitive equation general-circulation models incorporate baroclinic as well as barotropic effects, i.e. energy conversions in these models involve available potential energy as well as kinetic energy; this is no less true of the tropics than of midlatitudes. Apart from studies based on these complex models, however, few results have been reported of integrations carried out relating to the tropical atmosphere using simplified models. Only those of Rennick (1976) and Simmons (1977), both concerned with the normal modes of a flow resembling the west African jet, come into this category.

One of the reasons for this is that the simple criteria for baroclinic instability suggest that only very long waves are likely to have observable growth rates whereas synoptic-scale systems in low latitudes have wavelengths typically in the range 2000 to 4000 km. However, the existence of large vertical wind shears in low latitudes, of which that associated with the Asian easterly jet is an example, indicates that the atmosphere is often highly baroclinic. Disturbances which do occur, even though they may develop initially as a result of barotropic instability, must be expected to involve vertical motion and conversion of available potential energy, as well as kinetic energy exchange with the basic flow. The west African jet studies referred to above show that this is indeed the case, the vertical motion associated with the computed normal modes having a magnitude approaching that of midlatitude synoptic-scale systems; the computed wavelengths corresponding to maximum growth rates are near those observed but are clearly closer to those corresponding to barotropic than to baroclinic instability criteria.

The simplest model of baroclinic wave development is that of Eady (1949). The linearized forms of the vorticity (ζ) and thermodynamic equations are taken as

$$\frac{\partial \zeta}{\partial t} + U(z)\frac{\partial \zeta}{\partial x} = f\frac{\partial w}{\partial z}, \tag{25.4}$$

and

$$\frac{\partial(\ln\theta)'}{\partial t} + U(z)\frac{\partial(\ln\theta)'}{\partial x} + v\frac{\partial(\ln\theta)_0}{\partial y} = -w\frac{\partial(\ln\theta)_0}{\partial z}, \tag{25.5}$$

where $\theta = T(p_0/p)^{R/c_p}$ is the potential temperature; suffix $_0$ denotes an unperturbed value and a prime denotes a perturbation. This formulation neglects latitudinal variations in f and, as far as vertical density (ρ) variations are concerned, neglects multiplication by the reciprocal of the density scale height $-\partial(\ln\rho)_0/\partial z$ compared with $\partial/\partial z$ in the dynamical equations. The perturbation quantities u and v are approximated by their non-divergent parts and assumed to be in geostrophic balance. Introducing a perturbation stream function ψ, and using the approximate thermal wind relationship (assuming hydrostatic balance) yields

$$(\ln\theta)' = \frac{f}{g}\frac{\partial\psi}{\partial z}. \tag{25.6}$$

The vertical velocity component w is eliminated from (25.4), (25.5) and (25.6) which leads to an equation in ψ of the form

$$\left(\frac{\partial}{\partial t} + U(z)\frac{\partial}{\partial x}\right)q = 0, \tag{25.7}$$

where

$$q = \nabla^2\psi + \frac{f^2}{N^2}\psi_{zz} \tag{25.8}$$

is a simplified form of the quasi-geostrophic potential vorticity and

$$N^2 = g\frac{\partial}{\partial z}(\ln\theta)_0$$

is the square of the Brunt–Väisälä frequency.

Solutions of (25.7) subject to the boundary conditions

$$w = 0, \quad \text{at } z = H/2 \quad \text{and at } z = -H/2, \tag{25.9}$$

where H is the depth of the atmosphere, are obtained by putting

$$\psi = F(z)\exp[ik(x-ct)]\exp(ily)$$

and solving the resulting eigenvalue problem for the complex phase speed c. Two wave regimes are found to exist. Waves with wavenumbers (k, l) satisfying

$$k^2+l^2<\frac{f^2X_0^2}{N^2h^2},$$

where $X_0 \approx 2.4$ will amplify, while those with larger wavenumbers do not. Furthermore the growth rate is a maximum when

$$k^2+l^2=\frac{f^2X_m^2}{N^2h^2}$$

where $X_m \approx 1.68$.

If values typical of the atmosphere over south Asia at 20° N during summer are substituted into these expressions, i.e. $f = 5\times10^{-5}\ \mathrm{s}^{-1}$, $N^2 = 10^{-4}\ \mathrm{s}^{-2}$ and $H = 10^4$ m, then, according to this very simple theory, amplifying baroclinic waves will exist with $k^2 = l^2$ only for zonal wavelengths greater than about 7000 km and their growth rate is small.

If the theory is modified to include the β-effect, the steering level is lowered in westerly shear and raised in easterly shear. There are corresponding increases in the amplitudes of the waves at lower levels in westerly shear and upper levels in easterly shear. Furthermore the shorter waves are considerably destabilized (Fig. 25.2). These effects will be discussed again below, since they have particular significance for the

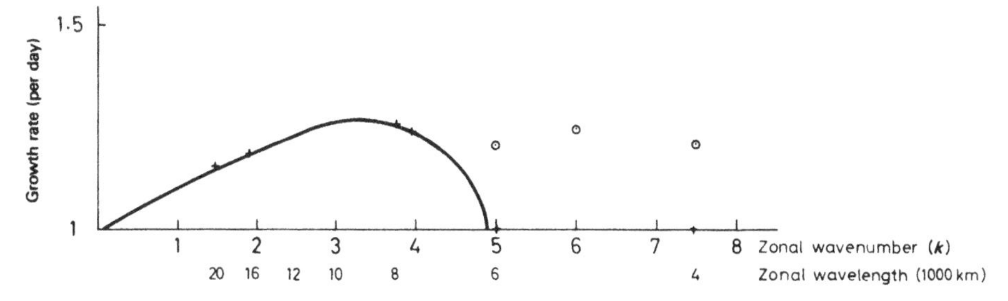

Fig. 25.2. Growth rate as a function of zonal wavelength for baroclinic waves on a basic flow with easterly shear at 20° N; $\pi/l = 5000$ km, $f = 5\times10^{-5}\ \mathrm{s}^{-1}$ and $U_z = -2.5\times10^{-3}\ \mathrm{s}^{-1}$. The full curve and crosses correspond to $\beta = 0, H_s \to \infty$; the circles correspond to $\beta = 2.15\times10^{-11}\ \mathrm{m}^{-1}\ \mathrm{s}^{-1}$ and $H_s = 7\times10^3$ m.

growth of baroclinic waves in an easterly shearing regime such as exists over south Asia during the summer months.

If the variation of density with height is included it increases the growth rate of some waves by as much as 50%, but does not much affect the shape of the growth rate curve. A further effect when $\beta \neq 0$ is to increase somewhat the phase speed (westward) of the longer waves.

25.3.2 *A simple numerical model*

This integrates over a region of the xz plane the equation expressing conservation of quasi-geostrophic potential vorticity

$$Dq + \psi_x\left\{\beta - \frac{f^2}{N^2}\left(U_{zz} + \frac{U_z}{H_s}\right)\right\} = 0, \tag{25.10}$$

where

$$D(z) = \frac{\partial}{\partial t} + U(z)\frac{\partial}{\partial x},$$

$$q = \nabla^2\psi + \frac{f^2}{N^2}\left(\psi_{zz} + \frac{1}{H_s}\psi_z\right) \tag{25.11}$$

and $U(z)$ is a specified zonal wind profile; the density scale height H_s is defined by

$$\frac{1}{H_s} = -\frac{\partial}{\partial z}(\ln \rho)_0.$$

Rigid lids are specified at the boundaries $z = \pm H/2$, i.e. the boundary conditions are taken as

$$D(-H/2)\psi_z = U_z\psi_x, \quad \text{at } z = -H/2, \tag{25.12}$$

$$D(H/2)\psi_z = U_z\psi_x, \quad \text{at } z = H/2. \tag{25.13}$$

The complex variable ψ is expressed in the form

$$\psi(x, y, z, t) = \sum_{k=0}^{K} \psi_k(z, t) \exp(ik\pi x/X) \cos(\pi y/Y) = \sum_{k=0}^{K} \psi_k, \tag{25.14}$$

with, correspondingly,

$$q(x, y, z, t) = \sum_{k=0}^{K} Q_k(z, t) \exp(ik\pi x/X) \cos(\pi y/Y). \tag{25.15}$$

Thus $K+1$ wave components are taken in the zonal direction but only one in the meridional direction; the length and width of the region considered are $2X$ and Y respectively. For consistency with (25.11),

$$Q_k = -\left(\frac{k^2\pi^2}{X^2} + \frac{\pi^2}{Y^2}\right)\psi_k + \frac{f^2}{N^2}\left(\psi_{kzz} + \frac{1}{H_s}\psi_{kz}\right), \qquad k = 0, 1, 2, \ldots, K. \tag{25.16}$$

An initial field $\psi(x, y, z, 0)$ is specified by a set of complex functions $\psi_k(z, 0)$. The initial set of functions $Q_k\ (z, 0)$ is given by (25.16) and is advanced forward a time Δt using (25.10), putting

$$D(z)Q_k = \left(\frac{\partial}{\partial t} + \mathrm{i}kU(z)\right)Q_k,$$

$$\psi_{kx} = \mathrm{i}k\psi_k$$

and specifying a finite-difference replacement for $\partial Q_k/\partial t$. The boundary values of ψ_{kz} are advanced in a similar manner using (25.12) and (25.13). The new $\psi_k(z, \Delta t)$ fields are then obtained by replacing the z-derivatives in (25.16) by finite differences and solving the resulting set of algebraic equations using matrix multiplication, i.e. the finite-difference analogue of (25.16) is solved as

$$\boldsymbol{\Psi}_k(z_i, \Delta t) = \boldsymbol{A}^{-1}\boldsymbol{Q}_k(z_i, \Delta t), \tag{25.17}$$

where, if $N+1$ levels are taken in the vertical (including the upper and lower boundaries), $\boldsymbol{A}$ is an $(N+1)\times(N+1)$ matrix and $\boldsymbol{\Psi}_k$ and $\boldsymbol{Q}_k$ are vectors of $N+1$ elements. (The first and last elements of $\boldsymbol{Q}_k$ are $\mathrm{i}k\boldsymbol{\Psi}_k(-H/2, \Delta t)$ and $\mathrm{i}k\boldsymbol{\Psi}_k(H/2, \Delta t)$.)

The process is then repeated; central time differences are used after the first time step.

25.3.3 Tests of the numerical model

Eady's analytical theory and in particular his growth-rate curve provides the obvious basis for a test of any simple baroclinic model of the atmosphere. An initial field $\Psi_1(z_i, 0)$ was specified, together with values of $U(z)$, N^2, H, and f typical of the Asian easterly jet; β was taken as zero and H_s made effectively infinite. First Y was taken as 5000 km and X specified as 10 000 km. The integration was allowed to proceed for 20 simulated days with time steps of 40 minutes, by which time the growth rate had settled down to a steady value. The process was repeated for further values of X down to below the cut-off. The growth rates are

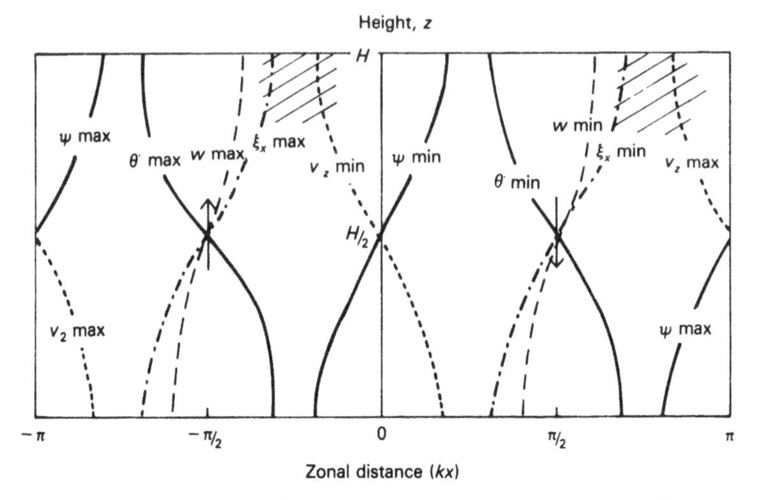

Fig. 25.3. The phases of ψ, θ', w, v_z and ζ_x as functions of zonal distance and height for the Eady wave with a meridional wavelength of 10 000 km and a zonal wavelength $(2\pi/k)$ of 8000 km at 20° N with $U_z = -2.5\times10^{-3}\ \mathrm{s}^{-1}$. The shaded regions indicate where ζ_x and v_z are large in magnitude and opposite in sign.

shown (as crosses) in Fig. 25.2 and are seen to agree very well with the theoretical values. So also do the phase speeds and structure (Fig. 25.3).

Further integrations were carried out with Y set equal to 2500 km, H_s set at 7×10^3 and β at $2.152\times10^{-11}\ \mathrm{m}^{-1}\ \mathrm{s}^{-1}$, its value for 20° N. These growth rates are shown in Fig. 25.2.

With H_s infinite and β zero, the pure Eady case, there is no amplification for this value of Y for any value of X, since the total wavenumber is higher than the cut-off value. The growth which is found in these integrations is associated with the β-effect and the vertical density gradient and is of a magnitude consistent with the theoretical calculations of Green (1960).

Since each growth rate has been calculated numerically by solving an initial value problem, the question arises as to whether the results are influenced to any extent by the form of $\Psi_1(z_i, 0)$. Although the initial stages of the development must be somewhat different for different initial vertical distributions of Ψ_1, the growth rate, if the disturbance eventually amplifies, will be unaffected.

25.4 The downstream development phenomenon in midlatitudes

Synoptic meteorologists were aware as far back as the 1940s that a major trough development over the eastern seaboard of North America was almost invariably followed two or three days later by a similar development over or near western Europe. A similar process has also been observed relating trough developments over the Pacific and over the western USA; the phenomenon is referred to as 'downstream development'. Between successive trough developments, ridges are observed to intensify.

Interest in this phenomenon was recently revived in the UK Universities' Modelling Group at Reading by the results of a numerical experiment in which a purely zonal flow, resembling the climatological mean, was perturbed in one sector only of the hemisphere. It had been anticipated that the baroclinically unstable zonal flow would soon break up into eddies round the whole hemisphere which would interfere with each other and create a disordered pattern. Instead, however, the development was extremely orderly, a baroclinic wave first developing near the original perturbation and, about two simulated days later, another downstream and so on. In addition a shallow small-scale disturbance developed upstream of the main disturbance. Fig. 25.4 shows the upper-level and surface flows after 13 days. Fuller details are contained in Simmons and Hoskins (1979).

Perhaps the most surprising feature of these results is the way in which the zonal flow, although baroclinically unstable, remains virtually undisturbed until the development process has progressed sufficiently far downstream. Clearly the baroclinic development in the model is initiated as a regular sequential process and does not break out in a random manner. It is reasonable to conclude that the same is likely to be true in the atmosphere, i.e. baroclinic waves do not in general develop in a random manner as apparently do some other kinds of instability features, e.g. small-scale convection plumes. The downstream development process seems therefore to be a major mechanism of baroclinic wave development in midlatitudes, representing an orderly release of baroclinic instability.

The simple model described above has also been used to study this process. An initial ψ distribution was chosen of the form

$$\psi(x, y, z, 0) = 100 \exp(-x^2/a^2) \cos^4(\pi x/2X) \cos(\pi y/Y), \quad |x| < a$$

with $a = 1500$ km, $X = 30\,000$ km and $Y = 5000$ km. $U(z)$ was specified (in m s^{-1}) as

$$U = -15 + 30z/H$$

with $H = 10$ km.

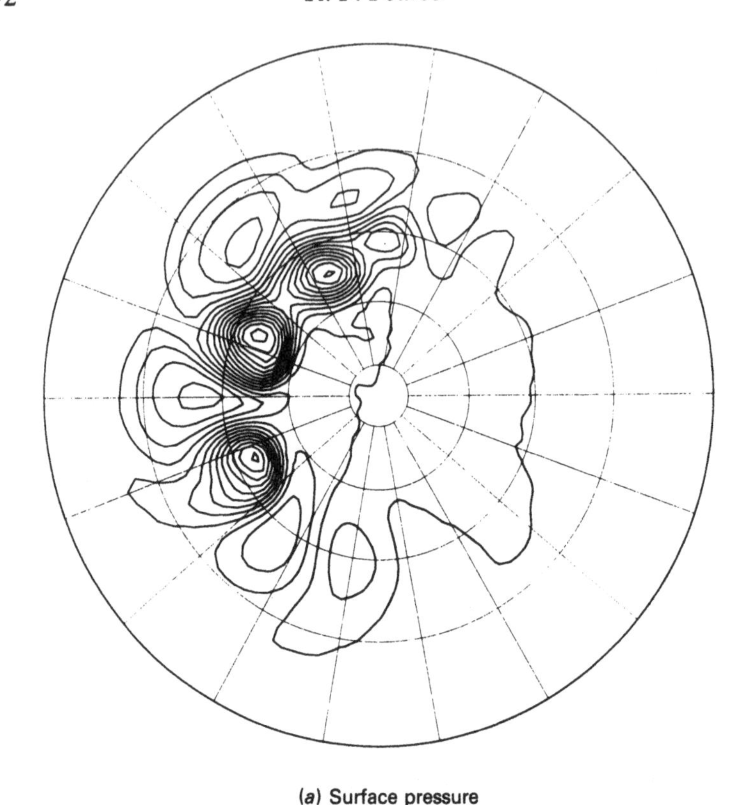

(*a*) Surface pressure

Fig. 25.4. (*a*) Computed surface isobars and (*b*) computed 500 mb (level 3) streamlines after 13 days of integration on a hemisphere using a 5-layer T42 truncation spectral model. An initial zonal flow configuration with no surface wind and a westerly jet maximum of approximately 35 m s^{-1} at about 35° N was slightly perturbed (day 0) in the top right-hand quadrant only. (After Simmons and Hoskins, 1979.)

Fig. 25.5*a* shows the computed ψ distribution after 4 simulated days; the maximum values have amplified by a factor of just over 4 and the pattern has assumed the backward tilt typical of developing midlatitude baroclinic waves. Perhaps the most striking feature of this result, like that of the hemispheric primitive equation experiment referred to above, is that the baroclinic development has evolved in a highly orderly and symmetrical manner with virtually no disturbance of the flow having yet occurred beyond about 7000 km from the axis of the original disturbance. The integration was continued to 10 days and the intensification and propagation continued in a similar manner with alternate positive and

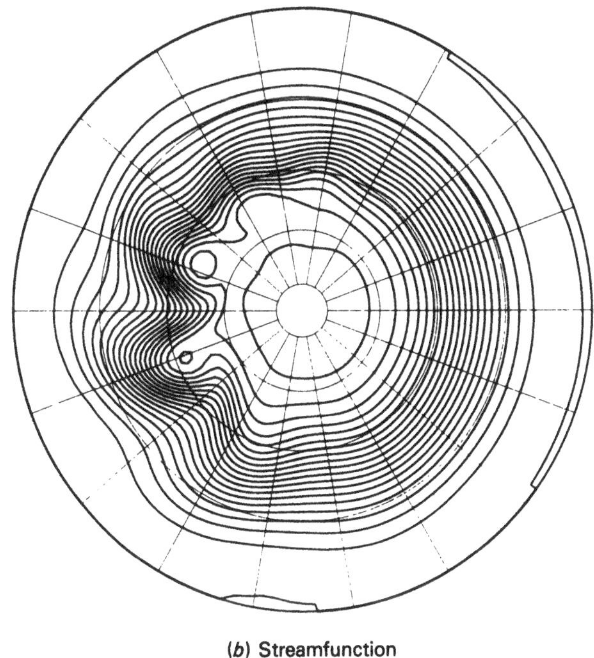

(*b*) Streamfunction

Fig. 25.4—*cont.*

negative regions extending to about 12 000 km on either side of the axis at the end of the period.

Fig. 25.5*b* shows the result obtained for the second experiment in which the only change made from the first was to set β equal to $1.6\times10^{-11}\ \mathrm{m}^{-1}\ \mathrm{s}^{-1}$. This has the effect of allowing Rossby wave dispersion to take place. The distribution of ψ still retains the features of amplification and propagation at a preferred wavelength seen in the first experiment, but there are some important modifications. The amplification rate is smaller and the development of the most rapidly amplifying waves mainly occurs downstream. There is development upstream as well, but this is associated both with long waves and shallow low-level short waves.

A third experiment was carried out in which a density scale height of 7×10^{3} m was used in place of the very large value used in the first two experiments, the value of β used for the second experiment being retained. The result, shown in Fig. 25.5*c*, was mainly to reduce the eastward phase speed of the primary amplifying wave components.

As a further experiment, designed to examine the rate of pure barotropic (Rossby wave) dispersion, the shear (U_z) was set to zero and H_s

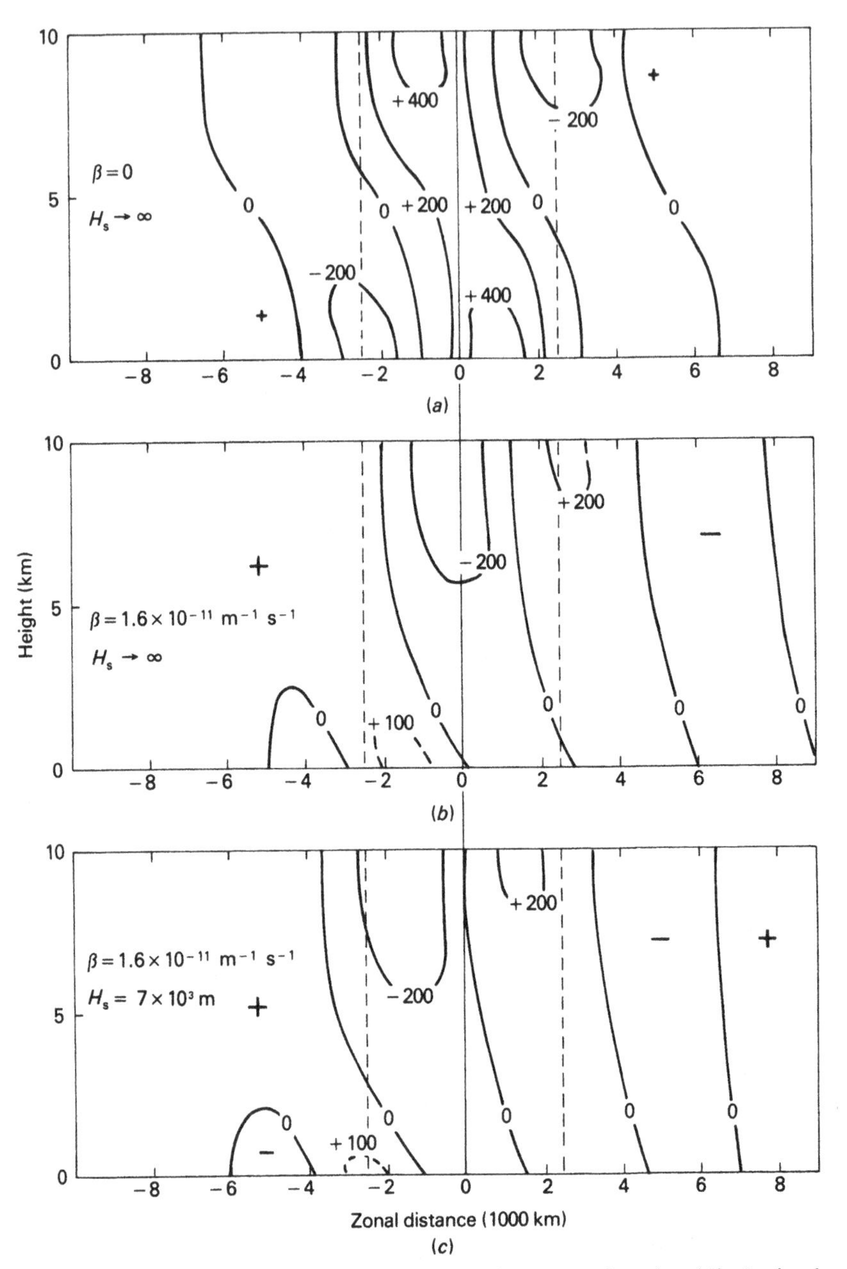

Fig. 25.5. Zonal cross-sections of the perturbation streamfunction (ψ) obtained after four simulated days of integration using the simple model described in § 25.3.2. The initial flow was a linear shear with $U = -15$ m s^{-1} at the surface and $+15$ m s^{-1} at $z = 10$ km, and the initial perturbation was independent of height and confined to the region between the hatched lines. The midlatitude value of 10^{-4} s^{-1} was taken for f. The values of β and H_s assumed for the three cases are indicated on the figures.

reset to a large value, f being retained at its midlatitude value. The resulting distribution of ψ for days 0 to 4 are shown in Fig. 25.6. As theory predicts, the initial wavepacket disperses throughout the region of integration (taken as 30 000 km in the x-direction), the longer wave components to the west and the shorter to the east.

One may therefore interpret Figs. 25.5*b* and 25.5*c* in two different ways; either as pure baroclinic development and propagation modified by Rossby wave dispersion, or as Rossby wave dispersion modified by baroclinic development. In this midlatitude case either interpretation is appropriate. The relative contributions of the β-effect and the other factors determining baroclinic development may be estimated theoretically by considering the simple form of the diagnostic equation for the vertical velocity obtained by eliminating the time-dependent terms from (25.4) and (25.5) and using (25.6):

$$N^2\nabla^2 w + f^2 w_{zz} = 2fU_z\zeta_x + f\beta v_z. \qquad (25.18)$$

This is a form of the well-known quasi-geostrophic ω-equation (where $\omega = \mathrm{d}p/\mathrm{d}t$).

The first term on the right represents the primary forcing of the vertical motion (with which 'development' is usually associated by synoptic meteorologists). It is essentially a combination of the terms associated with differential vorticity advection and the Laplacian of thermal advection (see for example Haltiner 1971). Thus positive vorticity advection ($U\zeta_x$ negative) at upper levels with smaller values at lower levels over a

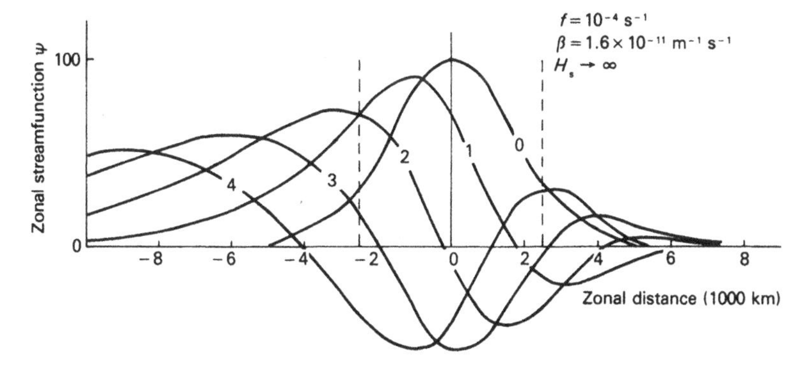

Fig. 25.6. Zonal streamfunction (ψ) distributions at days 0, 1, 2, 3 and 4 obtained using the simple model with zero initial vertical shear and the same initial ψ distribution as for the cases shown in Fig. 25.5.

given region implies that this term is negative and hence that w is generally positive in that region, and vice versa. The second term on the right represents the forcing of w associated with β – upward motion being associated with $v_z < 0$. With $U_z > 0$ as in midlatitudes, the terms reinforce each other where ζ_x and $v_z (= \psi_{xz})$ have the same sign and this is the case only at lower levels. They are of similar magnitudes for zonal wavelengths of about 8000 km. Hence, for wavelengths shorter than this one, the β-effect should be regarded as a secondary one, while for the very long waves it provides the primary forcing. For all wavelengths the β-effect has an important influence on the growth rate.

25.5 Downstream development and monsoon depressions

The observed mean flow and temperature distributions over southern Asia and the adjacent Indian Ocean for June, July, August and September are very similar (see for example Ramage and Raman, 1972), the main variation being associated with a gradual increase in the jet maximum near Sri Lanka from 30 m s^{-1} in June to 35 m s^{-1} in August falling to 27 m s^{-1} in September. Values of the variance of the westwards velocity component vary from a minimum of about 6 m s^{-1} near the jet axis to 12 m s^{-1} at 25° N (see Newell *et al.*, 1972). At low levels the monsoon westerlies also exhibit only small variations. The winds over the region during the summer months are therefore remarkably persistent and, compared with the winds in the extratropics, relatively steady. This is, therefore, a region for which studies of idealized flows, generally similar to the climatological mean, are likely to yield results having direct applicability to the flow observed over the region on any particular occasion during the summer months.

Krishnamurti *et al.* (1977) have carried out a study of over 30 occasions on which typhoons in the south China Sea were followed, 10 to 14 days later, by the development of a 'monsoon depression' in the Bay of Bengal, and have suggested that this could be a downstream development phenomenon similar to that of midlatitudes. They point out that the group velocity of Rossby waves in the midtropospheric flow would be consistent with this interpretation.

To test this hypothesis, two integrations were carried out using the simple baroclinic model described above. The first, analogous to that which produced the results for pure Rossby wave dispersion in Fig. 25.6, used the parameter values $U = 0$, $f = 5 \times 10^{-5}\,\mathrm{s}^{-1}$, $\beta = 2.15 \times 10^{-11}\,\mathrm{m}^{-1}\,\mathrm{s}^{-1}$ and $H_s = 7 \times 10^3$ m, appropriate for latitude 20° N, with

$$\psi(x, y, z, 0) = \begin{cases} -500 \cos^2(\pi x/2a) \cos(\pi y/Y), & |x| < a \\ 0, & |x| > a. \end{cases}$$

Y was taken as 1000 km and a as 1000 km. A high vertical resolution (32 levels) was used.

The result, shown in Fig. 25.7*a*, indicates that the amplitude of the initial disturbance is reduced by about $\frac{1}{3}$ after 10 days and that its maximum value is displaced about 1000 km to the west.

In the second experiment an easterly shear typical of that observed at 20° N and 90° W during the summer was specified with $U_z = -2.5 \times 10^{-3}$ s^{-1}, and $U = 0$ at $z = H/2$, the other parameters and the initial conditions remaining unchanged. The effect, as indicated in Figs. 25.7*b* and 25.8 was quite dramatic. After only 4 days, a train of baroclinic waves appear to the west of the origin at upper levels and to the east at lower levels. The Rossby wave dispersion can just be identified at middle levels, but is almost completely swamped by the dispersion associated with the vertical shear, i.e. by the advection of the initial disturbance westwards at upper levels and eastwards at lower levels. The main downstream effect is thus the generation of upper-level baroclinic disturbances, travelling at a speed somewhat less than that of the upper-tropospheric flow maximum of 12.5 m s^{-1}, i.e. at 8 m s^{-1} (about 7 degrees of longitude per day); any disturbance resulting from a tropical cyclone over the south China Sea would, if it behaved according to this result, reach the Bay of Bengal in about 3 days. Krishnamurti *et al* show a Hövmuller diagram of the surface-pressure wavenumber components 3 to 12 for one of the cases studied. This is reproduced in Fig. 25.9.

This shows waves between 60° E and 160° E which regularly move westwards at a rate of 5 to 8 degrees of longitude per day for the first part of the period. But, as well as moving westwards, they regularly intensify and decay with a period of 4 to 6 days. Such behaviour would be consistent with periodic interaction of the lower troposphere with westward-moving upper-tropospheric baroclinic waves. The concentration of the phenomenon between 60° E and about 160° E, where climatologically there is baroclinicity associated with easterly shear, adds support to this suggestion, although this baroclinicity is weak beyond about 120° E.

It seems, therefore, that downstream development, a process which has been shown to involve baroclinic development as well as Rossby wave dispersion, may be an important process in the low-latitude easterlies as

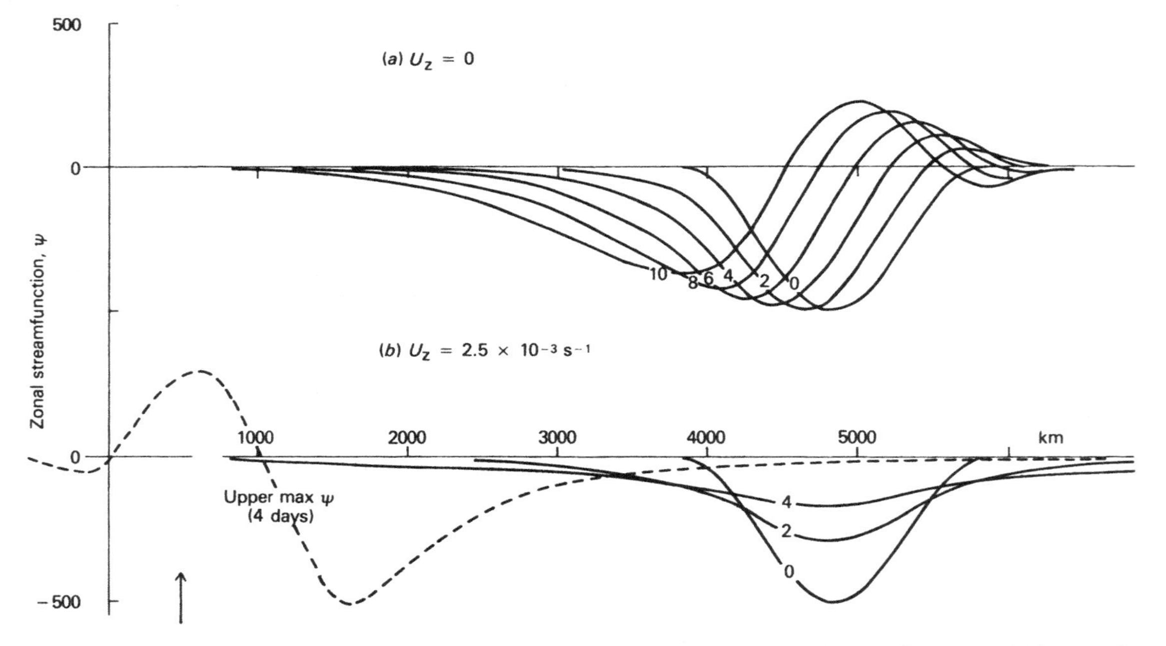

Fig. 25.7. (a) Zonal streamfunction (ψ) at days 0, 2, 4, 6, 8 and 10 obtained from a simple model integration with zero initial vertical shear and β set to a value of $1.215 \times 10^{-11}\ \mathrm{m^{-1}\ s^{-1}}$, corresponding to latitude 20°. The initial perturbation was taken as independent of height. (b) As (a) but with $U = +12.5\ \mathrm{m\ s^{-1}}$ at $z = 0$ and $U = -12\ 5\ \mathrm{m\ s^{-1}}$ at $z = 10$ km with linear shear. Only days 0, 2 and 4 are shown for the streamfunction at $z = 5$ km (full curves). The dashed curve shows the streamfunction at $z = 10$ km day 4.

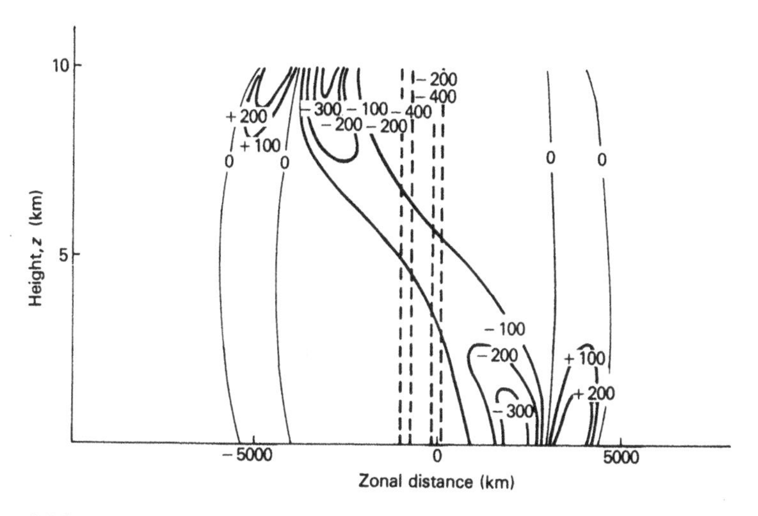

Fig. 25.8. Vertical cross-section of the zonal streamfunction (ψ) after 4 days integration of the simple baroclinic model with an easterly shear. $U = 12.5\ \mathrm{m\,s^{-1}}$ at $z = 0$ and $-12.5\ \mathrm{m\,s^{-1}}$ at $z = 10$ km. The dashed lines indicate the initial distribution of ψ. (The dashed curve in Fig. 25.7.*b* corresponds to the zonal variation of ψ at $z = 10$ km.)

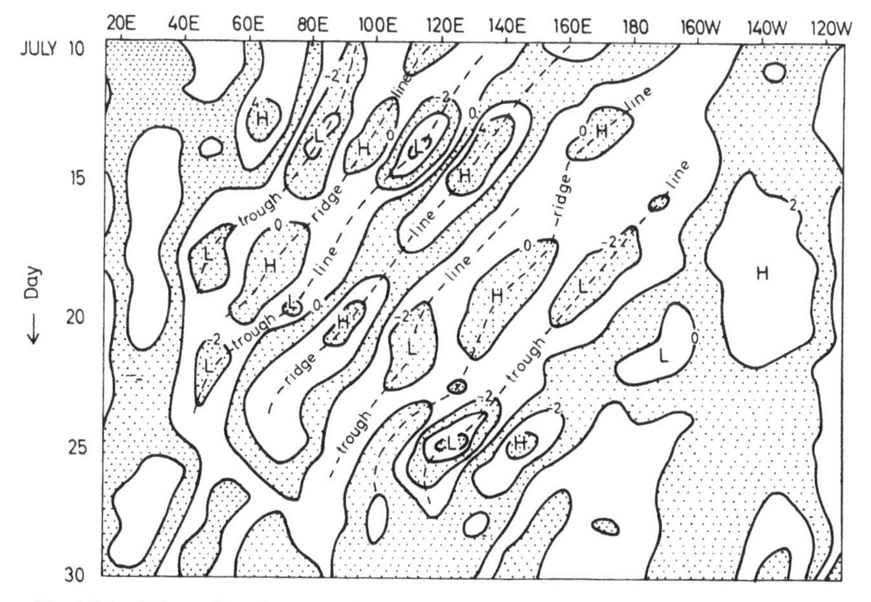

Fig. 25.9. Hövmuller diagram of the sum of zonal wavenumbers 3 to 12 based on mean sea-level pressure data at 20° N during July 1965. (After Krishnamurti *et al.*, 1977.)

well as in midlatitudes. However, it cannot manifest itself on the timescale of 10 to 14 days as a pure Rossby wave dispersion process as suggested by Krishnamurti *et al.* It is necessary to consider some other process, such as a regular modulation of wavenumbers 1 and 2, to account for changes on this timescale.

Finally, it is interesting briefly to consider the relative magnitudes of the forcing terms in (25.18) when low-latitude parameter values are inserted ($\beta = 2.15 \times 10^{-11}\ \text{m}^{-1}\ \text{s}^{-1}$ and $v_z = v/h$ where $h \approx H/2$). The terms are found to have similar magnitudes for wavelengths of about 4000 km. Furthermore, intensification occurs at upper levels since U_z is negative; this is indicated by the shaded regions in Fig. 25.3 where ζ_x and v_z have opposite signs and are large in magnitude.

In applying these idealized model results to the atmosphere it must be borne in mind that the model used here incorporates the quasi-geostrophic assumption, and that in latitudes below about 20° departures from geostrophy are likely to grow as the wave amplitude increases. However the general nature of the results is unlikely to be different in an integration of the full primitive equation set, although this has not yet been carried out.

25.6 Conclusions

The main conclusion of this study of monsoon depressions is that upper-tropospheric baroclinic development could be a significant factor in determining their occurrence or, at least, their variation in intensity over periods of a few days. This is only intended as a preliminary investigation of this phenomenon and to serve as an illustration of the way in which experiments with a simple numerical model can be carried out in order to study the basic dynamical processes which are likely to be involved. In this particular case the ability to use a high vertical (as well as horizontal) resolution is an advantage enabling the rather shallow baroclinic waves to be properly resolved. The next step is to carry out a study of observational data to see whether there exist westward-moving baroclinic waves in the upper troposphere over south Asia during the summer months. If this is confirmed to be the case, further model experiments need to be carried out incorporating dissipation by small-scale processes, both internally and in the surface boundary layer, and moisture. A simple form of convective parameterization must also be included.

The original hypothesis was that typhoon disturbances over the south China Sea occasionally provide the primary initiating mechanism for monsoon depressions. In the light of the simple model results, a modified form of this hypothesis is suggested, i.e. that upper-tropospheric baroclinic disturbances having wavelengths of about 4000 km travelling westwards over the west Pacific and south Asia have an important modifying influence on lower-level features and may contribute towards the initiation of both typhoons and monsoon disturbances.

The author wishes to acknowledge the invaluable assistance of Dr A. J. Simmons who wrote the original computer programme for the simple baroclinic model and made several very useful comments on the results; also the comments of Dr B. J. Hoskins who suggested the use of the omega-equation (25.18) to examine the importance of the β-effect in determining development.

References

Colton, D. E. (1973) Barotropic scale interactions in the tropical upper troposphere during the northern summer. *J. Atmos. Sci.*, **30**, 1287–1302.

Eady, E. T. (1949) Long waves and cyclone waves. *Tellus*, **1**, No. 3, 31–52.

Green, J. S. A. (1960) A problem in baroclinic stability. *Quart. J. Roy. Meteor. Soc.*, **86**, 237–51.

Haltiner, G. J. (1971) *Numerical Weather Prediction.* Wiley and Sons, New York.

Krishnamurti, T. N., Molinari, J., Pan, H. L. and Wong, V. (1977) Downstream amplification and formation of monsoon disturbances. Report 77–2, Department of Meteorology, Florida State University.

Kuo, H. L. (1949) Dynamic instability of two-dimensional non-divergent flow in a barotropic atmosphere. *J. Meteor.*, **6**, 105–22.

Kuo, H. L. (1951) Dynamic aspects of the general circulation and the stability of zonal flow. *Tellus*, **3**, 268–84.

Kuo, H. L. (1973) Dynamics of quasi-geostrophic flows and instability theory. *Advances in Applied Mechanics*, **13**, 247–330.

Lipps, F. B. (1962) The barotropic stability of the mean winds in the atmosphere. *J. Fluid Mech.*, **12**, 397–407.

Lipps, F. B. (1965) The stability of an asymmetric zonal current in the atmosphere. *J. Fluid Mech.*, **21**, 225–39.

Lipps, F. B. (1970) Barotropic stability and tropical disturbances. *Mon. Wea. Rev.*, **98**, 122–31.

Newell, R. E., Kidson, J. W., Vincent, D. G. and Boer, G. J. (1972) *The General Circulation of the Tropical Atmosphere.* Vol. 1. MIT Press, Cambridge, Mass.

Ramage, C. S. and Raman, C. V. R. (1972) *Meteorological Atlas of the International Indian Ocean Expedition.* Vol. 2, Upper Air. National Science Foundation, Washington DC.

Rennick, M. A. (1976) The generation of African waves. *J. Atmos. Sci.*, **33**, 1955–69.

Simmons, A. J. (1977) A note on the instability of the African easterly jet. *J. Atmos. Sci.*, 34, 1670–74.

Simmons, A. J. and Hoskins, B. J. (1979). The downstream and upstream development of unstable baroclinic waves. *J. Atmos. Sci.*, **36**, 1239–54.

Tupaz, J. B. (1977) A numerical study of barotropic instability of a zonally varying easterly jet. Ph.D. Thesis, US Naval Postgraduate School, Monterey.

Yanai, M. and Nitta, T. (1968) Finite difference approximations for the barotropic instability problem. *J. Meteor. Soc. Japan.*, **46**, 389–403.

26

The stability of the monsoon zonal flow with a superposed stationary monsoon wave

V. SATYAN, R. N. KESHAVAMURTY AND B. N. GOSWAMI

Using a two-level quasi-geostrophic model, the effect of the inclusion of a long stationary monsoon wave (basic meridional wind) on the combined barotropic–baroclinic instability of the monsoon current has been investigated. Both barotropic and baroclinic interactions of the perturbation with the monsoon wave and the zonal flow are included. This analysis, using realistic velocity profiles, yields fast-growing westward-moving upper-tropospheric waves corresponding to easterly waves, and lower-tropospheric modes corresponding to monsoon disturbances.

26.1 Introduction

The stability of the monsoon zonal flow has been studied by Shukla (1977) and Keshavamurty *et al.* (1977). They found that the monsoon atmosphere is not baroclinically unstable. Shukla found that the CISK mechanism can lead to a growth of disturbances of the same scale as the observed monsoon disturbances, but that there was no preferred scale with a fastest rate of growth. Keshavamurty *et al.* found that the monsoon zonal flow is barotropically unstable in the lower and midtroposphere and that this instability can yield disturbances of reasonable growth rate and scale. Mak (1975) studied the effect of the meridonal motion on baroclinic instability of a monsoon flow, but found that the meridonal components required for growth were very large compared to those obtained from observed winds. Lorenz (1972) studied the barotropic instability of Rossby waves and found that shorter waves can be barotropically unstable. The observed monsoon flow is seen to have an appreciable meridional component. Furthermore, in a global picture, the monsoon is found to be a stationary wave on the wavenumber 2 scale (Krishnamurti, 1971) induced by the land–ocean heating contrast of the Eurasian continent with respect to the Pacific Ocean as well as by the thermal and mechanical effects of the Himalayan massif. It is of interest to study the interaction of this global wave with the local monsoon. This chapter investigates the stability of a monsoon zonal flow with such a superposed stationary wave, taking into account both the barotropic and baroclinic interactions.

26.2 The basic equations and the model

The model is 2-level and uses the quasi-geostrophic and β-plane approximations. The perturbation potential vorticity equations at levels 1 and 3 (see Fig. 26.1) are, in the usual notation,

$$\left[\frac{\partial}{\partial t}+u_1\frac{\partial}{\partial x}+\overline{v_1^*}\frac{\partial}{\partial y}\right]\left[\zeta_1'+\frac{s_2}{f_0}(\phi_3'-\phi_1')\right]$$
$$+u_1'\left[\frac{\partial\overline{\zeta_1^*}}{\partial x}+s_2(\overline{v_3^*}-\overline{v_1^*})\right]$$
$$+v_1'\left[\beta-\frac{\partial^2 u_1}{\partial y^2}+\frac{\partial\overline{\zeta_1^*}}{\partial y}-s_2(u_3-u_1)\right]=-\frac{s_2}{f_0}\frac{R}{C_p p}\Delta p\,\dot{Q}_2', \tag{26.1}$$

Fig. 26.1. The vertical structure of the model.

$$\left[\frac{\partial}{\partial t}+u_3\frac{\partial}{\partial x}+\overline{v_3^*}\frac{\partial}{\partial y}\right]\left[\zeta_3'-\frac{s_2}{f_0}(\phi_3'-\phi_1')\right]$$

$$+u_3'\left[\frac{\partial\overline{\zeta_3^*}}{\partial x}-s_2(\overline{v_3^*}-\overline{v_1^*})\right]$$

$$+v_3'\left[\beta-\frac{\partial^2 u_3}{\partial y^2}+\frac{\partial\overline{\zeta_3^*}}{\partial y}+s_2(u_3-u_1)\right]-\frac{f_0\omega_4'}{\Delta p}$$

$$=\frac{s_2}{f_0}\frac{R}{C_p p}\Delta p\,\dot{Q}_2', \tag{26.2}$$

where $u_1(y)$, $u_3(y)$ are zonal winds at levels 1 and 3 and $\overline{v_1^*}$, $\overline{v_3^*}$ are stationary meridional winds at levels 1 and 3. They are functions of y which vanish at the northern and southern boundaries and are assumed to be described by the zonal waveform exp (ik_0x) in the x direction. ϕ' is the perturbation geopotential, u', v' are the perturbation velocity components and ζ' is the perturbation vorticity. Further, $s_2=f_0^2/(\Delta p^2\sigma)$ with $\Delta p=500$ mb and $\sigma=-(\alpha/\theta)(\partial\theta/\partial p)$ where $\alpha=-\partial\phi/\partial p$ and $\theta=T(p_0/p)^{R/c_p}$. The perturbation heating is denoted by $\dot{Q}_2'$ and f is the Coriolis parameter defined by $f=f_0+\beta y$. We assume wave solutions of the type

$$\phi_1'\approx\phi_1(y)\exp\left[ik(x-ct)\right],$$

$$\phi_3'\approx\phi_3(y)\exp\left[ik(x-ct)\right],$$

where $\phi(y)$ and c can be complex. Substituting these solutions in (26.1) and (26.2) gives

$$\left[ik(u_1-c)+\overline{v_1^*}\frac{\partial}{\partial y}\right]\left[\frac{1}{f_0}\left(-k^2\phi_1+\frac{\partial^2\phi_1}{\partial y^2}\right)+\frac{s_2}{f_0}(\phi_3-\phi_1)\right]$$

$$-\frac{1}{f_0}\frac{\partial\phi_1}{\partial y}\left[-k_{01}^2\overline{v_1^*}+s_2(\overline{v_3^*}-\overline{v_1^*})\right]$$

$$+\frac{ik}{f_0}\phi_1\left[\beta-\frac{\partial^2 u_1}{\partial y^2}+ik_{01}\frac{\partial\overline{v_1^*}}{\partial y}-s_2(u_3-u_1)\right]$$

$$=-\frac{s_2}{f_0}\frac{R}{C_p p}\Delta p\,\dot{Q}_2'\exp\left[-ik(x-ct)\right], \tag{26.3}$$

and

$$\left[ik(u_3-c)+\overline{v_3^*}\frac{\partial}{\partial y}\right]\left[\frac{1}{f_0}\left(-k^2\phi_3+\frac{\partial^2\phi_3}{\partial y^2}\right)-\frac{s_2}{f_0}(\phi_3-\phi_1)\right]$$

$$-\frac{1}{f_0}\frac{\partial\phi_3}{\partial y}[-k_{03}^2\overline{v_3^*}-s_2(\overline{v_3^*}-\overline{v_1^*})]$$

$$+\frac{ik}{f_0}\phi_3\left[\beta-\frac{\partial^2 u_3}{\partial y^2}+ik_{03}\frac{\partial\overline{v_3^*}}{\partial y}+s_2(u_3-u_1)\right]-\frac{f_0\omega_4}{\Delta p}$$

$$=\frac{s_2}{f_0}\frac{R}{C_p p}\Delta p\,\dot{Q}_2'\exp[-ik(x-ct)], \tag{26.4}$$

where k_{01} and k_{03} are the wavenumbers of the stationary monsoon wave at levels 1 and 3. The above equations are of third order and require three boundary conditions for their solution. Equations (26.3) and (26.4) are first converted into finite-difference form and use is made of the backward difference formula at the northern boundary and the central difference formula at the southern boundary. The boundary conditions used are: $\phi_1'=\phi_3'=0$ and $\partial\phi_1'/\partial y=\partial\phi_3'/\partial y=0$ at the lower latitudinal boundary; $\phi_1'=\phi_3'=0$ at the northern latitudinal boundary; $\omega=0$ at the top boundary; and $\omega=\omega_4$ at the bottom boundary (ω_4 being the value of ω at the top of the frictional boundary layer). As is well known, the difference scheme for the differential equations gives rise to a set of algebraic equations which can be solved by employing the standard eigenvalue technique.

There are $2n$ such algebraic equations in $2n$ unknowns (for a given grid of $(n+2)$ points at each level) and for a non-trivial solution the determinant of their coefficients must vanish. In matrix notation one has

$$(P-cQ)(\phi)=0,$$

or

$$(PQ^{-1}-cI)(Q\phi)=0,$$

where the matrices I, P, Q have their usual meanings (Keshavamurty *et al.*, 1977). It should be noted that in this case the matrix P is complex.

The analysis has been carried out for the mean zonal winds at the longitude of India (Fig. 26.2) and for meridional waves of wavenumber 2 around the globe at 700 mb (level 3) and 200 mb (level 1). The gradient of the potential vorticity of the July mean zonal wind is shown in Fig. 26.3. Profiles of the mean zonal winds at 200 mb and 700 mb are shown in Fig. 26.4. It is assumed here that the wavelength of the disturbance is small

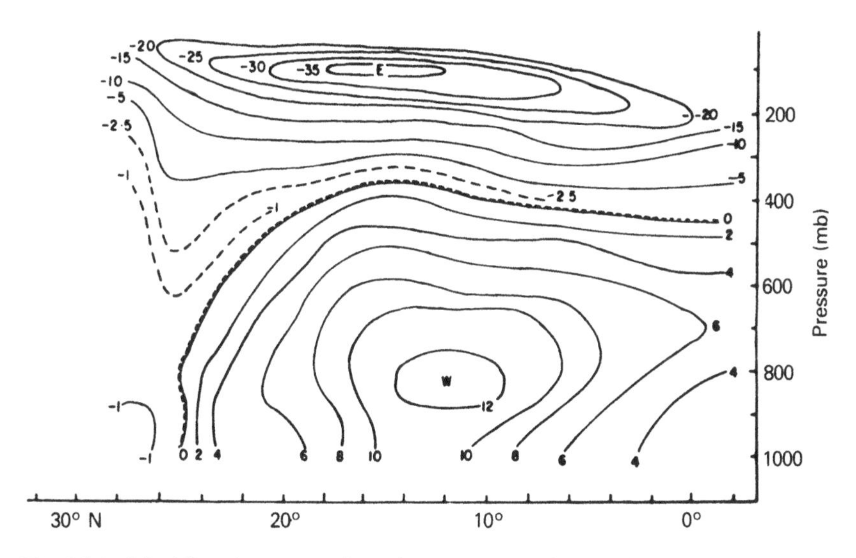

Fig. 26.2. Meridional cross-section of mean zonal winds at 80° E for July 1977. Wind speeds for the isotachs are given in m s^{-1}.

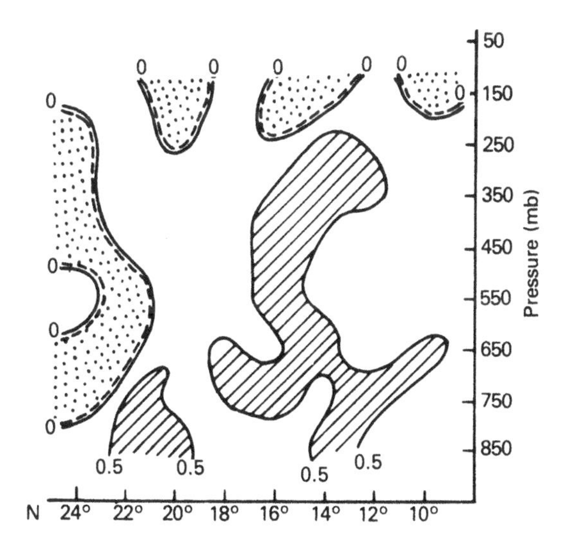

Fig. 26.3. Meridional gradient of the potential vorticity (P) of the July 197 man zonal wind at 80° E (shown in Fig. 26.2). The potential vorticity gradient has been computed (cf. (27.1)) s $B-(u_{j+1}+u_{j-1}-2u_j)/d^2+s_{k-1}(u_k-u_{k-2})+s_{k+1}(u_k-u_{k+2})$. Negative values are stipled.

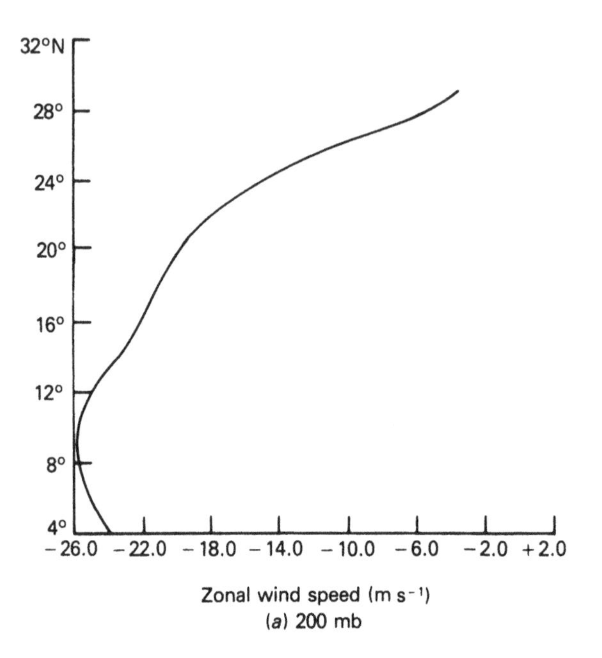

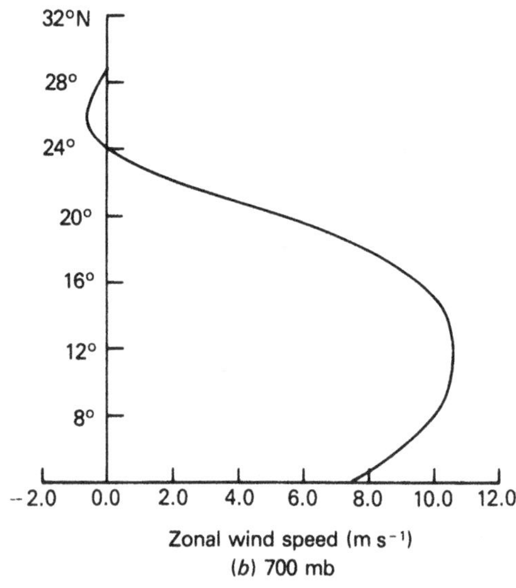

Fig. 26.4. Profiles of the mean zonal winds speeds (u) at (a) 200 mb and (b) 700 mb, obtained from Fig. 26.2.

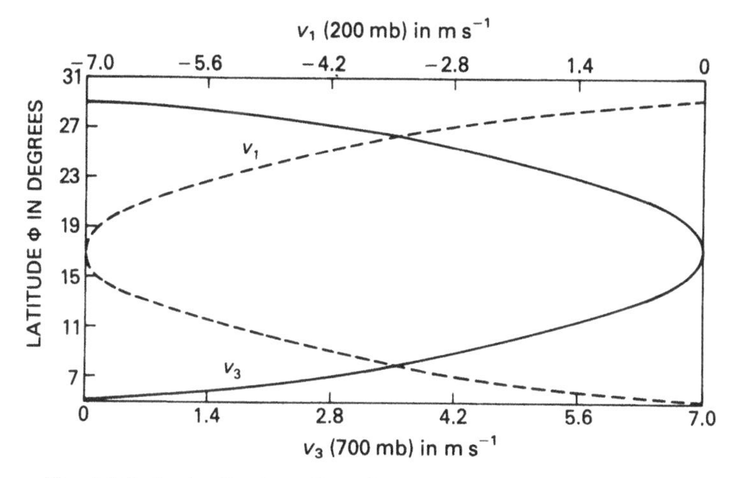

Fig. 26.5. Latitudinal profiles of the meridional wave amplitude.

compared with the wavelength of the stationary wave. Its energy exchanges with the zonal flow are both barotropic and baroclinic. The maximum amplitude of the wave is taken to be 7 m s^{-1} southerly in the lower level, and 7 m s^{-1} northerly in the upper level (Fig. 26.5). Each of the waves has a wavelength of 180 degrees which is the dominant mode of the observed global monsoon (Krishnamurti, (1971) and their amplitudes reduce to zero at the northern and the southern boundaries. The perturbation heating has not been included in this analysis.

26.3 Results and discussion

The most marked growing modes are indicated in Tables 26.1 and 26.2. In these tables, L denotes the scale length of the disturbance, c_r its phase speed, and τ_d its doubling time. ϕ_1 and ϕ_3 refer to the maximum values of the amplitude of the disturbance at levels 1 (200 mb) and 3 (700 mb) respectively. The latitude at which the maximum amplitude occurs is also shown below each ϕ value. It is seen from these results that there are growing modes in both the upper- and the lower-tropospheric layers. Furthermore, there are two distinct growing modes at each level. In the upper troposphere, we find a growing mode which moves westward with a phase speed of about 25 m s^{-1} and has a doubling time of 1.9 days. The wavelength of this mode is about 3000 km. We also see that there is another upper-tropospheric mode with a westward phase speed of 14 m s^{-1} and a doubling time of about 0.6 days around 3000 km. The

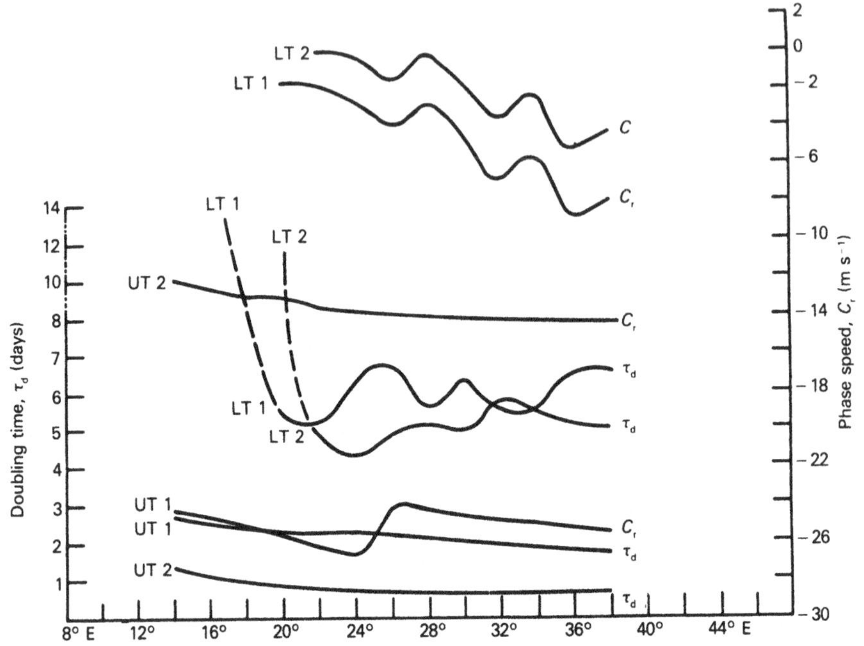

Fig. 26.6. Scale dependence of the growing modes. LT 1 and LT 2 correspond to the two lower-tropospheric modes, while UT 1 and UT 2 refer to the upper-tropospheric ones.

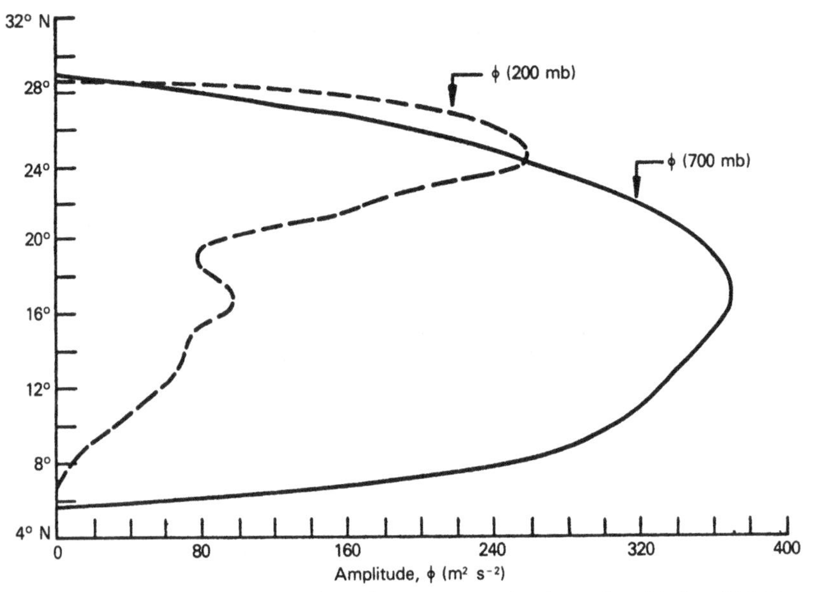

Fig. 26.7. Amplitude profile of the lower-tropospheric mode with $L = 3000$ km, $\tau_d = 5.6$ days and $C_r = 2.7$ m s^{-1}.

TABLE 26.1. *Description of the upper-tropospheric (UT) modes.*

	UT 1			UT 2		
L (km)	C_r (m s^{-1})	τ_d (days)	Max (ϕ_1) (m^2 s^{-2})	C_r (m s^{-1})	τ_d (days)	Max (ϕ_1) (m^2 s^{-2})
1400	−24.4	2.7	81 (18° N)	−12.0	1.3	220 (25° N)
1600	−24.8	2.5	87 (16° N)	−12.5	1.1	250 (25° N)
1800	−25.2	2.3	91 (16° N)	−12.9	0.91	280 (25° N)
2000	−25.6	2.26	96 (19° N)	−13.2	0.83	310 (25° N)
2200	−26.1	2.25	100 (19° N)	−13.5	0.78	340 (25° N)
2400	−26.6	2.26	110 (17° N)	−13.7	0.74	370 (25° N)
2600	−24.2	2.20	120 (19° N)	−13.9	0.71	400 (25° N)
2800	−24.4	2.08	120 (18° N)	−14.1	0.68	420 (25° N)
3000	−24.6	1.9	130 (18° N)	−14.1	0.67	450 (25° N)
3200	−24.9	1.9	130 (18° N)	−14.2	0.65	470 (25° N)
3400	−25.0	1.8	140 (18° N)	−14.3	0.62	500 (24° N)
3600	−25.4	1.7	140 (18° N)	−14.3	0.61	520 (24° N)
3800	−25.6	1.7	150 (18° N)	−14.3	0.59	530 (24° N)

doubling times and the phase speeds of the growing modes at both levels are shown as functions of wavelength in Fig. 26.6.

There are also interesting lower-tropospheric modes (Table 26.2). One such mode, whose amplitude structure is shown in Fig. 26.7, has a scale of about 3000 km with growth rate of 5–6 days moving westward with a speed of 2.7 m s^{-1}. It is the authors' that this corresponds to the monsoon disturbances forming over the north Bay of Bengal.

26.4 Conclusions

The inclusion of a stationary monsoon wave in a barotropic–baroclinic stability analysis has led to fast-growing westward-moving upper-tropospheric easterly waves of a scale of around 3000 km.

TABLE 26.2. *Description of the lower-tropospheric (LT) modes.*

	LT 1				LT 2			
L (km)	C_r (m s^{-1})	τ_d (days)	Max (ϕ_1) (m^2 s^{-2})	Max (ϕ_3) (m^2 s^{-2})	C_r (m s^{-1})	τ_d (days)	Max (ϕ_1) (m^2 s^{-2})	Max (ϕ_3) (m^2 s^{-2})
2200	−1.4	5.4	170 (26° N)	130 (20° N)				
2400	−1.6	5.2	220 (26° N)	200 (19° N)	0.05	4.8	190 (26° N)	250 (11° N)
2600	−2.5	6.3	240 (25° N)	250 (19° N)	−0.3	4.2	230 (26° N)	340 (12° N)
2800	−3.9	6.7	290 (24° N)	270 (20° N)	−1.3	4.8	240 (25° N)	370 (14° N)
3000	−2.7	5.6	260 (24° N)	370 (18° N)	−0.13	5.1	240 (25° N)	490 (13° N)
3300	−4.5	6.3	320 (24° N)	340 (19° N)	−1.6	4.9	270 (24° N)	540 (13° N)
3500	−6.8	5.5	390 (24° N)	290 (20° N)	−3.5	5.7	310 (24° N)	550 (14° N)
3700	−5.9	5.4	360 (24° N)	360 (19° N)	−2.5	5.4	310 (24° N)	620 (13° N)
4000	−8.7	6.4	460 (22° N)	350 (20° N)	−5.1	5.1	350 (24° N)	580 (14° N)
4200	−8.0	6.5	430 (22° N)	400 (19° N)	−4.3	5.1	350 (23° N)	680 (13° N)
4400	−7.3	6.3	390 (22° N)	540 (16° N)	−3.8	5.5	380 (23° N)	640 (13° N)

A lower tropospheric mode with a scale of 3000 km and a doubling time of 5–6 days is also obtained, which could correspond to monsoon disturbances forming over Bay of Bengal.

The authors wish to express their thanks to Professor D. Lal, Professor P. R. Pisharoty and Professor S. P. Pandya for their interest and encouragement. We also wish to record our grateful thanks to Professor J. G. Charney for enlightening discussions.

References

Keshavamurty, R. N., Asnani, G. C., Pillai, P. V. and Das, S. K. (1977) Some studies of growth of monsoon disturbances. Paper presented at the International Symposium on Monsoons, New Delhi, March, 1977.

Krishnamurti, T. N. (1971) Tropical east–west circulations during the northern summer, *J. Atmos. Sci.*, **28**, 1342–7.

Lorenz, E. N. (1972) Barotropic instability of Rossby wave motion. *J. Atmos. Sci.*, **29**, 2, 258–64.

Mak, M. K. (1975) The monsoonal mid-tropospheric cyclogenesis, *J. Atmos. Sci.*, **32**, 2246–53.

Shukla, J. (1977) Dynamics of monsoon disturbances. Paper presented at the International Symposium on Monsoons, New Delhi, March, 1977.

27

Growth of monsoon disturbances over western India

B. N. GOSWAMI, V. SATYAN AND R. N. KESHAVAMURTY

Active monsoon epochs over western India are often associated with cyclonic disturbances most marked in the midtroposphere. In order to understand the mechanism of their growth, linear barotropic, baroclinic and combined barotropic–baroclinic stability analyses of the mean flow in this region have been carried out. The potential vorticity of the mean flow shows extreme values in the region of the midtropospheric shear zone. The analyses yield barotropically unstable modes (i) at 700, 600, 500 mb and (ii) at 200 mb, with wavelengths of 20 to 30 degrees of longitude and doubling times of 3 to 4 days. Combined barotropic–baroclinic stability analyses using a 2-level quasi-geostrophic model yield essentially the same upper- and lower-tropospheric barotropically unstable modes, with only a marginal effect due to the presence of baroclinicity.

27.1 Introduction

There seem to be certain parts of the monsoon region from southeast Asia to western India which are particularly cyclogenetic. Shukla (1977) and Keshavamurty *et al.* (1977) have carried out a general stability analysis of the monsoon zonal flow. It turns out that the zonal wind profile over western India shows some slightly different features compared to that, say, over southeast Asia. In the former longitudes, the heat low over northwest India–Pakistan is overlain by anticyclonic flow in the midtroposphere, so that there is an east–west shear zone in the middle levels. This feature is not so well marked over other longitudes in the monsoon region. It seemed worthwhile, therefore, to study the stability characteristics of the zonal flow over western India.

During active monsoon epochs, disturbances form in the midtropospheric east–west shear zone over western India. Depressions forming over the north Bay of Bengal coupled with the above disturbances result in well-distributed rainfall over central and western India. One such midtropospheric disturbance was studied in detail by Miller and Keshavamurty (1968) with the help of IIOE aircraft and dropsonde data. Mak (1975) studied the baroclinic instability of the monsoon current with basic meridional motion also present. From Mak's study it appears that the meridional wind required for instability is rather large compared with reality. Carr (1977) made a detailed numerical study of the midtropospheric disturbance and found that condensation heating in association with orographic rain is important for its maintenance. In the present study, a linear stability analysis of the basic zonal flow in this region is carried out.

27.2 Zonal flow characteristics

The mean zonal flow over western India is depicted in Fig. 27.1 and shows the well-known westerly maximum in the lower troposphere and easterly jet in the upper troposphere. The midtroposphere is a shear zone with

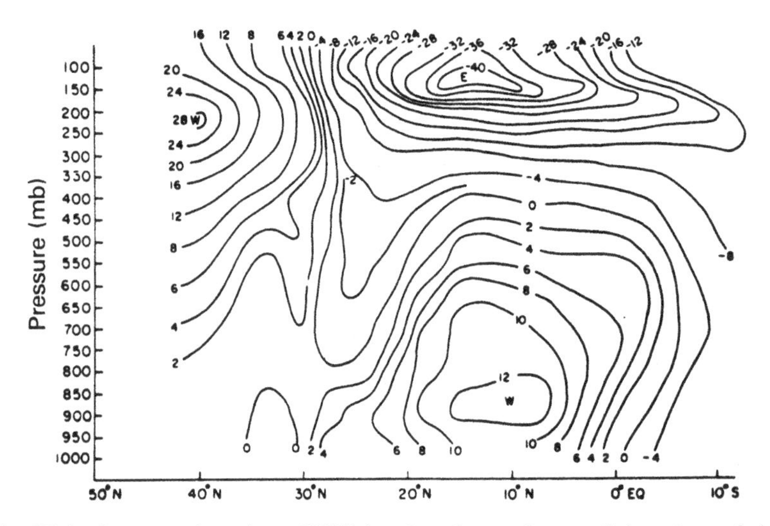

Fig. 27.1. Cross-section along 73° E for the observed mean July values of the zonal wind. The isotachs are labelled in m s^{-1}.

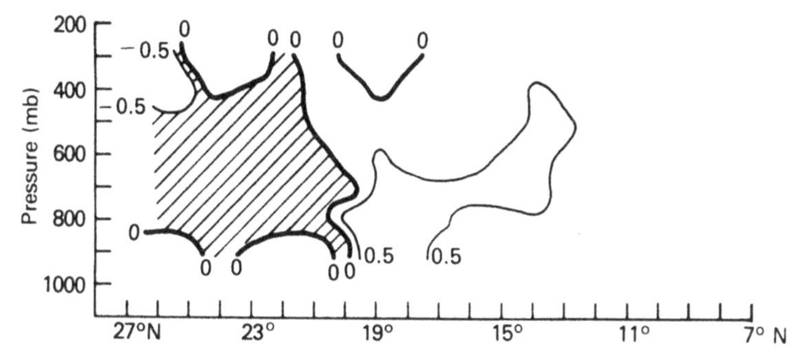

Fig. 27.2. Plot of the gradient of the potential vorticity of the basic flow shown in Fig. 27.1. The potential vorticity gradient has been computed from (27.1).

easterlies to the north. Charney and Stern (1962) and Pedlosky (1964) showed that, if the gradient of the potential vorticity of the basic flow vanishes somewhere within the channel, there is a possibility that the current may be unstable. This is a necessary but not a sufficient condition. The meridional gradient of the mean flow potential vorticity is computed as

$$\frac{\partial \bar{P}}{\partial y} = \beta - \frac{\partial^2 U}{\partial y^2} - f_0^2 \frac{\partial}{\partial p}\left(\frac{1}{\sigma}\frac{\partial U}{\partial p}\right)$$
$$= \beta - (U_{j+1} + U_{j-1} - 2U_j)/d^2 + S_{k-1}(U_k - U_{k-2})$$
$$+ S_{k+1}(U_k - U_{k+2}), \qquad (27.1)$$

where $S_k = f_0^2/(\Delta p^2 \sigma_k)$ and $\sigma = -(\alpha/\theta)(\partial\theta/\partial p)$. The basic U-profile shown in Fig. 27.1 and a standard profile of σ for the monsoon atmosphere were used for this purpose. This calculation shows (see Fig. 27.2) that $\partial\bar{P}/\partial y = 0$ at several levels around 20° N. Analysis of the behaviour of the different components in the equation for the gradient $\mathrm{d}\bar{P}/\mathrm{d}y$ shows that its vanishing is mainly due to the cancelling of the β and $\partial^2 U/\partial y^2$ terms.

This would suggest that it is worthwhile making computations to see if the basic current is barotropically or baroclinically unstable. This has been done using the eigenvalue technique and numerical methods.

27.3 Barotropic instability of the monsoon zonal flow

The potential vorticity of the mean monsoon zonal flow over western India has extreme values in the region of the east–west midtropospheric shear zone, and this is associated mainly with the barotropic term. This is therefore examined to establish whether or not the zonal current in the lower and middle troposphere is barotropically unstable.

Starting from the perturbation vorticity equation,

$$\left(\frac{\partial}{\partial t}+U\frac{\partial}{\partial x}\right)\nabla^2\psi'+\left(\beta-\frac{\partial^2 U}{\partial y^2}\right)\frac{\partial\psi'}{\partial x}=0, \tag{27.2}$$

and introducing a wave solution of the type

$$\psi'=\psi(y)\exp\left[ik(x-ct)\right],$$

gives

$$(U-c)\left(-k^2\psi+\frac{\partial^2\psi}{\partial y^2}\right)+\left(\beta-\frac{\partial^2 U}{\partial y^2}\right)\psi=0. \tag{27.3}$$

Now consider a channel between y_1 and y_n as shown in Fig. 27.3. Equation (27.3) for the jth latitude can be written in finite difference form as

$$(U_j-c)\left(\frac{\psi_{j+1}+\psi_{j-1}-2\psi_j}{\Delta y^2}-k^2\psi_j\right)+\left(\beta-\frac{U_{j+1}+U_{j-1}-2U_j}{\Delta y^2}\right)\psi_j=0. \tag{27.4}$$

Using the boundary conditions $\psi_0=\psi_{n+1}=0$ gives n equations in n unknowns, ψ_j $(j=1,2,\ldots,n)$, and for a non-trivial solution the determinant of the matrix of coefficients must vanish. In matrix notation

$$(P-cQ)(R)=0,$$

or

$$(PQ^{-1}-cI)(QR)=0, \tag{27.5}$$

where, P and Q are matrices consisting of the coefficients and R is the

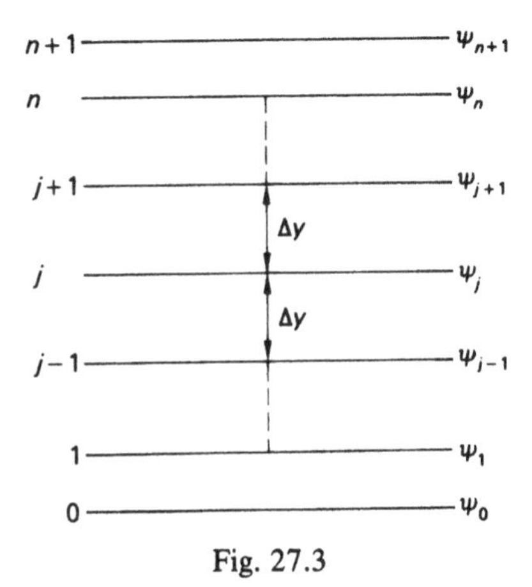

Fig. 27.3

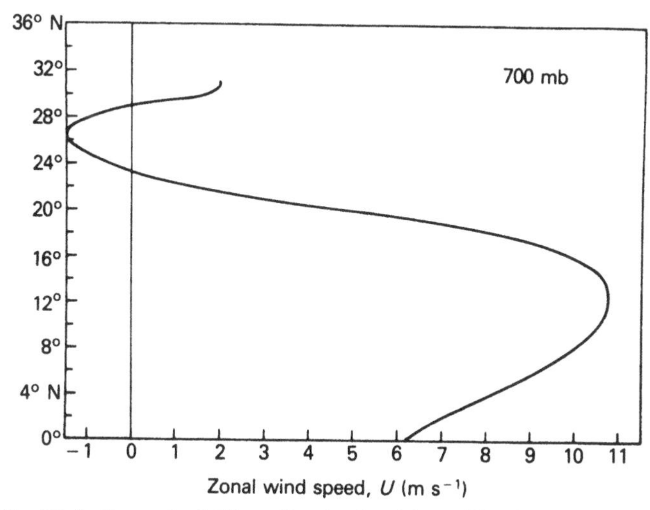

Fig. 27.4. Smoothed U-profile obtained from Fig. 27.1 at 700 mb.

column vector $\{\psi_1, \psi_2, \ldots, \psi_n\}$. The problem now reduces to finding the eigenvalues of the matrix PQ^{-1}, the eigenvectors QR, and hence R.

Such stability analyses have been carried out for the profiles between 850 and 200 mb along 73° E longitude (shown in Fig. 27.1) using a one-degree grid in the channel between the Equator and 31° N. The smoothed profiles used in the numerical computations for 700, 600 and 200 mb are shown in Figs. 27.4 to 27.6. It is found that the zonal current,

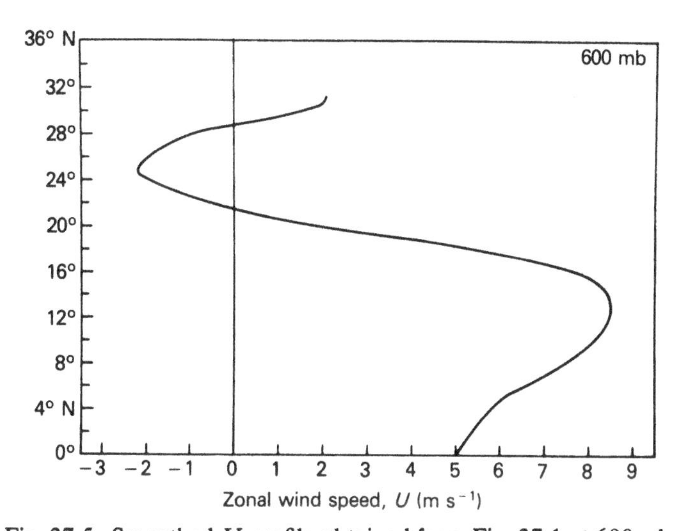

Fig. 27.5. Smoothed U-profile obtained from Fig. 27.1 at 600 mb.

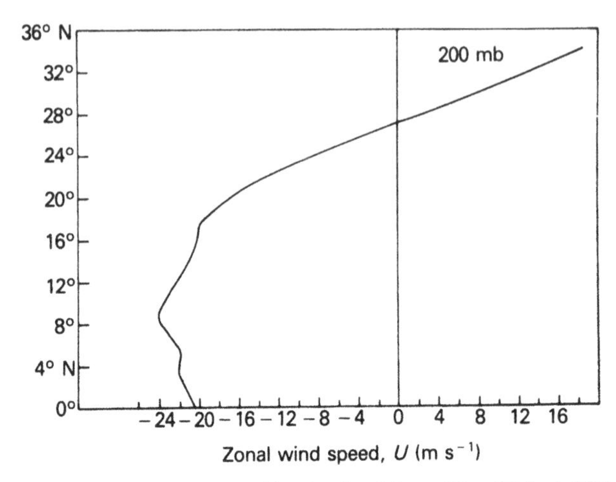

Fig. 27.6. Smoothed U-profile obtained from Fig. 27.1 at 200 mb.

both in the middle levels as well as in the upper troposphere, shows unstable modes with wavelengths from 20 to 30 degrees of longitude and doubling times of 3 to 4 days (Table 27.1). Figs. 27.7 and 27.8 show the scale dependence of the growth rates at 700 and 200 mb. The maximum growth rate at 700 mb occurs for disturbances of a scale around 2500 km. At 200 mb the scale of the disturbance with maximum growth rate is around 1800 km. The upper-tropospheric mode possibly corresponds to

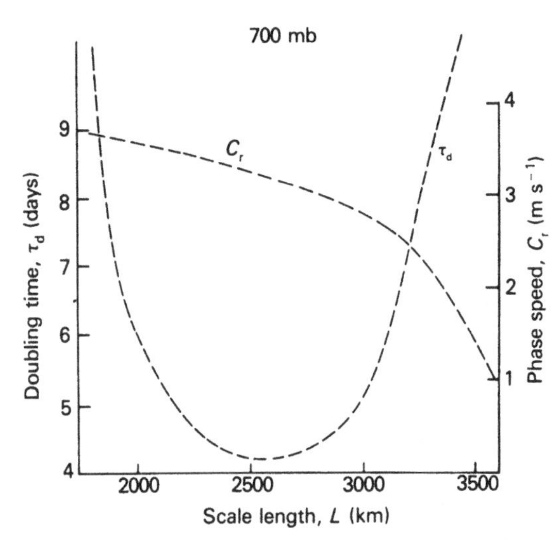

Fig. 27.7. Scale dependence of the growth rates at 700 mb from the barotropic analysis.

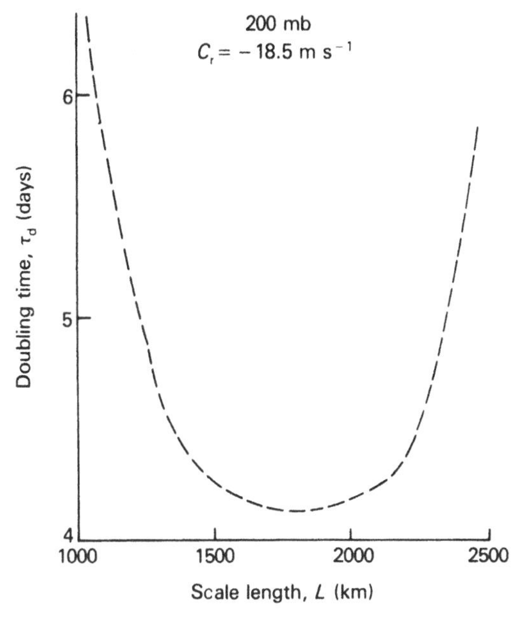

Fig. 27.8. Scale dependence of the growth rates at 200 mb from the barotropic analysis, with $C_r = 18.5$ m s^{-1}.

westward-moving easterly waves and the midtropospheric modes to the observed midtropospheric disturbance. The wavelength of the latter agrees well with that of the observed disturbance. However, an eastward phase speed of 2 to 3 m s^{-1} is found. The amplitude distributions for these modes show a realistic structure. As an example, the amplitude distributions of the most unstable modes at 700 mb and 200 mb are shown in Fig. 27.9. It was shown by Carr (1977) that the actual disturbance is stationary because of latent heat release in association with orographic rain.

TABLE 27.1. *Unstable modes found in barotropic stability analysis.*

Level (mb)	Wavelength (degrees of longitude)	C_r (m s^{-1})	Doubling time (days)
700	26	3.2	4.2
600	22	2.3	3.9
200	18	−18.7	4.1

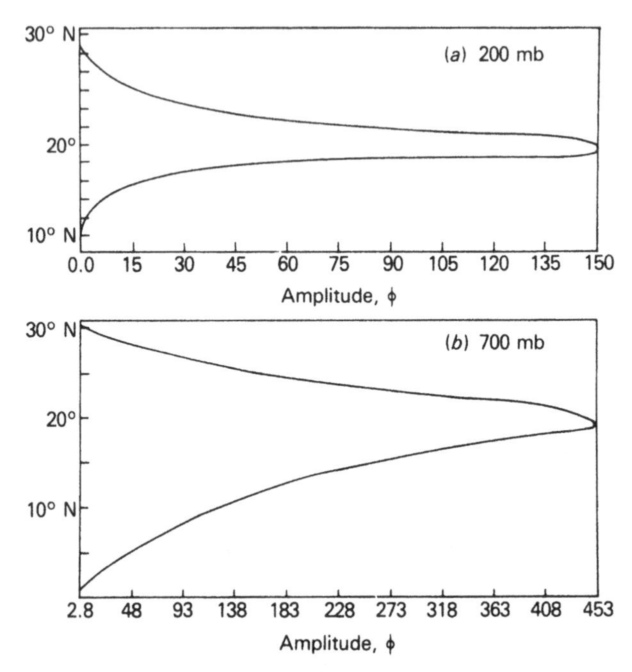

Fig. 27.9. The amplitude profiles of the most unstable modes obtained from the one-level barotropic stability analysis at: (*a*) 200 mb, $\tau_d = 4.1$ days, $C_r = 18.7\ \mathrm{m\,s^{-1}}$, $L = 18$ degrees of longitude; and (*b*) 700 mb, $\tau_d = 4.2$ days, $C_r = 3.2\ \mathrm{m\,s^{-1}}$, $L = 26$ degrees of longitude.

27.4 Stability analysis of a zonal flow with vertical shear but no horizontal shear

The stability characteristics of the zonal flow with only vertical shear is now examined, using a multi-level quasi-geostrophic model. The perturbation potential vorticity equation at level i can be written as

$$\left(\frac{\partial}{\partial t} + U_i \frac{\partial}{\partial x}\right)\left[\frac{1}{f_0}\nabla^2 \phi'_i + \frac{S_{i-1}}{f_0}(\phi'_i - \phi'_{i-2}) + \frac{S_{i+1}}{f_0}(\phi'_i - \phi'_{i+2})\right]$$

$$+ \frac{1}{f_0}\frac{\partial \phi'_i}{\partial x}[\beta + S_{i-1}(U_{i-2} - U_i) + S_{i+1}(U_{i+2} - U_i)] = 0.$$

Assuming wave solutions of the type $\phi' = \phi(p) \exp[ik(x - ct)]$ and introducing boundary conditions such that $\omega = 0$ at the top of the atmosphere and at 1000 mb, the above equation can be reduced to a set of n equations in n unknowns $(\phi_1, \phi_2, \ldots, \phi_n)$ and then solved numerically using the eigenvalue technique.

Experiments were carried out with the zonal wind (U) and σ-profile around 20° N with a 20-level model. In the lower levels the profile was modified to have the same speed as the maximum speed above the friction layer. In other words, the reduction of speed due to friction is not considered. No unstable modes were found. A zonal wind profile with speed increased to 1.5 times the normal speed at all levels yielded a slowly-growing mode with maximum amplitude in the upper troposphere.

27.5 Combined barotropic–baroclinic instability

As seen in § 27.2, the Charney–Stern (1962) condition for the possible existence of an internal jet instability is satisfied by the basic flow over western India. Hence, a combined barotropic–baroclinic stability analysis of the zonal flow has been carried out using a two-level quasi-geostrophic model.

The perturbation potential vorticity equations at levels 1 and 3 (Fig. 27.10) are

$$\left(\frac{\partial}{\partial t}+U_1\frac{\partial}{\partial x}\right)\left[\zeta_1'+\frac{S_2}{f_0}(\phi_3'-\phi_1')\right]+v_1'\left[\beta-\frac{\partial^2 U_1}{\partial y^2}-S_2(U_3-U_1)\right]$$
$$=-\frac{R}{C_p}\frac{\Delta p}{p}\frac{S_2}{f_0}\dot{Q}_2', \qquad (27.6)$$

and

$$\left(\frac{\partial}{\partial t}+U_3\frac{\partial}{\partial x}\right)\left[\zeta_3'-\frac{S_2}{f_0}(\phi_3'-\phi_1')\right]+v_3'\left[\beta-\frac{\partial^2 U_3}{\partial y^2}+S_2(U_3-U_1)\right]-\frac{f_0}{\Delta p}\omega_4'$$
$$=\frac{R}{C_p}\frac{\Delta p}{p}\frac{S_2}{f_0}\dot{Q}_2', \qquad (27.7)$$

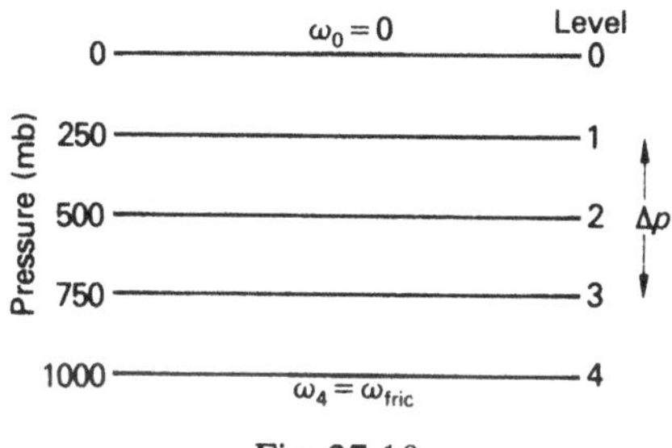

Fig. 27.10

where S_2 has the same definition as in § 27.3 and $\dot{Q}'_2$ is the rate of diabatic heating at level 2. If the channel is divided in the y-direction into n-latitudes (as was done in § 27.3) and solutions sought of the type $\phi' = \phi(y) \exp[ik(x-ct)]$, the equations corresponding to the jth latitude can be written as

$$(U_j^1 - c)\left[\frac{\phi_{j+1}^1 + \phi_{j-1}^1 - 2\phi_j^1}{\Delta y^2} - k^2\phi_j^1 + S_2(\phi_j^3 - \phi_j^1)\right]$$
$$+\phi_j^1\left[\beta - \frac{U_{j+1}^1 + U_{j-1}^1 - 2U_j^1}{\Delta y^2} - S_2(U_j^3 - U_j^1)\right]$$
$$= \frac{i}{k}\frac{R}{C_p}\frac{\Delta p}{p_2} S_2 \dot{Q}'_2 \exp[-ik(x-ct)], \tag{27.8}$$

and

$$(U_j^3 - c)\left[\frac{\phi_{j+1}^3 + \phi_{j-1}^3 - 2\phi_j^3}{\Delta y^2} - k^2\phi_j^3 - S_2(\phi_j^3 - \phi_j^1)\right]$$
$$+\phi_j^3\left[\beta - \frac{U_{j+1}^3 + U_{j-1}^3 - 2U_j^3}{\Delta y^2} + S_2(U_j^3 - U_j^1)\right]$$
$$= -\left(\frac{i}{k}\frac{R}{C_p}\frac{\Delta p}{p_2} S_2 \dot{Q}'_2 + \frac{i}{k}\frac{f_0^2}{\Delta p}\omega'_4\right) \exp[-ik(x-ct)]. \tag{27.9}$$

There are $2n$ equations in the $2n$ unknowns (ϕ_j^1, ϕ_j^3). For a non-trivial solution the determinant of the matrix of coefficients must vanish. In matrix notation,

$$(P - cQ)(\Phi) = 0$$

or

$$(PQ^{-1} - cI)(Q\Phi) = 0. \tag{27.10}$$

Once again the problem reduces to finding the eigenvalues of the matrix PQ^{-1} and eigenvectors $Q\Phi$ and hence Φ. This is done numerically.

The channel chosen for this numerical study is between 10° N and 30° N with a one-degree grid. The boundary conditions used are $\omega = 0$ at the top boundary, $\omega = \omega_4 = \omega_{\text{fric}}$ at the bottom boundary and $\phi^1 = \phi^3 = 0$ at the southern and northern boundaries. In this part of the study the cumulus heating is assumed to be absent.

The fastest growing modes obtained from this stability analysis are listed in Table 27.2. It yields the same two modes, one in the lower

TABLE 27.2. *Fastest growing modes obtained from combined barotropic-baroclinic stability analysis without heating.*

Mode	Position in the vertical	Wavelength (degrees of longitude)	C_r (m s^{-1})	Doubling time (days)
1	Lower troposphere	24	3.2	5.0
2	Upper troposphere	18	−18.6	4.2

midtroposphere and one in the upper troposphere earlier obtained in the barotropic stability analysis (see Table 27.1). Figs. 27.11 and 27.12 show the scale dependence of the growth rate of the lower- and upper-tropospheric modes. The amplitude distributions of the most unstable modes are shown in Fig. 27.13. Thus, these are essentially seen to be barotropically unstable modes which are not very much affected by the presence of the baroclinic effects.

This would remind one of Charney's (1969) conclusions from his scale analysis for the tropics, that there is no large vertical coupling in the tropics and the tropical atmosphere is quasi-barotropic in the absence of deep cumulus convection.

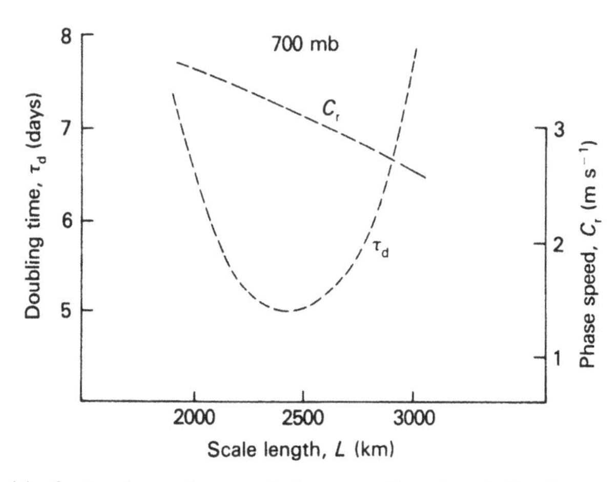

Fig. 27.11. Scale dependence of the growth rate of the lower-tropospheric 700 mb mode obtained with the combined barotropic–baroclinic model.

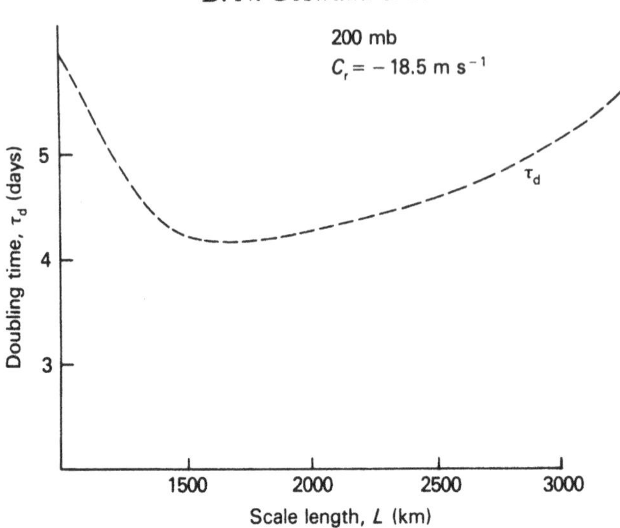

Fig. 27.12. Scale dependence of the growth rate of the upper-tropospheric (200 mb) mode obtained with the combined barotropic–baroclinic model, with $C_r = -18.5$ m s^{-1}.

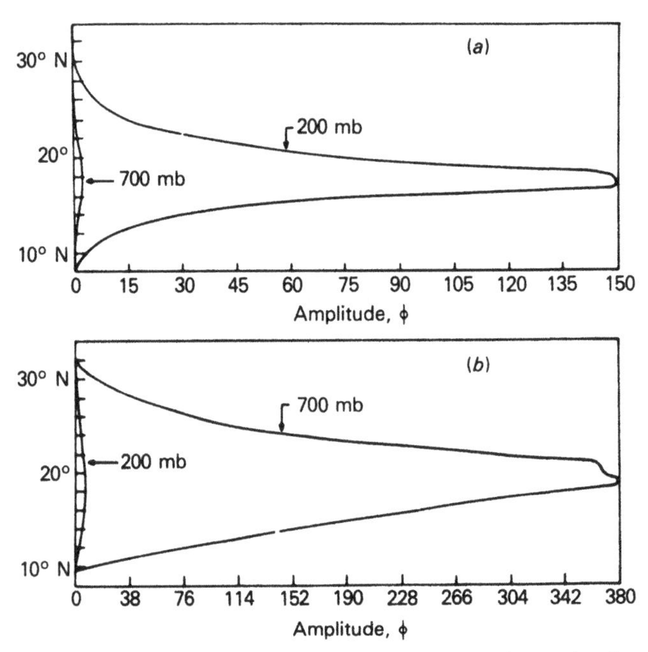

Fig. 27.13. The amplitude profiles of the most unstable modes in a 2-level combined barotropic-baroclinic analysis in the absence of heating with: (*a*) $L = 18$ degrees longitude, $\tau_d = 4.2$ days, $C_r = 18.5$ m s^{-1}; and (*b*) $L = 24$ degrees longitude, $\tau_d = 5$ days, $C_r = 3.2$ m s^{-1}.

27.6 Conclusions

Some experiments with stability analyses of the monsoon zonal flow over western India reveal the following interesting features:

(i) The zonal flow shows barotropically unstable modes, one in the upper troposphere with wavelength of about 18 degrees of longitude, a doubling time of 4.1 days and a westward phase speed of 18 m s^{-1}, and another in the lower troposphere with a wavelength of about 26 degrees of longitude and a doubling time of 4.2 days. The wavelength of these unstable modes agrees with that of the observed disturbances. The phase speed lies between the minimum zonal wind and the critical zonal wind speed (where $\beta - U_{yy} = 0$), which is in agreement with Kuo's (1949) theory. The phase of the disturbance at different latitudes has also been computed. The lower- and midtropospheric modes have a tilt of the trough (ridge) from north-northeast to south-southwest which would be associated with a northward transport of westerly momentum. The disturbance has largest amplitude north of the maximum westerly zonal wind. In the equation for the rate of change of eddy kinetic energy (K_E)

$$\frac{\partial K_E}{\partial t} = -\int \overline{u'v'} \frac{\partial U}{\partial y} \, dm, \tag{27.11}$$

where $\overline{u'v'}$ is the momentum transport by the wave and is positive in the present case in the region where $\partial U/\partial y$ is negative. Thus $\partial K_E/\partial t$ is positive. This would also suggest that the unstable modes obtained by this numerical method correspond to real disturbances.

(ii) The monsoon zonal flow does not yield any baroclinically unstable modes. Only when the vertical shear is increased to 1.5 times the normal value is a baroclinically growing mode obtained in the upper troposphere.

(iii) In the combined barotropic–baroclinic stability analysis the modes obtained in the barotropic analysis are still present with only marginal changes. In other words, the baroclinic effects do not significantly modify these modes.

Therefore, the midtropospheric disturbances over western India are likely to originate from the barotropic instability of the shear zone in this region.

The authors wish to thank Professor P. R. Pisharoty and Professor S. P. Pandya for their continued interest in this work.

References

Carr, F. H. (1977) The mid-tropospheric cyclone: Numerical experimentation. Paper presented at the International Symposium on Monsoons, New Delhi, March 1977.

Charney, J. G. (1969) A further note on large scale motions in the tropics. *J. Atmos. Sci.*, **26**, 182.

Charney, J. G. and Stern, M. C. (1962) On the stability of internal baroclinic jets in a rotating atmosphere. *J. Atmos. Sci.*, **19**, 159–72.

Keshavamurty, R. N., Asnani, G. C., Pillai, P. V. and Das, S. K. (1977) Some studies of the growth of monsoon disturbances, paper presented at the International Symposium on Monsoons, New Delhi, March 1977.

Kuo, H. L. (1949) Dynamic instability of two dimensional non-divergent flow in a barotropic atmosphere, *J. Meteor.*, **6**, 105–22.

Mak, M. K. (1975) The monsoon mid tropospheric cyclogenesis, *J. Atmos. Sci.*, **32**, 2246–53.

Pedlosky, J. (1964) The stability of currents in the atmosphere and ocean, Part I. *J. Atmos. Sci.*, **21**, 201–19.

Miller, F. R. and Keshavamurty, R. N. (1968) Structure of an Arabian Sea summer monsoon system, IIOE Meteorological Monograph 1 (East–West Centre Press, Honolulu, Hawaii, USA), p. 94.

Shukla, J. (1977) CISK–barotropic–baroclinic instability and the growth of monsoon depressions. Paper presented at the International Symposium on Monsoons, New Delhi, March 1977.

28

Topographic Rossby waves in the summer monsoon

SULOCHANA GADGIL

Stationary Rossby waves induced in a westerly flow over orography are investigated analytically using a quasi-geostrophic β-plane model. A radiation condition is imposed to obtain a steady solution as the limit of a transient problem in which the forcing has been built up from zero. The results agree well with observed mean monthly contours of the 850 mb surface for July over the Indian peninsula.

28.1 Introduction

Stationary Rossby waves induced in westerly flow over topographic features are investigated using a quasi-geostrophic β-plane model for homogeneous fluids. The model is of interest because of its application to the interaction between the topography of the Indian peninsula and the lower-tropospheric summer monsoon circulation. Earlier investigations (Gadgil and Sikka, 1976 and Gadgil, 1977; these references are hereafter referred to as GS) have suggested that the observed wavy pattern in the mean monthly low-level pressure charts with a trough around 87° E over the southern Bay of Bengal (in the latitudinal belt between 5° and 18° N) and a prominent ridge over Burma (around 100° E at the surface, as shown in Fig. 28.1*a* and 110° E at 850 mb, as shown in Fig. 28.1*b*) could be considered as a manifestation of a Rossby wave pattern induced by this topography.

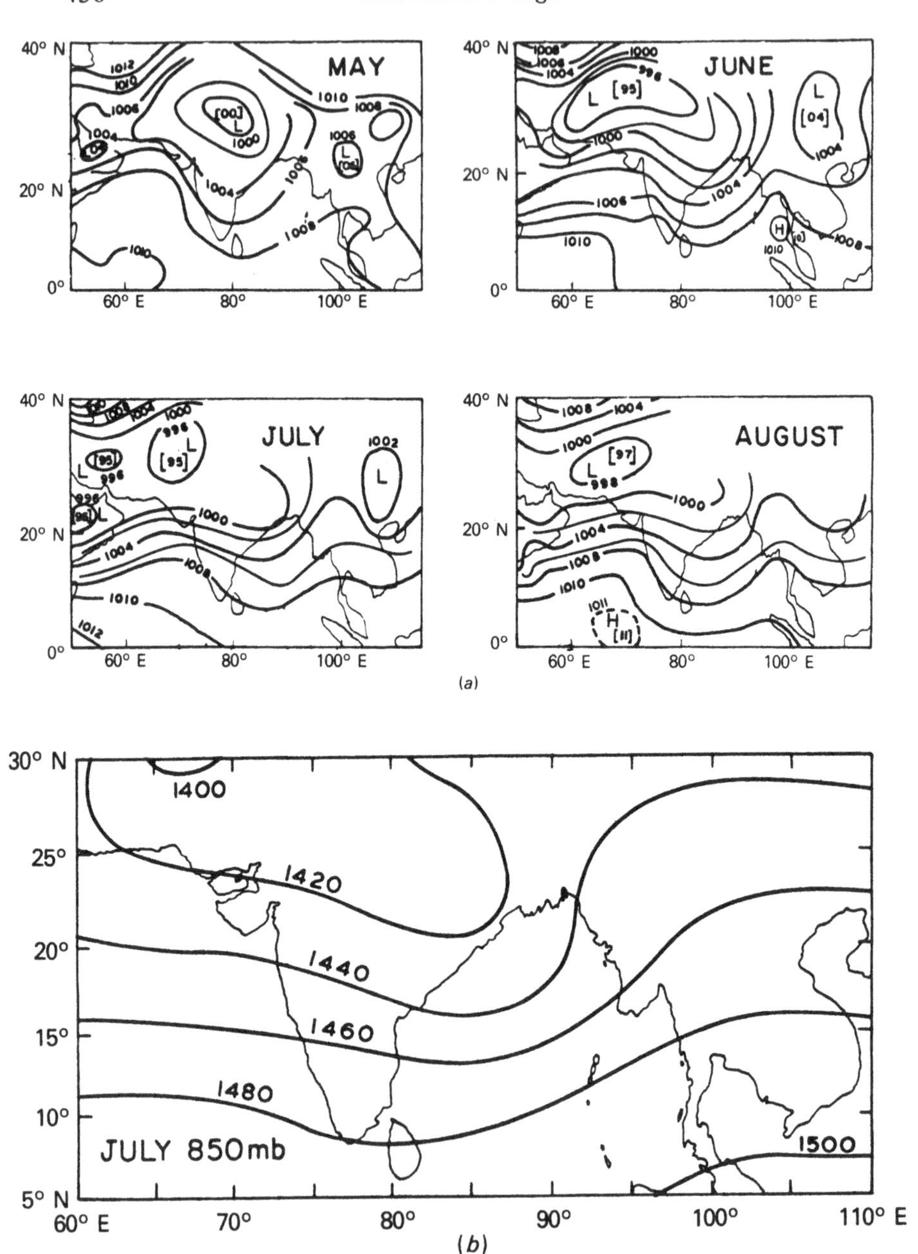

Fig. 28.1. (*a*) Mean monthly sea-level pressure charts for May to August. Isobars are labelled in mb and the numbers in square brackets indicate values of minima and maxima (after Ananthakrishnan *et al.*, 1968). (*b*) Mean monthly height contours of the 850 mb surface for July. Contours are labelled in geopotential metres.

In GS, an approximate solution for the flow past a finite obstacle on a β-plane was obtained by considering obstacles with a zonal extent smaller than a Rossby wavelength. For such obstacles, provided that the appropriate radiation condition for Rossby waves is imposed on the f-plane solution, the effect on a β-plane can then be deduced by combining the f-plane solutions over the region of the obstacle and matching with Rossby wave solutions in the region downstream of the obstacle. When such calculations were carried out, the results showed that the major effects of a meridionally elongated obstacle of a nondimensional height δ on a uniform westerly current characterized by a Rossby number ε, are:

(i) A *southward* flow component of magnitude δ/ε over the eastern part, and to the east of the obstacle (it should be noted that this component is in the direction opposite to that of the observed cross-isobaric flow);

(ii) Rossby waves with the zonal wavenumber $(\beta/\varepsilon)^{\frac{1}{2}}$ downstream of the obstacle. (The meridional wavenumber of the Rossby waves was assumed to be zero.) For appropriate values of the parameters, shown in Table 28.1, it was shown that these qualitative predictions were consistent with the observations.

TABLE 28.1. *Parameter values for the Rossby wave solution. H refers to the height of the lower troposphere.*

Parameter	Value	Parameter	Value
ε	0.5	L	500 km
δ	0.1	D	0.6 km
β	0.25	H	6 km
δ/ε	0.2	U	$10\ \mathrm{m\,s^{-1}}$
γ	0.447		

The solution is now considered for flow past obstacles of arbitrary zonal and meridional extent on a β-plane. The meridional wavenumber of the Rossby waves is not assumed to be zero, but the group velocity for wavenumbers corresponding to stationary waves is eastward (as in the case with zero meridional wavenumber), and the radiation condition derived by considering the linearized time-dependent problem is shown to be again that of no waves upstream i.e. westward of the obstacle.

The analysis of this general case yields the scales obtained earlier and, in addition, the detailed pattern of the streamlines over the entire region.

Comparison with the observations for 850 mb (Fig. 28.1*b*) indicates excellent agreement with the solution obtained, substantiating the hypothesis proposed in GS regarding the effect of the topographic feature. A new aspect of the solution for the general case presented here is the decay in amplitude of the Rossby waves with distance downstream from the obstacle.

The planetary waves thus induced in the westerly monsoon current of the lower troposphere may be expected to be restricted to the lower troposphere because the presence of strong easterly winds in the upper troposphere throughout the summer over the region would prevent their propagation into the upper troposphere (Charney and Drazin, 1961; Dickinson, 1968).

In the next section the quasi-geostrophic model is formulated; the radiation condition is derived in § 28.3, and the β-plane solution for flow past a rectangular plateau is given in § 28.4 and discussed in § 28.5.

28.2 A quasi-geostrophic model of flow over orography on a β-plane

For the quasi-geostrophic motion of an ideal fluid on a β-plane, with the coordinate axes chosen so that the x-axis points eastward, y-axis northward and the z-axis vertically upward, in the presence of bottom topography which is given in non-dimensional coordinates as

$$z = \delta h(x, y) \tag{28.1}$$

the governing equations are, in the usual notation,

$$v = p_x, \tag{28.2}$$

$$u = -p_y, \tag{28.3}$$

$$\phi = v_x - u_y \tag{28.4}$$

and

$$\varepsilon[\phi_t + (u\phi_x + v\phi_y)] + \delta(uh_x + vh_y) + \beta v = 0, \tag{28.5}$$

where

$$\varepsilon = U/f_0L, \qquad \delta = D/H, \qquad \beta = (L/A)\cot\theta_0.$$

Here p refers to the departure of pressure from the mean hydrostatic pressure over the region considered; this is non-dimensionalized by using the geostrophic scale. The other scaling factors are

f_0 Coriolis parameter,
L horizontal scale,

U scale of the velocity of the basic current,
D maximum height of the topography,
H total depth of the fluid,
θ_0 mean latitude,
A radius of the earth,
T time scale ($T = L/U$).

For steady flow, using the geostrophic relations (28.2) and (28.3), the vorticity equation (28.5) can be integrated to yield a potential vorticity integral of the form

$$\varepsilon\phi + \delta h + \beta y = F(p). \tag{28.6}$$

The function $F(p)$ is arbitrary and can be determined by the upstream vorticity under certain conditions (§ 28.3). If the current is uniform upstream of the topography, and parallel to the x-axis, i.e.

$$p = -y \quad \text{at } x = -\infty. \tag{28.7}$$

Then

$$F(p) = -\beta p,$$

and (28.6) reduces to

$$\varepsilon\phi + \delta h + \beta(p + y) = 0. \tag{28.8}$$

Introducing the pressure perturbation, P, defined by

$$p = -y + P,$$

(28.8) can be expressed as

$$\varepsilon(P_{xx} + P_{yy}) + \delta h + \beta P = 0. \tag{28.9}$$

28.3 The radiation condition

The solution to the steady state problem is not unique. The physically significant solution, which is the steady limit of a transient problem in which the forcing has been built up from zero, cannot be obtained until an appropriate radiation condition has been imposed. Lighthill (1967) has shown that for the general dispersive case such a solution can be obtained by taking the forcing to be proportional to $e^{\alpha t}$ and then letting α tend to zero. When this method is used in the treatment of the linearized version of the time-dependent vorticity equation (28.5) i.e.,

$$\phi_t + U\phi_x + \frac{\delta}{\varepsilon}Uh_x + \frac{\beta}{\varepsilon}v = 0, \tag{28.10}$$

$$h(x, y) = e^{\alpha t}h'(x, y),$$

the solution for the pressure field can be obtained by taking the two-dimensional Fourier transform. This yields

$$p(R, \psi) =$$

$$-\frac{\delta}{4\pi^2\varepsilon}\int_0^\infty a\,da \int_0^{2\pi} \frac{d\theta H(a\cos\psi, a\sin\psi)\exp[-iaR\cos(\theta-\psi)]}{(\beta/\varepsilon U) - a^2 - (ia\alpha/\varepsilon U\cos\theta)} \tag{28.11}$$

where

$$H(k, l) = \int_{-\infty}^{\infty} \exp[i(kx + ly)]h'(x, y)\,dx\,dy,$$

$R = (x^2 + y^2)^{\frac{1}{2}}$ and $\psi = \tan^{-}(y/x)$. The integrand has poles at the points

$$a = \frac{-i}{2\varepsilon U\cos\theta} \pm \frac{1}{2}\left(\frac{4\beta}{\varepsilon U} - \frac{\alpha^2}{\varepsilon^2 U\cos^2\theta}\right)^{\frac{1}{2}}.$$

These poles are located in the upper half-plane when $\cos\theta$ is negative. At large distances from the obstacle, the phase of the integral is stationary at

$$\theta_1 = \psi, \quad \theta_2 = \pi + \psi.$$

An approximate expression for the integral can be obtained by using the method of stationary phase (Chapter 6 in Carrier *et al.*, 1966), as

$$p = \frac{-\delta}{2\varepsilon(2\pi)^{\frac{3}{2}}}\int_0^\infty da\, a^{\frac{1}{2}}\left[\frac{H(a\cos\psi, a\sin\psi)\exp(iaR - i\pi/4)}{(\beta/\varepsilon U) - a^2 - (ia\alpha/\varepsilon U\cos\psi)}\right.$$

$$\left. + \frac{H(-a\cos\psi, a\sin\psi)\exp(-iaR + i\pi/4)}{(\beta/\varepsilon U) - a^2 + (ia\alpha/\varepsilon U\cos\psi)}\right]. \tag{28.12}$$

In the evaluation of these integrals using Jordan's lemma, we note that the poles contribute to these integrals only when $\cos\psi > 0$ i.e. only in the right semicircle or the eastern semicircle. Thus, the radiation of Rossby waves is restricted to the downstream semicircle and the appropriate condition for the steady state problem is that of no waves upstream of the obstacle. Hence, the determination of $F(p)$ in (28.6) by using upstream conditions is also justified.

28.4 Solution for flow over a rectangular plateau

We consider the solution of the equation for conservation of potential vorticity (28.9) for a rectangular plateau with zonal and meridional scales $\bar{x}$ and $\bar{y}$, i.e.,

$$h(x, y) = 1 \quad \text{for } |x| < \bar{x} \text{ and } |y| < \bar{y}$$
$$= 0 \quad \text{otherwise.}$$

Taking the Fourier transform in y,

$$\phi(x, \lambda) = \int_{-\infty}^{\infty} e^{i\lambda y} P(x, y)\, dy,$$

and

$$P(x, y) = \frac{1}{2\pi} \int_{-\infty}^{\infty} e^{-i\lambda y} \phi(x, \lambda)\, d\lambda,$$

the equation reduces to

$$\phi_{xx} + (\gamma^2 - \lambda^2)\phi = -\frac{\delta}{\varepsilon} H_1(x, \lambda), \tag{28.13}$$

where

$$\gamma^2 = \delta/\varepsilon.$$

For a given λ, the solution to (28.13) is exponential if $\lambda^2 > \gamma^2$ and sinusoidal otherwise. Imposing the radiation condition that there be no waves for $x < -\bar{x}$, and that the pressure should decay at sufficiently large distances in every direction, the solution for the transformed pressure and the meridional velocity which satisfies matching conditions at the edges of the plateau is given by

$$\phi = \frac{\delta}{2\varepsilon} \frac{H(\lambda)}{\lambda^2 - \gamma^2} \left\{ S(\lambda^2 - \gamma^2) \left[S(-x_1 x_2) + \frac{x_2}{|x_2|} e^{-\mu |x_2|} - \frac{x_1}{|x_1|} e^{-\mu |x_1|} \right] + S(\gamma^2 - \lambda^2) [S(-x_1 x_2) + S(x_2) \cos(\nu x_2) - S(x_1) \cos(\nu x_1)] \right\}, \tag{28.14}$$

$$v = \frac{\delta}{2\varepsilon} H(\lambda) \left\{ \frac{S(\lambda^2 - \gamma^2)}{\mu} (e^{-\mu |x_1|} - e^{-\mu |x_2|}) - S\frac{(\gamma^2 - \lambda^2)}{\nu} [2S(x_1) \sin(\nu x_1) - 2S(x_2) \sin(\nu x_2)] \right\}, \tag{28.15}$$

where

$$x_1 = x + \bar{x}, \qquad x_2 = x - \bar{x}, \qquad \mu = (\lambda^2 - \gamma^2)^{\frac{1}{2}},$$
$$\nu = (\gamma^2 - \lambda^2)^{\frac{1}{2}}, \qquad H_1(x, \lambda) = H(\lambda) S(-x_1 x_2),$$

and S is the Heaviside step function defined as

$$S(z) = 1 \quad \text{for } z > 0,$$
$$= 0 \quad \text{for } z < 0.$$

For the particular obstacle chosen, inversion of the Fourier transforms gives the following expressions for pressure (P) and meridional velocity (v)

$$P(x, y) = \frac{\delta}{2\pi\beta} \left\{ \int_0^\infty \frac{\sin(\gamma y_1 \cosh t) - \sin(\gamma y_2 \cosh t)}{\sinh t \cosh t} P_1(t)\, dt \right.$$
$$\left. + 2 \int_0^{\pi/2} P_2(t) \frac{\sin(\gamma y_1 \cos t) - \sin(\gamma y_2 \cos t)}{\sin t \cos t}\, dt \right\}, \tag{28.16}$$

$$v(x) = \frac{\delta}{2\pi(\varepsilon\beta)^{\frac{1}{2}}} \int_0^\infty \frac{\sin(\gamma y_1 \cosh t) - \sin(\gamma y_2 \cosh t)}{\cosh t} v_1(t)\, dt$$
$$+ \frac{\delta}{\pi(\varepsilon\beta)^{\frac{1}{2}}} \int_0^{\pi/2} \frac{\sin(\gamma y_1 \cos t) - \sin(\gamma y_2 \cos t)}{\cos t} v_2(t)\, dt, \tag{28.17}$$

where

$$P_1(t) = 2S(-x_1 x_2) + \frac{x_2}{|x_2|} \exp(-|x_2| \gamma \sinh t) - \frac{x_1}{|x_1|} \exp(-|x_1| \gamma \sinh t),$$

$$P_2(t) = -S(-x_1 x_2) + S(x_1) \cos(\gamma x_1 \sin t) - S(x_2) \cos(\gamma x_2 \sin t),$$

$$v_1(t) = \exp(-|x_1| \gamma \sinh t) - \exp(-|x_2| \gamma \sinh t),$$

$$v_2(t) = S(x_1) \sin(\gamma x_1 \sin t) - S(x_2) \sin(\gamma x_2 \sin t).$$

The first terms in (28.16) and (28.17) represent the contributions of meridional wavenumbers larger than γ, whereas the second terms represent those of the smaller wavenumbers. Also, the second term does not contribute to the solution in the region upstream of the obstacle because of the imposed radiation condition. (This is similar to that used by Miles, 1968.) The solutions (28.16) and (28.17) have been computed by evaluating the integrals numerically as well as by obtaining approximations by asymptotic methods.

28.5 Discussion

Those features of the solution which are pertinent to the interaction between the peninsular topography and the lower-tropospheric wind of the summer monsoon are considered first. The zonal profiles of the meridional velocity induced by a symmetric obstacle ($\bar{x} = \bar{y} = 1$) and a meridionally elongated obstacle ($\bar{x} = 0.5$, $\bar{y} = 2$) at $y = 0, 1, 2$ are shown in Fig. 28.2. Within the restricted range of variations of y shown, the meridional velocity is a slowly varying function of y and the waves resemble the one-dimensional waves induced by an infinite meridional ridge, considered in GS. The zonal profiles of the meridional component for different values of γ at $y = 0$ are shown in Fig. 28.3 for the symmetric obstacle. It is seen that a reasonable estimate of the scale of the meri-

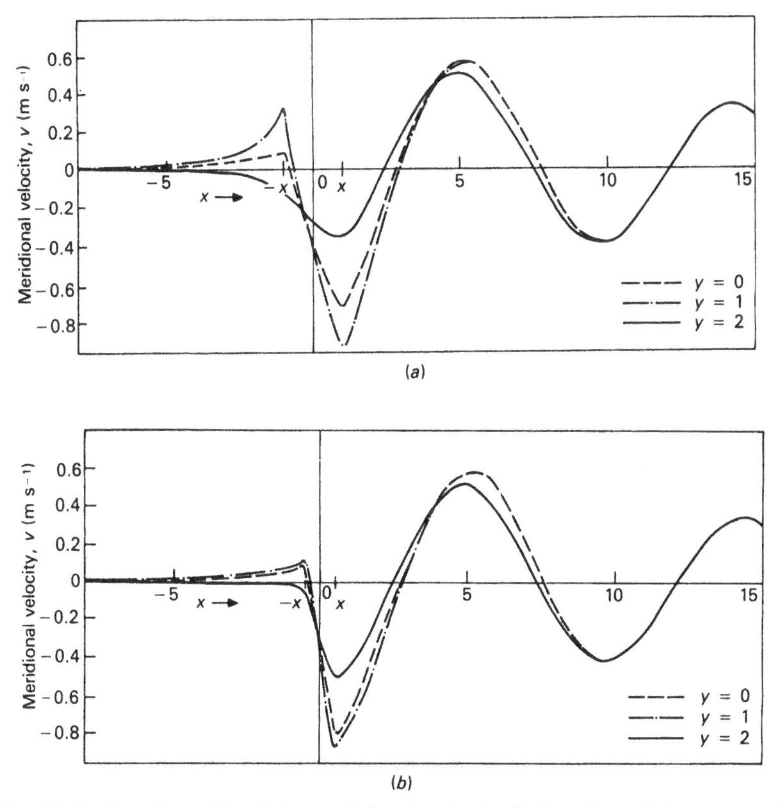

Fig. 28.2. Zonal profile of the meridional velocity induced by: (a) an obstacle with $\bar{x} = \bar{y} = 1$ at $y = 0$, 1 and 2; and (b) an elongated obstacle with $\bar{x} = 0.5$, $\bar{y} = 2$ at $y = 0$, 1 and 2.

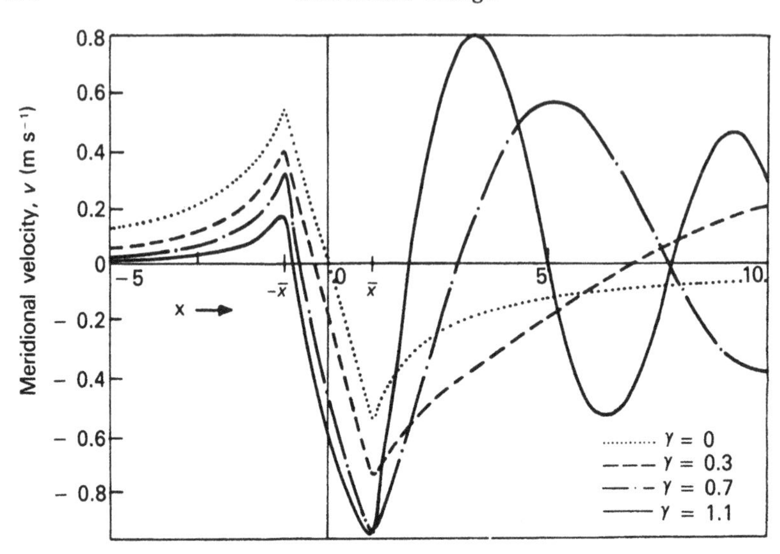

Fig. 28.3. Zonal profile of the meridional velocity $v(x, 0)$ for an obstacle with $\bar{x} = \bar{y} = 1$ for different values of γ.

dional velocity over the obstacle for the value of γ given in Table 28.1 can be obtained from the f-plane solution ($\gamma = 0$) as assumed in GS. Thus the qualitative predictions of GS are valid.

The streamlines for flow past the meridionally elongated obstacle are shown in Fig. 28.4 for the parameter values in Table 28.1. Since the model used is inviscid and quasi-geostrophic, the results have to be compared with observations above the frictional boundary layer, i.e. around 850 mb or so. Comparison of the observed mean monthly height contours of the 850 mb surface for July (Fig. 28.1*b*) with Fig. 28.4 indicates excellent agreement. Detailed comparison of various features of the solution with the observations is given in Table 28.2.

TABLE 28.2. *A comparison of values of various features yielded by the model and by observation. The observed values refer to the* 850 *mb level.*

Feature	Calculated	Observed
Maximum magnitude of v over the obstacle	$2\ \mathrm{m\,s^{-1}}$	$1\text{–}2\ \mathrm{m\,s^{-1}}$
Zonal distance of the first trough from the central longitude of the obstacle	1.3×10^3 km	1.2×10^3 km
Distance of the first ridge	3.9×10^3 km	3.5×10^3 km
Meridional displacement of the central streamline	300 km	200–300 km

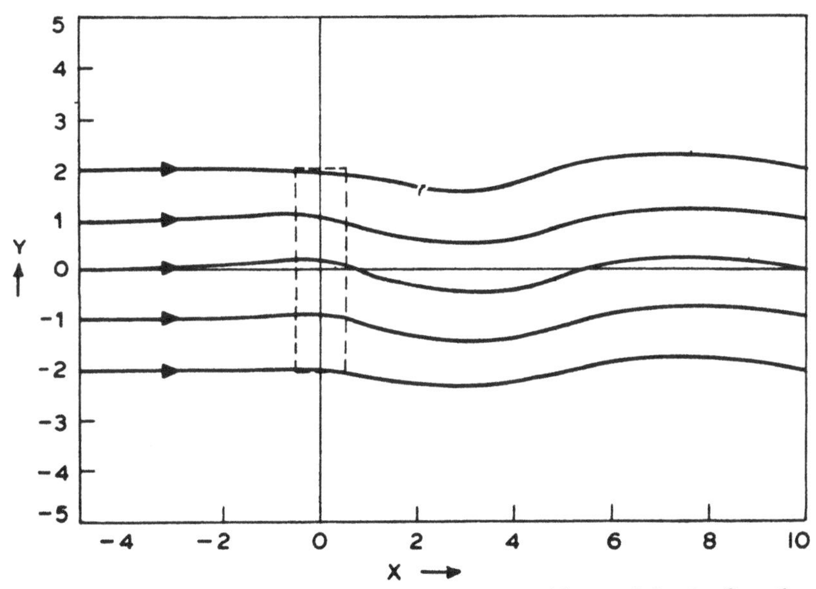

Fig. 28.4. Streamlines for flow past obstacles with $\bar{x}=0.5$, $\bar{y}=2$; other parameters are given in Table 28.1.

In contrast to the one-dimensional Rossby waves considered in GS, the amplitudes of the waves in (28.16) decay with distance downstream from the obstacle. The general nature of the solution at large distances from the obstacle can be determined by using standard asymptotic methods, i.e. Laplace's method and the method of stationary phase, which are ideally suited for the approximate evaluation of the integrals in (28.16) and (28.17). The asymptotic analysis reveals that the xy plane of flow splits naturally into the following three regions:

region I $\quad x \ll -\bar{x}, \quad \gamma|x|^2 \gg y,$

region II $\quad \gamma|x|^2 \ll |y|,$

region III $\quad x \gg \bar{x}, \quad \gamma|x|^2 \gg |y|.$

In the first two regions, only the first terms in (28.16) and (28.17) are significant, whereas in the third, only the second terms are important. The approximate expressions for the pressure field in these regions are

region I

$$p = \frac{\delta}{\beta}\left(\frac{2\sqrt{2}\,\bar{x}\sin(\gamma\bar{y})}{\pi x}\right)\sin(\gamma y - \pi/4),$$

region II

$$p = -\frac{\delta}{\beta}\left(\frac{2\gamma\bar{x}\sin(\gamma\bar{y})}{\pi y}\right)\sin(\gamma y - \pi/4),$$

region III

$$p = -\frac{\delta}{\beta}\left(\frac{2R^2}{xy}\sin(\gamma\bar{x}x/R)\sin(\gamma\bar{y}y/R)\right)\left(\frac{2\gamma}{\pi R}\right)^{\frac{1}{2}}\sin(\gamma R - \pi/4),$$

i.e. for $x \gg y$

$$p = -\frac{\delta}{\beta}(2\bar{y}\sin\gamma\bar{x})\left(\frac{2\gamma}{\pi x}\right)^{\frac{1}{2}}\sin(\gamma x - \pi/4).$$

In region I, upstream of the obstacle, the pressure and the zonal velocity decay as the inverse power of the distance, whereas the meridional velocity decays much faster. The waves in this region have crests parallel to the zonal direction and a wavelength of $2\pi/\gamma$. Note that waves with meridional wavenumbers smaller than γ, which can propagate westward of the obstacle, have been suppressed and hence only the more rapidly decaying waves remain in the solution. For a symmetric obstacle $(\bar{x} = \bar{y})$, the waves in regions II and III have circular wavecrests with the maximum velocity component at any point in the direction normal to the radius vector of length R. These waves fill the eastward hemisphere behind the obstacle, their amplitude decaying $(\alpha R^{-\frac{1}{2}})$ with distance as pointed out by Lighthill (1967). The solution in region II represents the contribution of meridional wavenumbers larger than γ, which will be generally excited by obstacles of meridional extent smaller than the Rossby wavelength. This region would collapse onto the y-axis in the limit of an obstacle with infinite meridional extent, and the waves in region III in this limit become one dimensional with crests parallel to the meridional direction. For an elongated obstacle (with $\bar{y} \gg \bar{x}$), with a zonal extent smaller than the Rossby wavelength, the velocity perturbation at a given distance from the origin is larger at a point on the x-axis than on the y-axis

$$\frac{v(y=0, x=R)}{u(x=0, y=R)} = \frac{\bar{y}\sin\gamma\bar{x}}{\bar{x}\sin\gamma\bar{y}} = \frac{\gamma\bar{y}}{\sin\gamma\bar{y}} \gg 1.$$

However, significant zonal motion which oscillates in the meridional direction is induced in the region north and south of the obstacle for all obstacles of finite meridional extent. The solution in region II for the parameters in Table 28.1, implies a maximum westerly velocity perturbation at the northern edge of the obstacle which becomes easterly at a distance of about 10^3 km to the north of the northern edge. One may speculate that this effect of the orography would contribute to the strengthening of the easterlies which normally occur at these latitudes.

It is a pleasure to thank Professor Sir James Lighthill for a stimulating discussion of the problem and for proving the radiation condition. I am indebted to Dr R. Sundar for help in the computations.

References

Ananthakrishnan *et al.* (1968) Monthly mean sea level isobaric charts. India Meteorological Department Forecasting Manual Report 1-1.

Carrier, G. F., Krook, M. and Pearson, C. E. (1966) *Functions of a complex variable*, McGraw-Hill, New York.

Charney, J. G. and Drazin, P. G. (1961) Propagation of planetary disturbances from lower into the upper atmosphere. *J. Geophys. Res.*, **66**, 83–110.

Dickinson, R. E. (1968) Planetary Rossby waves propagating vertically through weak westerly wind wave guides. *J. Atmos. Sci.*, **25**, 984–1002.

Gadgil, S. (1977) Orographic effect on the southwest monsoon – a review, *Pure and Applied Geophysics*, **115** (Special issue on monsoons).

Gadgil, S. and Sikka, D. R. (1976) The influence of the topography of the Indian peninsula on the low-level circulation of the summer monsoon. Fluid Mechanics Report 76FM 16, Indian Institute of Science, Bangalore, India.

Lighthill, M. J. (1967) On waves generated in dispersive systems by travelling forcing effects, with application to dynamics of rotating fluids, *J. Fluid Mech.*, **27**, 725–52.

Miles, J. W. (1968) Lee waves in a stratified flow, Part 1. Thin barrier. *J. Fluid Mech.*, **32**, 549–67.

Part III

The physics and dynamics of the Indian Ocean during the summer monsoon

Introduction

SIR JAMES LIGHTHILL

The Indian Ocean is of especial significance to oceanographers because of its unique feature of currents that switch direction on an annual cycle. This feature makes the Indian Ocean a natural 'laboratory' where oceanographers can seek to test their understanding of the dynamics of how ocean currents respond to changing wind patterns. In addition, Parts I and II of this book have suggested how the oceanography of the Indian Ocean may be of crucial importance to meteorologists. Charney and Shukla, in Chapter 6 of this book, indicated the *potential* 'predictability' of monsoon meteorology given a prediction of *boundary* values, including, especially, sea-surface temperatures. In this context, they emphasized that timescales of oceanic responses are all relatively slower than their atmospheric counterparts, so that longer-term forecasting of ocean boundary values (for use as the input to atmospheric models) might ultimately be attainable.

They are certainly right about timescales of response. Indeed, it has long been known for a midlatitude ocean that the timescale associated with the accumulation of the baroclinic part of its response to wind stress (including its western boundary current) may be as much as a decade (Veronis and Stommel, 1956). On the other hand, for a low-latitude system such as the Indian Ocean, the response is by no means as slow as that, and almost a decade ago Lighthill (1969) suggested that the corresponding timescale would be about a month rather than a decade. Lighthill's studies were particularly concerned with the Somali Current, and indicated the theoretical possibility that on a timescale of about one month a western boundary current *could* be generated in response to a distributed pattern of wind stress over the Ocean *even if* no local wind was blowing.

Since then, we have come to know all about the low-level jet over the Somali coast (especially due to the work of Findlater) and the likelihood that, in the main, the Somali Current is a direct response to forcing by the local winds. However, it remains possible that the Somali Current, like other western boundary currents, is further reinforced by the remoter influence of wind stresses distributed over the whole Ocean (Anderson and Rowlands, 1976).

Part III(*a*) begins with an up-to-date account of this field by John Swallow, whose unique contributions to our knowledge of time-variability of currents all over the world have been applied especially in this important area. This is followed by further important reviews of data, concerned with areas of the Indian Ocean closer to the Indian subcontinent itself.

Part III(*b*) begins with two chapters applying mathematical modelling to the Somali Current problem. These are followed by a chapter which reminds us of another modelling technique, using carefully designed laboratory experiments on appropriately shaped bodies of rotating fluid to elucidate the crucial features necessary to reproduce some of the more prominent features of ocean–current phenomena in different ocean systems.

The construction and use of combined atmosphere–ocean models to give realistic accounts of the dynamics of both media are fundamental to possible longer-term advances in a practical, useful understanding of monsoon dynamics. A study of the work in this Part shows some of the formidable complexity of the ocean processes, a complexity that must be properly allowed for in any realistic ocean–atmosphere model.

References

Anderson, D. L. T. and Rowlands, P. B. (1976) The Somali Current response to the S.W. monsoon: the relative importance of local and remote forcing. *J. Mar. Res.*, **34**, 395–417.

Lighthill, M. J. (1969) Dynamic response of the Indian Ocean to the southwest monsoon. *Phil. Trans. Roy. Soc. London*, **A265**, 45–92.

Veronis, G. and Stommel, H. (1956) The action of variable wind stresses on a stratified ocean. *J. Mar. Res.*, **15**, 43.

Part IIIa

Observational studies

29

Observations of the Somali Current and its relationship to the monsoon winds

J. C. SWALLOW

Observations of the Somali Current are presented and discussed, with particular emphasis on those aspects relating to the upwelling near the Somali coast and its generation by the monsoon winds.

29.1 Introduction

The Somali Current is the most dramatic feature of the seasonally reversing circulation in the northern part of the western Indian Ocean. At its peak, at the height of the southwest monsoon, surface currents of 3 to 4 m s^{-1} have been reported, and its total transport has been estimated at 70 million tonnes per second. The present state of knowledge of this current has been thoroughly reviewed in two recent papers, by Düing (1978) on observations and by Anderson (1978) on theoretical studies. From the point of view of monsoon dynamics, there are two particularly significant aspects. For monsoon meteorologists, the upwelling related to the Somali Current and its effect on sea-surface temperatures in the western Indian Ocean may be most important. For dynamical oceanographers, the onset of the Somali Current has been a particular concern: how do the monsoon winds generate such a strong current so quickly? The purpose of this chapter is to briefly review observations that bear on these two aspects of the Somali Current. First, though, for the benefit of those to whom it may be unfamiliar, a short description of the fully-developed current will be given.

29.2 The fully-developed Somali Current

It is convenient to think of the Somali Current as the seasonally reversing part of the boundary current along the east coast of Africa; that is, the part extending northwards from approximately 3° S latitude. A fairly detailed survey of the current at its maximum development was made by the research vessels *Argo* and *Discovery* in August 1964, during the International Indian Ocean Expedition, and described by Warren *et al.* (1966) and Swallow and Bruce (1966). Looking at the surface current arrows from that survey (Fig. 29.1), between 2° S and 2° N the boundary

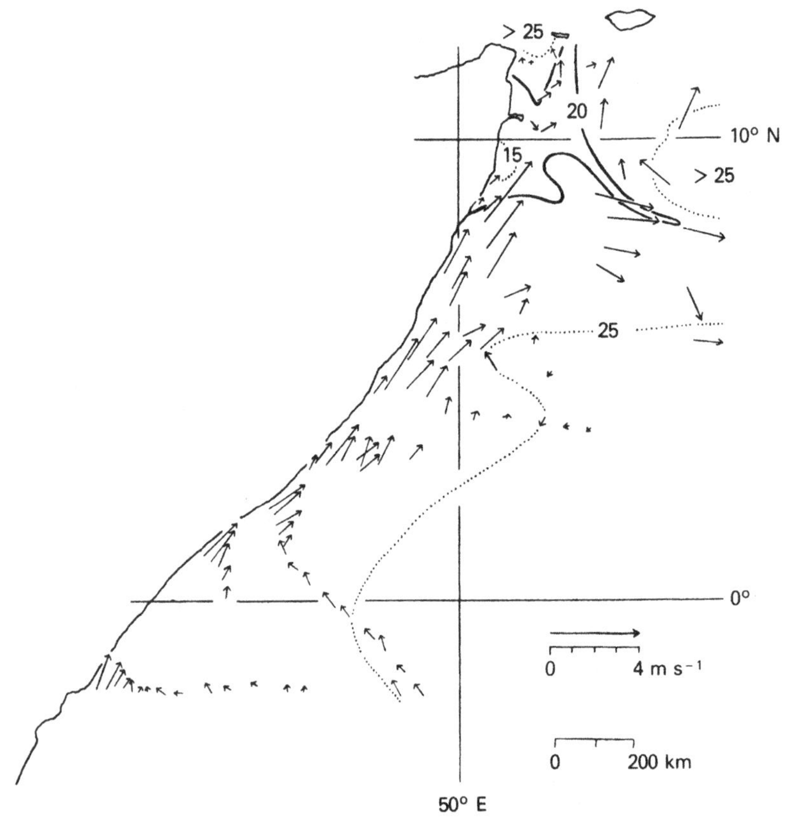

Fig. 29.1. Surface currents and selected isotherms (labelled in °C) observed off the Somali coast in August 1964. (Adapted from Stommel and Wooster, 1965; Swallow and Bruce, 1966; Warren *et al.*, 1966.)

current was quite narrow, with maximum speeds inshore of about 2 m s^{-1} decreasing to less than 0.5 m s^{-1} at 100 km offshore. Northward from 3° N, the boundary current increased in speed and width until near 9° N a maximum speed of 3.5 m s^{-1} was observed, and the current turned eastward, away from the coast. Some of the water was recirculated back into the boundary current by a clockwise eddy centred near 5° N, 52° E. This could be seen more clearly below the surface, at 100 m depth for example, being diminished near the surface by the offshore Ekman transport due to the strong local southwest winds. What appeared to be the western part of another clockwise eddy was found, with its centre near 10° N, 55° E. Surface temperatures over most of the area surveyed were between 24° C and 26° C, but between 9° N and 11° N, just north of where the boundary current turned eastward, much lower surface temperatures were observed, some as low as 14° C. Vertical profiles of temperature and current showed that the strong currents were confined to the near-surface layer, most of the shear being across the thermocline which was typified by temperatures between 20° C and 14° C. In the boundary current moving northeastwards from the Equator, the thermocline became progressively more tilted until it reached the surface near the coast where the boundary current turned offshore. The 20° C isotherm attained its greatest depth, more than 200 m, in the centres of the clockwise eddies. The surfacing of the thermocline and the turning offshore of the boundary current occurred quite close to the mean latitude in which, according to Findlater (1971), the low-level jet comes out over the sea in August.

To the extent that they could be compared, the pattern of surface currents observed in August 1964 was quite similar to the climatological mean for that month in the KNMI atlas (1952). That does not mean, of course, that the current is the same every year. In a survey made by the research vessel *Chain* in August and September 1970, reported by Bruce (1973), two regions of cold water were found near the coast at 6° N and 10° N, each being associated with a branch of current turning offshore. And there were relatively rapid changes during both of the surveys; in 1964, for example, part of the 20° C surface isotherm shifted by 50 km in less than 2 weeks.

29.3 The region of upwelling and its boundaries

As a first step towards understanding the relationship between the Somali Current and sea-surface temperatures in the western Indian Ocean, it seems useful to try to describe what appears to be happening at the boundaries of the upwelling region. Taking the 20° C surface isotherm in Fig. 29.1 as a convenient boundary for the upwelled water seen in August 1964, the southern edge coincided with the boundary current turning offshore and running approximately eastwards near 9° N. On its eastern side also, the upwelling region was bounded by a strong current, but distinguishable from the main Somali Current by its much higher salinity. In the boundary current turning offshore, surface salinities were below 35.2‰, suggesting a close connection to the relatively fresh water of the South Equatorial Current. At the centre of the 'clockwise eddy' near 10° N, 55° E, surface salinities exceeded 35.8‰. For such high salinities to exist in it, this eddy must have been generated and remained as a separate feature from the main boundary current. Between these two separate strong currents the upwelled surface water colder than 20 °C was drawn out into a thin tongue extending to at least 55° E, with strong cyclonic shear across it. On the northern edge of the upwelling region, outwards from the coast near 11° N, the warm high-salinity surface water of the Gulf of Aden was encountered. There was no large-scale strong current in the Gulf of Aden water, but some visual evidence of small-scale (a few kilometres) strong eddies near Cape Guardafui. Vertical profiles of current suggested that, in general, the cold upwelled water was moving slowly northwards and disappearing under the less dense Gulf of Aden surface water. The boundary there was very variable; observations only a few days apart could not be contoured together.

Similar-looking features can be seen in sea-surface temperature maps of the upwelling region produced by means of infrared radiometers in satellites. Two examples are shown in the review article by Düing (1978). In one of them, taken on 6 July 1966 (Szekielda and La Violette, 1970), a tongue of cold upwelled water extended eastwards from the coast near 10° N as far as 54° E and then continued southeastwards to about 8° N, 56° E, somewhat broader than but similar to the one observed in 1964. There was another patch of cold water, roughly triangular, with its southern edge running from the coast near 6° N northeastwards, then eastwards to 8° N, 53° E. One could readily imagine that the boundary current turned offshore near 6° N and that there was another clockwise eddy, or possibly a clockwise loop following a meander in the main current, on the southern side of the tongue of cold water near 10° N. The warm surface water of the Gulf of Aden met the cold upwelled water in a zone running northeast from Cape Guardafui.

The position and extent of the region of cold surface water seems to depend, then, on the latitude at which the boundary current leaves the coast, the positions and sizes of any offshore eddies (which are frequently present), and the extent to which the warm saline Gulf of Aden surface water can spread over the cold upwelled water.

There are other, less dramatic but perhaps more widespread, effects of the Somali Current on sea-surface temperatures. For example, the advection of cool fresh water from the South Equatorial Current northward across the Equator, and its subsequent distribution, together with the upwelled water, into the surface layers of the northwestern Indian Ocean may be important, but this will not be considered here.

29.4 The onset of the Somali Current

The few direct observations that have been made of the onset of the north-going boundary current show that it occurs within a few days of the local onset of southerly winds. Leetmaa (1972, 1973) found such reversals in the boundary current at latitudes 1° S and 2° S in April 1970 and in late March 1971, and ascribed them to the combined effects of direct acceleration by the local wind and a diversion of the north-going boundary current that is present throughout the year south of about 4° S. This part of the boundary current system is fed by the northern branch of the South Equatorial Current. During the northeast monsoon it meets the south-going Somali Current in the neighbourhood of 3° S and together they form the east-going Equatorial Countercurrent. With the onset of local southerly winds the south-going Somali Current diminishes, and the southern part of the boundary current continues northward beyond 3° S, instead of turning eastward. More recently, Düing and Schott (1978) observed the same process using an array of moored current meters, set off the coast of Kenya in January to July 1976. Northeastward flow was found throughout that period at 4° S, but north of 3° S it began above the thermocline at the beginning of May within a few days of the appearance of local southerly winds. Soon after that onset of the boundary current, however, strong variable surface currents were observed farther offshore in the same region (Fig. 29.2). These suggest that the water in the boundary current detected by the moored array may not have continued very far up the coast; much surface water of low salinity similar to that of the boundary current was being carried southeastwards or southward farther offshore. Surface transports in these offshore variable currents were 2 to 3 times as large as that of the boundary current. Satellite-tracked buoys released in that region at the same time (Regier and

Stommel, 1976) moved in clockwise ellipses with axes roughly 700 km and 300 km and a circulation time of 25 days. The deepening and slight northward turning of the boundary current as seen at the moored array in early June 1976 (Düing and Schott, 1978) is consistent with the appearance of a broad westward flow offshore in June (Fig. 29.2).

North of the Equator, no direct observations of the onset of the boundary current have yet been made, but indirect evidence was reported

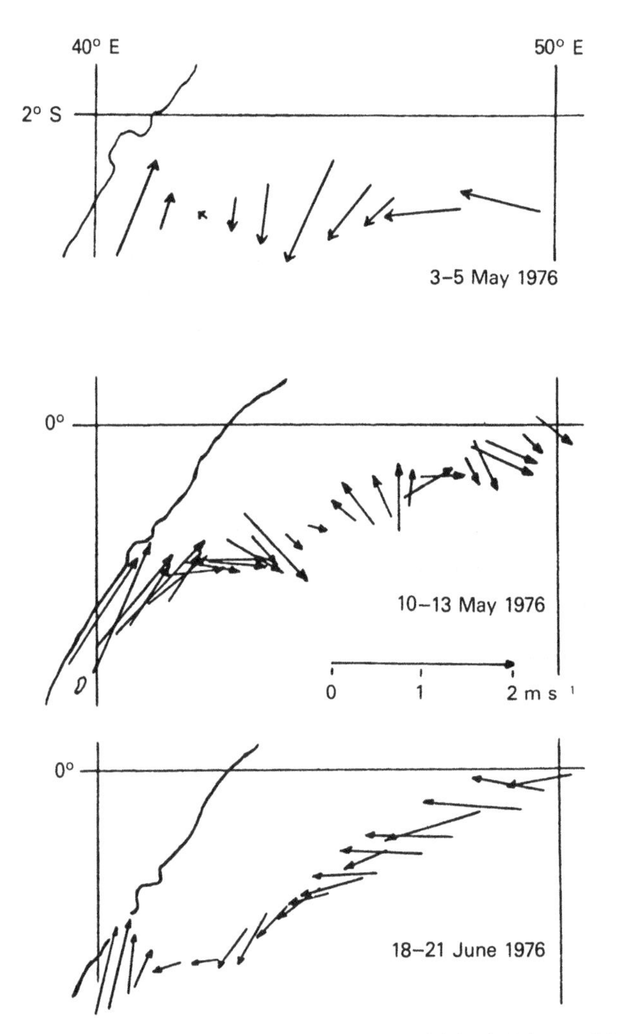

Fig. 29.2. Some surface currents observed off East Africa in May and June 1976. (Adapted from Swallow and Luyten, 1976.)

by Düing and Szekielda (1971). Inferring current speeds from the horizontal temperature gradients observed in infrared radiation maps from satellites, they concluded that the initial response followed the local southwest wind onset with a lag of about 12 days, but that subsequent growth of the current with a delay of about a month could not be accounted for by an increase of local wind. They inferred that both local wind forcing proposed by Cox (1970) and the remote forcing invoked by Lighthill (1969) were important, as indeed one might conclude from the observations south of the Equator.

One other observation, that may have some bearing on the onset of the Somali Current, needs to be mentioned. Although it is true that the strongest currents are near the surface, and most of the shear is found in the thermocline, there is some evidence for fairly vigorous motion (of order 0.5 m s^{-1}) at greater depths in some places. In particular, Bruce and Volkmann (1969) found a subsurface clockwise eddy near 9° N, 53° E in early 1965, that they interpreted as a remnant of the eddy seen near there in the previous summer. Leetmaa (1972) has surmised that the eddy may come to the surface again as the northeast monsoon weakens and reinforce the onset of the Somali Current, acting like a flywheel between one summer and the next.

References

Anderson, D. L. T. (1978) Low latitude western boundary currents. In *FGGE/INDEX/NORPAX Equatorial Workshop Review Papers,* Nova/NYIT, University Press, Austin, Texas.

Bruce, J. G. (1973) Large-scale variations of the Somali Current during the southwest monsoon, 1970. *Deep-Sea Res.*, **20**, 837–46.

Bruce, J. G. and Volkmann, G. H. (1969) Some measurements of current off the Somali coast during the northeast monsoon. *J. Geophys. Res.*, **74**, 8, 1958–67.

Cox, M. D. (1970) A mathematical model of the Indian Ocean. *Deep-Sea Res.*, **17**, 47–75.

Düing, W. (1978) The Somali Current: past and recent observations. In *FGGE/INDEX/NORPAX Equatorial Workshop Review Papers,* Nova/NYIT, University Press, Austin, Texas.

Düing, W. and Schott, F. (1978) Measurements in the source region of the Somali Current during the monsoon reversal. *J. Phys. Oceanogr.*, **8**, No. 3.

Düing, W. and Szekielda, K. H. (1971) Monsoonal response in the western Indian Ocean. *J. Geophys. Res.*, **76**, 18, 4181–7.

Findlater, J. (1971) Mean monthly airflow at low levels over the western Indian Ocean. *Geophys. Mem.* (HMSO, London), **16**, 115, 1–53.

Koninklijk Nederlands Meteorologisch Institut (1952) *Indische Oceaan. Oceanografische en Meteorologische gegevens.* 2nd edition. Publication No. 135: Vol. 1, 31 pp; Vol. 2, 24 charts.
Leetmaa, A. (1972) The response of the Somali Current to the southwest monsoon of 1970. *Deep-Sea Res.*, **19**, 319–25.
Leetmaa, A. (1973) The response of the Somali Current at 2° S to the southwest monsoon of 1971. *Deep-Sea Res.*, **20**, 397–400.
Lighthill, M. J. (1969) Dynamic response of the Indian Ocean to the southwest monsoon. *Phil. Trans. Roy. Soc. London*, **A265**, 45–92.
Regier, L. and Stommel, H. (1976) Trajectories of INDEX surface drifters. INDEX Occasional Notes 5 (unpublished document, Nova University Oceanographic Laboratory).
Stommel, H. and Wooster, W. S. (1965) Reconnaissance of the Somali Current during the southwest monsoon. *Proc. Nat. Acad. Sci. USA*, **54**, 8–13.
Swallow, J. C. and Bruce, J. G. (1966) Current measurements off the Somali coast during the southwest monsoon of 1964. *Deep-Sea Res.*, **13**, 861–88.
Swallow, J. C. and Luyten, J. R. (1976) Some observations of surface currents in the western Indian Ocean. *Ocean Modelling*, **1**, 4–5.
Szekielda, K. H. and La Violette, P. E. (1970) Variations in detailed sea surface temperature structure along the Somali coast as defined by NIMBUS II HRIR .data. (unpublished manuscript).
Warren, B. A., Stommel, H. and Swallow, J. C. (1966) Water masses and patterns of flow in the Somali Basin during the southwest monsoon of 1964. *Deep-Sea Res.*, **13**, 825–60.

30

Structure of currents and hydrographic conditions in the western equatorial Indian Ocean during the summer monsoon

L. V. GANGADHARA RAO, V. RAMESH BABU
AND V.V.R. VARADACHARI

This chapter presents the results of studies on the currents and hydrography of the waters of the western equatorial Indian Ocean during the period 28 May to 3 June 1973. Studies of hydrographic conditions in the area after about one month (i.e. 25 June to 2 July 1973) are also discussed. Time-series data on horizontal currents were gathered by recording current meters attached to a buoy anchored at the Equator near 60° E during the Indo-Soviet Monsoon Experiment 1973. Hydrographic data were also obtained from repeated Nansen casts at 3-hourly intervals around the buoy. The time variation of currents (3-hour averages) at different depths (75 m, 100 m, 150 m, 200 m, 300 m, 500 m and 800 m) and the vertical time sections of temperature, salinity and density from the surface to 1000 m depth are presented and discussed.

The studies show a complex vertical structure of horizontal currents with predominant easterly components at 75 m and 100 m depth and predominant westerly components at 200 m, 300 m, 500 m and 800 m depth, and a zone of strong current shear between 100 m and 200 m depth. At 150 m depth, the current has been found to be oscillatory with a period of oscillation of about 2 hours. The currents are rather strong even at deeper levels, particularly at 200 m and 800 m depths where westward jets with speeds of the order of 0.35 m s^{-1} and 0.4 m s^{-1} respectively have been encountered.

Studies of the hydrographic conditions have revealed the presence of internal waves in the area. A significant fall in temperature (about 2 °C) and a slight decrease in salinity in the surface layer were noticed during the one-month interval between the two series of hydrographic observations (May/June–June/July).

30.1 Introduction

Important among the unique features of the Indian Ocean is its exposure to the seasonally reversing monsoon winds. As a result, the surface currents, especially in the equatorial part of the Indian Ocean, differ markedly from those in the other oceans. A study of the structure of currents in the equatorial region of the Indian Ocean therefore assumes special significance.

Some investigations of the structure of currents in the equatorial Indian Ocean, aimed at achieving a better understanding of the behaviour, driving mechanism and method of maintenance of currents in this region, were carried out during the International Indian Ocean Expedition (1960–5) which used direct measurement of currents together with measurement of water properties (temperature, salinity, density, etc.). Results of the current measurements made from RV *Argo* of the Scripps Institution of Oceanography, USA during 1962 and 1963 (reported by Knauss and Taft, 1963 and 1964, and Taft, 1967) and those conducted from RRS *Discovery* of the UK in 1964 (reported by Swallow, 1964 and 1967), established the presence of an undercurrent in the Indian Ocean. This undercurrent showed some characteristics different from those of the undercurrents present in the Atlantic and the Pacific Oceans. These current measurements were, however, mostly confined to the upper 400 m. Some measurements of currents at greater depths were made in the equatorial region of the Indian Ocean from RV *Vityaz* of the USSR, during the southwest monsoon period of 1962, when a strong westward flow was found in the deeper regions (Neyman, 1964).

A joint Indo-Soviet Monsoon Experiment (ISMEX) was carried out from May to July 1973. During this experiment a vast amount of meteorological and oceanographic data was collected. The data was used for a preliminary exercise in the investigation of the atmospheric and oceanic processes associated with various phases of the Indian summer monsoon in preparation for the detailed studies to be taken up during the International Monsoon Experiment (MONEX) planned for 1979. The programme of ISMEX included collection of time-series data on horizontal currents and water properties from a few current-meter moorings and the associated research ships. Results of measurements made at a current-meter mooring positioned on the Equator in the western part of the equatorial Indian Ocean during the ISMEX are presented and discussed in this chapter.

30.2 Collection and treatment of data

During the ISMEX, *Okean*, a research vessel of the Hydrometeorological Services of the USSR laid a current-meter mooring on the Equator at 60° E from 27 May to 4 June 1973 to gather information on the structure of currents during the early phase of the summer monsoon. Time-series data on horizontal currents at seven depths (75 m, 100 m, 150 m, 200 m, 300 m, 500 m and 800 m) were collected with *Alekseev* recording current meters attached to the moored buoy at the respective depths. The ship was allowed to drift in the area and manoeuvered to keep herself within 3 miles (4.8 km) distance from the buoy which she had laid and which she subsequently recovered. Nansen casts were made down to about 1200 m depth at 3-hourly intervals and data on hydrographic parameters were collected for a study of the water properties. The ship returned to the same area after about a month and the observations were repeated from 25 June to 2 July 1973, to identify any changes.

Data on currents sampled at 20-minute intervals for different depths were resolved into zonal and meridional components. Three-hourly averages of currents (zonal component, meridional component and current vector) for the period 28 May to 3 June 1973 were calculated and presented in a chart describing their variation with time at different depths. Arithmetic and vectoral averages of currents for the entire period were also computed for different depths to distinguish the oscillatory type of current from the steady type.

Using the processed hydrographic data, vertical time sections of temperature, salinity and density from the surface to 1000 m depth, were prepared for the periods 28 May to 3 June and 25 June to 2 July. In these sections, the vertical scale is reduced by half for the depth range 500 to 1000 m as the variations of the parameters at these depths are less marked. The vertical time sections of temperature, salinity and density have been presented with isotherms, isohalines and isopycnals drawn at intervals of 1 °C, 0.1‰ and 0.2 kg m^{-3} (grammes per litre) respectively. However, in the region of strong density gradients ($\sigma_t = 23$ to 25 kg m^{-3}) the σ_t lines have been drawn at unit intervals.

30.3 Results and discussion

The western part of the equatorial Indian Ocean, with bathymetric contours of the region (shown in metres) and the location of the buoy station where the observations were taken, is shown in Fig. 30.1. It is seen that the station is located in an ideal deep-ocean environment free from any boundary effects of continents or islands. The water depth at the station was 4470 m.

Fig. 30.2 shows the time variation of currents (3-hourly averages) for the period 27 May to 4 June 1973 for different depths. Arithmetic and vectoral averages of the currents for the entire period are presented in Table 30.1. At 75 m depth, the current speeds vary between 0.3 m s^{-1} and 0.5 m s^{-1} with directions mostly between northeasterly and easterly exhibiting predominantly easterly components. The currents are rather weak at 100 m depth with speeds ranging between 0.2 m s^{-1} and 0.4 m s^{-1} and directions varying mostly between southeasterly and easterly. At 150 m depth, even though the individual values of speeds sampled at 20-minute intervals are of the order of 0.3 m s^{-1}, the vectorial averages over 3-hour periods are much less (less than 0.1 m s^{-1}) indicating that the currents at this depth are oscillatory. To depict the oscillatory nature of currents at this depth, a progressive vector diagram of currents for one day (28 May 1973) using 20-minute sampling has been prepared and presented in Fig. 30.3. It can be inferred from this diagram that the period of oscillation of currents at this depth is about 2 hours, and that

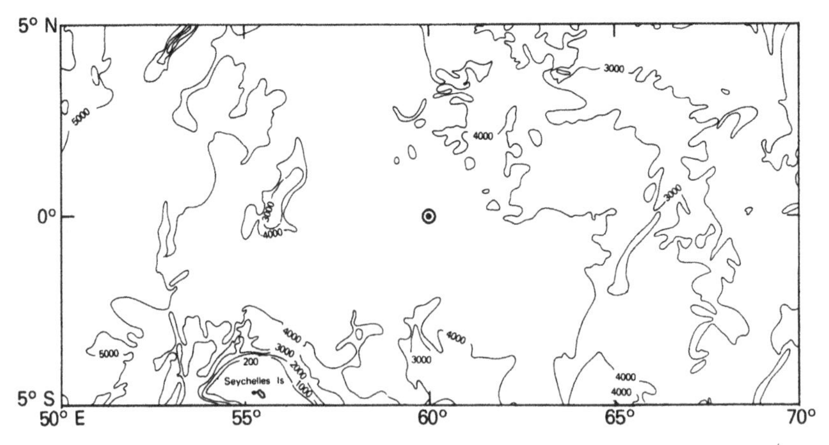

Fig. 30.1. Station location (at the centre of the region). The contours of ocean depth are labelled in metres.

TABLE 30.1. *Average currents for the period 28 May to 3 June 1973.*

Depth of sampling (m)	Average current (arithmetic mean) ($m s^{-1}$)	Average zonal component, $\bar{u}$ ($m s^{-1}$)	Average meridional component, $\bar{v}$ ($m s^{-1}$)	Average current (vectoral mean), $\bar{V}$ Speed ($m s^{-1}$)	Direction
75	0.41	0.36	0.09	0.37	76°
100	0.34	0.28	−0.13	0.31	115°
150	0.29	−0.03	−0.02	0.04	236°
200	0.37	−0.35	0.01	0.35	272°
300	0.33	−0.28	0.15	0.32	298°
500	0.37	−0.34	−0.01	0.34	268°
800	0.43	−0.32	−0.24	0.40	233°

there is a net southward movement at the rate of about 2.5 km per day. At 200 m depth, the current is again strong (speeds range from 0.3 $m s^{-1}$ to 0.45 $m s^{-1}$) and the directions vary between west-southwest and west-northwest exhibiting predominantly westerly components. At 300 m depth, the current speeds are slightly less than those at 200 m (speed variation between 0.2 and 0.43 $m s^{-1}$) and the directions vary between west and northwest. The current speeds vary between 0.25 and 0.46 $m s^{-1}$ at 500 m depth, and the directions vary between west-northwest and west-southwest, exhibiting predominantly westerly components. At 800 m depth, the current speeds and directions vary between 0.35 and 0.5 $m s^{-1}$ and west-southwest and south-southwest respectively.

These features agree fairly well with the findings of the recent observations on currents (time series of vertical profiles of horizontal velocity extending from ocean surface to the bottom, employing an acoustic dropsonde, a free-fall acoustically self-navigating vehicle) carried out from RV *Atlantis II* of the Woods Hole Oceanographic Institution, USA in the western Indian Ocean (along 53° E meridian) during May–June 1976, during the Indian Ocean Experiment (INDEX), reported by Luyten and Swallow (1976).

A critical study of the values of average currents (Table 30.1) reveals that except at 150 m depth, the currents are rather steady as no significant difference between the arithmetic mean and the vectoral mean could be found at these depths. A clockwise shift in the direction of currents from 75 m to 300 m depth and a counter-clockwise shift further down to 800 m depth is encountered (Table 30.1) suggesting that the complex variations

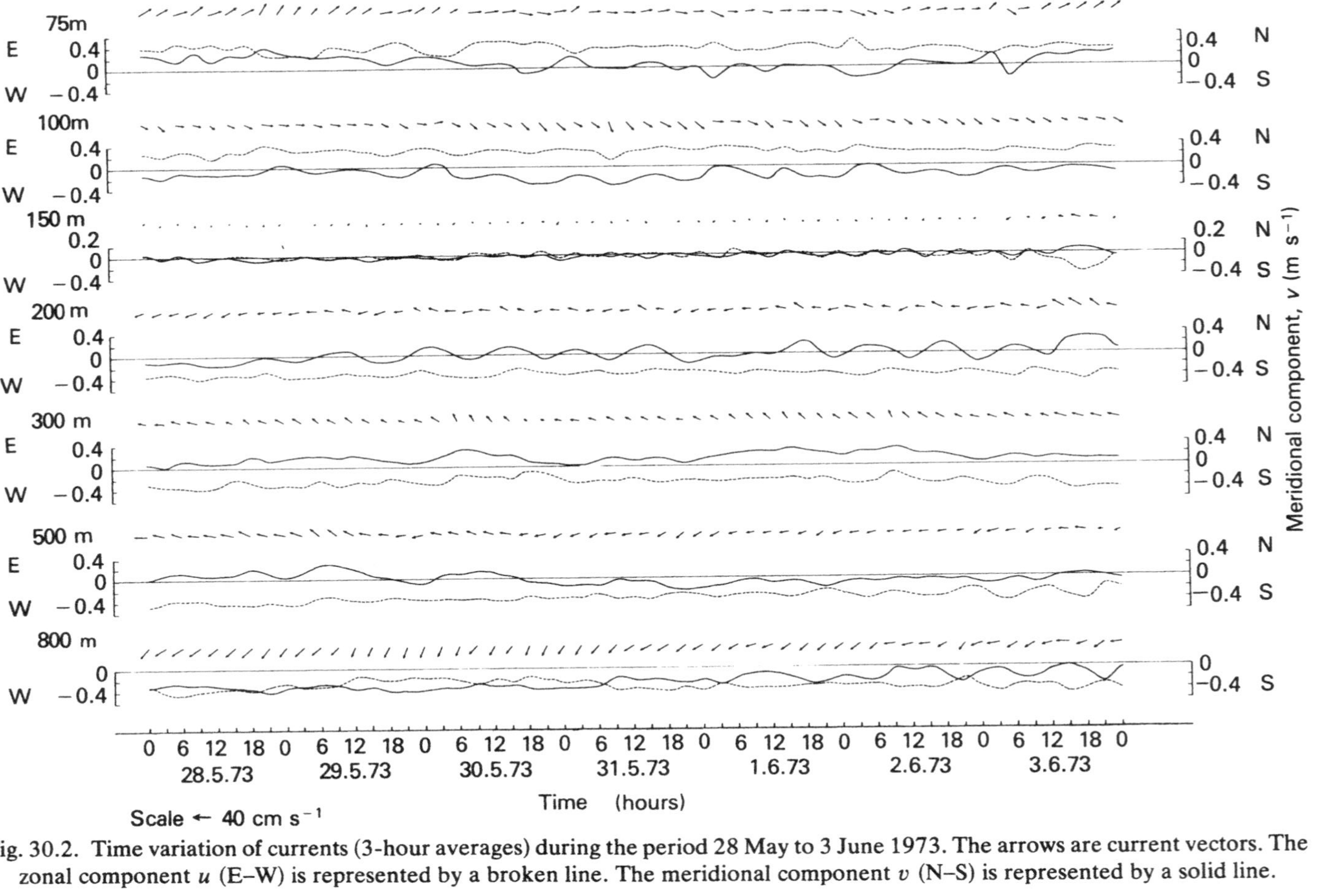

Fig. 30.2. Time variation of currents (3-hour averages) during the period 28 May to 3 June 1973. The arrows are current vectors. The zonal component u (E–W) is represented by a broken line. The meridional component v (N–S) is represented by a solid line.

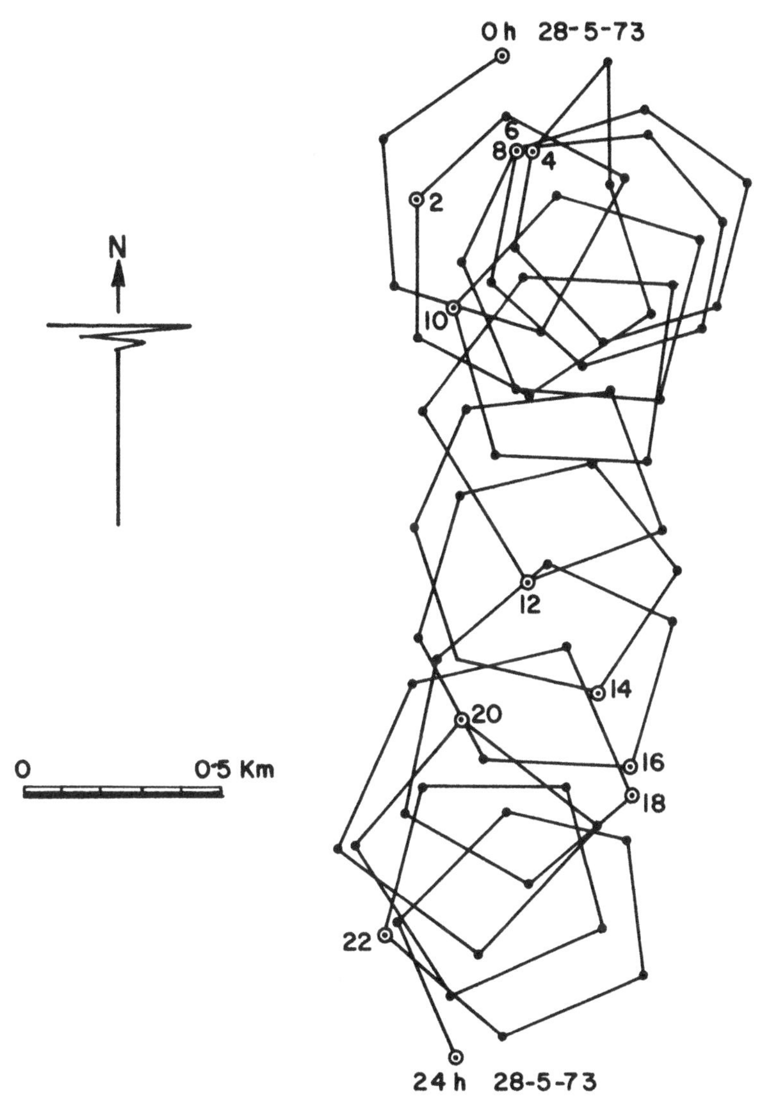

Fig. 30.3. Progressive vector diagram of currents at 150 m depth, using 20-minute sampling, for 28 May 1973.

of currents with depth, and the shear zones, are not confined to the upper layers, but extend perhaps to depths greater than 800 m. Rather strong currents with average speeds of the order of 0.35 m s^{-1} and 0.4 m s^{-1}, with predominant westward components encountered at depths of 200 m and 800 m respectively, could be referred to as westward jets since currents with such high magnitudes are not common in deeper regions. These findings agree well with those reported recently by Luyten and Swallow (1976) based on the current measurements carried out from RV *Atlantis II*.

Vertical time sections of temperature, salinity and density for the periods 28 May to 3 June and 25 June to 2 July are presented in Figs. 30.4 and 30.5 respectively, depicting the hydrographic conditions prevailing in the area at the time of current-meter mooring and a month later. During the former period, the temperature of the surface layer is more than 30 °C, the thermocline is very well developed with sharp vertical temperature gradients (10 °C per 25 m) and the isotherms exhibit a regular wavy pattern indicating the presence of internal waves. Within a month, the temperature of the surface layer fell by about 2 °C and the thermocline became slightly diffuse. (Fig. 30.5).

The density of surface waters is slightly higher during the latter period, periods, with slightly lower values being encountered during the latter period. An interesting feature of these sections is the presence of a series of cells of high and low salinity in the upper part of thermocline during the former period. This may be the result of local circulations associated with intense internal-wave activity (La Fond and Lee, 1962). The salinity structure during the latter period is comparatively simple with very few cells.

The density of surface waters is slightly higher during the latter period, reflecting the effect of relatively low surface temperatures encountered during this period. The presence of the strongly developed density discontinuity layer exhibiting a wavy pattern during the former period confirms the intense internal-wave activity revealed in the thermal structure.

The effect of local circulations associated with internal waves can be seen in the currents at depths of 75 m and 100 m, where slight variations in velocity have been observed periodically during May–June (Fig. 30.2). The period of oscillation (of about 2 hours) of currents at 150 m depth may as well, perhaps, be related to that of internal gravity waves present in this region as this feature cannot be attributed either to inertial oscillations or to tidal motions. These findings are in agreement with the

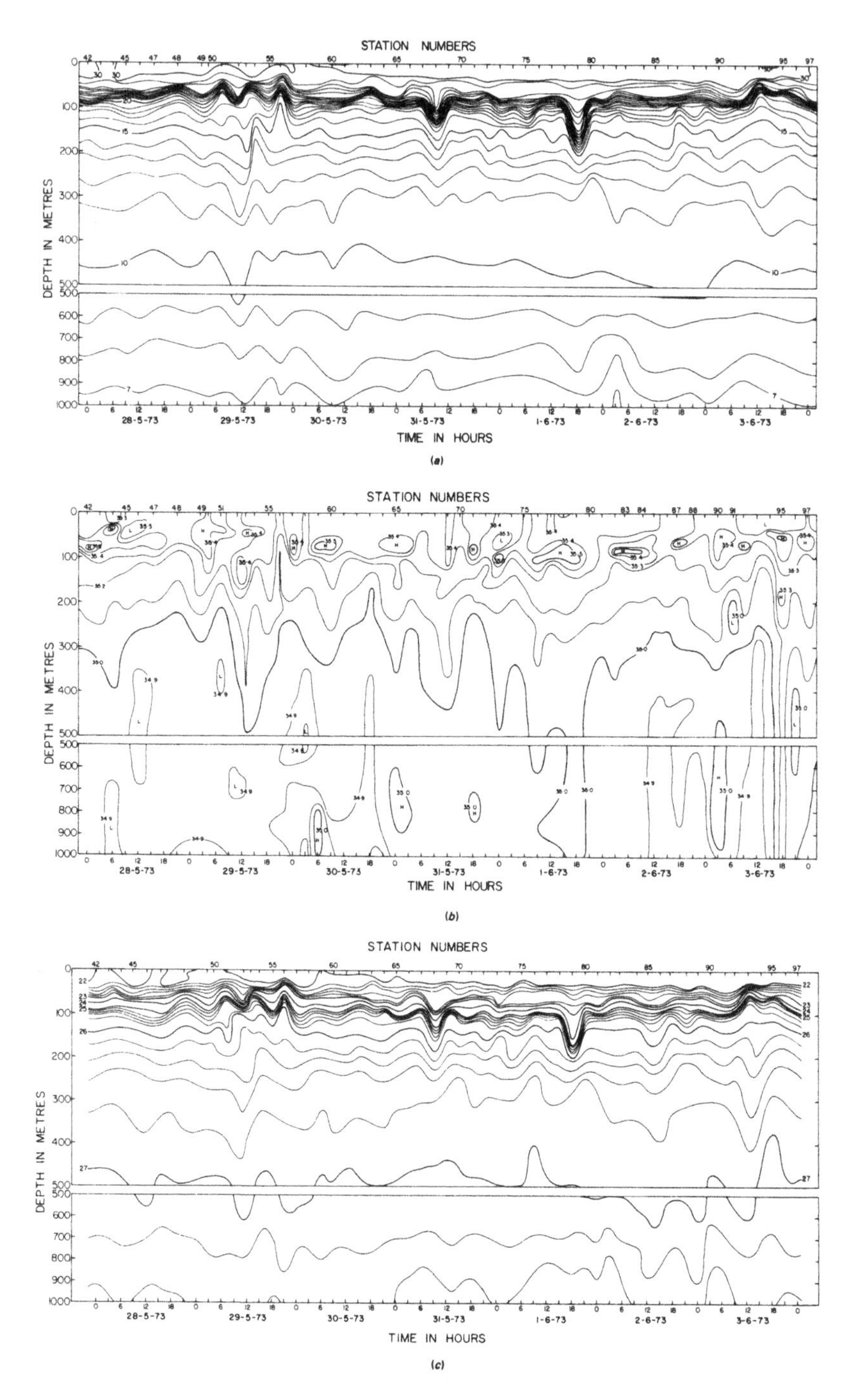

Fig. 30.4. Vertical time sections of: (*a*) temperature (contours labelled in °C); (*b*) salinity (contours labelled in ‰); and (*c*) density (contours labelled in kg m^{-3}) for the period 28 May to 3 June 1973.

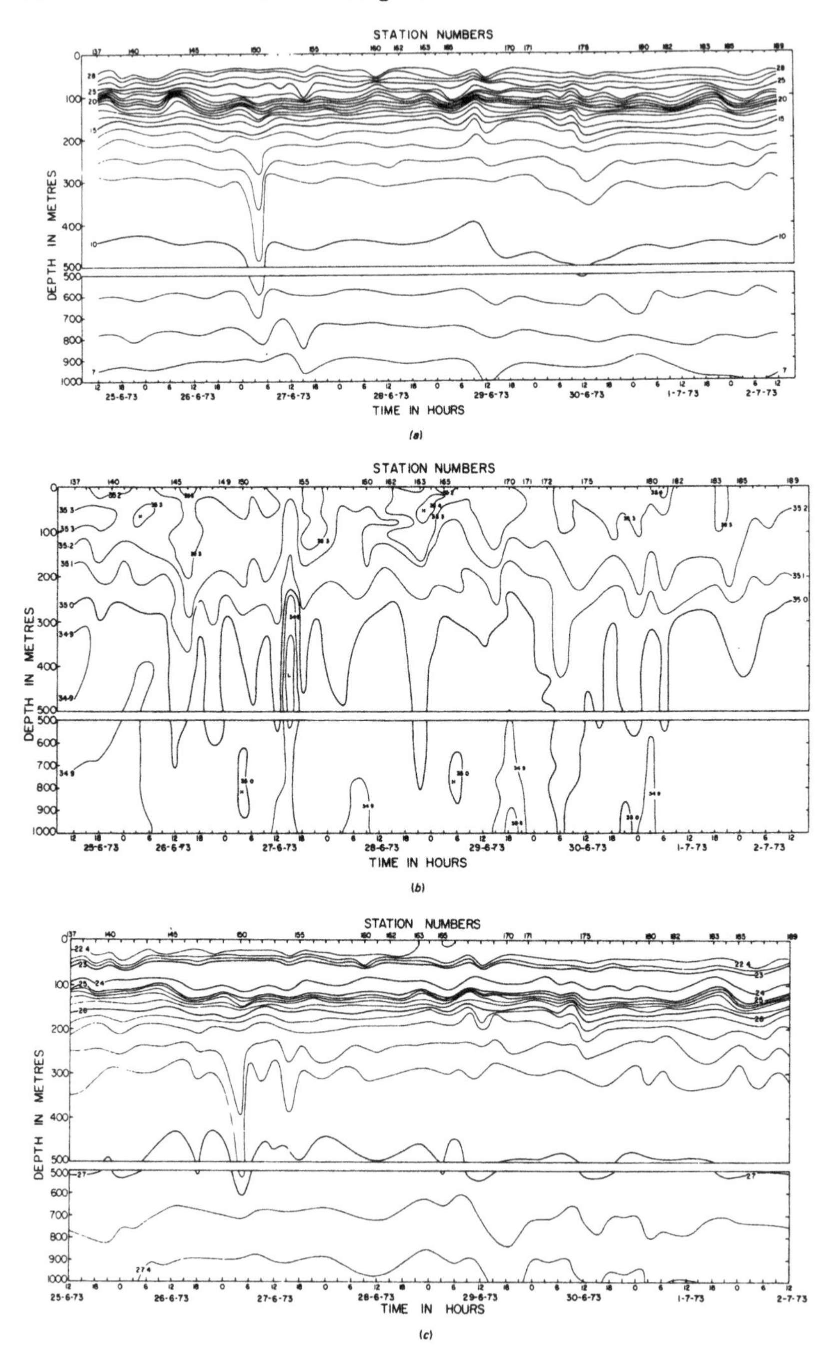

Fig. 30.5. Vertical time sections of: (*a*) temperature (contours labelled in °C); (*b*) salinity (contours labelled in kg m^{-3}); and (*c*) density (contours labelled in kg m^{-3}) for the period 25 June to 2 July 1973.

complexity of vertical structure of the horizontal velocity field with periods of oscillation of appreciably longer timescales, as encountered during the recent observations in the equatorial Indian Ocean (Luyten and Swallow, 1976).

The authors wish to express their thanks to Dr S. Z. Qasim, Director, National Institute of Oceanography for his keen interest in this study.

References

Knauss, J. A. and Taft, B. A. (1963) Measurement of currents along the Equator in the Indian Ocean. *Nature*, **198**, 376–7.

Knauss, J. A. and Taft, B. A. (1964) Equatorial undercurrent of the Indian Ocean. *Science*, **143**, 354–6.

La Fond, E. C. and Lee, O. S. (1962) Internal waves in the Ocean. *Navigation*, **9**, 231–6.

Luyten, J. R. and Swallow, J. C. (1976) Equatorial undercurrents. *Deep-Sea Res.* **23**, 1005–7.

Neyman, V. G. (1964) On the structure of zonal currents in the equatorial region of the Indian Ocean. *Okeanologiya*, **5**, 920.

Swallow, J. C. (1964) Equatorial undercurrent in the western Indian Ocean. *Nature*, **204**, 436–7.

Swallow, J. C. (1967) The equatorial undercurrent in the western Indian Ocean. *Studies in Tropical Oceanography* (Miami), **5**, 15–36.

Taft, B. A. (1967) Equatorial undercurrent of the Indian Ocean, 1963. *Studies in Tropical Oceanography* (Miami), **5**, 3–14.

31

Recent observations in the equatorial Indian Ocean[†]

J. R. LUYTEN

Recent velocity profile data for the western equatorial Indian Ocean are presented and discussed. These display a very low frequency variability of small vertical scales. Wunsch's (1977) theoretical model of a forced equatorial response at the annual period is shown to provide a rational explanation of the phenomenon.

31.1 Introduction and historical background

The oceanographic response of the western Indian Ocean to the atmospheric monsoon regime is a complicated one, both from the observational and theoretical points of view. Several phenomena have a clearly documented relationship to the monsoon forcing although many of the details are not clearly understood. The most dramatic of these monsoon phenomena, the Somali Current, has been discussed by Swallow in Chapter 29 of this book.

In the open, equatorial Indian Ocean two phenomena exist whose seasonal variability is clearly associated with the monsoon–the equatorial undercurrent and the equatorial surface jet. Recent observations suggest there may be other monsoon-related currents at greater depths.

The equatorial undercurrent is a permanent feature of both the Atlantic and Pacific Oceans (Philander, 1973) and is generally believed to be

† Contribution No. 4142 from the Woods Hole Oceanographic Institution.

associated with the persistent trade winds from the east. This westward stress at the surface, and corresponding difference in elevation of the sea surface, is balanced by a zonal pressure gradient between the boundaries. Because the fluid is stratified, there is a compensating pressure gradient in the thermocline which drives a flow down the pressure gradient. In the Indian Ocean such a zonal pressure gradient can only be established by the northeast monsoon. The observations (Taft and Knauss, 1967; Swallow, 1967; Knox, 1976; Kort *et al.*, 1975) have shown an undercurrent only during the latter part of the northeast monsoon, with large variations in strength from year to year. These variations are presumably related to variations in the strength of the westward wind stress. Recent observations along a section across the Equator north of the Seychelles (55° E) have shown a marked variation in the undercurrent between 1975 and 1976 (Leetmaa and Stommel, personal communication) which can be rationalized in terms of the strength of the winds. This is also true of Knox's (1976) observations at Gan Island. Nearly always, however, these arguments must be made in terms of the local winds, which may only partially be indicative of the large-scale wind field responsible for setting up the sea-surface and the pressure gradients.

The equatorial surface jet, identified in the historical data by Wyrtki (1973), is a narrow, approximately 500 km wide, equatorially-trapped flow to the east during the transition periods between the northeast and southwest monsoons. The historical ship-drift observations show a maximum strength ($\approx 0.5\ \mathrm{m\,s^{-1}}$) between 60° E and 90° E. Swallow (1967) observed a strong equatorially-convergent surface flow at 67°30′ E in mid-April 1964. This flow, shown in Fig. 31.1, extended to 100 m depth and appears to have been preceded (in March) by a surface flow to the west, during a well-developed northeast monsoon.

Knox's (1976) observations at Gan Island (75° E, 0°) extend over nearly two years. These data, consisting of bi-weekly profiles of horizontal velocity to 200 m depth, show a strong flow to the east, apparently due to the winds from the west in April–May and again in October–November, the transition periods in the monsoons.

Knox has estimated the local force balance and finds that either nonlinearities or pressure gradients are required to obtain a balance. The jets persist after the local winds have subsided, which is also consistent with non-local effects. A simple model of the equatorial jet was put forward by O'Brien and Hulburt (1974). In this model, a zonal eastward wind is applied impulsively, producing a rapidly accelerating zonal jet. This eastward flow becomes nonlinear and the net transport of water

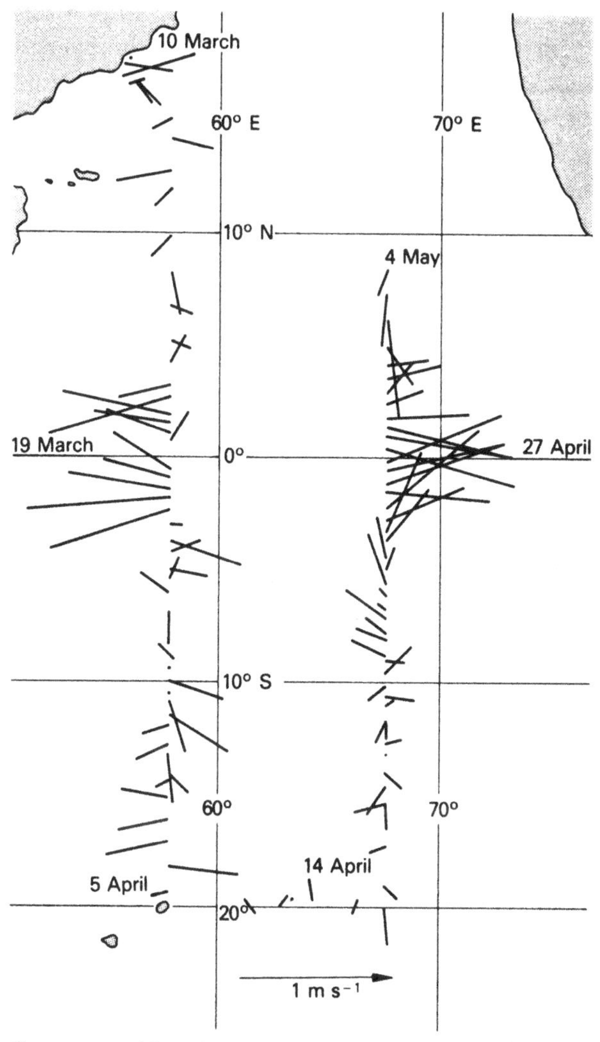

Fig. 31.1. Currents at 10 m depth relative to currents at 200 m depth. (After Swallow, 1967.)

across the basin results in an upward tilt of the sea surface from west to east and a corresponding zonal pressure gradient. Wyrtki has documented the variation of the sea level, apparently in response to the equatorial jet, so that the rudiments of the model are consistent with the observations.

There are considerably fewer observations of the flow below the thermocline. In 1964, Swallow (1967) observed a vigorous undercurrent at 58° E (in March) and 67°30′ E (late April) with westward flow ($\approx 0.2\ \mathrm{m\,s^{-1}}$) at 67°30′ E, although not at 58° E. In June, the 200 m westward flow was established at 58° E. In mid-June, sections at 58° E, 60°E and 67°30′ E all showed a narrow westward flow at 200 m still present near 61° E toward the end of August. Whether this flow extended to significantly greater depths was not known.

A section north of the Seychelles at 55° E from 3° S to 3° N was occupied repeatedly during the spring of 1975 and again in 1976 (Firing *et al.*, 1977). The data consist principally of relative-velocity profiles to 250 m depth with a reference velocity estimated from the ship's drift, using satellite navigation. In 1975, there was a persistent meridional circulation and a pronounced asymmetry across the Equator with a weak and intermittent undercurrent in the zonal flow. The region of westward flow was confined to the north of the Equator, in some cases extending from the surface to 250 m depth. The shear beneath the thermocline was weak with the eastward flow penetrating through the thermocline. In 1976 the structure was markedly different–the shear at the Equator was strong and began in April, and there was a symmetrical westward flow near the Equator at 200 m depth. This westward flow extended into the southwest monsoon well after the undercurrent had disappeared. North of the Equator this flow was significantly weaker. These data suggest that there may be considerable interannual variability, not only in the undercurrent, but in the deeper circulation as well. There is very little data on the flow below 200 m depth. Neyman (1964) has reported a narrow equatorial flow to the west between 700 and 800 m in the eastern basin during the summer Monsoon. Neyman also reports a broad eastward flow extending to 200–300 m depth, which he identified with the North Equatorial Current, perhaps similar to the data obtained by Leetmaa and Stommel in 1975 (personal communication). Neyman suggests that the apparent two-layer structure reverses with the season, although he does not report any data from the 700 m level in the winter monsoon season.

31.2 Recent observations

In 1976, a pilot experiment was undertaken near 53° E in which moored current-meter and profile observations were combined to examine the spatial and temporal scales of the equatorial current system. The positions of the observation stations are shown in Fig. 31.2. These observations were begun in May. The profiling extended over one month. The current-meter observations extended over 8 months, until January 1977.

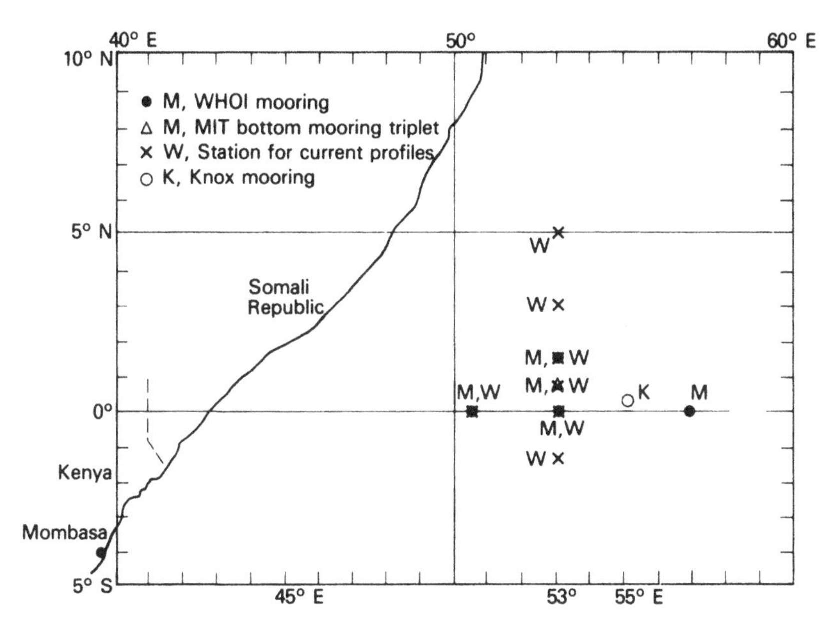

Fig. 31.2. Station positions for the INDEX-76 Equatorial Experiment. The figure shows the positions of the Woods Hole Oceanographic Institution (WHOI) moorings, the Massachusetts Institute of Technology (MIT) moorings, the stations at which current profiles were taken and the Knox mooring.

The current-meter array was exploratory and was designed with a 'midlatitude' intuition, essentially the dominance of large vertical scales at long timescales. The velocity profiles have shown that this view is inappropriate – there are periods of time of at least a month when vertical scales of a few hundred meters are dominant. The profile data indicate that the zonal flow in the equatorial region near 53° E is composed of several subsurface jets, similar in character to the undercurrent but generally flowing to the west (Luyten and Swallow, 1976). A time series of the zonal component of flow is shown in Fig. 31.3 from the equatorial station. The most energetic of these subsurface jets lie at 200 m and 750 m depth, with flow to the west of about 0.25 m s^{-1} (Luyten, 1978). The meridional component of flow in the jets varies over the month of observations, suggesting a meandering of the jet, although there is little change in the basic structure of the jets. The jet at 200 m depth was accelerating over most of the month, roughly doubling its amplitude in the interval, as shown in Fig. 31.3. No such acceleration was observed in the 750 m jet.

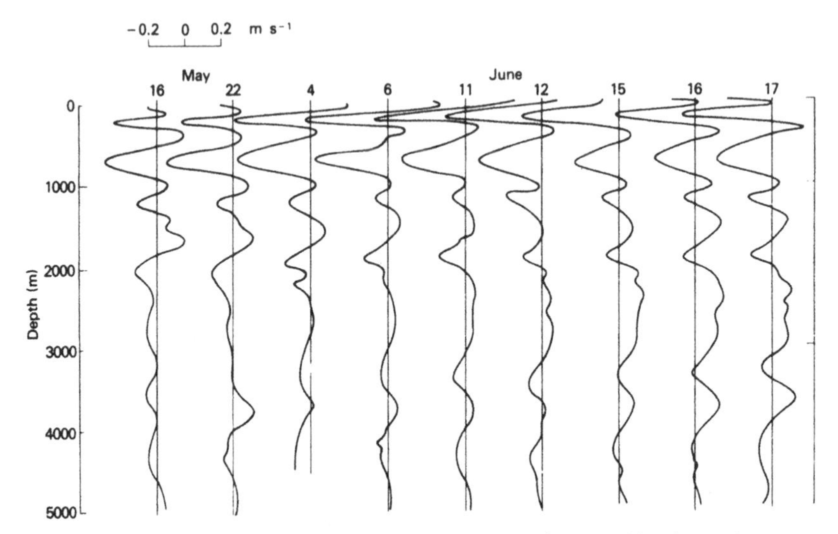

Fig. 31.3. Eastward components of velocity in vertical profiles from the equatorial site at 53° E. The data on which each profile was taken is given above the profile.

Representative profiles from each of the stations, shown in Fig. 31.4, demonstrate that the jets are trapped within 3° of the Equator. The jet at 750 m depth changes direction between 1°30′ N and 3° N, although it is clear that the main characteristic of the flow near the Equator, the dominance of small vertical scales, has disappeared by 5° N.

The principal overlap between the mooring array and the profile observations is at the 200 m level. Two of the equatorial moorings have complete records at 200 m; the mooring at 53° E, 1°30′ N had a sticky compass so only a speed record was obtained there. On the fourth mooring, at 57° E, the instrument was at 550 m depth. The low-passed data, shown in Fig. 31.5, illustrates the 200 m jet along the Equator. The flow is nearly identical to the profile data over the common interval. The current-meter data shows that the jet breaks down in mid-July. Fig. 31.6 compares the speeds from the three instruments at 200 m and shows that the breakdown is nearly simultaneous at the three locations. The wind data in this region does not show a strong event in the atmospheric circulation which might be responsible for this breakdown. There is no other data to suggest how, let alone why, this jet breaks down.

The interpretation of the instrument at 57° E, 550 m depth is ambiguous. By comparison with the profile data at 53° E and 50°30′ E in Fig. 31.4, the flow at 550 m is weakly to the east. Thus, if the velocity field at

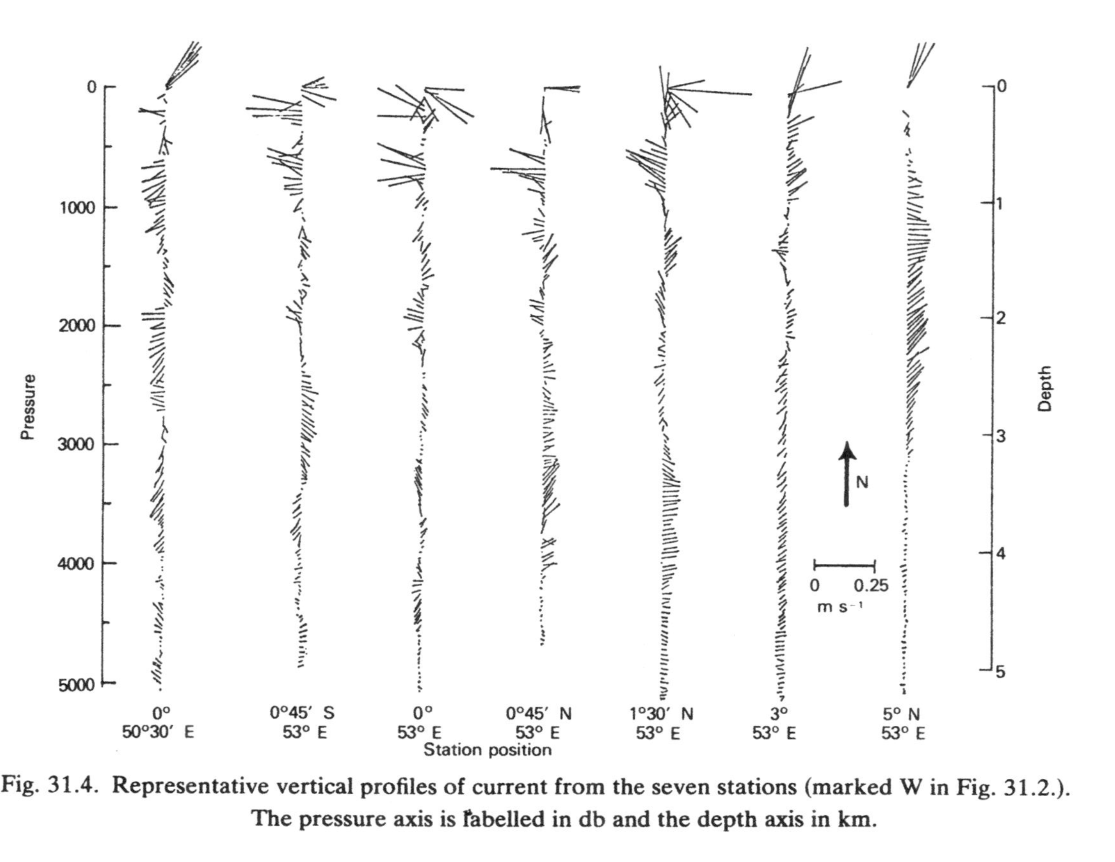

Fig. 31.4. Representative vertical profiles of current from the seven stations (marked W in Fig. 31.2.). The pressure axis is labelled in db and the depth axis in km.

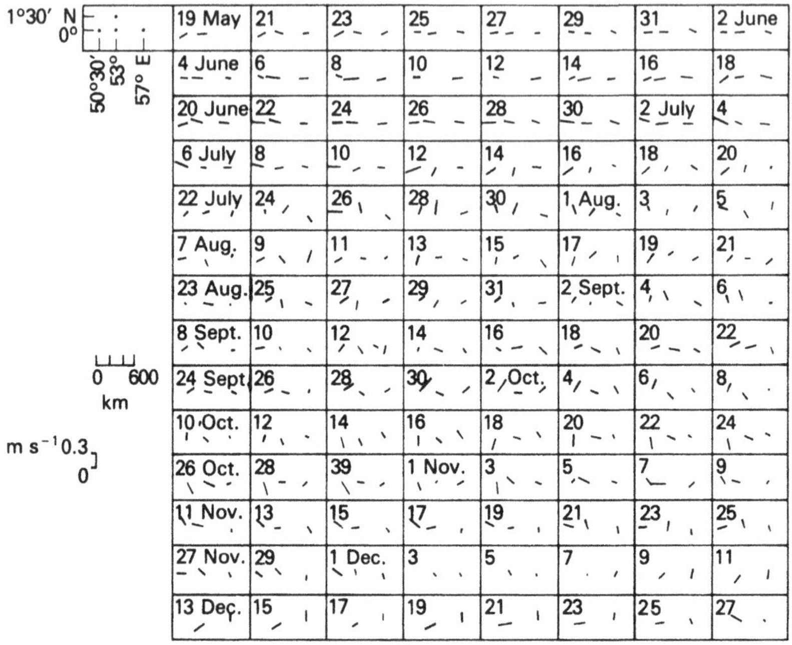

Fig. 31.5. Low-passed current from the moored current-meter array. The geographical position of each mooring is shown in the upper left panel. Each panel represents 2-day averaged current and the date on which values were recorded is given in the upper left corner of each panel. No current readings were obtained from the station at 53° E, 1°30′ N because of a faulty compass. Note that the instrument at the easternmost mooring, 57° E, is at 551 m depth. The equatorial stations at 50°31′ E and 53° E are at 203 m and 201 m depth, respectively.

these levels has a long zonal scale, the flow at 57° E should also be weakly to the east. The fact that it is to the west suggests a shorter zonal scale, or a significant change in the vertical structure – neither of which are indicated by the available historical data. An investigation of this question by a combined mooring and profile experiment is planned for 1979.

31.3 Theoretical considerations

The equatorial regions are dynamically unique due to the relative smallness of the Coriolis force, which gives rise to a number of model phenomena trapped in the vicinity of the Equator. Lighthill (1969) observed that energy and momentum can be redistributed rapidly in this region by the equatorial Kelvin and planetary waves. The response to the monsoon in the ocean interior, and specifically the rapid onset of the Somali Current, was rationalized by Lighthill as the accumulation of

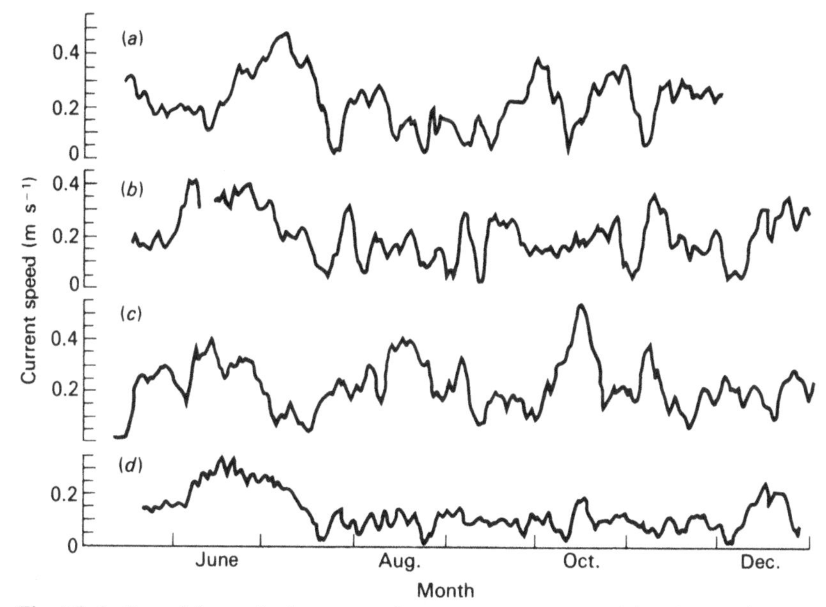

Fig. 31.6. Speed from the low-passed current-meter record for the stations at: (*a*) 0°, 50°30′ E (203 m depth); (*b*) 0°, 53° E (201 m depth); (*c*) 1°30′ N, 53° E (200 m depth); and (*d*) 0°, 57° E (551 m depth).

northward momentum carried to the western boundary by the free planetary waves. One can take the opposite point of view, and examine the forced equatorial response to periodic disturbances at the annual period. Such a model was constructed by Wunsch (1977) in order to rationalize the persistent small-vertical-scale variability seen in the profile data. The downward propagation of energy at the annual period is so slow, a few hundred meters per year, that one might argue that the energy would be dissipated by other processes before reflection from the bottom occurred. Thus, one might be justified in considering an infinitely deep ocean. Wunsch further assumes that the ocean beneath the surface mixed layer, which is directly forced by the wind, is driven by a convergence or divergence associated with the extra-equatorial Ekman flux.

The linear potential vorticity equation appropriate to the problem of the oceanic response to periodic forcing at a frequency ω, can be written

$$\mathrm{i}\omega\left[\nabla_1^2 p + (f^2-\omega^2)\frac{\partial}{\partial z}\left(\frac{1}{N^2}\frac{\partial p}{\partial z}\right) - \left(\frac{2\beta f}{f^2-\omega^2}\right)\frac{\partial p}{\partial y}\right] - \left(\frac{f^2+\omega^2}{f^2-\omega^2}\right)\beta\frac{\partial p}{\partial x}$$

$$= \mathrm{i}\omega\left[\boldsymbol{\nabla}\cdot\boldsymbol{\tau} - \left(\frac{2\beta f}{f^2-\omega^2}\right)\tau^{(y)}\right] - \left(\frac{f^2+\omega^2}{f^2-\omega^2}\right)\beta\tau^{(x)} - f\,\mathrm{curl}\,\boldsymbol{\tau},$$

where $p\,e^{-i\omega t}$ is the pressure, $f=\beta y$, $\nabla_1^2=\partial^2/\partial x^2+\partial^2/\partial y^2$, N is the Brunt–Väisälä frequency, and x, y, z have their usual meanings. The forcing due to the wind stress is modelled as a near-surface body stress, $\boldsymbol{\tau}=(\tau^{(x)}, \tau^{(y)})$. A complete derivation of this equation can be found in Wunsch (1977) and Moore and Philander (1977). For forcing at sufficiently long periods, the potential vorticity equation is elliptic in character, so that one expects solutions which are periodic in the vertical and trapped meridionally, and vice versa. The simple relation between the meridional and vertical scales can be found from a balance of the most highly differentiated terms

$$\frac{p_{yy}}{(f^2/N^2)p_{zz}}\approx\frac{p/y_0^2}{\beta^2 y_0^2 p/N^2h^2}\sim 1,$$

where y_0 is the meridional and h, the vertical scale. This balance suggests

$$y_0\approx\left(\frac{N^2h^2}{\beta^2}\right)^{1/4}\approx\left(\frac{gh_e}{\beta^2}\right)^{1/4},$$

where h_e is an equivalent depth and y_0 is known as the equatorial radius of deformation.

If one makes the assumption that both the pressure and the forcing function are zonally periodic† and that the meridional and vertical dependence can be separated, one obtains a homogeneous equation for the meridional dependence, and an inhomogeneous equation for the vertical structure, related through a separation constant. For the case of free waves (τ taken as zero) in a finite ocean depth, the vertical equation is homogeneous and the separation constant is determined by the conditions at the vertical boundaries, yielding discrete vertical modes. The separation constant is then related to the equivalent depth for the mode, giving a meridional scale, the equatorial radius of deformation. The dispersion relation for the free waves in a given vertical mode is determined by the meridional problem (homogeneous equation and boundary conditions).

For the forced problem the separation constant, or equivalently a radius of deformation, R_D, which defines the equatorial trapping scale, is fixed by the meridional problem, giving discrete relations between the imposed frequency and zonal wave length, and the meridional mode number, n. This radius of deformation defines an equivalent depth $R_D\approx(gh_e/\beta^2)^{1/4}$ which controls the vertical scale of the solution to the inhomogeneous vertical equation. Each meridional model is charac-

† An arbitrary zonal distribution of stress can be synthesized from its Fourier coefficients.

terized by a turning latitude, $y = (2n+1)^{\frac{1}{2}}(gh_n/\beta^2)^{1/4}$, beyond which the solution decays exponentially, and a meridional symmetry: for n odd, the meridional and zonal velocity are respectively odd and even in latitude, and vice versa for n even. Thus it is only for the relatively low meridional modes that the flow is equatorially trapped, with a correspondingly small vertical scale.

The inhomogeneity in the vertical equation represents the vertical velocity arising from the wind-induced convergence or divergence in the near-surface layer. It is only through this vertical velocity that wave energy is propagated into the interior. The zonal scales of symmetry of this vertical velocity determine zonal scale and meridional symmetry of the response. The character of the response depends upon whether the perturbations move to the east or the west–the westward-travelling disturbances can force a westward-propagating Rossby wave. For eastward-moving disturbances, the forced response consists of an equatorial Kelvin wave, internal waves of very small vertical scale and a variety of waves propagating energy away from the equatorial region. The equatorially-trapped response comes primarily from the westward-moving components of the disturbance.

Wunsch has examined the character of the solution for a particular choice of zonal wavenumber (approximately 3000 km) and a symmetric distribution of vertical velocity, varying slowly in the meridional direction. The stratification is chosen as exponential. The solution, given by Wunsch (1977), is shown in Fig. 31.7, and includes the odd meridional modes (appropriate to symmetric forcing) up to $n = 13$, although the structure is dominated by the first ($n = 1$) mode.

The propagation of energy into the ocean is associated with an upward propagation of phase, corresponding to a downward group velocity. Thus, the zonal jets shown in Fig. 31.7 will progress upward.

There is a tantalizing similarity between the solution presented by Wunsch, shown in Fig. 31.7, and the zonal velocity structure at the equatorial site, Fig. 31.3. Indeed the model rationalized persistent small-vertical-scale variability at depths well below the thermocline. The profiles from the meridional section, Fig. 31.4, also exhibit some similarity to the model, in that the structures, especially at 750 m depth, reverse sign near 3° N, and disappear by 5° N where larger vertical scales dominate the flow regime. No vertical phase propagation was found from the time series of profiles in the one month period from 16 May to 17 June 1976 (Fig. 31.3), however, although a shift of 360°/12 = 30° per month would be marginally detectable in such a complex field. Many of the

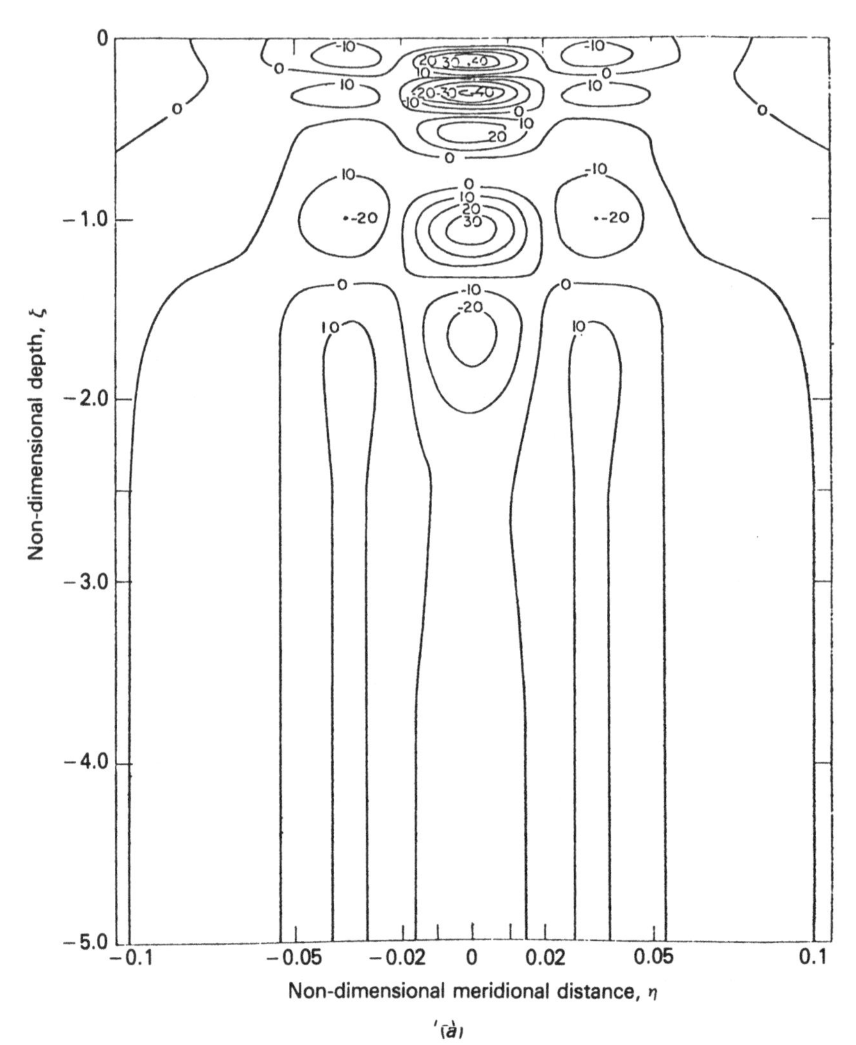

Fig. 31.7. Forced equatorial response to westward-travelling symmetric distribution of vertical velocity, corresponding to a zonal wavelength of approximately 3000 km, annual period and zero surface stress ($\tau = 0$), of (*a*) zonal velocity; (*b*) meridional velocity; and (*c*) zonal profile on $\eta = 0$. In dimensional terms, velocity contours are labelled in 10^{-2} m s^{-1}, the meridional coordinate runs from -6 to $+6$ degrees and the depth from 0 to 5000 km. (After Wunsch, 1977.)

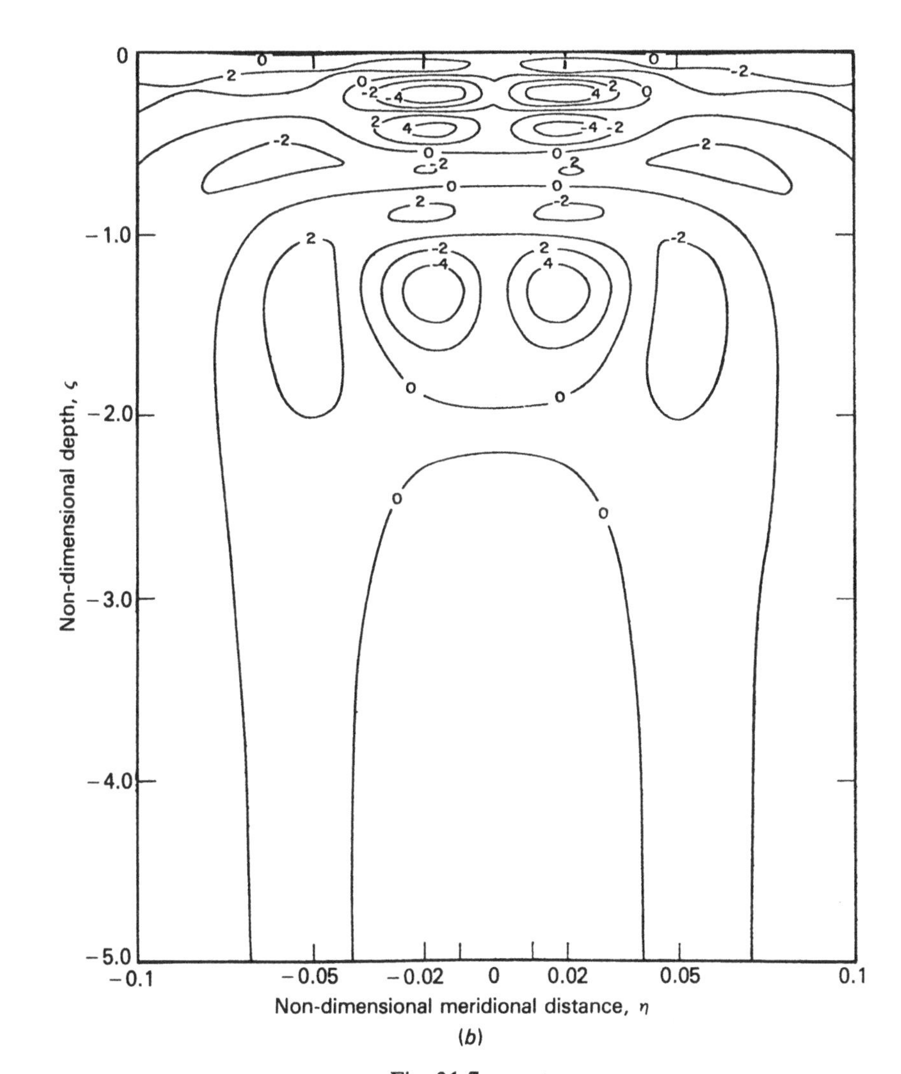
0
−1.0
−2.0
−3.0
−4.0
−5.0
Non-dimensional depth, ς
−0.1
−0.05
−0.02
0
0.02
0.05
0.1
Non-dimensional meridional distance, η
(b)

Fig. 31.7—*cont.*

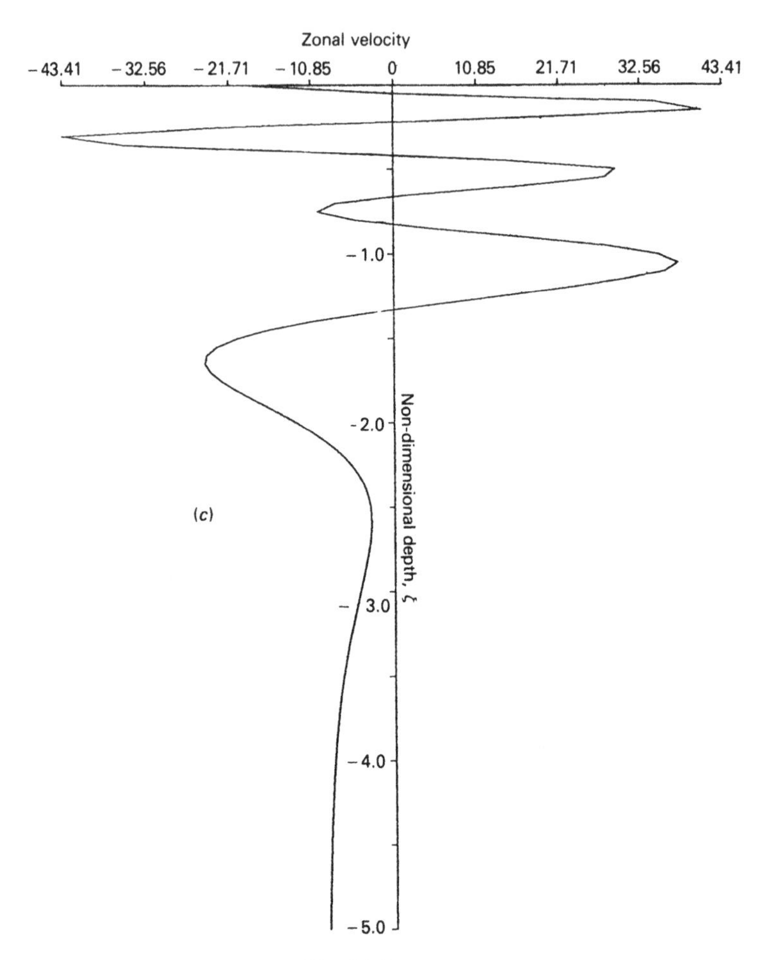
Zonal velocity
-43.41
-32.56
-21.71
-10.85
0
10.85
21.71
32.56
43.41
-1.0
-2.0
3.0
-4.0
-5.0
Non-dimensional depth, ζ
(c)

Fig. 31.7—*cont.*

theoretical limitations of the model were discussed by Wunsch, including the difficulties with the infinitely deep ocean and vanishingly small buoyancy frequency. Similarly, nonlinearities play an important role in equatorial models, and these must control the development of the flow with time. In addition, the actual forcing will contain a spectrum of zonal and temporal scales of variability giving rise to a correspondingly complicated response.

The most difficult point is, however, provided from the long-term mooring experiment, which showed that the 200 m jet, observed in both the profiles and current meters at 50°30′ E and 53° E, broke down in mid-July. This large zonal-scale phenomenon did not reappear in the current-meter data but was rather replaced by a highly variable smaller-scale eddy field. The underlying idea of the slow vertical propagation of a forced wave rationalizes the persistence of small-vertical-scale variability and may apply best to the deep water, with a more nonlinear model for the upper ocean driving the deep water.

31.4 Conclusions

The recent velocity profile data in the western Indian Ocean has demonstrated the existence of very low frequency variability of small vertical scales, trapped meridionally at the Equator. Observationally, there is as yet no clear understanding of the origin of these structures, their long-term recurrence or their stability, although previously there have been reports of the westward flow at 200 m and 750 m depth. Their role in the overall response of the Indian Ocean to the monsoon forcing is an open question which, along with the other questions, can only be answered by further observations. The model proposed by Wunsch, of the forced equatorial response at the annual period propagating down into the ocean, provides a theoretical rationalization for the small vertical scales at long timescales. This model is complementary to Lighthill's analysis of the oceanic response to the southwest monsoon. Because a spectrum of temporal and zonal scales is necessarily involved, together with significant nonlinear modifications, the oceanic response must be more complex than either extreme model suggests. The possible insights into the underlying mechanisms help in the design of further observational programmes to unravel the oceanic response to the monsoon. An experiment is at present being carried out to look for similar structures in the equatorial Pacific Ocean.

The observations from field programmes reported here were possible only through the generous and continuing support of the Ocean Science and Technology Division of the Office of Naval Research (contract N00014-76-C-0147 NR 083-400), and the Office of Climate Dynamics of the National Science Foundation (grant ATM-76-02196).

References

Firing, E., Leetmaa, A. and Stommel, H. (1977) INDEX Data Report: La Curieuse, 1976. Technical Report (Unpublished document).

Knox, R. A. (1976) On a long series of measurements of Indian Ocean Equatorial Currents near Addu Atoll. *Deep-Sea Res.*, **23**, 211–21.

Kort, J. G., Neyman, V. G. and Titov, V. B. (1975) Equatorial currents of the Indian Ocean in a period of the winter equinox. *Dokl. Akad. Nank.* (SSSR), **220**, 6, 1306–9.

Lighthill, M. J. (1969) Dynamic response of the Indian Ocean to the onset of the southwest monsoon. *Phil. Trans. Roy. Soc. London*, **A265**, 45–92.

Luyten, J. R. (1978) Subsurface jets in the western Indian Ocean. (Unpublished manuscript).

Luyten, J. R. and Swallow, J. C. (1976) Equatorial undercurrents. *Deep-Sea Res.*, **23**, 1005–7.

Moore, D. and Philander, S. G. H. (1977) Modelling of the tropical ocean circulation. *The Sea*, ed. E. D. Goldberg, Vol. 6, pp. 319–61, Wiley-Interscience, New York.

Neyman, V. G. (1964) On the structure of zonal currents in the equatorial region of the Indian Ocean (abstract). *Okeanologiya*, **4**, 5, 920.

O'Brien, J. J. and Hulbert, H. E. (1974) An equatorial jet in the Indian Ocean, theory. *Science*, **184**, 1075–7.

Philander, S. G. H. (1973) The Equatorial Undercurrent: Measurements and theories. *Rev. Geophys. Space Phys.*, **2**, 513–70.

Swallow, J. C. (1967) The equatorial undercurrent in the western Indian Ocean in 1964. *Studies in Tropical Oceanography* (Miami), **5**, 15–36.

Taft, B. A. and Knauss, J. A. (1967) The equatorial undercurrent of the Indian Ocean as observed by the Lusiad Expedition. *Bull. Scripps Inst. Oceanogr.*, **9**, 1–163.

Wunsch, C. (1977) Response of an equatorial ocean to a periodic monsoon. *J. Phys. Oceanogr.*, **7**, 497–511.

Wyrtki, K. (1973) An equatorial jet in the Indian Ocean. *Science*, **181**, 262–4.

32

Sea temperature variations in the northeastern Arabian Sea in relation to the southwest monsoon

V. RAMESH BABU, L. V. GANGADHARA RAO
AND V. V. R. VARADACHARI

The structure and variability of the temperature in the upper layers (0–200 m) of the northeast Arabian Sea have been studied in relation to the southwest monsoon over the Indian peninsula. Data from repeated bathythermograph (BT) lowerings at three-hourly intervals at a station around 18° N, 67° E during the periods 28 to 31 May 1973 and 29 June to 2 July 1973 (representing conditions before and after the onset of the monsoon) have been used for these studies. Mean temperature differences and autocorrelation coefficients have been computed and analysed for assessment of short-period variations.

The mixed layer was well developed with an average thickness of about 70 m during the monsoon period, while its thickness was about 40 m during the premonsoon period. The thermal structure reveals the presence of internal waves in the thermocline.

Studies of short-period variations indicate that changes of temperature occur in the upper part of the thermocline during the premonsoon and monsoon periods; however, the maximum temperature changes for the monsoon period are generally larger (ranging from 0.8 °C for a 3-hour interval to 1.2 °C for 9-hour and 12-hour intervals) than those for the premonsoon period, which are of the order of 0.8 °C for time intervals up to 12 hrs. Autocorrelation values reveal significant correlation (at the 1 % level) in the upper part of the mixed layer for intervals greater than 3 hours, the feature being more prominent during the monsoon period. Within the thermocline, significant correlation has been found around 170 m depth for intervals greater than 6 hours during the premonsoon period and around 80 m depth for intervals greater than 3 hours during the monsoon period. The magnitude of the temperature changes do not seem to bear any definite relationship to the degree of correlation.

32.1 Introduction

Sea temperature varies with depth and time as a result of several factors, such as insolation, wind mixing, internal waves, advection of waters of different temperatures, tidal currents, etc. (La Fond, 1954). Changes in sea temperature are generally significant in the upper few hundred metres only. Notable among the earlier works describing the nature of temperature variability in the upper layers of the sea, are those reported by La Fond and Moore (1960 and 1972) for the regions off the coasts of Southern California and Andhra.

Studies based on the data collected during the International Indian Ocean Expedition (IIOE) have revealed that the interaction of the southwest monsoon with the Arabian Sea is quite strong and significantly affects the thermal structure of the sea, especially in the upper layers (Colon, 1964; Sastry and D'Souza, 1970; Wyrtki, 1971). A detailed investigation of the variability of temperature in the upper layers of the Arabian Sea in relation to the southwest monsoon is, therefore, of special interest.

During May–July 1973, a joint Indo-Soviet Monsoon Experiment (ISMEX) was conducted in the northwestern Indian Ocean (including the Arabian Sea) to gather more information on the meteorological and oceanographic conditions associated with the various phases of the southwest monsoon as a prelude to the detailed study to be carried out during the International Monsoon Experiment (MONEX) planned for 1979.

In this chapter, a comparative study of the vertical temperature structure in general, and short-period temperature variations in particular, in the upper layers of the northeastern Arabian Sea is attempted using the bathythermograph (BT) data collected near a buoy positioned in the area during the ISMEX.

32.2 Data and methods

Data from repeated BT lowerings at three-hourly intervals, carried out on board the Soviet research vessel *Priliv* near a buoy station at 18° N,

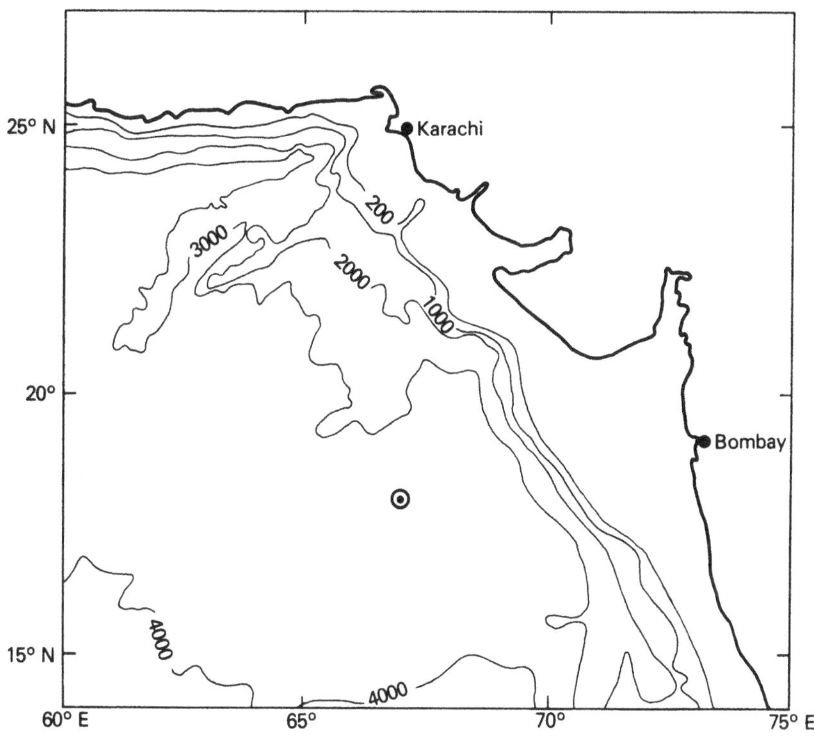

Fig. 32.1. Station location (lower left of centre). The contours of ocean depth are labelled in metres.

67° E (Fig. 32.1), are used in the present study. Data collected during the periods 28 to 31 May and 29 June to 2 July are chosen to represent the premonsoon and monsoon conditions respectively. With the processed BT data, vertical time sections of temperature up to 200 m depth are prepared for the two periods to compare and evaluate the gross changes in the vertical temperature structure.

Methods of power spectrum and harmonic analysis are considered to be appropriate for describing short-period variations when a long series of data is available. But, in the present study, as the data are for a short duration (3 to 4 days for each situation) two different methods (La Fond and Moore, 1960), employing mean differences and autocorrelation coefficients, have been used for assessing the short-period temperature variations. The computations have been carried out using the BT data processed for 10 m depth intervals.

In the differences method, the average change of temperature in a given time interval at a particular depth is computed from

$$D_\lambda = \frac{\sum_{i=1}^{N-\lambda} |(X_i - X_{i+\lambda})|}{N-\lambda},$$

where D_λ is the average magnitude of the temperature differences (°C) in a given time interval, X_i and $X_{i+\lambda}$ are the temperatures (°C) at the beginning and end of a given time interval, λ is the time interval used (in the present case 1, 2, 3 and 4 corresponding to 3-hours, 6-hours, 9-hours and 12-hours intervals respectively), and N is the total number of temperature readings at a given depth ($N = 28$ for the first period, $N = 31$ for the second period in the present case).

In the second method, the general dependence of the temperature values at one time on the values observed at another time for a particular depth, is evaluated by computing the autocorrelation coefficient, R_λ, given by

$$R_\lambda = \frac{(N-\lambda) \sum_{i=1}^{N-\lambda} X_i X_{i+\lambda} - \sum_{i=1}^{N-\lambda} X_i \sum_{i=1}^{N-\lambda} X_{i+\lambda}}{\left[(N-\lambda) \sum_{i=1}^{N-\lambda} X_i^2 - \left(\sum_{i=1}^{N-\lambda} X_i\right)^2\right]^{\frac{1}{2}} \left[(N-\lambda) \sum_{i=1}^{N-\lambda} X_{i+\lambda}^2 - \left(\sum_{i=1}^{N-\lambda} X_{i+\lambda}\right)^2\right]^{\frac{1}{2}}}.$$

32.3 Results and discussion

Vertical time sections of temperature for the premonsoon and monsoon periods are shown in Figs. 32.2*a* and *b* respectively. A comparison of the two sections reveals that the mixed layer† is well developed with an average thickness of about 70 m during the monsoon period, while its thickness is only about 40 m during the premonsoon period. A significant fall (about 1° C) of the temperature of the surface mixed layer from the premonsoon period to the monsoon period is noticed. These features, which could be attributed to the effects of strong winds, clouding and rain associated with the monsoon, are in agreement with those reported earlier by Ramesh Babu *et al.* (1976) and Gangadhara Rao *et al.* (1976). The spreading of thermocline in its lower part, resulting in relatively weak thermal gradients at 100 to 150 m depth, is significantly different during

† In the present study, a vertical temperature gradient of 2 °C/(25 m) is taken as the lowest limit of the thermocline, after Schott (La Fond and Rao, 1954) and the warm top layer of relatively low gradients of temperature above the thermocline is referred to as the mixed layer.

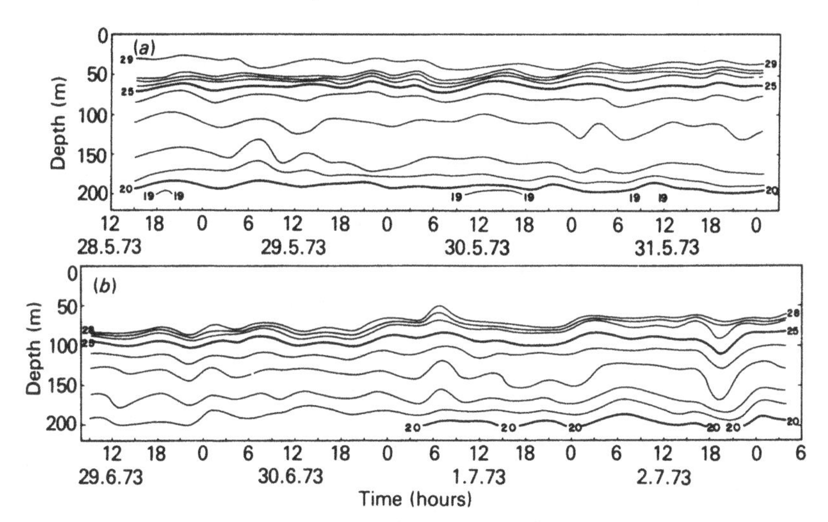

Fig. 32.2. Vertical time sections of temperature for: (*a*) 28 to 31 May 1973; and (*b*) 29 June to 2 July 1973. Contours are labelled in °C.

the premonsoon and monsoon periods. The presence of internal waves is indicated by the regular wavy pattern exhibited by the isotherms of the thermocline during both periods.

The results of a statistical analysis of the two sets of data describing the short-period temperature variations during the premonsoon and monsoon periods are presented in Figs. 32.3 and 32.4 respectively.

Figs. 32.3*a* and 32.4*a* show the average temperature differences (D_λ) for intervals of 3 hours, 6 hours, 9 hours and 12 hours. The effect of instrumental (and scaling) errors on the variation depicted by this method can be taken as very small because only the differences are considered here and not the actual temperatures. The width of the solid line is proportional to the average temperature difference, D_λ, at that depth for the time shown. The upper limit of shading represents the depth of the thermocline. The relatively wide range in the depth of the thermocline observed during the monsoon period is indicated in Fig. 32.4*a* by shading with alternate short and long diagonals.

In the mixed layer, the average temperature differences were of the order of 0.2 °C during the premonsoon period. They were less than 0.1 °C (which is the limit of accuracy of the instrument) during the monsoon period, reflecting rather steady environmental conditions which result in the persistence of a well-developed mixed layer for a longer duration during this period. Within the thermocline, the differences varied with

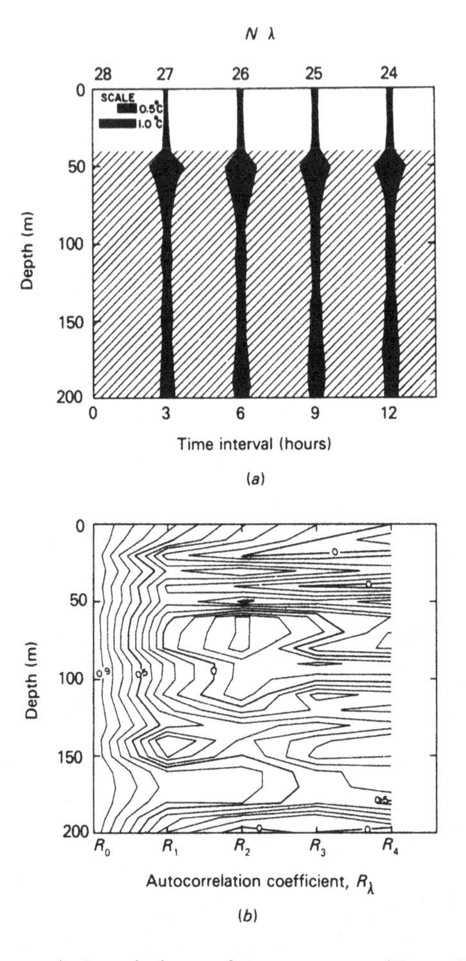

Fig. 32.3. Short-period variations of temperature 28 to 31 May 1973. (*a*) Average differences (D_λ); (*b*) autocorrelation coefficient (R_λ) for $\lambda = 0, 1, 2, 3, 4$.

depth, the maximum differences being encountered in the upper part of the thermocline (where stronger temperature gradients exist) during both periods. The maximum differences for the monsoon period were generally more (ranging from about 0.8 °C for a 3-hour interval to 1.2 °C for 9-hour and 12-hour intervals) than those for the premonsoon period, which were of the order of 0.8 °C for time intervals up to 12 hours.

A secondary maximum of average differences (of the order of 0.4 °C) was observed at around the 180 m depth, where the thermal gradients were relatively higher during both the periods (Fig. 32.2). These features, which are more prominent during the monsoon period, reflect the effect

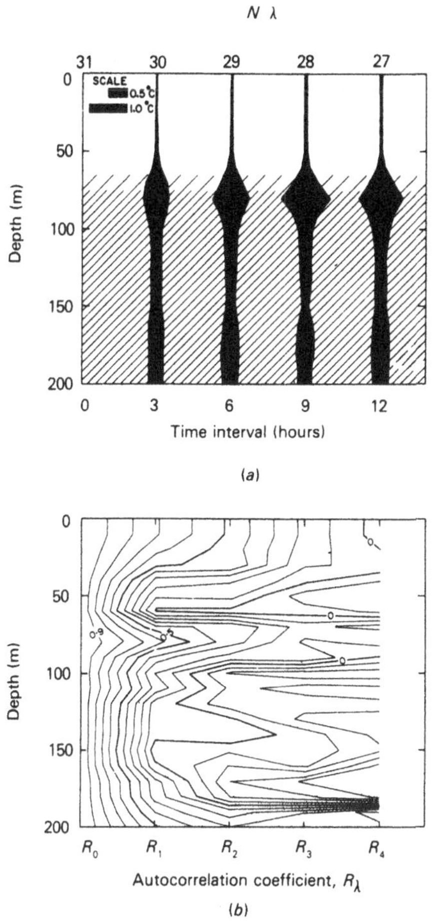

Fig. 32.4. Short-period variations of temperature 29 June to 2 July 1973. (*a*), Average differences (D_λ); (*b*) autocorrelation coefficient (R_λ) for $\lambda = 0, 1, 2, 3, 4$. $0, 1, \ldots, 4$.

of vertical oscillations associated with internal waves in the thermocline region (La Fond and Rao, 1954).

Analyses of autocorrelation coefficients (R_λ) for the two periods are shown in Figs. 32.3*b* and 32.4*b*. Autocorrelation values R_1 to R_4 have been contoured using linear interpolations with respect to time and depth. For a significant (1%) level of correlation, R_λ works out to be about 0.5 for both sets of data. The areas with less significant values are stippled in these figures.

In the upper part of the mixed layer, significant correlation between successive values was encountered for intervals of more than 3 hours, beyond which the correlation decreased with time. This feature is more prominent during the monsoon period than in the premonsoon period. A general increase in correlation was observed in the lower part of the mixed layer during the premonsoon period, while no such increase was noticed during the monsoon period. In the upper part of the thermocline, where the maximum average differences were encountered, an increase in correlation is noticed during both the periods, the feature being more prominent (with significant correlation for more than 3 hours) during the monsoon period. It is interesting to note that significant correlation was maintained for intervals much larger than 6 hours around the 170 m depth during the premonsoon period, suggesting a possible intense mixing in this layer.

The preceding discussion confirms that methods employing temperature differences and autocorrelations to describe temperature variability are quite useful, generating estimates which provide a basis for discussion of the physical processes responsible for the variability. The magnitudes of the temperature changes do not seem to be necessarily related to the degree of correlation.

The authors wish to express their thanks to Dr S. Z. Qasim, Director, National Institute of Oceanography for his keen interest in this study.

References

Colon, J. A. (1964) On interactions between the southwest monsoon current and the sea surface over the Arabian Sea. *Ind. J. Meteor. Geophys.* **15**, 183–200.

Gangadhara Rao, L. V., Ramesh Babu, V., Fernandes, A. A. and Varadachari. V. V. R. (1976) Studies on thermal structure of the northwestern Indian Ocean in relation to southwest monsoon over the Indian peninsula. *Proc. Symp. Tropical Monsoons*, Indian Institute of Tropical Meteorology, Poona, p. 219.

La Fond, E. C. (1954) Environmental factors affecting the vertical temperature structure of the upper layers of the sea. *Andhra University Memoirs in Oceanography*, **1**, 94–101.

La Fond, E. C. and Rao, C. P. (1954) Vertical oscillations of tidal periods in the temperature structure. *Andhra University Memoirs in Oceanography*, **1**, 109–16.

La Fond, E. C. and Moore, A. T. (1960) Short period variations in sea water temperatures. *Ind. J. Meteor. Geophys.*, **11**, 163–5.

La Fond, E. C. and Moore, A. T. (1972) Sea temperature variations off the east coast of India. *Ind. J. Mar. Sci.*, **1**, 63–5.

Ramesh Babu, V., Gangadhara Rao, L. V., Varkey, M. J. and Udayavarma, P. (1976) Temperature distribution in the upper layers of the northern and eastern Arabian Sea during Indo-Soviet monsoon experiment. *Ind. J. Meteor. Hydrol. Geophys.*, **27**, 291–3.

Sastry, J. S. and D'Souza, R. (1970) Oceanography of the Arabian Sea during southwest monsoon season, Part-I: Thermal structure. *Ind. J. Meteor. Geophys.*, **21**, 367–82.

Wyrtki, K. (1971) *Oceanographic atlas of the International Indian Ocean Expedition*, National Science Foundation, Washington, DC.

33

Heat budget of the north Indian oceanic surface during MONSOON-77

R. RAMANADHAM, S. V. S. SOMANADHAM AND R. R. RAO

The energy budget components at selected areas of the north Indian Ocean during different epochs of the summer monsoon period of 1977 are investigated. Marine meteorological data collected by the four-USSR-Ship-Polygon (MONSOON-77 Experiment) are used to evaluate net radiation, latent and sensible heat exchanges, and the net heat gain by the oceanic surface. Variations of these parameters are critically examined in relation to the fluctuations in surface pressure gradient force (∇p) and surface relative vorticity (ζ) which are taken as measures of atmospheric instability. Significant differences are found in the heat budget components during different monsoon periods over both the Arabian Sea and the Bay of Bengal.

33.1 Introduction

The large-scale flow patterns associated with the Indian summer monsoon have been described by many workers as being associated with the annual cycle of solar radiation and the differential heating of the land and sea surfaces. This approach broadly explains the onset and the existence of the monsoon, but it cannot explain the fluctuations and variability of monsoon activity. Attempts have recently been made to investigate the air–sea interaction over the north Indian Ocean along the track of the low-level monsoonal flow. Studies by Saha (1970, 1974) and Ellis (1952) indicated that the sea-surface temperature (SST) may have an important influence on the monsoonal flow and the associated rainfall.

Shukla (1975) suggested that an SST anomaly of a few degrees centigrade over the western part of the Arabian Sea could have a significant effect upon the monsoonal precipitation in the neighbourhood of the Indian subcontinent.

Few estimates have so far been made of fluxes from the sea surface into the atmosphere. The studies so far reported are limited to calm conditions over the sea surface. Fluxes over the Indian Ocean under disturbed conditions have not yet been estimated because of a lack of data. Rao *et al.* (1977) have attempted to evaluate fluxes over the north Indian Ocean using climatological data for normal and break-monsoon conditions. They reported that during July:

(i) the net radiation is smaller than that of the monsoon break throughout the north Indian Ocean;

(ii) higher energy losses occur over the northern Bay of Bengal than over the north Arabian Sea, the opposite being the case during the monsoon break;

(iii) there is a net loss of heat over an area of the Bay of Bengal which is larger than the corresponding area over the Arabian Sea. During the monsoon break, the northern Bay of Bengal gains heat while the northern Arabian Sea loses heat.

In the present study the energy budget components at selected areas of the northern Indian Ocean during different epochs of the summer monsoon of 1977 are estimated. Advantage is taken of the marine meteorological data collected by four Russian ships during fair and bad weather periods of MONSOON-77.

33.2 Data sources and computational procedures

The data are extracted from the Daily Weather Reports published by the India Meteorological Department. The surface parameters, (i) sea-surface temperature, (ii) dry bulb temperature, (iii) dew point temperature, (iv) wind speed and direction, (v) cloud amount and (vi) atmospheric pressure, are taken from ship observations for 0000 GMT and 1200 GMT. The stationary positions of the ships in the north Indian Ocean are shown in Fig. 33.1.

The empirical formulae developed by Laevastu and Hubert (1970) are used for calculating energy fluxes over the sea surface.

The energy balance equation is

$$Q = Q_i - (Q_b + Q_a + Q_e + Q_s),$$

which can be written as

$$Q = Q_N - (Q_e + Q_s),$$

where $Q_N = Q_i - (Q_b + Q_a)$. Here Q_i denotes the net downward flux of solar radiation, Q_b the effective upward flux of long-wave radiation, Q_a the reflected solar radiation, Q_s the sensible heat flux, Q_e the latent heat flux, Q_N the net radiation and Q the heat gain at the sea surface.

The various components of Q are generally complicated functions of time, cloud cover, latitude, wind speed, air–sea temperature difference and vapour pressure. They are computed (in W m^{-2}) from the empirical formulae:

$$Q_i = 0.0068 A_n t_d (1.0 - 0.0006 C^3),$$

$$Q_a = 0.0726 Q_i - (0.01 Q_i)^2,$$
$$Q_b = 143.7 - 0.90 T_s - 0.46 U_0 (1 - 0.0765 C),$$
$$Q_e = (0.126 + 0.037 V)(0.98 e_w - e_a), \quad \text{if } e_w - e_a > 0,$$
$$= 0.037 V (0.98 e_w - e_a), \quad \text{if } e_w - e_a \leq 0,$$

and

$$Q_s = 18.90\,(0.26 + 0.077 V)\,(T_s - T_a), \quad \text{if } T_s - T_a > 0,$$
$$= 1.45 V (T_s - T_a), \quad \text{if } T_s - T_a \leq 0.$$

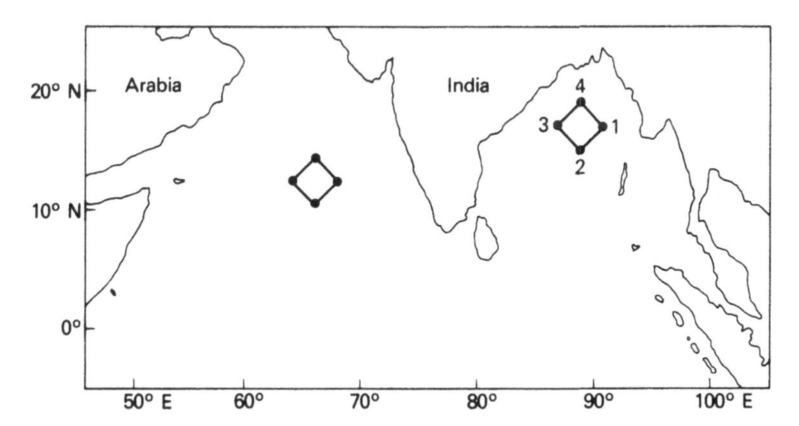

Fig. 33.1. The stationary positions of the ships in the north Indian Ocean: Bay of Bengal observation period – 13 to 19 August 1973; Arabian Sea observation period – 8 to 12 June 1977 and 30 June to 11 July 1977.

Here t_d denotes duration of daylight (minutes), A_n the noon altitude of the sun (degrees), C the total cloud amount (in tenths), e_a the water vapour pressure (mb), e_w the saturation vapour pressure at the sea surface (mb), U_0 the relative humidity (%), V the wind speed (m s^{-1}), T_s the sea-surface temperature (°C), and T_a the air temperature (°C).

33.3 Analysis and discussion

Fig. 33.2 shows the distributions of the various observed and computed parameters for the four stationary ships in the Bay of Bengal for the period 13 to 19 August 1977. On 18 August a monsoon low was reported to lie towards the east of the stationary ships. This low intensified into a depression on 19 August and moved northwest.

Fig. 33.2(*a*) refers to the ship *Umay* located at 17° N, 91° E. The net radiation, which was initially 181 W m^{-2}, started decreasing from 16 August until 18 August after which it remained constant. Simultaneously, the sea-surface pressure also decreased from 1004 mb to 1001 mb and the cloud amount increased from 9 to 10 tenths.

The latent heat flux Q_e showed an increase from 16 to 17 August, then a decrease from 17 to 18 August, followed by an increase from 18 to 19 August. These variations can be accounted for in terms of the sea-surface temperature, the dew point temperature, and the variations in the wind field. The sea-surface temperature showed an increase, whereas the dew point temperature decreased on 17 August. Simultaneously, the wind speed increased. This resulted in the increase of latent heat flux. Subsequently there is an increase of dew point temperature, decrease of sea-surface temperature and a fall in wind speed. This accounts for the decrease of flux from 17 to 18 August. From 18 to 19 August the latent heat flux showed an increase associated with the increase of wind speed and a steep fall of the dew point temperature. The sensible heat flux shows relatively little variation during this period.

The net gain of heat (Q) by the sea surface showed an increase up to 15 August and then a slow decrease until 18 August, subsequently steepening; its sign change to negative indicated that the sea surface had become an energy sink.

Figs. 33.2*b*, *c* and *d* represent the energy fluxes and the associated meteorological parameters observed during the period 13 to 19 August by the other three ships in stationary positions (*Unac* 17° N, 87° E; *Ereh* 19° N, 89° E; *Erei* 15° N, 89° E). At these locations, the net radiation showed an increase from 13 to 15 August and a decrease from 15 to 18

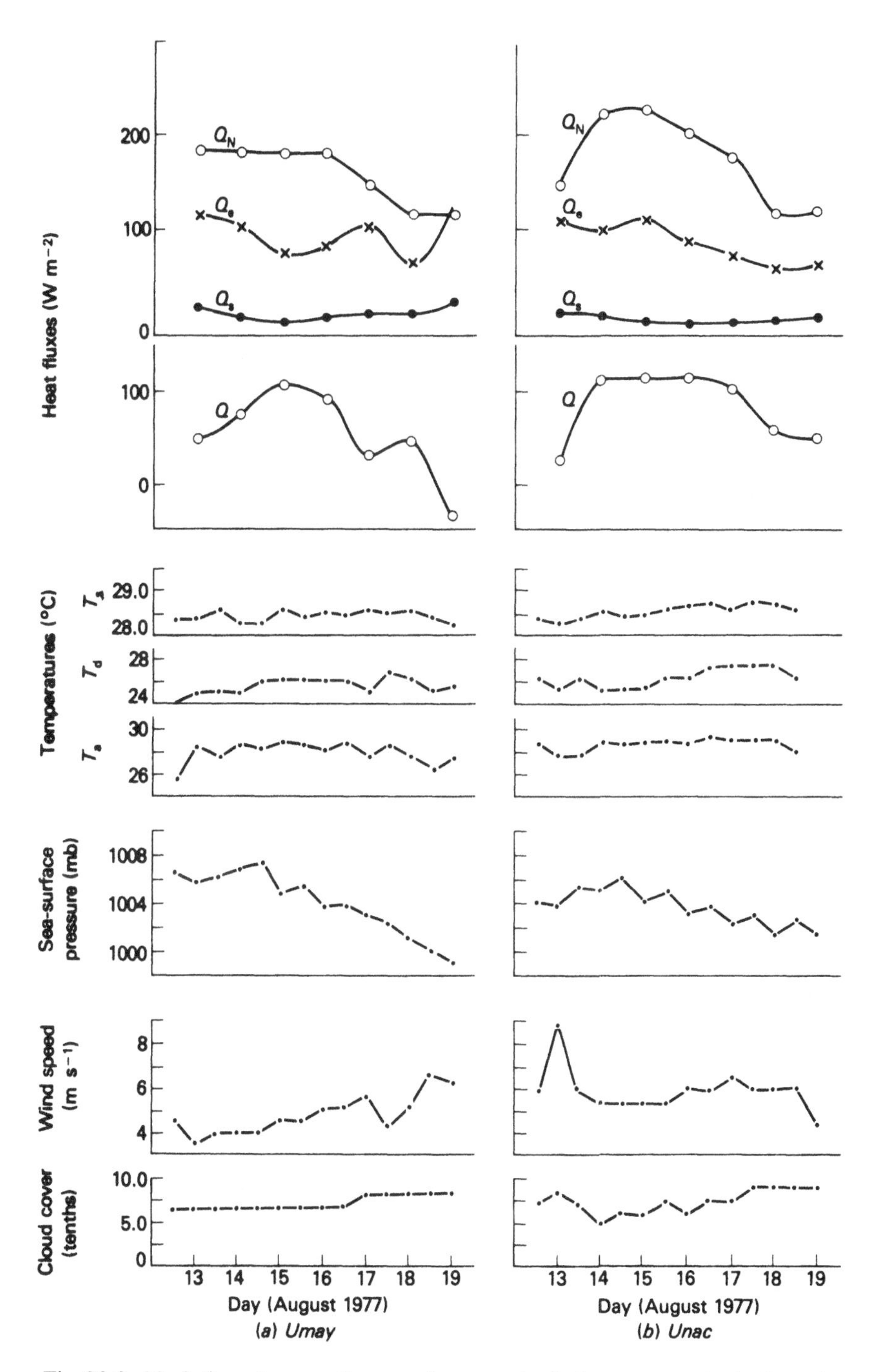

Fig. 33.2. Variation of energy fluxes and meteorological parameters in the Bay of Bengal, 13 to 19 August 1977, observed by (*a*) *Umay* (17° N, 91° E); (*b*) *Unac* (17° N, 87° E); (*c*) *Ereh* (19° N, 89° E); and (*d*) *Erei* (15° N, 89° E).

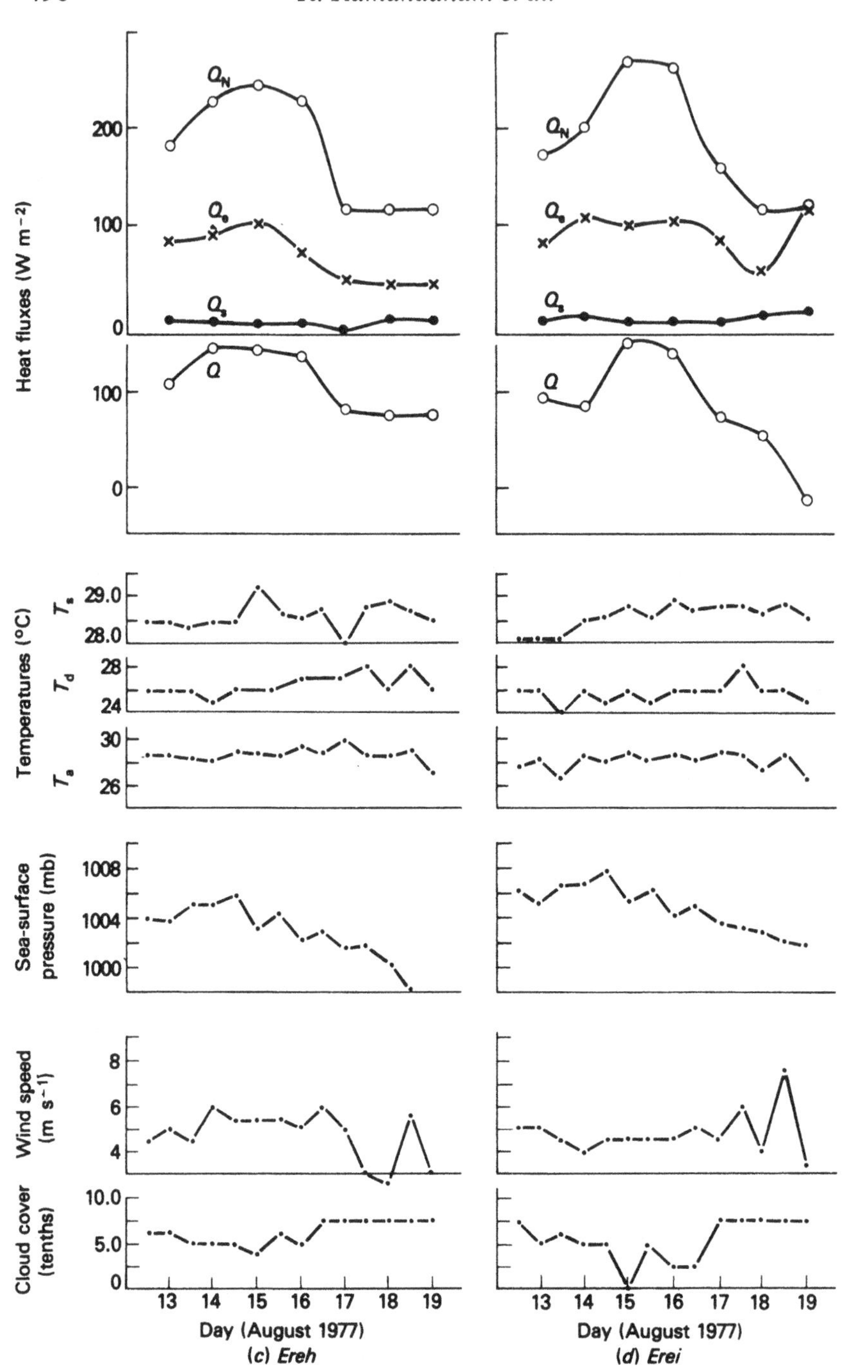

Heat fluxes (W m⁻²)
Q_N
Q_e
Q_s
Q
200
100
0
Temperatures (°C)
T_s
29.0
28.0
T_d
28
24
T_a
30
26
Sea-surface pressure (mb)
1008
1004
1000
Wind speed (m s⁻¹)
8
6
4
Cloud cover (tenths)
10.0
5.0
0
13 14 15 16 17 18 19
Day (August 1977)
(c) Ereh
(d) Erei

Fig. 33.2—*cont.*

August. The increase is associated with the increase of surface pressure and decrease of cloud amount, indicating fair weather conditions. After 15 August, a low-pressure area started developing and this resulted in an increase of cloud amount and a decrease in surface pressure.

The latent heat flux showed similar variations; whereas the sensible heat flux was more or less constant.

The net gain of heat (Q) by the sea surface also showed variations broadly similar to those at the first location. However, one important aspect to be noted is that at two locations, 17° N, 91° E and 15° N, 89° E, the net gain (Q) values were negative on 19 August, indicating that the sea surface supplied more energy to the atmosphere than it received from it. This seems to be associated with the presence of the depression lying very close to these two locations. The energy supply from the sea surface to the atmosphere is in the form of a latent heat flux which showed a sudden increase from 18 to 19 August.

The computed energy fluxes over the polygon area are presented in Fig. 33.3. Also, the surface vorticity and pressure gradient force are calculated and presented for this area. The net energy flux showed an increase from 13 to 15 August and a decrease from 15 to 19 August. The latent heat flux showed a similar variation, with a small dip on 15 August. An important aspect to be noted is that this flux increased from 18 to 19 August by about 3.4×10^{12} W. The variation of the sensible heat flux is very small and is not of much consequence. The vorticity over this area showed an increase after 17 August. This is associated with the depression lying near the polygon area. This increasing vorticity is responsible for the increasing vertical transport of heat flux from the sea surface to the atmosphere.

Table 33.1 shows the fluxes Q_N, $Q_e + Q_s$, Q, and also the percentage of outgoing to net incoming flux. Ship *Umay*, which showed the highest percentage of this ratio ($(Q_e + Q_s)/Q_N$), appears to be in the region where energy transport is high, leading to the intensification of the low-pressure system.

Figs. 33.4*a*, *b*, *c* and *d* illustrate the variation of energy fluxes and the associated meteorological parameters observed over the Arabian Sea from 8 to 12 June 1977 by the four Russian ships which were maintained at stationary positions in the form of a polygon as shown in Fig. 33.1. During this period a severe cyclonic storm is observed towards the northeast of the polygon. The system started as a low-pressure area on 8 June, intensified into a depression on 9 June and became a severe cyclonic storm from 10 June, moving towards the Oman coast.

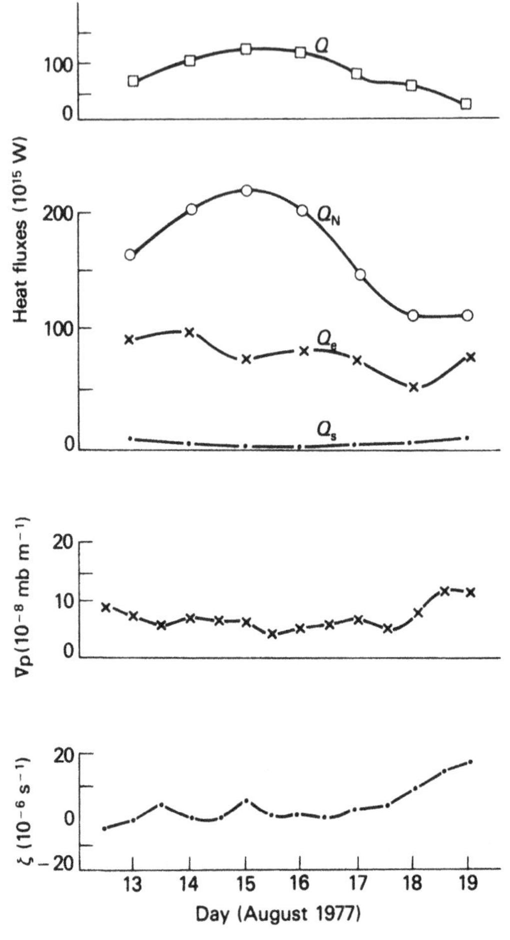

Fig. 33.3. Areal averages of the heat fluxes, the surface-pressure gradient force (∇p) and the surface relative vorticity (ζ) in the Bay of Bengal, 13 to 19 August 1977

TABLE 33.1. *Radiation fluxes at four locations in the Bay of Bengal* ($W m^{-2}$).

Ship and position	Q_N	$Q_e + Q_s$	Q	$(Q_e + Q_s)/Q_N$
Umay – 17° N 91° E	166	100	66	0.60
Unac – 17° N 87° E	185	98	87	0.53
Ereh – 19° N 89° E	184	71	112	0.39
Erei – 15° N 89° E	196	94	103	0.48

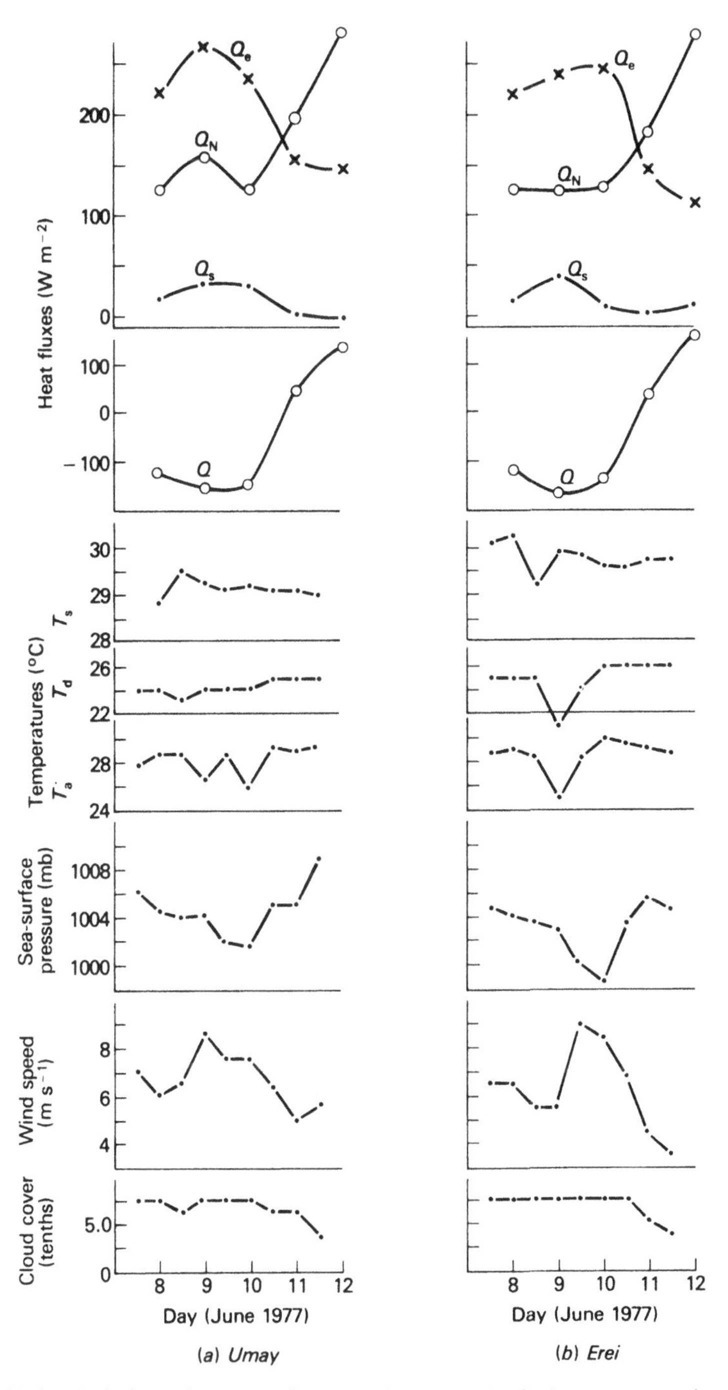

Fig. 33.4. Variation of energy fluxes and meteorological parameters in the Arabian Sea, 8 to 12 June 1977, observed by: (*a*) *Umay* (12°30′ N, 68° E); (*b*) *Erei* (14°30′ N, 66° E); (*c*) *Ereh* (12°30′ N, 64° E); and (*d*) *Erec* (10°30′ N, 66° E).

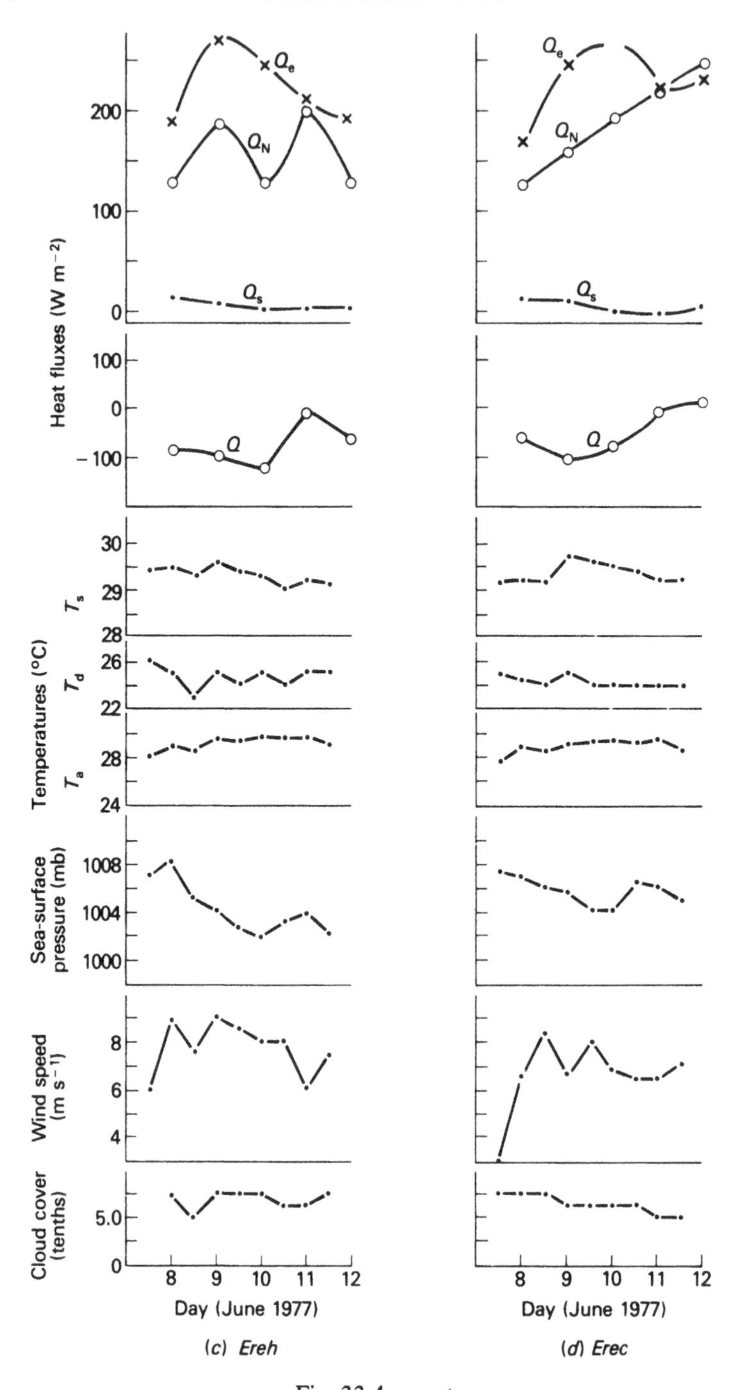

Heat fluxes (W m⁻²)
200
100
0
100
0
−100
Q_e
Q_N
Q_s
Q
Temperatures (°C)
T_s
30
29
28
T_d
26
22
T_a
28
24
Sea-surface pressure (mb)
1008
1004
1000
Wind speed (m s⁻¹)
8
6
4
Cloud cover (tenths)
5.0
0
8
9
10
11
12
Day (June 1977)
(c) Ereh
(d) Erec

Fig. 33.4—*cont.*

In general, the net heat gain by the sea surface was negative in the first four days and gradually changed its sign on the last day. This increase on the last day can be explained on the basis of the position of the cyclonic storm which has moved away from the polygon, thereby giving rise to fair weather conditions over the polygon.

Overall, the net heat gain over the sea surface is negative, indicating that the energy transported into the atmosphere by way of latent heat flux is nearly twice that of the transport during normal conditions.

In Fig. 33.4*b* the net gain by the sea surface is higher from 11 June onwards than at the other three locations. This peculiarity may be due to drifting of the ship *Erei* away from the system.

In the locations shown in Figs. 33.4*c* and *d*, the net gain by the sea surface, though varying, is negative throughout the period. These two locations seemed to be under the influence of the moving cyclone during the period under consideration.

Fig. 33.5 gives the areal averages of energy fluxes, vorticity and pressure gradient force at the central point of the polygon for the period 8 to 11 June discussed above. The energy flux is negative from 8 to 11 June and positive thereafter, and the values are much higher than the fluxes associated with the depression in the Bay of Bengal. The higher values of vorticity clearly indicate larger transports of heat flux from the sea surface to the atmosphere.

Figs. 33.6*a*, *b*, *c* and *d* illustrate the variation of energy fluxes from 30 June to 11 July 1977 computed from observations made by the Russian ships in the stationary positions shown in Fig. 33.1.

The left sides of Figs. 33.6*a*, *b*, *c* and *d* indicate the energy fluxes and the corresponding meteorological parameters associated with a normal monsoon regime; the right sides indicate the fluxes associated with a trough. Net radiation values are higher during the normal monsoon regime than during the regime with a trough because of the differences in cloudiness. Except at the northern ship of the polygon, latent heat losses are higher during the later regime causing an increase in cloudiness. On the left sides of Figs. 33.6*a*, *b*, *c* and *d*, the energy fluxes are greater than on the right sides. The decrease on the right sides can be explained on the basis of the weather associated with the trough, involving an increased cloud amount which reduces the net radiation received at the sea surface.

Fig. 33.7 shows the energy fluxes computed for the polygon area and the vorticity and pressure gradient force at the central point. The energy fluxes show a variation similar to that seen in Fig. 33.6. The negative

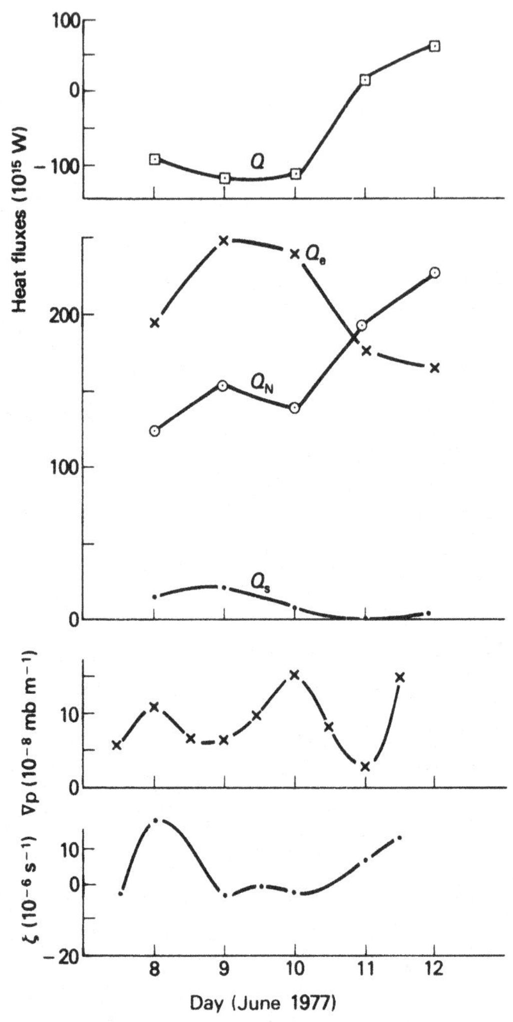

Fig. 33.5. Areal averages of the heat fluxes, the surface-pressure gradient force (∇p) and the surface relative vorticity (ζ) in the Arabian Sea, 8 to 12 June 1977.

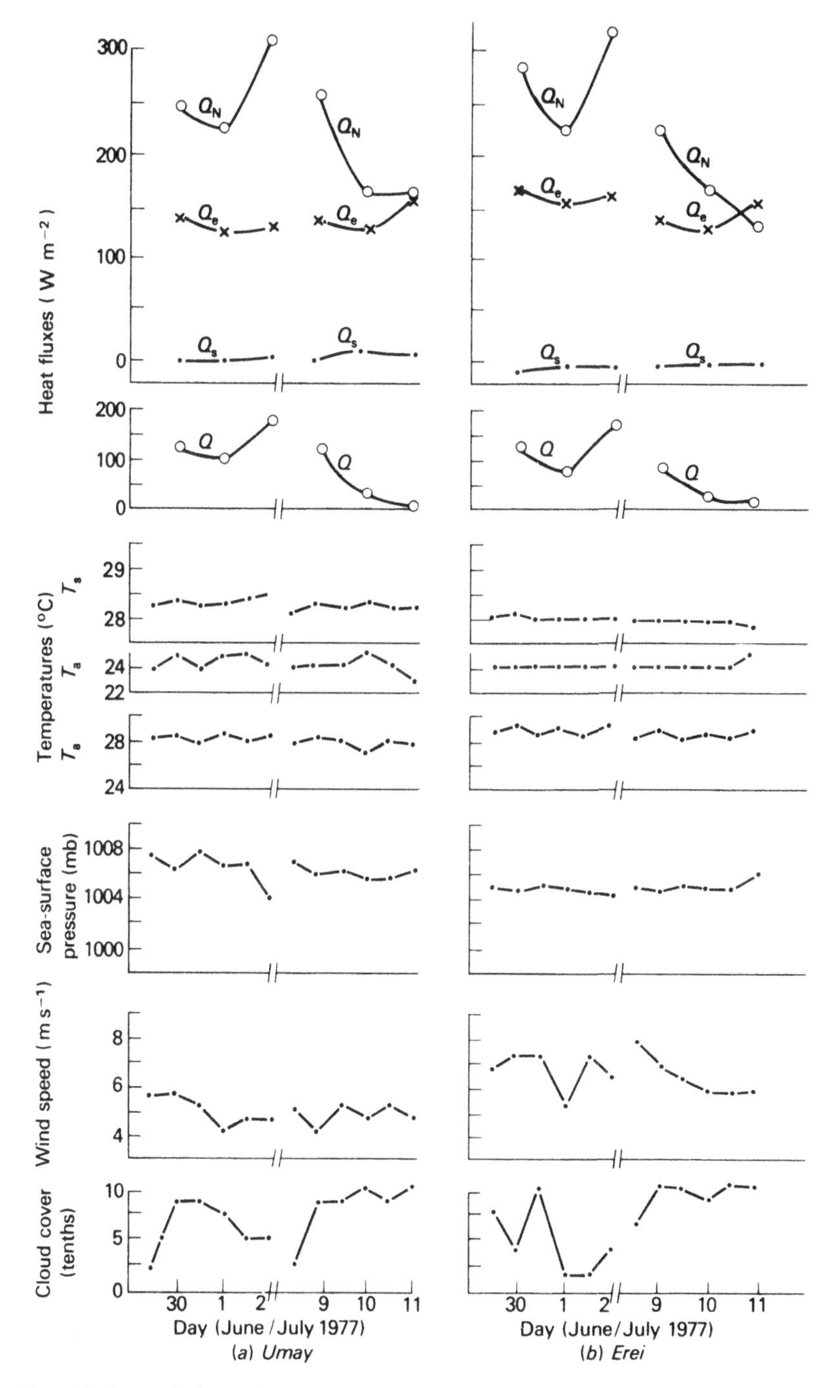

Fig. 33.6. Variation of energy fluxes and meteorological parameters in the Arabian Sea, 30 June to 11 July 1977, observed by: (*a*) *Umay* (12°30′ N, 68° E); (*b*) *Erei* (14°30′ N, 66° E); (*c*) *Ereh* (12°30′ N, 64° E); and (*d*) *Erec* (10°30′ N, 66° E).

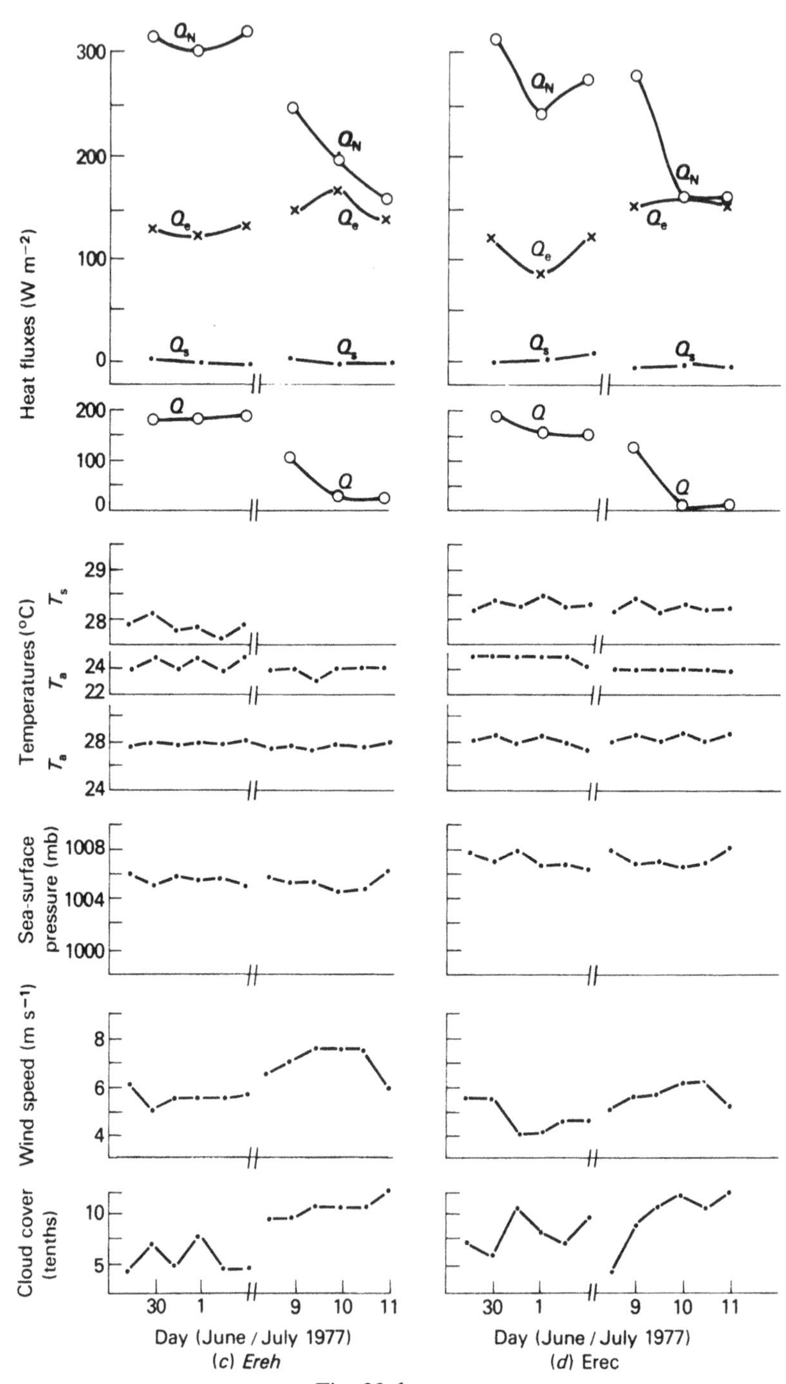
Heat fluxes (W m⁻²)
Q_N
Q_e
Q_s
Q
Temperatures (°C)
T_s
T_a
Sea-surface pressure (mb)
Wind speed (m s⁻¹)
Cloud cover (tenths)
Day (June / July 1977)
(c) Ereh
(d) Erec

Fig. 33.6—*cont.*

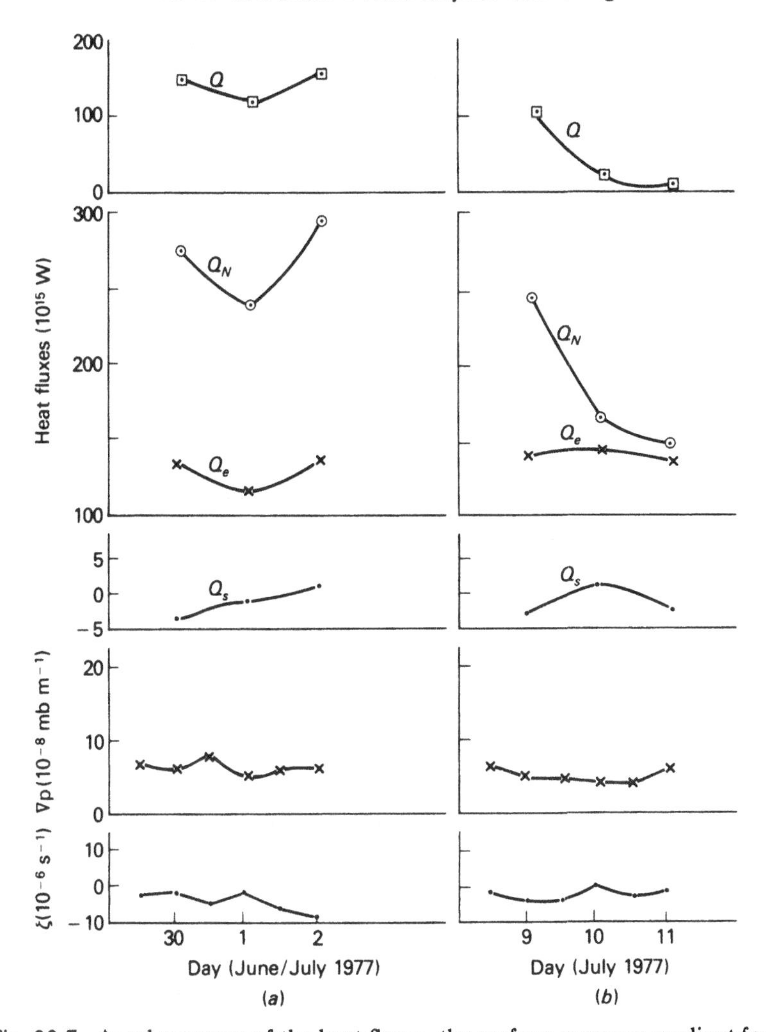

Fig. 33.7. Areal averages of the heat fluxes, the surface-pressure gradient force (∇p) and the surface relative vorticity (ζ) in the Arabian Sea from: (*a*) 30 June to 2 July 1977 (normal); and (*b*) 9 July to 11 July 1977 (trough).

vorticity indicates less transport of energy from the sea surface to the atmosphere.

Table 33.2 gives the average values of fluxes at different locations in normal monsoon conditions. The fluxes over the Arabian Sea are comparable to those at the Equator. Even under normal monsoon conditions, the latent heat flux from the Arabian Sea surface is much higher than that over the Bay of Bengal.

TABLE 33.2. *Average radiation fluxes* ($W m^{-2}$) *at different locations in the north Indian Ocean.*

Location	Q_N	Q_e	Q
Equator	268.7	124.0	136.0
Arabian Sea	271.6	128.4	128.7
Bay of Bengal	166.0	91.6	66.1

33.4 Conclusions

(1) The energy fluxes over the Arabian Sea ($Q_N = 266$ W m^{-2}, $Q_e =$ 145 W m^{-2}) are higher than the corresponding fluxes over the Bay of Bengal ($Q_N = 194$ W m^{-2}, $Q_e = 85$ W m^{-2}) as there is a northward shift in the monsoon trough from 11 to 18 August 1977 leading to a break-monsoon condition during the observation period in the Bay of Bengal.

(2) The latent heat fluxes associated with a cyclone (ranging from 145 to 290 W m^{-2}) are larger than those associated with a depression or a trough (50 to 120 W m^{-2}).

(3) The net heat gain by the atmosphere is larger in the case of a cyclonic storm (100 to 145 W m^{-2}) than in the case of a depression (25 to 50 W m^{-2}).

(4) The present studies indicate that accumulation of energy in the surface layers of the sea encourages evaporation leading to a fall in the pressure over the region.

Analysis of the data collected during MONSOON-77 over the north Indian Ocean, at the sea surface and in the upper air, may lead to a better understanding of monsoon phenomena in relation to air–sea interaction processes.

References

Ellis, R. S. (1952) A preliminary study of a relation between the surface temperature of the north Indian Ocean and precipitation over India. M.S. Thesis, Department of Meteorology, Florida State University.

Laevastu, T. and Hubert, W. E. (1970) The nature of sea surface temperature anomalies and their possible effects on weather. Technical Note No. 55, Fleet Numerical Weather Center, Monterey, California, 14 pp.

Rao, R. R., Somanadham, S. V. S. and Nizamuddin Syed (1977). Study of the influence of the surface energy budget of the north Indian Ocean on the behaviour of the Indian summer monsoon. Presented at the International Symposium on Monsoons held at New Delhi, March 1977.

Saha, K. R. (1970) Zonal anomaly of sea surface temperature in the equatorial Indian Ocean and its possible effect upon the monsoon circulation. *Tellus*, **22**, 403–9.

Saha, K. R. (1974) Some aspects of the Arabian Sea summer monsoon. *Tellus*, **26**, 464–76.

Shukla, (1975) Effect of Arabian Sea surface temperature anomaly on the Indian summer monsoon: A numerical experiment with the GFDL model. *J. Atmos. Sci.*, **32**, 3, 503–11.

34

The energy budget at selected stations over the north Indian Ocean during MONSOON-77

R. R. RAO, K. V. SUNDERARAMAM AND M. R. SANTA DEVI

The heat budget of the oceanic surface at selected stations over the north Indian Ocean is evaluated for the three phases of MONSOON-77 by computing the insolation, effective back radiation, latent heat flux and sensible heat flux using empirical expressions. The diurnal variations of wet bulb depression, sea-surface temperature, T, and wind speed are examined for each of three phases of the monsoon over the eastern Arabian Sea and the central Bay of Bengal. The so-called 'heat potential' values ($T \geqslant 27\,°C$) of the upper-oceanic layer at the stations located over the offshore trough along the west coast of India are computed and related to the net heat gain at the surface. Mean temperature and thickness values of the mixed layer are also analysed. The results are discussed in relation to the behaviour of the summer monsoon over India.

34.1 Introduction

The net energy accumulated over the tropics is transported to higher latitudes by the atmospheric and oceanic circulations in maintaining the thermal equilibrium of the earth–atmosphere system. Earlier studies have indicated that the atmosphere transports more energy than the oceans. However, the studies of Von der Haar and Oort (1973) based on satellite radiation data clearly indicate that the oceans play the larger role in transporting the surplus energy from the tropics. The relative contributions of the Indian, Atlantic and Pacific oceans must clearly differ. As the Indian Ocean is surrounded on three sides by the African, Asian and Australian continents, its efficiency in transporting the surplus net heat gain from the equatorial region towards the north pole is very much restricted and must lead to an accumulation of energy on its northern boundary.

The southeast trades of the southern hemisphere continually contribute to the accumulation of energy in the north Indian Ocean from March onwards. During the northern summer the land is heated and this results in the formation of the seasonal surface low-pressure area over north Africa and southern Asia. The associated pressure gradients imply an associated atmospheric circulation which constitutes the south Asian summer monsoon flow and this continuously interacts with the energy-rich tropical Indian Ocean. The general characteristics of the monsoon system are partially determined by the air–sea interaction processes which take place in the low-level south westerly flow over the Indian Ocean, Arabian Sea and Bay of Bengal. Several investigators (Ellis,

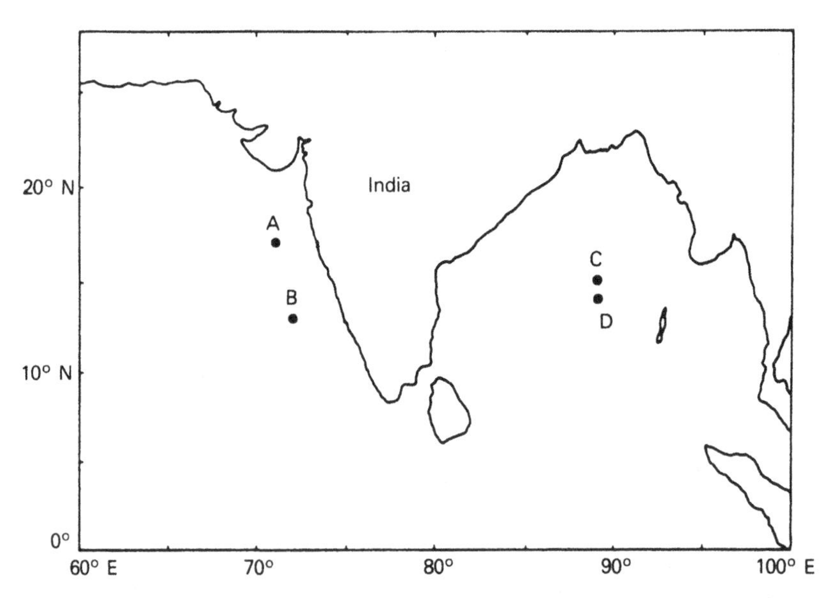

Fig. 34.1. Station location map (MONSOON-77). Station A: *Betwa*, 17° N, 71° E; phase I (26 May to 7 June), phase II (26 June to 15 July) and phase III (4 to 19 August). Station B: *Beas*, 13° N, 72° E, phase I (26 May to 7 June) and phase II (26 June to 15 July). Station C: *Beas*, 15° N, 89° E; phase III (4 to 11 August). Station D: *Beas*, 14° N, 89° E; phase III (11 to 17 August).

1952; Saha and Suryanarayana, 1972; Ramage *et al.*, 1972; and Shukla, 1975) have postulated the importance of the sea-surface temperature, which is a measure of the oceanic energy content, on the monsoonal flow and associated rainfall over southern Asia. Valuable data on the energy exchange processes were collected during the International Indian Ocean Expedition (1960–5). As a prelude to the MONEX experiment two preliminary experiments, the Indo-Soviet Monsoon Experiment (ISMEX) in 1973 and MONSOON-77 in 1977, were conducted with the collaboration of India and the USSR. The present study is based on data collected by two Indian ships during the period from the end of May to the middle of August, 1977 (3 phases of MONSOON-77). The location of the two Indian ships and the periods for the three phases are shown in Fig. 34.1.

34.2 Method of analysis

All the marine meteorological observations, collected at 3-hourly standard synoptic hours, are averaged for each day to compute the heat budget components using the empirical relations listed in Chapter 33 of this book (§ 33.2). For the computation of Q_i, the daytime average of cloudiness is used. The computed values of the parameters are shown in Figs. 34.2 and 34.3 and are discussed in relation to the behaviour of the monsoon over India for the three phases:

phase I 26 May to 7 June 1977;
phase II 26 June to 15 July 1977;
phase III 4 August to 19 August 1977.

The 'hurricane heat potential' H is defined (Whitaker, 1967) as the excess of heat in a given water mass over that of the same mass of water at 27° C, i.e.

$$H = \rho c_p \, \Delta T \, \Delta z, \tag{34.1}$$

where ρ is the density of sea water evaluated from its temperature and salinity, c_p is its specific heat at constant pressure, and ΔT is the average temperature excess above 27° C for a given depth increment Δz (5 m). H is evaluated from (34.1) for layers having mean temperatures greater than 27 °C for phases I and II at the southern location in the eastern Arabian Sea.

All the timings refer to the Indian Standard Time.

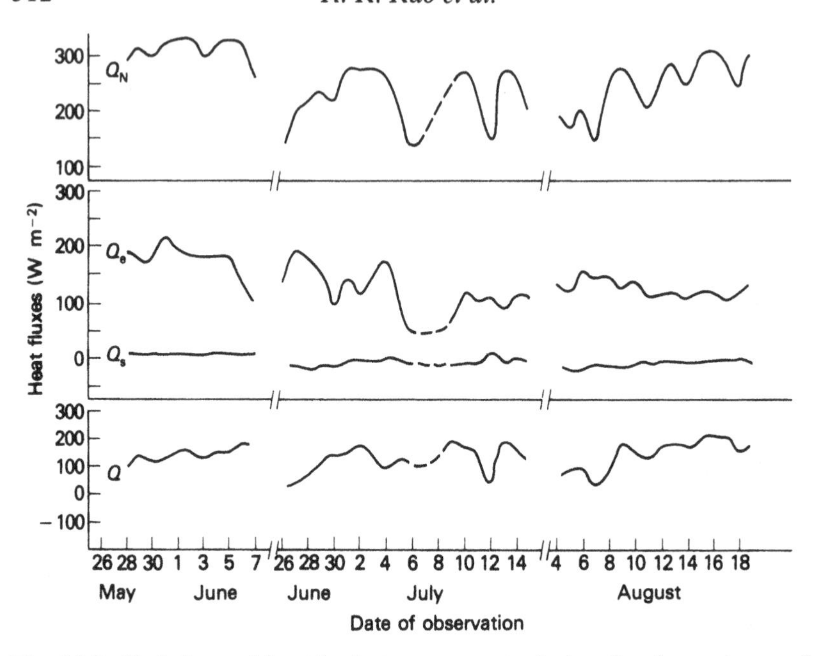

Fig. 34.2. Variations of heat budget components during the three phases of MONSOON-77 at the northern location (17° N, 71° E). Q_N is the net radiation absorbed by the ocean, Q_s is the sensible heat transfer, Q_e the latent heat flux across the air-sea interface (positive when the sea surface is gaining heat). Q is the net heat gain at the sea surface.

34.3 Analysis and discussion

Figs. 34.4 and 34.5 show the diurnal variations of wet bulb depression, sea-surface temperature (SST) and wind speed – the three important elements in the energy exchange processes at the air–sea interface. The data are presented as frequency distributions for each of the three phases.

The wet bulb depression shows a diurnal cycle at both locations (*Beas* – southern, *Betwa* – northern) over the eastern Arabian Sea during phase I. The variance is higher at the southern location due to the occasional development of cloud and precipitation over Lakshadweep and off the southwest coast of India. The maximum value occurred at around 1430 hours at both locations when the corresponding SST values were high. Using BOMEX data, Delnore (1972) reported the highest evaporation rates for a 6-hour period centred at the time of local sunset, and the lowest values around midnight and during mid-morning. This coincidence of maxima of wet bulb depression and SST suggests that an increase in SST enhances the dry bulb temperature through radiative and

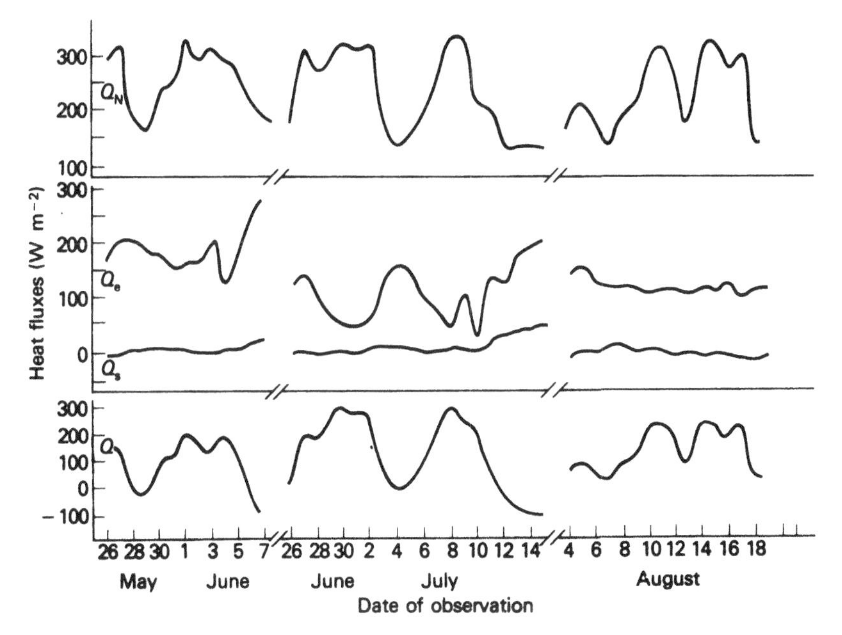

Fig. 34.3. Variations of heat budget components during the three phases of MONSOON-77 at the southern locations (13° N, 72° E for phases I and II; 15°/14° N, 89° E for phase III).

conductive processes more than the wet bulb temperature, causing the wet bulb depression to increase. After the onset of the southwest monsoon (phase II) the curvature of the wet bulb depression plots is reversed at both the locations, lower values occurring during the daytime. The average magnitudes fall by 2 °C after the onset of the monsoon associated with a corresponding fall in SST and the advection of moist monsoon air.

In phase III at both the locations (*Beas* over the Bay of Bengal) the curves again clearly show a diurnal variation with peak values around 1430 hours and minimum values around 0530 hours. This resemblance with phase I may be due to the prevailing break-monsoon conditions from 11 to 18 August.

During phase I the SST also shows a diurnal variation at both the locations with a maximum value around 1430 hours and a minimum value around 0530 hours. The variance is higher at the southern location with the occurrence of cloudiness associated with the pulsations in the trough of low pressure off the southwest coast. After the onset of the monsoon (phase II) the diurnal variation almost disappears. The average

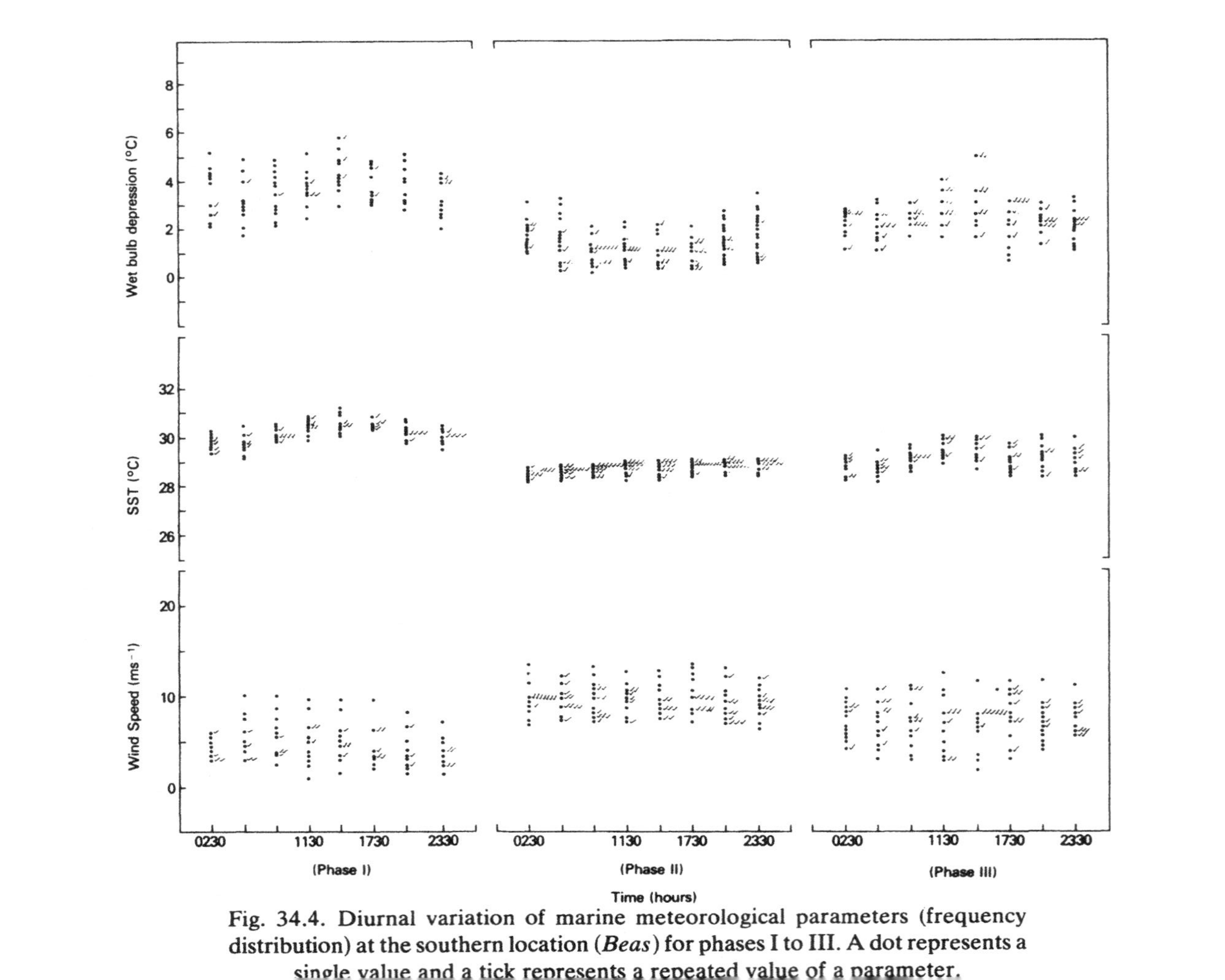

Fig. 34.4. Diurnal variation of marine meteorological parameters (frequency distribution) at the southern location (*Beas*) for phases I to III. A dot represents a single value and a tick represents a repeated value of a parameter.

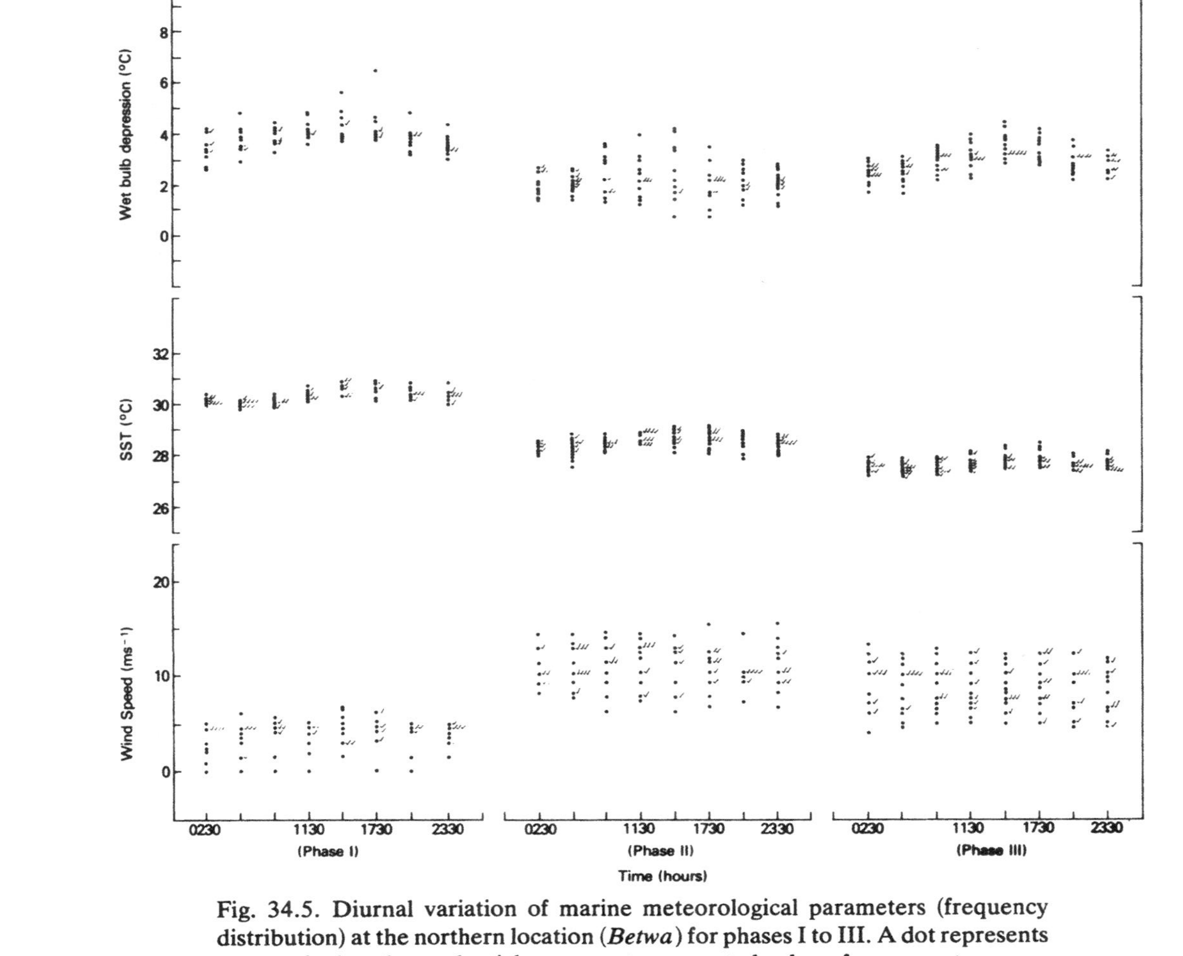

Fig. 34.5. Diurnal variation of marine meteorological parameters (frequency distribution) at the northern location (*Betwa*) for phases I to III. A dot represents a single value and a tick represents a repeated value of a parameter.

SST before the monsoon onset is 30.4 °C, and after the monsoon onset it falls by 2.1° C at both locations. During phase III the SST variance is high over the central Bay of Bengal and there is a fall of temperature by 1 °C from phase II to phase III over the east Arabian Sea. Except in phase II, the SST has higher dispersion at *Beas* compared to *Betwa*.

Wind speeds do not show any diurnal variation at either location during any of the three phases. During phase I at the southern location, magnitudes varied between 2 and 10 m s^{-1}, only on a few occasions exceeding 8 m s^{-1}. At the northern location the magnitudes are smaller, varying between 0 and 5 m s^{-1}. When the monsoon had completely set in, a well-marked increase in the wind speed is observed at both the locations (phase II), the values varying between 7 and 15 m s^{-1}. During phase III there is a decline in the average wind speed, the magnitudes varying between 5 and 12 m s^{-1}.

Table 34.1 (in § 34.4) shows the phase-averaged values of these three marine meteorological parameters at *Beas* and *Betwa* for all the three phases.

Figs. 34.2 and 34.3 show the variations of the main heat budget components during the three phases of MONSOON-77 for the northern and southern locations respectively.

Phase I. At the southern location the values of Q_N varied between 170 and 315 W m^{-2}, the corresponding range being somewhat smaller at the northern location; here the values are above 290 W m^{-2} throughout, except on the last day. The prominent dip in net radiation during the period 28 to 31 May at the southern location is associated with an increase in cloudiness with a well-marked trough of low pressure over Lakshadweep and off the southwest coast. Another large fall in Q_N occurred from 4 June at the southern location and from 5 June at the northern location as the offshore trough intensified with an associated low-level cyclonic circulation. The Q_e curves vary between values of 100 and 270 W m^{-2}. A steep rise in Q_e associated with the burst of the monsoon occurs from 4 June at *Beas*, and a steep fall from 5 June onwards at *Betwa*. The latter feature may be attributed to the northward transport of moist air along the trough at the end of phase I. Although the sensible heat exchange is small, the values remained positive (flux is from sea to atmosphere) throughout at the northern location, while the corresponding values at the southern location oscillated near zero. The net heat gain by the oceanic surface is positive at the northern location with a slight increasing trend. There are two occasions when the net heat gain by the oceanic surface became negative at the southern location with the decrease in net radiation, and in the latter case, increase in evaporation. At the time of the monsoon burst the oceanic heat loss is at a maximum of about 100 W m^{-2}.

Phase II. During this phase a well-marked dip in Q_N is observed from 3 to 7 July at the southern location associated with the formation of a cyclonic storm in the northwest Bay of Bengal and a simultaneous intensification of the trough off the southwest coast. During this period of rough weather a well-marked increase in evaporation occurred. Another significant fall in Q_N and rise in Q_e from 10 July lasted till the end of the phase. The corresponding pattern for Q_N at the northern location is similar, apart from a phase shift of two days, with an exception on 13 July. A conspicuous fall in Q_e at the northern location between 4 and 6 July may be a result of the horizontal advection of moist air from the south where intense evaporation took place around 4 July. On the whole, the evaporative loss is higher in the latter half of phase II at the southern location with the reverse at the northern location. The sensible heat exchange is again very small, apart from at the end of phase II at the southern location. The net heat gain curve at the southern location shows two large peaks during fair weather with negative values during disturbed weather. An increase in latent heat flux and a reduction in solar radiation resulted in the sea becoming an energy sink from 12 July onwards at *Beas*. There was a net heat gain at *Betwa* throughout the phase with minor fluctuations associated with changes in cloudiness.

Phase III. Over the eastern Arabian Sea the values of Q_N show an increasing trend with oscillations superposed (Fig. 34.2). The monsoon was initially weak, and the development of the trough off the southwest coast of India took place on 7 August causing a dip in Q_N. From 11 August onwards the axis of the seasonal monsoon trough shifted towards the north with break-monsoon conditions prevailing until 18 August. Day-to-day variations in the latent heat exchange during phase III are not significant, because of the absence of weather systems over the eastern Arabian Sea. The average value is of a magnitude comparable with the corresponding values in phase II. The sensible heat exchange is very small, but always from the atmosphere to the ocean. The resulting heat gain values are always positive with an increasing trend. The curve of Q very much resembles the curve of Q_N.

Over the central Bay of Bengal (*Beas*, phase III) Q_N has low values from 4 to 9 August in association with a monsoon depression over the northwest Bay of Bengal and adjoining areas (Fig. 34.3). On 7 August the depression intensified over central India and the winds strengthened. The cloudiness increased over the Bay of Bengal resulting in the reduction in the values of Q_N on 7 August. Again after 17 August another low-pressure area formed over the north Bay of Bengal and intensified the next day, this apparently resulting in a steep fall in Q_N values. Surprisingly, the Q_e values are free from violent changes, showing a slight decreasing trend. The sensible heat exchange is from the atmosphere to the sea throughout, except on 7 August. The heat gain curve follows a pattern similar to that of Q_N, the latent and sensible heat exchanges being relatively steady.

The subsurface thermal components. Fig. 34.6 shows the variation of the subsurface thermal components for the first two phases at 13 °N and 72° E (the southern location). The mean temperature of the mixed layer oscillated around 30 °C during phase I before the onset of the southwest monsoon. After the monsoon onset a temperature drop of 1.6 °C occurred. This fall of temperature may be mainly due to convective and wind-induced mixing extending to deeper layers. The average thickness of the mixed layer before the onset of the monsoon is 38.7 m and this increases by 15 m during phase II. This would appear to be mainly due to the increase of wind speed from 5 to 10 m s^{-1} enhancing the mechanical mixing. The average hurricane heat potential, *H*, during phase I is about 60 kJ and this is reduced by half during phase II. During phase I the hurricane heat potential began to fall from 4 June onwards as the solar input decreased in association with increased cloudiness. There is a broad agreement between the variations in the net heat gain at the surface and hurricane heat potential during phase I. The decreasing trend of hurricane heat potential from 26 to 30 May corresponds well with that of the net heat gain at the surface. Both parameters decrease again from 4 June, the date of onset of the monsoon. During phase II the correspondence between these parameters is weak. Both increase from 26 June to 2 July and decrease from 8 to 15 July. Differences of hurricane heat potential between successive days show violent changes in phase I, but these changes are smaller in phase II. Thus regimes of alternative heating and cooling seem to prevail in the upper layers of the ocean.

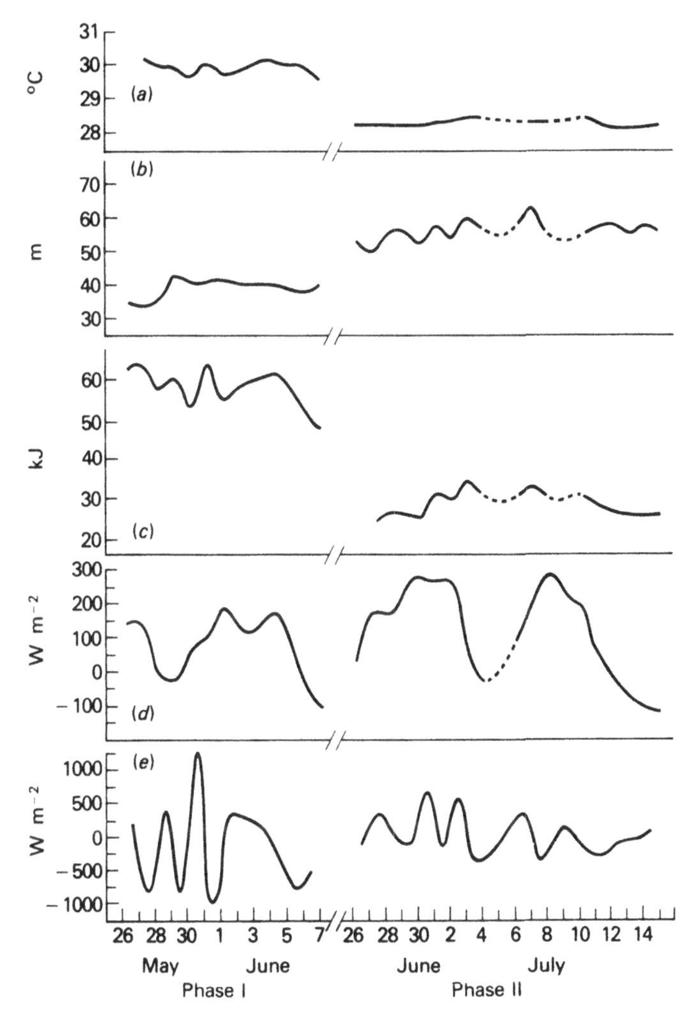

Fig. 34.6. Variation of the subsurface thermal components for phases I and II at the southern location. (*a*) The mean temperature of the mixed layer; (*b*) the thickness of the mixed layer; (*c*) heat potential, (*d*) net heat gain at the surface, and (*e*) day-to-day variations of hurricane heat potential.

34.4 Summary and conclusions

The main results of this study are presented in Tables 34.1 and 34.2

The analysis has shown well-marked diurnal variations during each of the three phases in wet bulb depression and SST, but not in winds. Following the monsoon onset the wet bulb depression and SST fell by 2 °C and the wind speeds doubled to 10 m s^{-1}. During August (phase III)

TABLE 34.1. *Phase-averaged values of wet bulb depressions, sea-surface temperatures and wind speed at* Beas *and* Betwa *for the three phases.*

Meteorological parameter	Phase I		Phase II		Phase III	
	Beas	*Betwa*	*Beas*	*Betwa*	*Beas*	*Betwa*
Wet bulb depression (°C)	3.8	3.8	1.3	2.0	2.3	2.7
Sea surface temperature (°C)	30.1	30.7	28.3	28.3	28.4	27.3
Wind speed (m s^{-1})	4.8	4.5	10.1	11.0	8.4	8.7

TABLE 34.2. *Ocean heat budget components* (W m^{-2}). (*Values in the brackets are percentages of net radiation.*)

	Beas			*Betwa*		
	Phase I	Phase II	Phase III	Phase I	Phase II	Phase III
Net radiation	253.3	222.6	220.2	310.6	224.2	238.7
Latent heat flux	182.6 (72)	104.4 (47)	110.3 (49)	171.6 (55)	118.2 (50)	120.2 (48)
Sensible heat flux	3.0 (1)	7.1 (3)	−6.8	4.5 (1)	−10.3	−13.9
Net heat gain	67.5 (27)	111.2 (50)	116.6 (51)	134.4 (43)	116.3 (50)	132.5 (52)
Mean temperature of the mixed layer (°C)	29.9	28.3				
Mixed layer depth (m)	38.7	55.8				
Hurricane heat potential (10^8 J m^{-2})	5.80	2.88				

the SST of the central Bay of Bengal was 1 °C higher than that of the eastern Arabian Sea. Over the eastern Arabian Sea, the net radiation values at the surface at the northern location are higher for all the three phases than at the southern location. The most rapid evaporation occurred before the onset of the monsoon, during phase I at both locations when the air was relatively dry. About half the net radiative energy was retained as net heat gain at the surface during phases II and III. With the drop of 1.8 °C in SST at the southern location of the east Arabian Sea, the hurricane heat potential of the upper layers of the ocean was reduced by half after the monsoon onset; also the mean temperature of the mixed layer was reduced by 1.6 °C and the mixed layer depth increased by 17 m.

These results support the view that surface energy exchanges and subsurface energy storage must be accurately quantified for the study of the behaviour of the summer monsoon over India.

The authors wish to thank Dr D. Srinivasan for his continuous inspiring encouragement throughout this investigation. Thanks are also due to Mr C. Ravikumaran Nair for his assistance with the computations.

References

Delnore, V. E. (1972) Diurnal variation of temperature and energy budget for the oceanic mixed layer during BOMEX. *J. Phys. Oceanogr.*, **2**, 3, 239–47.

Ellis, R. S. (1952) A preliminary study of a relation between surface temperature of the north Indian Ocean and precipitation over India. M.S. Thesis, Department of Meteorology, *Florida State University.*

Ramage, C. S., Miller, F. R. and Jeffries, C. (1972) *Meteorological Atlas of the International Indian Ocean Expedition.* Vol. 1. The surface climate of 1963 and 1964. National Science Foundation, Washington, DC.

Saha, K. R. and Suryanarayana, R. (1972) Mean monthly fluxes of sensible and latent heat from the surface of the Indian Ocean. *J. Mar. Bio. Ass. Ind.*, **14**, 663–70.

Shukla, J. (1975) The effect of the Arabian Sea surface temperature anomaly on the Indian summer monsoon. *J. Atmos. Sci.*, **32**, 503–11.

Von der Haar, T. H. and Oort, A. H. (1973) New estimate of annual poleward energy transport by northern hemisphere oceans. *J. Phys. Oceanogr.*, **3**, 2, 169–72.

Whitaker, W. D. (1967) Quantitative determination of heat transfer from sea to air during passage of hurricane Betsy. M.S. Thesis, Texas A & M University, USA.

35

Observations of coastal-water upwelling around India

A. V. S. MURTY

The seasonal locations of upwelling in the coastal waters around India and the importance of upwelling to the Indian fisheries are discussed.

35.1 Introduction

The monsoons and coastal-water upwelling have much in common. While the former improves the agricultural production, the latter enriches the living resources of the seas. Both phenomena have, therefore, tremendous impact on the economy of the region under their influence.

35.2 Materials and methods

The hydrographic data collected by the Central Marine Fisheries Research Institute of the Indian Council of Agricultural Research along the west coast of India have been utilized to present the thermal properties of the surface mixed layer and its associated thermocline. The average surface currents around the Indian subcontinent (Anonymous, A., 1976), published in the *U.S. Navy Marine Climatic Atlas of the World* (volume 3, Indian Ocean), are taken into consideration.

35.3 Results and discussion

Before dealing with the observations on coastal-water upwelling around India, it may be desirable to identify the areas and seasons wherein the process of upwelling might be expected.

By examining the general circulation around India, the coastal areas of upwelling can be speculatively located and the season of its occurrence identified. Upwelling is inferred to occur when the surface currents run parallel to the coastline with the land to the left; under these circumstances, the non-geostrophic component of the surface flow is directed away from the coast and it is likely that upwelling will be associated with it. This argument is highly simplified, however, and can only be used to provide a rough estimate of likely upwelling locations. The general system of currents around the Indian subcontinent is presented in Fig. 35.1 for the discrete seasons of the year. Fig. 35.1 indicates that upwelling is possible along the west coast of India during the southwest monsoon period and also during the summer transition. Currents at the head of the

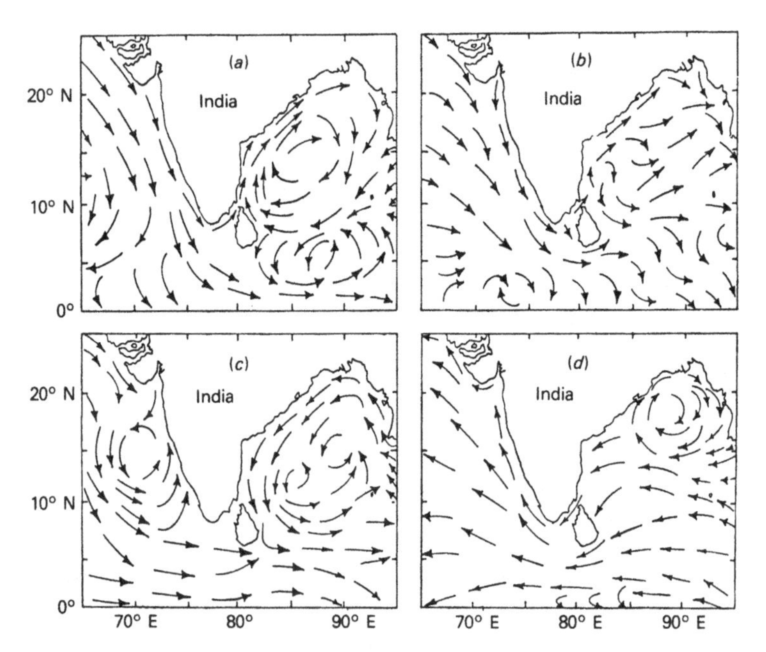

Fig. 35.1. The system of coastal currents around India for: (*a*) April (summer spring, winter and the monsoon seasons. The depths of the 3 blocks indicate the (*d*) January (winter). (From the *US Navy Marine Climatic Atlas of the World*, Volume 3, Indian Ocean, 1976.)

Bay of Bengal are favourable for upwelling all through the year, except for the short period of the winter transition. Upwelling is expected in the southeastern and central area off the east coast of India during the southwest monsoon season and also during the pre-southwest monsoon season.

Waters off the southwest and central west coast of India (between 7° N and 17° N) have been thoroughly explored over different years (Murty, 1965; Sharma, 1968; Sastry and D'Souza, 1972; Ramamirtham and Rao, 1973; and Anonymous, B., 1976). Upwelling in these waters is identified by the ascent of isolines of one or more parameters such as temperature, thermocline, density and dissolved oxygen. (The layer of minimum oxygen in the Arabian Sea is associated with the thermocline.)

The thermal conditions of the waters of the mixed layer in this region are presented in Fig. 35.2 for the three discrete seasons of the year.

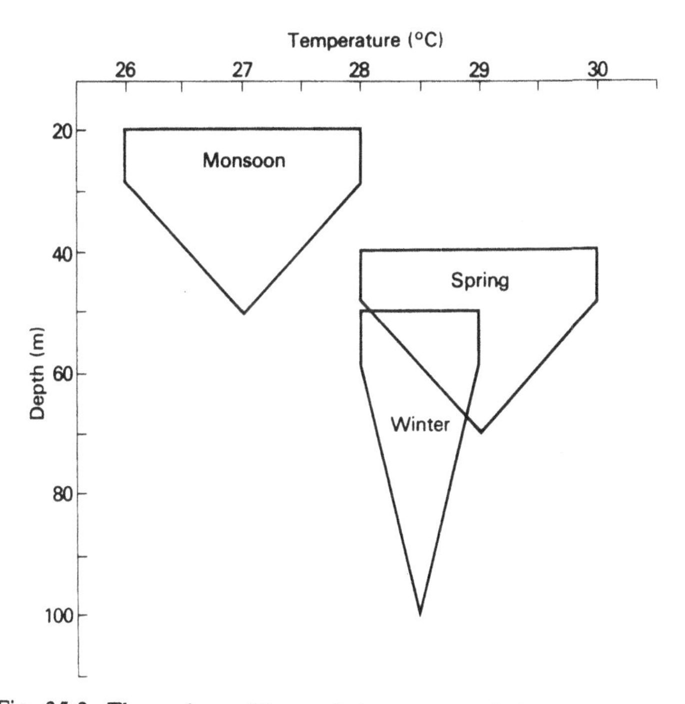

Fig. 35.2. Thermal conditions of the waters off the west coast of India for spring, winter and the monsoon seasons. The depths of the 3 blocks indicate the ranges of the depth of the lower limit of the mixed layer; their widths indicate the observed ranges of temperature of the mixed layer (upper horizontal axis). The sharpness of the thermocline in each case is: the monsoon season 0.17 °C m^{-1}; the winter season 0.15 °C m^{-1}; and the spring season 0.13 °C m^{-1}.

During winter, when there is no upwelling in the waters, the mixed layer extends to 50 m or even 100 m at some places and the associated thermocline is moderately strong (0.15 °C m^{-1}). By spring the mixed layer reduces in depth to between 40 m and 70 m and its associated thermocline is only as intense as 0.13 °C m^{-1}. Conspicuously enough, the mixed layer is very shallow (20 m to 50 m only) during the southwest monsoon period and in addition, the thermocline is strong (0.17 °C m^{-1}).

The duration of upwelling in these waters is identified, by the rise of the layer of minimum oxygen content, to be the pre-southwest monsoon and the southwest monsoon periods (Anonymous, B., 1976). The phenomenon was found to be more intense in the central part (Calicut–Karwar) of the southwest coast of India (Ramamirtham and Rao, 1973; Anonymous, B., 1976).

Taking winter as a reference, a conservative estimate of the average rate of upwelling in these waters by the middle of the southwest monsoon period is about (see Fig. 35.2) 0.35 m per day. Judging from the ascent of the layer of low oxygen content (Anonymous, B., 1976), the average rate of upwelling in these waters appears not to exceed 0.8 m per day.

Upwelling in the Bombay region seems to have a different origin (Jayaraman and Gogate, 1957). The offshore winds during the northeast monsoon appear to be favourable for local upwelling instead of the normal circulation off Bombay. Upwelling of the layer of minimum oxygen was, in fact, found during October/November 1958 by identifying the layer with an average oxygen content of as low as 0.7 ml l^{-1} occurring at an average depth of about 18 m some 30 kilometers out from the coast of Bombay (Carruthers *et al.*, 1959).

Studies on upwelling in the east coast waters of India are very limited. La Fond (1958) observed the upwelled waters in the surface area near Waltair during the months of March, April and May by following the layer of maximum salinity and moderate temperature.

Nevertheless, observations on seasonal variations of thermal structure in the western Bay of Bengal (Colborn, 1971) indicate that upwelling in the east coast surface waters, in general, prevails only during the transition periods of the two monsoons.

35.3.1 Significance of upwelling to Indian fisheries

As upwelling causes lifting of nutrients into the euphotic zone (i.e. the zone in which light penetration is sufficient for photosynthesis), thereby making the waters fertile, the areas of upwelling become pastures of phytoplankton with the consequent result of enhanced secondary and tertiary production.

The rate of primary production in the waters off the southwest coast during the southwest monsoon was found to be as high as five times the winter productivity. Based on the 75 m to surface vertical hauls over the shelf and adjacent waters of this region, the displacement volume of phytoplankton was found to be 9.7 ml m^{-2} during winter, 15.9 ml m^{-2} during spring and 25.1 ml m^{-2} during the southwest monsoon season (Subrahmanyan *et al.*, 1971).

Upwelling was found to have a significant effect on the fisheries. This effect was distinctly reflected in the distribution of benthic fish in the shelf area along the west coast of India (Banse, 1968). As the layer of low oxygen content ascends on to the shelf as a result of upwelling, a band of shelf bed along the coast is overlayed by badly aerated water. The demersal fish abandon this portion of the shelf and migrate either to the deeper or to the shallower areas of the shelf where sufficient aeration is found. Thus, demersal fish have been found to disappear from a rather broad belt parallel to the southwest coast during the southwest monsoon period, with the result that bottom trawling is profitable only in the deeper waters (below 35 m).

Similar observations on the operational fisheries were made off Bombay by Carruthers *et al.* (1959). Demersal fish in the Bombay region were forced to migrate to shallower areas (less than 15 m) of the shelf during the northeast monsoon period in order to escape the lethal low oxygen conditions of upwelled waters.

It is my pleasure to acknowledge with thanks the encouragement I received from Dr E. G. Silas, Director of the Central Marine Fisheries Research Institute, Cochin.

References

Anonymous, A. (1976) *U.S. Navy Marine Climatic Atlas of the World*, Vol. 3, Indian Ocean. NAVAIR 50-1C-530, Naval Weather Service Command, Washington, DC, p. 348.

Anonymous, B. (1976) UNDP/FAO Pelagic Fisheries Project, Cochin (IND 69/593) Progress Report No. 16, p. 39.

Banse, K. (1968) Hydrology of the Arabian Sea shelf of India and Pakistan and effects of demersal fishes. *Deep-Sea Res.*, **15**, 45–79.

Carruthers, J. N., Gogate, S. S., Naidu, J. R. and Laevastu, T. (1959) Shorewards upslope of the layer of minimum oxygen off Bombay, its influence on marine biology, especially fisheries. *Nature*, **183**, 1084–7.

Colborn, J. G. (1971) Thermal structure dynamics in the upper 500 metres of the Indian Ocean. NUC TP 266, p. 99. Naval Undersea Research and Development Centre, San Diego, California.

Jayaraman, R. and Gogate, S. S. (1957) Salinity and temperature variations in the surface waters of the Arabian Sea off the Bombay and Saurashtra coasts. *Proc. Ind. Nat. Acad. Sci.*, **B45**, 151–64.

La Fond, E. C. (1958) On the circulation of the surface layers of the east coast of India. *Andhra University Memoirs in Oceanography*, Series No. 62, **2**, 1–11.

Murty, A. V. S. (1965) Studies on the surface mixed layer and its associated thermocline off the west coast of India and the inferences thereby for working out a prediction system of the pelagic fisheries of the region. *Ind. J. Fish.*, **12**, 118–34.

Ramamirtham, C. P. and Rao, D. S. (1973) On upwelling along the west coast of India. *J. Mar. Bio. Ass. Ind.*, **15**, 306–17.

Sastry, J. S. and D'Souza, R. S. (1972) Upwelling and upward mixing in the Arabian Sea. *Ind. J. Mar. Sci.*, **1**, 17–27.

Sharma, G. S. (1968) Seasonal variation of some hydrographic properties of the shelf waters off the west coast of India. *Bull. Nat. Inst. Sci. Ind.*, No. 38, 263–76.

Subrahmanyan, R., Gopinathan, C. P. and Thankappan Pillai, C. (1971) Phytoplankton of the Indian Ocean: Some ecological problems. Presented to the Symposium on Indian Ocean and adjacent seas held at Cochin, 12–18 January 1971 by the Marine Biological Association of India.

Part IIIb

Modelling and theoretical studies

36

A numerical study of surface cooling processes during summer in the Arabian Sea

M. D. COX

One of the primary causes of surface cooling in the Arabian Sea during late spring and early summer is upwelling along the Somali and Arabian coasts. Along the Somali coast, upwelling apparently occurs as rather intense mesoscale phenomena associated with variations in the Somali Current. A three-dimensional numerical model of an equatorial ocean indicates that the formation and behaviour of these upwelling areas is sensitive to the orientation of the western boundary. Analysis of the local vorticity sources and sinks in the model solution indicates that the movement of the systems along the coast is caused by an imbalance between advective effects, producing northeastward motion, vortex stretching and the beta-effect producing southwestward motion, and the curl of the wind stress which is capable of producing either. These results suggest that surface cooling may be very sensitive to wind scales of the order of the baroclinic radius of deformation of the ocean near the coast.

36.1 Introduction

The effect of the sea-surface temperature (SST) of the Arabian Sea on rainfall over India has recently been studied by several investigators. While cause and effect is difficult to determine from observational data, Shukla (1975) has found, using a numerical atmospheric model, that SST anomalies of 1 to 3 °C during the Arabian Sea summer monsoon strongly affect the rainfall downstream. Fieux and Stommel (1976) have analysed data from heavily used shipping lanes in the Arabian Sea and found that such anomalies in the mean monthly SST do occur from year to year. The cause of summer cooling in the Arabian Sea is generally believed to be a combination of strong upwelling along the Somali and Arabian coasts, and evaporative cooling and forced mixing under the monsoon winds. These mechanisms (as well as others which could produce SST anomalies) must be represented accurately in any model which is used to study the air–sea interaction over the Arabian Sea. This chapter discusses the first of these processes by considering its representation in a numerical model.

Certainly near the coasts of the western Arabian Sea, summer cooling is predominantly caused by upwelling. However, the degree of intrusion of this water into the ocean interior is not well known. The presence of very strong surface currents near these coasts, with strong offshore components in many cases, indicates the potential of this mechanism for widespread cooling of the interior. On the other hand, dynamic topography charts indicate that such offshore currents follow a rather meandering path after leaving the coast, suggesting that eastward advection of cold water may have only a limited effect on the interior (Bruce, 1968). In any case, coastal upwelling is responsible for significant coastal surface cooling, which may have far-reaching effects on the south Asian climate. An understanding of the upwelling process along the Somali coast requires a study of the Somali Current system, of which it is a part. This chapter will attempt to present the most significant results so far obtained from three-dimensional numerical models designed for the study of the Somali Current. Two experiments will be described. The first involves a model of the meridional equatorial Western Boundary Current. The second simulates a sloping western wall, which models more realistically the boundary conditions felt by the Somali Current.

36.2 Review of Somali Current models

The development of an equatorial Western Boundary Current, driven by surface stresses, has been studied both analytically and numerically for the simplified, meridional boundary case. In a pioneering work, Lighthill (1969) used classical methods of analysis to investigate the coastal response to a wind-driven ocean circulation. Near the Equator, the westward-propagating planetary waves associated with the lowest wavenumber baroclinic modes, move at a speed sufficient to provide a response at the western boundary within a period of weeks after being excited at points as much as 500 km to the east. This mechanism explains the Western Boundary Current as an accumulation of northward fluxes, generated in the interior of the ocean and carried to the west as planetary waves. No local wind driving at the boundary itself is required. Using a wind stress of 0.4 N m^{-2}, Lighthill is able to account for some 40% of the observed flux of the Somali Current one month from onset. The maximum current occurs near 2° N. He suggests that advective terms, not included in his analysis, may serve to move the peak nearer the observed 8° to 9° N.

Cox (1970) used a three-dimensional numerical model with a realistic topography, driven by realistic seasonal winds, to study the temporal response of the Indian Ocean. The Lighthill mechanism, discussed above, was not evaluated in this solution, although it was clearly present. An analysis of the one-sided Ekman divergence from the coast, produced by alongshore stresses, indicated that the Somali Current can be thought of as a purely locally-driven current. The coarse resolution and high mixing coefficients used in this study prevented the simulation of any but its grossest features.

Anderson and Rowlands (1976), in an attempt to make a specific comparison of the effects of local and remote forcing, used an analytical, linearized model, similar to that of Lighthill, on which they applied both types of forcing. For local forcing, the asymptotic solution at the boundary is such that the alongshore velocity increases linearly with time, t. However, for remote forcing, the same velocity increases as t^2. Clearly, the relative importance of the two is determined by the magnitudes of the stresses involved. However, in the light of these findings, the idea that local driving dominates in the first stages, giving way to remote driving after one month or so, is not unreasonable.

Two numerical models with high resolution and low mixing coefficients have recently been used to study advective effects in the Somali Current.

Hurlburt and Thompson (1976) use a two-layer box model driven by an alongshore wind. A clockwise gyre forms at the equatorial western boundary and, as a result of inertial effects, progresses northward along the boundary. They relate this to the cyclonic eddy known to exist off the Somali coast during summer. However, the model gyre continues to move northward indefinitely, whereas stagnation occurs at 9° N or so in the real Somali Current. Wind stress, bottom topography, or other effects may explain the stagnation. The importance of this study is that it was the first to illustrate the critical role of the inertial effects in the Somali Current.

Cox (1976) considered a box model of similar shape but with 12 levels. This study, like that of Anderson and Rowlands, discussed above, sought to evaluate the relative roles of local and remote driving on the Somali Current. Further, the role of the inertia terms in altering the solutions of Lighthill, and Anderson and Rowlands, was investigated. In the first part of the study, results of the linearized numerical model were illustrated and compared with those of Lighthill. By analysing the numerical results in terms of the normal modes used by Lighthill, it was shown that the westward-moving equatorial waves are accurately described by the model, and the boundary current obtained was very similar to that of Lighthill. The model was then extended to include local forcing and inertial terms. The conclusion reached from the study is that local forcing dominates the boundary current development during at least the first month, with effects of remote forcing playing a significant role thereafter. As in the study of Anderson and Rowlands, the wind stress distribution is critical and, from year to year, the relative importance of the two may change. A further finding of the Cox study is that the Western Boundary Current formed by local forcing undergoes horizontal shearing instability and breaks into a northward progression of eddies. Some evidence for this type of behaviour has been reported by Bruce (1973) and is evident on some satellite-based SST charts.

A potentially important assumption, common to all the studies discussed here, is that the western boundary is meridional. Although speculation has been offered for the effect on these solutions of a sloping boundary, no one has studied the problem formally. This chapter discusses some possible effects of such a boundary.

36.3 Description of the model

The model used in this study is that of Bryan (1969) in which the equations of motion are taken to be the Navier–Stokes equations modified by the Boussinesq approximation and the hydrostatic assumption. Closure is obtained by replacing molecular viscosity by similar terms representing turbulent mixing. Temperature and salinity distributions are obtained by solving the conservation equations. A further description of the model may be seen in Bryan (1969) or Cox (1976). The model is fully three-dimensional with a one-third-degree grid spacing horizontally and 14 levels vertically. The spacing of the vertical levels is variable, allowing increased resolution near the surface. There are 5 levels in the upper 176 m of the model ocean. The horizontal regions considered are shown

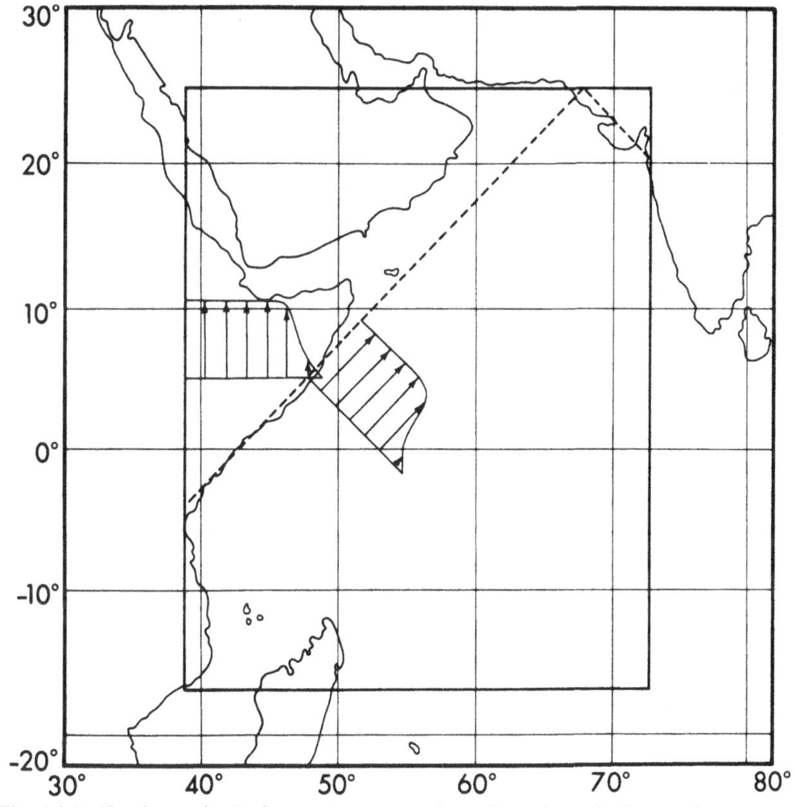

Fig. 36.1. Regions of solution and cross-section of applied wind stress for the two experiments, Case I with a meridional western boundary and Case II with a southwest-northeast boundary, superimposed on a map of the Arabian sea for reference. (From Cox, 1979.)

in Fig. 36.1, superimposed on a map of the Arabian Sea area for reference. The inner, bold rectangle is the boundary for the meridional western boundary case. The sloping boundary case is further bounded by the dashed line. In each case the bottom is taken to be flat at a depth of 4000 m. Unless stated otherwise, mixing coefficients are taken to be $2\times10^3\ m^2\ s^{-1}$ in the horizontal for both heat and momentum, $5\times 10^{-5}\ m^2\ s^{-1}$ in the vertical for heat and $10^{-2}\ m^2\ s^{-1}$ in the vertical for momentum. The gradients of temperature and salinity are taken to be zero at all boundaries. There is no-slip at the lateral boundaries and free-slip at the bottom and top. Surface stress is applied as a body force on the first layer of the model. The system of equations is solved as an initial value problem starting from conditions of a water mass at rest and horizontal homogeneity in temperature and salinity. The temperature and salinity, however, are given an initial vertical stratification which corresponds to the average stratification observed in the Arabian Sea area.

The system of equations is then numerically integrated in time, driven by the imposed wind stress shown in Fig. 36.1. In each case the stress is taken parallel to the western boundary, falling to zero about 1100 km from the boundary. It is applied along the entire western boundary. In order to match the temporal behaviour of the southwest monsoon, the stress is tapered at the beginning, rising from zero to its maximum in 6 days, and thereafter remaining constant. Since these experiments are designed to study the transient response of the western Indian Ocean to the southwest monsoon, the time integration is carried forward for only 3 to 4 months. In this relatively short period of time, the only effect of the fictitious wall at the eastern boundary of the model region is to convert any eastward-propagating equatorial Kelvin or Yanai waves into coastal Kelvin waves. These waves are produced by the driving used here and, at speeds of up to 3 m s^{-1}, would travel around the northern part of the basin quickly enough to affect the western boundary solution after approximately one month. In order to eliminate these waves, the so-called 'sponge' boundary condition is applied along the eastern wall, in which horizontal mixing of heat and salinity is assumed to increase strongly near the wall, causing rapid decay of any Kelvin waves which are produced there. The effect of the fictitious southern wall is very small in this short period of time.

36.4 Description of the results

Case I: Meridional western boundary

This experiment is very similar to the locally-driven case described in Cox (1976). The principal difference is the no-slip condition at the lateral boundaries, rather than the free-slip condition used then. In both cases the wind stress is taken to be 0.2 N m^{-2}. In Cox's (1976) experiment, it was found that a series of gyres formed at low latitudes and proceeded northward in sequence. An analysis of the eddy energy budget indicated that the energy for the gyres comes from the mean flow, via horizontal shearing instability. Results for the present case, with no-slip walls, are given in Fig. 36.2. This figure shows the mass transport streamfunction, integrated over the upper five levels (176 m) of the model. It is interesting to note that the instability occurs in the boundary current in a fashion very

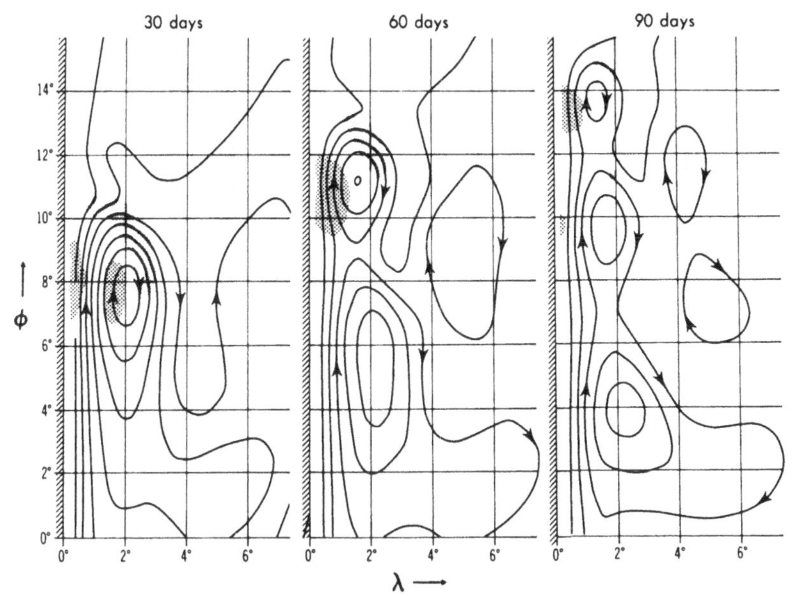

Fig. 36.1. Regions of solution and cross-sections of applied wind stress for the two circulation in the upper 176 m of the model for Case I. Each line represents 4×10^6 m^3 s^{-1}. Divergence is shown by stippling. Fine stippling: $w(z = 176\text{ m}) < -10^{-4}$ m s^{-1}. Coarse stippling: $w(z = 176\text{ m}) > +10^{-4}$ m s^{-1}. (From Cox, 1979.)

similar to its occurrence in Cox's (1976) experiment. The mean current retains an unstable profile offshore of its maximum in spite of the imposed zero value of longshore velocity at the meridional wall. Also of interest is the vigorous downwelling in the offshore segment to the north of the first gyre. It is present to a lesser extent in the later gyres. This feature has been observed in the two-layer model of Hurlburt and Thompson (1976), and some observational evidence for it has been presented by Düing (1978).

The corresponding temperature patterns at 45 m depth are shown in Fig. 36.3. Vigorous coastal upwelling associated with each gyre results in a succession of cold wedges similar in shape and size to those observed at the surface along the Somali coast during summer. The behaviour of these gyres (and therefore, cold water wedges) under varying conditions of boundary slope and wind stress is the primary concern of the remainder of this study.

Case II: Sloping western boundary

In order to study the effect of the sloping coast, the experiment described in Case I was re-run, changing only the slope of the western boundary and

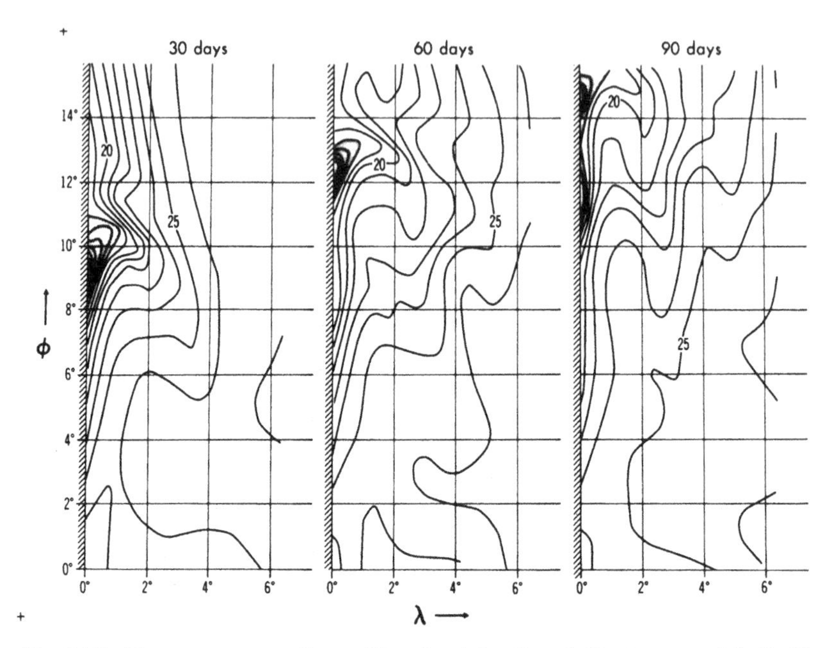

Fig. 36.3. Temperature profiles at 45 m depth for Case I. Contours are labelled in °C. Regions of cold water upwelling (temperature less than 18 °C) are stippled in this figure.

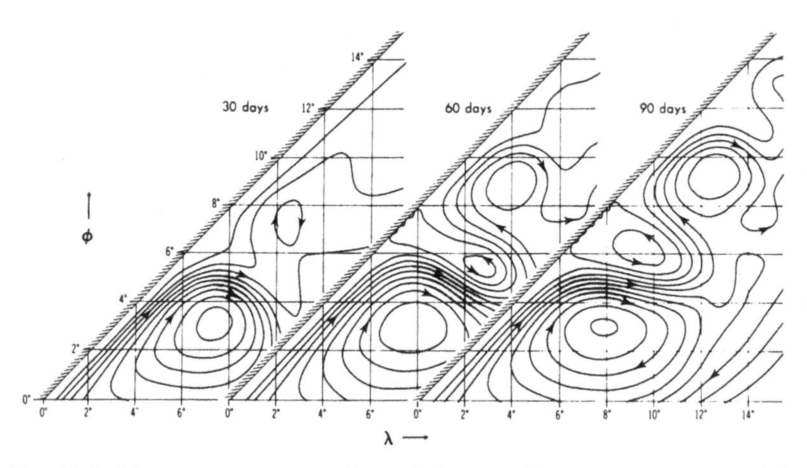

Fig. 36.4. Mass transport streamlines of the non-divergent component of the circulation in the upper 176 m of the model for Case II. Each line represents $4\times10^6\ \mathrm{m^3\,s^{-1}}$. Divergence is shown by fine stippling: $w(z=176\ \mathrm{m}) < -10^{-4}\ \mathrm{m\,s^{-1}}$. (From Cox, 1979.)

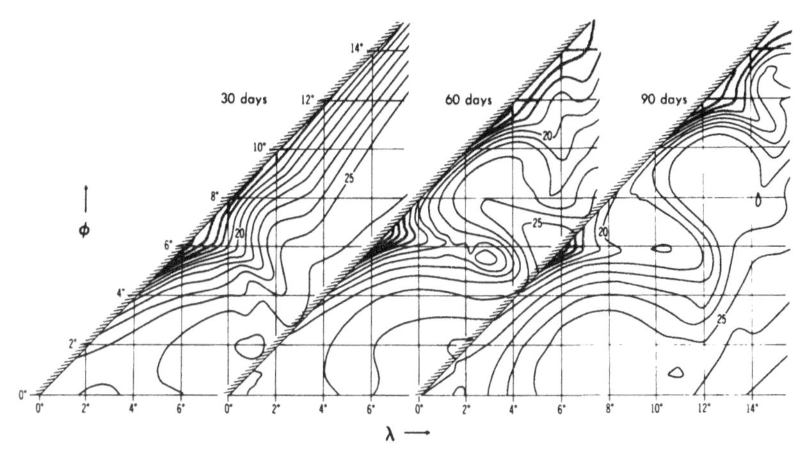

Fig. 36.5. Temperature profiles at 45 m depth for Case II. Contours are labelled in °C. Regions of cold water upwelling (temperatures less than 18 °C) are stippled in this figure.

the wind stress, as illustrated in Fig. 36.1. The streamfunction of the vertically-integrated flow over the upper 176 m is shown in Fig. 36.4 with areas of vigorous downwelling at 176 m indicated by stippling. Once again, the boundary current which is formed is unstable and breaks up into gyres. However, instead of being formed near the Equator and moving northward, they are now formed locally and tend to remain in place, with only a relatively slow movement along the boundary. The patterns of temperature near the surface are shown in Fig. 36.5. Wedge-shaped cold patches form along the boundary in association with each gyre, as in Case I.

36.5 Analysis of the results

It is clear from the preceding figures that surface cooling in this region can be understood only after the movement of the gyres is explained. The most convenient method for analysing highly nonlinear regimes such as this is the examination of the vorticity equation,

$$\nabla^2 \psi_t = -\beta \psi_x + f w_z - J(\psi, \nabla^2 \psi) + F + \text{curl}\,(\boldsymbol{\tau}), \tag{36.1}$$

where F represents frictional effects. A detailed analysis has been carried out for both Case I and Case II. The local rate of change of vorticity is partitioned into its various terms (the beta-effect, vortex stretching, advection, friction and wind stress), and each is analysed at several points of time during the development of the solution. It should be noted that the analysis is interpreted in terms of net forces on each gyre as a whole, rather than in relation to individual particles. Space prohibits the presentation of the detailed results here. Therefore, a descriptive account will be given, based upon the findings of the full analysis.

Vorticity sources occur in both cases, along the no-slip boundary (positive source) and in the region of negative curl of the wind stress some 900 km from the boundary (negative source), as well as from planetary vorticity. The observed flow patterns are a result of the propagation and advection of vorticity produced from these sources. Early in the experiments, westward propagation of negative vorticity takes place along the Equator in the form of Yanai waves. The result is the formation of a clockwise gyre at the Equator, intensifying against the boundary. Ignoring for the moment effects of rotation and friction, and considering the rules of vortex dynamics (e.g. Sommerfeld, 1950), the nonlinear interaction of the boundary vorticity and the westward-moving negative vorticity produces an advection of the gyre along the boundary. After 30 days (see Figs. 36.2 and 36.4), the westward propagation and subsequent advection northward is evident. As the negative vorticity is advected along the boundary, positive vorticity is advected out from the boundary and into its path, as indicated by the weak counterclockwise circulations after 30 days in Figs. 36.2 and 36.4. The result of the redistribution of positive vorticity is to shift the direction of advection of the main gyre from alongshore to somewhat offshore. The degree to which this occurs is determined by the magnitude of the new gyre in comparison with that of the main gyre. If the two become equal for instance, the advection becomes perpendicular to the boundary.

Of the remaining terms in the vorticity equation, the curl of the local wind stress is zero and frictional effects are very small. The beta-effect and vortex stretching contribute to $\nabla^2\psi_t$ in a sense opposite to that of advection. Of these two, the beta-effect is the stronger, and describes the tendency of the gyre to propagate westwards. However, as the vertical motion develops, particularly in the region of outflow, vortex stretching shifts the direction of their combined influence on the displacement of the gyre to the south of west. The movement of the gyres at any point of time is determined by the resultant of the influence of the advection and the rotation. The former results in a displacement of the gyre somewhat offshore, depending on the development of the counterclockwise gyre, and the latter is a displacement somewhat to the south of west, depending on the position and magnitude of the vertical motion. The difference between Cases I and II results mainly from the fact that, in general, the effect of vorticity advection on the gyres changes direction as the boundary slope changes, while the beta-effect does not. In Case I, the alongshore component of the advective effect is relatively unopposed and the gyres move northward at a high speed. In Case II, the alongshore component of the advective effect is more strongly opposed by rotational effects. The gyres move much more slowly and eventually stagnate. Two additional cases have been run in which the sloping boundary experiment is repeated with the magnitude of the stress halved (0.1 N m^{-2}) in the first case and doubled (0.4 N m^{-2}) in the second. While intuition may lead one to expect less tendency toward stagnation for stronger driving, the model results indicate that just the opposite is the case. The counterclockwise gyre ahead of the main gyre forms much more rapidly under stronger driving, the vorticity advection tending to cause the gyre to turn away from the coast and produce stagnation more rapidly, and at a latitude some three degrees further south, than in the case with a stress of 0.2 N m^{-2}. In the weak driving case, slow movement continues alongshore throughout the integration.

Both northward-moving and stationary cold water wedges have been reported in the Somali Current system. It is possible that the balance of effects present in this very simplified model may occur there. A potentially important effect, which has been neglected here, is that of local input of vorticity from wind stress. The vorticity balance of Case II could easily be upset by such an event, causing alongshore shifting of the gyres. It is possible that the behaviour of the gyres in the real Somali Current is very sensitive to this effect.

References

Anderson, D. L. T. and Rowlands, P. B. (1976) The Somali Current response to the S. W. monsoon: the relative importance of local and remote forcing. *J. Mar. Res.*, **34**, 395–417.

Bruce, J. G. (1968) Comparison of near surface dynamic topography during the two monsoons in the western Indian Ocean. *Deep-Sea Res.*, **15**, 665–77.

Bruce, J. G. (1973) Large-scale variations of the Somali Current during the southwest monsoon, 1970. *Deep-Sea Res.*, **20**, 837–46.

Bryan, K. (1969) A numerical method for the study of the circulation of the World Ocean. *J. Computational Phys.*, **4**, 347–76.

Cox, M. D. (1970) A mathematical model of the Indian Ocean. *Deep-Sea Res.*, **17**, 47–75.

Cox, M. D. (1976) Equatorially trapped waves and the generation of the Somali Current. *Deep-Sea Res.*, **23**, 1139–52.

Cox, M. D. (1979) A numerical study of Somali Current eddies. *J. Phys. Oceanogr.*, **9**, No. 2.

Düing, W. (1978) The Somali Current: Past and recent observations. FGGE/INDEX/NORPAX Equatorial Workshop Proceedings held at Scripps Institution of Oceanography, June 1977.

Fieux, M. and Stommel, H. (1976) Historical sea surface temperatures in the Arabian Sea. *Annals De L'institute Oceanographique*, **52**, 5–15.

Hurlburt, H. E. and Thompson, J. D. (1976) A numerical model of the Somali Current. *J. Phys. Oceanogr.*, **6**, 646–64.

Lighthill, M. J. (1969) Dynamic response of the Indian Ocean to onset of the southwest monsoon. *Phil. Trans. Roy. Soc. London*, **A265**, 45–92.

Shukla, J. (1975) Effect of Arabian Sea surface temperature anomaly on Indian summer monsoon: a numerical experiment with the GFDL model. *J. Atmos. Sci.*, **32**, 503–11.

Sommerfeld, A. (1950) *Mechanics of Deformable Bodies*. Academic Press, New York.

37

Maximum simplification of nonlinear Somali Current dynamics

L. B. LIN AND H. E. HURLBURT

A nonlinear model representing only the first baroclinic mode shows a remarkable ability to simulate observed features of the Somali Current system during the southwest monsoon. These features include the strength of the current and its highly nonlinear horizontal structure, a succession of warm and cold eddies just east of the main current, and appropriate patterns of upwelling. The response to the wind along the coast is rapid with large changes occurring in a period of a week. The maximum upwelling occurs near the coast northwest of the warm eddies where the Somali Current separates from the coast, forming the northern branch of an eddy. Associated with this upwelling, sharp cold tongues extend up to 500 km offshore separating the warm eddies. The eddies tend to stall as they move northward from the equatorial region. A comparison of open- and closed-basin solutions of the model equations demonstrates that the upwelling, the strength of the current and the eddies, and the movement of these features along the coast are very sensitive to the strength and distribution of the wind stress and to the boundary conditions. However, the basic structure is not. (The low-level jet is modelled on one of the wind patterns used.)

Simulation of the Somali Current system requires high resolution both longshore and offshore. Here, $\Delta y = 20$ km (longstream), and $\Delta x = 6$ km near the boundary to 192 km in the interior to cover a 5000×5000 km region with a grid of 250×61 points per variable. Using a 20-minute time step, only $7\frac{1}{2}$ minutes central processor (CP) time was consumed for each 60-day integration on the 2-pipe TI/ASC computer at the US Naval Research Laboratory. Because of the success of this simple model in simulating the Somali Current system and the great economy of its operation, it is suggested that a model with a single internal vertical mode is a logical choice for future air–ocean coupled models of the Indian monsoon system.

37.1 Introduction

As suggested by the work of Saha (1974), Shukla (1975), and others, the sea-surface temperature of the Arabian Sea provides a potentially important feedback to the southwest monsoon. During the northern summer the surface of the Arabian Sea is cold compared with most of the ocean surface at the same latitude. The Somali Current system is one source of cold sea-surface temperatures in the western Arabian Sea. It is a major oceanic response to the southwest monsoon in the northern Indian Ocean, flowing northeastward along the east African coast from about 4° S to 10° N each northern summer with speeds of up to 3 $\mathrm{m\,s^{-1}}$. In winter it reverses direction. Observational studies of this current are discussed by Swallow in Chapter 29.

In this chapter we discuss the simplest possible nonlinear model of the Somali Current system capable of representing its salient features, including the rapid response to the wind, the strength and highly nonlinear horizontal structure of the current, and the separation of the current from the coast forming strong warm eddies. These eddies, centred a few hundred kilometres offshore, move poleward. Also represented are the upwelling along the coast with maxima northwest of the eddies, and the sharp cold tongues which extend up to several hundred kilometres offshore north of the eddies. After discussing the dynamics of the numerical solutions, some aspects of ocean model design for a coupled air–ocean model of the southwest monsoon system are considered. A simple, properly designed ocean model is suggested as a logical starting point for a coupled model in this case.

The usefulness of the simple model described here was suggested by the results from a highly nonlinear two-layer model of the Somali Current system described by Hurlburt and Thompson (1976). They were able to explain most of the dynamics in their model using only the first internal mode.

37.2 The model

The numerical model uses the concept of reduced gravity and contains a single internal vertical mode. It consists of two homogeneous layers of different density with the lower layer infinitely deep and motionless. The interface represents the pycnocline. The resulting equations are those for shallow water waves with gravity replaced by the reduced gravity g'. In a right-handed Cartesian coordinate system with x increasing eastward, they are:

$$u_t + uu_x + vu_y - fv = -g'h_x + \tau^x/\rho h + A\nabla^2 u, \tag{37.1}$$

$$v_t + uv_x + vv_y + fu = -g'h_y + \tau^y/\rho h + A\nabla^2 v, \tag{37.2}$$

$$h_t + (hu)_x + (hv)_y = 0, \tag{37.3}$$

where (u, v) and (τ^x, τ^y) are components of velocity and wind stress, h is the layer thickness, f the Coriolis parameter, A the coefficient of lateral friction, t is time, and the subscripts denote partial derivatives. The model utilizes a basin of side 5000 km, about the width of the Indian Ocean at the Equator. The origin of coordinates is at the centre of the western boundary, and the Equator at $y = 0$ bisects the basin. The β-plane approximation is used with $\beta = 2.25 \times 10^{-11}\ \mathrm{m^{-1}\,s^{-1}}$. Other parameters are the mean layer thickness = 200 m, $g' = 2.85 \times 10^{-2}\ \mathrm{m\,s^{-2}}$ and $A = 10^3\ \mathrm{m^2\,s^{-1}}$.

In most cases the boundary conditions are the kinematic conditions assuming no-slip. In some, a variant of the open boundary condition developed by Hurlburt (1974) is applied along the northern and southern boundaries. The free-slip condition has also been tested, but yielded much less realistic results than those obtained using the no-slip condition along solid boundaries.

The model is driven from rest by a wind stress which is increased to its final magnitude over a period of 2 days. Because the Somali Current is such an overwhelming response to the wind, an initial state of rest is adequate for the simulation of the time-dependent response. In most of the cases only local forcing with a meridional wind component in the vicinity of the western boundary is considered. The remote forcing of the Somali Current proposed by Lighthill (1969) is not studied here. Lighthill suggested that the formation of the current is a result of westward packing of equatorial Rossby waves which are generated by the onset of the southwest monsoon in the interior of the northern Indian Ocean. His theory predicts a lag of about a month between the onset of the wind and the formation of the current. However, Leetmaa (1972, 1973) reported rapid response of the coastal current (2 m s^{-1} in 10 days) to the local wind, which started blowing northeastward along the coast about a month before the commencement of the monsoon in the ocean interior. In a linear, inviscid model Anderson and Rowlands (1976) found that the longshore velocity increases linearly with time under local longshore forcing, and quadratically with time under remote zonal forcing by the curl of the wind stress. They found that the Somali Current responds initially to the local forcing. About 2 months later the remote forcing becomes important. Numerical experiments by Cox (1976), which included nonlinear and viscous effects, indicate a similar conclusion, although the local forcing appears to remain important as well (see Chapter 36 for Cox's more recent work).

The finite difference scheme for the numerical model used here is explicit and conserves total energy except for explicit sources and sinks. The staggered grid and the numerical scheme are the same as those used by Holland and Lin (1975). The grid spacing is constant (20 km) in the meridional direction and is stretched in the zonal such that it increases from 5.5 km at the western boundary to 192 km in the central basin, then decreases to 15 km at the eastern boundary. The resulting grid is of 250×61 points per variable. The time step is 20 minutes. Despite the high resolution, only $7\frac{1}{2}$ minutes CP time is consumed for each 60-day integration on the 2-pipe TI/ASC computer at the US Naval Research Laboratory.

37.3 Results

37.3.1 Early development of the Somali Current system

The model Somali Current responds very rapidly to the longshore winds. The solution produced by the model for a wind stress of 0.2 N m^{-2} is shown in Figs. 37.1 and 37.2. The current speed attains its maximum value of 2.1 m s^{-1} in just over two weeks. The values of v after 10 and 30 days are illustrated in Figs. 37.1*b*, and 37.2*b*, showing the rapid growth of the current. The pycnocline anomaly (PA) distribution at day 10 shows weak pressure centres which have formed a few degrees on each side of the Equator and east of the boundary. Associated with these is an eddy centred on the Equator, exhibiting speeds of up to 0.2 m s^{-1} outside the boundary current. It is barely evident in the contours for u and v at day 10.

A series of eddies forms along the Equator as a Yanai wavefront excited by the onset of the meridional wind propagates eastward from the western boundary. With the passage of this front the equatorially-trapped inertial oscillation ceases at different phases at different longitudes. As a result, some of the energy of the oscillation is converted to short planetary waves which propagate westward. The eddies are centred on the Equator with antisymmetric pressure centres about 400 km on either side of the Equator. There are also mean currents (about 0.1 m s^{-1} for this case) centred about 400 km on either side of the Equator,

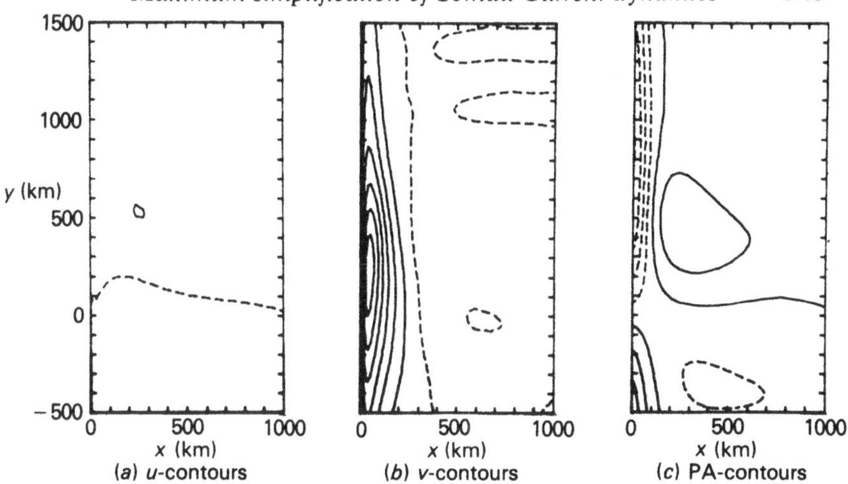

Fig. 37.1. Contours of: (*a*) velocity component u; (*b*) velocity component v; and (*c*) pycnocline anomaly (PA), predicted by the closed-basin model at 10 days. The model was driven from rest by a south wind with a uniform stress of 0.2 N m^{-2} except that it is brought smoothly to zero within 2000 km of the eastern boundary. The pycnocline anomaly is the deviation of the pycnocline from its initial uniform elevation. Dashed contour lines represent negative values. For the velocity components only zero is also dashed. Positive is eastward for u, northward for v, and downward for PA, i.e. upper-layer thickness greater than initial. Contour intervals are 0.2 m s^{-1} for velocity and 10 m for the pycnocline anomaly. The Equator is at $y = 0$ in all three cases. Only a portion of the modelled area within 1000 km of the western boundary is shown.

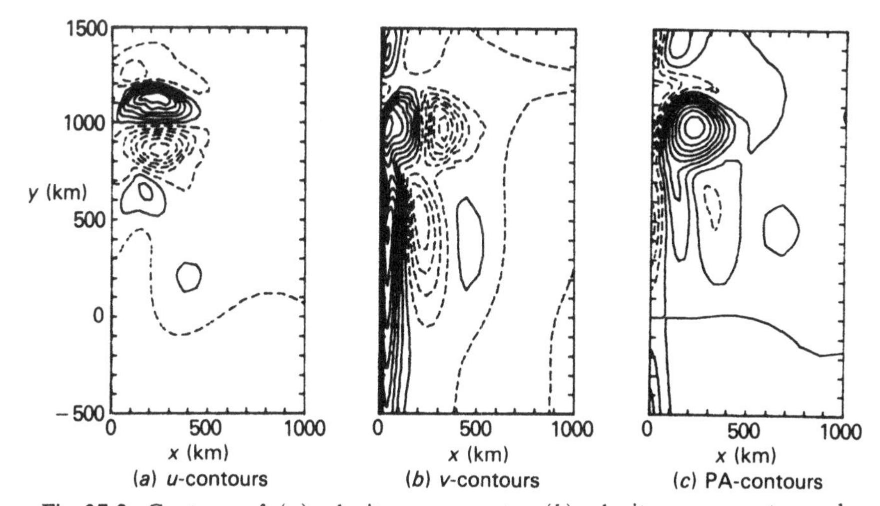

Fig. 37.2. Contours of: (*a*) velocity component u; (*b*) velocity component v; and (*c*) pycnocline anomaly (PA), predicted by the closed-basin model at 30 days. Contour intervals are 0.2 m s^{-1} for velocity and 10 m for the pycnocline anomaly.

eastward in the northern hemisphere, westward in the southern. Bretherton (1964) discusses the theory of equatorially-trapped inertial oscillations (wavenumber 0) in some detail, Moore and Philander (1977) review some aspects of equatorial wave theory for the ocean, and Anderson and Rowlands (1976) discuss the linear, inviscid response of the Somali Current to meridional forcing.

The PA shows that north of the Equator, cold water has upwelled along the western boundary. South of the Equator, downwelling has occurred. This pattern is a result of one-sided divergence driven by the longshore wind near the coast, a phenomenon common along eastern ocean boundaries. Associated with the midlatitude phenomenon of coastal upwelling or downwelling is a geostrophically balanced coastal jet (e.g. O'Brien and Hurlburt, 1972). The Somali Current is dynamically somewhat different, except at distances from the Equator much larger than the equatorial radius of deformation. On a scale governed by the equatorial radius of deformation, about 325 km in our case, the longshore pressure gradient and the wind stress act in the same direction so as to rapidly accelerate a northward coastal jet, with maximum effectiveness at the Equator (Hurlburt and Thompson, 1976).

At day 10 the solution is nearly linear. For the wind stress and geometry used here, the linear solution exhibits symmetry about the Equator, with the meridional flow symmetric, and the zonal flow and the pressure antisymmetric. Only a few days later the numerical solution loses its symmetric properties as it becomes strongly nonlinear.

Fig. 37.2 depicts the asymmetric properties of a strongly nonlinear solution in the western ocean at day 30. Most of the activity is in the northern hemisphere. The contours of v at day 30 show the current maximum displaced almost 1000 km north of the Equator. A well-developed countercurrent is also in evidence. East of the coastal upwelling a strong warm eddy (depressed pycnocline) has formed. To the southeast of that a cold eddy (raised pycnocline) is seen, and to the east of that another eddy. The first two develop because the coastal jet becomes nonlinear. The third is a linear phenomenon, and only it has a signature in both hemispheres (poorly visible in Fig. 37.2).

Hurlburt and Thompson (1976, p. 657) discuss the dynamical development of the strong warm eddy in some detail. Briefly, the longshore jet becomes supergeostrophic due to advection. Since the magnitude of the Coriolis force exceeds the pressure gradient force normal to the coast, the flow is accelerated offshore, and the Somali Current separates from the coast. This in turn increases the one-sided divergence near the coast and enhances the upwelling there. When the Somali Current turns offshore forming the eddy, it remains super-geostrophic with speeds at times over 1 m s^{-1}. Hence, the flow is continually accelerated to the right and spirals toward the centre across the pressure contours. An important dynamical characteristic of the eddy is that it is a forced anticyclonic inflow in the surface layer. This means it can entrain some of the cold coastal water. Satellite pictures shown at the October 1976 INDEX meeting by O. Brown of the University of Miami, indicate that while the warm eddy is warmer than the coastal water, it is cooler than the water further towards the interior of the Arabian Sea.

37.3.2 Later development of the Somali Current system

The more mature development of the Somali Current is now traced, especially the development of multiple eddies, a phenomenon also found by Cox (1976) in a 12-level model. Figs. 37.3 and 37.4 show the early development of a second major warm eddy for 3 different cases. The 0.1 N m^{-2} cases show the development process more clearly than the 0.2 N m^{-2}. Thus, in all three cases the wind stress is 0.1 N m^{-2}, but the boundary conditions and the wind stress distribution differ. The case

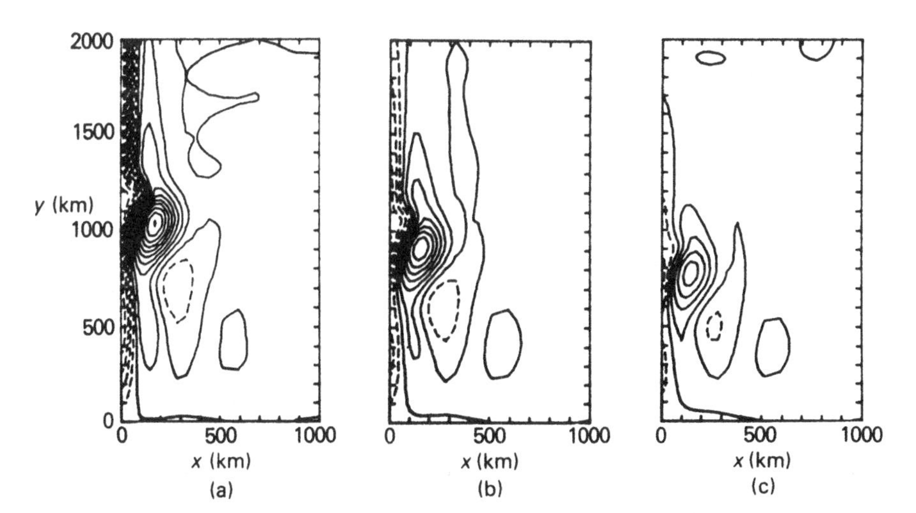

Fig. 37.3. Comparison of the pycnocline anomaly for three different cases at day 35: (*a*) open-basin model; (*b*) closed-basin model; and (*c*) closed-basin model with the wind applied only within 1500 km of the Equator. The wind stress is 0.1 N m^{-2} in all three cases and the contour interval is 10 m. Dashed contour lines represent negative values.

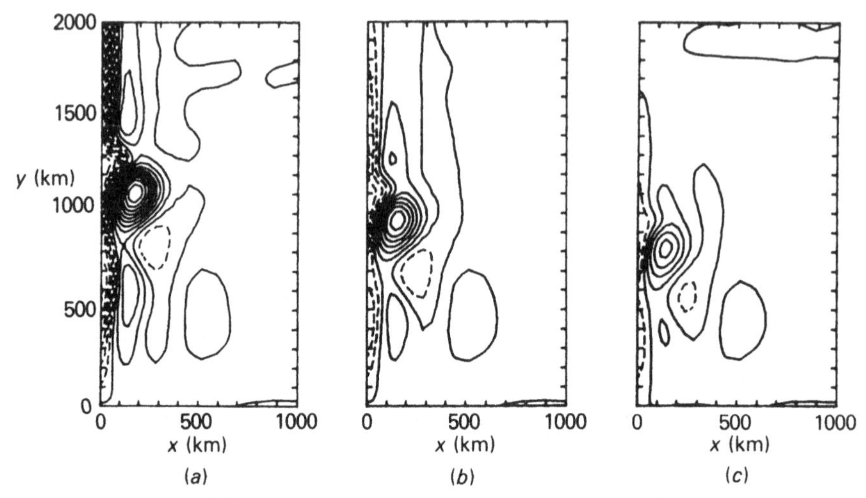

Fig. 37.4. Comparison of the pycnocline anomaly for three different cases at day 40: (*a*) open-basin model; (*b*) closed-basin model; and (*c*) closed-basin model with the wind applied only within 1500 km of the Equator. The wind stress is 0.1 N m^{-2} in all three cases and the contour interval is 10 m. In the open-basin case the solution near the western boundary is unaffected by the wind stress near the eastern boundary on a 60-day time scale.

shown in Figs. 37.3*a* and 37.4*a* was driven by a uniform south wind everywhere, and has open boundary conditions along the northern and southern boundaries. The case shown in Figs. 37.3*b* and 37.4*b* is the same except that the boundaries are closed and there is no wind within 200 km of the eastern boundary. It is also the same as the case in Figs. 37.1 and 37.2, but with half the wind stress amplitude. The case shown in Figs. 37.3*c* and 37.4*c* is the same as that shown in Figs. 37.3*b* and 37.4*b*, except that the wind stress is applied only within 1500 km of the Equator. The solution for the corresponding open-basin case is almost identical.

The substantial differences in the three solutions are primarily due to the effects of the wind stress distribution and the normal boundary conditions on the longshore structure of the pressure field near the western boundary. Despite the striking differences, the basic structural development of the current system is very similar, and the timing for the formation of the second warm eddy is almost the same. This suggests that the triggering mechanism is a linear phenomenon.

In all three cases there is a primary warm eddy and a sympathetic cold eddy. The third eddy is a pressure centre for a short planetary wave which has propagated westward from the interior. Comparison of Figs. 37.3 and 37.4 shows that as the wave propagates westward, a second warm eddy forms from the Somali Current at almost the same time in all three cases. Furthermore, the formation of the next major warm eddy is timed with the next incoming pressure centre of the wave. It is worth noting that the major warm eddies form locally and not from the wave pressure centres. The waves are only a perturbing mechanism. Other perturbing mechanisms, such as temporal variation in the wind, should also be capable of generating additional eddies. Figs. 37.5 and 37.6 show the more mature development of the Somali Current and the eddies for the 0.1 Nm^{-2} (Fig. 37.5) and 0.2 N m^{-2} (Fig. 37.6) cases shown in Figs. 37.1, 37.2, 37.3*b* and 37.4*b*. The first anticyclonic eddy slows down considerably in its northward movement with the development of the sharp cold tongue north of it. In the 0.1 N m^{-2} case it slows to 0.065 m s^{-1}, in the 0.2 N m^{-2} case to only 0.03 m s^{-1}, in agreement with the observations of Düing (1977). In the 0.1 N m^{-2} case the southern (second) warm eddy intensifies while the northern one decays. Since f increases with latitude and the eddy is in approximate geostrophic balance, the kinetic energy decays much more rapidly than the potential energy as the eddies progress up the coast. Near day 95 a third warm eddy forms south of the second one, with the approach of another planetary wave.

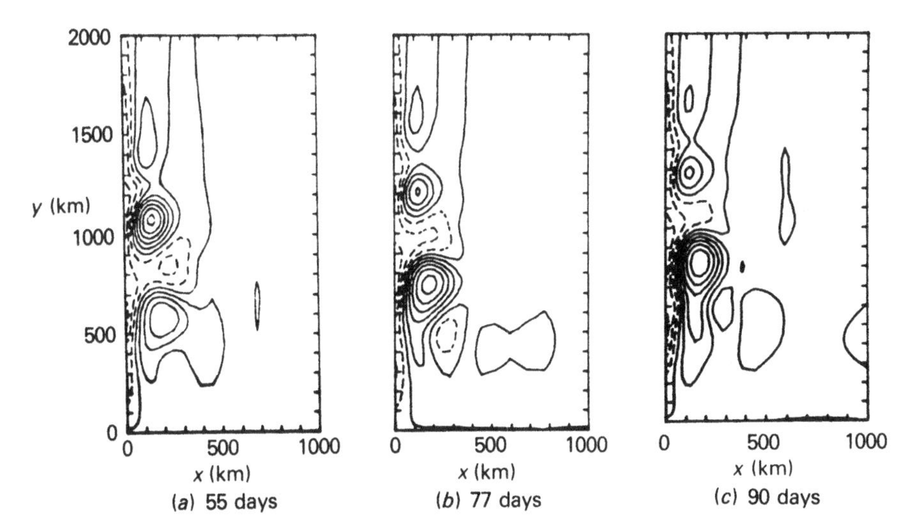

Fig. 37.5. The pycnocline anomaly in the closed-basin model with a wind stress of 0.1 N m^{-2} for: (*a*) 55 days; (*b*) 77 days; and (*c*) 90 days. The contour interval is 5 m.

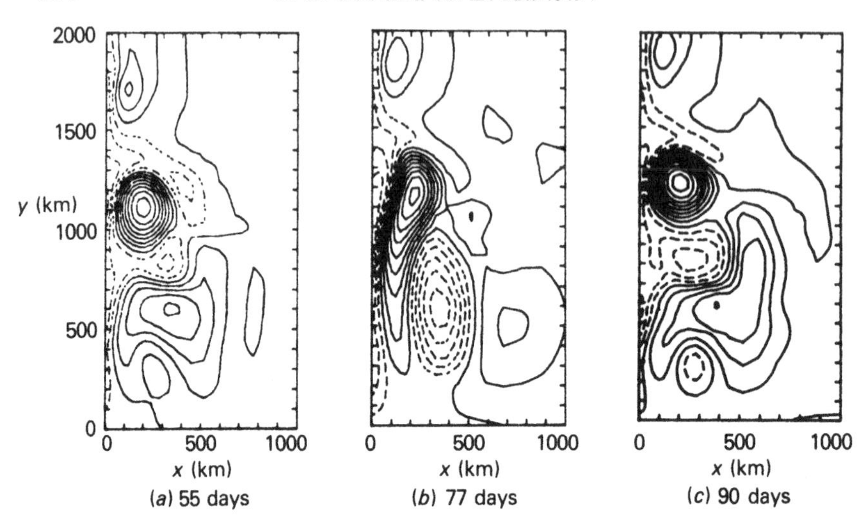

Fig. 37.6. The pycnocline anomaly in the closed-basin model with a wind stress of $0.2\,\mathrm{N\,m^{-2}}$ for: (*a*) 55 days; (*b*) 77 days; and (*c*) 90 days. The contour interval is 10 m.

In the $0.2\,\mathrm{N\,m^{-2}}$ case, the cold eddy takes longer to pinch off the second warm eddy in the pressure field, but it is observed in the velocity field about the same as in the $0.1\,\mathrm{N\,m^{-2}}$ case. The cold tongue is also much more developed, extending up to several hundred kilometres offshore. Unlike the $0.1\,\mathrm{N\,m^{-2}}$ case, the second eddy rejoins the first (see day 77), only to reform near day 85. In the process the cold tongue expands and contracts, shedding cold patches of water on contraction.

The translation speed of the inertial current and the eddies along the coast is determined by the residual of several large terms. Thus, it is sensitive to several external factors, including the boundary conditions (free-slip or no-slip, open or closed), the distribution and magnitude of the wind stress, the eddy viscosity and, as is shown by Cox in Chapter 36, the orientation of the coastline, to name but a few. There is a common factor in the internal dynamics. Both of the advective terms in (37.2) are large, implying that the Somali Current tends to become strongly inertial, the current maximum moving northward. Through the Coriolis force in (37.2), a well-developed and strongly supergeostrophic offshore flow north of the eddy provides a powerful negative feedback to the northward progression of the inertial current (Hurlburt and Thompson, 1976). The advective terms in (37.1) aid in slowing the northward translation of the system, but play their essential role in the development of the sharp cold tongues protruding offshore.

Although in (37.2) horizontal friction is a large term only within a few kilometres of the coast, it plays, in conjunction with the tangential boundary condition, an important role as the dissipative mechanism in the model. It greatly affects the strength, scales, and movement of the system. It is interesting that setting A equal to $10^3\,m^2\,s^{-1}$ gives much better agreement with observations in all respects than a value of 10^2, surprising skill for such a crude representation of friction. Only friction in (37.1) and the nonlinear variation of the layer thickness in (37.2) and (37.3) play no major roles in the model. Hence, with these exceptions the model contains only the minimum physics required to produce a remarkable simulation of the Somali Current system.

37.3.3 Low-level jet case

Fig. 37.7 shows the wind forcing and solution for a case driven by a crude representation of the low-level jet (Findlater, 1971). The longshore wind exerts a maximum stress of $0.4\,N\,m^{-2}$ at $y = 900$ km and an offshore component with the same amplitude at the same latitude. By day 10 a moderately strong eddy has developed just south of the wind maximum (compare with Fig. 37.1). This eddy develops further, moving northward, and later begins to decay. Near day 60 a second warm eddy forms, again just south of the wind maximum. It continues to intensify during the remainder of the 90 day integration, but *does not move*. By day

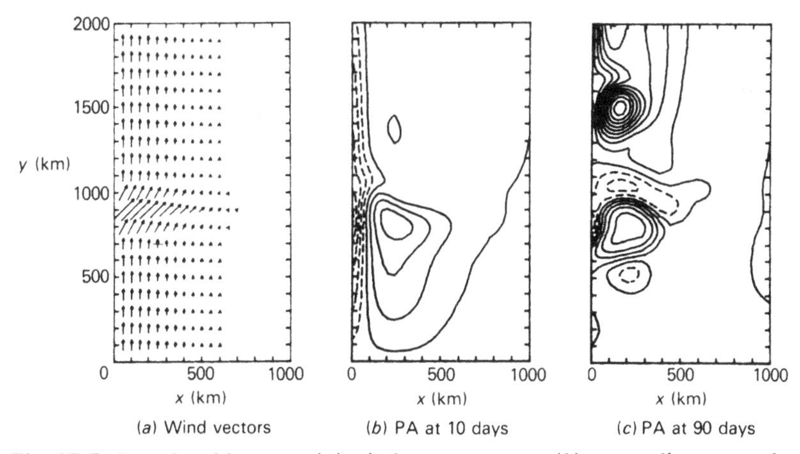

Fig. 37.7. Low-level jet case: (a) wind stress vectors; (b) pycnocline anomaly at 10 days, and (c) pycnocline anomaly at 90 days. The contour interval in (b) and (c) is 10 m. Dashed contour lines represent negative values.

90 its kinetic energy greatly exceeds that of the first warm eddy further north. Near day 90 a third warm eddy (not yet evident in the pressure field) forms 250 km north of the Equator. By this time the Somali Current has broken into a series of eddies with weak reversals in the longshore current between them.

37.3.4 Effects of open and closed boundaries

In Figs. 37.3*c* and 37.4*c*, where the wind stress was restricted in meridional extent, the open- and closed-basin solutions are virtually identical. Fig. 37.8 shows this is not the case when the wind stress extends to the northern and southern boundaries of the model region. In Fig. 37.8*a* the model has open boundaries and in Fig. 37.8*b* closed boundaries. Both were driven from rest by a uniform south wind. The open-basin case shows a raised pycnocline all along the western boundary north of the

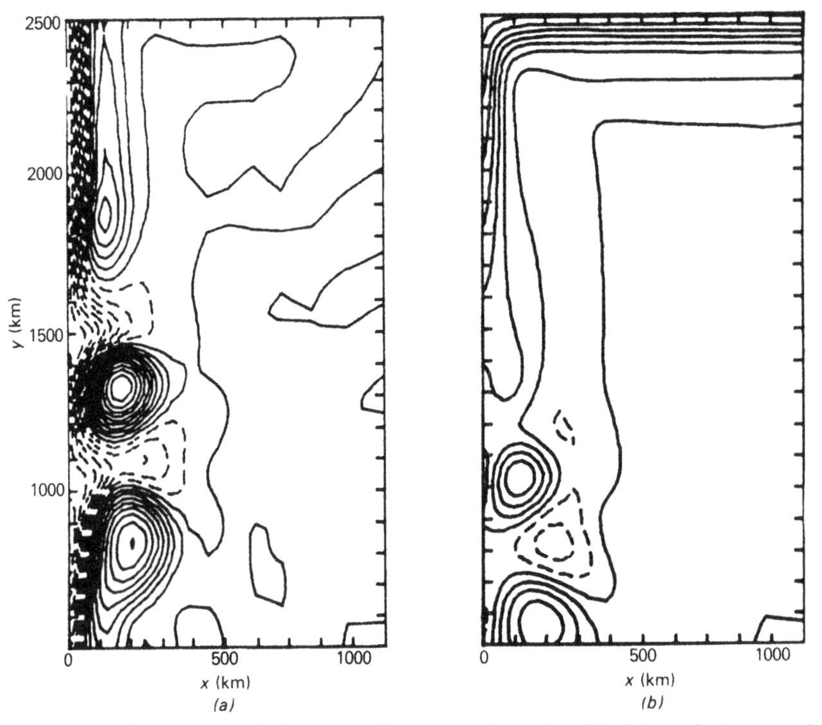

Fig. 37.8. Comparison of the pycnocline anomaly at day 60 with a wind stress of 0.1 N m^{-2} for: (*a*) the open-basin solution; and (*b*) the closed-basin solution. The northern boundary to the region is at $y = 2500$ km. The contour interval is 10 m. Dashed contour lines represent negative values.

Equator. The solution is much less vigorous in the closed-basin case and the pycnocline has actually been displaced downward along the northern part of the western boundary. South of this, the displacement of the pycnocline is almost zero along the western boundary. This situation with no cold water should definitely be avoided in a coupled air–ocean model. It resulted because a large amplitude Kelvin wave, originating from the eastern boundary, propagated along the northern boundary and into the region of interest. Eliminating the wind stress within a few hundred kilometres of the eastern boundary greatly reduces this problem, but the open- and closed-basin cases still differ in important respects (compare Figs. 37.3*a*, 37.4*a* and 37.3*b*, 37.4*b*).

37.4 Ocean model design for a coupled air–ocean model

Although a reduced gravity model is inadequate for many ocean circulation problems, it has demonstrated remarkable success in simulating the horizontal structure of the Somali Current system. This success and the great economy of its operation suggests that an ocean model with a single internal vertical mode is a logical initial choice for coupled air–ocean modelling of the southwest monsoon system. This simple model contains almost the minimum physics and resolution needed to simulate the strength of the Somali Current system, its highly nonlinear horizontal structure, the succession of warm and cold eddies, and appropriate patterns of upwelling, including the sharp cold tongues. It also models the very rapid response to the wind, with major changes occurring in a period of only a week. Because the Somali Current exhibits such an overwhelming response to the wind, it is not necessary to initialize the model using observed ocean data; nor is a costly multi-year spin-up of the ocean model climatology required.

High horizontal resolution is crucial, especially for the sharp cold tongues, the coastal upwelling, the Kelvin waves, and the strength and translation speed of the eddies. In a case with 60 km grid spacing in the meridional direction, the cold tongue extends only one third as far as in the 20 km resolution case. Also, the cold eddy which usually develops southeast of the first warm eddy is greatly reduced in strength, as is the second warm eddy.

As shown in Fig. 37.8, it is essential that the model should not allow unwanted large amplitude Kelvin waves to propagate into the region of interest. It is also important to note that no-slip boundary conditions are required on solid boundaries for realistic results with this model.

Perhaps the most difficult problem is that of coupling the model to the atmosphere. In its present hydrodynamical form, some functional relationship between the pycnocline anomaly and the surface temperature could be used. However, including the physics of the mixed layer should give better results. It has been suggested that wind mixing and evaporation by themselves play a very important role in the northern summer cooling of the Arabian Sea. A formulation suitable for layered models is needed to provide results superior to the simple convective adjustment currently used in most ocean models incorporating thermodynamical processes.

This work was supported by the Ocean Science and Technology Division of the Office of Naval Research on a subcontract from the Naval Research Laboratory as part of INDEX, an ONR/NSF sponsored project. Partial support was also provided by the Naval Oceanographic Laboratory of the Naval Ocean Research and Development Activity.

References

Anderson, D. L. T. and Rowlands, P. B. (1976) The Somali Current response to the southwest monsoon: the relative importance of local and remote forcing. *J. Mar. Res.*, **34**, 395–417.

Bretherton, F. P. (1964) Low-frequency oscillations trapped near the Equator. *Tellus*, **16**, 181–5.

Cox, M. D. (1976) Equatorially trapped waves and the generation of the Somali Current. *Deep-Sea Res.*, **23**, 12, 1139–52.

Düing, W. (1977) Large-scale eddies in the Somali Current. *Geophys. Res. Let.*, **4**, 4, 155–8.

Findlater, J. (1971) Mean monthly airflow at low levels over the western Indian Ocean. *Geophys. Mem.* (HMSO, London), **16**, 115, 1–53.

Holland, W. R. and Lin, L. B. (1975) On the generation of mesoscale eddies and their contribution to the oceanic general circulation. I. A preliminary numerical experiment. *J. Phys. Oceanogr.*, **5**, 642–57.

Hurlburt, H. E. (1974) The influence of coastline geometry and bottom topography on the eastern ocean circulation. Ph.D. Thesis, Florida State University, 103 pp.

Hurlburt, H. E. and Thompson, J. D. (1976) A numerical model of the Somali Current. *J. Phys. Oceanogr.*, **6**, 5, 646–64.

Leetmaa, A. (1972) The response of the Somali Current to the southwest monsoon of 1970. *Deep-Sea Res.*, **19**, 319–25.

Leetmaa, A. (1973) The response of the Somali Current at 2° S to the southwest monsoon of 1971. *Deep-Sea Res.*, **20**, 397–400.

Lighthill, M. J. (1969). Dynamic response of the Indian Ocean to the onset of the southwest Monsoon. *Phil. Trans. Roy. Soc. London*, **A265**, 45–92.

Moore, D. W. and Philander, S. G. H. (1977) Modelling of the equatorial oceanic circulation. *The Sea*, ed. E. D. Goldberg, Vol. 6, pp. 319–61, Wiley–Interscience, New York.
O'Brien, J. J. and Hurlburt, H. E. (1972) A numerical model of coastal upwelling. *J. Phys. Oceanogr.*, **2**, 14–26.
Saha, K. (1974) Some aspects of the Arabian Sea summer monsoon. *Tellus*, **26**, 464–76.
Shukla, J. (1975) Effect of Arabian sea-surface temperature anomaly on Indian summer monsoon: a numerical experiment with the GFDL model. *J. Atmos. Sci.*, **32**, 503–11.

38

Laboratory modelling of the oceanic response to monsoonal winds†

RUBY KRISHNAMURTI

In a laboratory model ocean, fluid in a rotating tank of varying depth is subjected to various stress patterns which simulate both steady and seasonally varying winds, including monsoonal winds. For a certain range of the governing parameters (Rossby number, Ekman number and Froude number), a homogeneous fluid displays steady westward intensified flow. For the same range of parameters a two-layer fluid can have baroclinic instabilities. The parameter range for these instabilities is mapped in a regime diagram. The northward transport of the western boundary current is measured as it varies with the curl of the imposed wind stress, and is compared with the corresponding values in a homogeneous fluid. The condition for surfacing of the lower layer is measured as it varies with Rossby number and Froude number.

38.1 Introduction

A laboratory model of the wind-driven ocean circulation is described in which various wind stress patterns are simulated which drive fluid in a tank of varying depth. Both homogeneous and two-layer fluids were used. The following three experiments are described:

(i) measurement of northward transport in the western boundary current as it varies with the magnitude of the wind stress;

† This is contribution no. 148 of the Geophysical Fluid Dynamics Institute, Florida State University.

(ii) conditions for surfacing of the lower layer and separation of the western boundary current from the boundary;

(iii) periodic forcing of a two-layer model.

The first of these was motivated by the observation that the oceans effect a major portion of the transport of heat from Equator to pole. Von der Haar and Oort (1973) have shown that the poleward transport by the oceans, when averaged over latitude, is comparable to the atmospheric transport. However, unlike the atmosphere which is primarily thermally driven, the upper oceans are to a large extent driven by the atmospheric winds. These winds change not only with the seasons, but also with the changing climate. For example T. N. Krishnamurti, in Chapter 2, shows that there are major changes in the airflow pattern in years of normal rainfall compared with years of drought over India and north Africa. It would be interesting to know the change in oceanic transport for a given change in the winds. This change in the poleward heat transport by the oceans not only affects the climate at a given latitude; it is also an important factor in the question of maintenance of the permanent Arctic sea ice.

The second experiment, relating to surfacing of the lower layer, was performed to test a theory due to Veronis (1973*a*, 1977). For a two-layer model with a quiescent lower layer, his argument proceeds as follows. In a region where the curl of the wind stress is anticyclonic (such as the subtropical North Atlantic), the Sverdrup transport is to the south. Mass balance across a latitude line requires an equal northward transport in the region near the western boundary. This would require, for the geostrophically balanced part of the transport alone, that the height of the interface at the western boundary be equal to that at the eastern boundary. In addition, one must also consider the Ekman transport. In the trade wind regions, the Ekman transport is to the north. Thus, the return transport on the west must be less than that required by the geostrophic flow alone, so the interface must be *lower* at the western than at the eastern boundary. If surfacing occurs at these latitudes, it would occur first on the eastern boundary. However, in the anticyclonic part of the Westerlies, the Ekman transport is to the south, adding to the geostrophic transport to the south. The interface at the western boundary must therefore be *higher* than at the eastern boundary. At these latitudes, surfacing would occur first on the western boundary. A similar argument has also been presented by Parsons (1969).

The third experiment was performed to simulate, in a very rough way, the response of a sea (such as the Arabian Sea) to time-varying winds (such as the southwest and northeast monsoons.) The spin-up time of a two-layer laboratory model can be adjusted to resemble the response time of a tropical ocean basin.

38.2 Laboratory model

Certain gross features of the general circulation of the oceans can be modelled in laboratory experiments, as has been shown by Stommel *et al.* (1958), Pedlosky and Greenspan (1967), Beardsley (1969), Baker and Robinson (1969), Veronis (1973*b*), Hart (1975) and others. For example, a constant curl of the wind stress can be simulated by a rotating lid in contact with the fluid, and the variation of the Coriolis parameter with latitude (β-effect) can be simulated by variation of fluid depth with location in the tank.

A schematic diagram of the apparatus is shown in Fig. 38.1. A circular cylindrical tank is divided into three separate basins: a 180° basin, a 120° basin, and a 60° basin. The tank is placed on a rotating table whose angular velocity is Ω. A cone-shaped lid rotates relative to the tank at an angular velocity $\Delta\Omega$. As in Beardsley's (1969) experiments this model possesses a western boundary. It differs from his homogeneous experiments in that these are two-layer experiments. In this respect, it is similar to Hart's two-layer models. However, Hart's studies, directed to polar ocean basins, had no western boundary.

The physical basis for the model lies in the analogy between the vortex stretching by flow across constant depth contours, and the relative vorticity produced by the oceanic flow between latitudes of different Coriolis force. Greenspan (1968) has elucidated the difference in the flows that may arise in closed containers having or not having closed geostrophic (constant depth) contours. It is seen in Fig. 38.1 that each of the three basins possesses no closed contours of constant depth. The nature of the steady driven flow is completely different from the case with closed geostrophic contours. The interior flow is a slow, order $E^{\frac{1}{2}}$ (where E is the Ekman number) drift across depth contours.

Since the speed on the top boundary is a linear function of radial distance, the top Ekman layer has a constant divergence with a vertical velocity w_E of order of $E^{\frac{1}{2}}$. Because the interior flow is of order of $E^{\frac{1}{2}}$, the bottom Ekman layer, for the case of a homogeneous fluid, has a much smaller flux of order of E. However, if the interior flow was across depth contours at a speed (v) such that the orographically induced vertical velocity ($w = v \tan \alpha$) balances the Ekman vertical velocity w_E, then the

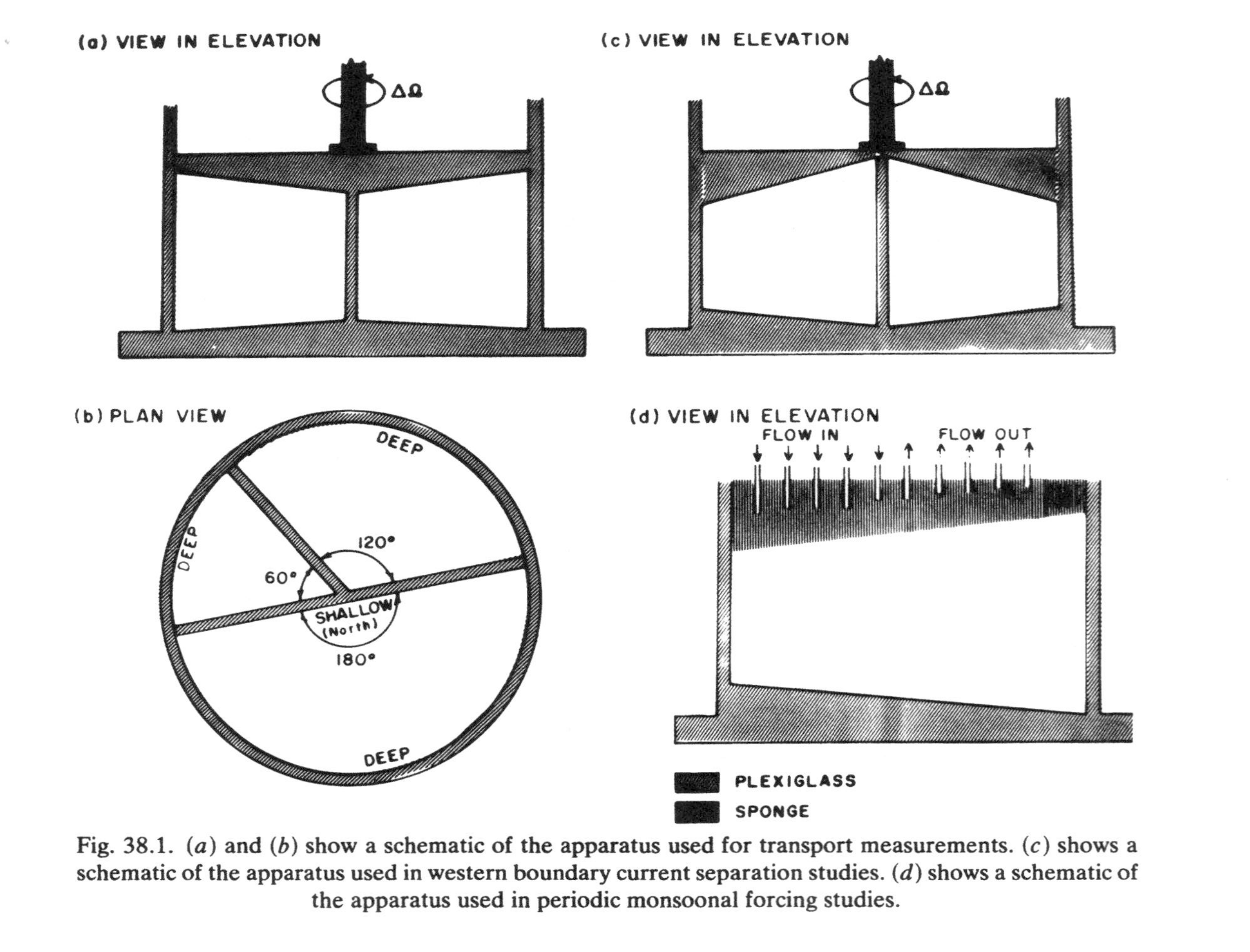

Fig. 38.1. (*a*) and (*b*) show a schematic of the apparatus used for transport measurements. (*c*) shows a schematic of the apparatus used in western boundary current separation studies. (*d*) shows a schematic of the apparatus used in periodic monsoonal forcing studies.

interior velocity could be depth independent. In the case where Ω is counterclockwise and $\Delta\Omega$ is clockwise, this is a constant flow to greater depths (or southwards flow) and is the equivalent of the Sverdrup balanced solution.

Mass is returned to the north by a fast current along the western wall. Beardsley (1969) shows that in the case where $\tan\alpha \ll E^{\frac{1}{4}}$, a Stommel (1948)-type geostrophic western boundary layer of thickness $E^{\frac{1}{2}}/\tan\alpha$ forms in which vortex shrinking is balanced by Ekman suction. When $\tan\alpha$ approaches $E^{\frac{1}{4}}$, the western boundary layer thins down towards the Stewartson $E^{\frac{1}{4}}$ layer. In this case, the diffusion of vorticity from the side wall is balanced by orographic vortex stretching as well as by Ekman-layer suction. When $\tan\alpha \gg E^{\frac{1}{4}}$ the flow is analogous to the Munk and Carrier (1950) model with lateral friction.

Rotating a rigid lid simulates a constant curl of the wind stress, with uniform vertical velocity out of the top Ekman layer. This effect could as well have been simulated by a uniform distribution of sources of mass at a non-rotating top boundary (Baker, 1971). Wind stress patterns are not always describable by a constant curl. Any jet such as the Findlater or 'low-level' jet over the Arabian Sea could better be described as a concentrated region of anticyclonic curl adjacent to another concentrated region of cyclonic curl. Such a wind stress pattern is simulated by a line of sources adjacent to a line of sinks in a porous upper boundary. Fig. 38.1*d* and Fig. 38.8 demonstrate this.

For the two-layer fluid case, these experiments showed that the lower layer has mean velocities approximately two orders of magnitude smaller than the upper-layer velocities. Thus, the assumption made in many theories that there is no stress transmitted across the interface appears to be most reasonable. A Sverdrup interior and a Stommel-type western boundary layer (Stommel, 1948) is then a possible upper-layer solution. The interface becomes distorted in such a way as to prevent the development of pressure gradients in the lower layer, the slope of the interface being proportional to the interior velocity of the upper layer in accordance with the Witte–Margules formula. The change in upper-layer depth produced by such distortion of the interface does not affect the vorticity of the upper layer since the geostrophic flow is perpendicular to the depth gradient. This being the case, distortions of the interface should not be thought of as contributing to the β-effect.

Lower-layer motions are, however, possible in the western boundary layer even for steady flow. Welander (1968) shows a recirculating flow confined to the western boundary lower layer if the northward gradient f/h is positive, where f is the Coriolis parameter, and h is the lower-layer depth.

For a two-layer laboratory model governed by Ekman-type dynamics, a convenient starting point is described in the next section.

38.3 Theoretical basis for the two-layer model

For steady linear flow, the vorticity equations in each layer can be shown to be (Hart, 1972)

$$\boldsymbol{V}_1 \cdot \boldsymbol{\nabla} H_{\mathrm{T}} - \tfrac{1}{2} E_1^{\frac{1}{2}} (\nabla^2 P_1 - 2) + \tfrac{1}{2} E_1^{\frac{1}{2}} \frac{\chi}{1+\chi} (\nabla^2 P_2 - \nabla^2 P_1) = 0, \quad (38.1)$$

$$\boldsymbol{V}_2 \cdot \boldsymbol{\nabla} H_{\mathrm{B}} - \tfrac{1}{2} E_2^{\frac{1}{2}} \nabla^2 P_2 - \tfrac{1}{2} E_2^{\frac{1}{2}} \frac{1}{1+\chi} (\nabla^2 P_2 - \nabla^2 P_1) = 0, \quad (38.2)$$

where $\boldsymbol{V}_1$ and $\boldsymbol{V}_2$, are the upper- and lower-layer velocities respectively, P_1 and P_2 are the upper- and lower-layer geostrophic streamfunctions respectively and H_{T}, H_{B} are the shapefunctions of the top and bottom lids respectively. Here, $\chi = \nu_1/\nu_2$ and we shall take

$$H_{\mathrm{T}} = -y \tan \alpha_1,$$

$$H_{\mathrm{B}} = y \tan \alpha_2.$$

The first term in (38.1) represents the orographically induced vertical velocity resulting from geostrophic flow along the top boundary. The second term is the Ekman suction at the top associated with interior vorticity, the third is the vertical velocity resulting from Ekman layer divergence. The fourth and fifth terms represent Ekman suction at the interface associated with the interior vorticities of the lower and of the upper layers. The Froude number is assumed to be small and vortex stretching as columns of fluid move over the parabolic mean interface is neglected. For a 10 cm radius tank rotating at 20 rpm, the interface is only approximately 1 mm higher at the rim than at the centre. This may be compared to the change in height of 2 cm due to the slope of the lid.

If we take $\nu_1 = \nu_2$ the above equations become

$$(\nabla^2 + \beta_1 \partial/\partial x) P_1 - \tfrac{1}{3} \nabla^2 P_2 = \tfrac{4}{3}, \quad (38.3)$$

$$(\nabla^2 + \beta_2 \partial/\partial x) P_2 - \tfrac{1}{3} \nabla^2 P_1 = 0, \quad (38.4)$$

where we have defined

$$\beta_1 = \tfrac{4}{3} \tan \alpha_1 / E_1^{\frac{1}{2}},$$

$$\beta_2 = \tfrac{4}{3} \tan \alpha_2 / E_2^{\frac{1}{2}}.$$

For a flat-bottomed ocean we would choose a model with $\beta_1 = \beta_2$. However, β_2 could be varied to represent different bottom topographies. For example, if β_2 is chosen as zero, we find the upper-layer equation reduces to Stommel's equation for a homogeneous fluid, while the lower layer is motionless everywhere.

The interior balance is given by

$$\beta_1 \partial P_1 / \partial x = \tfrac{4}{3},$$

$$\beta_2 \partial P_2 / \partial x = 0,$$

which is the analogue of the Sverdrup interior flow.

The Laplacian terms in (38.3) and (38.4) become important in the western boundary layer. Equations for P_1 and P_2 alone can be obtained from (38.3) and (38.4), of the form

$$LP_1 = 0,$$

$$LP_2 = 0,$$

where

$$L = \nabla^4 + \tfrac{9}{8}(\beta_1 + \beta_2)\nabla^2 \frac{\partial}{\partial x} + \tfrac{9}{8}\beta_1\beta_2 \frac{\partial^2}{\partial x^2}.$$

If $\tan \alpha_1$ is of the order of E^b and $\tan \alpha_2$ is of the order of E^c, various types of western boundary layers are possible for various magnitudes of b, c. For the upper layer these are generally of the kind discussed by Beardsley (1969), and include a Stommel-type western boundary current. In the lower layer, the western boundary region has been shown by Welander (1968) to have various forms depending upon the bottom topography. The present laboratory models include studies with $\tan \alpha_2 =$ 0, 0.04, and 0.84. The second corresponds to $\beta_1 = \beta_2$ for the fluid layer depths chosen.

Table 38.1 gives the definitions and the ranges of values of the basic parameters of the laboratory model. It also shows some dimensionless numbers (not all independent). The Ekman number, Rossby number and internal Froude number are the main parameters that are varied in the experiments. The table lists the approximate magnitudes of these numbers for large-scale oceanic flows. For the laboratory model a range

TABLE 38.1 *Values of parameters and dimensionless numbers used in the laboratory model.*

Basic parameters		
d	Mean depth of upper layer	1–3 cm
D	Mean depth of tank	10 cm
L	Radius of tank	10 cm
$\tan \alpha_1$	Slope of lid	0.0, 0.1, 0.2, 0.3
$\tan \alpha_2$	Slope of bottom	0.0, 0.04, 0.84
Ω	Angular velocity of tank	10–30 rpm
$\Delta\Omega$	Angular velocity of lid	0.04–1.5 rpm
g	Acceleration of gravity	
$g^* = g\Delta\rho/\rho$	Reduced gravity	1–10 cm s^{-2}
ρ	Density of upper-layer fluid	
$\rho + \Delta\rho$	Density of lower-layer fluid	
ν_1	Kinematic viscosity of upper layer	0.95 cm^2 s^{-1}
ν_2	Kinematic viscosity of lower layer	1.37 cm^2 s^{-1}

Dimensionless numbers and length scales			
		Ocean	Laboratory model
$E_1 = \nu_1/2\Omega d^2$	Ekman no.	10^{-3}	10^{-3}–10^{-5}
$R = \Delta\Omega/\Omega$	Rossby no.		10^{-1}–10^{-3}
$\varepsilon = U/\Omega L$	Rossby no.	10^{-2}–10^{-4}	10^{-1}–10^{-5}
$F = \Omega^2 L/g$	Froude no.	10^{-3}	10^{-2}–2×10^{-3}
$f = \Omega^2 L/g^*$	Internal Froude no.	1–10	1–10^2
$\pi = \Delta\rho/\rho$		$\sim 2\times10^{-3}$	10^{-2}–10^{-3}
$\delta = \sqrt{\nu_1/\Omega}$	Ekman layer depth	3×10^1 m	10^{-1} cm
d	Upper layer depth	10^2 m	10^0 cm
$\lambda = \sqrt{gD/4\Omega^2}$	Rossby radius of deformation	10^3 km	2×10^1–5×10^1 cm
$\lambda^* = \sqrt{g^* d/4\Omega^2}$	Baroclinic radius of deformation	30 km	10^{-1}–10^0 cm

of values was chosen to span the oceanic values. In order to keep the Froude number small, the tank radius L was chosen as 10 cm. Other useful parameters are listed at the bottom of the table.

38.4 Experimental procedure

For the homogeneous experiments, thymol blue prepared according to the Baker (1966) formula was used as the working fluid. For the two-layer experiments thymol blue was used as the upper layer, and a starch or sugar solution for the lower.

To obtain transport in the western boundary current, the tank shown in Fig. 38.1a was used. The velocity profile $v(x)$ was determined by photographing at fixed time intervals the northwards progress of a dye line. The areal transport A is defined as

$$A = \rho_1 \int_0^{x_0} v(x)\,\mathrm{d}x,$$

where ρ_1 is the density of the thymol blue, and where the western wall is at $x = 0$. The value $x = x_0$ is at the centre of the gyre. The volume transport T is defined as

$$T = \rho_1 \int_0^{x_0} v(x)h(x)\,\mathrm{d}x.$$

The depth $h(x)$ of the upper layer was determined as it varied with x by suspending fine vertical wires which were marked with a colour-coded scale.

The separation experiments were prepared in a similar manner, but using the tank shown in Fig. 38.1*c*.

The periodic forcing experiments were performed using the tank shown in Fig. 38.1*d*. To simulate the monsoons over the Arabian Sea, the line of sources and sinks was at a thirty degree angle to the east–west direction. The sources (pumping into the interior) were to the south, the sinks (pumping out of the interior) to the north for the southwest monsoons of summer. The pumping direction was reversed and of smaller magnitude for the northeast monsoons of winter. No attempt was made to simulate the complication of the double branching of the Findlater jet of the summer monsoon (shown in Fig. 38.2). The summer condition was maintained for 'three months', corresponding to June, July, and August. One day corresponds to one rotation of the table. The pumps were turned off for 'two months', then the reversed pumping corresponding to the winter condition was continued for the following 'four months' of November to February. The pumping was stopped for the 'three months' of March to May. Then the cycle was repeated for 'five years'.

38.5 Observations

38.5.1 *The first (northward transport) experiment*

Fig. 38.3*a* shows streaklines of the steadily driven flow of a homogeneous fluid, with $\tan\alpha_1 = 0.21$, $\tan\alpha_2 = 0.0$ and $R = -7.35\times10^{-2}$. (The negative value of R indicates that $\Delta\Omega$ was in the opposite sense to Ω.) It shows the following typical features:

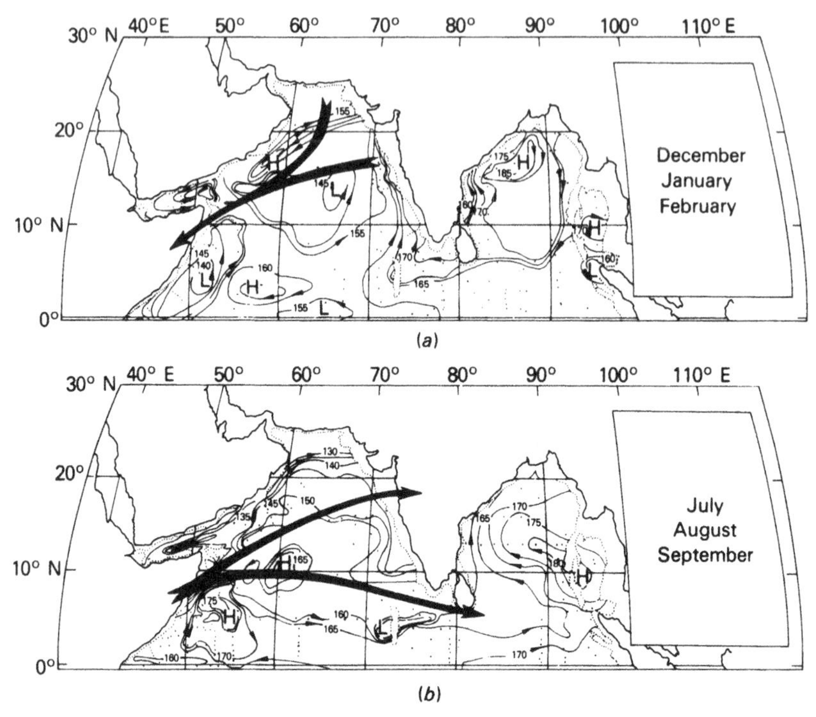

Fig. 38.2. The axis of the Findlater (1971) jet is superposed on the dynamic topography for: (*a*) the November jet on the dynamic topography of winter; and (*b*) the jet of June on the dynamic topography of late summer. The contours are labelled in 10^{-7} N m. The topographic contours are at intervals of 5 geodynamic centimetres.

(i) A divergence of this flow to the west across constant depth contours over much of the tank, from shallow to deep for clockwise rotation of the lid. This is analogous to the Sverdrup interior flow to the south.

(ii) This flow is diverted to the west in the rim boundary layer which is several millimetres thick at the eastern part of the rim and becomes several centimetres thick near the western part of the 180° basin.

(iii) A return flow to the north occurring in a fast narrow current along the western boundary. The centre of the gyre is far to the west of the centre of the basin.

(iv) A stationary topographic Rossby wave where the western boundary current enters the interior. Associated with this is a region of weakly cyclonic flow. There are no instabilities and no time dependence. All of these features are in agreement with Beardsley's homogeneous flows.

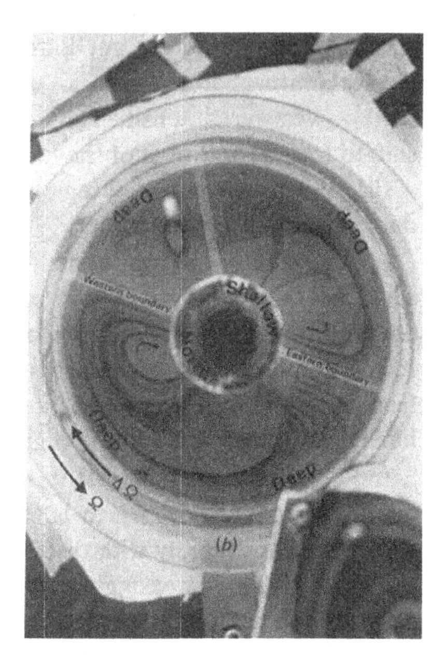

Fig. 38.3. Streaklines showing the flow patterns for different experiments: (a) with the homogeneous fluid ($E-2.6\times10^{-5}$); (b) and (c) with the two-layer fluid ($\Delta p/p=2.0\times10^{-3}$, internal Froude number $f=20$). In all cases the Rossby number $R=-7.35\times10^{-2}$, the Froude number $F=4\times10^{-2}$, tan $\alpha_1=0.21$ and tan $\alpha_2=0$. The arrows in (c) indicate the countercurrent.

For the same slopes and Rossby number, a two-layer fluid (Fig. 38.3*b*) shows similar features, but baroclinic instabilities are now possible. Typical features are:

(i) A westward intensified anticyclonic gyre, with the centre of the gyre shifted to the southwest of the location for the homogeneous case.

(ii) A more pronounced region of cyclonic flow as the western boundary current enters the interior.

(iii) Instability and time dependence in primarily two scales. There are small waves of wavelength approximately 5 mm, corresponding with the baroclinic radius of deformation for this experiment. There is also a larger scale instability, with wavelengths of several centimetres. This is evident from time-lapse film sequences, and the conditions for its occurrence have been summarized in the regime diagram below (Fig. 38.4).

(iv) A countercurrent in the rim boundary layer. This is a highly depth-dependent flow which is strongest just above the interface and which vanishes rapidly with distance above the interface. This countercurrent becomes increasingly stronger to the east. It continues along the eastern wall as a northward current and finally enters the interior by flowing cyclonically around the low formed by the topographic Rossby wave.

A summary of the observed instabilities is shown in the regime diagram in Fig. 38.4. The curves are taken from Hart's (1972) theory of baroclinic instability for a two-layer fluid in a circular cylindrical tank. Since the tank used here has western boundaries and boundary currents, different modes may be excited. Also, since Hart had computed the stability curves for only one value of tan α_1, these curves were used on the plane defined by tan $\alpha_1 = 0.21$ as well as the plane defined by tan $\alpha_1 = 0.32$. On the plane defined by tan $\alpha_1 = 0.105$, Hart's curves for tan $\alpha_1 = 0$ were used. In spite of all these differences, his curves still give a good structure upon which to display and understand the data from these experiments.

The measured northwards volume transport T in the western boundary current is plotted against Rossby number in Fig. 38.5. The data has been made dimensionless by dividing by Ω and the volume of upper-layer fluid in the basin. The lower curve is the transport in a homogeneous fluid, again made dimensionless by dividing by the volume and Ω.

It is seen that as R is increased, the transport by the upper-layer fluid is considerably greater than the transport by an equal volume of fluid over a rigid bottom. The curves of measured areal transports by homogeneous and two-layer fluids, seen in Fig. 38.5*b*, show only a small divergence of the two curves at higher Rossby number where the two-layer fluid

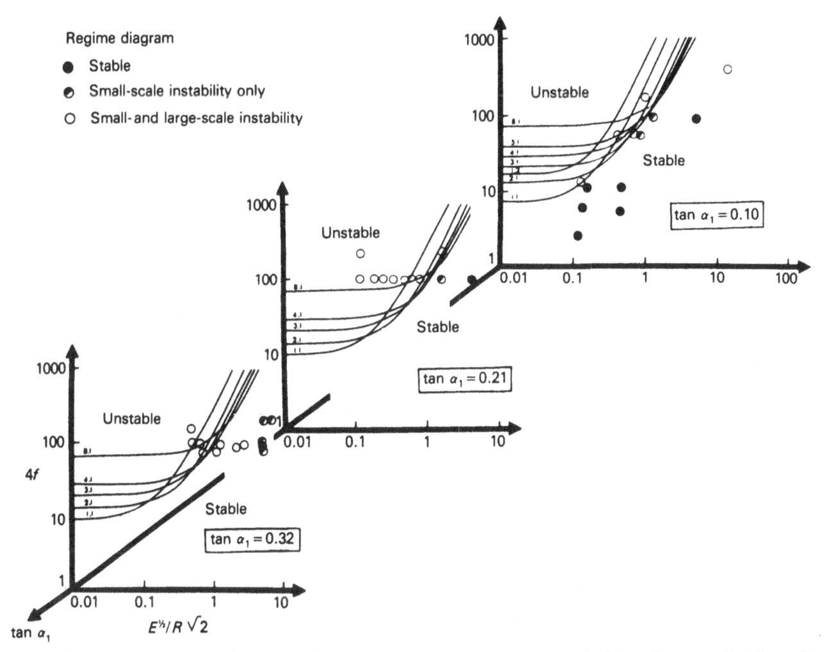

Fig. 38.4. Regime diagram in the parameter space f (the internal Froude number), $E^{\frac{1}{2}}/R$ (where E is the Ekman number, R the Rossby number), and $\tan \alpha_1$. $\tan \alpha_2 = 0$. The curves are from a theory of baroclinic instability by Hart (1972).

appears to have a somewhat larger areal transport. (The large error bar represents different values obtained in repeated observations, due to the time dependence of the two-layer flow at these large values of R.) It is primarily the increased depth of the upper layer in the particular region of the fast western boundary flow that results in the greatly increased transport compared with that of the homogeneous case.

38.5.2 *The second (surfacing and separation) experiment*

The following observations were made:

(i) With the tank shown in Fig. 38.1*a*, and with the lid rotating in a clockwise sense, as in the trade wind regions, surfacing first occurs on the eastern boundary near the weak cyclonic flow associated with the stationary topographic Rossby wave.

(ii) With the tank shown in Fig. 38.1*c*, and with the lid rotating in a clockwise sense, as in the anticyclonic part of the westerlies, surfacing of the lower layer occurs on the western boundary, and the fast western

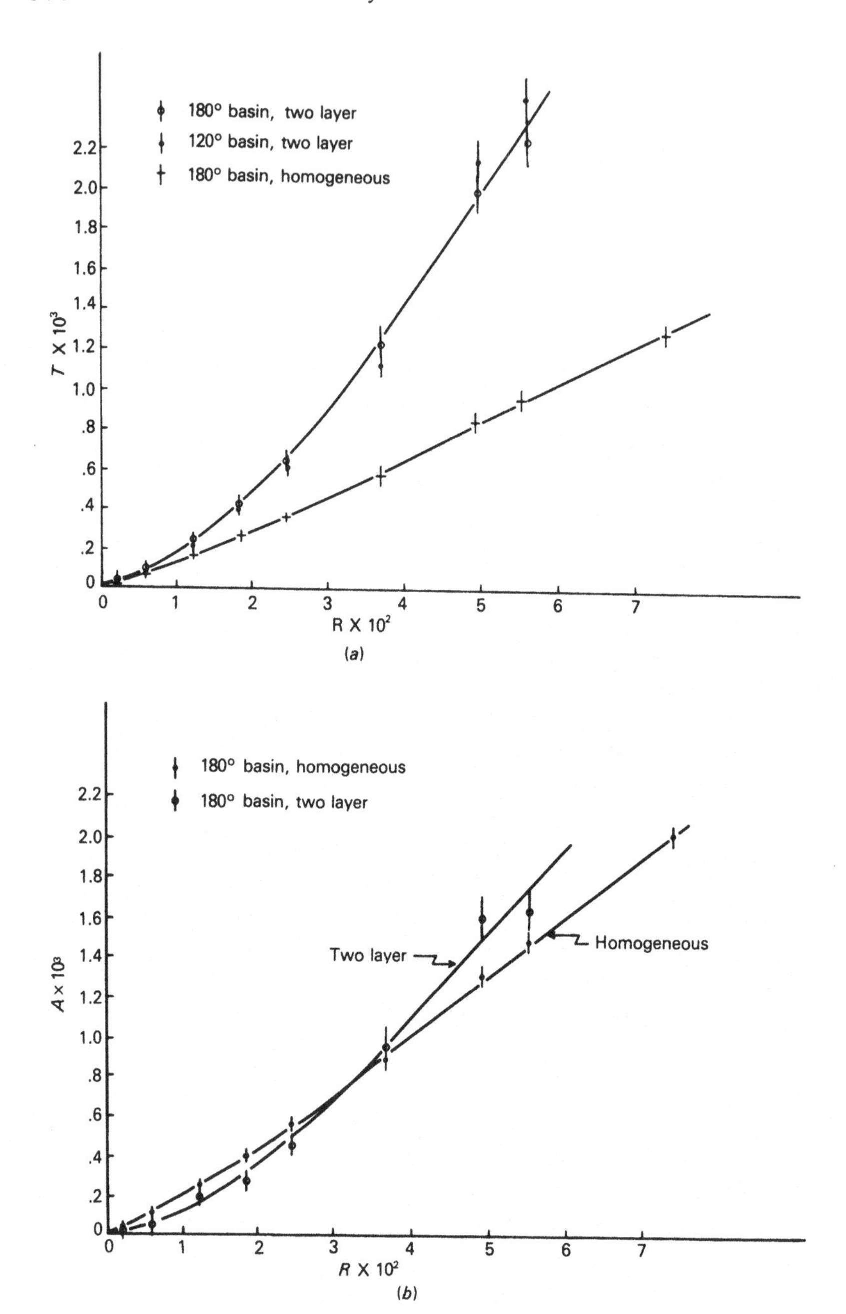

Fig. 38.5. (*a*) Dimensionless northward volume transport T of the western boundary current as a function of the Rossby number R. (*b*) Dimensionless northwards areal transport A as a function of Rossby number R.

boundary current separates from the western boundary before it is forced to do so by the rim. This is shown in Fig. 38.6. The conditions for this surfacing appear to be in agreement with Veronis' (1973*a*) theory, but these experiments are not yet completed.

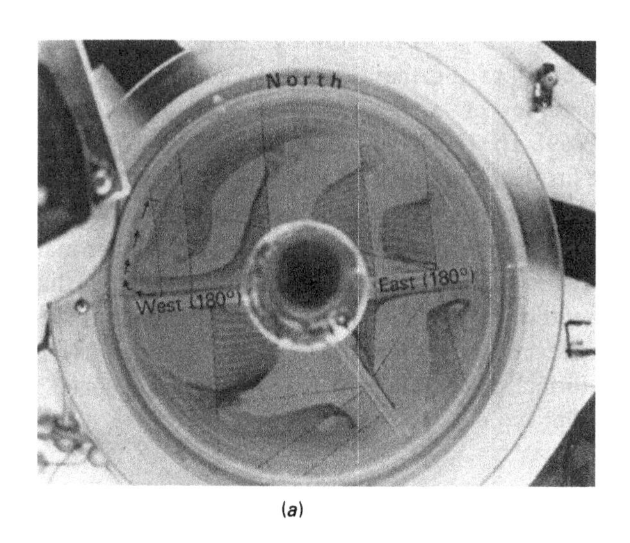

(*a*)

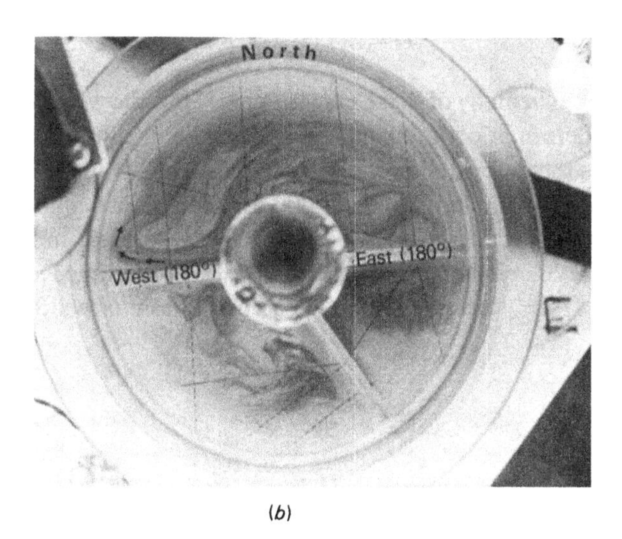

(*b*)

Fig. 38.6. (*a*) Non-separated flow and (*b*) separated flow at the western boundary. All parameters are the same in (*a*) and (*b*) except that the upper-layer depth is slightly smaller in (*b*). The tank used is shown in Fig. 38.1*c*.

38.5.3 *The third (periodic forcing) experiment*

The two-layer laboratory model responds to changes in rotation rate or to applied stress in a time short compared to 'decades' which applies for a midlatitude ocean (Veronis and Stommel, 1956). The tank shown in Fig. 38.1*a* containing 10 cm basins spins up in about 4 minutes or 80 'days'. The tank shown in Fig. 38.1*d* with a 20 cm basin spins up in about 8 minutes or 160 'days'. Furthermore, they can be made to respond in a time on the order of 'a month' corresponding to a tropical ocean (Lighthill, 1969). The rapid spin-up of the laboratory model is due to two factors. The first is the presence of a meridional boundary. The second is also related to the geometry of the model; the depth to width ratio of the oceans can never be modelled in a geometrically similar manner in the laboratory. In modelling a steady situation, we could argue that the depth is unimportant since the flow is depth independent (in each layer). However in the time-dependent situation, the layer depths enter into the radius of deformation and the baroclinic Rossby wave speed. The radius of deformation

$$\lambda = \frac{\sqrt{(g^* H)}}{2\Omega},$$

where

$$H = d(D-d)/D,$$

is about 0.5 cm in many of these experiments. Thus the width of the tank is just 40 deformation radii across and can be adjusted at will. In midlatitudes the deformation radius is about 30 km, while in the tropics this may be about 300 km. Thus a 6000 km ocean basin in midlatitudes is 200 deformation radii across, while a similar sized basin in the tropics would be just 20 deformation radii across. At the baroclinic Rossby wave speed of $\beta\lambda^2/4\pi^2$, the travel time (τ) across a basin of width L will be given by

$$\tau = \frac{4\pi^2 f^2}{\beta g^*}\frac{L}{H},$$

where f is the Coriolis parameter. Using $\beta = \tan \alpha_1/E^{\frac{1}{2}}$, this gives $\tau = 80$ 'days' for the model in Fig. 38.1*a*, and $\tau = 160$ 'days' for the model in Fig. 38.1*d*. The response time of 160 'days' or approximately 5 months is too slow for an effective simulation of the Arabian Sea. However, interesting effects might be expected since the forcing varies with a period of 1 year.

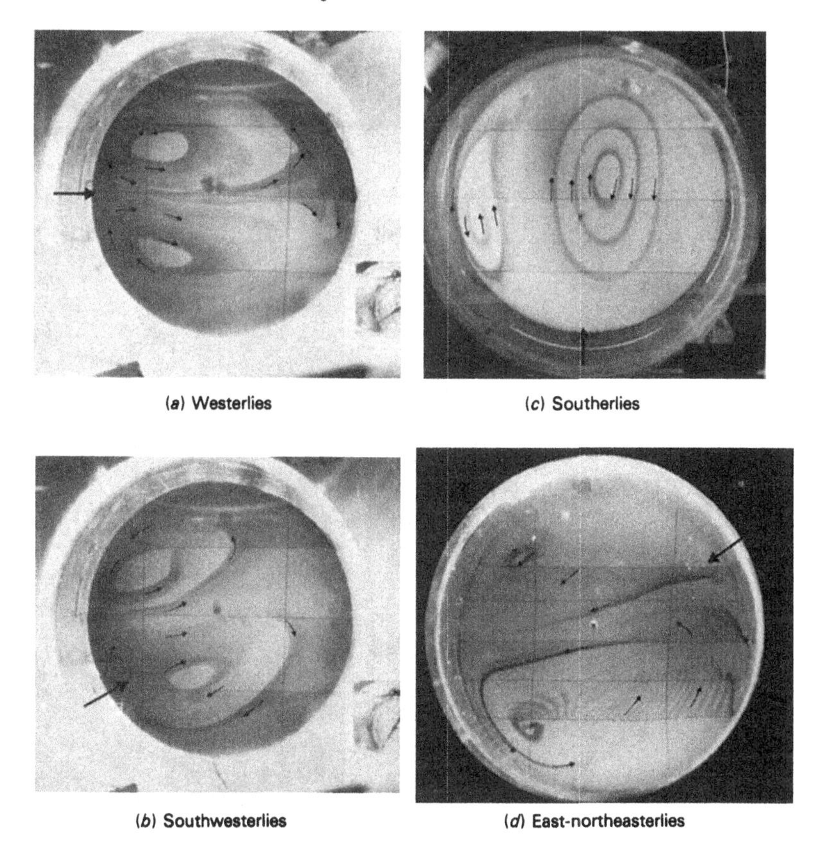

Fig. 38.7. Homogeneous β-plane model showing the oceanic response to various winds: (*a*) westerlies; (*b*) southwesterlies; (*c*) southerlies; and (*d*) east-northeasterlies.

The response of a homogeneous model to constant forcing is shown in Fig. 38.7 for four different directions of 'winds', all of the same magnitude. The response of a two-layer model to a steady westerly jet of the same strength as in Fig. 38.7 is shown in Fig. 38.8. It consists of an anticyclonic gyre to the south, a cyclonic gyre to the north where there is near-surfacing of the lower layer. It is noted that the response is now time dependent, and there is an interesting relationship between the time dependence in the two gyres. Some of the changing features are indicated by arrows.

The two-layer model, whose response time is 5 months, is subject to a southwest jet and its response is shown after 5 months and after 17 months in Fig. 38.9. A similar situation for a northeast jet of wind is also shown in Fig. 38.9.

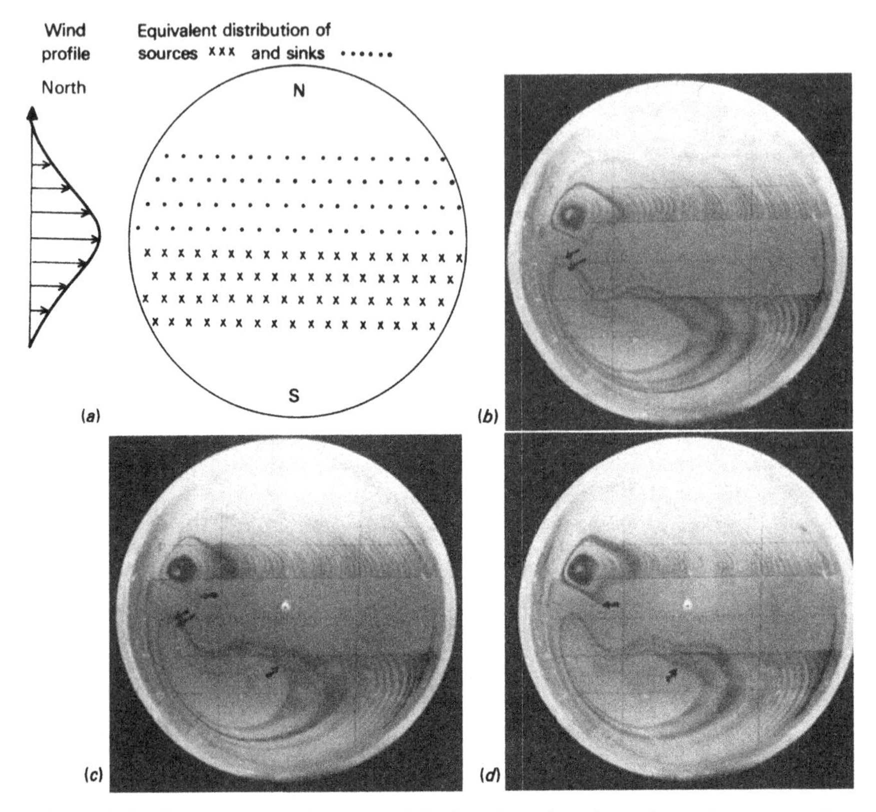

Fig. 38.8. Two-layer β-plane model showing the time-dependent oceanic response to steady westerlies. (*a*) shows schematically the assumed wind profile and the equivalent distribution of sources and sinks. (*b*), (*c*) and (*d*) are at successive times.

38.6 Conclusion

This two-layer laboratory model of a wind-driven ocean circulation appears to be capable of simulating some of the gross features observed or expected from theories of the oceanic general circulation. It also produced some unexpected features such as the deep countercurrent along the eastern boundary, which is reminiscent of the Davidson Current, and the California Undercurrent. Certainly the model allows features not observable in a homogeneous model of ocean circulation.

It is a pleasure to acknowledge the support of this research by the Fluid Dynamics Division of the Office of Naval Research.

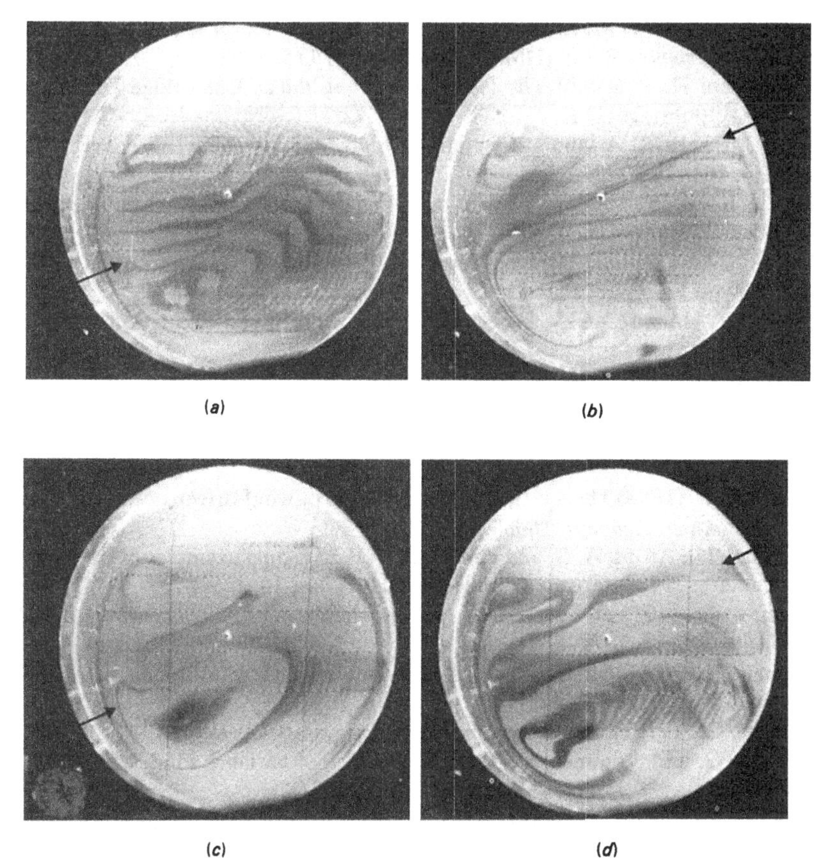

Fig. 38.9. Two-layer β-plane model showing the response at: (*a*) 3 months, and (*b*) 17 months to steady southwesterlies; (*c*) 5 months, and (*d*) 17 months to steady northeasterlies.

References

Baker, D. J. (1966) A technique for the precise measurement of small fluid velocities. *J. Fluid Mech.*, **26**, 573–6.

Baker, D. J. (1971) A source–sink laboratory model of the ocean circulation. *Geophys. Fluid Dyn.*, **2**, 17–29.

Baker, D. J. and Robinson, A. R. (1969) A laboratory model for the general ocean circulation. *Phil. Trans. Roy. Soc. London*, **A265**, 533–66.

Beardsley, R. C. (1969) A laboratory model of the wind-driven ocean circulation. *J. Fluid Mech.*, **38**, 255–72.

Düing, W. (1970) *The Monsoon Regime of the Currents in the Indian Ocean.* East–West Center Press, Honolulu.

Findlater, J. (1971) Mean monthly airflow at low levels over the western Indian Ocean. *Geophys. Mem.* (HMSO, London), **16**, 115, 1–53.
Greenspan, H. P. (1968) *The theory of rotating fluids.* Cambridge University Press, Cambridge, U.K.
Hart, J. E. (1972) A laboratory study of baroclinic instability. *Geophys. Fluid Dyn.*, **3**, 181–209.
Hart, J. E. (1975) The flow of a two-layer fluid over topography in a polar ocean. *J. Phys. Oceanogr.*, **5**, 615–24.
Lighthill, M. J. (1969) Dynamic response of the Indian Ocean to onset of the southwest monsoon. *Phil. Trans. Roy. Soc. London*, **A265**, 45–92.
Munk, W. H. and Carrier, G. F. (1950) The wind-driven circulation in ocean basins of various shapes. *Tellus*, **2**, 158–67.
Parsons, A. T. (1969) A two-layer model of Gulf Stream separation. *J. Fluid Mech.*, **39**, 511–28.
Pedlosky, J. and Greenspan, H. P. (1967) A simple laboratory model for the ocean circulation. *J. Fluid Mech.*, **27**, 291–304.
Stommel, H. (1948) The westward intensification of wind-driven ocean currents. *Trans. Amer. Geophys. Union.*, **29**, 202–6.
Stommel, H., Arons, A. B. and Faller, A. J. (1958) Some examples of stationary planetary flow patterns in bounded basins. *Tellus*, **10**, 179–87.
Veronis, G. (1973*a*) Large scale ocean circulation. *Advances in Applied Mechanics.*, **13**, 1–92.
Veronis, G. (1973*b*) Model of world ocean circulation: 1. wind-driven, two-layer. *J. Mar. Res.*, **31**, 228–89.
Veronis, G. (1977) Personal communication.
Veronis, G. and Stommel, H. (1956) The action of variable wind stresses on a stratified ocean. *J. Mar. Res.*, **15**, 43–75.
Von der Haar, T. H. and Oort, A. H. (1973) New estimate of annual poleward energy transport by northern hemisphere oceans. *J. Phys. Oceanogr.*, **3**, 169–72.
Welander, P. (1968) Wind-driven circulation of one- and two-layer oceans of variable depth. *Tellus*, **20**, 1–16.

Part IV

Some important mathematical modelling techniques

Introduction

R. P. PEARCE

In the previous three sections, the results of applying mathematical modelling techniques to the study of various aspects of both the atmospheric and oceanic components of the monsoon have been described. These studies exploit a wide range of mathematical methods ranging from those of classical mathematical analysis, for example wave perturbation theory, to those of numerical analysis. The development of the techniques themselves is an activity which is necessary if models are to be improved, and the five chapters in this section deal with particular areas of importance. The incorporation of orography into numerical prediction models is one of the major technical problems in the field of numerical modelling. Substantial advances have been made, particularly through the device of computing the meteorological parameters on surfaces which generally follow the orographic features. However, difficult problems still remain, associated with the inability of these models to properly resolve small-scale circulations in the neighbourhood of mountain ranges. Nowhere is this problem more acute than over the Indian subcontinent where the Himalayan range has such an enormous influence. A review of the technical problems associated with this phenomenon is given in Chapter 39. The estimation of vertical motion in the monsoon circulation, a dynamical feature to which the rainfall is intimately related, is described in Chapter 40. In Chapter 41, the use of a turbulence closure scheme in the context of the planetary boundary layer is described: the use of this new technique, or some modification of it, may well prove to be of importance in the difficult problem of dealing with the unresolved small scales of motion in numerical models. In Chapter 42, the use of empirical orthogonal functions in the technique of estimating rainfall is described, and in Chapter 43 an account is given of an interesting application of the concepts of functional analysis in the general context of the simulation of atmospheric processes.

39

On the incorporation of orography into numerical prediction models

H. SUNDQVIST

This chapter begins with a superficial review of examples of orographic effects on atmospheric motion. The major part of the chapter is a review of various ways of incorporating orography into prediction models. Advantages and disadvantages of different vertical coordinate systems are discussed in relation to the calculation of the pressure force and advection terms. The representation of very steep mountains as blocking walls is also considered and some viewpoints on initialization problems are presented.

39.1 The principal methods in current use

The results of many numerical model experiments have indicated that when the Earth's topography is taken into account an improved simulation of the atmospheric circulation is achieved (see for example, Hahn and Manabe 1975; Hahn and Manabe, 1976; Kasahara *et al.*, 1973; Manabe and Terpstra, 1974; Sadler and Ramage, 1976). Generally, it is necessary to ensure that model results, intended to elucidate physical processes, are contaminated as little as possible by spurious effects that may be caused by the numerical treatment. This view appears to require special attention in conjunction with the incorporation of orography.

Orography is commonly described in the same way and with the same resolution as the other horizontally varying quantities of the numerical model. Most of this chapter is devoted to methods of that type. It has also been suggested that mountain ranges be regarded partly as vertical walls and such approaches also deserve attention and will be referred to.

Consideration is restricted to motions for which the hydrostatic approximation is valid, implying that the pressure (p) or functions of p may be used as the vertical coordinate. The modelling of orography and the choice of a vertical coordinate system are closely connected with each other. Methods for taking the height variation of the Earth's surface into account may be divided into three principal groups, which differ either with respect to the choice of the vertical coordinate or to their generality.

Before defining these groups, it is first noted that the inclusion of the surface topography is basically a matter of describing the kinematic boundary condition at the surface. The condition that the flow normal to the Earth's surface is everywhere zero leads to the well-known relation (in the z-system):

$$w_H = \mathbf{v}_H \cdot \nabla H = u_H \frac{\partial H}{\partial x} + v_H \frac{\partial H}{\partial y}, \tag{39.1}$$

where $H(x, y)$ is the height of the orography and (u, v, w) are the velocity components. All approaches employ condition (39.1), but there are still basic differences. These may be grouped as follows:

(i) the z-or p-system is used and the model atmosphere is assumed to extend all the way down to mean sea level (msl), at which the boundary condition (39.1) is applied;

(ii) the z-or p-system is used, the model atmosphere extending only to the actual ground surface;

(iii) the vertical coordinate used is a transformation of the basic coordinate (z or p) such that the Earth's surface is a coordinate surface, a typical example being the so-called σ-system introduced by Phillips (1957), where $\sigma(=p/p_s)$ is the pressure normalized by dividing by the surface pressure, in which the wind component normal to the coordinate surfaces vanishes identically at the ground, implying that condition (39.1) is implicitly fulfilled.

Only the basic z-, p-, or σ-systems are considered here to bring out the characteristic features of these groups. A more comprehensive review of various coordinate systems used for numerical weather prediction is found in a paper by Kasahara (1974).

The main approximation models of group (i) suffer is that condition (39.1) is applied at msl instead of at the surface. This can lead to unsatisfactory effects, especially on thermal and condensation processes since the forcing does not appear at the proper altitude. There are no special numerical problems and therefore this type of model will not be considered further.

Groups (ii) and (iii) differ primarily with regard to numerical and programming techniques, the former affecting the behaviour of the model atmosphere during a time integration.

In the case of group (ii), coordinate systems using z and p as vertical coordinate are natural choices because they harmonize so well with the stratification of the atmosphere. However, the presence of mountains causes irregular holes in the coordinate surfaces, leading not only to problems in calculating partial derivatives at constant z or constant p (e.g., the pressure gradient force), but also to complexities in the programming. Also, because of these holes, it is not possible to employ a spectral representation of variables in the horizontal in this group of models.

Group (iii) (i.e., σ-type) models are characterized by a convenient lower boundary condition, $\dot{\sigma}_H = 0$. In contrast to the models described in the preceding paragraph, these models have unbroken coordinate surfaces. Some concern about truncation errors in σ-system models has often been expressed. However, as will be seen later, these problems are of much the same dimensions even in the models of group (ii). One important distinction between groups (ii) and (iii) is that the models in group (iii) have coordinate surfaces that are tilted (in mountainous regions) with respect to the natural stratification, and therefore part of this is projected onto the surfaces $\sigma = \text{constant}$. This is seen by considering the adiabatic equation applied near the ground on a mountain slope; thus $\dot{\sigma} = 0$ and the relatively large temperature changes due to a non-vanishing w_H (39.1) have to be accomplished by the horizontal advection $\mathbf{v}_H \cdot \nabla_\sigma \theta$, implying that $|\nabla_\sigma \theta|$ is greater than $|\nabla_z \theta|$ and that the former is enhanced by the projection of $\partial\theta/\partial z$ onto the surface $\sigma = \text{constant}$.

Models of a fourth type may also be considered, i.e. those which try to take into account unresolvably steep mountain slopes by treating them as vertical walls, where a condition of vanishing normal component of the horizontal wind is imposed (Egger, 1972). Any of the vertical coordinates z, p or σ may be used. Note, however, that this type of approach involves the same kind of complications in the programming as those in group (ii) even if the σ-system is adopted.

39.2 Individual features of the z-, p- and σ-systems

When designing finite-difference analogues of the governing differential equations, one of the basic considerations is that the difference equations should possess the same integral properties as the original equations. Such requirements are sometimes relaxed in certain respects depending on the character of the problem. However, regarding the formulation of finite differences in the vertical, it seems to be especially important that such conditions are satisfied. In that way a guide to a consistent coupling between temperature, pressure and geopotential via the hydrostatic relation is obtained. This is of great importance, especially when orographic effects are included. Only an outline of these derivations will be given here.† The z-system is used to exemplify the principle; the method should be applied in an analogous way in any other vertical coordinate system.

Conventional notation is used and it is sufficient to define only the following symbols:

$$\pi = (p/p_0)^\kappa, \qquad \kappa = R/c_p,$$

$$\delta_k A = A_{k+\frac{1}{2}} - A_{k-\frac{1}{2}},$$

where A is any scalar dependent variable, and

$$\hat{w} = \rho w.$$

Fig. 39.1 shows where the dependent variables are defined. The horizontal pressure gradient force in the momentum equation at a grid-point k is not explicitly available at the level at which the momentum equation is solved. Thus, it is necessary to find

$$\rho_k^{-1} (\nabla p)_k, \tag{39.2}$$

at the altitude $z = z_k$, with

$$\rho_k = -\frac{1}{g} \frac{\delta_k p}{\delta_k z}. \tag{39.3}$$

We shall use θ in the thermodynamic equation, T and θ being related by

$$T_k = \pi_k \theta_k. \tag{39.4}$$

From the set of difference equations (written in flux form) the energy equation of the system is derived and the terms collected so that they can be identified with their analogues in the differential equation (see Kasahara, 1974). Hence, putting

$$E_k = \tfrac{1}{2} \mathbf{v}_k \cdot \mathbf{v}_k + c_p T_k,$$

† A detailed analysis will be presented in the forthcoming Volume II on *Numerical methods used in atmospheric models* in the GARP Publication Series. The reader may also refer to Phillips (1974).

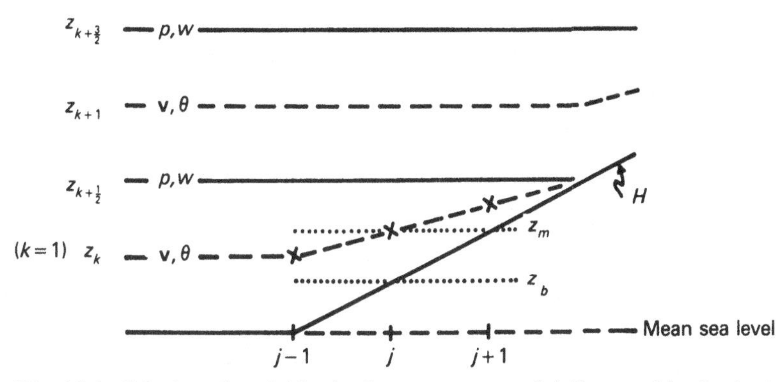

Fig. 39.1. Display of variables in the z-system model discussed in the text.

gives

$$\frac{\partial}{\partial t}(\rho_k E_k) + \nabla\cdot[\rho_k \mathbf{v}_k(E_k+\phi_k)] + \frac{1}{\delta_k z}\delta_k[\hat{w}(E+\phi)]$$

$$= \frac{1}{\delta_k z}\delta_k\left(z\frac{\partial p}{\partial t}\right) + \frac{1}{\delta_k z}\delta_k[\hat{\omega}(c_p T+\phi)]$$

$$-\frac{\pi_k}{\delta_k z}\delta_k(\hat{w}c_p\theta) + \phi_k\nabla\cdot(\rho_k\mathbf{v}_k) + c_p\rho_k T_k\frac{\partial \ln \pi_k}{\partial t}$$

$$+\mathbf{v}_k\cdot(\nabla \ln \pi_k) - \mathbf{v}_k\cdot(\nabla p)_k. \tag{39.5}$$

In order to ensure that (39.5) has the same properties as its differential analogue, all terms but the first on the right-hand side of (39.5) have to vanish. This leads to the following relations:

$$c_p\rho_k T_k(\nabla \ln \pi)_k = (\nabla p)_k, \tag{39.6a}$$

$$\frac{\partial \ln \pi_k}{\partial t} = \frac{\phi_k}{c_p T_k}\frac{1}{\delta_k p}\frac{\partial}{\partial t}\delta_k p, \tag{39.6b}$$

$$\phi_{k+1} - \phi_k = c_p\theta_{k+\frac{1}{2}}(\pi_k - \pi_{k+1}), \tag{39.6c}$$

$$(\phi + c_p T)_{k+\frac{1}{2}} = \tfrac{1}{2}(\phi_k + \phi_{k+1}) + \tfrac{1}{2}c_p\theta_{k+\frac{1}{2}}(\pi_k + \pi_{k+1}). \tag{39.6d}$$

(The fourth term on the right-hand side of (39.5) is first rewritten with the aid of the continuity equation. Relations (39.6c) and (39.6d) are obtained by the requirement that the coefficients of $\hat{w}_{k+\frac{1}{2}}$ and $\hat{w}_{k-\frac{1}{2}}$ are zero.) Only (39.6a) and (39.6b) are discussed here. Relation (39.6a) shows how to evaluate $(\nabla p)_k$ and relation (39.6b) relates the rate of change of $\ln \pi_k$ to the pressure tendencies at $(k+\frac{1}{2})$ and $(k-\frac{1}{2})$, implying that an initial π_k-field is required. Relations (39.6) are valid for this particular choice of vertical coordinate and relative vertical distribution of variables. For other coordinate systems and/or variable distributions, similar but not identical relations result.

39.2.1 The z-system

The general-circulation model of the National Center for Atmospheric Research (NCAR) is probably the most extensively used z-system model (see Kasahara and Washington, 1971; Washington, 1974).

To make the discussion concrete the vertical structure of that model is considered and the calculation of the pressure gradient force in a layer which intersects a mountain slope, as shown in Fig. 39.1, will be carried out.

It may be seen from Fig. 39.1 that z_k of the first layer above the ground is situated halfway between $z = H$ and $z = z_{k+\frac{1}{2}}$; that is:

$$z_k = \tfrac{1}{2}(H + z_{k+\frac{1}{2}}). \tag{39.7}$$

Now, regarding evaluation of the pressure gradient at altitude z_k and point j in Fig. 39.1, it might appear natural either to take the average of the gradients along z_{k+1} and z_b or to find proper π-values along z_m. In both alternatives, interpolation and/or extrapolation (into the ground) would be necessary. However, this is not a satisfactory method, because the lapse rate within the layer (and especially in the ground) is not available. This circumstance may be demonstrated by the following example.

Consider standard atmosphere conditions at all grid-points (i.e., no pressure gradients exist). Then values of p are defined on $z = z_{k+\frac{1}{2}}$ and on $z = H$, and values of T on $z = z_k$; from these we shall evaluate the pressure gradient centred around j and representative at height $z_k (= z_m)$. We do this averaging of the gradients along $z = z_{k+\frac{1}{2}}$ and $z = z_b$, the former of which is zero. In order to obtain a pressure value at $z = z_b$ we now integrate the hydrostatic equation from $z = H$ to $z = z_b$ adopting isothermal conditions at temperature T_k since this is the only temperature available. This gives

$$\ln \pi_b = \ln \pi_H + \frac{g}{c_p T_k}(H - z_b)$$

and (see (39.6*a*)),

$$\rho_k^{-1}\left(\frac{\partial p}{\partial x}\right)_k = \tfrac{1}{2}\left(c_p T_k \frac{\partial}{\partial x} \ln \pi_H + g\frac{\partial H}{\partial x}\right). \tag{39.8}$$

For an isothermal atmosphere the two terms on the right cancel and the result is exact. However, inserting values for a standard atmosphere it is found that, corresponding to a slope of 2000 m in 300 km between $(j+1)$ and $(j-1)$, (39.8) yields a fictitious pressure gradient corresponding to a geostrophic wind amounting to 7.5 m s^{-1} at 45° latitude.

Part of this error is due to the extrapolation into the ground in the presence of an atmosphere which is not isothermal. Therefore, it appears that such procedures should be abandoned. Actually, (39.8) gives a hint of a way of dealing with this problem. Namely, let us regard $z = z_k(x, y)$ as a coordinate surface, $\zeta = \text{constant}$, not only above flat ground but above mountain slopes as well, implying that we introduce a hybrid z–σ-system in the lowest layer. Transforming (39.6*a*) accordingly, yields the following expression which is valid for all points on the surface z_k:

$$\rho_k^{-1}(\nabla p)_k = [c_p T_k(\nabla_\zeta \ln \pi)_k + g(\nabla_\zeta z)_k], \tag{39.9}$$

where ∇_ζ implies differentiation along the surface $z = z_k(x, y)$. Definition (39.7) gives

$$(\nabla_\zeta z)_k = \tfrac{1}{2}\nabla H. \tag{39.10}$$

The problem is now reduced to finding a proper initial π_k-field. (During a subsequent integration, π_k is then obtained from (39.6*b*).) In real-data applications, the initial π_k-distribution should be taken principally from the analyses, but some kinds of tests probably need to be performed to find out if adequate accuracy is being achieved. (A corresponding procedure in the above example would be to evaluate π_k according to an isothermal atmosphere; then of course (39.9) yields a vanishing pressure gradient.)

In order to demonstrate the care with which this matter has to be handled, two approximate ways of obtaining π_k are suggested for the above example. In one case, the mean with respect to geometrical height is used:

$$\begin{aligned}\pi_k &= \frac{1}{\delta_k z}\int_{k-\frac{1}{2}}^{k+\frac{1}{2}} \pi \,\mathrm{d}z \\ &= \frac{1}{\delta_k z}\left[\int_{k-\frac{1}{2}}^{k+\frac{1}{2}} \mathrm{d}(\pi z) - \int_{k-\frac{1}{2}}^{k+\frac{1}{2}} z \,\mathrm{d}\pi\right] \\ &= \tfrac{1}{2}(\pi_{k+\frac{1}{2}} + \pi_{k-\frac{1}{2}}),\end{aligned} \tag{39.11}$$

and in another case, the average with respect to $\ln p$ is adopted:

$$\pi_k = \frac{1}{\delta_k \ln p}\int_{k-\frac{1}{2}}^{k+\frac{1}{2}} \pi \,\mathrm{d}\ln p = \frac{1}{\kappa}\frac{\delta_k \pi}{\delta_k \ln p}. \tag{39.12}$$

Inserting definition (39.11) in (39.9) the resulting spurious pressure gradient force now corresponds to a geostrophic wind of $-3.77\ \mathrm{m\,s^{-1}}$. Using definition (39.12) instead, the corresponding figure becomes $+3.74\ \mathrm{m\,s^{-1}}$. Hence, these choices yield individually an error that is half as large as the one from (39.8) and, at least in this particular example, an average of the gradients resulting from relations (39.11) and (39.12) produces a highly satisfactory result.

So far interest has been focussed on the pressure gradient term. But the same kind of problem naturally arises with all horizontal advection (or flux divergence) terms. Consequently, these should also be treated in the hybrid z–σ-coordinate system. A proper handling in this respect is probably most important of all in the case of the thermodynamic equation. The boundary condition (39.1) has also to be appropriately adjusted to the sloping coordinate surface. The apparent vertical velocity is reduced by one half according to (39.10). However, the total vertical advection effect is still accounted for, because part of this is then included in the 'horizontal' advection term $\mathbf{v} \cdot (\nabla_\zeta \theta)_k$.

39.2.2 The p-system

The boundary condition at the surface in the p-system is given by

$$\omega_H = \frac{\partial p_H}{\partial t} + \mathbf{v}_H \cdot \nabla p_H \qquad (39.13)$$

which corresponds to (39.1) in the z-system.

The calculation of the pressure gradient force and advecting terms in the p-system is very similar to that described for the z-system in § 39.2.1. Let us therefore merely note one peculiarity of the p-system, pointed out by Katayama *et al.* (1974). Since a p-surface changes altitude in the course of the integration, it may happen that grid-points disappear or new ones appear in the vicinity of mountain slopes. The latter situation, especially, poses a difficult problem since values have to be assigned to physical variables at the new points. There is no obvious means of handling such situations. They therefore constitute a source of spurious orographic effects.

39.2.3 The σ-system

Truncation in σ-systems has been reviewed by Sundqvist (1975*a*).

Consider an atmosphere with a vertical temperature distribution given by

$$T = T_0 + \sum_{n=1}^{N} \gamma_n z^n, \tag{39.14}$$

where

$$z = \ln (p/p_0).$$

It is assumed that both T_0 and γ_n are independent of x and y, implying that no horizontal pressure gradients exist. To simulate a realistic temperature distribution, it is generally necessary to take $N \geq 2$. Then a simple orographic field,

$$z_s = \ln (p_s/p_0) = -\hat{P}_m\left(1 - \cos \frac{2\pi m}{D} x\right), \tag{39.15}$$

is considered where m is the number of waves across the domain D (or $D/(m\ \Delta s)$ is the number of grid intervals per wavelength and subscript s denotes the Earth's surface). Expressing T in the σ-system gives, since $z = \ln \sigma + z_s$,

$$\begin{aligned} T &= T_0 + \sum_{n=1}^{N} \gamma_n (\ln \sigma + z_s)^n \\ &= T_0 + \sum_{n=1}^{N} \gamma_n \left(\ln \sigma - \hat{P}_m + \hat{P}_m \cos \frac{2\pi m}{D} x\right)^n. \end{aligned} \tag{39.16}$$

Thus, along $\sigma =$ constant, T will have an x-variation that is composed of all wavenumbers up to mN.

If the model mountain has pronounced amplitudes in the shorter wave components (e.g., $3\ \Delta s$ and $4\ \Delta s$), (39.16) shows that a significant truncation, and possible aliasing as well, will be attached to the description of T. These effects will consequently affect the accuracy of the calculations of terms in the thermal equation and of the pressure gradient force. It is difficult to make a quantitative estimate of the effect on the former, but the following example gives some insight into the impact on the pressure force.

The pressure force in the σ-system has the form

$$F_s = F_{s1} + F_{s2}, \tag{39.17}$$

where

$$F_{s1} = \nabla_\sigma \phi, \quad F_{s2} = RT\nabla z_s,$$

which is zero in the present example. Integrating the hydrostatic equation to a level σ with T computed using (39.16), the difference form of the x-component of F_{s1} becomes

$$\phi_x = -\frac{R}{2\,\Delta s}\left(T_0(z_{s,j+1}-z_{s,j-1}) + \sum_{n=1}^{N}\frac{\gamma_n}{n+1}[(\ln\sigma+z_s)_{j+1}-(\ln\sigma+z_s)_{j-1}]\right). \tag{39.18}$$

Tentatively we write F_{s2} as

$$RT(z_s)_x = \frac{R}{2\,\Delta s}\bar{T}(z_{s,j+1}-z_{s,j-1}). \tag{39.19}$$

For an isothermal atmosphere, (39.18) and (39.19) add to exactly zero regardless of the width of the mountain. Furthermore, for $N=1$ (i.e., if T varies linearly with $\ln p$) the two terms still cancel each other exactly if

$$\bar{T} = \tfrac{1}{2}(T_{j+1}+T_{j-1}).$$

This was also pointed out by Corby *et al.* (1972).

When γ_n is a function of x and y, and/or N is greater than 1 it is impossible even in this simple case to find a finite-difference formulation which generally ensures an exact cancelling of the two terms. So an error in the pressure gradient due to horizontal truncation in the presence of orography has to be accepted. A quantitative test (Sundqvist, 1975*a*) indicates however, that it may be possible to keep such errors smaller than a magnitude corresponding to a geostrophic wind of about 0.5 m s^{-1}.

In general, the wave numbers associated with the orography will extend beyond those describing synoptic scale variations of T_0 and γ_n. The implication of this is that a higher resolution is usually needed to describe synoptic scale systems when orography is incorporated.

Although this article focusses on grid-point models, it should be noted that the above truncation problems are likely also to appear if a spectral representation of the variables is employed. A closer examination of the problems in this case is needed.

In the previous sections attention has been focussed on fictitious effects that are repeated at each timestep during an integration of the model equations. Another effect likely to be introduced into the σ-system is associated with interpolation of initial analyses from the p-system to σ-levels. This aspect is closely analogous to the problem of finding a proper initial π_k, which was discussed in § 39.2.1. The following example again indicates the importance of establishing consistent relations between temperature, pressure and geopotential. (For details see Sundqvist, 1976.)

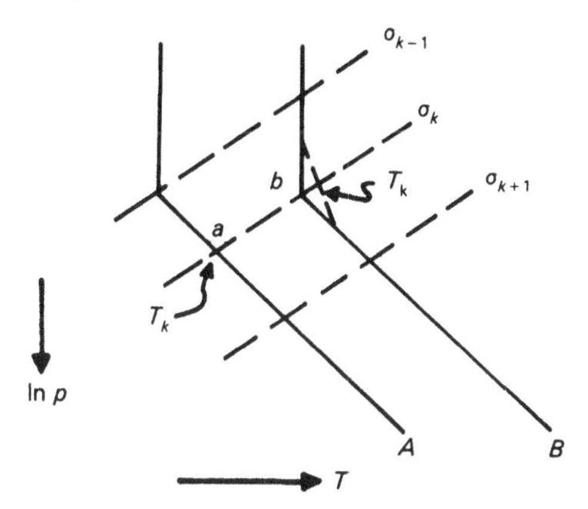

Fig. 39.2. Illustration of truncation error arising when T is calculated from the finite-difference form of the hydrostatic equation (39.19) using $\phi(\sigma)$ as obtained from an analytic expression for $T(p)$.

Assume the lapse rates at points A and B in Fig. 39.2 to be identical. Also assume that the σ-surfaces are sloping as shown in the figure and that the p-system geopotentials (hydrostatically consistent with the lapse rates) have been interpolated to the σ-levels. (Such an interpolation may be carried out to any arbitrarily assigned accuracy.) In order to calculate the pressure gradient (39.17), the temperature along $\sigma = \sigma_k$ is needed. To obtain this a commonly used finite-difference form of the hydrostatic equation with second-order accuracy, is applied, namely

$$T_k = \sum_{j=k-1}^{k+1} s_j \phi_j, \tag{39.20}$$

where s_j is a function of $\ln \sigma$. Then, in this example, $T_k(A)$ becomes the same as $T(a)$ of the p-system, whilst $T_k(B) \neq T(b)$ as illustrated in Fig. 39.2. Consequently, using an average of $T_k(A)$ and $T_k(B)$ in the pressure gradient calculation, introduces a fictitious force.

To obtain a quantitative estimate of these errors the above simple state (i.e., T varying linearly with $\ln p$ to the tropopause at 200 mb, above which the atmosphere is isothermal) was introduced as the p-system 'analysis' into a numerical prediction system. Interpolation from p- to σ-surfaces was performed and then a balance equation in the σ-system (Sundqvist, 1975*b*) was solved. The resulting streamfunction in the Tibet–Himalaya region is shown in Figure 39.3. At the lower levels the magnitude of the fictitious winds is about 0.5 m s^{-1}, but at the two upper levels the magnitude is 10 to 15 m s^{-1}. This circulation is of synoptic-scale extent and it changes its sense from one level to the next.

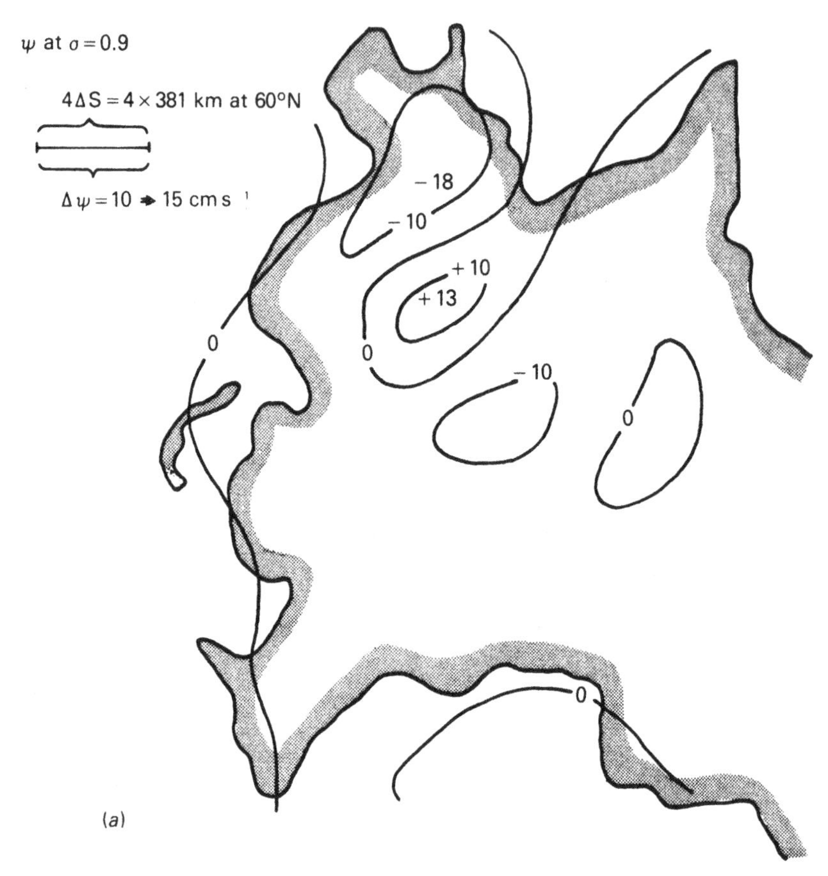

Fig. 39.3. Streamfunction obtained from the solution of the σ-system balance equation for the case where ϕ had been interpolated from the p-system, in which the pressure force is zero. (a), . . . , (e) show σ-levels 0.9, 0.7, . . . , 0.1, respectively. The Asian land mass is outlined with India and SE Asia at the top of the diagram. (After Sundqvist, 1975b.)

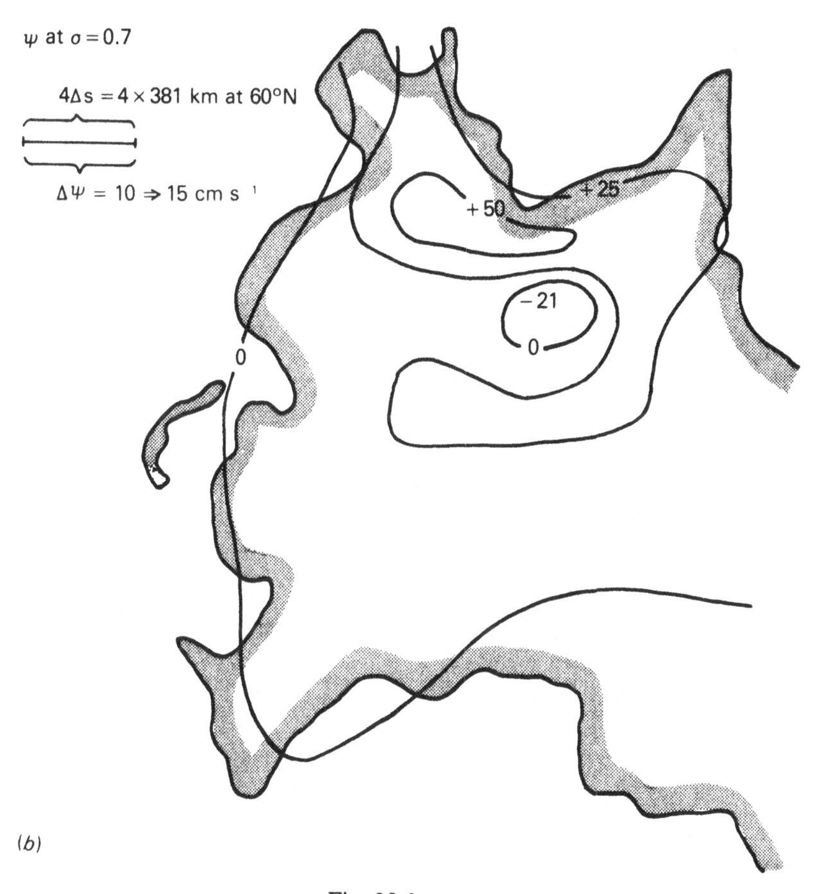

Fig. 39.3—*cont.*

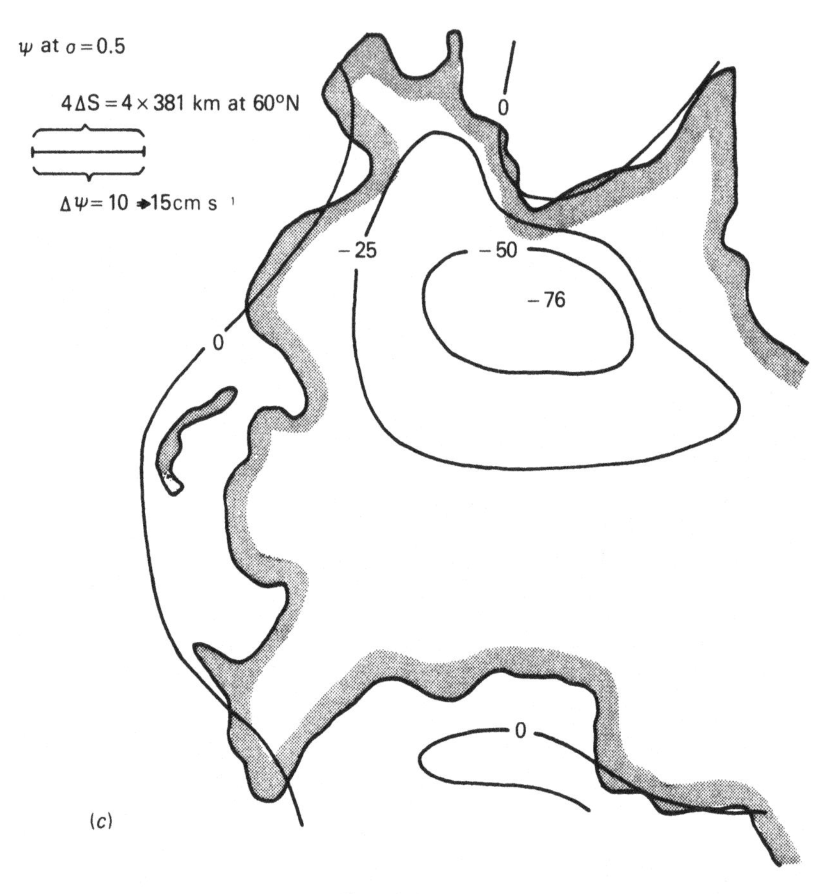

Fig. 39.3—*cont.*

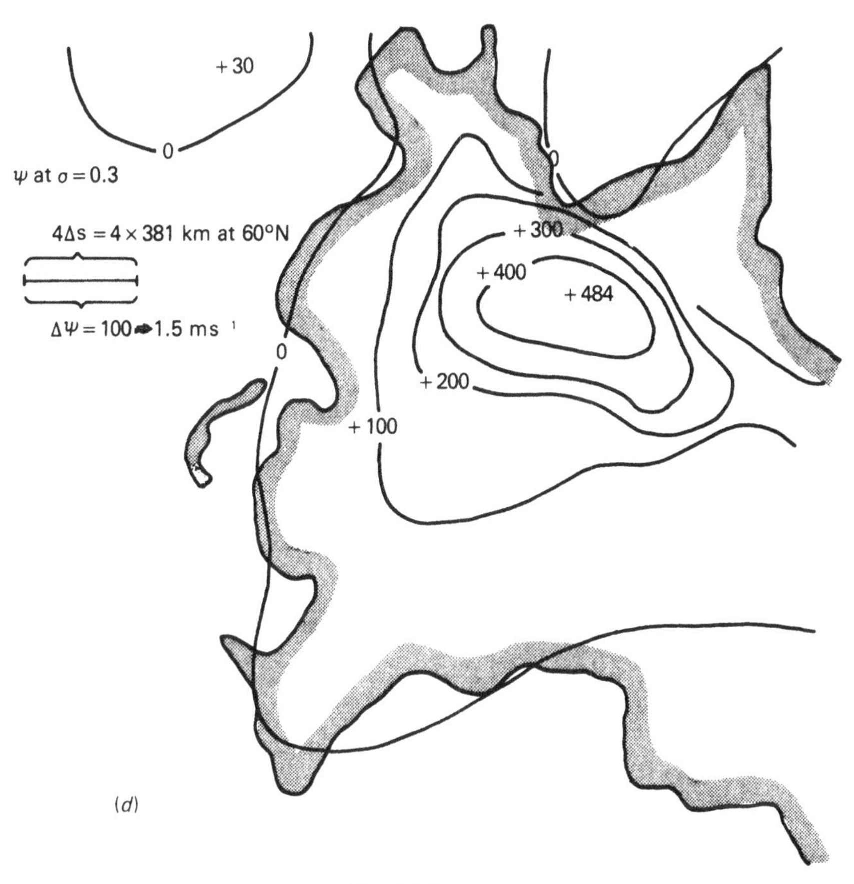
+30
0
ψ at σ = 0.3
4Δs = 4 × 381 km at 60°N
ΔΨ = 100 ⇔ 1.5 ms⁻¹
0
0
+300
+400
+484
+200
+100
(d)

Fig. 39.3—*cont.*

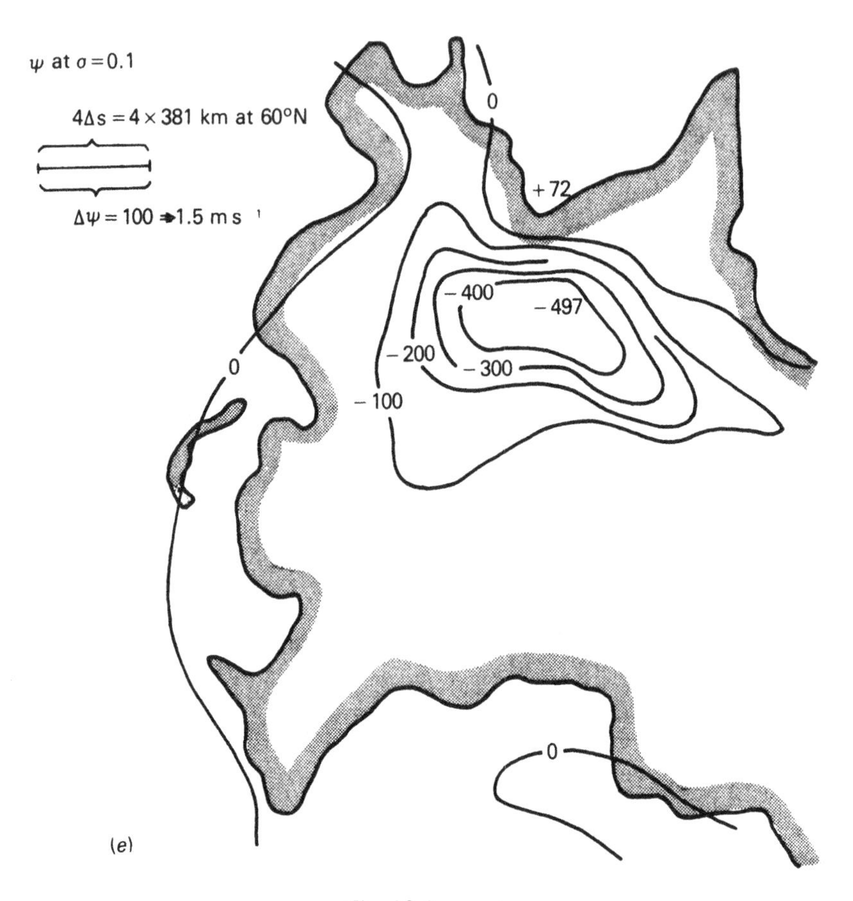
ψ at σ=0.1
4Δs = 4 × 381 km at 60°N
Δψ = 100 ⇛ 1.5 m s⁻¹
0
+72
−400
−497
−200
−300
0
−100
0
(e)

Fig. 39.3—*cont.*

The following method, used to avoid such false pressure gradients, was investigated. The pressure gradient force, $\boldsymbol{F}_p$, is evaluated in the p-system analysis and is then interpolated to σ-surfaces where the divergence along constant σ-surfaces is calculated. This is then put equal to the divergence of the pressure gradient force expressed in σ-coordinates, thus yielding a differential equation for the geopotential on a σ-surface (level k, say):

$$\nabla_\sigma \cdot \left[\nabla_\sigma \phi_k + \left(\sum_{j=k-1}^{k+1} s_j \phi_j \right) \nabla \ln p_s \right] = (\nabla_\sigma \cdot \boldsymbol{F}_p)_k, \qquad (39.21)$$

where $\boldsymbol{F}_p = \nabla_p \phi$, and relation (39.20) is substituted, consequently imposing the hydrostatic relation of this discrete system. Equation (39.21) was solved for the case described in the previous paragraph, and the subsequent solution of the balance equation for this state yielded fictitious winds that nowhere exceeded 0.5 m s^{-1}.

39.3 Steep mountain slopes considered as vertical walls

Egger (1972) presented an interesting numerical study in which an attempt to incorporate steep mountain slopes was made. These slopes are regarded as vertical walls at which the normal component of **v** vanishes. Two types of barrier were considered. Type (a) is a single wall along a row of grid points (the Pyrenees and the Alps) and type (b) is a massif having steep outer boarders and covering an area containing several grid-points (Greenland, for example).

The model used by Egger has σ as vertical coordinate. For type (a) barriers, the level $\sigma = 1$ is at mean sea level. The treatment for type (b) is illustrated in Fig. 39.4, which shows a cross-section of Greenland in the

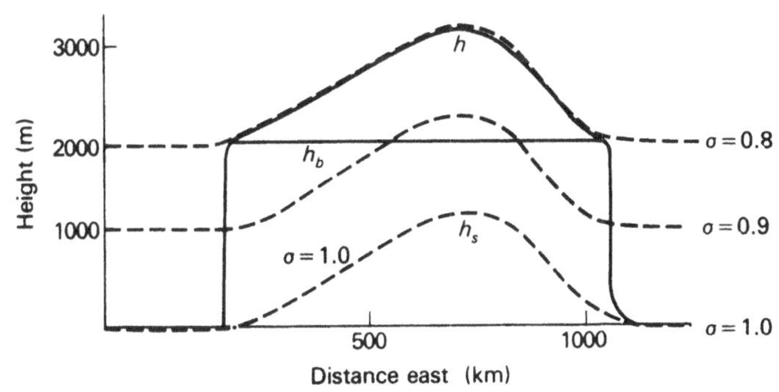

Fig. 39.4. West–east cross-section of Greenland and the lower levels of the model. The full line denotes the assumed land mass elevation. (After Egger, 1972.)

model. The wind is defined at $\sigma = 0.9$, 0.7 etc. and $\dot{\sigma}$ at $\sigma = 1.0$, 0.8, etc. The vertical walls are assumed to reach $\sigma = 0.8$, above which level a smooth topography (constituting the surface $\sigma = 0.8$) is superposed. At all grid-points inside the barrier, $\dot{\sigma} = 0$ along $\sigma = 0.8$ and $\mathbf{v} = 0$ along $\sigma = 0.9$.

Starting from an artificial initial field, consisting of a depression in a westerly current, Egger performed two 48-hour forecasts, one with no mountains and one where the Pyrenees and the Alps were considered. The resulting surface pressure distribution for the two cases is shown in Fig. 39.5. The difference between the two cases is quite clear and the result of the case in which the mountains are included agrees qualitatively with synoptic experience.

Fig. 39.6 shows the results of a corresponding experiment on the flow around Greenland. When a conventionally smoothed topography of Greenland is used, the 48-hour integration yields the pattern displayed in Fig. 39.6*a*. If the blocking technique described above is used, the resulting pattern is as shown in Fig. 39.6*b*. Egger infers that his suggested approach yields flow patterns that have more resemblance to real situations than have those patterns produced by smooth topography (or no topography).

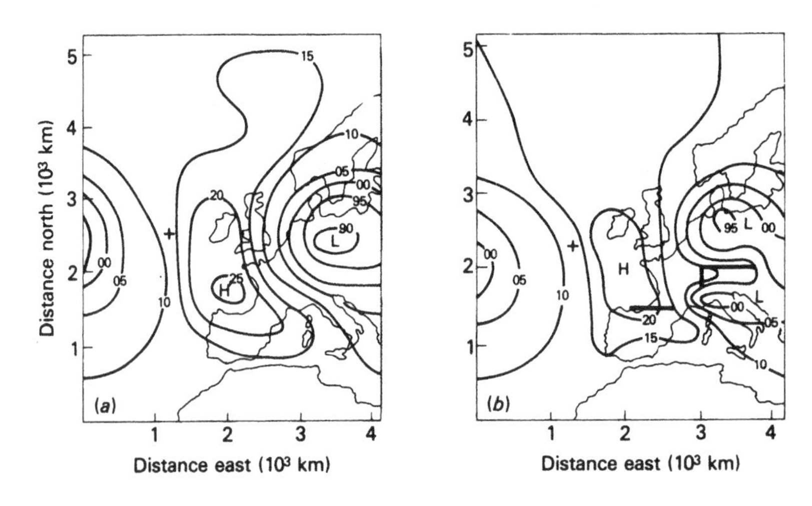

Fig. 39.5. 48-hour forecast of surface pressure: (*a*) without incorporation of orography; and (*b*) with the Pyrenees and the Alps considered as vertical walls along the straight lines indicated in the figure. A + in the figure shows the initial position of the low. (After Egger, 1972.)

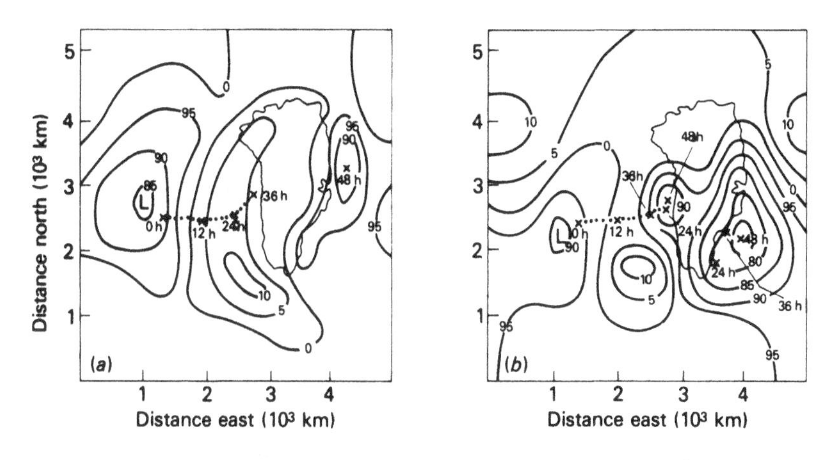

Fig. 39.6. 48-hour forecast of surface pressure: (*a*) with conventionally smoothed topography of Greenland incorporated; and (*b*) with Greenland having vertical walls from mean sea level up to 2000 m. (After Egger, 1972.)

It is quite likely that this blocking technique introduces substantial forcing on the very shortest scales–which we usually attempt to suppress by diffusion terms. For a fair appraisal of this method, the results of which are highly intriguing, these aspects need further investigation (see Egger, 1974).

39.4 Concluding remarks

Fictitious pressure gradient forces and temperature changes of appreciable magnitudes may readily occur in model calculations when orographic effects are incorporated. This has been demonstrated with examples in the preceding sections. These false effects appear in the shape of truncation errors although they are, to a large extent, the result of slight inconsistencies in the model relations for pressure, temperature and geopotential. Truncation and aliasing effects–which appear in both spectral and grid-point models–may be kept within tolerable limits provided that the model mountains are essentially composed of wave-components of wavelengths greater than 6 to 8 Δs. So the obvious way to incorporate mountains of relatively small horizontal extent or steep mountain slopes, seems to be to increase the horizontal resolution of the model. Possible approaches include the use of limited area models, or horizontal variation of the grid distance (see for example, Vergeiner and Ogura, 1972).

The above-mentioned aspects are generally applicable regardless of the type of vertical coordinate system employed.

From a technical point of view there is a clear distinction between the z- and the p-system on one hand, and the σ-type systems on the other. The former does not allow the use of spectral models and introduces, an extra programming burden compared with the latter type.

Verification problems that are specific for models including orography have not been taken up here. However, one point should be noted. When comparing model results with real data, one should probably avoid reducing the model results to mean sea level (or standard pressure levels) in mountainous regions, because of possible contamination resulting from extrapolation through a fictitious atmosphere below ground.

In this chapter, attention has mainly been given to the incorporation of orographic scales that are compatible with the resolution of the model in question. Steep slopes and possibly influential small-scale mountain components are smoothed considerably by the commonly-employed resolutions in prediction models. An adequate knowledge of how the atmospheric motion of synoptic and larger scales is affected by steep mountain barriers is lacking. The work of Egger (1972)–although an extreme approach in the sense that the air is, by-and-large, forced to flow *around* the mountain barriers–shows that it may be necessary to account for very steep slopes. Studies that can shed additional light on the need for parameterization of such subgrid-scale effects ought, therefore, to be pursued.

Theoretical studies show that orographically induced gravity waves of relatively small scales may be responsible, under certain conditions, for considerable vertical transfer of momentum and energy. These (nonlinear) processes and their relation to the main motion are not, as yet, fully understood. An appreciable amount of numerical experimentation with global or hemispheric models remains to be done in order to improve our insight into these interactions and their importance for the evolution of the large-scale atmospheric flow. A large portion of the wave-spectrum that may be important in this context is probably not resolvable in ordinary numerical models, implying that parameterization has to be employed. This was recognized as an important problem and was given a good deal of attention at the JOC Study Group Conference on parameterization of subgrid-scale processes (GARP, 1972). Yet, it seems that the subject has attracted relatively modest research efforts so far.

References

Corby, G. A., Gilchrist A. and Newson R. L. (1972) A general circulation model of the atmosphere suitable for long period integrations. *Quart. J. Roy. Meteor. Soc.*, **98**, 809–32.

Egger, J. (1972) Incorporation of steep mountains into numerical forecasting models. *Tellus*, **24**, 324–35.

Egger, J. (1974) Numerical experiments on lee cyclogenesis. *Mon. Wea. Rev.*, **102**, 847–60.

GARP (1972) Parameterization of sub-grid scale processes. GARP Publications Series No. 8 (ICSU/WMO), pp. 52–61.

Hahn, D. G. and Manabe, S. (1975) The role of mountains in the South Asian monsoon circulation. *J. Atmos. Sci.*, **32**, 1515–41.

Hahn, D. G. and Manabe, S. (1976) Reply to comments by Sadler and Ramage (1976). *J. Atmos. Sci.*, **33**, 2258–62.

Kasahara, A. (1974) Various vertical coordinate systems used for numerical weather prediction. *Mon. Wea. Rev.*, **102**, 509–22. See also Corrigendum in *Mon. Wea. Rev.* (1975), **103**, 664.

Kasahara, A. and Washington W. M. (1971) General circulation experiments with a six-layer NCAR model, including orography, cloudiness and surface temperature calculations. *J. Atmos. Sci.*, **28**, 657–701.

Kasahara, A., Sasamori, T. and Washington W. M. (1973) Simulation experiments with a 12-layer stratospheric global circulation model. I. Dynamical effect of the Earth's orography and thermal influences of continentality. *J. Atmos. Sci.*, **30**, 1229–51.

Katayama, A., Kikuchi, Y. and Takigawa, Y. (1974) MRI global 3-level model. GARP Publications Series No. 14, (ICSU/WMO) pp. 174–88.

Manabe, S. and Terpstra, T. B. (1974) The effects of mountains on the general circulation of the atmosphere as identified by numerical experiments. *J. Atmos. Sci.*, **31**, 3–42.

Phillips, N. A. (1957) A coordinate system having some special advantages for numerical forecasting. *J. Meteor.*, **14**, 184–5.

Phillips, N. A. (1974) Application of Arakawa's energy-conserving layer model to operational numerical weather prediction. Office Note 104, US Department of Commerce, NMC.

Sadler, J. C. and Ramage, C. S. (1976) Comments on 'The role of mountains in the South Asian monsoon circulation'. *J. Atmos. Sci.* **33**, 2255–8.

Sundqvist, H. (1975*a*) On truncation errors in sigma-system models. *Atmosphere*, **13**, 81–95.

Sundqvist, H. (1975*b*) Initialization for models using sigma as the vertical coordinate. *J. Appl. Meteor.*, **14**, 153–158.

Sundqvist, H. (1976) On vertical interpolation and truncation in connection with use of sigma system models. *Atmosphere*, **14**, 37–52.

Vergeiner, I. and Ogura, Y. (1972) A numerical shallow-fluid model including orography with a variable grid. *J. Atmos. Sci.*, **29**. 270–84.

Washington, W. M. (1974) NCAR global circulation model. GARP Publications Series No. 14, (ICSU/WMO) pp. 61–78.

40

Vertical motion in the monsoon circulation

M. C. SINHA AND O. P. SHARMA

A review is presented of methods for solving the omega-equation by (*a*) finite differences, and (*b*) finite elements. The different types of forcing terms that arise when the wind vector is resolved into a rotational and a solenoidal part are described. A computational procedure for evaluating omega by successive approximations is outlined. Computations of omega by the method of finite elements, using prismatic elements, is described. Results of omega computations made by finite differences and by finite elements are presented and discussed.

40.1 Introduction

A problem of considerable importance in meteorology is the estimation of vertical velocity. As its magnitude is very small, direct measurements are not possible. Consequently, the pattern of vertical velocity is usually inferred from measurements of horizontal velocity, the pressure and temperature distribution in the atmosphere. A commonly used technique is to derive a solution of the diagnostic omega-equation. An excellent review of vertical velocity computations has been provided by Pearce (1974). In this paper we wish to consider solutions of the omega-equation by (*a*) finite differences, and (*b*) by the method of finite elements.

40.2 Basic equations

The basic equations express the first law of thermodynamics, the conservation of vorticity and conservation of mass. Pressure coordinates are generally used to derive the omega-equation but, as we shall see later, this gives rise to difficulties in the vicinity of steep orographic features.

Assuming that the geostrophic approximation holds for the horizontal advective flow, it follows that:

$$\boldsymbol{V}_g = \boldsymbol{k} \times \nabla\phi, \tag{40.1}$$

with

$$\nabla \cdot \boldsymbol{V}_g = 0, \tag{40.2}$$

where $\boldsymbol{V}_g$ denotes the geostrophic wind, ϕ the geopotential, and $\boldsymbol{k}$ the unit vector along the vertical.

The vorticity equation and the first law of thermodynamics may be written as:

$$\zeta_t + \boldsymbol{V}_g \cdot \nabla(\zeta + f) = f_0\omega_p, \tag{40.3}$$

and

$$(-\phi_{pt}) + \boldsymbol{V}_g \cdot \nabla(-\phi_p) - \sigma\omega = \frac{R\dot{Q}}{c_p p}, \tag{40.4}$$

where the notation used is

f Coriolis parameter,
f_0 Coriolis parameter at a fixed latitude,
ζ vertical component of relative vorticity,
σ $-\alpha(\ln\theta)_p$, the static stability parameter,
α specific volume,
θ potential temperature,
p pressure,
∇ horizontal gradient operator in an isobaric surface,
ω $\mathrm{d}p/\mathrm{d}t$,
$\dot{Q}$ diabatic heating rate per unit mass,
c_p specific heat at constant pressure.

Partial derivatives are denoted by subscripts.

From (40.3) and (40.4) the omega-equation is derived as:

$$\sigma\nabla^2(\omega) + f_0^2\omega_{pp} = [J(\phi, \zeta + f)]_p + \nabla^2 J(\phi, -\phi_p)/f - R\nabla^2(\dot{Q})/c_p p, \tag{40.5}$$

where J stands for the Jacobian operator. The static stability σ is assumed to be independent of horizontal space coordinates.

It should be noted that the forcing terms of (40.5) depend on: (i) the geopotential; (ii) the vorticity; and (iii) the diabatic heating. If these are specified, then omega may be obtained by solving (40.5) with suitable boundary conditions. The boundary conditions at the lower boundary include the effects of ground terrain and surface friction. The value of omega at the lower boundary of the model may be written as

$$\omega_s = (p_s)_t + \boldsymbol{V}_s \cdot \nabla p_s, \tag{40.6}$$

where p_s is the surface pressure.

One of the difficulties of this formulation is that $\boldsymbol{V}_s$, the surface wind, cannot be determined near steep mountains with much accuracy.

Many techniques have been used for the inclusion of frictional effects, but as the free atmosphere above the planetary boundary layer is the main concern here, a simple formulation, which has been used in many other models, is adopted:

$$\tau_x = C_D \rho u |\boldsymbol{V}|, \tag{40.7a}$$

$$\tau_y = C_D \rho v |\boldsymbol{V}|, \tag{40.7b}$$

where τ_x and τ_y denote the components of frictional stress, u and v are the zonal and meridional components of $\boldsymbol{V}$, ρ is the air density and C_D is a representative drag coefficient.

An improvement on the geostrophic assumption may be made by expressing the velocity vector in terms of a streamfunction ψ and a velocity potential χ:

$$\boldsymbol{V} = \boldsymbol{k} \times \nabla \psi - \nabla \chi. \tag{40.8}$$

The relative vorticity is then

$$\zeta = \nabla^2 \psi,$$

and

$$\omega_p = \nabla^2 \chi.$$

Using (40.8), the vorticity equation becomes

$$\nabla^2 \psi_t = -J(\psi, \nabla^2 \psi + f) - g(\nabla \times \boldsymbol{\tau})_p + \nabla \chi \cdot \nabla(\nabla^2 \psi + f) + (\nabla^2 \psi + f)\nabla^2 \chi$$
$$- \nabla \omega \cdot \nabla \psi_p - \omega(\nabla^2 \psi)_p. \tag{40.9}$$

From this the omega-equation may be written as

$$\nabla^2(\sigma\omega) + f^2 \omega_{pp} = F(x, y, p), \tag{40.10}$$

where $F(x, y, p)$, the forcing function, now contains eleven terms. In (40.10), the Coriolis parameter (f) has not been treated as a constant.

Similarly, the horizontal variations of σ have been taken into consideration. In (40.10)

$$\begin{aligned}F(x, y, p) = {} & f[J(\psi, \nabla^2\psi + f)]_p + \pi\nabla^2[J(\psi, \theta)] - 2[J(\psi_x, \psi_y)]_{pt} \\ & -f[J(\nabla^2\psi, \nabla^2\chi)]_p + fg(\nabla \times \tau)_{pp} - R(\nabla^2\dot{Q})/C_p p \\ & +f[\omega\nabla^2\psi_p]_p + f(\nabla\omega \cdot \nabla\psi_p)_p - f[\nabla\chi \cdot \nabla(\nabla^2\psi + f)]_p \\ & -\pi\nabla^2(\nabla\chi \cdot \nabla\theta) - \beta(\psi_{ytp}), \end{aligned} \tag{40.11}$$

where $\pi = RT/p\theta$ and $\beta = f_y$.

Several experiments have been conducted for the monsoon regions with each of the versions (40.5) and (40.10) of the omega-equation. Thus, Das (1962) and Saha (1968) utilized modified versions of (40.5) to compute the mean vertical motion during the monsoon. An interesting result of these investigations was that computations of mean vertical motion provided an estimate of diabatic heat sources and sinks. Rao and Rajamani (1970) and Sinha (1973) calculated vertical velocities in the tropics using (40.5). Rajamani (1976) later used (40.5) to estimate energy conversions during the monsoon. On the other hand, Krishnamurti (1968) used (40.11) to compute vertical motion.

40.3 The method of finite elements

In recent years, there has been increasing interest in using the method of finite elements to solve boundary value problems. It is of some interest to see how this method could be applied to obtain a solution of (40.5).

Introduce a finite number of levels, say N, in the vertical and divide each level into a number of triangles. In this way the domain of interest, say Ω, is partitioned into a number of discrete prismatic elements. If a basis function is chosen for these prismatic elements having the form

$$\phi(x, y, p) = u(p)v(x, y), \tag{40.12}$$

then it is possible to carry out a finite-element formulation along the vertical and horizontal axes. The dependent variable ω may be expressed in terms of the basis given in (40.12). Choose the function $u(p)$ to be

$$u^l(p_k) = \begin{cases} 1, & l = k \\ 0, & l \neq k. \end{cases} \tag{40.13}$$

As the vertical and horizontal discretizations are independent, the finite-difference approximations may be applied in the vertical as well. This leads to

$$\nabla^2 \omega_k + c_1 A_{kl} \omega_l = F_k, \tag{40.14}$$

where

$$c_1 = f_0^2 / 4(\Delta p)^2$$

$$A_{kl} = \begin{vmatrix} -2/\sigma_1 & 1/\sqrt{(\sigma_1\sigma_2)} & 0 & \dots & 0 \\ 1/\sqrt{(\sigma_1\sigma_2)} & -2/\sigma_2 & 1/\sqrt{(\sigma_2\sigma_3)} & \dots & 0 \\ \vdots & \vdots & \vdots & \vdots & \vdots \\ 0 & 0 & 0 & 1/\sqrt{(\sigma_{N-1}\sigma_N)} & -2/\sigma_N \end{vmatrix}$$

Let G be the matrix whose columns are the eigenvectors of the matrix A. The following transformations may be then introduced

$$\omega_k = G\hat{\omega}_l,$$

$$F_k = G\hat{F}_l. \tag{40.15}$$

As the matrix A is symmetric, its eigenvalues are defined by

$$\Lambda = \{\lambda_k\}_{k=1}^{N},$$

and

$$GG^T = I,$$

where I is the unit matrix.

With these transformations the following decoupled set of equations is obtained:

$$\nabla^2 \hat{\omega}_l - c_1 \lambda_l \hat{\omega}_l = \hat{F}_l, \quad l = 1, 2, \dots, N, \tag{40.16}$$

while, for simplicity, the boundary condition may be expressed as

$$\hat{\omega}_l = 0.$$

To obtain a weak formulation of the system (40.16) multiply it by a test function v, which vanishes on the boundary. Integrating by parts, the set of algebraic equations:

$$(\nabla \hat{\omega}_l, \nabla v) + c_1 \lambda_l (\hat{\omega}_l, v) = -(\hat{F}_l, v), \tag{40.17}$$

is obtained where

$$(\hat{\omega}_l, v) = \int_\Omega \hat{\omega}_l v \, dx \, dy.$$

This system of algebraic equations may be now solved for each level.

40.4 Computation procedure

40.4.1 Finite differences

The computational algorithm is shown schematically in Fig. 40.1. Using wind data as input, the vorticity is first computed at each grid point. The streamfunction is next obtained by inverting the boundary value problem

$$\nabla^2 \psi = \zeta, \tag{40.18}$$

with suitable boundary conditions on the lateral walls. Using values of ψ as input, the geopotential is then obtained from the balance equation

$$\nabla^2 \phi = f\zeta + 2J(-\psi_y, \psi_x).$$

From this field of geopotential, the potential temperature (θ) and the static stability (σ) are obtained. The tendency ψ_t is thereafter computed by solving the vorticity equation (40.9), with the assumption that χ is everywhere zero. When this is inserted in the omega-equation (40.10), it provides a first guess for omega. Using this estimate of ω, the divergence is computed and the velocity potential χ is obtained by inverting $\nabla^2\chi = -D$ using $\chi = 0$ along the boundaries of the domain.

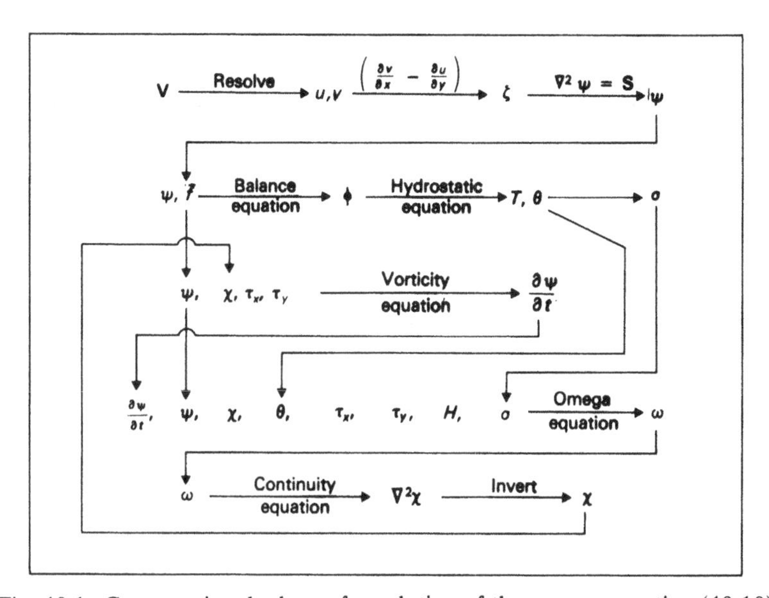

Fig. 40.1. Computational scheme for solution of the omega-equation (40.10).

The entire sequence is then repeated, and after 5 iterations the process is found to converge sufficiently to give reasonably accurate estimates of omega. In these computations nine-point finite-difference replacements are used for the Laplacian and Jacobian operators.

40.4.2 Finite elements

The domain is split up into a number of discrete elements in the form of prismoids. This is illustrated in Fig. 40.2. The main problem is to

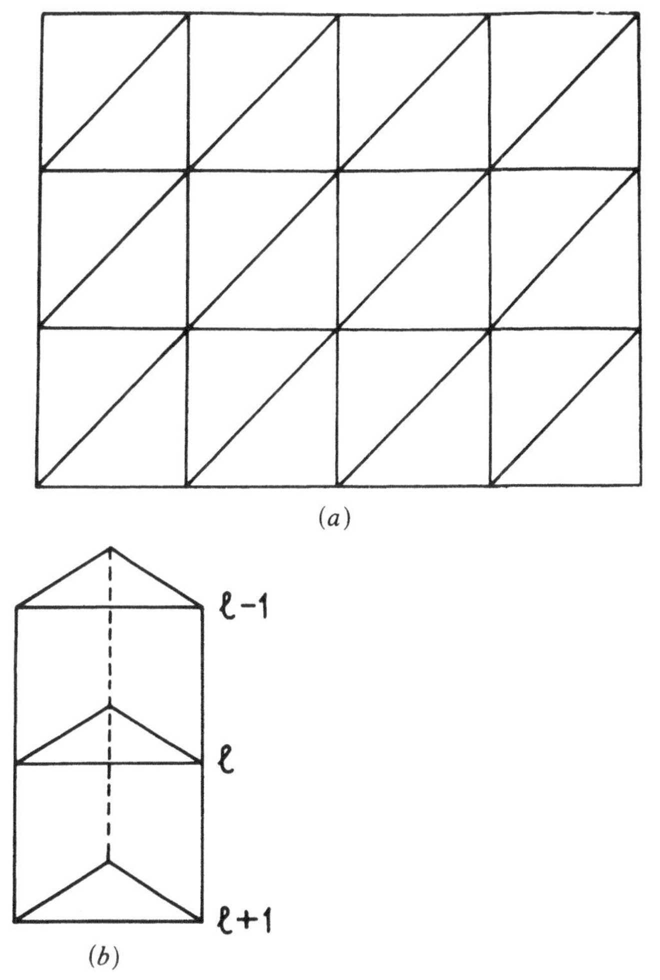

Fig. 40.2. The discrete elements employed in the domain of integration: (*a*) in the horizontal, and (*b*) in the vertical.

determine in which classes $\hat{\omega}_l$ and v should be allowed to lie. In accordance with the definition of Sobolev spaces, $\hat{\omega}_l$ and v are considered to be in the space of scalar functions which are square integrable over the domain Ω, and whose first derivatives are also square integrable over Ω. The approximate solution, as a linear combination of the basis functions, is

$$\hat{\omega}_l = \sum_i \hat{\omega}_{li} v_i.$$

When the above is inserted in (40.17), the problem is reduced to the evaluation of a series of integrals over a triangular domain. This can be achieved by integration by parts. Consider for example,

$$\int_\Omega J(z, \zeta) v \, dx \, dy = \int_\Omega (z_x \zeta_y - z_y \zeta_x) v \, dx \, dy.$$

Integrating by parts,

$$\int_\Omega J(z, \zeta) v \, dx \, dy = -\int_\Omega \zeta (z_x v_y - z_y v_x) \, dx \, dy.$$

z and v are interpolated by polynomials of degree one, so that ∇z and ∇v are constants and

$$\int_\Omega J(z, \zeta) v \, dx \, dy = -\sum_{i=1}^{M} \zeta (z_x v_y - z_y v_x) A(T_i),$$

where M is the number of triangles and $A(T_i)$ is the area of the ith triangle. It is important to note that the test function v is unity at each node and vanishes away from the node. Omega can now be evaluated over the domain.

40.5 Case studies

40.5.1 Finite differences

Results are presented of a case study for August 16, 1976 (00 GMT). This is a monsoon situation when a depression lay over north India, near the state of Rajasthan. The cyclonic circulation associated with the depression extended up to 500 mb. The sea-level isobars are shown in Fig. 40.3.

The horizontal grid was a portable latitude–longitude mesh with an adjustable size. In this study for a limited area, a grid of only 16×8 points

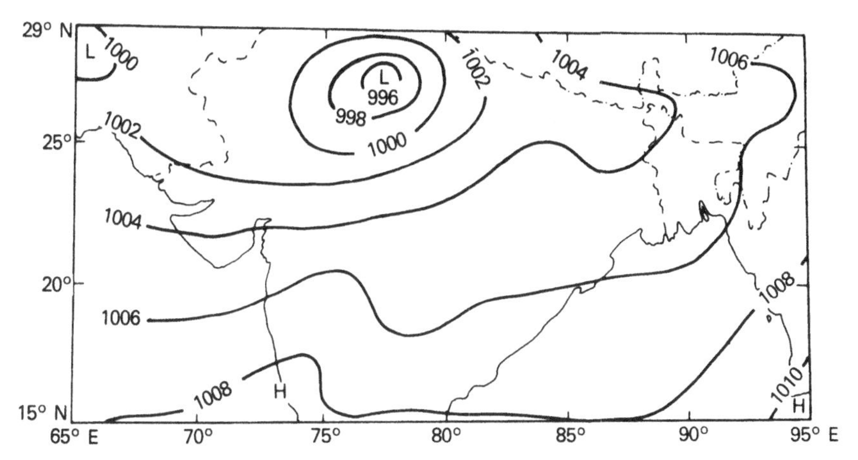

Fig. 40.3. Sea-level chart for 00 GMT, 16 August 1976. Isobars are labelled in mb.

was used with a horizontal resolution of 2 degrees. The domain covered the region from 15° N to 29° N and from 65° E to 95° E. The vertical grid was staggered. The details of the grid with levels for input are shown in Fig. 40.4.

The distribution of omega, evaluated by the finite-difference formulation, is shown in Fig. 40.5. The satellite observed cloud pictures are depicted in Fig. 40.6. It is seen that there is reasonable agreement between areas of ascending motion and heavy cloud. The computed omega fields at the surface, 775, 600 and 400 mb agree quite well with the cloud distribution. Similarly, the cloudless region over the eastern sector of Uttar Pradesh, Madhya Pradesh and adjoining Orissa agrees with the zone of descending motion.

40.5.2 *Finite elements*

The results of a similar case study using the method of finite elements is shown in Fig. 40.7 for 800 mb. Computations were also carried out for 600 and 400 mb, but they are not reproduced here. This refers to a monsoon depression after it had just crossed the coast. The sea-level isobars are shown in Fig. 40.8. A region of ascending motion is noticeable just ahead of the depression up to 800 mb. The present computations were, however, carried out assuming flat terrain and ignoring friction. Further work is in progress on the incorporation of more realistic boundary conditions.

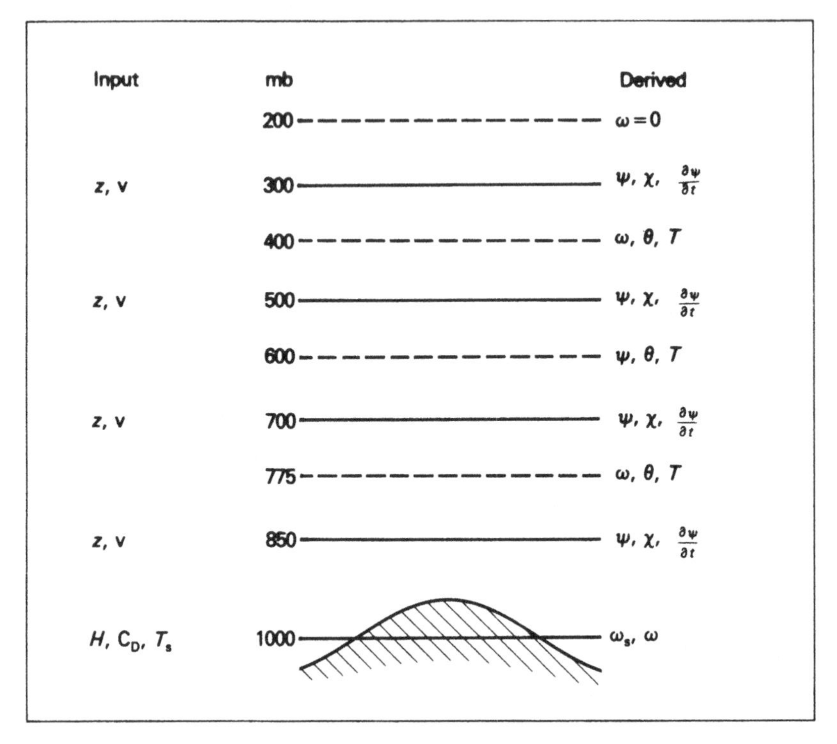

Fig. 40.4. Vertical grid with staggered input and working levels.

The finite-element method, which has been outlined here, is based on separation of variables with respect to the vertical and horizontal coordinates. Its main advantage is that it could be used with irregular boundaries. Thus, if the free surface at the top of the atmosphere was an undulating one, or if the Earth's surface had irregular terrain, suitable elements could be used to locate nodal points wherever needed, depending on the geometry of the domain. This would not be possible with finite differences, because the nodes have to be located at the intersection of horizontal and vertical lines. Such a restriction is not necessary with finite elements. But, as we can see, the finite-element method could create difficulties at the upper boundary of the domain ($p = 0$), because the static stability is not well defined at the top of the atmosphere. It is also not clear whether separation of variables, which leads to different forms of horizontal and vertical discretization, would be valid if the nonlinear omega-equation were used. It appeared from the trials that we have made so far, that a simple formulation in terms of finite elements could lead to

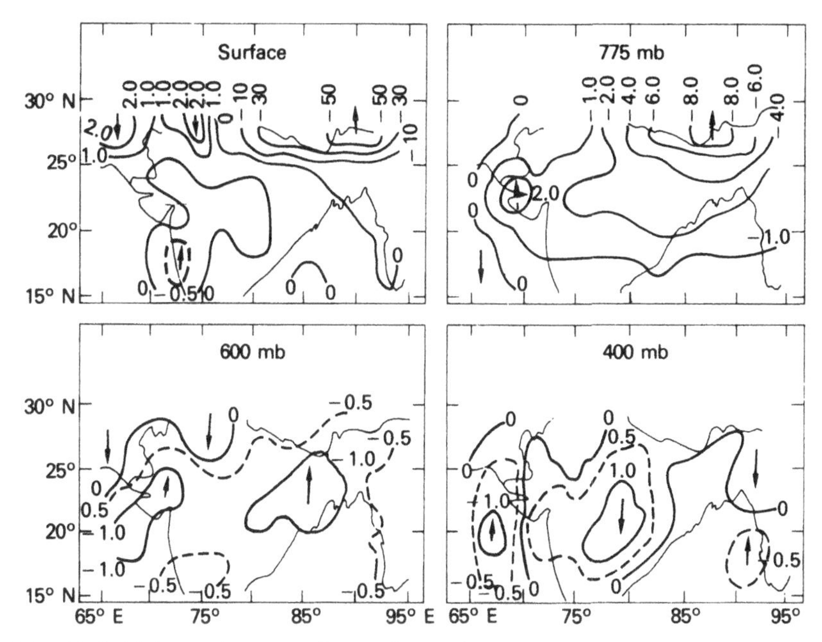

Fig. 40.5. Values of omega (in 10^{-4} mb s^{-1}) at different levels for 00 GMT, 16 August 1976.

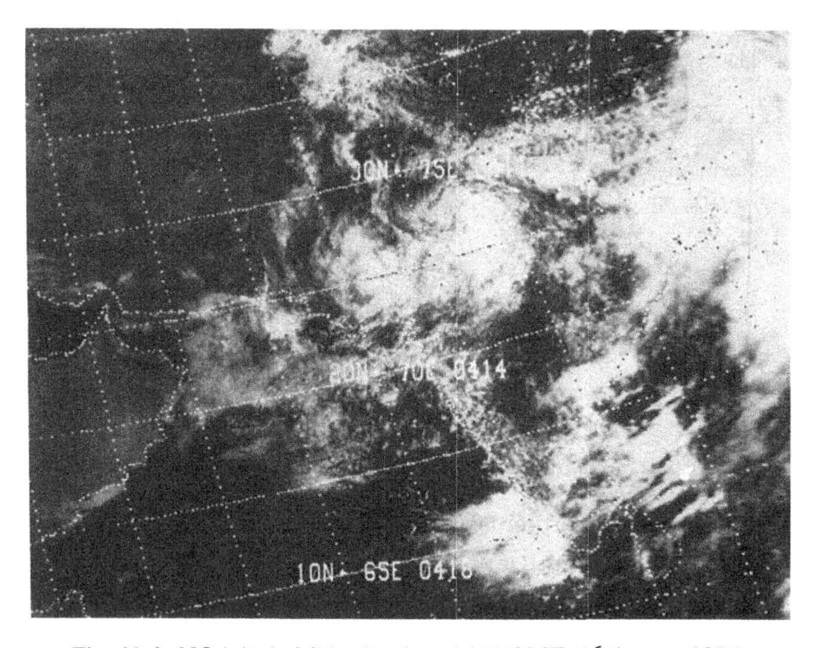

Fig. 40.6. NOAA-4 visible clouds at 0414 GMT, 16 August 1976.

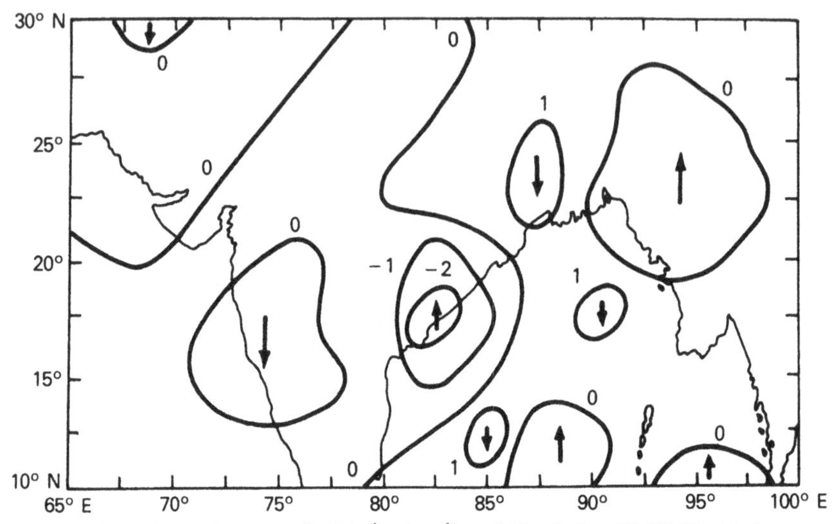

Fig. 40.7. Values of omega (in 10^{-4} mb s^{-1}) at 800 mb for 00 GMT, 10 August 1977. Arrows indicate regions of ascent and descent.

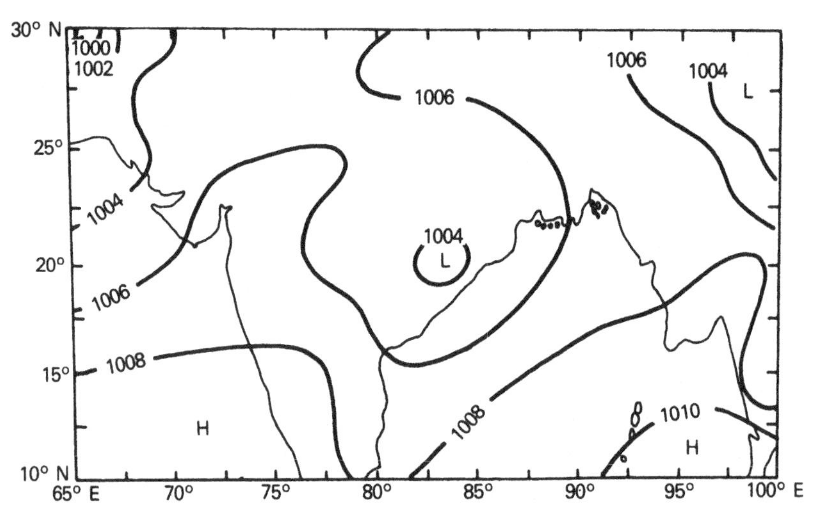

Fig. 40.8. Sea-level chart for 00 GMT, 10 August 1977. Isobars are labelled in mb.

some economy in speed of operation and storage in the computer memory, but these trials have been limited to linear models with very simple boundary conditions.

The authors are indebted to Dr P. K. Das, Director General of Meteorology for many helpful discussions. One of us (O. P. Sharma) is also indebted to Dr R. Sadourny and Dr O. Pironneau for their assistance in the formulation of the finite-element method.

References

Das, P. K. (1962) Mean vertical motion and non-adiabatic heat sources and sinks over India during the monsoon. *Tellus*, **14**, 212–20.

Krishnamurti, T. N. (1968) A diagnostic balance model for studies of weather systems of low and high latitudes, Rossby number less than 1. *Mon. Wea. Rev.*, **96**, 197–207.

Pearce, R. P. (1974) The design and interpretation of diagnostic studies of synoptic scale atmospheric systems. *Quart. J. Roy. Meteor. Soc.*, **100**, 265–85.

Rajamani, S. (1976) Energy aspects of the monsoon circulation. *Proc. Tropical Monsoons*, Indian Institute of Tropical Meteorology, Pune, India, pp. 87–96.

Rao, K. V. and Rajamani, S. (1970) Diagnostic study of a monsoon depression by geostrophic baroclinic model. *Ind. J. Meteor. Geophys.*, **21**, 187–94.

Saha, K. R. (1968) On the instantaneous distribution of vertical velocity in the monsoon field and structure of monsoon circulation. *Tellus*, **20**, 601–19.

Sinha, M. C. (1973) Computations of vertical velocities in the Indian region by quasi-geostrophic diabatic omega equation of four level model. Scientific Report No. 194, India Meteorological Department, New Delhi.

41

A one-dimensional model of the planetary boundary layer for monsoon studies

SAVITA VARMA

This chapter presents results obtained by numerical integration of a one-dimensional model of the planetary boundary layer above an oceanic surface. Wind hodographs are constructed for: (i) a constant eddy coefficient; and (ii) an eddy coefficient which depends on the mixing length. Starting with a set of given initial conditions, results are presented to compare this model with (*a*) a three-dimensional model constructed by Sommeria (1977) and (*b*) the results of the Puerto Rico experiment of 1972 (Pennell and LeMone, 1974).

41.1 Introduction

It is recognized that boundary-layer studies are likely to play an important role in improving our understanding of the monsoon. In recent years, experimental data have been collected by the NCAR 1972 Puerto Rico experiment (Pennell and LeMone, 1974) for the mixed layer over a tropical ocean. A three-dimensional model by Sommeria (1976 and 1978) has shown good agreement with the observed data. In this chapter, a similar one-dimensional model is presented which may be used for similar experiments for the monsoon region.

41.2 Basic equations

Atmospheric turbulence is assumed to be statistically homogeneous. The mean values of the dependent variables, namely, the components of the wind vector, potential temperature and the pressure gradient are assumed to be independent of horizontal space coordinates. This implies that the horizontal variation of the mean flow is assumed to be small compared with its vertical variation. With this assumption, the basic equations are taken as

$$\bar{u}_t = -(\overline{w'u'})_z + f(\bar{v} - v_g), \tag{41.1}$$

$$\bar{v}_t = -(\overline{w'v'})_z - f(\bar{u} - u_g), \tag{41.2}$$

$$\bar{\theta}_t = -(\overline{w'\theta'})_z + \frac{1}{c_p}\dot{Q} \tag{41.3}$$

where $\bar{u}$, $\bar{v}$ are components of the mean wind towards the east and north respectively, $\bar{\theta}$ is the mean potential temperature, f is the Coriolis parameter, and $\dot{Q}$ represents the rate of diabatic heating per unit mass.

Primed symbols have been used to denote deviations from the mean, which is defined by

$$\bar{u}(x, y, z, t) = \frac{1}{\Delta x\, \Delta y\, \Delta z\, \Delta t} \int_{t-\Delta t/2}^{t+\Delta t/2} \int_{z-\Delta z/2}^{z+\Delta z/2} \int_{y-\Delta y/2}^{y+\Delta y/2} \int_{x-\Delta x/2}^{x+\Delta x/2} u(x, y, x, t)\, \mathrm{d}x\, \mathrm{d}y\, \mathrm{d}z\, \mathrm{d}t. \tag{41.4}$$

Each dependent variable may be thus expressed as the sum of its mean value and a deviation from the mean; u_g and v_g represent the components of the geostrophic wind

$$u_g = -\frac{1}{f\rho_0}\frac{\partial p}{\partial y}, \qquad v_g = \frac{1}{f\rho_0}\frac{\partial p}{\partial x}. \tag{41.5}$$

The eddy stresses are now parameterized in terms of the mean flow, i.e.

$$\overline{w'u'} = -K\frac{\partial \bar{u}}{\partial z}, \tag{41.6a}$$

$$\overline{w'v'} = -K\frac{\partial \bar{v}}{\partial z}, \tag{41.6b}$$

$$\overline{w'\theta'} = -K_\theta\frac{\partial \bar{\theta}}{\partial z}, \tag{41.6c}$$

where K, K_θ represent eddy coefficients which, although varying in space and time, are assumed to remain positive. This study will be concerned with: (i) a constant K; and (ii) a value of K which depends on the vertical gradients of $\bar{u}$ and $\bar{v}$. For this purpose we put

$$K = l^2\left[\left(\frac{\partial \bar{u}}{\partial z}\right)^2 + \left(\frac{\partial \bar{v}}{\partial z}\right)^2\right]^{\frac{1}{2}} \tag{41.7}$$

where the mixing length (l) is defined as

$$l = \frac{k_0(z+z_0)}{1+k_0(z+z_0)/\lambda}, \tag{41.8}$$

where k_0 is Von Karman's constant, z_0 is the roughness parameter and λ is the maximum value of l at the top of the planetary boundary layer. This formulation of the eddy coefficient was suggested by Blackadar (1962). The value of λ is determined by external factors and, following Blackadar, we put

$$\lambda = 4\times 10^{-4}\ u_g/f. \tag{41.9}$$

It should be noted that in this definition, the dependence of K on thermal stratification has not been considered.

For the coefficient of thermal diffusivity we put

$$K_\theta = +ClE,$$

where C is a constant (1.0), l is the mixing length and $\frac{1}{2}E^2$ is the mean turbulent kinetic energy per unit mass, i.e.

$$\tfrac{1}{2}E^2 = \tfrac{1}{2}(\overline{u'^2} + \overline{v'^2} + \overline{w'^2}). \tag{41.10}$$

The time rate of change of the turbulent energy is

$$\frac{1}{2}\frac{\partial E^2}{\partial t} = \frac{\partial}{\partial z}\left[\tfrac{5}{3}\lambda_1 E \frac{\partial}{\partial z}(\tfrac{1}{2}E^2)\right] - \overline{u'w'}\frac{\partial \bar{u}}{\partial z} - \overline{v'w'}\frac{\partial \bar{v}}{\partial z} + g\beta\,\overline{w'\theta'} - \frac{E^3}{\Lambda_1}. \tag{41.11}$$

The above expression may be derived from a second-order closure approximation of turbulence. The details have been described by Mellor and Yamada (1974) and will not be repeated here. Λ_1 and λ_1 are two length scales proportional to the mixing length, and β is defined by

$$\beta = -\frac{1}{\rho}\left(\frac{\partial \rho}{\partial T}\right)_p = \frac{1}{T}.$$

In deriving (41.11) Mellor and Yamada (1974) used approximations for (i) viscous dissipation and (ii) third-order moments (i.e., terms containing products of u', v', w'). Viscous dissipation was considered to be inversely proportional to Λ_1, in agreement with Kolomogoroff's hypothesis for small-scale isotropic turbulence. The third-order moment was expressed as being proportional to another parameter λ_1. In addition, it was assumed that there was no correlation between pressure and the velocity fluctuations. Experimental evidence suggests that this correlation is indeed small.

In order to include moisture but not phase changes, the potential temperature (θ) is replaced by a virtual potential temperature (θ_v) defined by

$$\theta_v = \theta(1 + 0.61q), \tag{41.12}$$

where q is the specific humidity. The rate of change of the variance of q may be expressed by an expression similar to (41.11):

$$\frac{1}{2}\frac{\partial}{\partial t}(\bar{q}'^2) = \frac{\partial}{\partial z}\left[\lambda_2 E \frac{\partial}{\partial z}(\tfrac{1}{2}q'^2)\right] - \overline{w'q'}\frac{\partial q}{\partial z} - \frac{Eq'^2}{\Lambda_2}. \tag{41.13}$$

In (41.13), λ_2 and Λ_2 represent two other length scales which are both proportional to the mixing length. The vertical turbulent moisture flux is expressed as

$$\overline{w'q'} = -ClE\frac{\partial \bar{q}}{\partial z}, \tag{41.14}$$

which is consistent with the form assumed for K_θ. It will be noted that the appropriate expression for E during the integration of (41.13) includes the effect of moisture on the buoyancy by using

$$\overline{w'\theta_v'} = -ClE\frac{\partial \bar{\theta}_v}{\partial z}. \tag{41.15}$$

41.3 Numerical integration

Equations (41.1), (41.2) and (41.3) are integrated from the surface to a height of 2 km. This layer is divided into increments of 50 m depth. Denoting the vertical grid-points by the suffix j and timesteps by n, (41.1) is replaced by the following difference equation:

$$\bar{u}_j^{n+1}\left(1 + \frac{2\Delta t}{(\Delta z)^2}K\right) = \bar{u}_j^{n-1}\left(1 - \frac{2\Delta t}{(\Delta z)^2}K\right)$$

$$+ \frac{2\Delta t}{(\Delta z)^2}K\left(\bar{u}_{j+1}^n + \bar{u}_{j-1}^n\right) + 2\Delta t f(\bar{v}_j^n - v_g), \tag{41.16}$$

with a similar equation for $\bar{v}$.

It was found that while (41.16) was adequate when K was constant, a different scheme was necessary when K was variable.

For the latter, the equation used was

$$\bar{u}_j^{n+1}\left[1+\frac{\Delta t}{(\Delta z)^2}(k_{j+\frac{1}{2}}^n+K_{j-\frac{1}{2}}^n)\right]$$

$$=\bar{u}_j^{n-1}\left[1-\frac{\Delta t}{(\Delta z)^2}(K_{j+\frac{1}{2}}^n+K_{j-\frac{1}{2}}^n)\right]$$

$$+\frac{2\Delta t}{(\Delta z)^2}(K_{j+\frac{1}{2}}^n\bar{u}_{j+1}^n+K_{j-\frac{1}{2}}^n\bar{u}_{j-1}^n)+2\Delta tf(\bar{v}_j^n-v_g). \qquad (41.17)$$

Similar equations were used for $\bar{v}$, $\bar{\theta}$, $\bar{\theta}_v$, q and for $\frac{1}{2}E^2$, $\frac{1}{2}(\bar{q}')^2$.

The constants were taken as

$$z_0=1.0 \text{ cm},$$

$$f=7.29\times10^{-5}\text{ s}^{-1},$$

$$(u_g, v_g)=(10, 0)\text{ m s}^{-1}.$$

The diabatic heating rate $\dot{Q}$ was assumed to have a value corresponding to a rate of cooling of 2×10^{-5} K s^{-1}. This implies a radiational cooling approximately balancing the vertical advection of potential temperature by turbulence in a mixed layer. Thus, it is assumed that the flux of heat over an oceanic surface, such as the Arabian Sea, remains constant during the integration of the model.

The following initial and boundary values were assumed for obtaining the wind hodographs in Fig. 41.1:

$$\bar{u}=\bar{v}=0\text{ m s}^{-1}\quad\text{at } z=z_0,$$

$$\left.\begin{aligned}\bar{u}&=u_g=10\text{ m s}^{-1}\\ \bar{v}&=v_g=0\text{ m s}^{-1}\end{aligned}\right\}\text{at } z=2\text{ km}.$$

The integration was performed with a time step of 5 s.

In order to compare the results with (*a*) the three-dimensional model of Sommeria (1977), and (*b*) case (ii), i.e. with a variable eddy coefficient, of the Puerto Rico experiment, the initial conditions were subsequently changed to those adopted by Sommeria (1978). These were:

(*a*)

$$\bar{u}=16\text{ m s}^{-1}\qquad z_0\le z\le500\text{ m},$$

$$16\le\bar{u}\le11.3\text{ m s}^{-1}\qquad 500\le z\le1250\text{ m},$$

$$\bar{u}=11.3\text{ m s}^{-1}\qquad 1250\le z\le2000\text{ m},$$

(*b*)	$2 \leq \bar{v} \leq 4 \text{ m s}^{-1}$	$z_0 \leq z \leq 800$ m,
	$\bar{v} = 4 \text{ m s}^{-1}$	$800 \leq z \leq 2000$ m,
(*c*)	$298.8 \leq \bar{\theta} \leq 298.5\ K$	$z_0 \leq z \leq 50$ m,
	$\bar{\theta} = 298.5\ K$	$50 \leq z \leq 500$ m,
	$298.5 \leq \bar{\theta} \leq 300.1\ K$	$500 \leq z \leq 1100$ m,
	$300.1 \leq \bar{\theta} \leq 307\ K$	$1100 \leq z \leq 2000$ m,
(*d*)	$\bar{q} = 0.015 \text{ kg m}^{-3}$	$z_0 \leq z \leq 500$ m,
	$0.015 \leq \bar{q} \leq 0.0115 \text{ kg m}^{-3}$	$500 \leq z \leq 1450$ m,
	$0.0114 \leq \bar{q} \leq 0.005 \text{ kg m}^{-3}$	$1450 \leq z \leq 2000$ m.

The initial $\bar{u}$ component was selected to have a constant value in the upper part of the boundary layer, with a slight increase downward and becoming constant again below 500 m. The increase in the layer 500 to 1250 m represents a cloud layer. The $\bar{v}$ component was taken to be constant with height, except for a small veering in the lowest 800 m.

The initial potential temperature was taken to be constant throughout a mixed layer of 500 m depth. Above it, the lapse rate becomes conditionally unstable up to 1100 m, and later strongly stable. This simulates case (ii) of the Puerto Rico experiment.

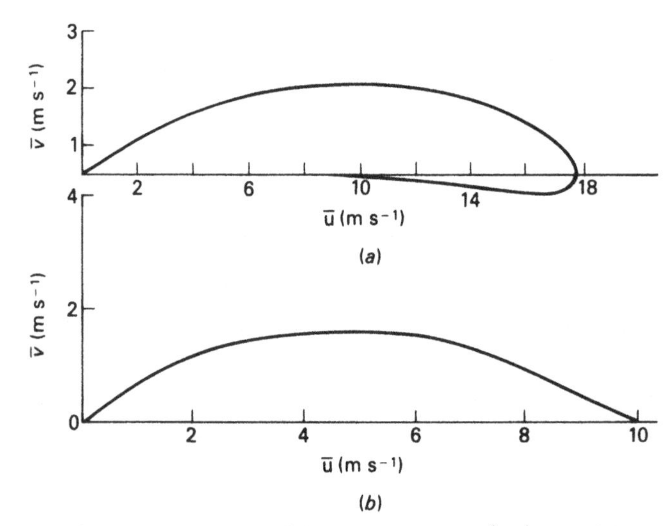

Fig. 41.1. Wind hodographs for (*a*) constant K (5 $\text{m}^2\,\text{s}^{-1}$) and (*b*) Blackadar's formula for K.

The specific humidity shows a slight decrease in the mixed layer (500 to 1450 m) and later decreases rapidly in the stable conditions in the upper part of the boundary layer.

41.4 Results

Fig. 41.1 shows wind hodographs for a constant K (5 m^2 s^{-1}), and a variable K using Blackadar's model for the mixing length. The results were obtained after 12.5 hours of model time. The hodographs are similar except for the fact that with Blackadar's model the reversal of the wind was not observed.

Fig. 41.2 illustrates the growth of $\bar{u}$ and $\bar{v}$ in a period of 3 hours. The results are compared with (*a*) case (i) of the Puerto Rico experiment, and

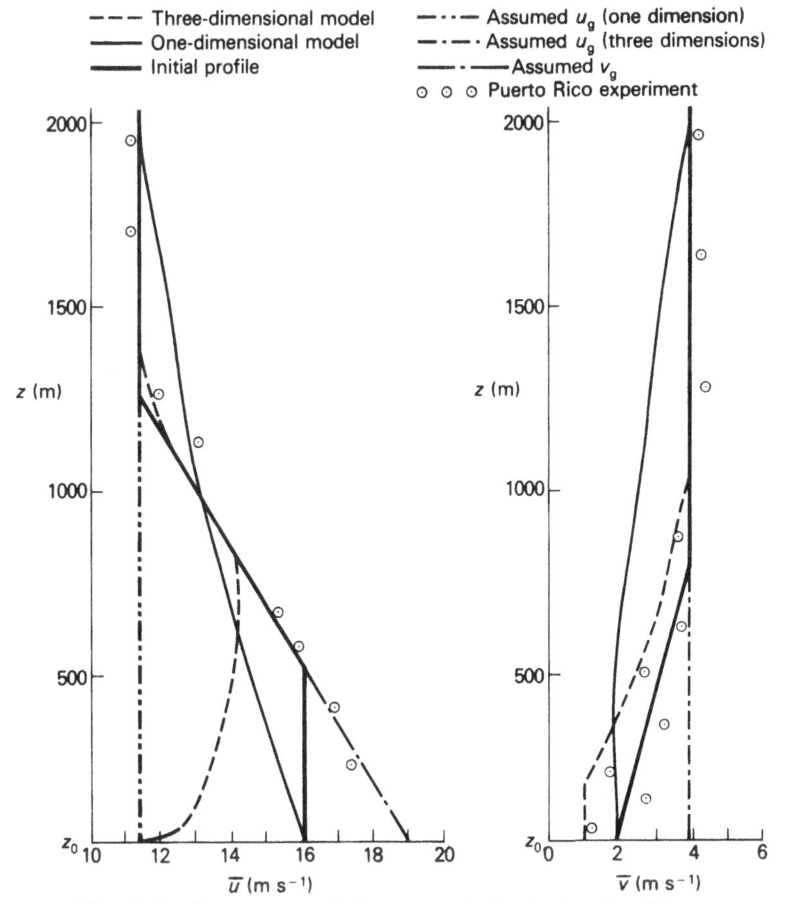

Fig. 41.2. Components of the mean wind velocity after 3 hours.

(b) the three-dimensional model of Sommeria. Case (i) of the Puerto Rico experiment refers to a region where convective activity was suppressed by a stable upper part of the planetary boundary layer, with cumulus clouds only a few hundred metres deep in the form of streets. We note that the $\bar{u}$, $\bar{v}$ profiles do not show good agreement with the three-dimensional model, or the Puerto Rico experiment. In fact, the gradients of $\bar{u}$, $\bar{v}$ obtained with a one-dimensional model appear to be different from the observations of the Puerto Rico experiment. This difference is attributed to the neglect of space derivatives of $\bar{u}$, $\bar{v}$ in our model.

The profiles of $\bar{\theta}$ and $\bar{\theta}_v$ (Fig. 41.3) are in reasonable agreement both with the observations from the Puerto Rico experiment and the three-dimensional model. Similar agreement is observed for the profile of $\bar{q}$, in Fig. 41.4. However, the region of super-adiabatic lapse rate near the surface is too exaggerated.

Figs. 41.5 and 41.6 show the vertical profiles of E^2 and $\frac{1}{2}\overline{q'^2}$ after 3 hours. They show a maximum of kinetic energy between 500 and 1000 m, but a similar maximum was not observed in the variance of q. The maximum in E indicates rapid production of turbulent kinetic energy

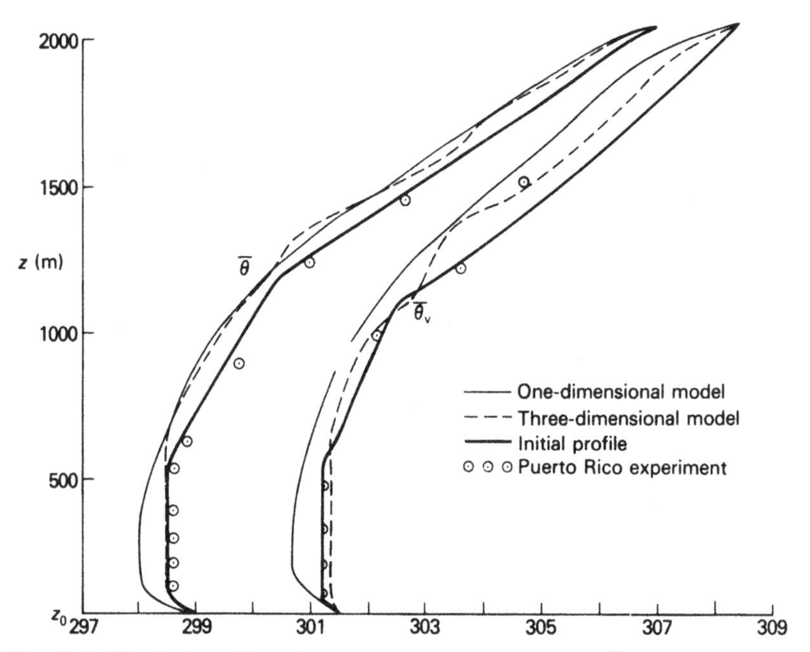

Fig. 41.3. Vertical profiles of mean potential temperature ($\bar{\theta}$) and mean virtual potential temperature ($\bar{\theta}_v$) after 3 hours.

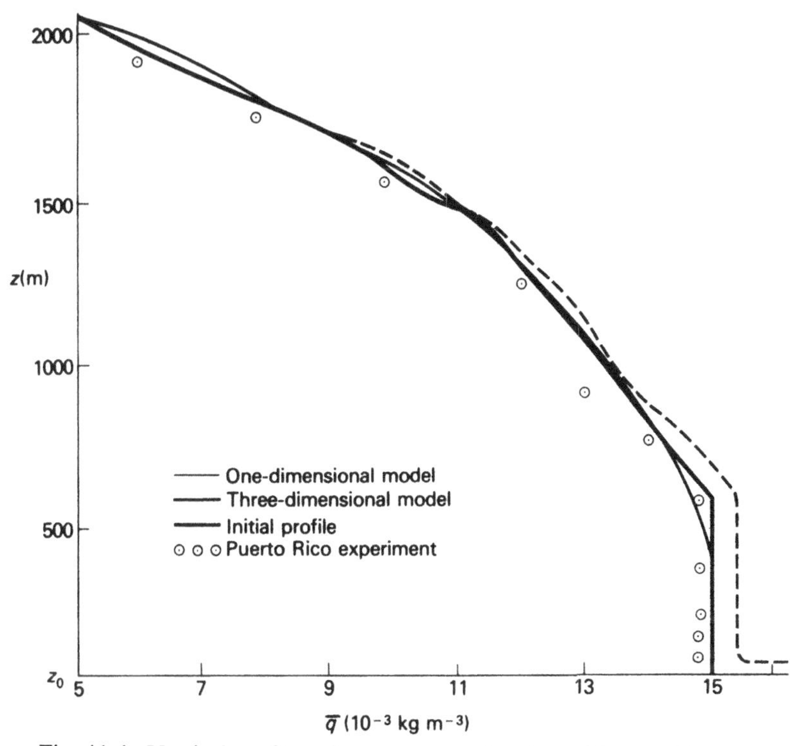

Fig. 41.4. Vertical profiles of mean specific humidity ($\bar{q}$) after 3 hours.

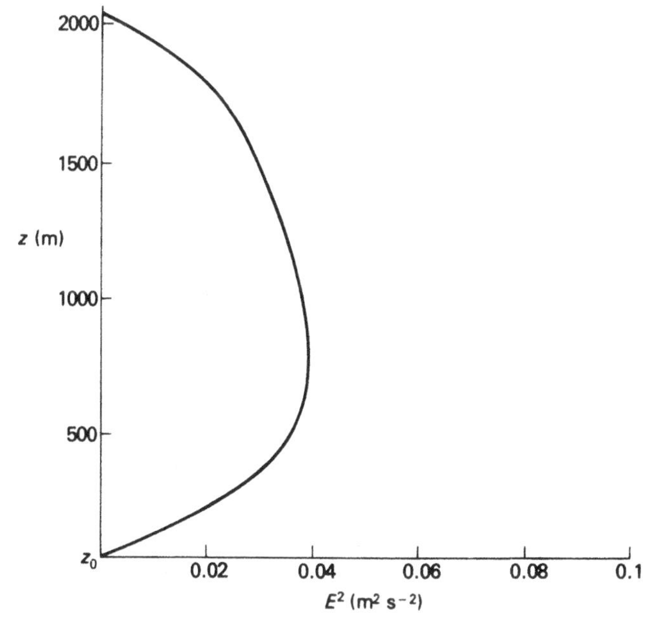

Fig. 41.5. Vertical profiles of computed mean turbulent kinetic energy ($\times 2$) after 3 hours.

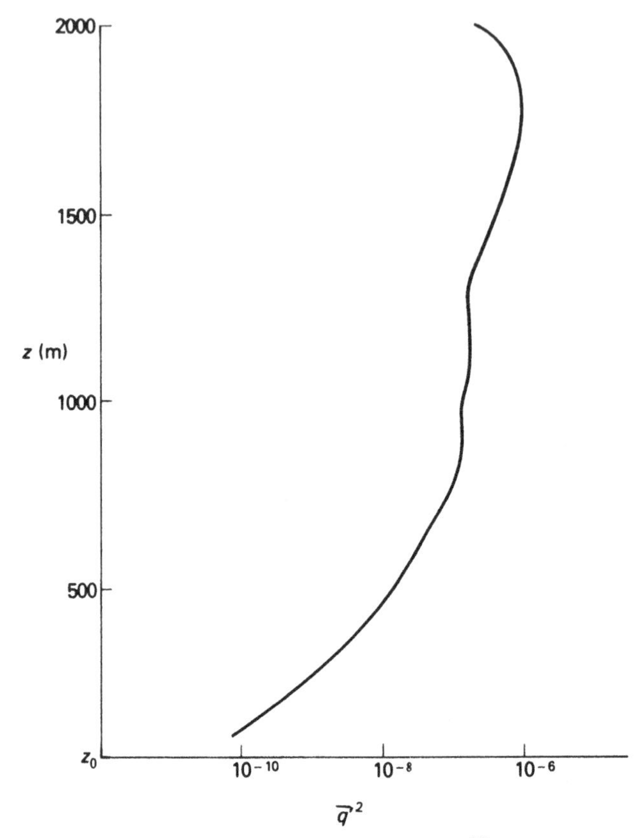

Fig. 41.6. Variance of the specific humidity ($\overline{q'^2}$) after 3 hours.

near the middle of the boundary layer, approximately coinciding with the cloud layer.

It is proposed to extend this study with representative profiles for the monsoon circulation, after the field data for MONSOON-77 and MONEX-79 have been processed. There appear to be interesting possibilities in this study. In particular, it seems likely that in the region of a low-level wind maximum as observed off the east coast of Africa, the turbulence kinetic energy profile (Fig. 41.5) may show a maximum at a lower level because of cooling due to oceanic upwelling and the absence of clouds. On the other hand, near the Indian coastline or in the eastern sector of the Arabian Sea, the profile could be quite different because of low-level clouds and a warmer sea.

This study was started with Dr G. Sommeria, Laboratoire de Météorologie Dynamique, France, during his visit to India. I wish to thank Professor M. P. Singh, IIT Delhi and Dr P. K. Das, IMD Delhi for their subsequent guidance.
I gratefully acknowledge help given by Dr P. K. Das in providing computer facilities at the India Meteorological Department, Delhi.

References

Blackadar, A. K. (1962) The vertical distribution of wind and turbulent exchange in a neutral atmosphere. *J. Geophys. Res.*, **67**, 3095–102.

Mellor, G. L. and Yamada, T (1974) A hierarchy of turbulence closure models for atmospheric boundary layer. *J. Atmos. Sci.*, **31**, 1791–1806.

Pennell, W. T. and LeMone, M. A. (1974) An experimental study of turbulence structure in the fair-weather trade wind boundary layer. *J. Atmos. Sci.*, **31**, 1308–23.

Sommeria, G. (1976) Three-dimensional simulation of turbulent processes in an undisturbed trade wind boundary layer. *J. Atmos. Sci.*, **33**, 216–41.

Sommeria, G. (1978) Direct testing of a three-dimensional model of the planetary boundary layer against experimental data. *J. Atmos. Sci.*, **35**, 25–39.

42

The use of empirical orthogonal functions for rainfall estimates

P. R. RAKHECHA AND B. N. MANDAL

In the present study the empirical orthogonal function or 'eigenvector' approach is used to determine the dominant rainfall patterns from normal seasonal rainfall records over Rajasthan. Two contrasting years (1917 and 1918) in which rainfall was in excess and deficient are also examined separately to see what anomalies, if any, exist in the associated patterns. Empirical orthogonal functions or eigenvectors are derived from the sets of 12-monthly rainfall values of 40 stations in Rajasthan. In the years of normal rainfall the first eigenvector is found to account for 99% of the variance in the original 12×40 matrix of rainfall data, thus indicating that the entire area is homogeneous as far as the normal seasonal variation of rainfall is concerned. However, in years of excessive or deficient rainfall, 3 or 4 vectors are needed to account for 99% of the variance. The first eigenvector in practically all cases largely resembles the seasonal variation of rainfall over the area, while the higher-order eigenvectors arise mainly as adjustment vectors to account for the balance of the variance. The eigenfunctions are used to estimate the mean monthly rainfall for places having no rainfall records. It is found that a reasonably good estimate of the normal seasonal distribution of rainfall over Rajasthan is given by just one vector.

42.1 Introduction

Empirical orthogonal functions, Tchebycheff polynomials and simple mathematical functions have been used to dissect two-dimensional fields of meteorological data. The primary object is to synthesize a large quantity of data into a much smaller number of components that still convey all the essential information contained in the original data. Empirical orthogonal functions have certain advantages over the conventional orthogonal functions since they are not of any predetermined form but are developed as unique functions from the data matrix. This is particularly useful if nothing is known in advance about the existence or nature of the component patterns. Several investigators have used the technique of empirical orthogonal functions to objectively describe the characteristics of meteorological data. Grimmer (1963) used this technique to derive characteristic patterns of surface temperatures in Europe. Kutzbach (1967) derived empirical orthogonal functions from sea-level pressure, surface temperature and precipitation in North America. Stidd (1967) used empirical orthogonal functions to represent the seasonal variation of rainfall over Nevada and found that the first three terms in order of importance account for 93% of the variance in the original 12×60 matrix of data. He found that these have features in common with the three natural cycles of precipitation.

Because of their advantages, empirical orthogonal functions, here referred to as eigenvectors, are used to determine the dominant precipitation patterns from the normal seasonal rainfall data over the Rajasthan region. Two contrasting years, 1917 when the rainfall over Rajasthan was very much in excess and 1918 when it was highly deficient, are also examined separately to see what anomalies, if any, exist in the associated patterns. The eigenvectors are derived from measurements of monthly rainfall at 40 stations in Rajasthan pertaining to normal rainfall averaged over 50 years and for the years 1917 and 1918. The method of derivation of the eigenvectors is outlined in § 42.2.

42.2 Eigenvectors of a climatic data matrix and their computation

Let $R = \{R_{ij}\}$ be an $m \times n$ climatic data matrix, where the elements in the ith row are the n station values corresponding to the ith sample (say the ith month). If $N(N < m, n)$ is the rank of the matrix, the spatial and temporal variation of the field can be described as a linear function of N independent vectors as

$$R_{ij} = \sum_{k=1}^{N} e_{ik} m_{kj},$$

where e_{ik} $(k = 1, 2, \ldots, N)$ are the elements of time vectors and m_{kj} are empirical space coefficients. Thus, with the aid of N independent vectors, the time variation of the field can be specified, and similarly the space variation at any point of time can also be specified as a linear function of N independent vectors. These vectors are termed eigenvectors and contain m elements if variation between the rows is sought, or n elements if variation between the columns is determined. The rank of the matrix is equal to the rank of the covariance matrix obtained as the product of the matrix and its transpose (R') i.e. $R'R$ or RR'. Since the covariance matrix obtained is symmetrical, the derived eigenvectors are real and mutually orthogonal.

If one considers the climatic data matrix of 12×40 elements consisting of 12 monthly values of an element for 40 stations and the variation between rows is determined, there will be 12 eigenvectors, each of 12 elements. The 12 eigenvalues are obtained by solving for λ the characteristic equation

$$|RR' - \lambda I| = 0. \tag{42.1}$$

The corresponding eigenvectors $\boldsymbol{e}$ are then obtained by solving the equation

$$(RR' - \lambda I)\boldsymbol{e} = 0. \tag{42.2}$$

Here I is the unit matrix. All eigenvalues and eigenvectors are real and all the eigenvalues are positive.

The eigenvalues are proportional to the variances associated with each of the corresponding eigenvectors. If the eigenvalues are arranged in decreasing order of magnitude, the first few terms, say K, may account for a substantial part of the total variance (95% and more) so that the residual is small and can be neglected as representing contributions within observational and sampling error bounds.

42.3 Discussion of results

The eigenvalue and the variance associated with each eigenvector are given in Table 42.1 in decreasing order of magnitude. These correspond to the normal seasonal rainfall and to the particular years 1917 and 1918. It is seen that in the case of normal seasonal rainfall the first eigenvector accounts for over 99% of the variance thus indicating that the entire area is homogeneous so far as the normal seasonal variation of rainfall is concerned, while in the two years of excessive and deficient rainfall 3 or 4 vectors are needed to account for 99% of the variance, indicating that the area becomes heterogeneous in such years.

TABLE 42.1. *Eigenvalues of the rainfall matrix in decreasing order of magnitude and the percentage variance associated with the corresponding eigenvectors.*

	Normal		1917		1918	
Eigenvalue Number	Eigenvalues	Variance (%)	Eigenvalues	Variance (%)	Eigenvalues	Variance (%)
1	5163.99	99.3	14776.91	90.0	1560.59	90.7
2	18.25	0.4	625.73	3.9	81.20	4.7
3	12.14	0.2	396.95	2.4	49.85	2.9
4	4.07	0.1	266.40	1.6	16.83	1.0
5	2.74		153.04	0.9	8.84	0.5
6	0.43		36.93	0.2	1.74	0.1
7	0.36		7.07		0.82	
8	0.17		3.59		0.41	
9	0.09		1.52		0.09	
10	0.06		0.43		0	
11	0.05		0.13		0	
12	0.03		0		0	

The 12 components of the significant eigenvectors for normal seasonal rainfall are given in Table 42.2 and are represented graphically in Figs. 42.1 and 42.2. The space fields associated with the first two eigenvectors are shown in Fig. 42.3.

It can be seen from Fig. 42.1 that the first eigenvector largely represents the annual cycle of rainfall, and in particular exhibits the predominant monsoon pattern over the whole area. The remaining eigenvectors are not significant in the normal case. The analysis clearly shows that the first eigenvector brings out the predominant features of the time variation.

TABLE 42.2. *Values of the first two significant eigenvectors for the normal seasonal rainfall.*

	Jan.	Feb.	Mar.	Apr.	May	Jun.
e_1	0.022	0.018	0.013	0.010	0.030	0.191
e_2	0.008	0.056	0.033	0.014	0.081	−0.403
	Jul.	**Aug.**	**Sep.**	**Oct.**	**Nov.**	**Dec.**
e_1	0.671	0.645	0.305	0.037	0.013	0.014
e_2	−0.436	0.724	−0.324	−0.045	−0.075	−0.007

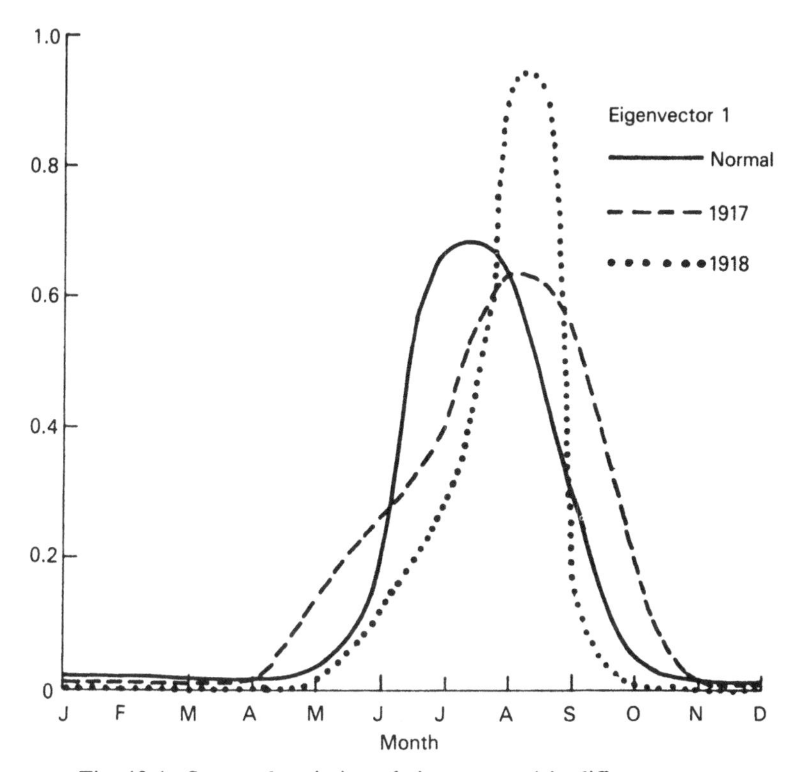

Fig. 42.1. Seasonal variation of eigenvector 1 in different years.

42.4 Physical significance of eigenvectors

Various investigators, e.g. Grimmer (1963), Stidd (1967), reported that it was possible to provide a physical explanation for the horizontal distributions described by eigenvectors. Stidd (1967) used eigenvectors to represent the seasonal variation of rainfall over Nevada and found that the first 3 eigenvectors in order of importance account for 93% of the mean square rainfall in the original 12×60 data matrix, and stated that they have features in common with the three natural cycles of precipitation. He identified the first eigenvector with Houghton's (1964) 'Pacific' component, the second vector with the 'Gulf' component, and the third with a composite of two of the several distributions that Houghton assigned to his spring and fall continental components after removing the effects of the first two. An attempt was, therefore, made to determine if such a separation of the components could be effected in respect of the rainfall of Rajasthan.

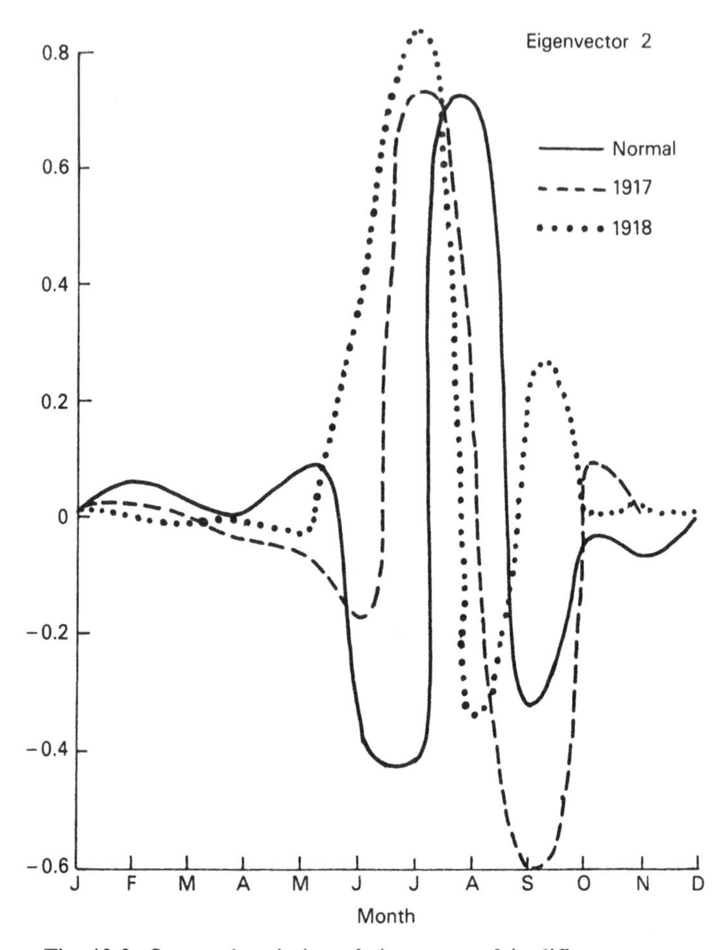

Fig. 42.2. Seasonal variation of eigenvector 2 in different years.

42.4.1 Synoptic features associated with rainfall over Rajasthan

It is well known that very little rain occurs over most of Rajasthan during the monsoon season unless some trough or depression moves towards it from the east or southeast. When a depression moves over Rajasthan, rain is well distributed over this region. However, there are occasions when some districts receive rain even in the absence of a depression. This rain can be associated with an active monsoon current, either from the Bay of Bengal or from the Arabian Sea, or with purely local thunderstorm activity. Therefore, the basic synoptic features associated with rainfall over Rajasthan are: (i) monsoon depressions; (ii) active monsoonal

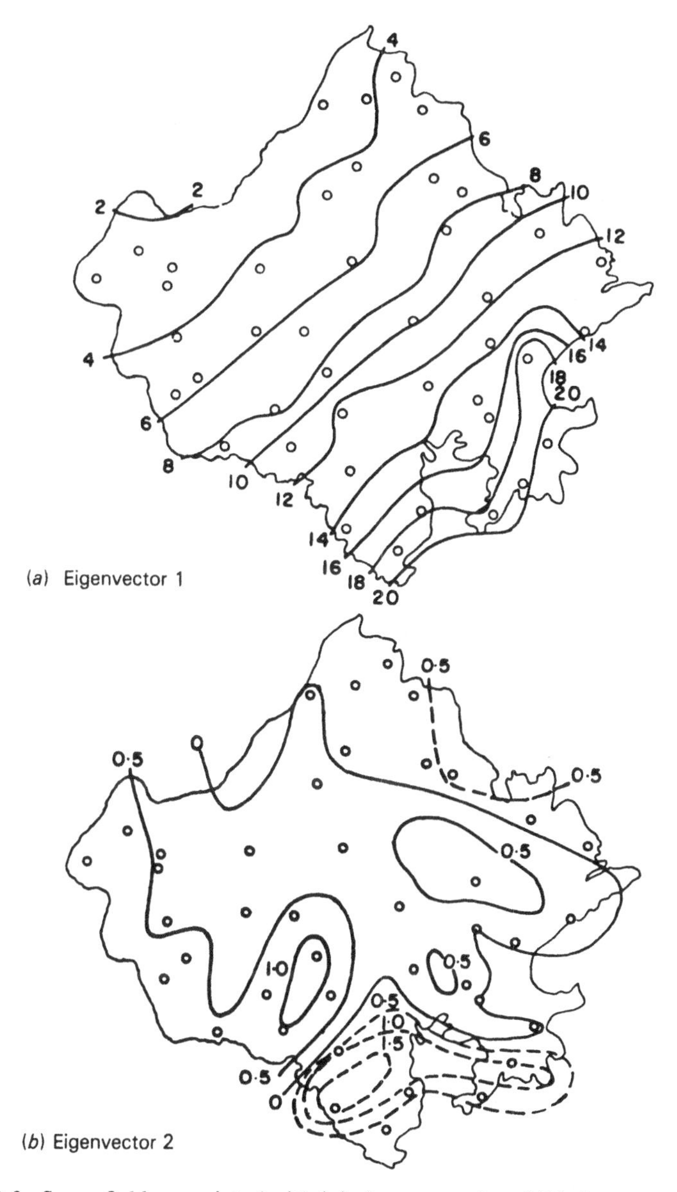

Fig. 42.3. Space fields associated with (*a*) eigenvector 1 and (*b*) eigenvector 2 for normal seasonal rainfall.

airflow from the Arabian Sea; (iii) active monsoonal airflow from the Bay of Bengal; or (iv) an area of local thunderstorm activity. Furthermore, these systems become modified as a result of mutual interaction. Thus, the pattern of distribution of rainfall over Rajasthan derives from different rain-producing systems which have occurred over the area. If these patterns could be explained in their entirety, then it would be possible to specify the precise factors which contributed to excessive rainfall in one year or less rainfall in another year. Jagannathan (1968) computed rainfall contributed by the different rain-producing systems over Rajasthan from an examination of daily rainfall data during the monsoon months of July and August for the two contrasting years 1917, when the rainfall over the area was very much in excess, and 1918 when it was very deficient. He found that in August 1917 the rainfall contributed by monsoon depressions, the Arabian Sea current and the Bay of Bengal current were about 50%, 33% and 5% respectively.

An attempt was, therefore, made to separate out the components of the rainfall distribution of Rajasthan for which Jagannathan (1968) showed the respective contributions. In order to determine the distributions of rainfall associated with different rain-producing systems, the actual rainfall distribution was resolved into component distributions and their physical meanings ascertained as far as possible. For this purpose, eigenvectors were derived from the seasonal rainfall data matrix of the two contrasting years 1917 and 1918.

The 12 components of the significant time variation eigenvectors are given in Table 42.3 and shown graphically in Figs. 42.1 and 42.2 for the

TABLE 42.3. *Significant eigenvectors for the two contrasting years of 1917 and 1918.*

		Jan.	Feb.	Mar.	Apr.	May	Jun.
e_1	1917	0.008	0.011	0.006	0.020	0.142	0.270
	1918	0.008	0	0.018	0.002	0.025	0.145
e_2	1917	0.017	0.016	0.006	−0.030	−0.046	−0.169
	1918	0.031	0	−0.015	−0.002	−0.033	0.353
		Jul.	**Aug.**	**Sep.**	**Oct.**	**Nov.**	**Dec.**
e_1	1917	0.395	0.632	0.557	0.202	0	0.001
	1918	0.269	0.936	0.169	0.001	0.020	0.001
e_2	1917	0.733	0.156	−0.633	0.081	0	0.006
	1918	0.831	−0.339	0.259	−0.003	0.018	0

two contrasting years 1917 and 1918. The space fields associated with the first eigenvector are shown in Fig. 42.4. Figs. 42.1 and 42.2 show that the first eigenvectors in both cases largely resemble that describing the seasonal variation of rainfall and bring out the predominant features of time variation over the area. The second and third vectors do not show any systematic seasonal patterns, as can be seen from Fig. 42.2. By analogy with regression analysis, the first eigenvector was fitted such that the sums of squares of different normals from the points represented by the observation vectors in 12-dimensional space were minimized. It must then resemble the seasonal pattern characteristic of the area. As already stated, in the normal seasonal case the first vector accounts for over 99% of the variance and about 91% in the individual years 1917 and 1918. The analysis clearly shows that the first significant eigenvector brings out the predominant features of the time variation while the other vectors do not have any systematic seasonal pattern. Hence the eigenvectors as such are not capable of providing physical insight into the precise factors determining rainfall over Rajasthan.

42.5 The use of eigenvectors for estimating rainfall

The eigenvector approach, which facilitates the representation of a large quantity of data in terms of a small number of orthogonal components accounting for a large fraction of the total variance in the original data, certainly has great advantages for analytical work. In the example presented here it can be seen that a reasonably good estimate of the seasonal distribution of rainfall over Rajasthan as a whole is given by just one or two eigenvectors.

The space fields m_{kj} associated with k different eigenvectors were obtained as follows:

Let $E = (e_{ik})$ be the 12×12 matrix of eigenvectors. Then

$$E'R = M, \tag{42.3}$$

where

$$M = (m_{kj}),$$

and

$$m_{kj} = \sum_{i=1}^{12} e_{ik} R_{ij},$$

is the sum of the 12 products of the elements of e'_k (an eigenvector e_k

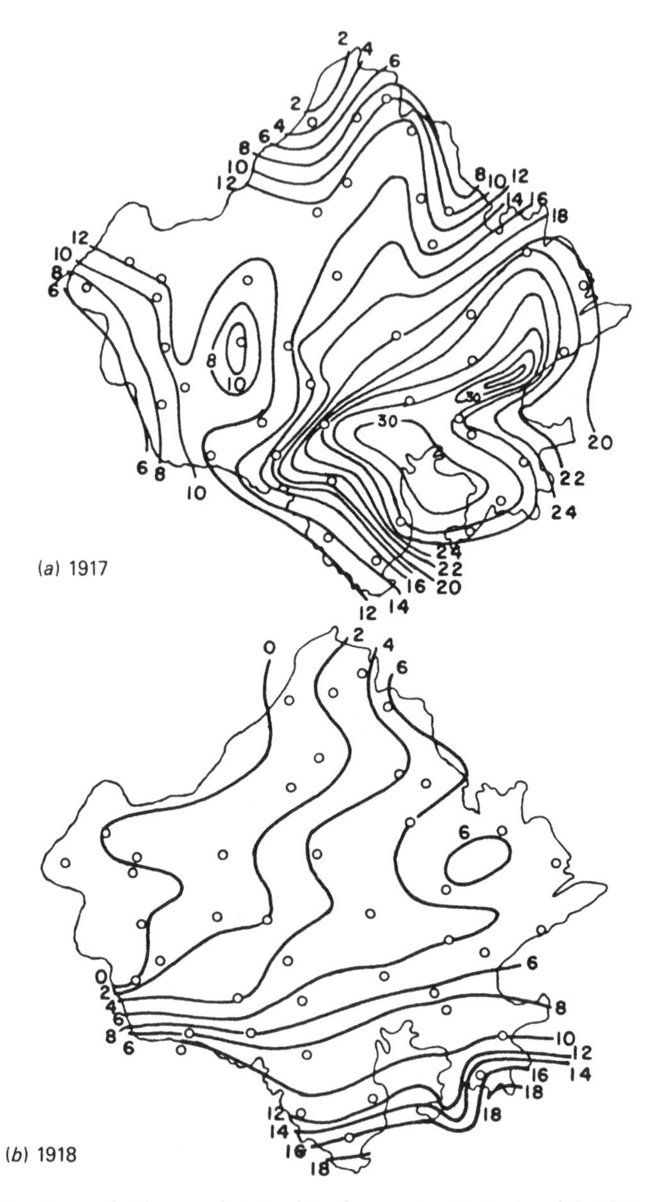

Fig. 42.4. Space fields associated with eigenvector 1 during (*a*) 1917 and (*b*) 1918.

transposed) and the corresponding elements of the 12 monthly values of the jth station. (Here i runs from 1 to 12, j from 1 to 40 and k from 1 to 12.)

If only the N significant eigenvectors are included, ($N < 12$), then there will be only N significant space fields associated with N eigenvectors and the matrix M of space components will be of order $N \times 40$. Since the eigenvectors are orthogonal, $\boldsymbol{e}_k \boldsymbol{e}'_k = I$ and $\boldsymbol{e}'_k = \boldsymbol{e}_k^{-1}$ i.e. the inverse exists. The rainfall in the ith month at the jth station can be obtained using the relationship

$$R = EM, \tag{42.4}$$

giving

$$R_{ij} = \sum_{k=1}^{N} e_{ik} m_{kj},$$

where m_{kj} is the value of the space field corresponding to the kth vector and jth location (not necessarily coinciding with the observational location), and e_{ik} is the ith element corresponding to the ith month in the kth eigenvector.

As already stated above, in the case of normal seasonal rainfall the first eigenvector accounts for over 99% of the variance; thus the first eigenvector only was used to estimate the mean monthly rainfall for places having no rainfall records. The first significant space field associated with the first eigenvector is shown in Fig. 42.3. The space field value m_{1j}, together with the first eigenvector $\boldsymbol{e}_1$, helps to estimate the seasonal variation of rainfall for the jth station.

As an illustration, the monthly rainfall values computed by the eigenvector method for Siwana (25°39′ N, 72°25′ E), which was not utilized in the derivation of eigenvectors, was compared with the actual mean rainfall based on data for 1901–50. The spacefunction related to Siwana was interpolated from Fig. 42.3. The rainfall for the year 1917 which was in excess and required 3 eigenvectors to account for the space–time variance was also calculated on the basis of interpolated values of spacefunctions. The computed and the actual values of rainfall in the case of Siwana for the normal rainfall are given in Table 42.4.

Table 42.4 shows a reasonable agreement between the values obtained by the eigenvector method and the actual values for the normal seasonal rainfall. Thus, it can be seen that an adequate estimate of the seasonal distribution of rainfall over Rajasthan as a whole is given by just one

TABLE 42.4. *Comparison of actual and computed rainfall values (in inches) for Siwana.*

	Jan.	Feb.	Mar.	Apr.	May	Jun.
Computed by eigenvector method	0.15	0.15	0.10	0.07	0.21	1.33
Actual rainfall (1901–50)	0.15	0.16	0.12	0.07	0.28	1.18
	Jul.	Aug.	Sep.	Oct.	Nov.	Dec.
Computed by eigenvector method	4.69	4.32	2.13	0.23	0.09	0.09
Actual rainfall (1901–50)	4.14	5.16	2.12	0.15	0.03	0.04

vector. This indicates that the normal seasonal variation of rainfall is homogeneous over the entire area.

The authors are grateful to Dr O. N. Dhar, Assistant Director (Hydrometeorology) for encouragement and guidance in the preparation of this paper.

References

Grimmer, M. (1963) The space-filtering of monthly surface temperature anomaly data in terms of pattern, using empirical orthogonal functions. *Quart. J. Roy. Meteor. Soc.*, **89**, 395–408.

Houghton, J. G. (1964) Prediction of long-term average rainfall at Nevada stations from short-term periods of observation. Desert Research Inst. Univ. Nevada, 21 pp.

Jagannathan, P. (1968) Climatic environment and its fluctuations in Rajasthan. IMO Scientific Report, 65, 21 pp.

Kutzbach, J. E. (1967) Empirical eigenvectors of sea-level pressure, surface temperature and precipitation complexes over North America. *J. Appl. Meteor.*, **6**, 5, 791–802.

Stidd, C. K. (1967) The use of eigenvectors for climatic estimates. *J. Appl. Meteor.*, **6**, 2, 255–64.

43

Applications of perturbation theory to problems of simulation of atmospheric processes

G. I. MARCHUK AND V. V. PENENKO

The applications of perturbation and control theories to numerical simulation of the atmosphere and ocean are discussed.

It is a characteristic of mathematical simulation of the atmosphere and ocean that input parameters of the models and initial fields are defined fairly roughly and often only the range of parameters is known. Therefore, a problem arises in defining the influence upon a simulation of (*a*) input parameter variations, and (*b*) the method of estimation of parameter values from observational data.

Methods are considered for the construction of discrete models and for numerical analysis and prediction of meteorological fields, as well as for the investigation of the model sensitivity to input parameter variations. The computational algorithms are based on a variational principle in combination with a splitting method. The latter provides stability and economy of computation.

Methods based on perturbation and control theories make it possible to carry out a qualitative analysis of a particular numerical model to determine the influence of different factors, and to aid rational design of numerical experiments.

In particular, application of sensitivity methods and control theory to initialization allows one to adjust the scales of processes described by the atmospheric model to correspond to the scales of the processes described by the input observational data.

One of the applications of the sensitivity theory is to the design of numerical experiments for evaluation of the effect of man's activity upon the climatic system.

43.1 Introduction

The theory of modern dynamical models of the atmosphere and ocean is founded on the laws of conservation of mass, momentum and energy which, along with thermodynamic laws, simulate interaction processes occurring between the atmosphere, ocean and land masses. Mathematically this is a system of multi-dimensional nonlinear differential equations solved under the assumption that the external energy source is solar radiation. Initial conditions are defined from the observed data in the actual atmosphere–ocean–land system. The systems of equations include a number of parameters, namely coefficients in the equations, initial fields, characteristics of integration fields, etc. In solving particular problems one can describe, to some extent reliably, a certain set of input parameter values. Determination of the range of parameters is based on *a priori* data of the model and observed values of fields of hydrometeorological elements, as well as on the intuition and experience of researchers. Solutions of each problem are found not only as functions of space variables and time, but also as functions of the model's input parameters. To find out whether any particular solution is reliable, one must investigate its behaviour for a variety of input parameters. This constitutes the problem of the model's sensitivity to variations of its input data.

Another aspect of the sensitivity problem is the effect of man's activity upon the climatic system. This is represented by variations of basic parameters describing the state and the regime of the atmosphere. This means that man's activity should be considered as one of the factors in the climatic system. One application of the sensitivity theory is to determine the effect of man's activity upon the climatic system.

There is a wide class of sensitivity problems, for whose solution a special mathematical apparatus has been designed. It is based on variational principles, methods of perturbation theory, and system identification. Since this investigation deals with the model's behaviour and its stability to input parameter variations, the practicality of the implementation of computational algorithms is of great importance. These computational algorithms are at present based on variational principles in combination with splitting methods.

Of no less importance is the formulation of a method to solve the inverse problem, i.e. to find values of input data which yield simulated fields of meteorological elements similar to those which are observed. Splitting techniques give economical and stable computational algorithms for implementation in the models, particularly in experiments carried out to investigate their sensitivity.

Perturbation and control theory methods, described e.g. in Eykhoff (1974) and Moiseev (1971), make it possible to conduct a qualitative analysis of the model and hence determine the involvement in it of different factors. This, in its turn, allows the rational design of numerical experiments, the formulation of problems for analysis and the forecasting of hydrometeorological fields.

43.2 The formulation of a generic problem and methods for its solution

Suppose that the structure of the mathematical model is given and is represented by a system of nonlinear differential equations which describe the basic hydrodynamic laws of simulated processes. This approach is similar to existing approaches to the construction of mathematical models for the class of problems under consideration.

For the sake of simplicity, the mathematical model will be expressed in the form

$$L(\boldsymbol{\phi}) \equiv B\frac{\partial \boldsymbol{\phi}}{\partial t} + G(\boldsymbol{\phi}) = 0, \qquad (43.1)$$

where $\boldsymbol{\phi}$ is a vector function representing a state of the system, B is a diagonal matrix (some of its diagonal elements may equal zero), and $G(\boldsymbol{\phi})$ is a nonlinear matrix differential operator. The structure of the operator is represented by a system of model equations and boundary conditions. The components of $\boldsymbol{\phi}$ may include the velocity vector, temperature, pressure or geopotential, density, etc. The range of variations of the space variable $\mathbf{x}$ and the time t is denoted by $\mathscr{D}_t$, the range of variations of the space variable $\mathbf{x}$ only by $\mathscr{D}$, and a set of functions $\boldsymbol{\phi}$, satisfying boundary and initial conditions, by $Q(\mathscr{D}_t)$. In addition to the vector $\boldsymbol{\phi}$, the model has a set of parameters denoted by $\boldsymbol{Y}$. The parameters are understood to be coefficients in equations, and in particular those representing turbulence; the parameters include also initial values of the state vector $\boldsymbol{\phi}$, the characteristic of the underlying Earth's surface, external sources and so on. Let the vector of input parameters belong to a certain range of values, i.e. $\boldsymbol{Y} \in R(\mathscr{Y})$.

Mathematical simulation consists of: investigation of solvability of the system of equations (43.1), construction of discrete analogues and development of computational algorithms, investigation of the model's behaviour in the domain $\{(\mathbf{x}, t) \in \mathscr{D}_t, \ \boldsymbol{Y} \in R(\mathscr{Y})\}$, and its sensitivity to variations of vector $\boldsymbol{Y}$. In fact the computational algorithm yields

$$\boldsymbol{\phi} = \boldsymbol{\phi}(\mathbf{x}, t, \boldsymbol{Y}), \tag{43.2}$$

which defines the state vector of the model as a function of independent variables and input parameters. A functional specifying the deviation of computed and observed values of the state vector is a criterion for a model's quality and methods of its implementation. Let $\Phi(\boldsymbol{\phi}, \boldsymbol{\phi}_m)$ be such a functional where $\boldsymbol{\phi}$ is a value of the state vector computed using the model and $\boldsymbol{\phi}_m$ a value obtained by measurements in actual conditions. Components of $\boldsymbol{\phi}_m$ are determined on a discrete set of points $\mathscr{D}_t^m \subset \mathscr{D}_t$. In practice, several criteria are usually used to estimate a model's quality.

At present there are a number of approaches to the construction and use of discrete models (Marchuk, 1967 and 1974; Döös, 1970; and GARP, 1974). A detailed review of the approaches is presented in Penenko (1975) and GARP (1974). For our purpose, the most convenient approach is that stated in Marchuk (1967 and 1974) and Penenko (1975 and 1977) which is based on the splitting method and a variational principle.

An integral identity, subsequently used to find a solution in a weak, generalized sense to a problem corresponding to problem (43.1), is determined. The corresponding problem is written as

$$I(\boldsymbol{\phi}, \boldsymbol{Y}, \boldsymbol{\phi}^*) = 0, \tag{43.3}$$

where $\boldsymbol{\phi}^*$ is an arbitrary, sufficiently smooth vector function. Let its structure correspond to that of $\boldsymbol{\phi}$. The integral identity (43.3) is chosen so that representations of the model as (43.1) and (43.3) are equivalent in a class of sufficiently smooth functions $\boldsymbol{\phi}$ and $\boldsymbol{\phi}^*$. To correctly define the functional of the left-hand side (43.3) is chosen so that

$$I(\boldsymbol{\phi}, \boldsymbol{Y}, \boldsymbol{\phi}) = 0 \tag{43.4}$$

represents an equation for the energy balance of the system.

An example of the construction of such an integral identity is as follows. Consider a non-adiabatic model of the atmosphere on a spherical Earth using an isobaric coordinate system. In this case the vector function $\boldsymbol{\phi}$ has the form $\boldsymbol{\phi} = (u, v, T, H, w)$ where (u, v, w) are components of the velocity vector $\boldsymbol{u}$, and T and H are deviations of temperature and geopotential from their standard values $\bar{T}$ and $\bar{H}$, respectively. In a similar way $\boldsymbol{\phi}^* = (u^*, v^*, T^*, H^*, w^*)$ and $\boldsymbol{u}^* = (u^*, v^*, w^*)$ are defined.

For the model under study the integral identity can be defined as follows (Penenko, 1977)

$$
\begin{aligned}
&I(\boldsymbol{\phi}, Y, \boldsymbol{\phi}^*)\\
&\quad= \int_{\mathscr{D}_t} \Big\{ [(\Lambda u, u^*) + (\Lambda v, v^*) + \sigma(\Lambda T, T^*)]\\
&\qquad + (\boldsymbol{u}^* \cdot \operatorname{grad} H - \boldsymbol{u} \cdot \operatorname{grad} H^*) + \left(f + \frac{\tan\theta}{a} u\right)(u^* v - v^* u)\\
&\qquad + \frac{R}{p}\left(T\omega^* - \sigma \frac{(\gamma_a - \gamma)}{g} \bar{T}\omega T^*\right) - \frac{\sigma\tilde{\varepsilon}}{c_p} T^* \Big\} \, \mathrm{d}D \, \mathrm{d}t\\
&\qquad + I_{\mathscr{D}}(\boldsymbol{\phi}, \boldsymbol{\phi}^*) + \int_{\mathscr{S}_t} \tfrac{1}{2}\rho\left(\frac{\partial H}{\partial t} H^* - H \frac{\partial H^*}{\partial t}\right)\Big|_{p=p_a} \mathrm{d}s \, \mathrm{d}t\\
&\qquad + \tfrac{1}{2}\left[\int_{\mathscr{D}} (uu^* + vv^* + \sigma TT^*) \, \mathrm{d}D + \int_{\sigma} \rho HH^* \Big|_{p=p_a} \mathrm{d}s\right]\Bigg|_0^{\bar{t}} = 0, \qquad (43.5)
\end{aligned}
$$

$$
\begin{aligned}
&I_{\mathscr{D}}(\boldsymbol{\phi}, \boldsymbol{\phi}^*)\\
&\quad= \int_{\mathscr{D}_t} \Big\{ \mu_1 [D_T(\boldsymbol{u}_s) D_T(\boldsymbol{u}_s^*) + D_s(\boldsymbol{u}_s) D_s(\boldsymbol{u}_s^*)]\\
&\qquad + \chi_1 \left(\frac{\partial u}{\partial p}\frac{\partial u^*}{\partial p} + \frac{\partial v}{\partial p}\frac{\partial v^*}{\partial p}\right) + \sigma\chi_2 \frac{\partial T}{\partial p}\frac{\partial T^*}{\partial p}\\
&\qquad + \sigma \frac{\mu_2}{a^2}\left(\frac{1}{\sin^2\theta}\frac{\partial T}{\partial \psi}\frac{\partial T^*}{\partial \psi} + \frac{\partial T}{\partial \theta}\frac{\partial T^*}{\partial \theta}\right)\Big\} \, \mathrm{d}D \, \mathrm{d}t\\
&\qquad + \int_{\sigma_t} (u^* \tau_u + v^* \tau_v + \sigma T^* \tau_T)|_{p=p_a} \, \mathrm{d}s \, \mathrm{d}t, \qquad (43.6)
\end{aligned}
$$

$$
\begin{aligned}
(\Lambda\phi, \phi) = \tfrac{1}{2}\bigg[&\left(\frac{\partial \phi}{\partial t}\phi^* - \frac{\partial \phi^*}{\partial t}\phi\right)\\
&+ (\phi^* \boldsymbol{u} \cdot \operatorname{grad} \phi - \phi \boldsymbol{u} \cdot \operatorname{grad} \phi^*)\bigg], \qquad (43.7)
\end{aligned}
$$

where

$$\boldsymbol{u}_s = (u, v), \qquad \boldsymbol{u}_s^* = (u^*, v^*).$$

The meanings of the symbols introduced here are:

$\mathrm{d}D$	element of volume, $\mathrm{d}D = \mathrm{d}s \, \mathrm{d}p$,
$\mathrm{d}s$	element of area, $\mathrm{d}s = a^2 \cos\theta \, \mathrm{d}\theta \, \mathrm{d}\psi$,
ψ	longitude,
θ	latitude,

p	pressure,
p_a	pressure at the lower boundary of the atmosphere,
ω	dp/dt
γ	temperature lapse rate
γ_a	dry adiabatic lapse rate
ρ	standard value of density,
a	Earth's radius,
f	Coriolis parameter,
c_p	specific heat at constant pressure,
g	acceleration due to gravity,
R	gas constant,
$\bar{\varepsilon}$	heat influx,
σ	the constant $[g/(\gamma_a-\gamma)T]_0$, chosen such that the summation is possible,
$\mu_\alpha, \chi_\alpha (\alpha=1,2)$	coefficients of turbulence,

The functions $\tau_d (d=u, v, T)$ represent the dynamic and thermal interactions of the underlying surface with the atmosphere. A boundary-layer model is used to define these functions.

The asymmetric form of $(\Lambda\phi, \phi)$ is obtained by symmetrizing and reducing the displacement operator Λ.

Forms of the functional $I_{\mathscr{D}}(\boldsymbol{\phi}, \boldsymbol{\phi}^*)$ and the expressions $D_T(\boldsymbol{u}_s)$, $D_s(\boldsymbol{u}_s)$ depend on the structure of the turbulent exchange model; in particular,

$$\left.\begin{aligned} D_T(\boldsymbol{u}_s) &= \frac{1}{a\cos\theta}\left(\frac{\partial u}{\partial\psi}+\frac{\partial(v\cos\theta)}{\partial\theta}\right), \\ D_s(\boldsymbol{v}_s) &= \frac{1}{a\cos\theta}\left(\frac{\partial v}{\partial\psi}-\frac{\partial(u\cos\theta)}{\partial\theta}\right). \end{aligned}\right\} \quad (43.8)$$

The expressions $D_T(\boldsymbol{u}_s^*)$ and $D_s(\boldsymbol{u}_s^*)$ are derived from (43.8) by identical formulations of $\boldsymbol{u}_s \equiv \boldsymbol{u}_s^*$.

The integration domains in (43.5) and (43.6) are

$$\mathscr{D}_t = \{\mathscr{D}\times[0, \bar{t}]\}, \quad \mathscr{D} = \{0\le\psi\le 2\pi, -\tfrac{1}{2}\pi\le\theta\le\tfrac{1}{2}\pi, p_T\le p\le p_a(p_a(p_T\ge 0)\}, \quad (43.9a)$$

$$\mathscr{S}_t = \{\mathscr{S}\times[0, \bar{t}]\}, \qquad \mathscr{S} = \{0\le\psi\le 2\pi, -\tfrac{1}{2}\pi\le\theta\le\tfrac{1}{2}\pi\}. \quad (43.9b)$$

For an input parameter vector one may take

$$\boldsymbol{Y} = (\boldsymbol{\phi}^0, \bar{\varepsilon}, \mu_1, \mu_2, \chi_1, \chi_2, \tau_u, \tau_v, \tau_T), \quad (43.10)$$

where $\boldsymbol{\phi}^0 = (u^0, v^0, T^0, H^0(p_a))$ is the vector of the system's initial state.

The integral identity (43.5) includes differential equations, boundary and initial conditions and external sources. The upper and lower atmospheric boundary conditions are the natural ones. Conditions of periodicity and the conditions at the poles are included in the definition of a functional class to which the solution sought belongs. A direct formulation without additional operations of differentiation and integration by parts yields an equation of energy balance.

On the basis of the identity (43.3), discrete models can be constructed. These can be finite-difference, spectral, or finite-element models depending on the method of discretization used.

For finite-difference models the construction of discrete approximations by means of identity (43.3) proceeds as follows:

(i) A grid domain $\mathscr{D}_t^{\mathrm{h}}$ is introduced and spaces of grid-functions $\boldsymbol{\phi}^{\mathrm{h}} \in Q^{\mathrm{h}}(\mathscr{D}_t^{\mathrm{h}})$ and $\boldsymbol{\phi}^{*\mathrm{h}} \in Q^{*\mathrm{h}}(\mathscr{D}_t^{\mathrm{h}})$ are defined.

(ii) Integrals in (43.5) are approximated by quadrature formulae.

(iii) Derivatives are substituted by difference relations, terms of the same type being approximated uniformly, i.e. to the same order.

(iv) Integrals in time are approximated by fractional steps.

As a result, the identity

$$I^{\mathrm{h}}(\boldsymbol{\phi}^{\mathrm{h}}, \boldsymbol{Y}^{\mathrm{h}}, \boldsymbol{\phi}^{*\mathrm{h}}) = 0, \tag{43.11}$$

is obtained which approximates (43.3) and possesses similar properties.

Systems of difference equations can now be obtained formally from the condition of stationarity for the functional $I^{\mathrm{h}}(\boldsymbol{\phi}^{\mathrm{h}}, \boldsymbol{Y}^{\mathrm{h}}, \boldsymbol{\phi}^{*\mathrm{h}})$ for arbitrary and independent variations of the functions $\boldsymbol{\phi}^{*\mathrm{h}}$ and $\boldsymbol{\phi}^{\mathrm{h}}$ in grid-points of the domain $\mathscr{D}_t^{\mathrm{h}}$, i.e.

$$\frac{\partial}{\partial \boldsymbol{\phi}^{*\mathrm{h}}} I^{\mathrm{h}}(\boldsymbol{\phi}^{\mathrm{h}}, \boldsymbol{Y}^{\mathrm{h}}, \boldsymbol{\phi}^{*\mathrm{h}}) = 0, \tag{43.12}$$

$$\frac{\partial}{\partial \boldsymbol{\phi}'^{\mathrm{h}}} [\lim_{\zeta \to 0} I^{\mathrm{h}}(\bar{\boldsymbol{\phi}}^{\mathrm{h}} + \zeta \boldsymbol{\phi}'^{\mathrm{h}}, \boldsymbol{Y}^{\mathrm{h}}, \boldsymbol{\phi}^{*\mathrm{h}})] = 0, \tag{43.13}$$

where ζ is a real parameter and $\boldsymbol{\phi}'^{\mathrm{h}}$ is the deviation of the state vector from the defined vector $\bar{\boldsymbol{\phi}}^{\mathrm{h}}$.

System (43.12) is a system of the model's basic equations approximating (43.1), and (43.13) is a dual set of hydrodynamic equations. Equations (43.12) and (43.13) are always energy-balanced systems. The method chosen for approximation in time is such that these are splitting schemes. The above two properties ensure stability, economy and simplicity of computation. The property of energy balance is invariant with respect to the choice of the grid domain and approximation of integrals and integrands.

Suppose that the statement of problem (43.1) is correct. In this case it may be assumed that the class of the vector functions $\boldsymbol{\phi}$ defined by expression (43.2) depends continuously on $\boldsymbol{Y}$. This means that to the small perturbations $\delta\boldsymbol{Y}$ of vector $\boldsymbol{Y}$ there correspond the small perturbations $\delta\boldsymbol{\phi}$ of vector $\boldsymbol{\phi}$. Taking into account the definition (43.10), $\delta\boldsymbol{Y}$ is defined as

$$\delta\boldsymbol{Y}=(\delta\boldsymbol{\phi}^0,\ \delta\tilde{\varepsilon},\ \delta\mu_1,\ \delta\mu_2,\ \delta\chi_1,\ \delta\chi_2,\ \delta\tau_u,\ \delta\tau_v,\ \delta\tau_T), \qquad (43.14)$$

where the symbol δ determines the variation. The variation $\delta\boldsymbol{\phi}$ of the vector $\boldsymbol{\phi}$ is determined in a similar way. In the discrete finite-difference model the vectors $\boldsymbol{\phi}$, $\boldsymbol{Y}$, $\delta\boldsymbol{\phi}$ and $\delta\boldsymbol{Y}$ are given at the points of the grid $\mathscr{D}_t^{\mathrm{h}}$. The variation symbol δ of the vector $\delta\boldsymbol{\phi}$ shows that at least one of the components receives a perturbation at any point of the domain $\mathscr{D}_t$ (or $\mathscr{D}_t^{\mathrm{h}}$). It is evident from (43.10) that the components of the vector $\boldsymbol{Y}$ are of various kinds. Therefore, for notational convenience, all the components are numbered sequentially and the following definitions are introduced

$$\boldsymbol{Y}=\{Y\}, \qquad \delta\boldsymbol{Y}=\{\delta Y_i\}, \qquad (i=1,\ldots,N), \qquad (43.15)$$

where N is the total number of the components determined in a typical case by the number of the points in $\mathscr{D}_t^{\mathrm{h}}$ and by the number of different kinds of components of the vector $\boldsymbol{Y}$.

The variations are considered to be small if the relations

$$|\delta Y_i|\ll|Y_i|, \qquad i=(1,\ldots,N) \qquad (43.16)$$

are satisfied. Relation (43.2) shows that the variations $\delta\boldsymbol{Y}$ and $\delta\boldsymbol{\phi}$ are interconnected. The sensitivity problem may be defined as that of evaluating the influence of the variations $\delta\boldsymbol{Y}$ on the state vector of the system.

As a measure for the estimation of the influence of perturbations $\delta\boldsymbol{Y}$ of the vector of input parameters on the vector $\boldsymbol{\phi}$ a particular functional is chosen. There is considerable freedom of choice available. Therefore, such a functional of a general type is first considered. Call it $\Phi(\boldsymbol{\phi})$ and assume that it depends explicitly on $\boldsymbol{\phi}$ and is bounded, continuous and differentiable on the set of the functions $\boldsymbol{\phi}\in Q(\mathscr{D}_t)$ and $\boldsymbol{Y}\in R(\mathscr{Y})$.

From these definitions it follows that the vectors

$$\mathrm{grad}_{\boldsymbol{\phi}}\,\Phi\equiv\left\{\frac{\partial\Phi(\boldsymbol{\phi})}{\partial\boldsymbol{\phi}}\right\}, \quad \mathrm{grad}_{\boldsymbol{Y}}\,\Phi\equiv\left\{\frac{\partial\Phi(\boldsymbol{\phi})}{\partial\boldsymbol{Y}}\right\}, \qquad (43.17)$$

are determined. Due to the explicit dependence of the functional on $\boldsymbol{\phi}$ the calculation of the vector $\text{grad}_{\boldsymbol{\phi}}\, \Phi$ is reduced to simple differentiation formulae of the scalar function in the vector arguments. Any functional defined on the functions $\boldsymbol{\phi}$ always implicitly depends on the components of the vector $\boldsymbol{Y}$. Thus the calculation of the vector $\text{grad}_{\boldsymbol{Y}}\, \Phi$ is a difficult mathematical problem. It is connected with the fact that the relationship between $\boldsymbol{\phi}$ and $\boldsymbol{Y}$, represented formally by relation (43.2), is in fact defined by the algorithm.

Definitions are next given of variations of the functional which is obtained by variations of the vector functions $\boldsymbol{\phi}$ and $\boldsymbol{Y}$

$$\delta\Phi(\boldsymbol{\phi}) = (\text{grad}_{\boldsymbol{\phi}}\, \Phi, \delta\boldsymbol{\phi}), \tag{43.18}$$

$$\delta\Phi(\boldsymbol{\phi}) = (\text{grad}_{\boldsymbol{Y}}\, \Phi, \delta\boldsymbol{Y}). \tag{43.19}$$

The inner products in (43.5) and (43.6) are defined by the structures of the functional $\Phi(\boldsymbol{\phi})$ and relation (43.2). Equation (43.19) connects the variations of the functional with the input parameter variations. This is the basic functional relation in the sensitivity theory of the model. Some applications of this relation to the forecast problem are discussed in Marchuk (1974) and Penenko (1975). The components of the vector $\text{grad}_{\boldsymbol{Y}}\, \Phi$ are the sensitivity functions of the model with respect to the functional $\Phi(\boldsymbol{\phi})$. They characterize the contribution of the variations of each component of the input parameter vector to the functional variations.

Omitting the intermediate arguments and following Penenko (1975), the algorithm for the calculation of the functional variations by formula (43.6) and the components of the vector $\text{grad}_{\boldsymbol{Y}}\, \Phi$, may be obtained through the following procedure:

(i) An unperturbed value of the vector $\boldsymbol{Y}$ is given.

(ii) The system of the basic equations of the model (43.12) is solved. As a result, the unperturbed value $\boldsymbol{\phi}$ of the state vector is determined.

(iii) The system of adjoint equations (43.13) is solved for the function $\boldsymbol{\phi}^*$ with a source term $\text{grad}_{\boldsymbol{\phi}}\, \Phi$ and under the condition $\boldsymbol{\phi}^*(t) = 0$. Formally this problem is written as

$$\frac{\partial}{\partial \delta\boldsymbol{\phi}} \left\{ \lim_{\zeta \to 0} \frac{\partial}{\partial \zeta} [\Phi(\boldsymbol{\phi} + \zeta\delta\boldsymbol{\phi}) + I^{\text{h}}(\boldsymbol{\phi} + \zeta\delta\boldsymbol{\phi}, \boldsymbol{Y}, \boldsymbol{\phi}^*)] \right\} = 0, \quad \boldsymbol{\phi}^*(t) = 0, \tag{43.20}$$

where ζ is a real parameter and the functional I^{h} is defined by the left-hand side of the summation identity (43.11).

(iv) The desired functional variation of the values of $\boldsymbol{\phi}$ and $\boldsymbol{\phi}^*$, calculated in steps (ii) and (iii), is determined by

$$\delta\Phi(\boldsymbol{\phi}) = \lim_{\zeta \to 0} \frac{\partial}{\partial \zeta} I^{\mathrm{h}}(\boldsymbol{\phi}, \boldsymbol{Y} + \zeta \delta \boldsymbol{Y}, \boldsymbol{\phi}^*). \tag{43.21}$$

(v) The vector components $\mathrm{grad}_{\boldsymbol{Y}}\, \Phi$ are determined by the formulae

$$\mathrm{grad}_{\boldsymbol{Y}}\, \Phi = \frac{\partial}{\partial \delta \boldsymbol{Y}} \left[\lim_{\zeta \to 0} \frac{\partial}{\partial \zeta} I^{\mathrm{h}}(\boldsymbol{\phi}, \boldsymbol{Y} + \zeta \delta \boldsymbol{Y}, \boldsymbol{\phi}^*) \right]. \tag{43.22}$$

From (43.21) it follows that the functional expressions in the left-hand side of the integral identity (43.3) and in its discrete analogue (43.11) give the inner products in (43.19) which connect the functional and the parameter variations. The vector function $\boldsymbol{\phi}^*$ in the integral and summation identities is arbitrary and has smooth components. In the above algorithm $\boldsymbol{\phi}^*$ is a Lagrange multiplier vector taking into account the implicit dependence of the functional on the parameter vector. The adjoint problem (43.20) is constructed in such a way that the relation for the variations (43.21) is likely to be the most simple. The algorithm for the solution of the adjoint problem (43.20) is reduced to the splitting scheme consistent with (43.12) for the solution of the basic problem by the summation functional.

43.3 Specification of model parameters from the observed data

The method used is based on system optimization and identification theory. It allows feedback between the changes of the parameters and the functionals defining quality criteria. For the diagnostic investigation of the model let us choose the quality criterion in the form of the functional which determines the root-mean-square deviations between the computed and measured values of the state vector. Assume that the measured values $\boldsymbol{\phi}_{\mathrm{m}}$ of the state vector are given on the discrete set of the points $\mathscr{D}_t^{\mathrm{m}}$ and define this functional as

$$\Phi(\boldsymbol{\phi}) = \|\boldsymbol{\phi} - \boldsymbol{\phi}_{\mathrm{m}}\|_{\mathscr{D}_t^{\mathrm{m}}}. \tag{43.23}$$

To calculate the value of functional (43.23) it is necessary to have the operator of the transformation of the information from the grid $\mathscr{D}_t^{\mathrm{h}}$ onto the set $\mathscr{D}_t^{\mathrm{m}}$, since the vector $\boldsymbol{\phi}_{\mathrm{m}}$ is given on the latter. The definition of the quality functional on the values of the state vectors given on the set $\mathscr{D}_t^{\mathrm{m}}$ is more convenient and reasonable than that on the grid $\mathscr{D}_t^{\mathrm{h}}$. Let us use the idea of the optimum gradient method for the construction of the feedback and introduce the following metric in the domain of the values of the parameters:

$$dS^2 = \sum_{i=1}^{N} \eta_i dY_i^2, \tag{43.24}$$

where the η_i are scale factors, and dY_i are components of $\boldsymbol{Y}$. The derivative $dS/dt \equiv V_P$ can be considered as the velocity of the movement of the model in the parameter space. If we take as a hypothesis that the velocity V_P is proportional to the length of the gradient vector of the functional in the direction of the vector $\boldsymbol{Y}$, i.e.

$$V_P = \kappa \left[\sum_{i=1}^{N} \frac{1}{\eta_i} \left(\frac{\partial \Phi(\boldsymbol{\phi})}{\partial Y_i} \right)^2 \right]^{\frac{1}{2}} \tag{43.25}$$

where κ is a proportionality coefficient, we can obtain the system of differential equations:

$$\frac{dY_i}{dt} = -\left(\frac{\kappa}{\eta_i} \right) \frac{\partial \Phi(\boldsymbol{\phi})}{\partial Y_i}, \qquad i = (1, \ldots, N). \tag{43.26}$$

Every equation of (43.26) connects the velocity of the i-component of the parameter vectors with the corresponding component of the vector $\text{grad}_{\boldsymbol{Y}}\, \Phi$ which is defined in the neighbourhood of the unperturbed value of the vector $\boldsymbol{Y}$.

The finite-difference approximation of equations (43.26) can be constructed in the following way:

$$Y_i^{j+1} = Y_i^j - \left(\frac{\kappa \Delta t}{\eta_i} \right) \frac{\partial \Phi(\boldsymbol{\phi})}{\partial Y_i}, \qquad i = (1, \ldots, N), \tag{43.27}$$

where Δt is a time step with which the integrations of the basic and the adjoint equations of the model are made.

The algorithm for the calculation of the vector components $\text{grad}_{\boldsymbol{Y}}\, \Phi$ was described earlier when the method of the evaluation of the functional variations was discussed. Formula (43.22) represents this vector. The integration of equations (43.27) is carried out step-by-step together with the integration of the equations of the model (43.12). As is seen from (43.21) and (43.26) the solution of the adjoint problem is a necessary link connecting the changes of the quality criteria with the changes of the vector $\boldsymbol{Y}$ in the feedback regime.

The gradient of the functional (43.23) defined on grid $\mathcal{D}_t^h$ is used as the right-hand part of the adjoint problem.

The specific feature of the problem is that the deviations between the calculated and measured values are defined on set $\mathscr{D}_t^{\mathrm{m}}$ and $\mathrm{grad}_{\boldsymbol{\phi}}\, \Phi$ is defined on grid $\mathscr{D}_t^{\mathrm{m}}$. It is reasonable to define the proportionality coefficient κ in (43.25) on each time step according to the changes of the functional (43.23) or from the minimum condition of this functional in the direction of the vector $\boldsymbol{Y}^{j+1} = \boldsymbol{Y}^j + \kappa\, \mathrm{grad}_{\boldsymbol{Y}}\, \Phi$.

As a particular case of the method of the estimation of the parameters, consider the problem of the definition of the initial state of the system with the conditions of (43.12) as constraints. This problem is solved on the class of functions satisfying the system of equations

$$\min_{\boldsymbol{Y} \in R(\mathscr{Y})} \Phi(\boldsymbol{\phi}) = \min_{\boldsymbol{Y} \in R(\mathscr{Y})} \|\boldsymbol{\phi} - \boldsymbol{\phi}_{\mathrm{m}}\|_{\mathscr{D}_t^{\mathrm{m}}}. \tag{43.28}$$

43.4 The sensitivity problem

To illustrate the method used, two examples will be considered. The first demonstrates the calculation of the sensitivity function to initial data of an atmospheric dynamic model for the northern hemisphere. This model is described by an integral identity in the domain $\mathscr{D}_t$ which corresponds to the northern hemisphere. The boundary conditions on the Equator are natural. The sensitivity is investigated with respect to the linear functional

$$\Phi(\boldsymbol{\phi}) = \int_{\mathscr{D}_t} \boldsymbol{\eta}\boldsymbol{\phi}\, \mathrm{d}D\, \mathrm{d}t, \tag{43.29}$$

where $\boldsymbol{\eta}$ is a Dirac-type non-negative weight function which is chosen in such a way that the value of the functional is equal to a single value of the geopotential at the point $\{\psi = 80°,\ \theta = 54°,\ p = 500\ \mathrm{mb},\ t = 188\ \mathrm{hours}\}$.

The grid domain with steps $\Delta\psi = 10°$, $\Delta\theta = 5°$ and with six standard levels in the vertical coordinate is used to discretize the model. Fig. 43.1 shows the isolines of the component H^* of the solution of the adjoint problem (43.20) at 500 mb, and the vertical cross-section at $\theta = 54°$, with $t = (7 - \kappa) \times 24$ hours ($\kappa = 1, \ldots, 7$). In this particular case the solution of the adjoint problem (43.20) plays the role of a sensitivity function to variations of the input data with respect to the functional (43.29). The analysis of the behaviour of this function shows that the domain of influence of the initial data gradually displaces from the East westwards and from the top downwards with increasing time.

Thus, the influence domain corresponding to each point or subdomain in $\mathscr{D}_t$ can be defined beforehand and the influence of the input data variations on the value of the prognostic at the given point can be

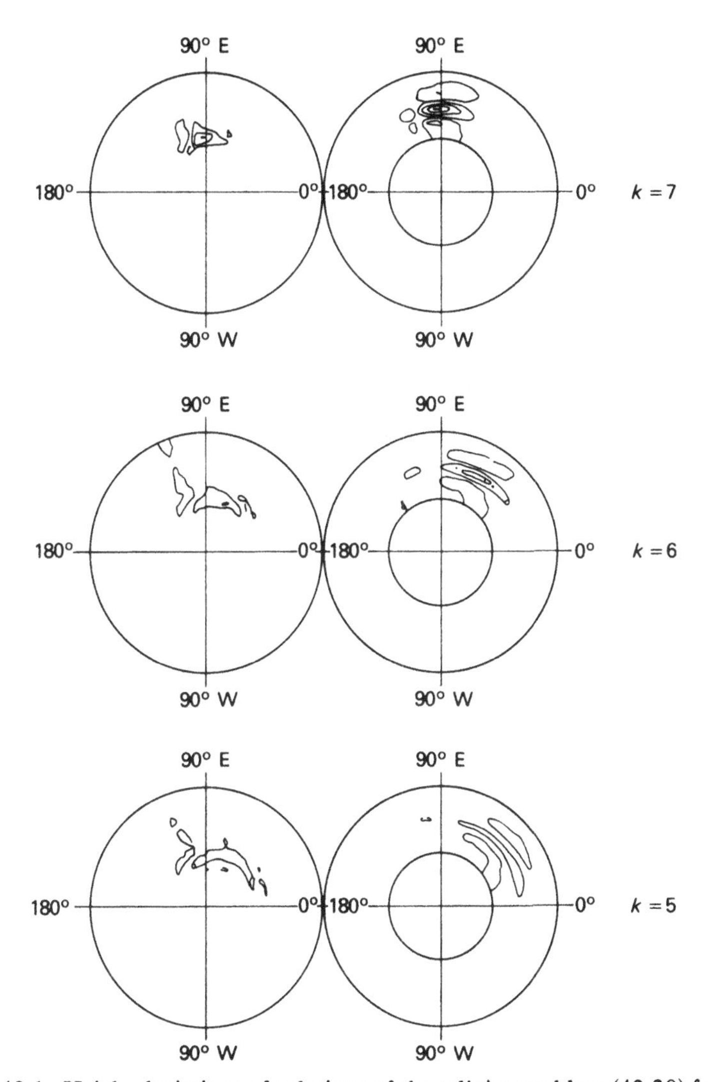

Fig. 43.1. Height deviations of solutions of the adjoint problem (43.20) for the northern hemisphere at times $t=(7-k)\times 24$ hours ($k=1,\ldots,7$). The contours are at decametre intervals. The left-hand patterns show the solutions at 500 mb over the northern hemisphere. The right-hand patterns show the solutions at latitude 54° as a function of height and longitude.

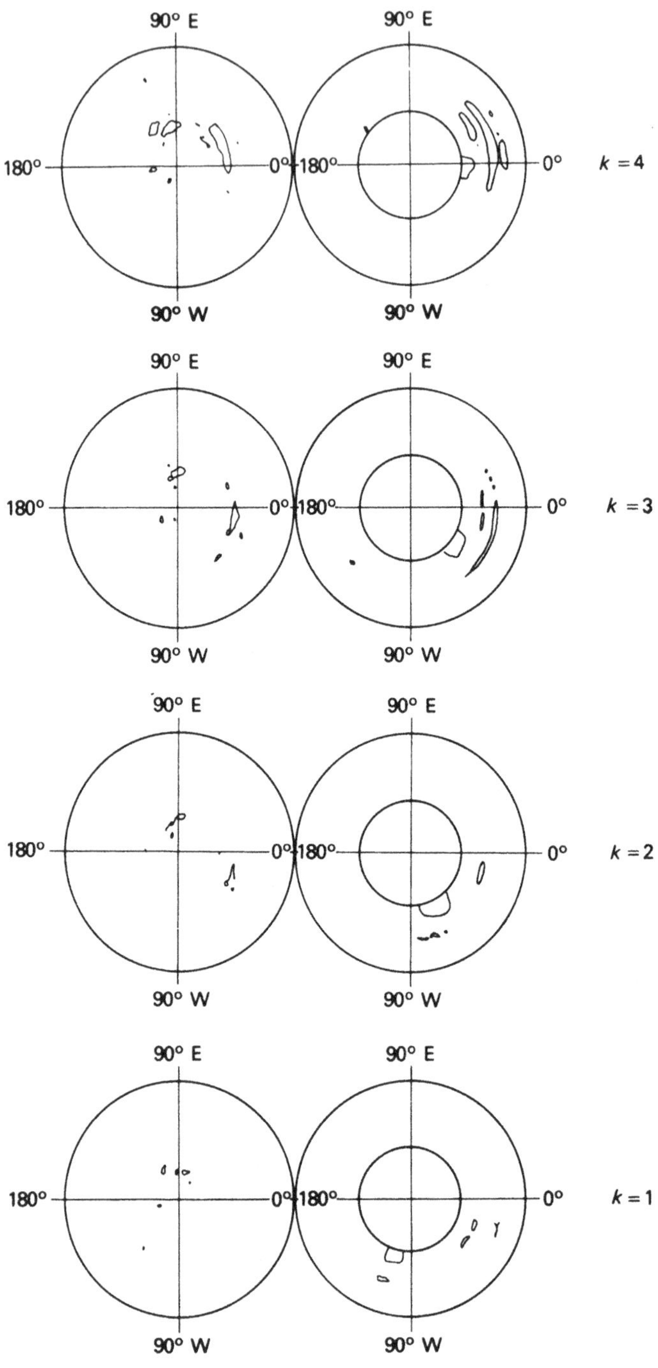
90° E
90° E
180°
0°
180°
0°
k = 4
90° W
90° W
90° E
90° E
180°
0°
180°
0°
k = 3
90° W
90° W
90° E
90° E
180°
0°
180°
0°
k = 2
90° W
90° W
90° E
90° E
180°
0°
180°
0°
k = 1
90° W
90° W

Fig. 43.1—*cont.*

evaluated. This example shows, in addition, that the solution of a problem of atmospheric dynamics over a limited area can make no sense because the influence domains can be beyond the domain of integration of the problem. In this case, the model proves to be insensitive to the perturbation, which in fact determines the behaviour of the simulated processes. This means that the forecast problem for a limited area is improperly posed.

The second example shows the result of the analysis and the reconstruction of the initial fields of meteorological elements using the variational principle (43.28). The linear discrete model described by integral identity (43.5) is used as a constraint for the class of admissible functions. The nonlinear terms such as $(\phi^* \boldsymbol{u} \cdot \text{grad}\, \phi - \phi \boldsymbol{u} \cdot \text{grad}\, \phi^*)$ where $\phi = (u, v, T)$ and $\phi^* = (u^*, v^*, T^*)$, which are in antisymmetric form $(\Lambda\phi, \phi^*)$, are omitted. All the components of the vector function $\boldsymbol{\phi}^0$ considered as the vector input parameters in the points of the grid $\mathscr{D}^h$ at $t = 0$ are subjected to the analysis and adjustment. For convenience of numerical experimentation, the functional in (43.5) is determined in the following way:

$$\Phi(\phi) = \tfrac{1}{2} \sum_{im\kappa} R_{im}(H_{im\kappa} - \tilde{H}_{im\kappa})^2 \Delta D_{im\kappa}, \qquad i = (1, \ldots, 36),$$
$$m = (1, \ldots, 18), \kappa = (1, \ldots, 6), \qquad (43.30)$$

where $\Delta D_{im\kappa} = a^2 \sin \theta_m \Delta\psi \Delta\theta \Delta p_\kappa$, $H_{im\kappa}$, $\tilde{H}_{im\kappa}$ are calculated and known geopotential values at the points $\boldsymbol{x} = (\psi_i, \theta_m, p_\kappa)$; R_{im} is a weight function depending on ψ and θ. Two variants of the choice of the weight function in the functional are made. In the first variant $R_{im} = 1$ $(i = 1, \ldots, 36;$ $m = 1, \ldots, 18)$ and in the second variant

$$R_{im} = \begin{cases} 1, & \text{if } i = (1, \ldots, 10, 22, \ldots, 27);\ m = (1, \ldots, 17), \\ 0, & \text{at the other points.} \end{cases}$$

The second variant is typical for the weather forecast problem when the measured data are sufficiently dense for two regions of the northern hemisphere, i.e. America and Eurasia, but are absent over the Atlantic and the Pacific Oceans.

The minimization problem of the functional (43.30) on the set of values of the vector $\boldsymbol{\phi}_0$ with the model equations as constraints are solved by means of an iteration procedure (43.14). Fig. 43.2 shows the calculated geopotential fields at 1000 mb, 500 mb and 100 mb. The fields shown on the left-hand side of Fig. 43.2 are obtained using all the available data, and are essentially a check on those shown on the right, which were

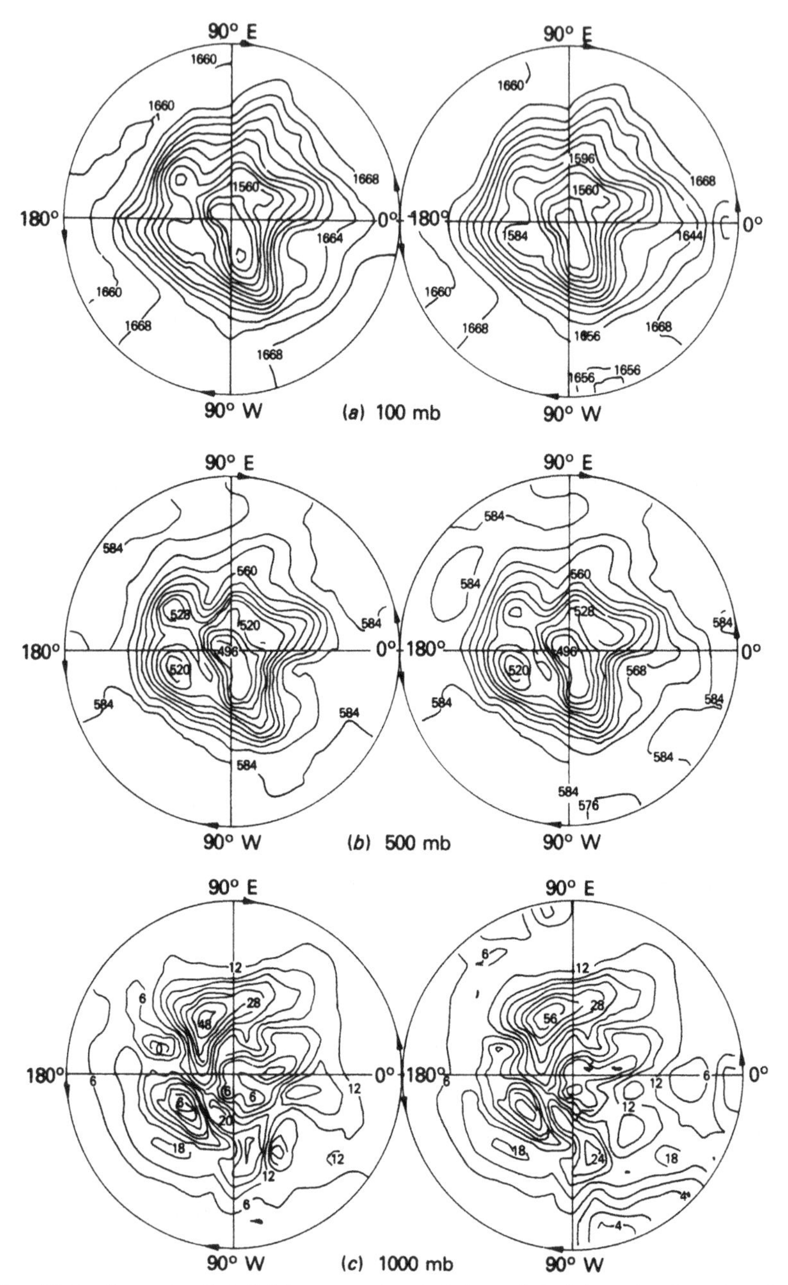

Fig. 43.2. Geopotential fields reconstructed by minimizing the functional (43.20): (*a*) at 100 mb; (*b*) at 500 mb; and (*c*) at 1000 mb. The contours are labelled in decametres. The patterns on the left are obtained using all available data. Those on the right are obtained with data omitted.

obtained with data omitted. Comparison shows that the fields on the right-hand side of Fig. 43.2 correctly describe the main features of the reconstructed geopotential fields. This is achieved because the discrete model itself acts as an interpolation operator while solving the optimization problem (43.28). As the influence domains for each grid-point, using the model operator, are sufficiently large, satisfactory reconstructions of meteorological fields, including those for regions with no observed data, are obtained. It is quite remarkable that results of this quality are obtained with such a large volume of oceanic data missing.

The suggested methods can be applied in many problems and lead to new methods for mathematical modelling of the ocean and atmosphere. At the same time, these methods suggest new ways of making more complete use of the available data for modelling.

References

Döös, B. R. (1970) Numerical experimentation related to GARP, GARP Publication Series No. 6, p. 68.

Eykhoff, P. (1974) *System Identification–Parameter and State Estimation.* John Wiley, London.

GARP (1974) Modelling for the first GARP Global Experiment. GARP Publication Series No. 14, VI, p. 261.

Marchuk, G. I. (1967) Numerical methods in weather forecasting. *Gidrometeoizdat.*

Marchuk, G. I. (1974) The numerical solution of problems in the dynamics of the atmosphere and oceans. *Gidrometeoizdat.*

Moiseev, N. N. (1971) *Numerical Methods in the Theory of Optimal Systems.* Nauka, Moscow, p. 424.

Penenko, V. V. (1975) Computational aspects of modelling the dynamics of atmospheric processes and of estimating the influence of various factors on the dynamics of the atmosphere. In *Some Problems of Computational and Applied Mathematics*, Nauka, Novosibirsk.

Penenko, V. V. (1977) Energetically balanced discrete models of the dynamics of atmospheric processes. *Meteor. Hydrol.*, No. 10, 3–20.

Part V

Storm surges and flood forecasting

Introduction

SIR JAMES LIGHTHILL

The final part of this book is devoted to the last part of the *cycle* of important fluid-dynamical processes which constitutes monsoon dynamics. The analysis of these concluding processes gains interest primarily from the hope that it can lead to practical methods for producing adequate and reliable advance warning of flooding dangers.

First, we have two chapters by leading experts on the prediction of marine flooding associated with storm surges. The numerical modelling involved allows for the combined input from atmospheric processes and astronomical tide-raising forces, together with a comprehensive treatment of depth-distribution effects and boundary conditions. These chapters describe both the successful present use of these techniques in connection with flood prediction for the British Isles, and their potential for application in the Bay of Bengal to forecast flooding surges resulting from monsoon wind stresses acting upon the shallow water at the mouth of the Ganges delta.

Although in these chapters the storm-surge experts concentrate on flooding dangers associated with monsoons, they were, of course, consulted during their stay in Delhi on the implications of the tragic events of the month (November, 1977) immediately preceding the Symposium upon which this book is based. Flooding of great severity was produced by a tropical cyclone in the state of Tamil Nadu. This was followed, within a few days, by a still more devastating, and indeed unprecedented, degree of flooding in Andhra Pradesh caused by another cyclone. The India

Meteorological Department had, in fact, given a creditable 2-day forecast of severe flooding dangers. Much active discussion among the participants at the Symposium was inevitably devoted to the question of what methods of data acquisition would be needed (as the input to storm-surge models) to extend still further the period of reliable warning and to make it still more quantitatively specific.

From these discussions a consensus emerged among the participants from outside India that it would be of immense help for all purposes of giving reliable warning of dangers from intense cyclones if a suitable aircraft were available for use by the India Meteorological Department. When weather satellites indicated a strong cyclone in the Bay of Bengal, such an aircraft could fly directly into it to determine on the spot its actual strength (as indicated by the wind-velocity pattern, which could be quickly determined by an aircraft fitted with inertial navigation) and other parameters relevant to its subsequent path. This information could be immediately communicated to the synoptic and numerical forecasting teams to help them make an early start on their predictions of the cyclone's effects. Acquisition of such an aircraft can additionally be recommended because, at seasons free of the danger, it could be of great value for meteorological research purposes, as experience in other countries has shown.

Another type of flood forecasting of vital importance in India is that associated with flooding dangers produced by runoff from heavy monsoon rainfall into India's great rivers. The final chapter, by a distinguished member of the institution which hosted the Symposium, Professor Subhash Chander, provides an authoritative review of one valuable technique of flood forecasting for those enormous river basins.

The joint IUTAM/IUGG Symposium on Monsoon Dynamics not only increased the knowledge of those attending, but also established new contacts between workers in their own fields and in closely related areas. In particular, contacts were established between participants from geographically remote areas and from different spheres of activity (especially operational departments and universities). The future impact of material so multidisciplinary and multinational as that involved in the studies of monsoon dynamics and forecasting depends crucially upon the maintenance and growth of such new contacts.

44

Storm-surge prediction using numerical models

R. A. FLATHER

In recent years, a new system for the prediction of storm surges in the North Sea has been under development at the Institute of Oceanographic Sciences (IOS) Bidston. The scheme is based on the use of dynamical finite-difference models of the atmosphere and of the sea. The atmospheric model, the Bushby–Timpson 10-level model on a fine mesh, used in routine weather prediction at the Meteorological Office, provides the essential forecasts of meteorological data which are then used in sea-model calculations to compute the associated storm surge. The basic sea model, having a coarse mesh, covers the entire sea areas surrounding the British Isles. Additional models giving improved resolution in areas of special interest are also under development. The scheme has been operational since the autumn of 1978.

This chapter outlines the scheme and indicates those factors found to be of particular importance. Some points of difference between the storm surge problem in the North Sea and in the Bay of Bengal are also discussed.

44.1 Introduction

Since the pioneering work of Richardson in the 1920s, numerical methods have been used increasingly to solve the equations governing the motion of the atmosphere and the sea. During the 1960s atmospheric models were put to work in operational weather prediction, and their use and development has continued up to the present. Similar developments have taken place in oceanography, where models have been used to investigate motion in oceans and shelf seas. There has been less emphasis on the use of sea models for forecasting, perhaps because of the success of the harmonic method in tidal prediction. However, in the last twenty years, numerical-hydrodynamical methods have been used to study both tides and meteorologically-driven motions and there is now considerable interest in predicting storm surges by these means. In Europe, with its extensive shallow continental shelf affected by storm surges, work is in progress in several national centres involving the use of models for forecasting water levels along the coasts of Britain, Belgium, Holland, Germany and Denmark (e.g. Duun-Christensen 1971, 1975; Timmerman 1975).

At the Institute of Oceanographic Sciences, we have been working in cooperation with the Meteorological Office on a system for predicting storm surges using models of the atmosphere and of the sea. The essence of the proposed scheme is to extract data from routine numerical weather predictions carried out by the Meteorological Office using a fine-mesh 10-level model of the atmosphere (Benwell *et al.*, 1971), then to process the data in order to determine, in advance, the changing distribution of wind stress and atmospheric pressure over the sea surface, finally using sea models with the processed data as input to compute the storm surge. An advantage of this procedure is that the calculations can all be carried out on a computer with minimal intervention from a forecaster. Various aspects of the procedure and its development have been described in a series of reports and papers (Flather and Davies, 1975, 1976, 1978; Davies and Flather, 1977, 1978; Flather 1976*a*, *b*; Davies 1976*b*).

The basic sea-model component in the system covers the whole of the continental shelf seas surrounding the British Isles, and is shown in Fig. 44.1 with grid-points of the 10-level model superimposed. Further sea models giving improved resolution in areas of special interest have also been developed. Thus, a model of the North Sea with one-third finer grid than the shelf model has been established by Davies (1976*a*) and the problem of incorporating it into the surge-prediction scheme considered (Davies and Flather, 1977). Several experiments carried out with the shelf model have also been run using the North Sea model to study the influence of resolution on the results. A model covering the Southern Bight of the North Sea and the eastern part of the English Channel, incorporating a one-dimensional representation of the River Thames, has been developed by Prandle (1975), specifically to give warnings at the site of a moveable barrier being constructed across the river to protect London from flooding due to surges.

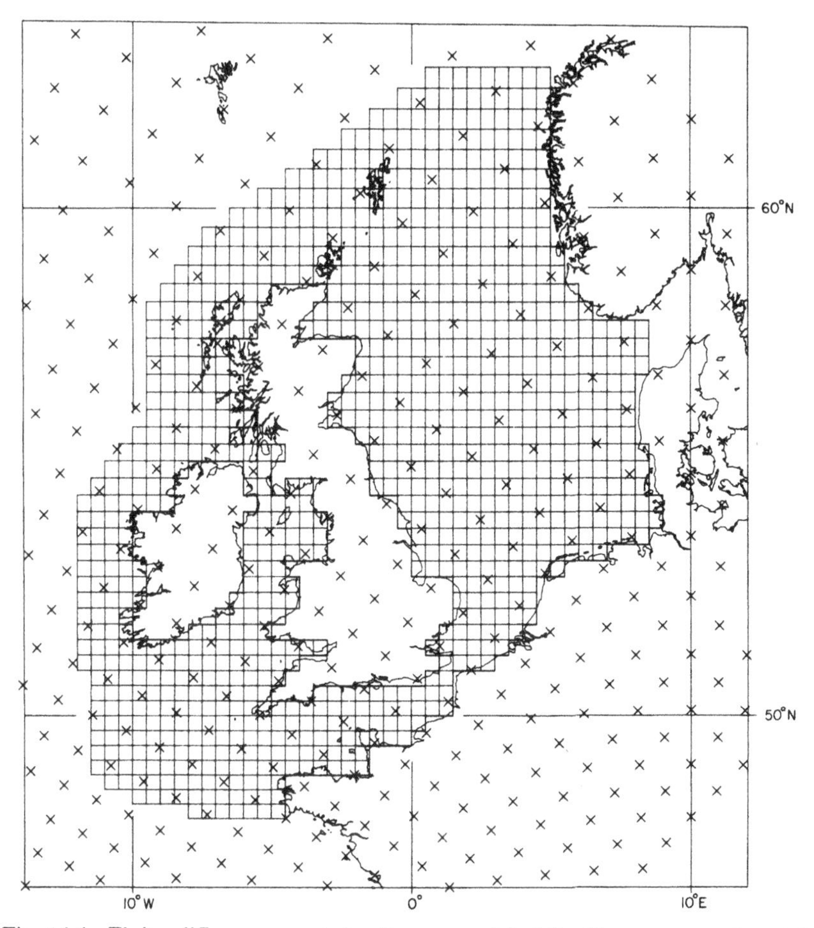

Fig. 44.1. Finite-difference mesh for the sea model of the European continental shelf with grid-points (×) of the 10-level atmospheric model.

The atmospheric data and the means by which the information required for the sea-model calculation of storm surges is derived from the numerical weather forecasts, are described in § 44.3. Only the basic method is given and the reader is referred to other reports and papers providing more complete details of some of the alternatives.

In § 44.4 a short account is given of a number of experiments aimed at determining those factors which most influence the calculated surges. Specifically, effects associated with: (i) changes in the sea model; (ii) changes in the meteorological interface; (iii) the influence of the tides; and (iv) forecasting errors in the atmospheric data are considered. The discussion, although somewhat incomplete, nonetheless represents a distillation of experience gained in experiments over some years, involving surge simulations covering several months of model time. The work has culminated in a scheme which has been operational since the autumn of 1978.

Although the present system is intended primarily for the North Sea, there is no reason, in principle, why it should not be used in other regions. In the context of this symposium, interest is centred on the storm surge problem in the Bay of Bengal. With a view to the possible implementation of a numerical prediction scheme for the Bay of Bengal, some points of difference between the Bay of Bengal and North Sea storm surges are discussed in § 44.5.

44.2 The sea model

The hydrodynamic equations which form the basis of the sea models employed are

$$\frac{\partial \zeta}{\partial t}+\frac{1}{R\cos\phi}\left(\frac{\partial}{\partial \chi}(Du)+\frac{\partial}{\partial \phi}(Dv\cos\phi)\right)=0, \qquad (44.1)$$

$$\frac{\partial u}{\partial t}+\frac{u}{R\cos\phi}\frac{\partial u}{\partial \chi}+\frac{v}{R\cos\phi}\frac{\partial}{\partial \phi}(u\cos\phi)-fv$$

$$=-\frac{g}{R\cos\phi}\frac{\partial \zeta}{\partial \chi}-\frac{1}{\rho R\cos\phi}\frac{\partial p_a}{\partial \phi}+\frac{1}{\rho D}(F^{(S)}-F^{(B)}), \qquad (44.2)$$

$$\frac{\partial v}{\partial t}+\frac{u}{R\cos\phi}\frac{\partial v}{\partial \chi}+\frac{v}{R}\frac{\partial v}{\partial \phi}+\frac{u^2\tan\phi}{R}+fu$$

$$=-\frac{g}{R}\frac{\partial \zeta}{\partial \phi}-\frac{1}{\rho R}\frac{\partial p_a}{\partial \phi}+\frac{1}{\rho D}(G^{(S)}-G^{(B)}), \qquad (44.3)$$

where the notation is:

χ, ϕ	east-longitude and latitude, respectively,
t	time,
ζ	elevation of the sea surface,
u, v	components of the depth mean current,
$F^{(S)}, G^{(S)}$	components of the wind stress on the sea surface,
$F^{(B)}, G^{(B)}$	components of the bottom stress,
p_a	atmospheric pressure,
D	total depth of water ($D=h+\zeta$),
h	undisturbed water depth,

ρ	density of water, assumed uniform,
R	the radius of the Earth,
g	the acceleration due to gravity,
f	Coriolis parameter ($f = 2\omega \sin \phi$),
ω	the angular speed of the Earth's rotation.

Equations (44.1) to (44.3) are depth-averaged equations written in polar coordinates. The component directions are those of increasing χ, ϕ; i.e. to the east and north respectively.

Adopting a quadratic law, relating bottom stress to depth mean current, gives

$$F^{(B)} = k\rho u(u^2+v^2)^{\frac{1}{2}}, \qquad G^{(B)} = k\rho v(u^2+v^2)^{\frac{1}{2}}, \tag{44.4}$$

where k is a coefficient of bottom stress.

Equations (44.1) to (44.4) are to be solved starting from a prescribed initial state

$$\zeta = \zeta(\chi, \phi, t), \qquad u = u(\chi, \phi, t), \qquad v = v(\chi, \phi, t), \qquad \text{at} \quad t = t_0, \tag{44.5}$$

subject to boundary conditions of vanishing flow across coastal boundaries

$$q = 0, \tag{44.6}$$

where q denotes the component of depth mean current along the outward directed normal to the boundary. Along open sea boundaries a generalized radiation condition is employed, taking the form

$$q = \hat{q}_M + \sum_i \hat{q}_i + \frac{c}{h}\left(\zeta - \hat{\zeta}_M - \sum_i \hat{\zeta}_i\right), \tag{44.7}$$

where $c = (gh)^{\frac{1}{2}}$,

$\hat{\zeta}_M$, $\hat{q}_M$	are contributions to elevation and current associated with the meteorologically-driven motion, and
$\hat{\zeta}_i$, $\hat{q}_i$	are similar contributions associated with the ith constituent of the tide.

With $\hat{\zeta}_M$, $\hat{q}_M$, $\hat{\zeta}_i$, and $\hat{q}_i$ specified, (44.7) relates the total elevation ζ to the normal component of the total current, q, seeking to prevent the artificial reflection of internally generated disturbances from the boundary.

Equations (44.1) to (44.4) are now approximated by finite-difference forms, which allow the numerical solution defined on the spatial grid (Fig. 44.1) to be developed through time from the prescribed initial state. For each box element shown, u is defined at the midpoint of longitudinal sides, v at the midpoint of latitudinal sides and ζ at the box centre. The approximation follows that introduced by Flather and Heaps (1975) for similar equations expressed in Cartesian coordinates. 'Angled derivatives' (Roberts and Weiss, 1966; Piacsek and Williams, 1970) are used to achieve a stable representation of the nonlinear advective terms suitable for long-term computations. The resulting equations are explicit, so that, by solving them, grid-point values of ζ, u, v at time $t+\tau$ are deduced from known values at time t, thereby advancing the solution by one timestep, $\tau = 180$s, in the shelf model. Repetition of this procedure allows the numerical solution to be built up step-by-step through time. A complete description of the method is given in Davies and Flather (1978).

44.3 The meteorological data and interface with the sea model

In this section we consider the specification of wind stress, gradients of atmospheric pressure and the meteorologically-driven contributions, $\hat{\zeta}_M$, $\hat{q}_M$, to the input along the open sea boundaries. Since the only forecast information available comes from the atmospheric model prediction runs, all these quantities must be derived or estimated from the model data. From this point of view, it is perhaps unfortunate that atmospheric pressure p_a, wind stress $\boldsymbol{\tau}$, and surface wind $\boldsymbol{w}$, are not calculated directly in the 10-level model. In common with a number of atmospheric models, pressure is used as vertical coordinate, so that winds defined on surfaces of constant pressure, and vertical distances between the pressure surfaces, are the basic dependent variables. The surface boundary condition allows the height, H, of the 1000 mb surface to be calculated. Grid-point values of H at hourly intervals constitute the basic data supplied by the Meteorological Office. However, additional data comprising 6-hourly grid-point values of h', the 1000 mb to 900 mb layer thickness, and w'_χ, w'_ϕ, components of the wind at 1000 mb, have also been obtained for use in surge-prediction experiments.

Unfortunately no unique method exists for deriving p_a and $\boldsymbol{\tau}$ from the available data; rather, many alternatives incorporating different assumptions and empirical results in combination are possible. A comprehensive account of all procedures examined is given in Flather and Davies (1978). Here, only the basis of the derivation is indicated.

It is assumed, first, that, expressed in mb,

$$p_a = 1000 + \rho_a g H, \tag{44.8}$$

where ρ_a, the density of air, is taken to be uniform and constant. The resulting atmospheric pressures defined at grid-points of the 10-level model, are then interpolated onto the sea-model grid, where the gradients $\partial p_a/\partial \chi$ and $\partial p_a/\partial \phi$ are estimated using centred finite differences. It is also assumed that the surge elevation input, $\hat{\zeta}_M$, follows the hydrostatic law, so that

$$\hat{\zeta}_M = (\overline{p_a} - p_a)/\rho g, \tag{44.9}$$

where $\overline{p_a}$ is the mean atmospheric pressure taken to be 1012 mb. The associated current $\hat{q}_M$ is taken to be zero.

Geostrophic wind components, $\tilde{w}_\chi$, $\tilde{w}_\phi$, are now calculated from the pressure gradients:

$$f\tilde{w}_\phi = \frac{1}{\rho_a R \cos\phi}\frac{\partial p_a}{\partial \chi}, \quad f\tilde{w}_\chi = -\frac{1}{\rho_a R}\frac{\partial p_a}{\partial \phi}. \tag{44.10}$$

At this point an empirical relationship between geostrophic and surface wind is applied. Since this relationship is dependent on the Coriolis parameter and hence on latitude, it is important to use an appropriate form. We use linear expressions derived from measurements taken in the North Sea by Hasse and Wagner (1971) and Hasse (1974). These take the form

$$w = a\tilde{w} + b, \tag{44.11}$$

where w and $\tilde{w}$ are magnitudes of surface and geostrophic wind, respectively; a and b are empirically derived values, which Hasse (1974) related to air–sea temperature difference ΔT (see Flather and Davies, 1978, for an account of the estimation of air temperature from h'). A further assumption about the relative directions of surface and geostrophic wind is also required. Hasse (1974) defines the cross-isobar angle, δ, as a numerical function of ΔT. If ΔT is unknown, then a constant value $\delta = 20°$ is used.

The wind stress is then calculated using a quadratic law

$$F^{(S)} = c_D \rho_a w_\chi w, \quad G^{(S)} = c_D \rho_a w_\phi w, \tag{44.12}$$

where c_D is a drag coefficient. Recent reviews (Bunker, 1976; Garratt, 1977) suggest that over the sea c_D increases with increasing wind speed w. We have used a law adopted by Heaps (1965), in which c_D increases linearly with w between 5 m s^{-1} and 19 m s^{-1} and remains constant above and below this range. Thus

$$c_D \times 10^3 = \begin{cases} 0.565, & \text{for } w \leq 5, \\ -0.12 + 0.137w, & \text{for } 5 < w \leq 19.22, \\ 2.513, & \text{for } w > 19.22, \end{cases} \quad (44.13)$$

where w is in m s^{-1}. The linear variation, derived by Smith and Banke (1975) from measurements on Sable Island, Nova Scotia,

$$c_D \times 10^3 = 0.63 + 0.066w, \quad (44.14)$$

has also been used.

44.4 Experiments to determine the sensitivity of the predictions to changes in various aspects of the procedure

A number of experiments have been carried out to determine the extent to which the predictions are influenced by the various stages in the prediction procedure. The experiments were carried out by using the method outlined in §§ 44.2 and 44.3 with data from a sequence of 10-level model forecasts covering storm-surge periods in April and November 1973 in the North Sea. Certain modifications to the method were introduced in order to investigate the influence of the changes on the results. Aspects of the scheme examined can be summarized under the headings: (1) the sea model; (2) the meteorological interface; (3) the influence of the tides; (4) forecasting errors in the atmospheric model – predictability.

44.4.1 The sea model

The original shelf model (Flather and Davies, 1976) incorporated the representation of coastal boundaries and depth distribution introduced by Heaps (1969). Using this model, rather poor results were obtained for a negative storm surge of 4 April 1973 on the east coast of England. Examination of weather charts suggested that the surge was associated with the passage of a frontal trough across the region, see Fig. 44.2, with strong offshore winds in shallow coastal areas reducing levels. Extensive changes to the model were introduced in order to give a better representation of the shape of the coastline. Barriers simulating islands and peninsulas, previously omitted, were introduced and the entrance to the Skaggerak, which Heaps treated as a coastal boundary, was changed into

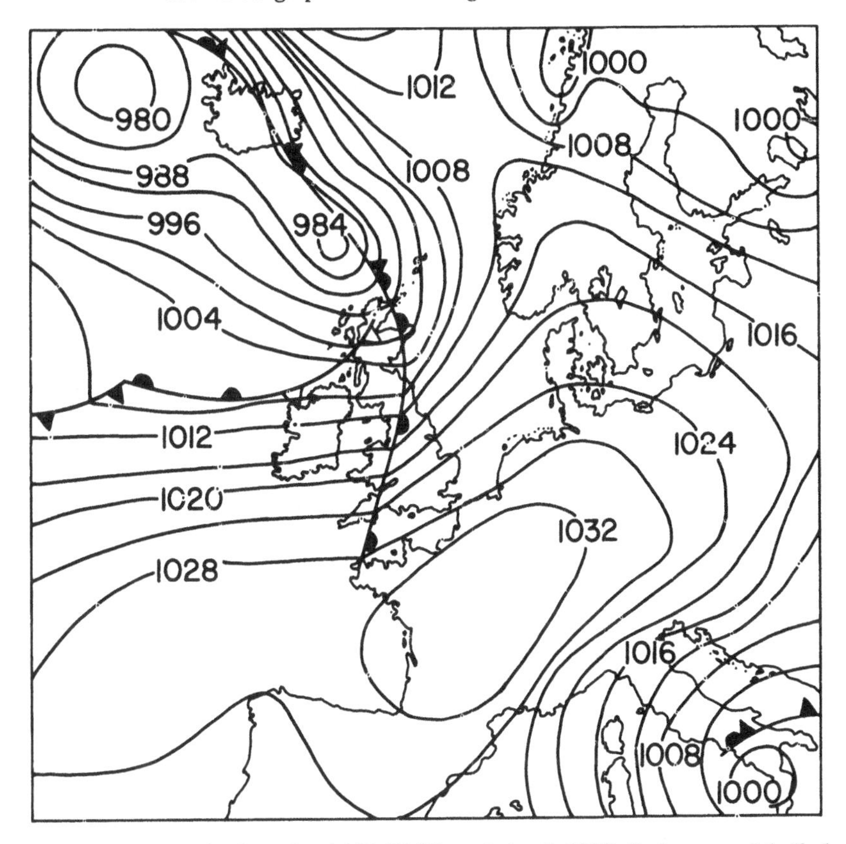

Fig. 44.2. Synoptic chart for 0600 GMT on 4 April 1973. Isobars are labelled in mb.

an open sea boundary. Advective terms in the equations of motion, previously omitted for simplicity, were now included.

As a measure of the accuracy of the results, root-mean-square errors based on comparisons between model calculated residuals and hourly values derived by subtracting the predicted tide from the observed water level were calculated for a number of ports around the North Sea. The magnitude of changes in these errors resulting from modifications to the scheme indicates the importance of the modification. In column (*a*) of Table 44.1, the differences generated by the sea-model changes just described are presented. These range from +4.7 cm at Esbjerg, a deterioration, to −10.2 cm at Southend, a considerable improvement. It appears therefore, that extensive changes in the sea model can have a significant effect on the results. Subsequent experiments showed that small or localized changes were unlikely to make much difference overall.

TABLE 44.1. *Change in rms error (cm) for the storm-surge period 2 to 6 April 1973 due to: (a) extensive revision of the sea model; (b) increasing the cross-isobar angle from 0° to 20°; and (c) taking account of the influence of the tide. The final rms error obtained with the revised sea model, taking $\delta = 20°$ and including tide–surge interaction is given in column (d).*

	(a)	(b)	(c)	(d)
Wick	0.1	−0.2	−4.6	13.5
Aberdeen	0.4	−2.3	−4.4	15.9
North Shields	−1.4	−6.1	−5.5	17.0
Inner Dowsing	−5.0	−7.1	−5.7	21.2
Immingham	−3.0	−7.2	−7.1	19.5
Lowestoft	−6.8	−9.3	−4.8	25.0
Walton-on-Naze	−5.3	−7.5	−7.1	27.2
Southend	−10.2	−3.1	−9.0	33.9
Ostende	−4.0	−7.5	−9.0	25.1
Ijmuiden	−7.2	−6.3	−7.7	34.6
Terschelling	−7.4	−10.1	−5.1	27.2
Cuxhaven	−2.4	−13.0	−10.6	37.4
Esbjerg	4.7	−11.8	−9.3	23.3
Bergen	−1.2	0.2	−2.2	11.1

44.4.2 *The meteorological interface*

In tests using the same storm period 2 to 6 April 1973 the effect of a change in wind direction brought about by increasing the cross-isobar angle, δ, from 0 to 20° was determined. The resulting change in rms errors is given in column (*b*) of Table 44.1. Quite substantial improvements in accuracy resulted from the change, suggesting that the surge prediction is strongly dependent on wind direction. Because of the square law of wind stress (44.12) and the form of the drag coefficient (44.13) or (44.14), the stress and hence the calculated surge will also be strongly influenced by the wind strength. Consequently, the relationship between geostrophic and surface wind is of fundamental importance within the overall scheme.

Related experiments were carried out to examine surface wind and stress estimates during the severe storm of 2 to 4 January 1976. Fig. 44.3 shows four estimates of the distribution of wind stress over the continental shelf at 0000 GMT on 3 January. The distributions shown in Figs. 44.3*a* and 44.3*b* were derived from forecast 10-level model data by the procedure outlined in § 44.3, with the geostrophic-surface wind relationships due to Hasse and Wagner (1971) using $\delta = 20°$ and Hasse

Fig. 44.3. Distribution of wind stress over the continental shelf at 0000 GMT 3 January 1976: (*a*) derived from predicted 1000 mb heights using the Hasse and Wagner (1971) relationship between geostrophic and surface wind; (*b*) derived from predicted 1000 mb height and 1000–900 mb thickness using the Hasse (1974) geostrophic to surface wind relationship; (*c*) derived from predicted 1000 mb winds assumed equal to surface winds; (*d*) derived from analysed information used to initialize the 10-level model. All derivations use the drag coefficient due to Heaps (1965), equation (4.13).

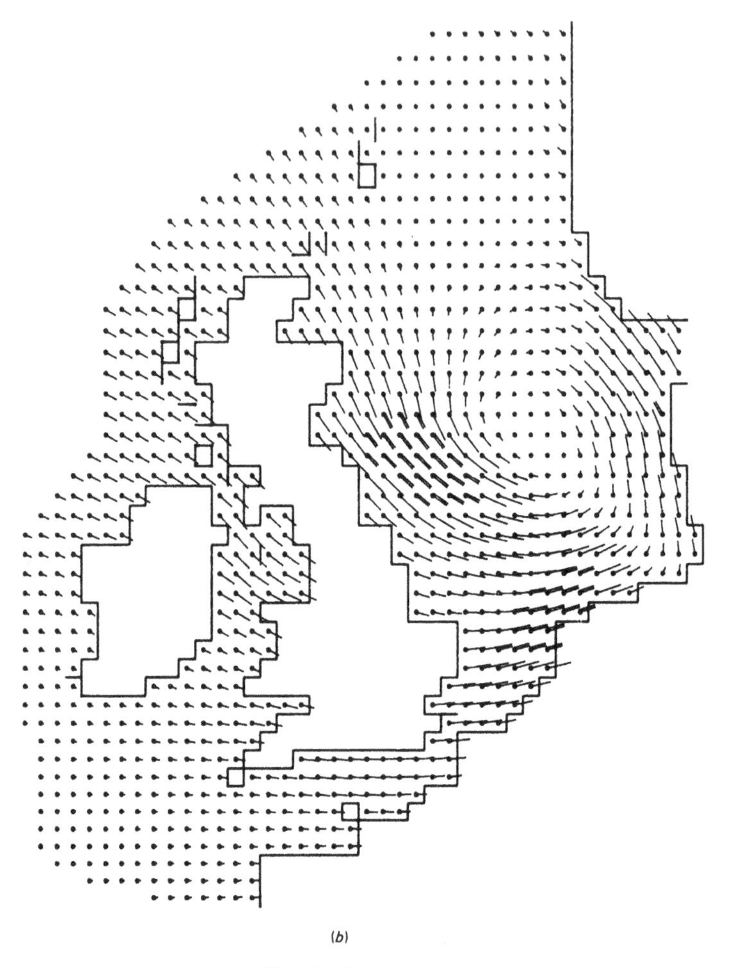

Fig. 44.3—*cont.*

(1974), respectively. Although similar, being based on the same 1000 mb height distribution, important differences are apparent in stress magnitude. Some differences in direction, associated with the dependence of cross-isobar angle on air–sea temperature difference, taken into account by Hasse (1974), do occur but are generally small and difficult to detect in the figures. The stress distribution calculated from 1000 mb winds is shown in Fig. 44.3*c*, and suggests that they cannot reasonably be accepted as approximations for the surface wind. Finally, Fig. 44.3*d* shows the

Fig. 44.3—*cont.*

stress computed from diagnosed surface winds derived by the Meteorological Office from the initial conditions used in the 10-level model forecast. Since observations are incorporated in the analysis, Fig. 44.3*d* should represent most closely the actual situation. Differences between the distributions shown in Figs. 44.3*a*, *b*, *c* and *d* reflect both errors introduced in deriving wind stress from the meteorological data and forecasting errors in the atmospheric model solutions, which are discussed later.

Fig. 44.3—*cont.*

44.4.3 Tidal influence

Offshore tidal measurements near the shelf edge between Ireland and Norway (Cartwright, 1976) made it possible to define tidal input to the continental shelf model and to obtain a satisfactory solution for the principal M_2 constituent (Flather, 1976*c*). Subsequently, the second largest constituent, S_2, was also introduced, giving a spring–neap cycle in the model tide.

An experiment was carried out to determine tidal influence on the storm surges of 2 to 6 April 1973. This required three separate model runs: one to predict the tide alone; the second to predict tide and surge together; and the third to predict the surge alone. The difference between the second and first solutions gave a surge prediction including the effects of tide–surge interaction, which could be compared with the surge solution. The reduction in rms error brought about by taking into account the tidal influence is given in column (*c*) of Table 44.1. Clearly, the effect is substantial and tide–surge interaction must be taken into account.

An important effect of interaction as far as flood prediction is concerned is that, in shallow water, maximum surges tend to avoid tidal high water (Rossiter, 1961; Banks, 1974). The effect can be seen even in the shelf model, which does not have a fine enough mesh to properly resolve estuaries and inlets. Fig. 44.4 (reproduced from Flather and Davies, 1978) shows a comparison of observed surges with calculated results using the shelf model and the meteorological forces derived from 1000 mb height predictions as described in § 44.3 and illustrated in Figure 44.3*a*. Observed surge heights at the times of maximum water level are circled. At Southend, in the outer Thames Estuary, it can be seen that surge peaks occur some 3 to 5 hours before high water in both the observations and model results. Further, the surge at high water, to be added to the predicted tide to give the total water height, does not exceed 0.7 m whereas the largest surge peak reaches 2.7 m two hours after low tide. Clearly this effect, due to interaction, is of fundamental importance in determining the maximum total water level, and hence the danger of flooding. At Lowestoft, where the tidal range is much smaller than at Southend and where consequently interaction is weaker, the effect does not occur. Prandle and Wolf (1978) have investigated the interaction in the Southern Bight of the North Sea, demonstrating its dynamical causes.

44.4.4 Forecasting errors in the atmospheric model – predictability

It is well known that the precision of atmospheric model forecasts decreases with increasing length of warning, so that weather conditions more than a few days ahead can only be predicted with much uncertainty. Some indication of the forecasting error can be seen by comparing Fig. 44.3*a*, a stress distribution derived from predictions 12 hours ahead, with Fig. 44.3*d*, based on diagnosed information. In particular, the different locations of the centre of the circulation mainly reflect forecast errors.

Experiments have been carried out to investigate the effect of errors in the 10-level model forecasts on the surge predictions, the aim being to determine whether reasonable accuracy could be retained from the initial data time up to 36 hours, when each meteorological prediction is terminated. Fine-mesh 10-level model predictions are carried out by the Meteorological Office twice each day, starting at midday and midnight, but because of the time required for initialization and the model runs, the forecast is issued some 3 to 4 hours later. This leaves more than 30 hours of data for possible use in surge prediction. For a period of 10 days from 11 to 20 November 1973, these data were extracted and divided into four subsets containing respectively: (*a*) hours 7 to 19; (*b*) hours 12 to 24; (*c*) hours 18 to 30; and (*d*) hours 24 to 36 of every forecast. Since each subset contains a sequence of 12-hour periods of data, starting 12 hours apart, complete coverage of the whole 10 days is obtained in each of the cases (*a*) to (*d*). However, the forecast error should, in general, be greater in case (*d*) than in case (*c*), and so on. Surges during the period were calculated using each of the four data sets, and results at some North Sea ports are plotted in Fig. 44.5 together with observed values. The meteorological situation associated with the surge on 19 to 20 November is shown in Fig. 44.6. It is apparent that the main surges are present in all solutions, though there are occasions when the observed levels fall well outside the range of the predictions. An example is the negative surge on the English coast on 12 November and the following positive surge on 13 November, both of which are underpredicted. At other times the observed levels lie within or close to those predicted. A further point is that the spread of predicted elevations varies considerably from time to time, perhaps reflecting the degree of uncertainty in the meteorological forecast. From the evidence of the rms errors, which give a measure of the overall accuracy, solution (*a*), based on data from hours 7 to 19, is best (Flather, 1976*b*), in accord with the expected behaviour. However, this does not mean that solution (*a*) is most accurate at all times and at all locations. More important, it appears that useful surge predictions can be obtained using data from the whole of each 10-level model forecast, so that predictions up to about 30 hours in advance are possible.

Fig. 44.4. Variation with time of computed surge residual elevations from grid-points of the shelf model approximating tide-gauge locations, compared with the equivalent observed variations. Observed residuals coinciding with maximum and minimum total water levels during each tide are indicated. The vertical scale indicated from the lowest (Dover) case applies also to the other cases. (From Flather and Davies, 1978.)

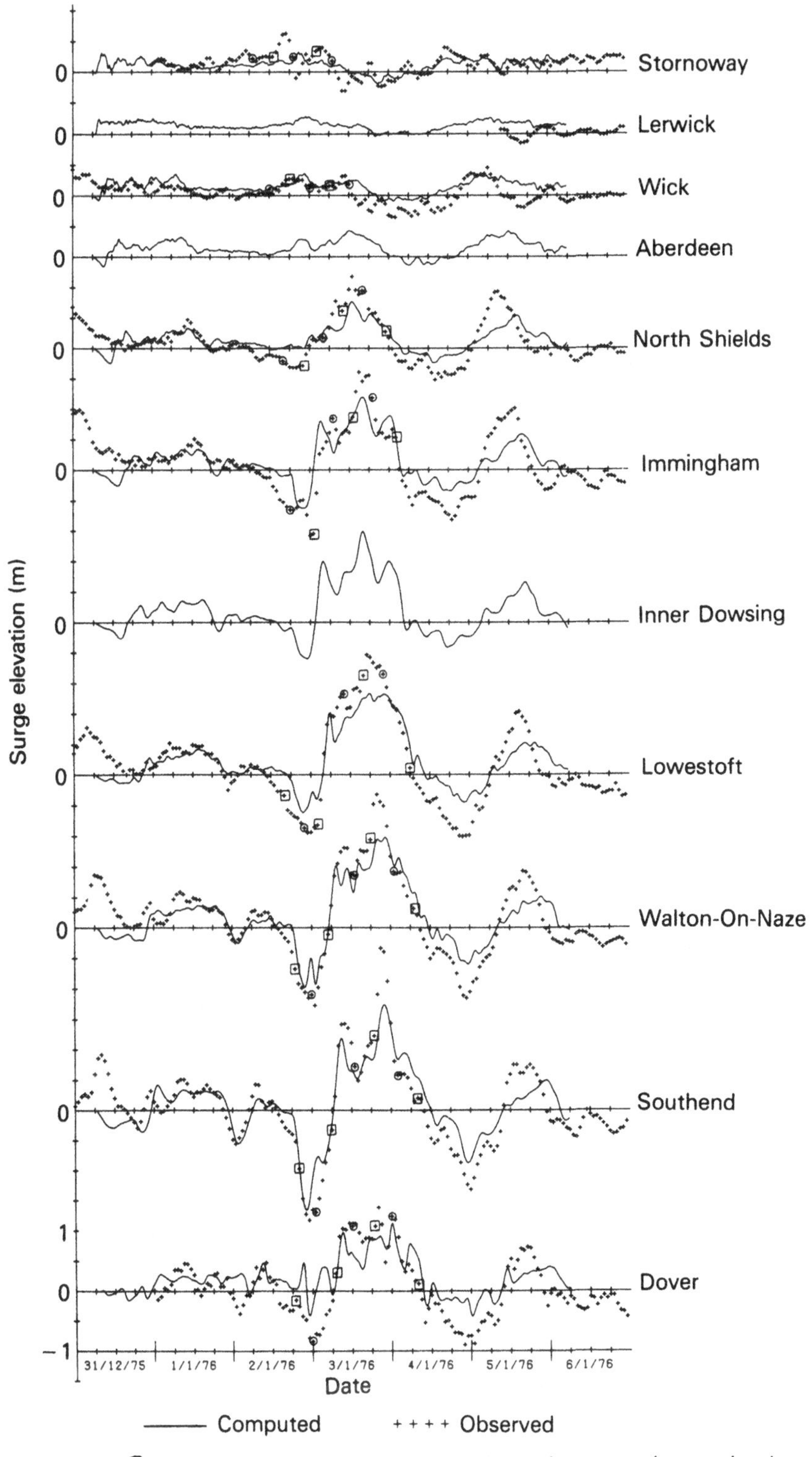
Stornoway
Lerwick
Wick
Aberdeen
North Shields
Immingham
Inner Dowsing
Lowestoft
Walton-On-Naze
Southend
Dover
Surge elevation (m)
0
1
−1
31/12/75
1/1/76
2/1/76
3/1/76
4/1/76
5/1/76
6/1/76
Date
Computed
+ + + + Observed
Observed residual coinciding with maximum total water level
Observed residual coinciding with minimum total water level

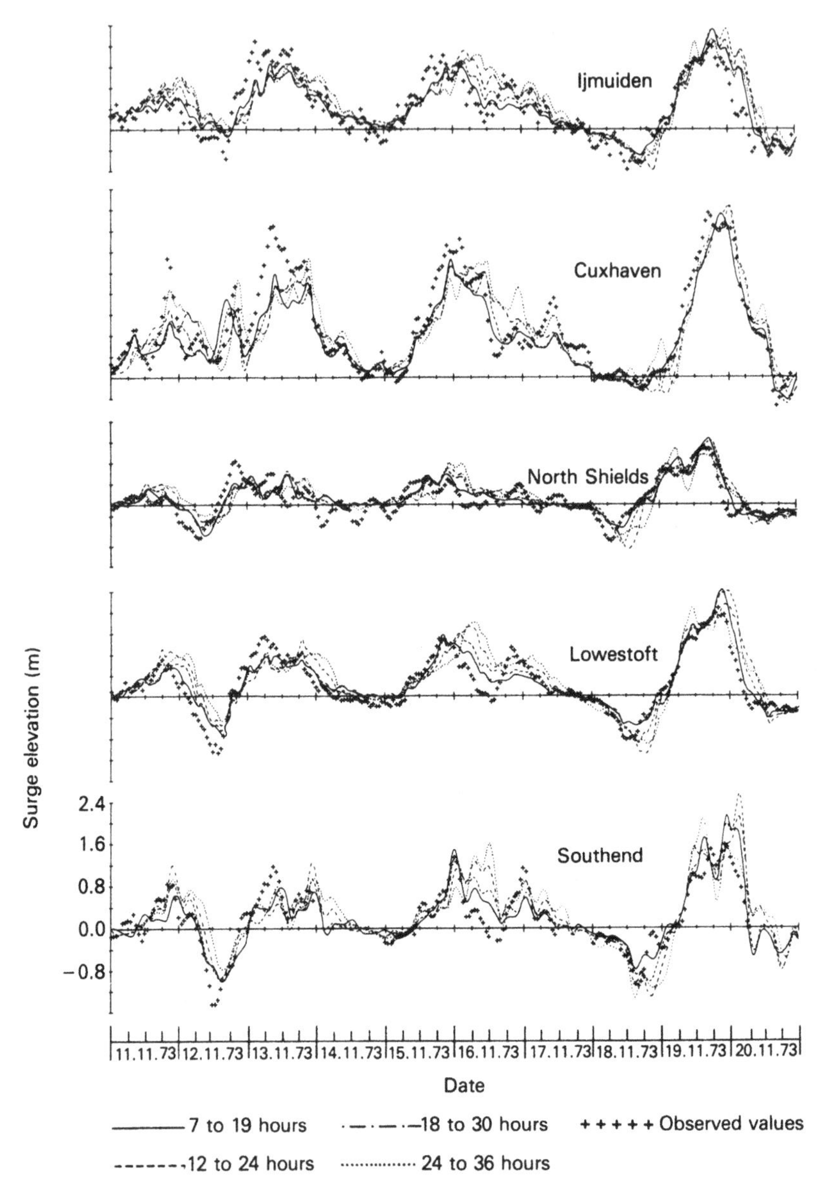

Fig. 44.5. Comparison of observed and predicted surge elevations. The predicted elevations are calculated using different sections of the meteorological forecasts (described in § 44.4), namely; 7 to 19 hours: 12 to 24 hours; 18 to 30 hours; and 24 to 36 hours. The vertical axis scale indicated for the lowest (Southend) case applies also to the other cases.

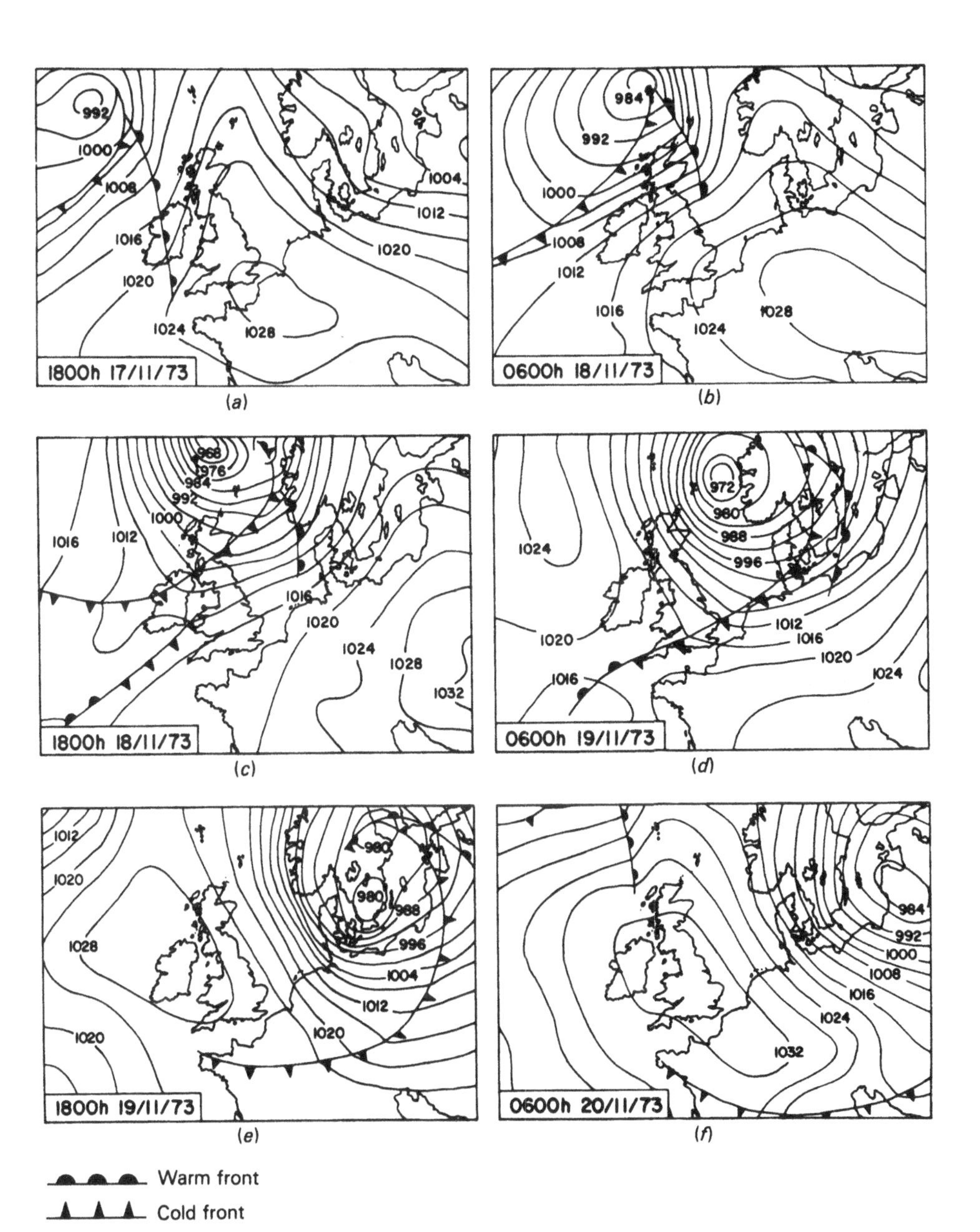

Fig. 44.6. Synoptic charts for 17 to 20 November 1973. Isobars are labelled in mb.

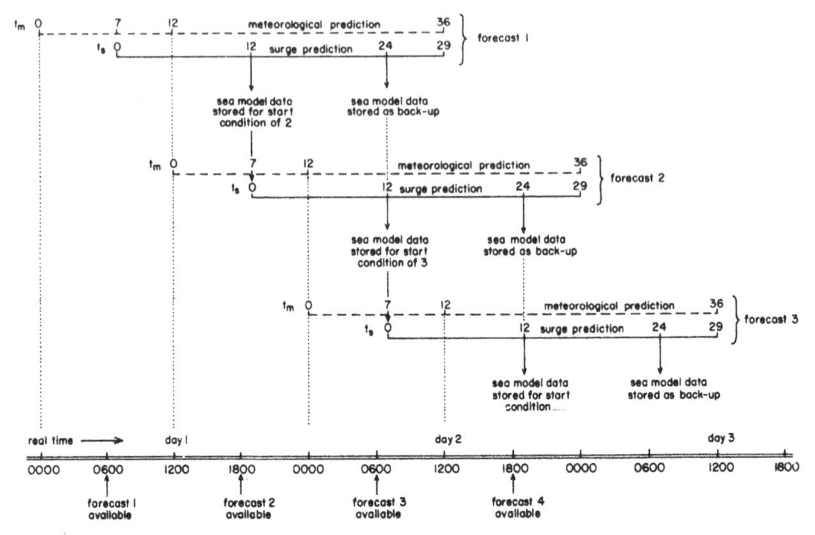

Fig. 44.7. Scheme for operational surge forecasting providing predictions up to 30 hours ahead issued twice each day. t_s denotes time in the sea-model calculation and t_m time in the atmospheric model.

Fig. 44.7 indicates the way in which predictions would be obtained in the real-time practical scheme, surge forecasts being run in parallel with the weather forecasts. As is well known, the level of frictional dissipation in shallow seas is such as to reduce rapidly the influence of initial conditions in sea-model calculations. This contrasts with the situation in atmospheric models where initial conditions are all-important and special initialization techniques must be employed: for the sea model it is sufficient simply to take as initial conditions data generated in an earlier integration. An important practical consequence is that by virtue of the overlap between successive forecasts, the proposed system could continue in the event of one forecast being lost, perhaps because of computer failure. The effect of the resulting additional error in the initial condition for the next forecast quickly becomes negligible in comparison with the disturbance generated by the meteorological forcing (Flather, 1976*b*).

A sequence of forecasts for the period 18 to 20 November 1973, generated using the operational scheme, is shown in Fig. 44.8. For each

Fig. 44.8. A sequence of predictions covering the period 18 to 20 November 1973 obtained by running the forecasting scheme in an operational mode. The vertical axis scale indicated for the lowest (Southend) case applies also to the other cases.

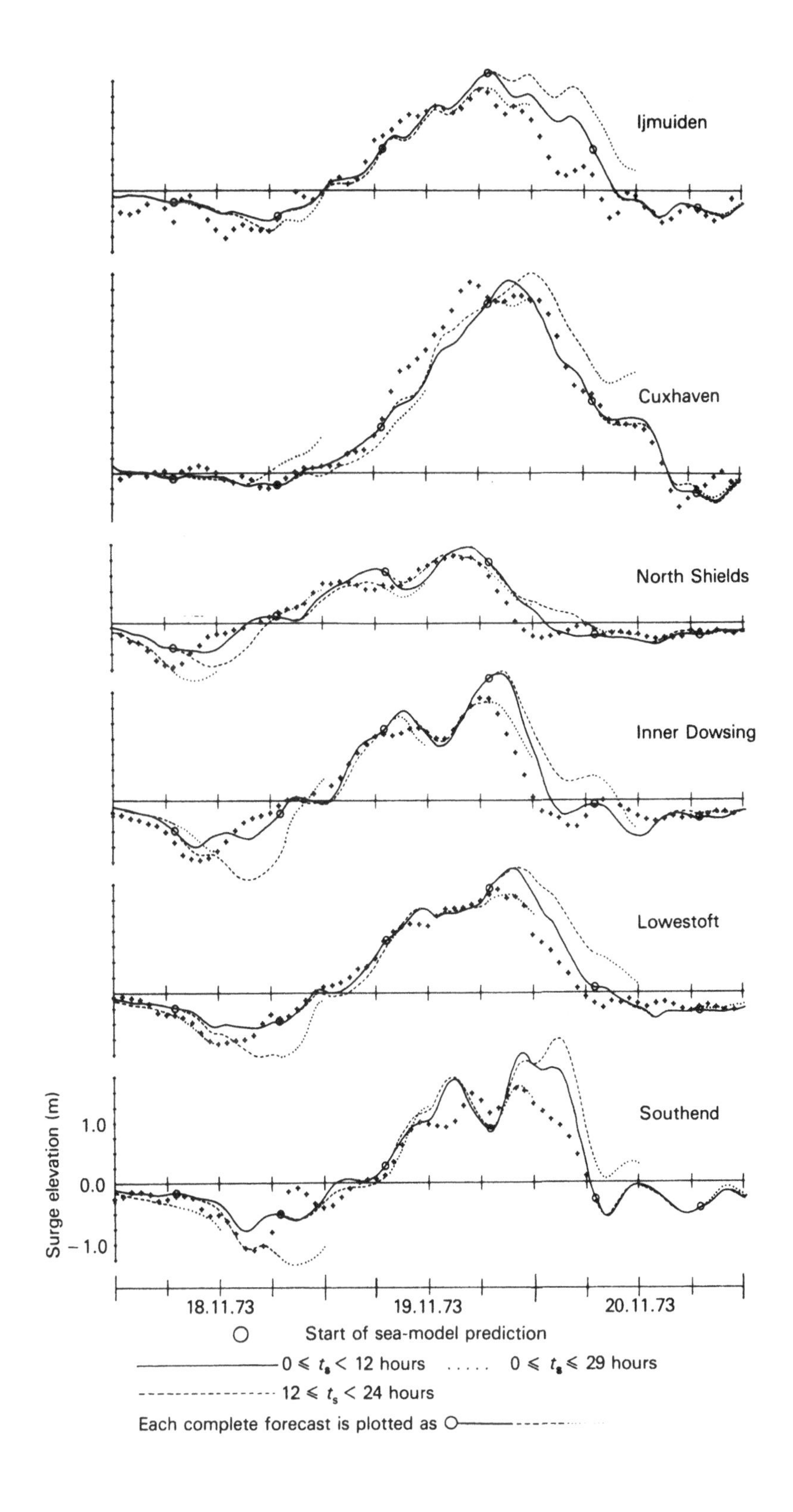
Ijmuiden
Cuxhaven
North Shields
Inner Dowsing
Lowestoft
Southend
Surge elevation (m)
1.0
0.0
− 1.0
18.11.73
19.11.73
20.11.73
Start of sea-model prediction
0 ≤ t_s < 12 hours
0 ≤ t_s ≤ 29 hours
12 ≤ t_s < 24 hours
Each complete forecast is plotted as

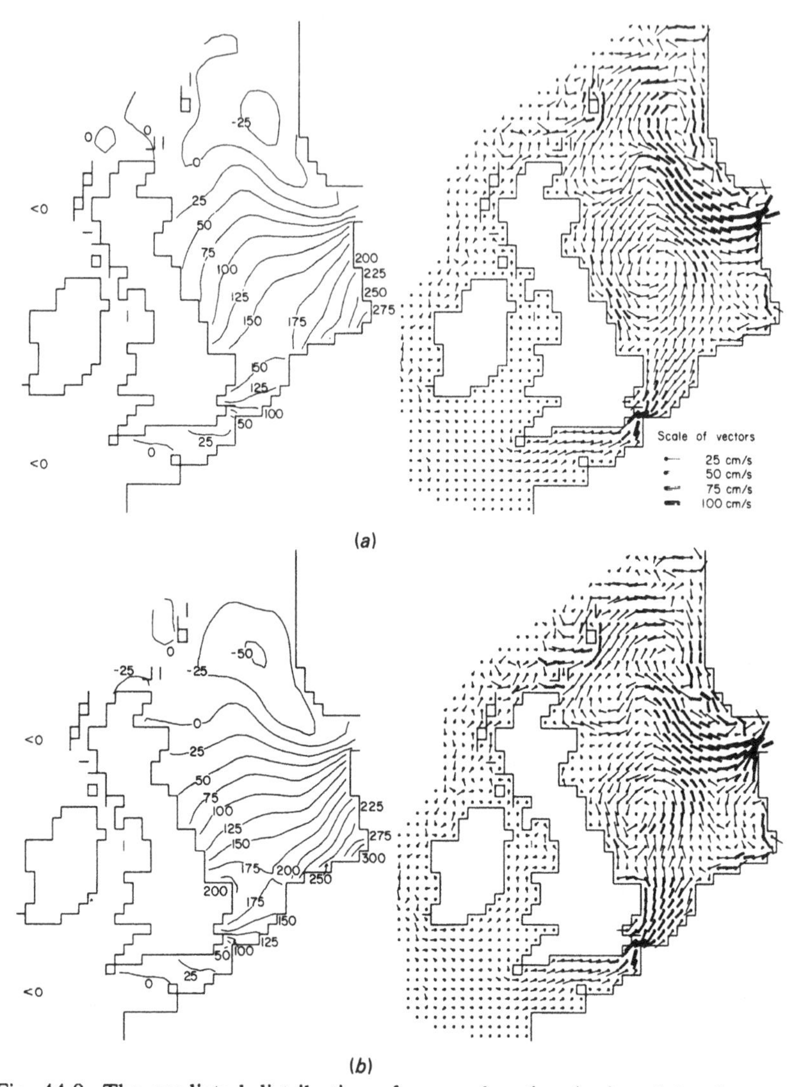

Fig. 44.9. The predicted distribution of surge elevation (cm) and depth mean current at 2100 GMT on 19 November 1973 obtained from different forecasts: (*a*) sea-model forecast starting at 1900 GMT on 18 November 1973; (*b*) sea-model forecast starting at 1900 GMT on 19 November 1973.

forecast run of the sea model, surge elevation at each port indicated during the period $0 \leq t_s < 12$ hours is plotted as a continuous line, for the period $12 \leq t_s < 24$ hours as a dashed line, and for $24 \leq t_s \leq 29$ hours as a dotted line. The whole sequence of prediction runs thus provides either two or three estimates of surge height at any time – two estimates between

0000 GMT and 0700 GMT; three between 0700 GMT and 1200 GMT; two between 1200 GMT and 1900 GMT and three from 1900 GMT to 2400 GMT (see Fig. 44.7). The earliest possible warning of any event, provided in real-time operation by the forecast issued more than 24 hours in advance, is indicated by the dotted lines in Fig. 44.8. This would be updated by part of the next forecast, issued 12 hours later and drawn as a dashed line. The final short-term prediction, issued 12 hours later still, would come from the section drawn as a continuous curve. Predicted spatial distributions of surge elevation and depth mean current at 2100 GMT on 19 November, extracted from two of the forecasts plotted in Fig. 44.8, are shown in Fig. 44.9. Despite the fact that the distribution in Fig. 44.9*a* would have been available 24 hours earlier than the distribution in Fig. 44.9*b*, and about 27 hours ahead of the event, the two patterns are closely similar, confirming the usefulness of predictions giving more than one day's warning of a developing storm surge.

44.5 Some points of difference between the storm-surge problems in the Bay of Bengal and in the North Sea

The storm-surge prediction procedure outlined in §§ 44.2 and 44.3 was developed for use in the North Sea. However, in principle, the same techniques could be adapted to other areas. In the context of this book, storm surges in the Bay of Bengal are of particular interest, and here we compare some of their characteristics with typical North Sea surges. Such a comparison is an important first step in the possible modification of the present method for use in the Bay of Bengal.

The North Sea is a shallow, partly enclosed, marginal sea of average depth 65 m (see Fig. 44.10). A number of reviews of North Sea surges exist, one of the most recent of which is that by Heaps (1967). The surges are associated with North Atlantic depressions, approaching from directions between southwest and northwest. Over waters of oceanic depth, atmospheric pressure gradients predominate over the forces due to wind stress, the latter occurring with a factor (1/depth) in (44.2) and (44.3). Timmerman (1975) has shown that atmospheric pressure effects off the shelf to the north west of Scotland produce a significant contribution to surges in the northern North Sea. Southwesterly winds over the shelf west of Scotland as the depression approaches land generate an 'external surge' component, which propagates, trapped against the coast by the Coriolis force, into the North Sea taking more than 12 hours to reach the Southern Bight. The same southwesterlies acting over the shallow coastal

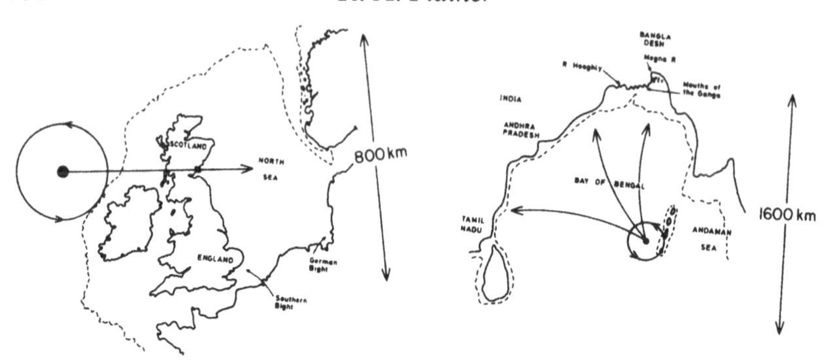

Fig. 44.10. Maps of the North Sea and Bay of Bengal showing the 200 m depth contours (dashed) and typical tracks of depressions.

waters along the east coast of England, as the depression enters the North Sea, also reduce levels locally, producing a negative surge. Subsequently, as the depression moves east the winds veer through west to northwesterly or northerly, blowing over the shallow waters of the southern North Sea and the German Bight and piling up water against the coastlines. Almost all North Sea surges contain these basic ingredients in varying proportion. The surges also enter rivers draining into the North Sea, in particular the Thames, which can be affected to a point upstream of London, and the Elbe, where the city of Hamburg experiences surges. Within the rivers the surges are modified importantly by shallow water nonlinear effects and the resulting dynamical interaction with the tide.

The Bay of Bengal (Fig. 44.10) is an area very different from the North Sea, being essentially a northward extension of the Indian Ocean, with depths decreasing from 4000 m in the south to 2000 m in the north. Bounding the deeper water is a narrow continental shelf, varying in width from less than 50 km along the coasts of Tamil Nadu and Andhra Pradesh to 300 km in the north and northeast of the Bay. The surges are associated with tropical cyclones which originate in the central or southern parts of the Bay or in the Andaman Sea, moving towards the west or northwest before curving to the north and eventually to the north east. There are, however, frequent and considerable deviations from this pattern of cyclone movement.

A number of general reviews and descriptions of individual cyclones and the associated surges have been published; see, for example, Eliot (1900), Dunn (1962), Ratnam and Nayar (1966), the Bay of Bengal Pilot (1966) and Frank and Husain (1971). However, many questions remain concerning their generation and subsequent behaviour. In view of the large water depth, atmospheric pressure effects may dominate over the deeper part of the Bay. Disturbances generated there would propagate much more quickly than the speed at which the cyclone moves and reach the coasts some time ahead of the storm. Very strong wind generated currents also occur (Bay of Bengal Pilot, 1966) and could influence sea levels where they impinge on the continental shelf. The main surge effect is generated on the continental shelves. Ratnam and Nayar (1966) suggested that the resonant coupling mechanism, shown by Proudman (1929) to occur when the speed of a travelling atmospheric pressure disturbance matches that of free gravity waves, could be significant there. Wind acting over shallow water is the main generating mechanism. The maximum onshore winds, and hence also the maximum surge levels, occur on the right of the storm track as it crosses the coast. A negative surge associated with offshore winds on the left of the storm centre may also appear. The surge maximum occurs near to the time at which the storm centre crosses the coast. Coastal geometry and bottom topography are important in determining the surge response on the open coast (Flierl and Robinson, 1972).

The extreme shallowness of the water in some near-shore areas, the focussing effect of inlets and estuaries and the absence of shore defences to protect the low-lying coastal lands leads to the formation of a 'storm wave', frequently described by eye witnesses as 'a wall of water', which can travel several miles inland flooding affected areas to a depth of as much as 13 metres (see the account of the Backergunge cyclone of October 1876 in Eliot, 1900). Some accounts suggest that the accumulating water associated with the surge may be held back by the river flow and ebbing tide in estuaries such as the Hooghly, eventually advancing upstream as a bore, overtopping the river banks and inundating low-lying land as a storm wave (e.g. the Calcutta cyclone of October 1864; Eliot, 1900). Clearly, nonlinear effects play a vital part in this stage in the development of the surge.

Tidal influence is important in the north of the Bay, especially in the Hooghly and the Ganga, where tidal currents are strong and the range of the tide is large. The greatest danger of flooding exists when the surge coincides with high tide. Thus, according to Eliot (1900), the surges associated with both the Calcutta cyclone of 1864 in the Hooghly and the Backergunge cyclone of 1876 in the mouth of the Meghna occurred close to the time of high water on spring tides. Conversely, the effects of the Midnapore cyclone of November 1874 were minimized by the coincidence of the storm wave with low tide in the mouth of the Hooghly. The apparent importance of nonlinear effects suggests that dynamical tide–surge interaction must also play a part here. The tidal range decreases to the south so that along much of the east coast the tides are small.

Several calculations of Bay of Bengal surges have been carried out. Janárdhan (1967) estimated some surge heights at Saugor Island on the basis of the static wind set up, assuming a balance between wind stress and surface slope terms in (44.2) and (44.3), apparently with some success. The long narrow continental shelf along the coasts of Tamil Nadu and Andhra Pradesh, affected by the recent cyclone of November 1977, suggests that an analytical model might be usefully applied. Thus, for example, neglecting longshore variations and linearizing the governing equations, analytical time-dependent solutions can be obtained using Laplace transforms. Such models have proved capable of reproducing open coast surges on the west coast of the British Isles (Heaps, 1965). In general, however, the problem demands numerical methods of solution. Numerical model studies of the surges on the northern shelf have been carried out by Das (1972) and Flierl and Robinson (1972) in response to the November 1970 Bangladesh cyclone, and subsequently by Das *et al.* (1974). The models used were based on simplified equations and the storms were represented by idealized circular cyclones moving with uniform speed towards the coast. From the results of their experiments Flierl and Robinson (1972) constructed graphs which allow estimates to be made of the maximum surge height as a function of the maximum wind speed, the speed of translation along the track, the direction of the track and the distance of the landfall point from Chittagong. Similar graphs giving the highest water level at the time of landfall as a function of storm intensity, as defined by the pressure difference between the centre and edge of the cyclone, and the storm speed were constructed by Das *et al.* (1974).

In considering the implementation in the Bay of Bengal of a prediction scheme similar to that for the North Sea, the importance of all stages in the procedure must be considered. Thus, for example, a sophisticated sea model will not produce accurate surge predictions from inaccurate meteorological forecasts. The difficulty of predicting the intensities and tracks of cyclones may at present justify only the use of graphs or simple models to estimate the surge. However, improvements in weather forecasting and the operational development of numerical models of the atmosphere will make a more sophisticated approach to flood prediction worthwhile.

The design of sea models to be used could follow broadly that adopted elsewhere. The first component would be a model covering the whole of the region affected by the cyclones, which, in view of the variability in their tracks, would need to be the entire Bay of Bengal, such that all the main features of coastline and topography – especially the continental shelves – were resolved. This model could then be used to specify boundary conditions for further shelf models, representing the shallower regions accurately and probably incorporating one-dimensional models of estuaries. One obvious model would cover the northern shelf, as in the models of Das *et al.*, but extending into the river mouths and estuaries. The great length of the continental shelf along the east coast of India suggests that a whole family of models would be required; only that model covering the shelf extending on either side of the expected cyclone track being actually used. These secondary models might have sufficiently fine resolution to include a rough representation of the flooding of coastal land using the moving boundary procedures proposed by Sielecki and Wurtele (1970), Leendertse and Gritton (1971) or Flather and Heaps (1975). Additional more detailed models of coastal inlets and estuaries with the surrounding low-lying land would eventually be required. The need to simulate the development of bores and storm waves in these models would present some formidable but interesting problems. All models would take account of nonlinear terms, tides and river flow where appropriate.

The author is indebted to Dr N. S. Heaps for valuable comments on a first draft of this paper and to the Meteorological Office for their continuing cooperation.

The work described in this chapter was funded by a Consortium consisting of the Natural Environment Research Council, the Ministry of Agriculture Fisheries and Food, and the Departments of Energy, the Environment and Industry.

References

Banks, J. E. (1974) A mathematical model of a river-shallow sea system used to investigate tide surge and their interaction in the Thames-Southern North Sea region. *Phil. Trans. Roy. Soc.*, **A 275**, 567–609.

Bay of Bengal Pilot (1966) London: The Hydrographer of the Navy, N.P. 21.

Benwell, G. R. R., Gadd, A. J., Keers, J. F., Timpson, M. S. and White, P. W. (1971) The Bushby–Timpson 10-level model on a fine mesh. Science Paper No. 32, Meteorological Office, London, 23 pp.

Bunker, A. F. (1976) Computations of surface energy flux and annual air–sea interaction cycles of the North Atlantic Ocean. *Mon. Wea. Rev.*, **104**, 1122–40.

Cartwright, D. E. (1976) Shelf boundary tidal measurements between Ireland and Norway. *Mém. Soc. R. Sci. Liège*, Series 6, **10**, 133–9.

Das, P. K. (1972) Prediction model for storm surges in the Bay of Bengal. *Nature*, **239**, 211–13.

Das, P. K., Sinha, M. C. and Balasubramanyam, V. (1974) Storm surges in the Bay of Bengal, *Quart. J. Roy. Meteor. Soc.*, **100**, 437–49.

Davies, A. M. (1976*a*) A numerical model of the North Sea and its use in choosing locations for the deployment of offshore tide gauges in the JONSDAP'76 oceanographic experiment. *Dt. hydrogr. Z.*, **29**, 11–24.

Davies, A. M. (1976*b*) Application of a fine mesh numerical model of the North Sea to the calculation of storm surge elevations and currents, Report No. 28, Institute of Oceanographic Sciences.

Davies, A. M. and Flather, R. A. (1977) Computation of the storm surge of 1 to 6 April 1973 using numerical models of the north-west European continental shelf and the North Sea, *Dt. hydrogr. Z.*, **30**, 139–62.

Davies, A. M. and Flather, R. A. (1978) Application of numerical models of the north-west European continental shelf and the North Sea to the computation of the storm surges of November–December 1973. *Dt. hydrogr. Z. Erg.-H.* A, No. 14.

Dunn, G. E. (1962) The tropical cyclone problem in East Pakistan. *Mon. Wea. Rev.*, **90**, 83–6.

Duun-Christensen, J. T. (1971) Investigations on the practical use of a hydrodynamic numeric model for calculation of sea level variations in the North Sea, the Skaggerak and the Kattegat. *Dt. hydrogr. Z.*, **24**, 210–27.

Duun-Christensen, J. T. (1975) The representation of the surface pressure field in a two-dimensional hydrodynamic numeric model for the North Sea, the Skaggerak and the Kattegat. *Dt. hydrogr. Z.*, **28**, 97–116.

Eliot, J. (1900) *Handbook of cyclonic storms in the Bay of Bengal.* Meteorological Department of the Government of India, Calcutta, 2 vols.

Flather, R. A. (1976*a*) Results from a storm surge prediction model of the north-west European continental shelf for April, November and December 1973. Report No. 24, Institute of Oceanographic Sciences.

Flather, R. A. (1976*b*) Practical aspects of the use of numerical models for surge prediction. Report No. 30, Institute of Oceanographic Sciences.

Flather, R. A. (1976*c*) A tidal model of the north-west European continental shelf, *Mém. Soc. R. Sci. Liège*, Series 6, **10**, 141–64.

Flather, R. A. and Davies, A. M. (1975) The application of numerical models to storm surge prediction. Report No. 16, Institute of Oceanographic Sciences.

Flather, R. A. and Davies, A. M. (1976) Note on a preliminary scheme for storm surge prediction using numerical models, *Quart. J. Roy. Meteor. Soc.*, **102**, 123–32.

Flather, R. A. and Davies, A. M. (1978) On the specification of meteorological forcing in numerical models for North Sea storm surge prediction, with application to the surge of 2–4 January 1976. *Dt. hydrogr. Z. Erg.-H.* A, No. 15.

Flather, R. A. and Heaps, N. S. (1975) Tidal computations for Morecambe Bay. *Geophys. J. Roy. Astron. Soc.*, **42**, 489–517.

Flierl, G. R. and Robinson, A. R. (1972) Deadly surges in the Bay of Bengal: Dynamics and Storm tide tables, *Nature*, **239**, 213–15.

Frank, N. L. and Husain, S. A. (1971) The deadliest tropical cyclone in history? *Bull. Amer. Meteor. Soc.* **52**, 438–44.

Garratt, J. R. (1977) Review of drag coefficients over oceans and continents, *Mon. Wea. Rev.*, **105**, 915–29.

Hasse, L. (1974) On the surface to geostrophic wind relationship at sea and the stability dependence of the resistance law. *Beit. Phys. Atmos.*, **47**, 45–55.

Hasse, L. and Wagner, V. (1971) On the relationship between geostrophic and surface wind at sea. *Mon. Wea. Rev.*, **99**, 255–60.

Heaps, N. S. (1965) Storm surges on a continental shelf, *Phil. Trans. Roy. Soc. London* **A 257**, 351–83.

Heaps, N. S. (1967) Storm surges. *Oceanogr. Mar. Biol. Ann. Rev.*, **5**, 11–47.

Heaps, N. S. (1969) A two-dimensional numerical sea model. *Phil. Trans. Roy. Soc. London*, **A 265**, 93–137.

Janardhan, S. (1967) Storm induced sea-level changes at Saugor Island situated in the north Bay of Bengal. *Indian J. Meteor. Geophys.*, **18**, 205–12.

Leendertse, J. J. and Gritton, E. C. (1971) A water quality simulation model for well-mixed estuaries and coastal seas: Volume 2, Computational procedures. Memorandum R-708-NYC, The Rand Corporation, New York.

Piacsek, S. A. and Williams, G. P. (1970) Conservation properties of convection difference schemes. *J. Computational Phys.*, **6**, 392–405.

Prandle, D. (1975) Storm surges in the southern North Sea and River Thames. *Proc. Roy. Soc. London*, **A 344**, 509–39.

Prandle, D. and Wolf, J. (1978) The interaction of surge and tide in the North Sea and River Thames. *Geophys. J. Roy. Astron. Soc.*, **55**, 203–16.

Proudman, J. (1929) The effects on the sea of changes in atmospheric pressure. *Mon. Not. Roy. Astron. Soc. Geophys. Suppl.*, **2**, 197–209.

Ratnam, V. and Nayar, P. S. (1966) Floods due to storm surges associated with cyclonic storms. *Ind. J. Meteor. Geophys.*, **17**, 363–8.

Roberts, K. V. and Weiss, N. O. (1966) Convective difference schemes. *Maths. Comput.*, **20**, 272–99.

Rossiter, J. R. (1961) Interaction between tide and surge in the Thames. *Geophys. J. Roy. Astron. Soc.*, **6**, 29–53.

Sielecki, A. and Wurtele, M. G. (1970) The numerical integration of the non-linear shallow-water equations with sloping boundaries. *J. Computational Phys.*, **6**, 219–36.

Smith, S. D. and Banke, E. G. (1975) Variation of the sea surface drag coefficient with wind speed. *Quart. J. Roy. Meteor. Soc.*, **101**, 665–73.

Timmerman, H. (1975) On the importance of atmospheric pressure gradients for the generation of external surges in the North Sea. *Dt. hydrogr. Z.*, **28**, 62–71.

45

Numerical simulation of storm surges in the Bay of Bengal

B. JOHNS

Using fairly standard techniques, a vertically-integrated model is formulated for simulating the interaction between surge and tide in the Bay of Bengal. The model includes a simplified representation of the river system at the head of the Bay that permits the incorporation of the effect of fresh water input into the system. The equations of the model are nonlinear and their numerical solution, corresponding to prescribed surface wind stress, tidal forcing and fresh water discharge, enables the interactions between the three processes to be evaluated. Some early results are given that correspond to the passage of an idealized tropical cyclone across the analysis area.

An important part of the formulation of the model relates to the way in which nonlinearity and bottom friction are included. The questionable nature of this is emphasized and an alternative scheme is proposed. This involves the use of an energy-based closure scheme to parameterize the Reynolds' stresses together with a transformed set of coordinates to facilitate the numerical solution of the free surface problem. These ideas are introduced in connection with the simulation of tide and surge in an elongated channel which could be readily adapted to represent the river system at the head of the Bay of Bengal. This reformulation, together with a choice of parameters representative of conditions in the River Thames, England, is used to assess the validity of the assumptions made in the earlier formulation of the storm-surge model. The conventional representation of bottom friction is found to be especially weak, and a proposal is made to abandon earlier approaches and to develop the method advocated in this presentation.

45.1 Introduction

Abnormal increases in the elevation of the sea surface above its mean level have led to well-documented cases of flooding with consequential loss of life and property in the low-lying land areas adjacent to the Bay of Bengal. These increases can be produced by a combination of the effects of the astronomical tide, wind-generated surge (forced by tropical cyclones) and fresh water outflow from the many river systems. A full representation of the interaction between these three processes is essential in a predictive scheme.

An account of numerical storm-surge models of continental-shelf seas (mainly in northwestern Europe) is given by Flather in Chapter 44 of this book and a further review is not given here. In the present contribution, we concentrate upon the development of a model for numerical storm-surge simulation in the Bay of Bengal. The formulation of this is considered carefully and, after making clearly defined hypotheses, a nonlinear vertically-integrated model is derived for the numerical simulation of the three processes. This applies to a region including the entire Bay of Bengal and an idealization of the Ganges, Brahmaputra and Meghna river system in Bangladesh.

A preliminary numerical experiment is described that relates to the use of the model with idealized coastlines and bathymetry. The sea-surface elevation is evaluated corresponding to the astronomical tide and the surge generated by the passage of an idealized cyclone across the analysis area. Particular attention is paid to the nonlinear interaction between these processes and an example is given of the temporal variation in the water elevation in the river system as the cyclone approaches the land. This example illustrates effectively the need to incorporate nonlinear interactions in the model; a linear superposition of surge and tide leads to substantial error.

It is pointed out that the formulation of a nonlinear depth-integrated model of the type used in this experiment involves several possibly weakening assumptions. The implications of these are considered within the framework of a multi-level model for unidirectional flow in an elongated channel. This model incorporates a mean turbulent energy closure scheme similar to that used by Johns (1977). It yields a representation of the stress at the floor of the channel based upon a greatly reduced empirical input. An assessment is therefore possible of the assumptions implicit in the depth-integrated model and conclusions are drawn concerning the adequacy of the representation of nonlinear interactions in this.

Finally, suggestions are made for the future development of numerical storm-surge simulation in the Bay of Bengal. These relate to the possible use, where valid, of a depth-integrated model and its matching to multi-level models in the shallow sea regions where nonlinearity is so important.

45.2 Formulation of the model

A system of rectangular Cartesian coordinates is used in which the origin, O, is in the equilibrium level of the sea-surface and is located at the most southerly point of the Bay of Bengal included in the analysis area. Ox points towards the east, Oy towards the north and Oz is measured vertically upwards. The displaced level of the free surface is given by $z = \zeta(x, y, t)$ and the floor of the bay corresponds to $z = -h(x, y)$. In the present idealized model, the analysis area consists of two rectangular regions bounded in the south by an open-sea boundary. Other lateral boundaries consist of vertical walls approximating the east coast of India, the head of the Bay of Bengal and the west coast of Burma and Thailand. In the northeast corner of the bay model an idealized river system connects with the main analysis area. This is an approximation to the Ganges, Brahmaputra and Meghna river system in Bangladesh. In the numerical applications, the details of surge and tide will be computed on fine and coarse meshes as indicated in Fig. 45.1. This will allow the necessary fine resolution in the northernmost shelf area and, at the same time, will enable the surge-generating capacity of a tropical cyclone to be recorded from the time that it first enters the Bay of Bengal.

Within the bay area, the Reynolds'-averaged components of velocity (u, v, w) then satisfy

$$\frac{\partial u}{\partial t} + u\frac{\partial u}{\partial x} + v\frac{\partial u}{\partial y} + w\frac{\partial u}{\partial z} - fv = -g\frac{\partial \zeta}{\partial x} + \frac{1}{\rho}\frac{\partial \tau_x}{\partial z}, \tag{45.1}$$

$$\frac{\partial v}{\partial t} + u\frac{\partial v}{\partial x} + v\frac{\partial v}{\partial y} + w\frac{\partial v}{\partial z} + fu = -g\frac{\partial \zeta}{\partial y} + \frac{1}{\rho}\frac{\partial \tau_y}{\partial z}, \tag{45.2}$$

and

$$\frac{\partial u}{\partial x} + \frac{\partial v}{\partial y} + \frac{\partial w}{\partial z} = 0. \tag{45.3}$$

Here, the pressure is hydrostatic, ρ is the density of homogeneous water, f the Coriolis parameter and (τ_x, τ_y) the components of horizontal Reynolds' stress. The effects of barometric forcing and the direct action of the tide-generating forces have been omitted.

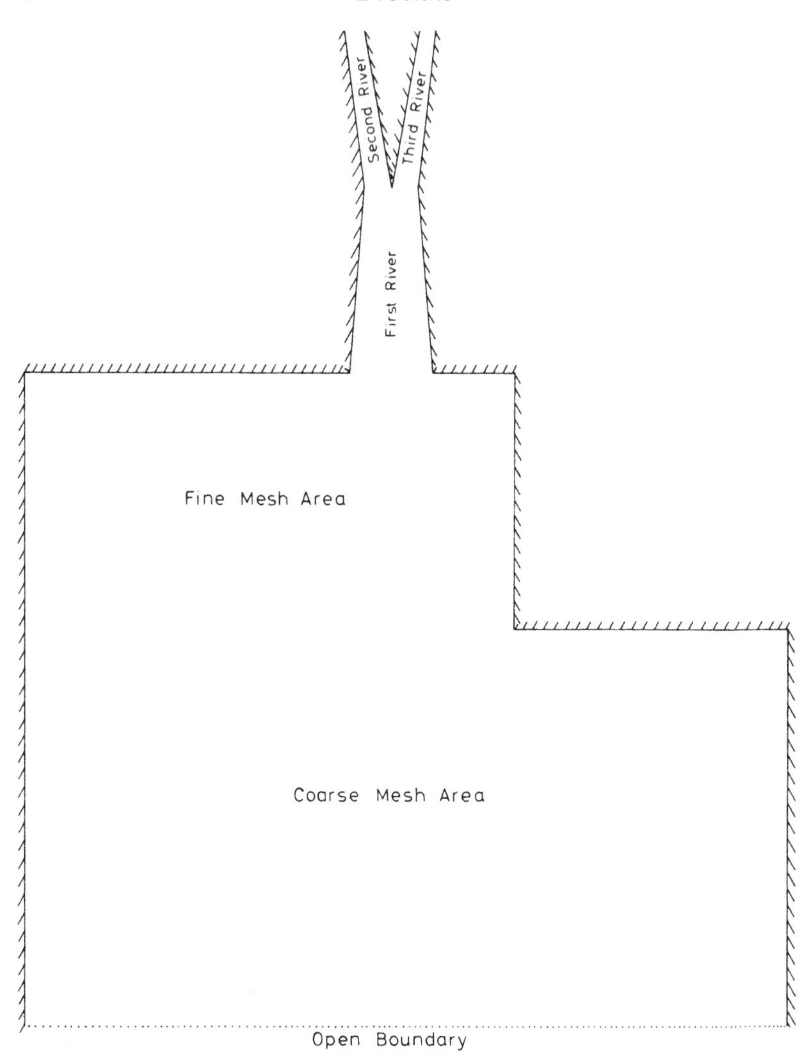

Fig. 45.1. Diagrammatic sketch of the analysis area showing the fine- and coarse-mesh areas.

In the river system, the flow in each of the three channels is assumed to be undirectional and to satisfy

$$\frac{\partial v}{\partial t}+v\frac{\partial v}{\partial y}+w\frac{\partial v}{\partial z}=-g\frac{\partial \zeta}{\partial y}+\frac{1}{\rho}\frac{\partial \tau_y}{\partial z}, \tag{45.4}$$

and

$$\frac{\partial}{\partial y}[b(y)v]+b(y)\frac{\partial w}{\partial z}=0, \tag{45.5}$$

where $b(y)$ is the prescribed breadth of the river at each of its rectangular cross-sections.

Applications of (45.1) to (45.5) have usually depended upon the use of derived equations for the depth-averaged velocity defined by

$$(\bar{u}, \bar{v}) = \frac{1}{\zeta + h} \int_{-h}^{\zeta} (u, v)\, dz. \tag{45.6}$$

Straightforward manipulation of (45.1) to (45.3) then leads to

$$\frac{\partial \bar{u}}{\partial t} + \bar{u}\frac{\partial \bar{u}}{\partial x} + \bar{v}\frac{\partial \bar{u}}{\partial y} - f\bar{v} + \frac{1}{\zeta + h}\left(\frac{\partial}{\partial x}[(\zeta + h)(\overline{u^2} - \bar{u}^2)] + \frac{\partial}{\partial y}[(\zeta + h)(\overline{uv} - \bar{u}\bar{v})]\right) = -g\frac{\partial \zeta}{\partial x} + \frac{1}{\rho(\zeta + h)}(\tau_x^{\zeta} - \tau_x^{-h}), \tag{45.7}$$

$$\frac{\partial \bar{v}}{\partial t} + \bar{u}\frac{\partial \bar{v}}{\partial x} + \bar{v}\frac{\partial \bar{v}}{\partial y} + f\bar{u} + \frac{1}{\zeta + h}\left(\frac{\partial}{\partial x}[(\zeta + h)(\overline{uv} - \bar{u}\bar{v})] + \frac{\partial}{\partial y}[(\zeta + h)(\overline{v^2} - \bar{v}^2)]\right) = -g\frac{\partial \zeta}{\partial y} + \frac{1}{\rho(\zeta + h)}(\tau_y^{\zeta} - \tau_y^{-h}), \tag{45.8}$$

and

$$\frac{\partial \zeta}{\partial t} + \frac{\partial}{\partial x}[(\zeta + h)\bar{u}] + \frac{\partial}{\partial y}[(\zeta + h)\bar{v}] = 0. \tag{45.9}$$

In these equations, the overbars specify depth-averaged values, $(\tau_x^{\zeta}, \tau_y^{\zeta})$ denotes the applied surface wind stress and $(\tau_x^{-h}, \tau_y^{-h})$ the bottom stress. Equations (45.4) and (45.5) for the unidirectional channel flow may be treated in the same way.

These equations may be used in numerical prediction schemes by prescribing the surface wind stress, forcing the tide along the open-sea boundary and prescribing the fresh water discharge at the head of the river system. Their numerical solution will then yield the free surface elevation as surge and tide propagate into the head of the bay and eventually into the river system.

Before such an application can be made, however, further assumptions must be introduced into (45.7) and (45.8). In the shallow water region at the head of the Bay of Bengal, and in the river system, the nonlinear terms are of especial importance and must be retained in the formulation. However, the retention in (45.7) of terms such as $\overline{u^2} - \bar{u}^2$ leads to a fundamental difficulty as they cannot be evaluated within the framework of a vertically-integrated model. In the many well-documented applications of (45.7) to (45.9), it is usual to make assumptions typified by

$$\overline{u^2} - \bar{u}^2 = 0. \tag{45.10}$$

Additionally, a parameterization of the bottom stress must be made in terms of the depth-averaged current. This is frequently done by writing

$$(\tau_x^{-h}, \tau_y^{-h}) = k\rho(\bar{u}^2 + \bar{v}^2)^{\frac{1}{2}}(\bar{u}, \bar{v}), \tag{45.11}$$

where k is an empirical friction coefficient which is often taken to be 2.6×10^{-3}.

Evidently, (45.10) is equivalent to neglecting the vertical structure in the flow and this may be quite unjustifiable in shallow water. A consequence of this would be a misrepresentation of the interactive effects with all the implications relating to flood prediction.

The viability of (45.11) should also be considered. The bottom stress is determined by the turbulent velocity correlation leading to the Reynolds' stress and it is evident that this is not a local function of the Reynolds'-averaged flow. At a given locality, turbulence in the flow is produced by transfer of energy from the averaged flow and by advection and diffusion from adjacent regions. Consequently, (45.11) may be criticized on the grounds that it attempts to relate the bottom stress to purely local conditions in the averaged flow.

If, however, (45.10) and (45.11) are incorporated into the model, (45.7) and (45.8) reduce to

$$\frac{\partial \bar{u}}{\partial t} + \bar{u}\frac{\partial \bar{u}}{\partial x} + \bar{v}\frac{\partial \bar{u}}{\partial y} - f\bar{v} = -g\frac{\partial \zeta}{\partial x} + \frac{1}{\zeta + h}\left[\frac{\tau_x^{\zeta}}{\rho} - k\bar{u}(\bar{u}^2 + \bar{v}^2)^{\frac{1}{2}}\right], \tag{45.12}$$

and

$$\frac{\partial \bar{v}}{\partial t} + \bar{u}\frac{\partial \bar{v}}{\partial x} + \bar{v}\frac{\partial \bar{v}}{\partial y} + f\bar{u} = -g\frac{\partial \zeta}{\partial y} + \frac{1}{\zeta + h}\left[\frac{\tau_y^{\zeta}}{\rho} - k\bar{v}(\bar{u}^2 + \bar{v}^2)^{\frac{1}{2}}\right], \tag{45.13}$$

together with (45.9).

After a similar process of reduction, (45.4) and (45.5) become

$$\frac{\partial \bar{v}}{\partial t} + \bar{v}\frac{\partial \bar{v}}{\partial y} = -g\frac{\partial \zeta}{\partial y} + \frac{1}{\zeta + h}\left(\frac{\tau_y^{\zeta}}{\rho} - k\bar{v}|\bar{v}|\right), \tag{45.14}$$

and

$$\frac{\partial \zeta}{\partial t} + \frac{1}{b(y)} \frac{\partial}{\partial y} [b(y)(\zeta + h)\bar{v}] = 0. \tag{45.15}$$

With a prescribed distribution of surface wind stress, these equations may be used to predict the ensuing pattern of sea-surface elevation.

45.3 A numerical experiment

The main analysis area extends from about 10° N to 20° N. The dimensions of the fine- and coarse-mesh areas are respectively 1080 km (east–west) × 684 km (north–south) and 1440 km (east–west) × 648 km (north–south). The lengths of the three rivers are respectively 342 km, 522 km and 414 km. The fine- and coarse-mesh areas consist of 61 × 39 and 21 × 11 grid-points respectively. The three rivers contain 20, 30 and 24 grid-points. Thus, for the fine-mesh area, $\Delta x = \Delta y = 18$ km and, for the coarse-mesh area, $\Delta x = \Delta y = 72$ km. For each of the river systems, $\Delta y = 18$ km.

In the coarse-mesh area, the equilibrium depth of the water is 500 metres. In the fine-mesh area, the depth varies exponentially in both directions equalling 10 metres along the coasts and 500 metres at the points of juncture with the coarse-mesh area. The first river has a uniform depth of 8 metres whilst the other two each have a uniform depth of 7 metres.

The first river has a breadth of 144 km at its point of communication with the sea. Over the first landward 90 km, the breadth is adjusted to simulate crudely the effect of islands and it then reduces exponentially from 18 km to 8 km at the point of juncture with the other river sections. The breadth of each of these then varies exponentially from 4 km to 2 km where the flow is prescribed as being due to fresh water discharge alone.

With this grid, (45.12) to (45.15) may be discretized and a numerical procedure used in which ζ, $\bar{u}$ and $\bar{v}$ are evaluated at staggered points. The solutions in the various regions are matched together at the positions of juncture. This enables a numerical solution to be obtained that describes the interaction effects between surge, tide and fresh water flow. It is first necessary, however, to obtain a solution that corresponds to tide and fresh water flow alone. This is done by preseribing an initial state and the fresh water discharge and by forcing the tidal solution along the open-sea boundary. In the experiment, an M_2 tidal constituent having a vertical amplitude of 1 metre was prescribed, this being in phase at all points along the open-sea boundary. After 24 cycles of integration, using a timestep of 155 s, the transients in the initial response are dissipated by friction and an effectively oscillatory solution is obtained that corresponds to the oscillatory forcing. The solution at the end of the 25th cycle is then taken as the initial condition for the surge calculation.

In the numerical experiment described here, the fresh water discharge is omitted and we consider the interaction between tide and the surge generated by an idealized cyclone tracking across the analysis area. The pressure distribution in this is given by

$$p = p_\infty - \Delta p \, e^{-r/R}, \tag{45.16}$$

where p_∞ is the ambient pressure, Δp the difference between the central and ambient pressure, R the radius of the cyclone and r the radial distance of any point from the centre of the cyclone.

In this application, we take $\Delta p = 50$ mb, $R = 350$ km and consider the initial position of the centre of the cyclone to be 432 km north of the open-sea boundary and 504 km from the western boundary. The cyclone moved at a constant speed of 14 km per hour and tracked along a straight line so that its central point struck the mouth of the Meghna after about three days. The associated wind speeds are determined from a gradient formula, with variable Coriolis parameter, and the surface wind stress is calculated by a quadratic drag law.

In Fig. 45.2, the time variation of the surface elevation is shown at a point midway along the first river. In this ζ_S denotes the elevation that results from surge alone (the initial condition being one of rest), ζ_T denotes the elevation due to tide alone (there being no wind stress forcing) and $\zeta_{S+T+IST}$ denotes the elevation due to the simultaneous presence of both surge and tide. Thus, $\zeta_{S+T+IST}$ contains the nonlinear interaction between surge and tide. The effect of this may be seen in ζ_{S+IST} which is the elevation due to surge alone plus this interactive effect. The effect of the surge is at its greatest about 36 hours before landfall and we see that

$$\zeta_{S+T+IST} - (\zeta_S + \zeta_T) = \zeta_{S+IST} - \zeta_S < 0. \tag{45.17}$$

Hence, the interactive effect reduces the surface elevation below the value derived from a purely linear superposition of surge and tide.

It is also of interest to note that the surge-generated component of the elevation is negligible at the time of landfall and that subsequently a small negative surge develops in the river.

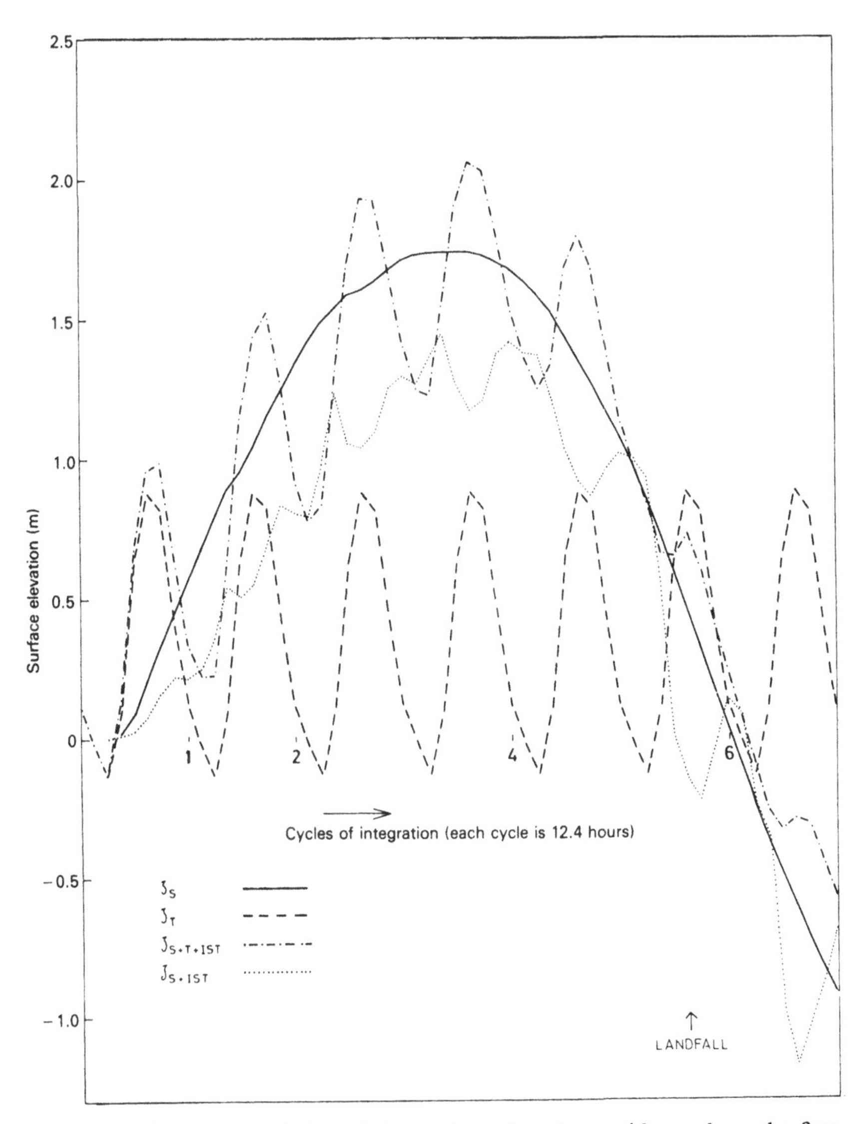

Fig. 45.2. Temporal variation of the surface elevations midway along the first river.

In Fig. 45.3, the same information is given for the surface elevation at the mouth of the first river. The situation here is different from that midway along the river. The interactive effect now reduces the surface elevation at times of high tide below the value derived from a linear combination of surge and tide. At times of low tide, the interactive effect

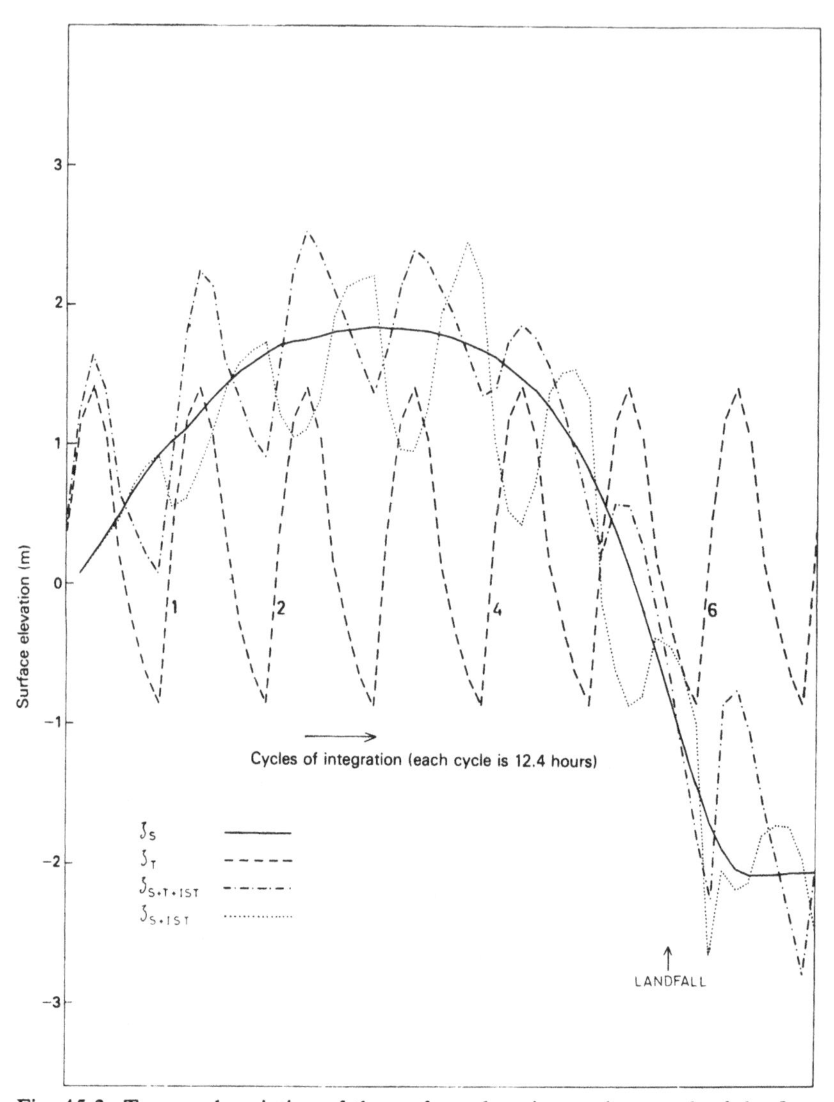

Fig. 45.3. Temporal variation of the surface elevation at the mouth of the first river.

increases the elevation. 9 hours before landfall, the surge-generated elevation is approximately zero and a substantial negative surge has developed about 12 hours after landfall.

In Fig. 45.4, contours of equal sea-surface elevation are given for the fine-mesh area. These again result from the interaction of surge and tide and correspond to about 1.5 hours before landfall. It is clear that a

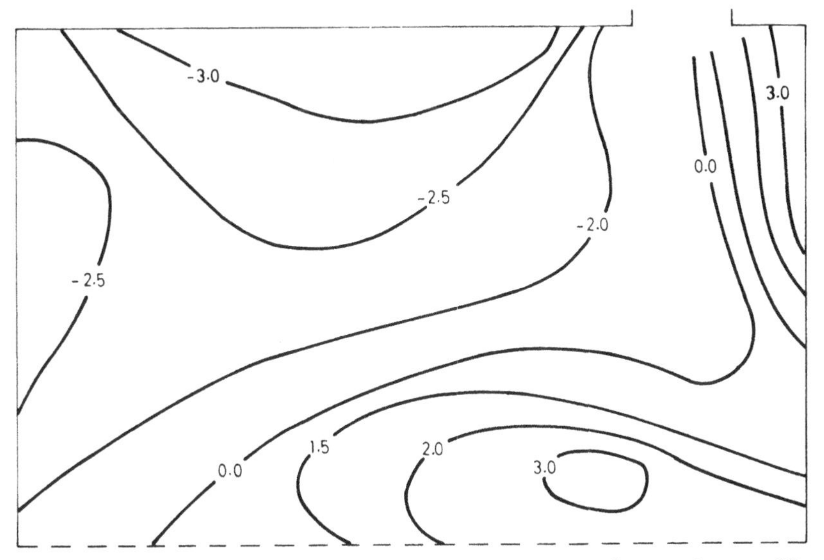

Fig. 45.4. Contours of equal sea-surface elevation in the fine-mesh area. The contours are labelled in metres.

positive surge (in excess of three metres when combined with the tide) occurs to the east of the estuary and a strong negative surge develops in the northwestern extremity of the bay. The surge is positive everywhere on the eastern side of the basin and negative on the western side.

45.4 Formulation of a multi-level model

The numerical experiment described in § 45.3 has been performed with the assumptions (45.10) and (45.11). The validity of these has been questioned on physical grounds and an outline account is now given of an alternative formulation. Although capable of generalization, this applies to the undirectional flow satisfying (45.4) and (45.5). Fuller details of this are given by Johns (1978).

Depth averaging is no longer used and to facilitate the development of a multi-level computational model, a new vertical coordinate is defined by

$$\sigma = \frac{z+h}{\zeta+h}. \tag{45.18}$$

Then, $\sigma = 0$ at $z = -h$ and $\sigma = 1$ at $z = \zeta$ and the floor and free surface of the channel correspond to fixed coordinate surfaces. This is reminiscent of the sigma-coordinates introduced into meteorology by Phillips (1957) and transformations of this type have been considered in marine modelling by Freeman *et al.* (1972).

Using x, σ and t as new independent coordinates, and prescribing a gradient transfer law for momentum in which

$$\tau_y = K\rho \frac{\partial v}{\partial z}, \tag{45.19}$$

(45.4) becomes

$$\frac{\partial v}{\partial t} + u\frac{\partial v}{\partial y} + \omega\frac{\partial v}{\partial \sigma} = -g\frac{\partial \zeta}{\partial y} + \frac{1}{(\zeta+h)^2}\frac{\partial}{\partial \sigma}\left(K\frac{\partial v}{\partial \sigma}\right). \tag{45.20}$$

Here, K is a vertical exchange coefficient and

$$\omega = \frac{\partial \sigma}{\partial t} + v\frac{\partial \sigma}{\partial y} + \omega\frac{\partial \sigma}{\partial z} \tag{45.21}$$

Within the river system, the surge-generating capacity of the surface wind stress is negligible. Tide and surge therefore penetrate into the river from the connecting bay area. By prescribing an absence of water slippage at the floor of the channel together with an absence of applied surface wind stress, it is readily found that

$$\left.\begin{aligned} v = \omega = 0, &\quad \text{at } \sigma = 0, \\ \frac{\partial v}{\partial \sigma} = 0, &\quad \text{at } \sigma = 1. \end{aligned}\right\} \tag{45.22}$$

The kinematical surface condition requires that

$$\frac{\partial \zeta}{\partial t} + v\frac{\partial \zeta}{\partial y} = w, \quad \text{at } z = \zeta \tag{45.23}$$

which, by use of (45.21), yields

$$\omega = 0, \quad \text{at } \sigma = 1. \tag{45.24}$$

From (45.5), we may then show that

$$(\zeta+h)b(y)\omega = \sigma\frac{\partial}{\partial y}\left(b(y)(\zeta+h)\int_0^1 v\,\mathrm{d}\sigma\right) - \frac{\partial}{\partial y}\left(b(y)(\zeta+h)\int_0^\sigma v\,\mathrm{d}\sigma\right), \tag{45.25}$$

and ω is determined diagnostically.

Turbulence closure is achieved by use of the equation for the Reynolds'-averaged turbulent energy density, E. This has the form

$$\frac{\partial E}{\partial t}+v\frac{\partial E}{\partial y}+\omega\frac{\partial E}{\partial \sigma}=\frac{K}{(\zeta+h)^2}\left(\frac{\partial v}{\partial \sigma}\right)^2+\frac{1}{(\zeta+h)^2}\frac{\partial}{\partial \sigma}\left(K\frac{\partial E}{\partial \sigma}\right)-\varepsilon. \tag{45.26}$$

The terms on the right-hand side of (45.26) refer respectively to the production of turbulent energy from the averaged flow, the redistribution of turbulent energy by the turbulence itself and the loss of turbulent energy through dissipation. The redistribution of turbulent energy by the turbulence is assumed to be diffusive and to follow the same gradient transfer law as for momentum.

Accompanying boundary conditions prescribing an absence of flux of turbulent energy across the floor and free surface are

$$\frac{\partial E}{\partial \sigma}=0, \quad \text{at } \sigma=0 \text{ and } \sigma=1. \tag{45.27}$$

Following Launder and Spalding (1972) and Johns (1977) we write

$$\left.\begin{aligned} K &= c^{\frac{1}{4}} l E^{\frac{1}{2}}, \\ \varepsilon &= c^{\frac{3}{4}} E^{\frac{3}{2}}/l. \end{aligned}\right\} \tag{45.28}$$

where $c=0.08$. The length scale l is determined from a local similarity hypothesis in which

$$l=-\frac{\kappa E^{\frac{1}{2}}/l}{\partial/\partial z\,(E^{\frac{1}{2}}/l)}, \tag{45.29}$$

which may be interpreted as a relation between the local mixing scale and frequency scale of the turbulence. Here, κ is Von Karman's constant, $l=\kappa z_0$ at $z=-h$ and z_0 is the roughness length of the bottom elements.

Numerical analysis has been applied to these equations to compute the M_2 tidal flow in the River Thames, England, between Southend and Richmond. The extension of this to determine the interaction between tide and surge is trivial.

A tidal regime is chosen having a length of 96.54 km with a depth decreasing exponentially from 9.75 m at Southend to 3.06 m at Richmond. The breadth at Richmond is about 1.2% of that at Southend. The roughness length of the bottom elements is taken as 5 mm and the tidal elevation at Southend is prescribed in the form

$$\zeta=a\sin(2\pi t/T), \tag{45.30}$$

with $a = 2.59$ m and $T = 12.4$ hours. The tidal current vanishes identically at Richmond. A discretization scheme is used with 18 grid-points along the channel and 10 computational levels through its depth. These levels are arranged so as to provide increased resolution near $\sigma = 0$ in comparison with that near $\sigma = 1$.

In Fig. 45.5, the profile of the tidal current is shown at four successive instants during the tidal cycle, the evaluation being midway along the channel where the depth is 5.28 m. From this, it is clear that there is an intense shear layer adjacent to the floor of the channel. In this, the tidal current is increased from zero to its near-surface value. This suggests that (45.10) may indeed be a good approximation for use in vertically-integrated tidal models and a quantitative evaluation of this conjecture may be made by using the numerical solution to compute both $\bar{v}^2$ and $\overline{v^2}$.

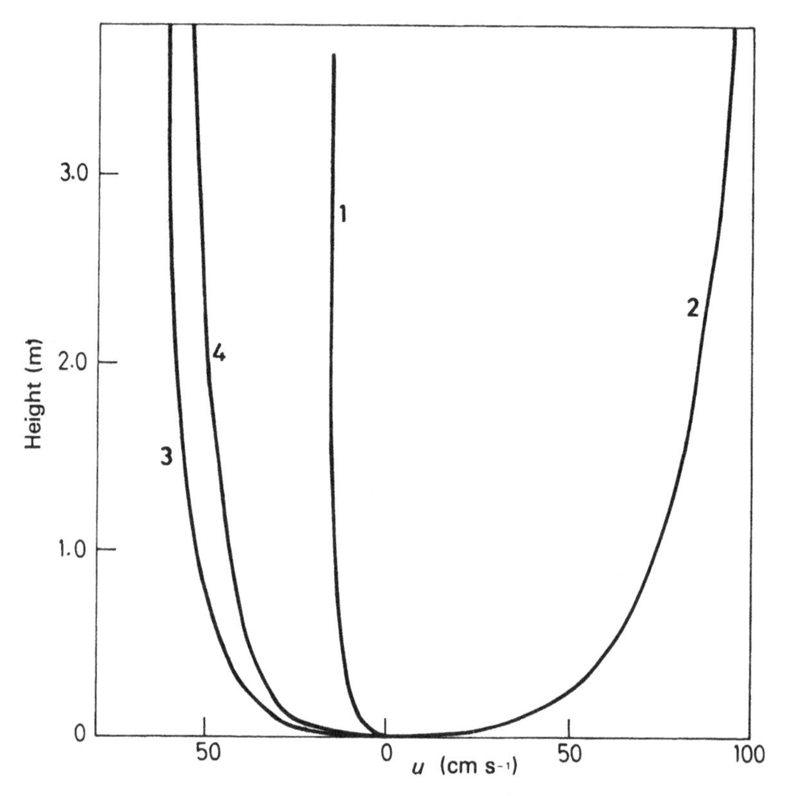

Fig. 45.5. Profile of the tidal current as a function of height above the bottom of the channel.

At all points along the channel, we find that

$$\bar{v}^2/\overline{v^2}<1.04, \tag{45.31}$$

which provides excellent support for (45.10). It must, however, be emphasized that (45.31) applies only for a tidal flow, or a surge-generated flow for which the direct effect of the applied surface wind stress is negligible. If this cannot be neglected, we must expect a vertical structure to evolve throughout the depth as momentum is transferred from the atmospheric winds to the lowermost layers of the flow. In this case, it seems unlikely that (45.10) will yield a good approximation.

The multi-level model may also be used to estimate the appropriateness of the empirical bottom stress parameterization (45.11). Near the floor of the channel, (45.26) reduces to an approximate balance between the local production and dissipation of turbulence and (45.19) then yields

$$\frac{\tau_b}{\rho}=c^{\frac{1}{2}}\ E_b \operatorname{sgn}\left[\left(\frac{\partial v}{\partial z}\right)_b\right], \tag{45.32}$$

where the subscript b refers to an evaluation at $\sigma=0$. Hence, on using (45.11) for unidirectional flow,

$$k=c^{\frac{1}{2}}\frac{E_b \operatorname{sgn}[(\partial v/\partial z)_b]}{\bar{v}|\bar{v}|}. \tag{45.33}$$

The evaluation of k during a tidal cycle shows that the friction coefficient has a complex variation with time. However, we may define an optimized coefficient, for which $\int_0^T (\tau_b/\rho - k\bar{v}|\bar{v}|)^2\,dt$ is a minimum, by writing

$$k=\frac{\int_0^T (\tau_b/\rho)\bar{v}|\bar{v}|\,dt}{\int_0^T \bar{v}^4\,dt}. \tag{45.34}$$

When evaluated midway along the channel, this yields

$$k=4.63\times10^{-3}. \tag{45.35}$$

This value of k has been used to compute the temporal variation of the bottom stress from the empirical formula and the result compared with an evaluation based upon (45.32). The results are given in Fig. 45.6. From this, it is clear that the empirical formula will not provide a good approximation to τ_b throughout the tidal cycle. It may be shown that the empirical formula is at its weakest during the maximum flood and ebb tide. At these times, the errors in the empirical estimates of $|\tau_b|$ are found to be about 80% and 23%.

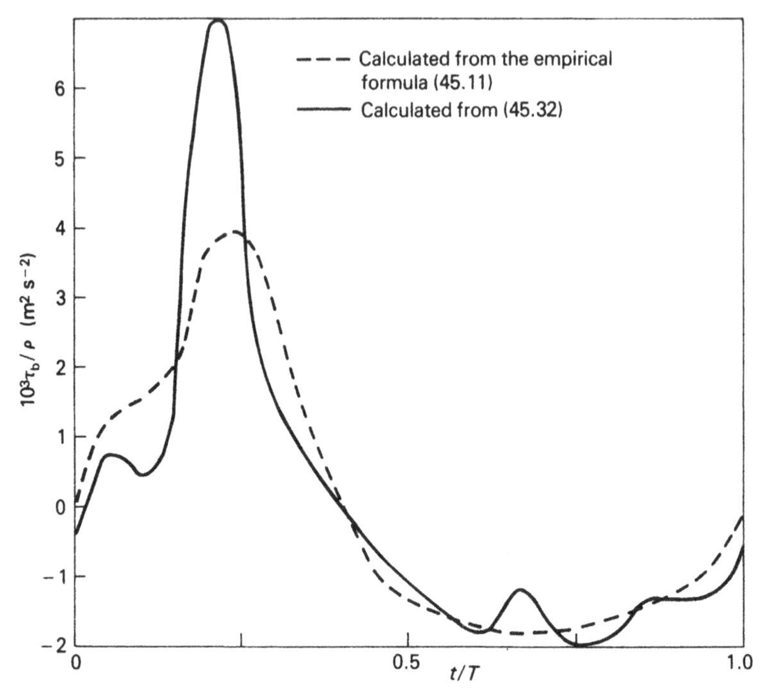

Fig. 45.6. Temporal variation of the bottom stress (τ_b) calculated from the empirical formula (45.11) with $k = 4.63 \times 10^{-3}$, and from (45.32). The tidal period, T, is 12.4 hours.

When used in a vertically-integrated model, misrepresentations of the bottom stress resulting from the use of (45.11) may be expected to lead to significant error. In such models, (45.11) is an important source of nonlinearity and contributes towards the interactions between surge, tide and fresh water flow. Consequently, on the basis of deductions from the multi-level model, these interactions will be represented incorrectly with the obvious implications that this has on the prediction of sea-surface elevation.

45.5 Conclusions

The use of an idealized numerical model for the Bay of Bengal and the Ganges, Brahmaputra, Meghna river system has shown the importance of nonlinear interactions between surge and tide in shallow water regions. By prescribing a hypothetical cyclone tracking across the Bay of Bengal, the model yields the value of the surge and tidally generated sea-surface elevation and its time of occurrence relative to the position of the cyclone. The idealized geometry of the present model must be emphasized and future work should be directed towards a more realistic representation of the coastlines and bathymetry (see e.g. Johns and Ali, 1980). Improved tidal data must be obtained along the open-sea boundary in order to refine the tidal input in the model. Appropriate hydrological data is required for the fresh water discharge in the river system.

Even with these refinements, however, the weaknesses in the vertically-integrated formulation may invalidate the prediction of interactive processes. Hence, in the longer term, the development of a multi-level model must be considered. This need not necessarily cover the entire analysis area. In the deep water region, where nonlinear interactions are insignificant, a vertically-integrated model is still appropriate, although this may require the addition of barometric forcing. This model could then be used to yield sea-surface elevation boundary conditions for a multi-level model in the shallow water shelf regions. In this, nonlinearity and bottom friction would be more correctly represented and a consequential improvement in the prediction of the interaction between surge and tide might then be expected.

An acknowledgement is made to Dr Anwar Ali for his contribution to the developent of the Bay of Bengal model.

References

Freeman, N. G., Hale, A. M. and Danard, M. B. (1972) A modified sigma equation approach to the numerical modelling of Great Lakes hydrodynamics. *J. Geophys. Res.*, **77**, 1050–60.

Johns, B. (1977) Residual flow and boundary shear stress in the turbulent bottom layer beneath waves. *J. Phys. Oceanog.*, **7**, 733–8.

Johns, B. (1978) The modelling of tidal flow in a channel using a turbulence energy closure scheme. *J. Phys. Oceanog.*, **8**, 1042–9.

Johns, B. and Ali, M. A. (1980) The numerical modelling of storm surges in the Bay of Bengal. *Quart. J. Roy. Meteor. Soc.*, **106**, 1–18.

Launder, B. E. and Spalding, D. B. (1972) *Mathematical Models of Turbulence.* Academic Press, New York, 169 pp.

Phillips, N. A. (1957) A coordinate system having some special advantages for numerical forecasting. *J. Meteor.*, **14**, 184–5.

46

Flood stage forecasting in rivers

S. CHANDER, S. K. SPOLIA AND A. KUMAR

Two case studies of flood stage prediction are described using a model based on state variables. In the first case, flood stages on the Brahmaputra river are predicted using stage data as input to the model with its parameters estimated by the extended Kalman filter algorithm. In the second case, flood stage forecasting on the Wainganga river is carried out using rainfall as the input. The parameters of the model in this case are estimated by the recursive least-square algorithm.

46.1 Introduction

Forecasting is, by definition, the art of estimating the probable behaviour of a phenomenon. Its accuracy depends upon the nature of the phenomenon, the nature of the available data and the adequacy of the fitted model. Hydrologists are commonly required to forecast the time development of hydrologic phenomena such as runoff, floods and rainfall. These phenomena are complex in nature and their distribution both spatial and temporal.

In India, where the terrain is highly irregular, the rainfall distribution during the year is highly skewed, 75% of rainfall being received during the monsoon months June to September. The rainfall of the country as a whole is less variable from year to year (±10% of the annual mean) but it is extremely variable from month to month. This seasonal variability in Indian rainfall is responsible for causing floods or droughts in some parts of the country almost every year.

Stochastic models have been widely used for generating synthetic data needed for optimal working in both design and operation of water resources systems (Lawrence and Kottegoda, 1977; Chander and Spolia, 1976). However, use of these models in hydrological forecasting has been limited.

Stochastic models for forecasting annual rainfall may give accurate predictions, but any suitable model for quantitatively forecasting the highly variable rainfall during the monsoon months for short time periods when the flood danger is a maximum, requires a large amount of meteorological data collected at very close intervals of time from densely distributed stations. Such predictions could not be carried out in the absence of such data.

Chander and Spolia (1976) and Chander *et al.* (1975) have used autoregressive moving average (ARMA) models with stages recorded at upstream stations on the rivers or their tributaries as inputs for forecasting flood stages at downstream stations. These forecasts are of great importance as they allow some time for evacuation of human as well as animal life. Stage data has been used in these studies as these are the only data which can be easily and economically collected during floods.

Two case studies of flood stage prediction are described in this chapter. In the first, flood stages on the river Brahmaputra are predicted using upstream stage data from three tributaries as a multiple input AR model. The parameters of this model are estimated using the extended Kalman filter algorithm. In the second case, flood stage forecasts on the Wainganga river are worked out using rainfall as an input to the system. The parameters of the model in this case are estimated by a recursive least-square algorithm.

46.2 Case study 1

In this case study, forecasting of stages on the river Brahmaputra at Dibrugarh has been attempted using the observed gauge data upstream on the three major tributaries, namely Dihang, Debang and Lohit. The hourly gauges during floods are measured at Passighat on Dihang, Jiagaon on Debang, and at Teju on Lohit. The river bed in the tributaries and the main river may suffer aggradation or degradation depending on the flow. A model based on gauges needs to be changed from flood to flood. Chander and Spolia (1976) overcame this difficulty in their multiple-input autoregressive model by making use of the differences in gauges at the three upstream tributary stations as inputs to the autoregressive model of differences in gauges at Dibrugarh.

The discrete linear, time-invariant model proposed in Chander and Spolia (1976) is given by

$$g_{i+T,i} = A_1 g_{i,i-T} + \sum_{j=1}^{m} A_{2j} h^{(J)}_{i-T_j+T,i-T_j} + \sum_{j=1}^{m} A_{3j} h^{(J)}_{i-T_j,i-T_j-T}, \qquad (46.1)$$

where the symbols denote the following

m number of multiple inputs (tributaries in this case),
T forecasting time,
T_j lag time between the upstream station on the jth tributary, and the forecasting station ($T < T_j$),
$h_{i-T_j+T,i-T_j}$ difference in gauges at the upstream station on the jth tributary between the $(i - T_j + T)$th and the $(i - T_j)$th instants,
$g_{i,i-T}$ difference in gauges at the ith and the $(i - T)$th instants of time at the forecasting station,

A_1, A_{2j}, and A_{3j} are parameters to be determined.

Equation (46.1) can also be written as

$$y(k) = A_1 y(k-1) + \sum_{j=1}^{m} A_{2j} w(k - T_j) + \sum_{j=1}^{m} A_{3j} w(k - 1 - T_j), \qquad (46.2)$$

where $y(k)$ and $y(k-1)$ are the outputs at times k and $k-1$ respectively and $w(k - T_j)$ and $w(k-1-T_j)$ are the inputs from tributaries at times $(k - T_j)$ and $(k-1-T_j)$ respectively.

46.2.1 Estimation of parameters

The parameters A_1, A_{2j} and A_{3j} can be estimated by a least-square method using an appropriate T_j for each of the m tributaries and observed input and output data. The observed input and output data were plotted and an estimate of delay time T_j for each tributary was made on the basis of the time difference between the peak of the jth tributary and the peak at the output station. Delay times T_j of 12, 12 and 14 hours were estimated for Passighat, Tezu and Jiagaon respectively on this basis.

Knowing the appropriate T_j, the optimal estimates of the parameters are given by

$$\hat{\boldsymbol{\theta}}(k) = |H'(k)H(k)|^{-1} H'(k)\boldsymbol{Y}(k), \tag{46.3}$$

where $\hat{\boldsymbol{\theta}}(k)$ is the column vector $[A_1, A_{21}, \ldots, A_{2m}, A_{31}, \ldots, A_{3m}]'$ of $2m+1$ elements, and $H(k)$ is the $N \times (2m+1)$ matrix of observed data; $\boldsymbol{Y}(k)$ is the vector $[y_k, y_{k-1}, \ldots, y_{k-N+1}]'$ of outflow values at Dibrugarh.

If $P(k) = |H'(k)H(k)|^{-1}$, then

$$\hat{\boldsymbol{\theta}}(k) = P(k)H'(k)\boldsymbol{Y}(k). \tag{46.4}$$

The parameter estimates generated by (46.3) for the model defined by (46.1) need to be updated as new observations become available.

System updating can be carried out using recursive least-squares (Mendel, 1973) or an extended Kalman filter (Jazwinski, 1970).

46.2.2 *Parameter estimation using recursive least-squares*

Let the updated estimates of the parameters be denoted by $\hat{\boldsymbol{\theta}}(k+1)$. Then

$$\hat{\boldsymbol{\theta}}(k+1) = P(k+1)H'(k+1)\boldsymbol{Y}(k+1). \tag{46.5}$$

But,

$$\boldsymbol{Y}(k+1) = \begin{bmatrix} y(k+1) \\ \hline \boldsymbol{Y}(k) \end{bmatrix},$$

an $(N+1) \times 1$ vector, and

$$H(k+1) = \begin{bmatrix} \boldsymbol{h}(k+1) \\ \hline H(k) \end{bmatrix},$$

an $(N+1) \times (2m+1)$ matrix. Thus, (46.5) can be rearranged as

$$\hat{\boldsymbol{\theta}}(k+1) = \hat{\boldsymbol{\theta}}(k) + P(k+1)\boldsymbol{h}'(k+1)[y(k+1) - \boldsymbol{h}(k+1)\hat{\boldsymbol{\theta}}(k)]. \tag{46.6}$$

Equation (46.6) can be used for updating parameters, knowing $\hat{\boldsymbol{\theta}}(k)$, $\boldsymbol{h}(k+1)$ and $y(k+1)$, but needs two inversions to obtain $P(k+1)$. The computation can be further simplified using the matrix lemma (Mendel 1973)

$$\hat{\boldsymbol{\theta}}(k+1) = \hat{\boldsymbol{\theta}}(k) + P(k)\boldsymbol{h}'(k+1)[\boldsymbol{h}(k+1)P(k)\boldsymbol{h}'(k+1) + 1]^{-1} \times [y(k+1) - \boldsymbol{h}(k+1)\hat{\boldsymbol{\theta}}(k)]. \tag{46.7}$$

Equation (46.7) reduces the computations as $(\boldsymbol{hPh}'+1)$ is a scalar quantity and can be easily used for upgrading the parameters.

46.2.3 Parameter estimation using the extended Kalman filter

To use extended Kalman filtering, (46.2) is written in the standard form as an input–output noise system as

$$\boldsymbol{X}(k+1)=\alpha(k)\boldsymbol{X}(k)+\beta(k)\boldsymbol{U}(k)+\boldsymbol{W}(k),$$
$$\boldsymbol{Y}(k+1)=H(k+1)\boldsymbol{X}(k+1)+\boldsymbol{V}(k+1), \qquad (46.8)$$

where $\boldsymbol{U}(k)$ is an accessible input vector of data from tributaries and $\boldsymbol{W}(k)$ is the n-vector noise sequence which takes into account any modelling error or the random inputs to the system. $\boldsymbol{V}(k)$ is the measurement error vector and $\boldsymbol{Y}(k)$ is the noisy observation of $\boldsymbol{X}(k)$.

Assuming the parameters are randomly varying, $\boldsymbol{\theta}$ can be written in the form

$$\boldsymbol{\theta}(k)=\boldsymbol{\theta}(k-1)+\boldsymbol{\xi}(k-1), \qquad (46.9)$$

where $\boldsymbol{\theta}(k)$ forms a vector of unknown elements of matrices $\beta(k)$ and $\alpha(k)$. $\boldsymbol{\xi}$ is a white Gaussian random sequence having zero mean with a covariance *S*, which is unknown to begin with but is subsequently estimated. (In the absence of a complete knowledge of all the processes involved, the choice of a simple random walk structure for the parameter variation seems the most sensible one.) The parameters $\boldsymbol{\theta}(k)$ were estimated using the standard extended Kalman filter algorithm (Jazwinski, 1970). The covariance estimates of noisy sequences were estimated using the adaptive estimation algorithm given by Sage and Husa (1969).

The parameters updated with every new observation using (46.2), were used for three floods shown in Figs. 46.1*a*, *b* and *c*. These forecasts were made at 4-hour and 8-hour intervals. The results were compared with the observed hydrographs. It is concluded that 4-hour forecasts are nearer to the observed hydrograph than the 8-hour forecasts.

46.4 Case study 2

In this study stage forecasting of the river Wainganga is attempted using rainfall data. Rainfall is measured at seven different stations, namely: (i) Seoin Chapra; (ii) Kunda; (iii) Mohgaon; (iv) Karabdol; (v) Bijna; (vi) Gopalganj; and (vii) Seoni, in this catchment. The weighted average

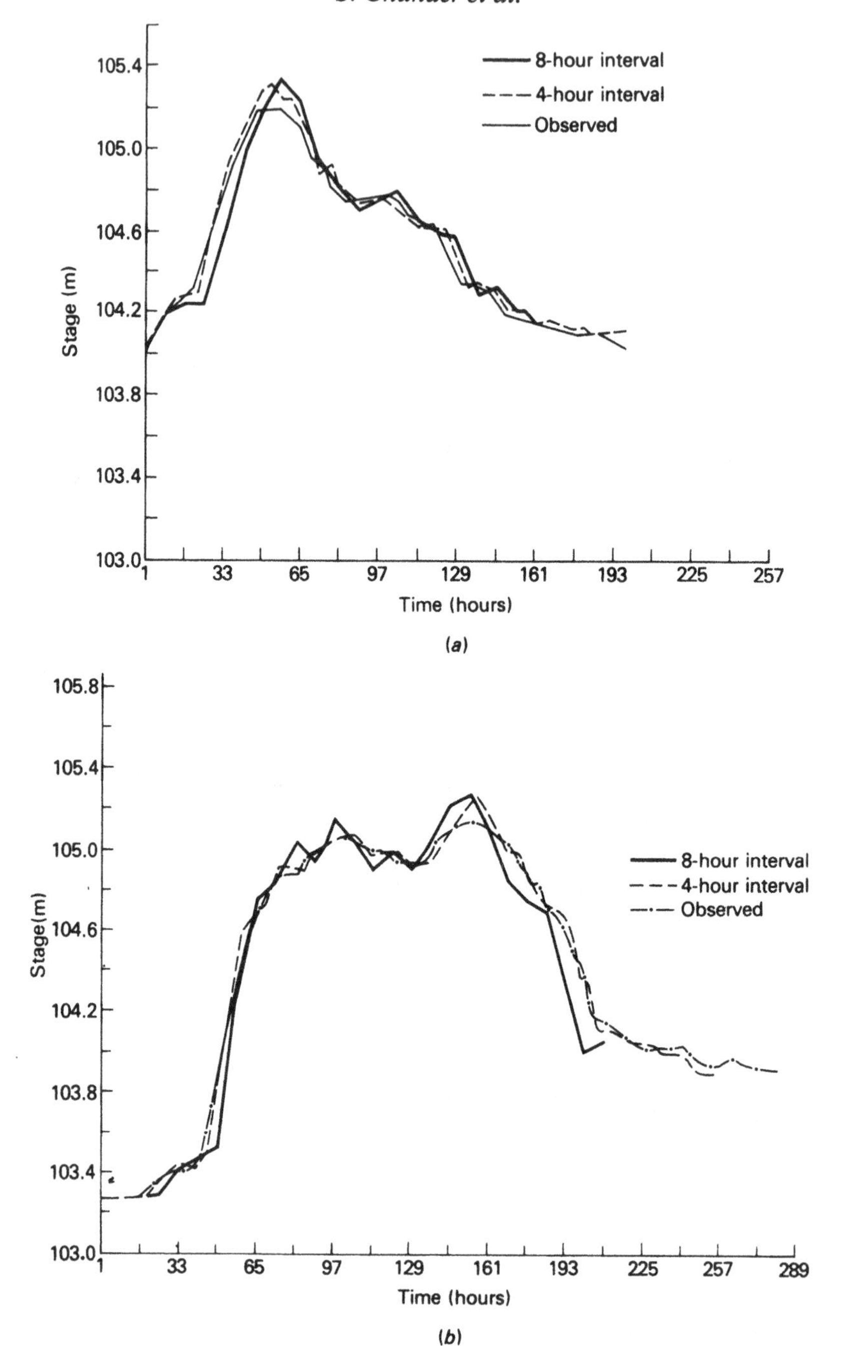

Fig. 46.1. Comparisons of 8-hour and 4-hour interval forecasts with the observed hydrographs for three floods at Dibrugarh.

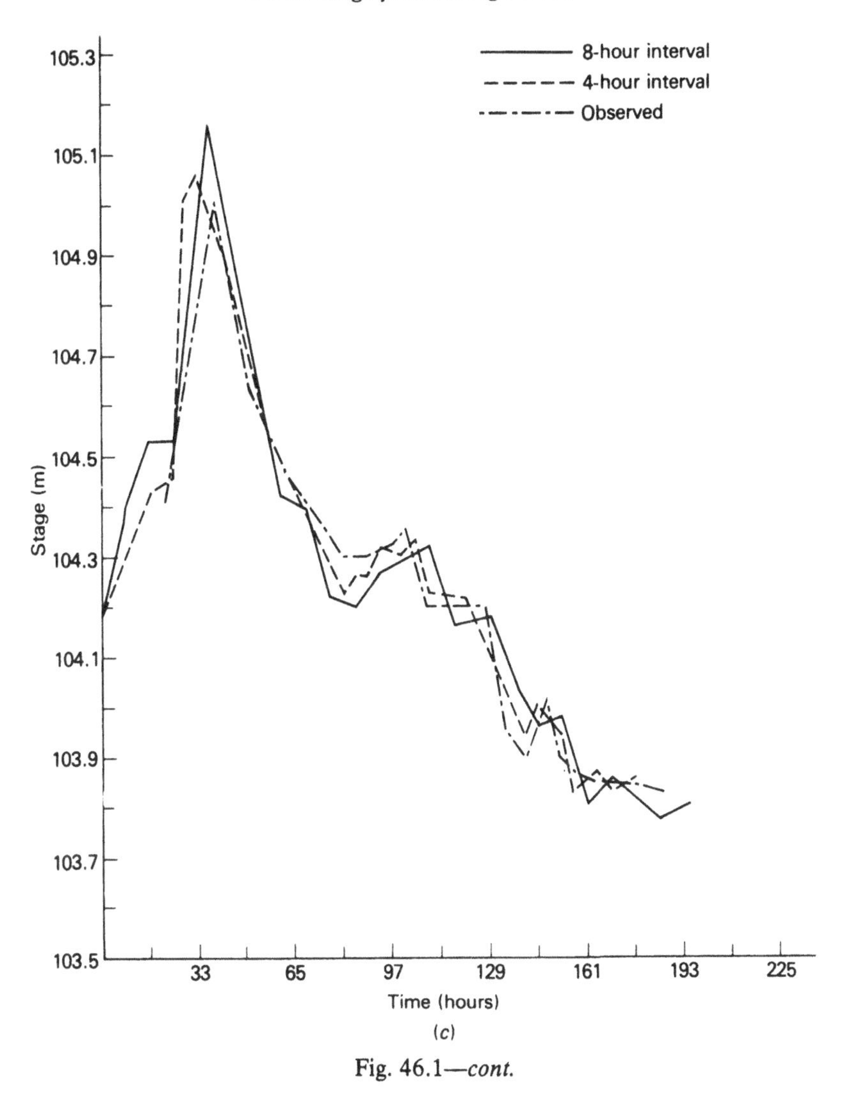

Fig. 46.1—*cont.*

rainfall was calculated using Theissen's method and was used as input to the model, i.e.

$$y(t)=\phi_1 y(t-T)+\phi_2 y(t-2T)+\theta_0+\theta_1 w(t-T)$$
$$+\theta_2 w(t-2T)+\theta_3 w(t-3T), \quad (46.10)$$

where $y(t)$ denotes the river stage at time t and $w(t-T)$ is the weighted average rainfall at time $(t-T)$. T was assumed to be 3 hours, and 3-hour forecasts were made in this case.

The parameters were initially estimated using the first 25 observations of a multipeaked flood observed during the period 29 July to 6 August 1968 (Fig. 46.2*a*). The model was then used to forecast the stages for the remaining period of the flood.

The observed and the predicted stage hydrographs using the same parameters are also shown in Fig. 46.2*a*.

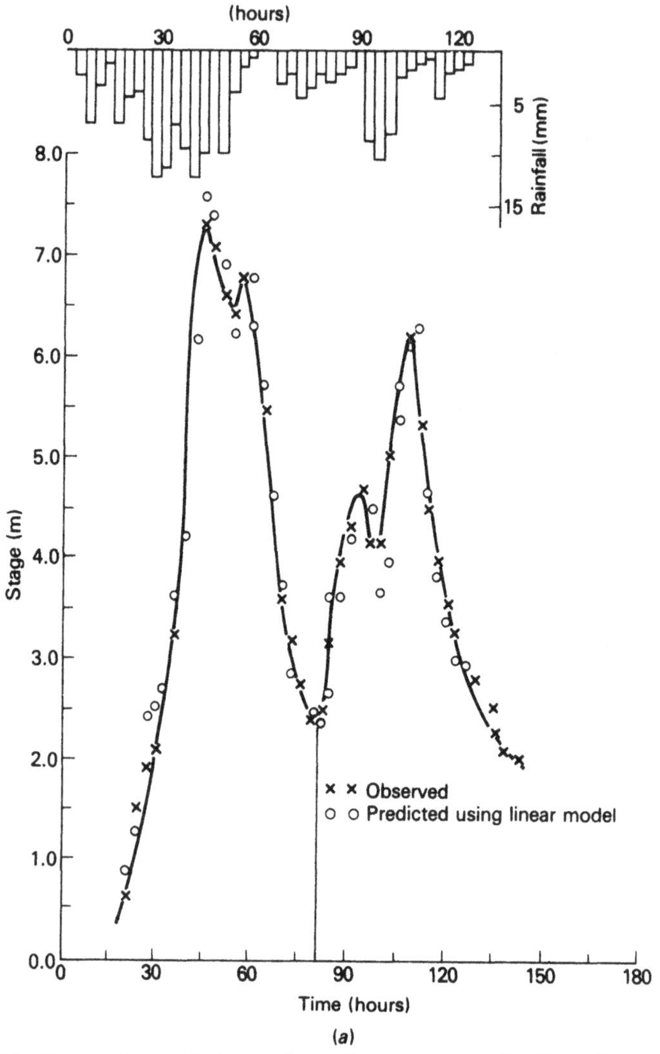

Fig. 46.2. Comparisons of observed and forecast stages for three floods on the river Wainganga. Flood dates are: (*a*) 30 July to 6 August 1969; (*b*) 28 August 1972; and (*c*) 28 August 1973.

In order to discover whether the basin was best represented by a time-variant or time-invariant model with the structure given in (46.10), two floods observed during the years 1972 and 1973 were analysed and 3-hour forecasts were computed using constant parameters and the time-invariant model with updated parameters (46.7). The computed and observed hydrographs are shown in Figs. 46.2*b* and *c* for both cases. The mean-square criterion was used for evaluating the models, and it was found that the recursive least-square algorithm reduced the mean-square error considerably; for 1972 it was reduced from 0.1663 to 0.0866 and

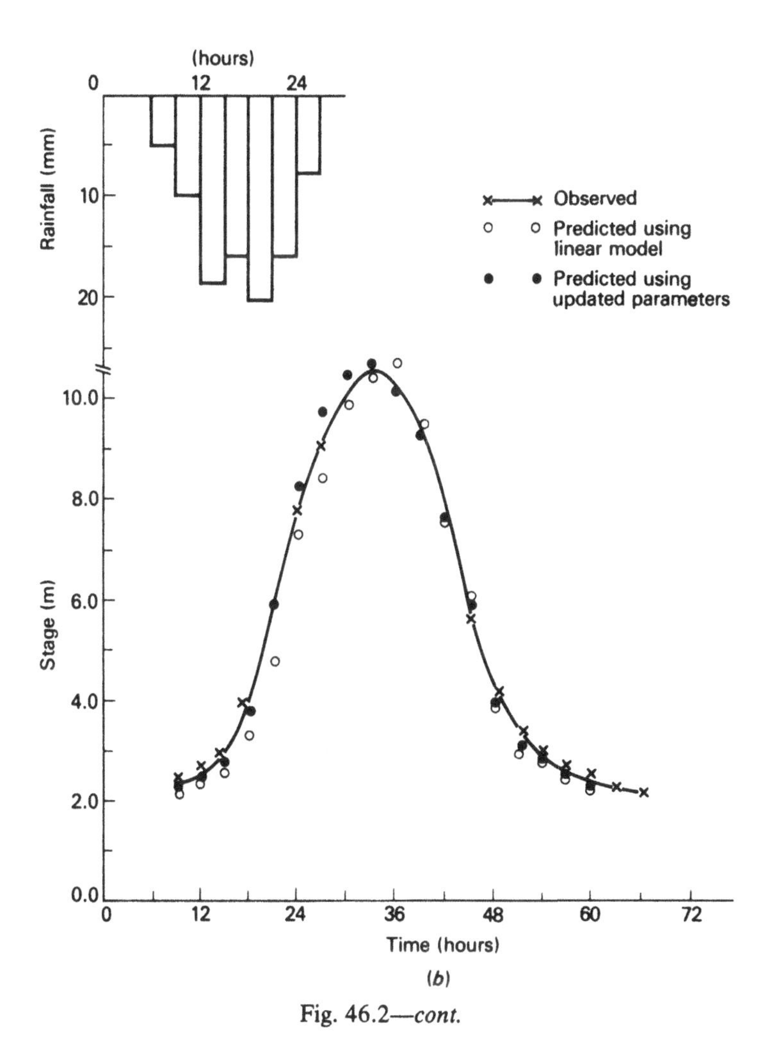

Fig. 46.2—*cont.*

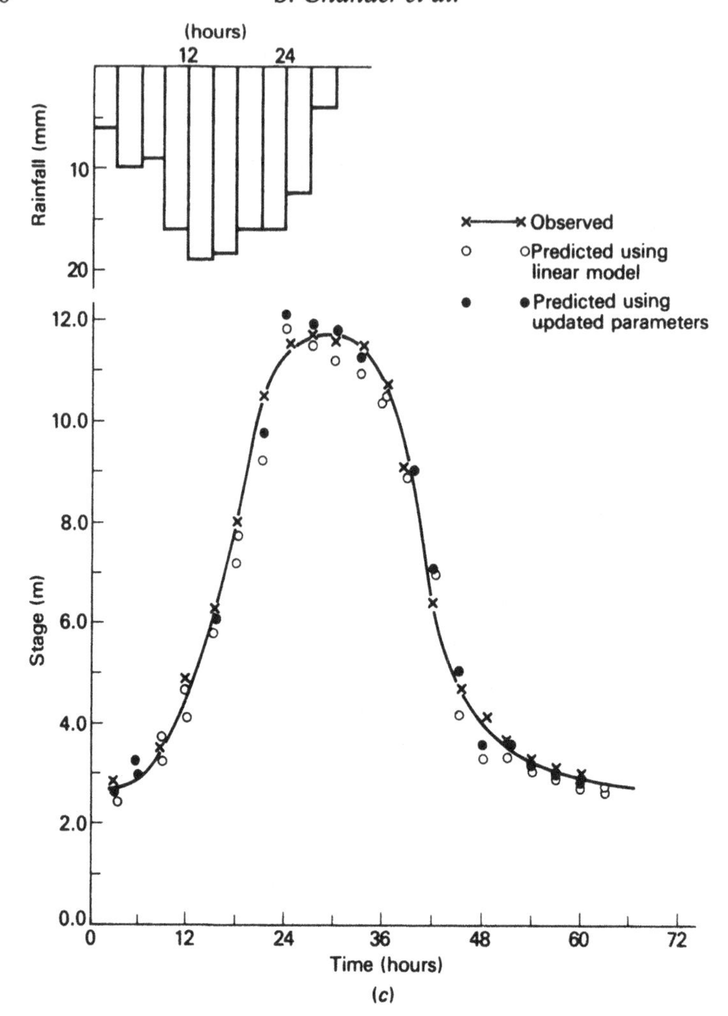

(c)

for 1973 from 0.245 to 0.01395, thus implying that the forecasts are better using updated parameters in the model.

46.5 Conclusions

It has been shown in this chapter that recursive least-square and extended Kalman filters can be used for the estimation of parameters of the models used in flood-stage forecasting. These methods are particularly useful as they do not require data storage and thus facilitate implementation of such models on mini-process control computers.

References

Chander, S., Das, C. J. and Padhi, S. (1975). An approach to flood stage forecasting. Proceedings of the 2nd World Congress of the International Water Resources Association, New Delhi, pp. 125–30.

Chander, S. and Spolia, S. K. (1976) Flood stage forecasting on Brahamputra river. (Unpublished paper.)

Jazwinski, A. H. (1970) *Stochastic processes and filtering theory*. Academic Press, New York.

Lawrence, A. J. and Kottegoda, N. T. (1977) Stochastic modelling of river flow time series. *J. Roy. Stat. Soc.*, A **140**, 1–47.

Mendel, J. M. (1973) Discrete Techniques of parameter estimation. Marcel Dekker, Inc. New York, pp. 90–101.

Sage, A. C. and Husa, G. W. (1969) Adaptive filtering with unknown prior statistics. Proc. JACC, pp. 760–7.

Index

Page numbers in parentheses indicate pages on which references are listed